TEACHING
PRIMARY SCIENCE
CONSTRUCTIVELY
4th edition
KV-621-263
Edited by Keith Skamp

Teaching Primary Science Constructively
4th Edition
Keith Skamp

Publishing manager: Alison Green
Publishing editor: Ann Crabb
Senior project editor: Nathan Katz
Developmental editor: Kylie McInnes
Cover design: Miranda Costa
Text design: Danielle Maccarone
Editor: Paul Smitz
Indexer: Russell Brooks
Permissions research: Georgina Wober
Proofreader: James Anderson
Cover: Images © Shutterstock
Typeset by Cenveo Publisher Services

First published in 1998 by Harcourt Australia Pty Ltd
Second edition published in 2004 by Nelson Australia Pty Ltd
Third edition published in 2008 by Thomson Learning

This fourth edition published in 2012

For product information and technology assistance,
in Australia call **1300 790 853**;
in New Zealand call **0800 449 725**

For permission to use material from this text or product, please email
aust.permissions@cengage.com

National Library of Australia Cataloguing-in-Publication Data
Title: Teaching primary science constructively / edited by Keith Skamp.
Edition: 4th ed.
ISBN: 9780170191746 (pbk.)
Notes: Includes index.
Subjects: Science--Study and teaching (Primary)
Other Authors/Contributors: Skamp, Keith.
Dewey Number: 372.35044

Cengage Learning Australia
Level 7, 80 Dorcas Street
South Melbourne, Victoria Australia 3205

Cengage Learning New Zealand
Unit 4B Rosedale Office Park
331 Rosedale Road, Albany, North Shore 0632, NZ

For learning solutions, visit **cengage.com.au**

Printed in China by RR Donnelley Asia Printing Solutions Limited.
2 3 4 5 6 7 15 14 13 12

Brief Contents

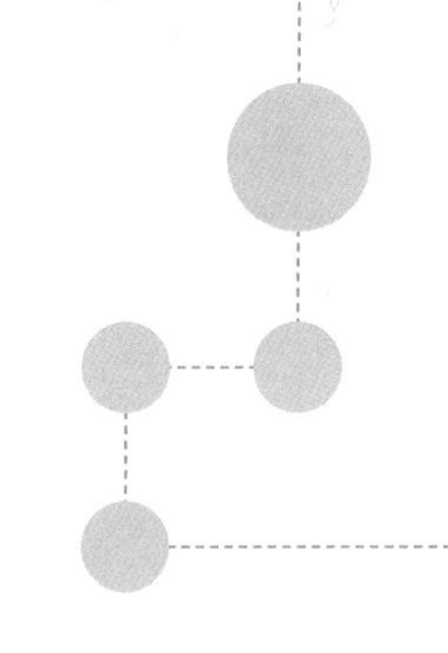

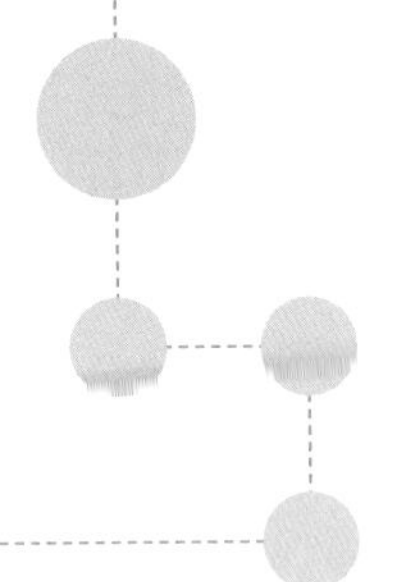

Contents

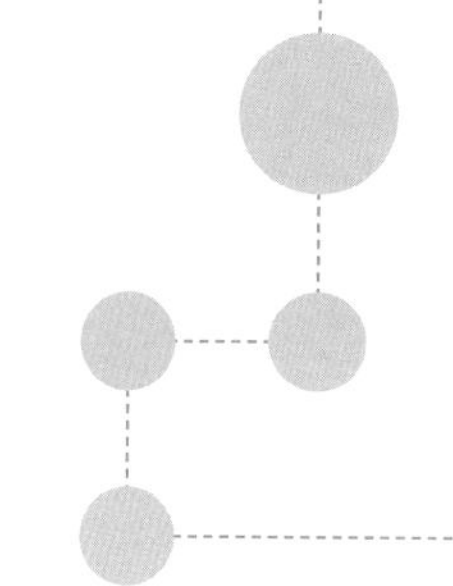

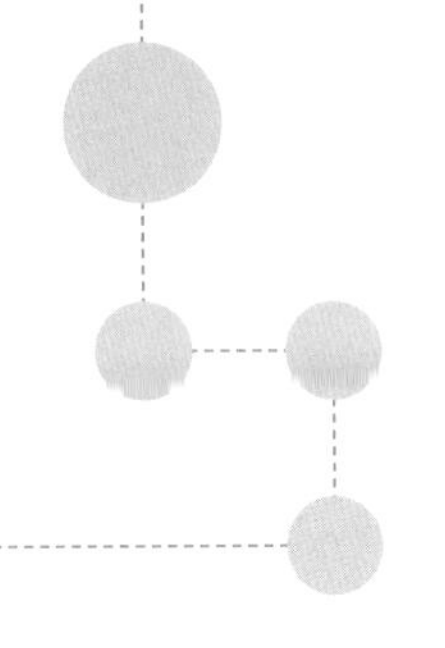

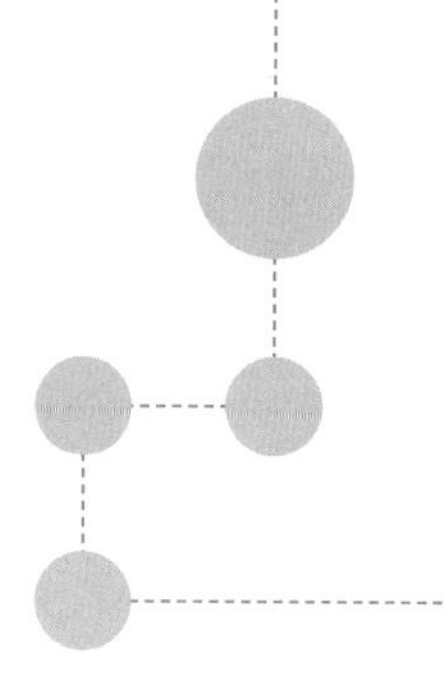

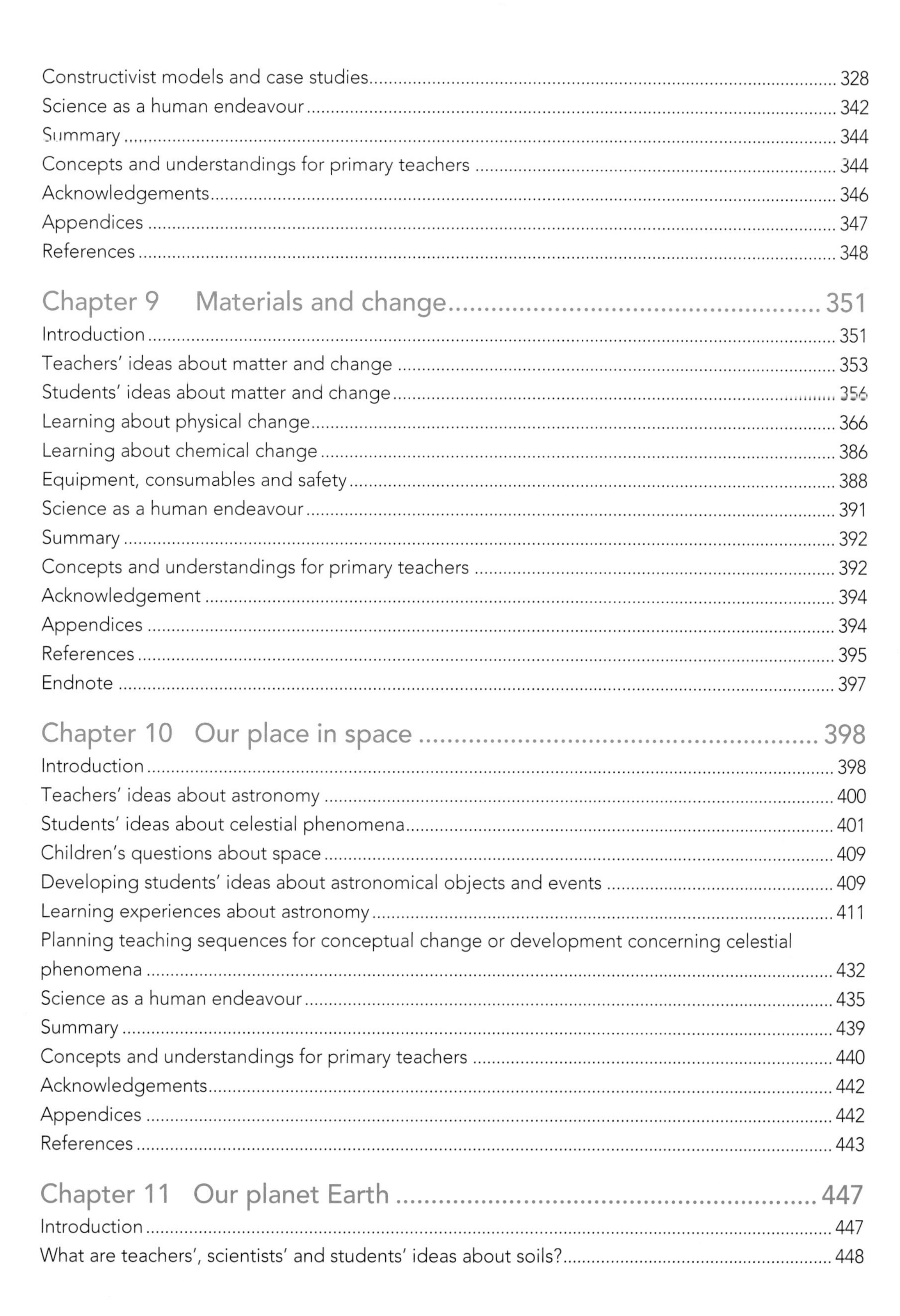

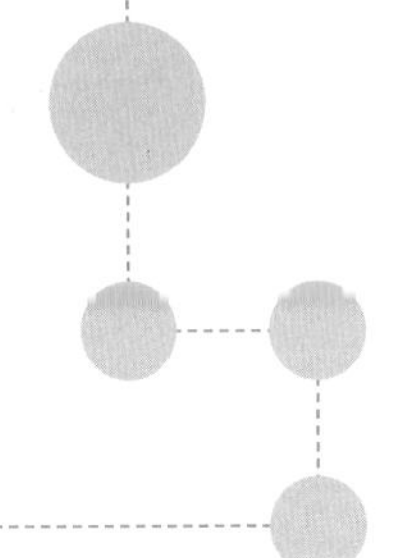

Preface

A fundamental belief of the authors of this book is that effective primary science teaching will support students in the development, formation and modification of their ideas so that they can make better sense of the way in which their world works. Further, effective primary science teaching will enhance students' understanding of how science ideas are derived, and will assist their agency in individual and social decision making in areas underpinned by scientific ideas and processes.

This interpretation of learning science, which underpins this book, considers a significant component of learning as conceptual development, modification and change. The intention is that students use stepping stones in learning concepts and understandings currently accepted by the scientific community. Learning science, in this book, is also interpreted as students becoming members of a classroom (scientific) learning community in which they are continually developing a range of scientific competencies and learning what it means to engage with the practices of the scientific community. These different dimensions to learning science are seen as commensurate.

To achieve these aims, particular approaches to teaching primary science are espoused. These approaches are based on the principle that we should teach in a way that is consistent with what we know about how students learn. Most of the research evidence related to students' science learning suggests that students continually construct their own ideas about how their world works and can be assisted in developing their competencies and identities within the practices of a scientific (classroom) community as these ideas are tested against the available evidence. This construction of ideas involves the interaction of the learner's existing ideas with further experiences and ideas from their peers, teachers and other sources. The sociocultural context, therefore, in which scientific ideas, competencies, practices and identities develop is an ever-present consideration in how students learn. Recent thinking strongly suggests that teachers need to be conscious of these interlinked aspects if they are to facilitate the movement of student thinking from an 'everyday' to a 'scientific' way of looking at the world, and assist them in making decisions about how to live sustainably in it. Consequently, there has been an increasing emphasis on the scaffolding of student learning by the teacher. This, in broad terms, is a constructivist view of learning, a view that has personal, social and cultural dimensions. A goal of this book is that it will encourage teaching which is consistent with this broad constructivist view of learning. On one level this requires teachers to adopt a 'constructivist' mindset about how learning occurs; on another level it encourages teachers to consider using particular teaching strategies which support this view of learning. The title of the book, *Teaching Primary Science Constructively*, endeavours to capture all of these ideals.

Consistent with this view of learning, several new and different emphases are apparent in this fourth edition. Social constructivism and the influence of the sociocultural context on learning are more prominent, while the focus on conceptual change that is strongly linked with a personal constructivist perspective is presented within a wider framework of factors that can influence change. 'Hot' conceptual change that acknowledges the impact on learning of a multitude of affective and related factors, for example, is now a key factor in thinking

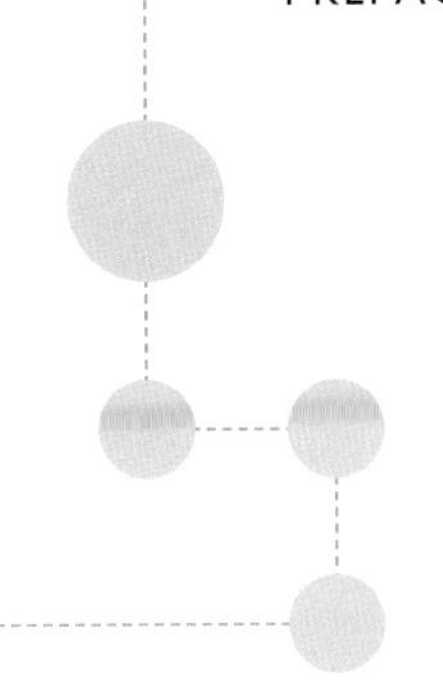

about how students learn science ideas and practices. The research evidence that effective science learning is enhanced through student-generated representations, and teachers using related pedagogy, is growing and integral to this edition. With reference to student agency in decision making in everyday life, the pedagogy associated with primary students engaging with socioscientific issues is now included. These facets of learning science are further exemplified by highlighting numerous examples of 'science as a human endeavour'.

The book has been written for two audiences. The first is preservice teachers. The content will encourage them to reflect upon their existing conceptions about how the world works; that is, their science ideas, as well as their current beliefs about what is science, science learning and science teaching. Consequently, most preservice teachers will be required to rethink what it means to facilitate primary students' science learning and hence the teaching approaches that they use.

The second audience is practising teachers, for whom the book will be a challenge to reconsider current teaching methods. For those teachers familiar with constructivist learning, the book will provide the most recent research findings related to teaching from a constructivist perspective and how teaching emphases have changed over the last few years within this way of looking at learning science. For these teachers, numerous (including several new) examples of this approach in action will assist their reflections on their own practices. The book's strength for practising teachers is that it provides teaching ideas on how to implement a constructivist approach within specific content strands of primary science syllabuses.

This book addresses the implementation of the science understandings, inquiry processes and 'science as a human endeavour' strands that comprise most international primary science syllabuses, including the most recent national curriculum changes in Australia. In so doing it is a unique text in that it provides suggestions for teaching specific concepts and understandings across a wide range of content areas, as well as explicit advice for facilitating an appreciation of the nature of science and humankind's interaction with science through these various content strands. This book, therefore, markedly differs from texts that only provide principles and advice related to general pedagogical issues such as using different teaching approaches, assessing students, teaching students from different backgrounds and so on. These teaching considerations are not overlooked but are raised within the context of teaching specific science concepts and/or science inquiry processes. A very rich source of 'pedagogical content knowledge', specific to the major content areas of primary science frameworks or syllabuses, is consequently readily available to teachers.

To illustrate the approach taken, teachers might be seeking advice about ways to teach, say, chemical change from a constructivist perspective to their middle and upper primary classes. This book directly addresses this need by focusing on what primary students think about chemical change and the conceptual difficulties they may encounter, and then provides specific teaching strategies and learning experiences related to chemical change, including cooking, burning and rusting. Further, it challenges teachers with research evidence that argues primary students can generate their own scientific representations of how materials interact with each other, and asks them to consider innovative ways to engage students with this type of scientific thinking. Chemical change case studies are also included: they illustrate how students have learned within a constructivist and representation-focused learning framework and provide examples of the espoused pedagogy in action.

To learn from a constructivist perspective implies that the ways in which teachers challenge and support students to further develop and modify their ideas are critical issues. A wide range of research-based models, strategies, procedures and techniques has been suggested to encourage learning. One key implication of the research which underpins this book is that one-off science lessons do not facilitate effective conceptual development and so, in this book, there is a major focus on planning and implementing sequential lessons. Numerous case studies are included to illustrate how teachers have applied constructivist learning sequences in their classrooms. Conceptual development, the acquisition of inquiry processes and skills, and the formation of identity within a scientific (classroom) learning community are evident in these exemplars of practice. The authors also believe that when primary students use the processes associated with technological capability, and practices involving the use of scientific and generic literacies, conceptual development is enhanced. There are case studies included that highlight, for example, fair testing, the symbiotic relationship that can exist between scientific and technological learning, multiple and multimodal representations of science ideas, and the literacy aspects of learning science.

Primary teachers are sometimes concerned about their own science knowledge background as well as what science encompasses. These issues are discussed in the initial chapter, with suggestions for ways forward. In the introductory sections of Chapters 3 to 12 there are activities for readers that encourage them to check their own understanding of science concepts and the nature of science related to the learning experiences with which they will engage their students. To further assist teachers, there is, at the end of each of these conceptual chapters, a listing of key concepts and understandings that the authors believe primary teachers will find of value. These concepts and understandings are derived from the research literature, which has investigated students' science ideas over the last three decades. They are for teachers and not necessarily students.

As implied, another feature of this book is that it draws on an extremely wide research and professional literature related to primary science education. To the authors' knowledge, no one book has attempted to draw together the findings of international research studies on the ideas primary students hold about concepts included in most primary science syllabuses. Apart from updating these research studies, this fourth edition emphasises how the development of students' ideas (about how their world works) is now being interpreted differently to the way it was at the time of the third edition. Learning science is now considered to be influenced by a wider range of factors and conditions than previously thought. This has implications for teaching science. Consequently, the latest research and thinking about the most effective ways to help students grow their understanding of particular science concepts, science inquiry processes, competencies and practices are included, and explicit indications are provided in this edition as to how teachers might incorporate these findings into their own practices. At all times, however, the commentary is worded on the assumption that readers are teaching professionals and that they will make their own pedagogical decisions based upon their own teaching experiences as well as the research and practices exemplified in this book. Allied to these advances, there are a growing

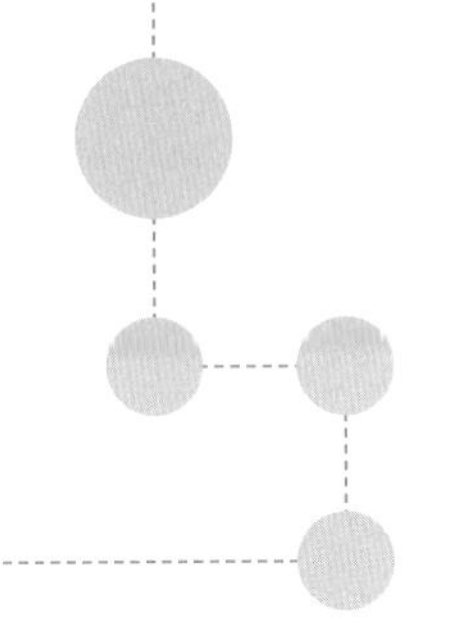

number of studies which suggest that teachers who blend literacy learning, in its various forms, with science learning will have a positive impact on students' conceptual and process-oriented understanding of science. Recent thinking acknowledges that learning science involves the student in accessing a range of multimodal literacies in science. Increased reference is made to the implications for teaching from this research. Overall, every effort has been made to be accurate in reporting the current state of research about primary students' learning in science and to describe and interpret this research so that it is meaningful for primary teachers.

It is appreciated that primary teachers are busy people. In that light, having these research findings interpreted and summarised, we believe, will be of invaluable assistance. Even so, we wish to emphasise that the research base related to teaching primary science has continued to expand with every edition of this book. References to the research and professional literature are therefore extensive. Primary science teaching is now underpinned by a significant and more refined research base, but some approaches still remain contested and this is acknowledged. To reiterate, one purpose of this book is to highlight how the suggested teaching ideas are built on a solid research base while also being accessible to testing by practising and future teachers.

This book can be used by an individual or with colleagues. It certainly encourages the reader to interact with the text and has numerous reader activities, which vary in their content. Some of them include hands-on and minds-on tasks using simple materials, while others are activities that require access to primary students, analysis and reflection; several activities ask questions relating to decisions made by teachers in the case studies. The overall range is extensive. These reader activities could be readily used as exercises and/or assignments in university and professional development courses.

One way to use the book in a structured course would be to focus on a conceptual area in a primary science syllabus or framework and work through the particular activities, reading and discussion in the relevant chapter. Supplementary reading and activities could readily be developed from the numerous questions, tasks and references suggested. There are, however, many other ways in which to productively access the book's content. As a guide, it would be most helpful if you read sections – or all – of Chapter 1 first. It acts as an advance organiser for the remainder of the chapters and sections of it are referred to in all later chapters. You could revisit this chapter several times and find that you glean more meaning from it each time you return. As all the conceptual chapters assume a knowledge of Chapter 2, which focuses on scientific inquiry (thinking and working scientifically), then this should be read next. The remaining chapters can be read in any order, although Chapter 8 should precede Chapter 9.

The content of this book could readily encompass two science education units in a university preservice primary teacher education degree. If it is used for a one-semester unit, in a postgraduate diploma or degree, or in limited professional development courses, it is suggested that the focus be on selected chapters or text activities. Further, the book could be of great value to a teacher using particular chapters to guide teaching about specific concepts.

Apart from accessing the research and professional literature, the authors, who are practising teachers and/or teacher educators, have drawn on their own experience and research and that of many others in writing this book. The assistance of these people, mainly

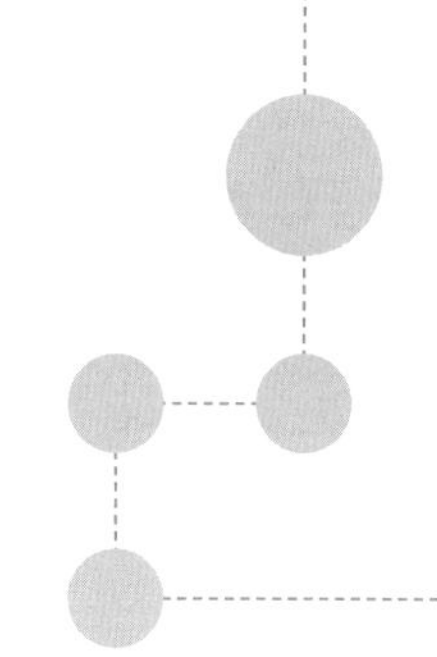

practising teachers, has been duly acknowledged in various places within the text, but we again thank them for it. Their advice, in particular their case study material, is an integral component of most of the chapters. As the editor, I would especially like to thank my co-writers, including the several new contributors to this edition, for the most cooperative manner in which they have written this book and their willingness to embrace the developments in learning science since the third edition and hence make significant revisions to most chapters. Finally, there have been several reviewers of the third edition and draft chapters of this fourth edition; the feedback provided by these specialist science educators was incisive and helpful, and their contributions are acknowledged. For this fourth edition we would like to thank Sally Birdsall (University of Auckland); Sue Wilson (Australian Catholic University); Kathy Paige (University of South Australia); Yvonne Zeegers (University of South Australia); Kathy Stewart (Macquarie University); Mitch O'Toole (University of Newcastle); Donna Satterthwait (University of Tasmania); Christine Redman (University of Melbourne); Gregory Smith (Charles Darwin University); and David Geelan (University of Queensland). Several of their suggestions are included in this final version.

If this book is instrumental in assisting some teachers to implement more effective science education for primary students, we shall all have achieved our purpose.

Keith Skamp

Using this book

This book is about helping preservice and practising teachers facilitate the learning of science by primary students. It has been structured in a particular way to assist your learning. If you understand how it has been organised, then you will be able to navigate your way through it far more effectively. Further, an appreciation of the wide range of ideas it includes will encourage you to look for aspects that might otherwise be overlooked.

The argument is advanced that we should teach science based on what we know about how students learn science. Most science educators accept that research findings indicate that constructivism, set within a sociocultural context and with appropriate emphases on learning through representation and the impact of factors such as students' affective engagement and an appreciation of their own agency, is the theory which underpins how students learn science. Constructivism, seen within such a landscape, is a learning theory, which describes what we understand about how students construct their ideas.

How students construct ideas and develop scientific competencies and practices in order to understand their world also applies to how teachers construct their ideas about science, teaching science and learning science. Therefore, we have applied many constructivist and related learning principles in the book's interaction with you, the reader. We take as a given that you, the reader, have your own ideas about what science is, what it means to learn science and, consequently, the role(s) you would adopt to teach it. So you are constantly challenged about your existing ideas related to how your world works and how you believe we determine those ideas; that is, the concepts, generalisations and inquiry processes and practices underpinning science, as well as your ideas about learning science and teaching science.

As stated in the Preface, Chapter 1 introduces constructivism and is an advance organiser for the other chapters. You are advised to keep returning to this chapter to obtain a broad overview of constructivism as a learning theory and its teaching implications. We see science being characterised as, firstly, a way of knowing that embraces various competencies and practices; secondly, an evolving body of knowledge and understandings; and, thirdly, a human endeavour. Each of these attributes is interdependent. Scientific inquiry, or thinking and working scientifically (Chapter 2), focuses on science as a way of knowing, and how young students can be introduced to its values, practices and processes. Explicit teaching of the nature of science is integral to this book's approach and is stressed in many chapters and their various case studies, such as chapters 3 and 8. Chapters 3 to 12 are the conceptually oriented chapters related to the content strands in most recent primary science syllabuses, irrespective of their country of origin. They have been written with a basic schema in mind. There is usually:

- a section on *your ideas*, as teachers, about the topic
- a significant discussion about, and summary of, the *research related to primary students' ideas* about the topic
- a clear description of (or references to) *scientists' ideas* on aspects of the topic, which are either integrated with the primary students' ideas or are separate parts of the text. These

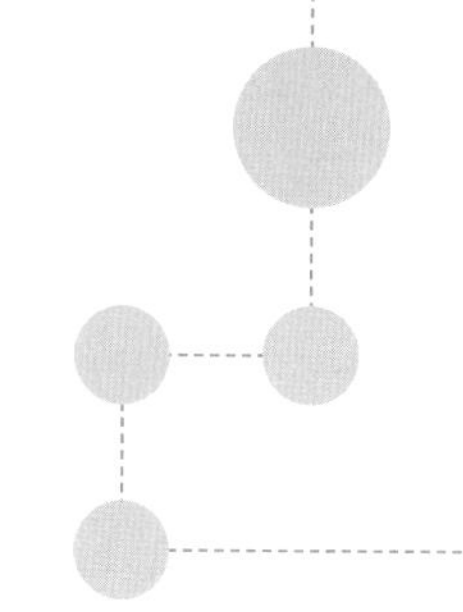

inclusions clarify the alternative and non-scientific conceptions many students and adults, including teachers, hold
- an indication of *what children want to know* about the topic
- a discussion of key *constructivist and related teaching principles* and selected *strategies* relevant to the topic
- one or more *teaching case studies or vignettes* of lesson sequences based on constructivist and related teaching approaches; some chapters (for example, 3, 4 and 6) are completely structured around one or more detailed case studies. These are reports of actual classroom teaching and often are based on the teachers' own descriptions
- reference to influential research and professional development initiatives such as the Australian Academy of Science's *Primary Connections* project
- innovative and specific examples of *science as a human endeavour*
- inclusion of *other teaching and learning considerations* such as the effective use of information and communication technology (ICT) related to the topic and sometimes primary science in general
- *concepts and understandings for primary teachers*
- *reader activities* to encourage interaction with the text.

All of these common elements in each chapter are introduced in various sections in Chapter 1. As a major focus of this book is the advocacy of various constructivist and related teaching approaches, each of chapters 3 to 12 then includes examples of constructivist models and schema in action, and exemplifies representation-focused learning and novel ways in which to prompt students to be engaged by their learning.

As suggested earlier, many issues, apart from specific constructivist learning approaches, influence teaching and learning. Conceptual development theories based upon constructivist premises acknowledge that these factors will also influence students' conceptual growth. These other considerations are integrated into the text of each chapter. Teaching and learning issues regularly referred to in this book, with reference to chapters where they are particularly emphasised, are:
- integration in general (3, 11, 12), especially science-literacy connections (2, 3, 6, 9, 12) but also with mathematics (2, 3); the symbiotic relationship between technology and science is a major focus at times (3, 6, 12)
- assessment for learning (3, 4, 6, 7)
- teaching using analogies (5)
- affective teaching (7, 8, 9)
- socioscientific issues, including sustainability (2, 6, 8, 9, 11, 12)
- equipment, consumables and safety (9)
- culturally appropriate teaching and learning, including different ways of knowing about our world (6, 10, 12)
- learning in informal settings such as science centres (3) and outdoors (12)

There are many other pedagogical issues raised in various chapters, such as the effective use of small groups, teaching for creative outcomes, and using appropriate and engaging contexts for learning with young students.

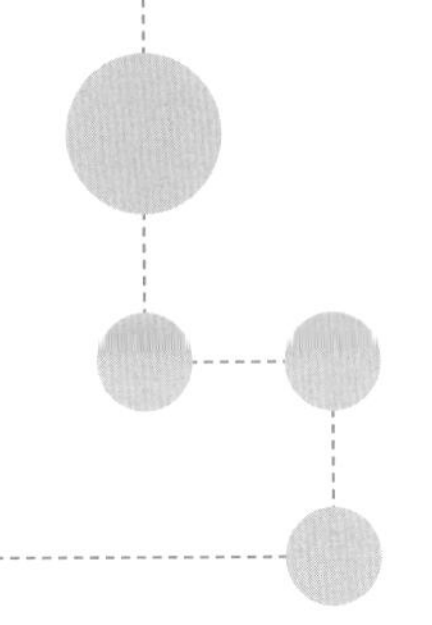

All chapters have a range of analytical, hands-on and/or reflective *activities* to encourage you, the reader, to engage with the content and the overall themes as outlined in the Preface and in Chapter 1. For this to occur, you will need to interact with the written text and think about how you are learning; that is, be metacognitive. You could do the activities by yourself, but in many instances it would be beneficial if you could share your thoughts with another teacher or a small group. Such an approach is consistent with you being an active learner and a member of a science learning community. More effective learning often occurs when you interact with others and are required to articulate your ideas and beliefs about science and learning or teaching science. Learning with others also fulfils some of the social constructivist conditions for learning which, in part, underpin the theory upon which this book is based.

Several of the activities can, of course, be done mentally or simply in discussion with others, but you are urged to keep a written and running record of your interactions with the book's contents and its reader activities. There are studies that strongly suggest that we learn far more effectively when we think and write about what and how we are learning, and diaries and portfolios have been found to be most helpful in this regard.

As authors we invite you to interact with this book as a 'deep' rather than a 'surface' learner. Deep learners are intrinsically motivated and seek meaning in the ideas with which they interact. We hope you accept the invitation!

Resources Guide

FOR THE STUDENT

As you read this text you will find numerous features to assist you with your learning.

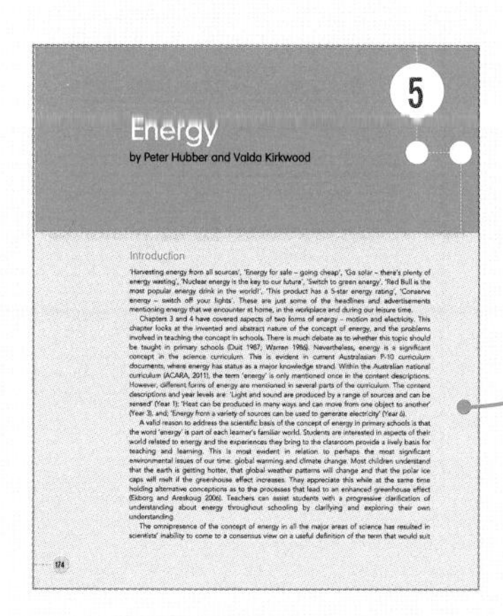

Each chapter opens with an **introduction**, giving you a clear idea of what the chapter will cover.

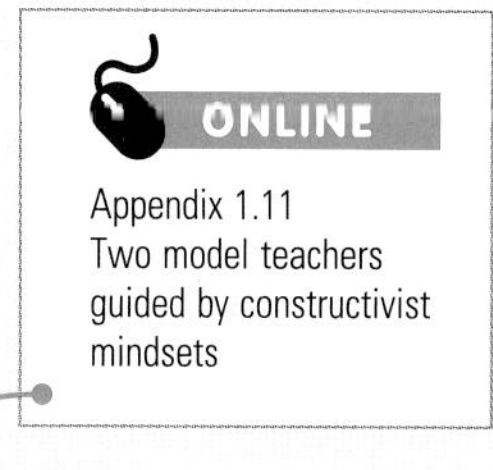

When you see this icon, go to the student companion website at **http://login.cengage.com** to view the relevant online Appendix.

Icons in the margins highlight the discussion of key and emerging pedagogical ideas that span across chapters such as:

- **ICT** the use of ICT
- **NOS** the nature of science
- **R** focus on representation
- **A** assessment of learning
- **M&E** motivation and engagement
- **SSI** socioscientific issues; and
- **S&C** science and culture.

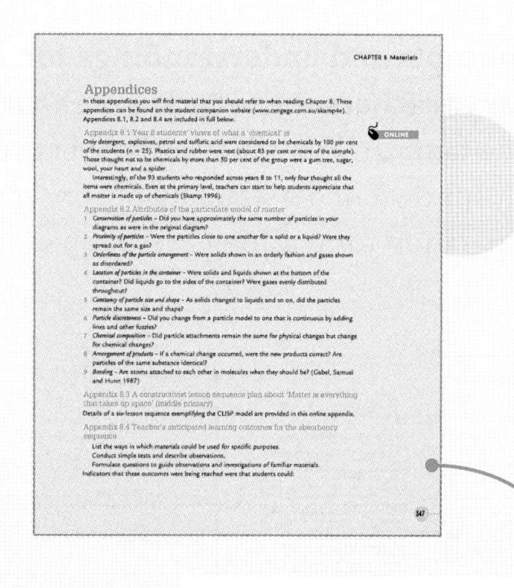

These online Appendices contain material that you should refer to when reading the chapters. A description of the online Appendices is provided at the end of each chapter.

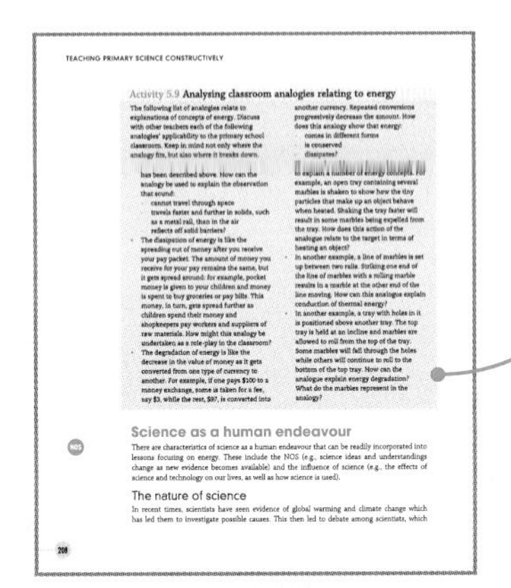

Activity boxes contain practical activities that encourage you to interact with the chapters and think about how you are learning.

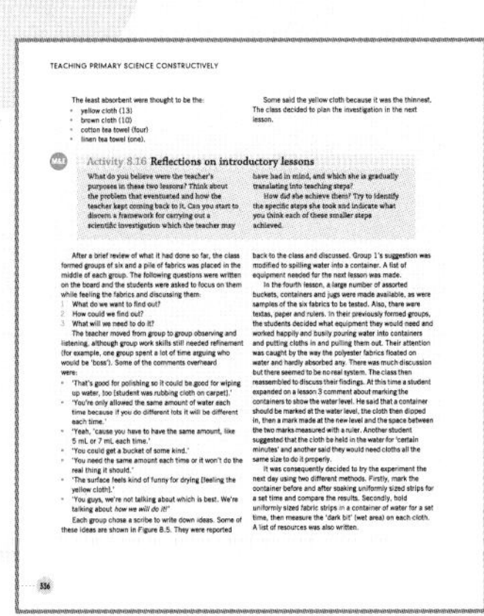

Case studies illustrate how teachers have applied constructivist learning sequences in their classrooms.

Chapters 3 to 12 include checklists of the key **concepts and understandings for primary teachers** with which you should become familiar. You can use these lists to audit your own understanding of concepts in preparation for teaching these topics.

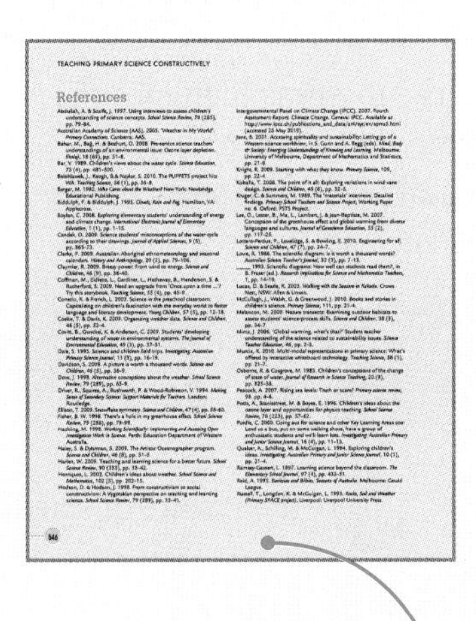

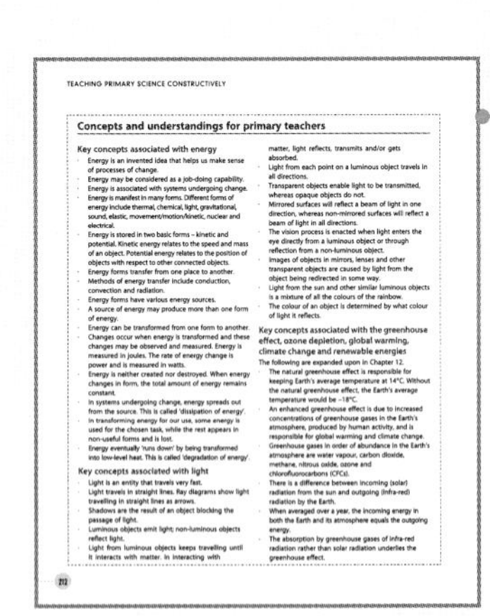

Each chapter includes a comprehensive list of references that you can use for further reading or research.

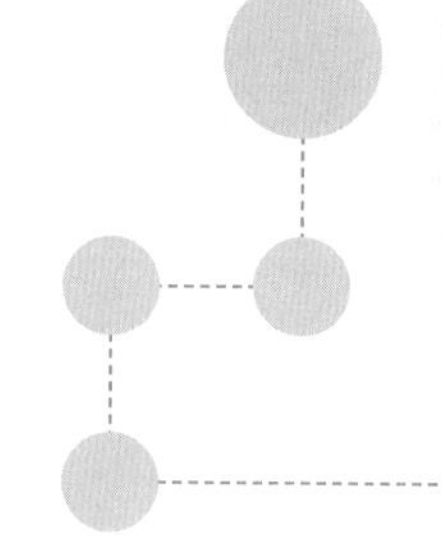

New copies of this book come with an access code that gives you a 12-month subscription to **Search me! science education**. Fast and convenient, this resource is updated daily and provides you with 24-hour access to full-text articles from hundreds of scholarly and popular journals, e-books and newspapers, including *The Australian* and *The New York Times*.

Use the **Search me! science education keywords** at the end of each chapter to explore topics further and find current references. These terms will get you started; then try your own search terms to expand your knowledge.

Online

Visit **http://login.cengage.com** and use the access code that comes with this book to get a 12-month access to:

- an **ebook** version of this text
- all material on the **student companion website**. There are online appendices, revision tools, information on teaching primary science in New Zealand, useful weblinks and more; and
- **Search me! science education**.

FOR THE INSTRUCTOR

Cengage Learning is pleased to provide you with a selection of resources that have been developed to supplement the fourth edition of *Teaching Primary Science Constructively*. These resources are available on the Instructor's companion website accessible via **http://login.cengage.com**

PowerPoint™ presentations
Chapter-by-chapter PowerPoint presentations cover the main concepts addressed within the text and can be edited to suit your own requirements. Use these slides to enhance your lecture presentations and to reinforce the key principles, or for student handouts.

Artwork from the text
The figures and diagrams from the text are available for you to use in your own PowerPoint presentations and student materials.

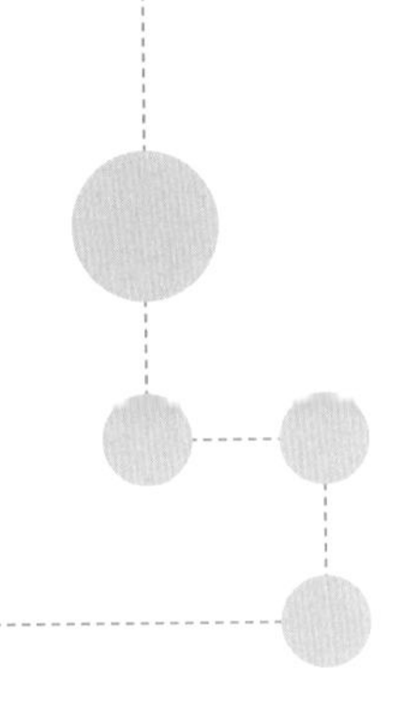

About the authors

Editor

KEITH SKAMP (Chapters 1, 8, 9 and 10) is an Adjunct Professor in the School of Education at Southern Cross University. He has held visiting scholar and professorial positions at Teachers College, Columbia University, New York; Durham University and the University of Liverpool, England; Universities of British Columbia and Lethbridge, Canada; and Flinders University, South Australia. Keith has been a lecturer in undergraduate and graduate units in science education, environmental education and research methodology for many years, and has led and been involved in professional and curriculum development initiatives in primary science at state, national and international levels. His research and consultancy in science and environmental education has included working with the Australian Academy of Science's *Primary Connections* initiative, as well as for State and federal governments, and has resulted in numerous refereed journal articles and other publications; currently his research focuses on students' responses to global warming. Keith has received distinguished awards for university teaching and service to professional science education associations.

Contributors

GRAHAM CRAWFORD (Chapters 11 and 12) is an Adjunct Senior Research Fellow at the University of South Australia, where he taught primary science education and geology for over 20 years. He has coordinated numerous projects in primary science, technology and mathematics in Australia and South-East Asia and is a former editor of *Investigating: Australian Junior and Primary Science Journal*. Graham is co-author of four books on classroom-level assessment in primary science as well as journal articles on curriculum and teacher development. His research interests include the development of key earth science concepts by adult and child learners, and teachers' understandings of how their students learn best.

LINDA DARBY (Chapter 3) has been involved in primary and secondary science teacher education since 2002, at the University of Ballarat and Deakin University, and since 2008 as Lecturer in Science Education at RMIT University where she has developed and coordinated a number of primary science education courses. Her research interests centre on pedagogical theory and teacher identity. Her research is school- and classroom-based and includes involvement in the evaluation of the Department of Education and Early Childhood Development's Teaching and Learning Coach (TALC) initiative and the Improving Middle Years Maths and Science Project, as well as research into students' perceptions of effective pedagogy in science, teachers' experience of the culture of teaching science, and the influence of teaching out-of-field on professional identity.

ROSEMARY FEASEY (Chapter 2) is a leading expert in primary science. She is currently a freelance consultant working in the UK and elsewhere. Rosemary was the first person from a primary science background to become Chair of the Association for Science Education (ASE) in its 100-year history. She is passionate about primary science, has always been proactive in primary science education, and has worked to change the nature of primary science and

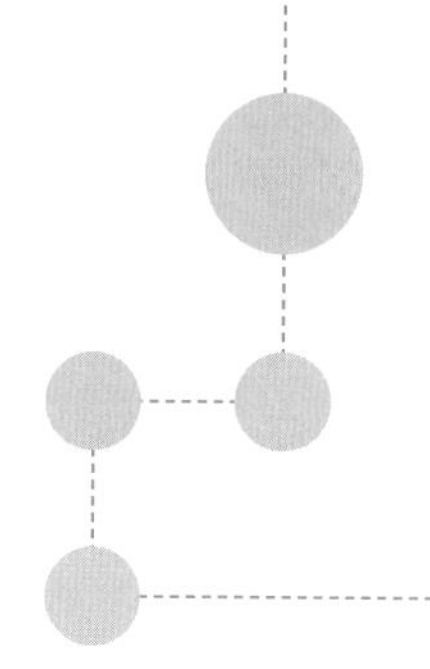

develop the quality of science education. Rosemary has produced many publications on primary science and been involved in programs for the BBC.

FILOCHA DE MELO (Chapter 7) spent more than 30 years in classrooms as a science, mathematics and horticulture teacher in Western Australia and Victoria, before joining the tertiary sector. She has a passion to support student learning in science and her research expertise is in qualitative research. Her record of helping schools in the successful implementation and evaluation of innovative projects spans many State and federal initiatives. She is currently working at Deakin University and is involved in research on the role of representation in learning science.

PETER HUBBER (Chapters 4 and 5) spent 27 years in the classroom as a science and mathematics teacher before arriving at Deakin University in 2000. He has a strong record in professional development, working with teachers and schools in local, State and federal initiatives. Such initiatives include several Australian Government Quality Teaching Program (AGQTP) projects, Australian School Innovation in Science, Technology and Mathematics (ASISTM) projects, Middle Years Pedagogy Research and Development (MYPRAD) projects, and the Science and Technology Education Leveraging Relevance (STELR) program. He regularly presents at national and international conferences and has several publications in the area of student learning in science. His current research is an Australian Research Council (ARC) funded project related to the role of representation in learning science.

BEVERLEY JANE (Chapters 6 and 11) began teaching as a biology teacher in Victorian schools and Malaysia. After 20 years of university teaching in science and technology education with preservice primary and early childhood teachers, Beverley is now an ordained deacon in the Anglican Church. Her passion for education continues as she preaches to multigenerational congregations and runs science activities in informal settings, such as Kids Club. One of the books she is currently writing focuses on intergenerational relationships and how grandparents can foster their grandchildren's sense of wonder and understanding of science.

WENDY JOBLING (Chapter 6) taught in Victorian State schools for many years and joined Deakin University in 2006. Since then, Wendy has taught undergraduate and postgraduate teacher education students in science and design, creativity and technology units. During 2008–09, Wendy was involved in a Malaysian project that focused on the development and implementation of a teacher education degree. In addition to her teaching commitments, she has also recently been involved in collaborative research related to effective teaching and learning, including the following two projects: Association of Independent Schools Victoria Project: The Development of Thinking Skills through ICT; and Promoting Effective Small Schools' Science: Maximising Student and Teacher Learning.

VALDA KIRKWOOD (Chapters 4 and 5) taught science and mathematics in New Zealand secondary schools from the late 1960s to the mid 1980s before moving into the tertiary system as a Project Officer for the Learning in Science Project (Energy) at the University of Waikato. Positions at the Canberra College of Advanced Education, Waikato University, the

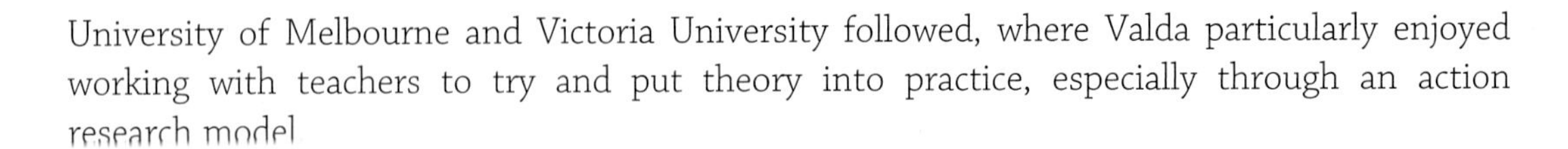
University of Melbourne and Victoria University followed, where Valda particularly enjoyed working with teachers to try and put theory into practice, especially through an action research model.

SUZANNE PETERSON (Chapters 3 and 7) was an exemplary teacher of primary science. Through her interaction with colleagues and records of her science pedagogy she inspired others. Her 'story', as partly recorded in sections of this book, will encourage a deeper understanding of the power of primary school science to educate. Her untimely death in 2009 was regarded, by those who knew her, as a tragic loss to education.

CHRISTINE PRESTON (Chapters 11 and 12) is a Lecturer in Early Childhood and Primary Science Education at the University of Sydney. She concurrently teaches kindergarten science at Abbotsleigh Junior School in Sydney. She taught secondary and K-12 science before specialising in primary science. Christine regularly presents at state and national conferences for science teachers and teacher educators in the areas of educational research and science teaching and learning. Her current research interests are primary children's interpretations of science diagrams, teaching science using toys and educating children about the marine environment.

RUSSELL TYTLER (Chapters 3 and 7) is Professor of Science Education at Deakin University. He has been involved over many years in system-wide curriculum development and professional development initiatives. He was the principal researcher for the School Innovation in Science project, which developed a framework for describing effective science teaching, and a strategy for supporting school and teacher change. He has researched and written extensively on student learning and reasoning in science, science investigations, learning and literacy in science, pedagogy, and school and teacher change. With Suzanne Peterson, he conducted a seven-year longitudinal study of children's learning in science. With Peter Hubber and Filocha De Melo, he has been involved in the ARC project The Role of Representation in Learning Science (RiLS). Russell has held visiting professor positions in Europe and Asia and has been involved in a number of international collaborations.

1

Teaching primary science constructively

by Keith Skamp

Introduction

Primary students enjoy science when it is student-centred, their 'voices' are heard, and there is a focus on investigation (Goodrum, Hackling and Rennie 2001; Logan and Skamp 2008). This interest in science starts at a very young age. Even preschool children's play can be focused on conceptual connections and kindergarten children can hold sophisticated scientific understandings (Fleer 2009; Fleer and Robbins 2003a). If this is the case, then teachers need to retain that interest by teaching in ways that assist effective learning.

We should teach science in a way that is consistent with what we know about how students learn science. To not do so would be a disservice to learners. There is not complete consensus about how primary students learn science but there is a growing body of research that indicates they learn by constructing ideas and developing ('constructing') competencies. What does this mean? What are the teaching implications?

When we talk about students learning science by constructing ideas and competencies we are referring, in broad terms, to the learning theory called 'constructivism'. It relates to students' 'sense' or 'meaning making'; that is, how they make sense or meaning of their world and how it works. This theory has been interpreted in specific ways in relation to teaching science. Numerous studies have investigated the effectiveness of teaching science based on constructivist premises. This book draws on this research to support the approaches advocated.

A deliberate decision has been made to cite these research studies in order to emphasise that the way in which students learn primary science is informed by an expanding international research base. This is important, as many teachers are not aware of what the research and professional literature say about the effective teaching, and learning, of science (Duit, Treagust and Widodo 2008). Teachers, teacher educators and researchers comprise the science education research community who have contributed to this research base. You, too, can become part of this community by testing out the ideas in this book with your students and reporting on them to colleagues.

Structure of the chapter

This chapter provides the background to, and the research bases for, learning and teaching primary science from a constructivist perspective. The generic pedagogical implications of constructivism are outlined and later chapters provide specific and concrete examples related to inquiry in primary science and its content areas.

Your navigation through this chapter will be assisted by an appreciation of how it is structured. After initially asking you to reflect on where you are starting from, the chapter gives an overview of various reasons why students should experience science at the primary level (section a) and why the emphases have changed in recent years (c). Between these two sections you are required to stop and think about your views on this subject called 'science' – what it is and how it works; this section (b), as will be explained, is best introduced at an early stage. The chapter's three major sections begin with how students *learn* science constructively (d and e), which leads into the *teaching* implications (f) that are consistent with the research on how students learn science. The concluding sections examine science curriculum/syllabus/framework requirements and related pedagogical issues, such as assessment, through a constructivist lens (g and h); they also compare the learning opportunities advocated in this book with accepted criteria for effective primary science practice (i). Finally, a fresh slant is given to the embracing and enhancement of your current science background (j). The following is a synopsis of these sections:

a *Changing emphases in learning primary science* – An interpretation of scientific literacy as the overall aim of teaching science is presented. The interdependence of conceptual understanding and 'thinking and working scientifically' is introduced.

b *The nature of science: how it works* – It is explained why teachers need to address this question before progressing further. Several tasks are suggested, leading to what it's generally agreed characterises science.

c *Why the change in emphasis? Alternative conceptions* – Students, in general, complete schooling with many 'alternative', rather than scientific, conceptions about how the world works. This research finding was the impetus for pedagogical changes in primary science education. The characteristics of 'alternative conceptions' are identified.

d *Learning science* – The implications of the 'alternative conceptions' research led to a revised view of how students learn science. Your own views on what it means to 'learn science', and then to 'teach science', are now sought. How constructivism became a focus for learning and teaching primary science is then described.

e *Constructivist views on how students learn science* – Personal and social constructivist perspectives on learning are compared and contrasted. The importance of context in understanding how students learn is emphasised by looking at learning through a sociocultural lens. Recognition that the perceptual, associative, expressive, aesthetic, emotional and/or affective nature of students' thinking may be as, or more, significant in appreciating how students learn science, rather than merely focusing on more formal and logical thinking patterns (personal constructivism has centred on the latter dimension), has resulted in revised views of how the learning of science occurs. These revised views and a new emphasis on developing ('constructing') competencies in the discursive practices of science are explained.

f *Teaching science constructively* – Some of these constructivist perspectives have resulted in models/schemata of how lessons, in general, could be sequenced. The major phases within these sequences are overviewed and then described in detail. Other learning perspectives, such as learning through representation, have suggested different emphases in constructivist sequences. These are also outlined and then expanded upon. Finally, research evidence evaluating the effectiveness of teaching from these different perspectives is summarised. It indicates that students learn more effectively when constructivist, rather than traditional, pedagogy is used.

g *Implementing science curriculum requirements with a constructivist mindset* – An interpretation of learning outcomes, assessment for learning, curriculum integration, and information and communication technology (ICT), using a constructivist mindset, is presented.
h *Content areas for constructivist learning* – An overview is given of how chapters 2 to 12 provide substantive coverage of 'science as understandings' and 'science as inquiry' in primary science curricula. 'Thinking and working scientifically' is especially developed in Chapter 2. The notion of 'science as a human endeavour' is discussed. This is related to all the content areas, such as energy and living things (chapters 3 to 12), and references Indigenous and cultural knowledge, careers in science and related areas, and socioscientific issues.
i *Conditions for effective primary science: an overview* – The pedagogical directions taken in this book to advance science learning opportunities for primary students are compared with generally accepted research-based characteristics of effective primary science practice.
j *Your own science background* – Various ways to address this issue, a concern for some primary teachers, are suggested.

Before progressing further, it is time to challenge you with an issue that is integral to this book. Talking this matter over with colleagues will help you to self-assess your current mindset about helping students learn when science is the focus.

Activity 1.1 Teaching science: getting you thinking

What do you think of the idea, held by a preservice teacher before she started her practicum, that 'a teacher should continue to teach a scientific concept until all the children show that they understand it' (Russell and Martin 2007, p. 1163)?

What do some of your colleagues think? Do you agree with them? Why?

You may find it valuable to revisit this activity at a later time and reflect on your initial responses.

Changing emphases in learning primary science

Many reasons have been advocated as to why science should be in the primary curriculum. During the 1970s and 1980s there was a strong push for teachers to emphasise the processes of science, such as observing, inferring and predicting, and to ensure students had 'hands-on' experiences. 'Discovery learning' was a catchcry with an almost implicit belief that students would learn simply by handling materials. Some primary teachers still adhere to these views and nominate their peers as exemplary teachers of science because children in their classes are manipulating materials, just as Goodrum (1987) reported many years ago.

Although effective primary science must include hands-on experiences as well as students using science processes, numerous research studies have led to the conclusion that conceptual understanding must also be a fundamental emphasis in science learning. Basic science concepts and understandings, not just about well-established content areas such as 'living things' but also 'nature of science' concepts such as 'evidence', help us to make sense of our world and they are relevant to learners of all ages. If primary students start to appreciate some of the major scientific ideas that help us to understand the objects, events and phenomena that comprise our day-to-day lives, as well as start to develop competencies in making informed decisions about everyday issues that involve scientific ideas, then they will be taking a step towards scientific literacy.

Scientific literacy is the key goal of science education in many countries (Murcia 2009): your students are to progress towards becoming scientifically literate people. The interpretation of scientific literacy is an important consideration when thinking about why we teach science (Rennie 2005). Although interpretations of scientific literacy vary (Murcia 2009; Tytler 2007), that which evolves in Chapter 2 – namely, thinking and working scientifically – has five components. Each of these components, which may be compared with Rennie's depiction (see Figure 1.1), is consistent with the structure of the remaining chapters in this book.

Thinking and working scientifically requires, in part, a focus on understanding science ideas in order to make sense of our world, as well as an appreciation of the way in which science derives those ideas and the forms of evidence it accrues to substantiate them. Part of understanding this latter feature of scientific literacy is being able to apply science processes, skills and attitudes (such as a respect for empirical evidence), while at the same time working with the ideas that help us make more sense of our world; this is emphasised by the three prongs in Figure 2.1 (see p. 65).

The teaching of scientific inquiry and its component science processes and dispositions, therefore, should not be separated from students' learning about science concepts and understandings. 'Learning concepts and understandings' and 'learning how to inquire scientifically' are interdependent and teaching science needs to integrate them for the most effective learning (Pratt 2009; Traianou 2006; Tytler and Peterson 2003, 2005).

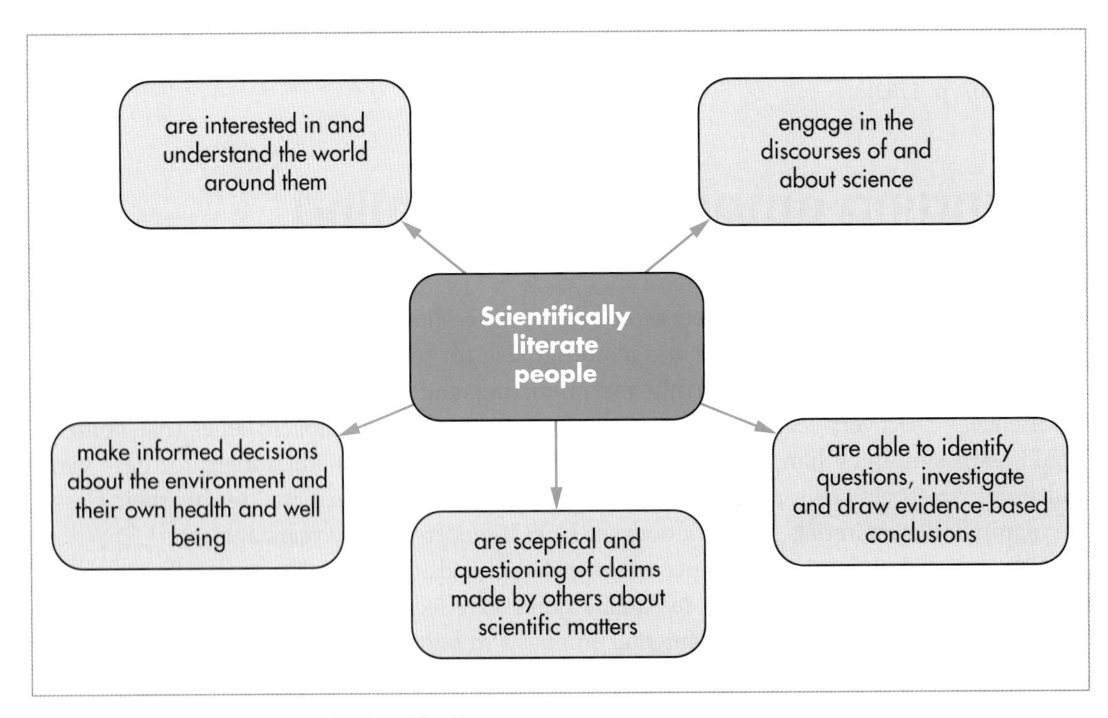

FIGURE 1.1 A definition of scientific literacy

Source: Rennie 2005, p. 11

Scientific literacy, as depicted in Figure 1.1, also refers to students making informed decisions about the environment and their health and wellbeing. These socioscientific issues have a place in the primary curriculum (Dolan, Nichols and Zeidler 2009; Tytler 2007; Zeidler and Nichols 2009) as responsible citizenship requires participation in resolving such issues; further, they add interest and relevance to the study of science. To effectively engage with socioscientific issues, students will need to draw upon all the other components of scientific literacy.

The pursuit of scientific literacy as a goal is the antithesis of focusing primary school science purely on the subject matter. School science needs to be seen as making a significant contribution to the overall general educational development of students and, in particular, helping them develop 'a sense of personal agency in engaging with science' (adapted from Tytler 2007, pp. 12, 20). This is a marked shift from earlier decades.

The nature of science: how it works

Implicit in the above discussion about scientific literacy is an understanding of what is 'science' and 'how it works'. A growing appreciation of the characteristics of science can assist students in constructing science ideas and competencies; it can also help them engage more effectively in everyday socioscientific issues.

Can primary students understand aspects of the nature of science (NOS)? If this is seen as a concomitant of developing conceptual understanding and/or thinking and working scientifically, then, provided the attributes of science are made explicit, primary students can learn and apply them (Lui and Lederman 2007). The NOS is now an integral component of many primary science curricula (e.g., Australian Curriculum, Assessment and Reporting Authority 2010). To explicitly teach the NOS you must have some understanding of it yourself.

'What is the nature of science?' is a 'big' question. We need to stop and consider it, not just for the above reasons, but because your views on the NOS will influence your day-to-day teaching and learning decisions, even down to the words you use when you are talking with your students (Park and Lee 2009). These decisions may impact on the interests and attitudes of your students towards science and their agency while involved in it. Only one text activity on this subject (see Activity 1.2) is included here so that the coherency of this chapter is not impeded, but this is not to devalue the significance of this topic. It underpins notions of scientific literacy and what is included in science curricula; additional tasks requiring you to reflect on the NOS are in the online Appendix 1.1.

Appendix 1.1
What do you think science is?

Activity 1.2 The nature of science: your current views

Your current conceptions of the NOS should become clearer as you respond to the following:

- What characteristics of science (how it 'works' and how scientists 'work' – the NOS derives from both) do you think should be evident as you teach your primary class? Try to identify several characteristics.
- Justify your responses to yourself and then to other colleagues. Add further characteristics if convinced by a colleague.
- Think back to teachers of science you have had. What characteristics of science did they convey either implicitly or explicitly? How did they do this?

- As a challenge, think about the following: From what you know about climate change, what further characteristics of science and how scientists work would you add to the above?

Your conceptions of the NOS will have been formed from your previous experiences with school science, science in the media and many other sources. As you engage with the text and activities in this book, as well as having further professional development experiences, your conceptions may change further. Hands-on and other activities in this book will ask you what aspects of the NOS were, or could have been, exemplified in the activities – this will remind you to keep revisiting this issue. At this stage, recognising your existing views and discussing them with colleagues may be a starting point for later refinement of your thinking.

You need to be clear about the key aspects of how science works and the ways in which to exemplify these in your teaching. However, you should appreciate that these can vary in meaning. The nature of scientific inquiry, for example, is not necessarily defined by one 'scientific method' (Williams 2007). Just stop and think: Do, for instance, astronomers, geologists, chemists and biologists go about investigating problems in the same way? Activity 1.2 and other tools in the online Appendix 1.1 aim to challenge or extend your thinking about the NOS so that you are clearer about the subject of 'science' that you are teaching. You will need to revisit this appendix when you attempt later activities.

Appendix 1.1 What do you think science is?

Seven 'non-contentious' aspects/characteristics of science suitable for primary students to learn have been proposed (Lui and Lederman 2007, p. 1284; see the online Appendix 1.1). There are many ways to help students learn about these characteristics apart from 'telling' them. Encouraging students to reflect on why they did various things during an investigation, as suggested in Chapter 2, could be a way forward. This is necessary because students do not learn about the NOS by simply engaging in a scientific investigation (Appleton 2007; also see the 'Science and literacy' section of this chapter on p. 38). Helping students to appreciate how models are used in science to help test ideas is another example. Models can be mental (internal) and physical (external) and may be generated by the teacher or students (Jonassen 2008; Treagust 2007), and they resemble, in some way, whatever is being studied (e.g., water cycle, causes of day and night).

Activity 1.3 encourages you to consolidate your thinking to this point about what school science might encompass. It refers to a categorisation that distinguishes various emphases in primary science. You may find it helpful.

Activity 1.3 Learning science, doing science and learning about science

Some writers have drawn the distinction between 'learning science', 'doing science' and 'learning about science' (Harlen 1998).

- What do you think could be the differences between these three ways of thinking about primary school science?

- Reference has been made to 'hands-on' science. We can also talk about 'hearts-on' and 'heads-on' science. How would you interpret these approaches to school science? In what ways might they relate to 'learning science', 'doing science' and 'learning about science'? How might they be connected to how students learn to become scientifically literate?

Discuss these broad ways of looking at school science with colleagues. Compare your thinking with the online Appendix 1.2 (Skamp 2007).

Appendix 1.2 Learning science, doing science and learning about science

Why the change in emphasis? Alternative conceptions

Reflections on the nature of primary science education and subsequent changes predominantly came about because research studies in several countries found that, despite 10 or more years of formal schooling, many students were not acquiring more scientific ideas about how to interpret their world. Rather, at the end of secondary school, these students still held ideas or conceptions that were not in accord with the way scientists currently understand our world.

These scientifically incorrect conceptions are usually called 'alternative conceptions'. This term has been selected because it not only 'refers to experience-based explanations constructed by the learner to make a range of natural phenomena and objects intelligible, but it also confers intellectual respect on the learner who holds those ideas' (Wandersee, Mintzes and Novak 1995, p. 176). It is the most accepted description, although it has its limitations (Fleer 1999; Fleer and Robbins 2003a; Leach and Scott 2002). Other terms encountered in this book include 'misconceptions' as well as 'everyday', 'intuitive' and sometimes 'naive' ideas and 'preconceptions';[1] the last mentioned usually signifies the ideas students hold before commencing a topic. When these various terms are used, their meaning will be clear from the text.

Appendix 1.12 Another view about 'alternative conceptions'

Some researchers (e.g., Novak 2002; Vosniadou 2008) distinguish between students' conceptions and their mental or cognitive structure(s) and/or framework(s). This is because they believe that students have coherent but not necessarily scientific ways of looking at the world that could account for many of their alternative conceptions. These 'ways of looking at the world', it is hypothesised, are underpinned by the manner in which various ideas are linked in each learner's mind (called their cognitive structure(s), framework(s) and so on). An example is how numerous people believe that movement is always associated with an applied force in the direction of movement (the so-called 'impetus' framework). This *non-scientific* framework has been useful in explaining students' alternative conceptions about the directions of forces and related phenomena. Not all researchers hold this view about students' cognitive structures, though, with some, such as diSessa (2008), believing that students' knowledge is more like 'knowledge in pieces' in their minds, possibly layered and more contextually based. Some recent research interpreting students' science learning with an emphasis on affective and similar traits, rather than rational thinking processes, orients to this latter theory (Tytler 2007; Tytler and Prain 2010)

Numerous research studies have reported students' alternative conceptions about most of the concepts and understandings found in the content strands of primary science curriculums. From these studies, the general characteristics of students' alternative conceptions have been identified (Bell, Cowie and Jones 2009; Duit 2006; Tytler 2002a; Wandersee, Mintzes and Novak 1995). These are listed below with little elaboration, as they are expanded upon later in this chapter and exemplified in chapters 3 to 12.

General characteristics of students' alternative conceptions

Alternative conceptions are held by all learners. Students' heads are not empty vessels; rather, students come to science lessons with a diverse range of everyday or alternative ideas. These alternative ideas:

- cut across age, ability, gender and cultural boundaries; that is, they have often been found to be similar, irrespective of these factors
- are influenced by direct everyday experiences, including direct observation and perception, peer culture, language, teachers' explanations and resource materials
- are context- or domain-specific; that is, students use different ideas to explain events that occur in different situations even though the events have the same scientific explanations
- are held by students without them being aware that they are the ideas that they use to explain how their world works; in other words, they are implicit and tacit to their thinking and hence students may, for example, not appreciate that they can be tested against the available evidence
- can be tenaciously held and resistant to change by conventional teaching strategies; students may
  - reject any new ideas
  - simply add new ideas encountered in science lessons to their existing ideas; that is, still retain their existing ideas
  - modify their existing ideas, but not necessarily in line with a more scientific interpretation of phenomena or events,
- which means that the outcomes of formal science learning can therefore lead to a range of unintended conceptual learning outcomes
- are not automatically replaced by the accepted scientific idea when students are confronted with conflicting evidence – students' ideas may change in a range of ways, as just mentioned
- may be more difficult to change when there is a greater qualitative difference between the everyday view and the scientific view. This difference has been labelled the 'learning demand' of a scientific idea (Scott, Asoko and Leach 2007).

These attributes of students' ideas need to be qualified. Although, as stated, students' ideas can be context-dependent, this list may give the impression that students' alternative conceptions are stable and consistent. In many instances this appears to be the case; however, there is evidence that younger students may acquire knowledge in a piecemeal fashion as they develop their conceptions of their world (Tytler 2007; Tytler and Prain 2010). These general attributes, therefore, are only a guide to thinking about alternative conceptions.

From where do students get these alternative conceptions? Many derive them from their everyday experience of the world – for example, the idea that objects only move because there are forces acting on them in the direction of movement seems like commonsense, but it is not always the case. Some alternative conceptions are picked up from 'the general misinformation in society' (peers, family, the media) and 'much of this "lay science" may be wrong or misunderstood'. Teachers themselves also can be a source of alternative conceptions (Taber 2001, p. 163).

Knowledge of students' alternative conceptions is critical in determining teaching approaches. Summaries of primary students' alternative conceptions are in all the conceptually oriented sections of this book (chapters 3 to 12). Students also have existing beliefs about 'how science works' (that is, the NOS), which are referred to in some chapters, and these views can influence their learning of other science ideas (e.g., about whether 'heat' is a substance or a process).

Activity 1.4 Alternative conceptions: How could this be?

In what ways can you identify with these findings about alternative conceptions? Which characteristics of alternative conceptions make sense to you? Why?

How could so many students go through school and still hold, for example, the following alternative conceptions about basic everyday objects and phenomena?

- A tree is not a plant and a spider is not an animal.
- The moon appears to change its shape because the Earth casts a shadow on it.
- Objects always have a force acting on them in the direction in which the objects are moving.
- When substances burn they always lose weight.
- A pebble is not a rock.
- Plants receive food through their roots.
- Heat is a substance.
- A light bulb in an electric circuit works because the current flows from both ends of the battery and clashes in the bulb.

Why do you think some students hold on to these ideas so tenaciously?

You may hold some of these alternative conceptions. This is not uncommon. The main issue for you as a teacher is to be aware that this could be the case. When teaching students, you need to clarify your own conceptions about the everyday objects and phenomena associated with the topic being taught. Each chapter of this book has text and activities devoted to analysing your own conceptions about the science content areas in your science curriculum. These should assist you in clarifying your science ideas as you plan for your teaching.

There are many factors involved in why students may or may not change their ideas (Taber 2001). For example, a young child may have a particular experience that markedly impacts on their interpretation of a science concept, such as their familiarity with decaying leaves and hence what they think 'decomposition' entails. Several longitudinal studies following the same students over many years (e.g., Hellden 2005; Tytler, Prain and Peterson 2007) are revealing more about the nature of young students' reasoning and thinking and how their ideas/conceptions are influenced in complex ways by, for example, associative, perceptual and analogical reasoning. Several chapters provide examples of this (e.g., Chapter 3).

Progression of conceptions with age

Several researchers have argued that students' alternative conceptions about some phenomena (e.g., gravity and chemical change), show a progression with age – see, for example, the relevant chapters in Vosniadou (2008). Examples of such progressions are given later in this book. Where progressions appear to exist, they tend to apply to the overall cohorts of students of varying ages and not to individual students. Several studies (e.g., Hellden 2005; Tytler 2002a; Tytler and Prain 2010) have shown that school students across a wide age range (and adults) can hold ideas of varying degrees of scientific sophistication and acceptability. Conceptual progressions with age are therefore contested by some (see Tytler and Prain 2010). We cannot make assumptions about the conceptual progression of individual learners; therefore, summaries of the progression of conceptual understandings across the school years must be used with care.

Learning science

Determining the alternative conceptions that learners (including ourselves) often hold is critical to further learning in science, and to the teaching of science. Along with obtaining student interest, it is usually the initial step that teachers should take in planning and implementing any science lesson sequence. However, what are teachers to do once they have determined – at least to some extent – the ideas that their learners hold? This question does not have definitive answers, but there are many research-based recommendations on how to proceed.

Your views

The research findings about students' alternative conceptions caused science educators (teachers and researchers) to rethink what it means to 'learn science'. Before proceeding, you need to think about your own views on what it means for primary students to learn science. Why is there a need to do this?

Just as many students are generally unaware of their own ideas about how the world works and often do not change these ideas while at school, some preservice teachers are unaware of their existing beliefs about 'learning science' (and 'teaching science') and, even after completing their degrees, have not changed their views about learning and teaching science (Russell and Martin 2007; Skamp 2001). Also, there are many practising teachers who do not have an explicit view on learning science (Duit, Treagust and Widodo 2008).

A starting point for changing our ideas is to first identify them. Activity 1.5 builds on the previous activities by asking you to identify your current thinking about what it means to *learn* science effectively.

Activity 1.5 Your current beliefs about 'learning science'

Develop a personal mind map around the bubble '*Learning* science means'.

- How does your mind map fit with your own experiences as a science learner? What are the positives and negatives?
- If you have taught science, how consistent is your mind map with how you think about learning when you are teaching?
- If not already included, then for each connection you made to your '*Learning*

science means' bubble, try to identify one or more roles for the learner. Write these in a different colour next to the connection.

- Refer back to the ideas concerning 'learning science', 'doing science' and 'learning about science', as well as 'heads-on', 'hands-on' and 'hearts-on' science: Does your mind map relate in any way to these notions?

Now try to complete the following statement: 'Currently, my explicit view of learning science is …'

If possible, share your statement with several others. What views do they hold? Do you want to expand or otherwise alter your statement? Why?

What do you think might be going on inside your own and young students' heads when each of you say you are learning science?

It is appropriate to now briefly divert from this theme to ask you to identify your current thinking about what it means to '*teach* science' (see Activity 1.6). When this chapter turns to the implications for teaching science of the findings about students' alternative conceptions, then you will be able to reflect back on your thoughts. This may assist you in modifying your thinking.

Activity 1.6 'Teaching science': Where are you at?

Having considered what science is (Activity 1.2) and what school science could involve (Activity 1.3), as well as having reflected on the meaning of 'learning science' (Activity 1.5), now try to encapsulate what you believe 'teaching science' to primary students is all about. You might like to list the various teaching roles associated with what you think 'teaching science' encompasses. Why would you take on those roles? Keep this list.

At a later point in this chapter, various teaching roles associated with particular aspects of science teaching will be introduced (e.g., Harlen 2009). You should compare them with your list.

You could consider activities 1.5 and 1.6 as assisting you to prepare a statement on your overall learning and teaching philosophy. You often need to do this when preparing for interviews for teaching positions, and teachers may be asked to express such views if seeking a promotion. The difference here is that we are focusing on science.

Appendix 1.3 Teachers' conceptions of 'learning science', 'teaching science' and 'science'

Questions about what 'science', 'learning science' and 'teaching science' mean have been asked of teachers in several countries, including Australia, as reported in the online Appendix 1.3. After completing activities 1.2, 1.5 and 1.6, as well as some of the tasks in the online Appendix 1.1, compare your responses with these findings. Consequently, reflect further on your perceptions of 'learning science' and 'teaching science'.

Your curriculum and this book

After you have attempted the aforementioned activities, you might find it useful to scan your science curriculum to ascertain its take on 'learning science'. It probably will not be stated explicitly but its learning outcomes would be one indication. Some curriculums may actually refer to a particular learning theory or theories, but this is not common.

This book would agree, in general terms, with what most recent primary science curriculums state or imply about what it means to learn science, and the interpretation of scientific literacy as shown in Figure 1.1 (see p. 4). The interpretation that underpins this book is: *students should be refining their ideas in line with those currently accepted by the scientific community so that they can make better sense of the way in which their world works, and also have an increasing understanding of how those ideas are derived and hence some agency in the individual and social decision making that is underpinned by them.* This interpretation regards a significant component of learning science as conceptual development, formation and change towards more scientifically acceptable concepts and understandings; it also embraces learning about how science derives these ideas and applying them in a range of everyday situations. This is consistent with a more conceptual approach to science topics required at the primary level (Jarvis, Pell and McKeon 2003).

The ways in which teachers encourage students to develop, form and possibly change their ideas then become critical issues. Your responses to activities 1.5 and 1.6 may have identified some of these ways. Duit and Treagust (1998, p. 19) state that 'learning science is only successful if learning pathways are designed to lead from certain facets of preinstructional knowledge towards the science perspective'. This view does not preclude a consideration of several ways of knowing about our natural world. It also implies that there are multiple 'pathways'. In this book, the view is taken that the research on how students learn favours a broad constructivist position and that the learning pathways selected need to be consistent with the various interpretations of learning from constructivist and related perspectives.

Constructivist views on how students learn science

The research findings about the characteristics of students' alternative conceptions resulted in what is now called a constructivist view of learning science. This may be distinguished from some constructivist interpretations of learning in other subjects (Gil-Perez et al. 2002). This means that different constructivist approaches may be needed to facilitate learning in science compared to, say, mathematics.

One constructivist view of learning focuses on concept development. Concepts, here, may be considered as 'basic units of knowledge that can be accumulated, gradually refined, and combined to form ever richer cognitive structures', although the notion of cognitive structures may be problematic. Another perspective on learning is that of a process of becoming a member of a 'learning [here, scientific] community'. These two perspectives – one of acquisition, the other of participation – are not a continuum but different dimensions of learning (Stard, in Leach and Scott 2008, p. 649). Learning 'competencies' associated with the discursive practices of science (e.g., ways of representing ideas), as described later in this chapter, have an alignment with the notion of scientific learning communities but are also helping students refine their conceptual understanding. This book acknowledges both positions. These different views of learning also help us to see that learning science may be considered from an individual and a social perspective, and that each offers insights as to how students learn and the subsequent teaching implications.

Personal constructivism

Constructivist learning argues that students construct rather than absorb new ideas. Learning is not the transmission of knowledge from the head of the teacher (or another source, such as a textbook or a web page) to the head of the learner. Learners actively generate meaning from experiences on the basis of existing ideas; each individual, when changing ideas, has to construct their own meaning from their experiences. This view of learning could be termed 'personal constructivism' – change takes place in the learner's head (Leach and Scott 2000). This was the focus of constructivist learning in the 1980s and early 1990s. It is underpinned by a generative model of learning (Osborne and Wittrock 1983, 1985), which has as its fundamental premise that

> people tend to generate perceptions and meanings that are consistent with their prior learning. These perceptions and meanings are something additional both to the stimuli and the learner's existing knowledge. To construct meaning requires effort on the part of the learner and links must be *generated* between stimuli and stored information.
>
> **Osborne and Wittrock, 1985, p. 63; emphasis in original**

This view of learning tends to focus on the role of internal logical and rational thinking processes in prompting conceptual change towards accepted scientific concepts. It is important that students develop understandings about how their world works, but the learning of these concepts needs to be in contexts that are relevant to the learner (Tytler 2007). The learning of abstract context-free scientific propositions has inherent problems associated with it, as it often does not engage learners; it also is not consistent with the growing view that the ways in which students learn about and understand phenomena is context-bound or 'situated'.

Personal constructivist approaches, in summary, emphasise that students' beliefs about their world and how it works are constructed, not received, and that such beliefs strongly influence later learning in science. These approaches tend to focus on individual internal cognitive processes and may undervalue the influence of sociocultural factors – these are discussed next.

Social constructivism

The social and cultural settings within which students learn also impact on the students as they alter their ideas. Learning, therefore, is situated: activities and their contexts,[2] which are provided for students, cannot be separated from the learning associated with them. Interactions with teachers and others are integral to these contexts and to understanding how learning occurs (Fleer and Robbins 2003b). 'An individual's [student's] *understanding* of the [scientific community's] concepts, theories and ideas is [therefore] a dynamic process resulting from acting in situations and negotiating with other members of that ["school classroom science"] community' (Traianou 2006, p. 835, parentheses added). This sociocultural viewpoint places emphasis on students' experiences and the settings in which they take place, rather than on the individual's mental structures. In schools, students will come to understand the ideas of science through their interactions with the community of learners in their class who are also thinking about scientific ideas. The teacher is part of that community, too.

From this perspective, sometimes referred to under the umbrella of 'situated cognition' (Bell, Cowie and Jones 2009; Scott, Asoko and Leach 2007), science 'learning means change from one sociocultural context, usually the everyday science context, to a new, science [often in-school] context' (Duit and Treagust 1998, p. 16). This change, from an everyday culture to a science culture, has implications for the learner. When taking the 'border crossing' (Aikenhead 1996; Hodson 2002; Traianou 2006) to a science culture, the 'cultural changes' that a student may experience would include language use, how meaning is derived from various contexts, and learning in a collaborative setting, because these are the characteristics of a science culture. In other words, constructivist learning has a social and a cultural aspect. Learning science, therefore, is not just about individual students interacting with the material world; it also involves students sharing in the 'practices, conventions, and modes of expression' that embody how science makes advances in knowledge about our world (Scott, Asoko and Leach 2007, p. 41). These could be referred to as the discursive practices of science.

NOS

It needs to be appreciated that science does not work by manipulating known concepts as school science may portray it. Classroom science learning communities need to reflect a more flexible view of how science works (the NOS) and hence what learning science concepts entails. A focus on learning science through multiple representations is one outcome of this view and is expanded upon below.

Related to this social and participative dimension of science learning is the hypothesis that 'all higher order psychological processes and structures (such as science concepts) originate on the social plane', which derives from the work of Vygotsky (1978) (Leach and Scott 2000, p. 44). What this means is that your students encounter science concepts through the '[science] talk' and '[science] writing' of others. Acquiring the social language of (school) science is therefore part of learning science. Social constructivism recognises these aspects of learning and knowledge construction (Duit and Treagust 1998; Leach and Scott 2000). It is 'social' in that learning has social origins, but also because the scientific community advances knowledge through social conventions and contexts.

Social constructivist approaches, in summary, emphasise that science learning:

- moves from social to personal planes
- is mediated by the sociocultural context; that is, the signs and symbols that comprise our environment, especially language
- requires that learners be introduced to the social language and context of science, usually a role for teachers to play
- is a dynamic and transformative process.

It needs to be noted, though, that most teachers would accept that the individual learner still must engage in steps of 'personal sense-making' (Scott, Asoko and Leach 2007, p. 40). There is clearly an overlap between personal and social constructivist views of learning, but with different emphases on, and interpretations of, how learning occurs. You will find research reported in later chapters which has a personal constructivist emphasis, such as students' conceptions about various concepts. However, many of the learning (and consequently, teaching) aspects that have been emphasised are underpinned by this sociocultural perspective of how learning occurs; the scaffolding role of the teacher in supporting the development of scientific concepts and competencies is special in this regard.

Constructivism and learning through participation in discursive practices

The above social constructivist and sociocultural perspectives have resulted in a renewed appreciation of how the scientific community really works, and a refined reinterpretation of how younger students learn science or engage in 'meaning making' about their world.

These fresh insights have provided a different perspective on the nature of concepts, conceptions and conceptual development to that generally accepted within the world of school science. Conceptual development, derived from this sociocultural position, is seen as students 'constructing competence' in particular practices that help them articulate/represent their views of how the world works. For particular fields of knowledge (science, history etc.), these have been termed 'discursive practices', which are, in part, ways in which students can engage and participate in the knowledge, conventions and practices of (here) the science community. In the primary school environment, the 'science community' can also refer to the 'community of classroom learners', who, through various forms of communication related to the social and symbolic world of science, are transforming their own microscience (classroom) community, its knowledge and themselves, in a similar way to the scientific community.

Discursive practices in the learning of science, therefore, would embrace language use and various other forms of representation (e.g., pictorial) as well as ways of being with others and the identities of learners as they participate in their own classroom science community (Kelly 2008). As this 'competence' in discursive practices develops and expands, conceptual development forges a complex and usually non-linear path for each individual. It involves a lifelong learning journey (Harlen 2009) and recognises that 'concepts' are, after all, human constructions and subject to change. This may be contrasted with the personal constructivist perspective in which students are seen as building mental models and frameworks and reach, at some time, a defined 'scientific conceptual endpoint' in relation to particular ideas (adapted from Jonassen 2008; Scott, Asoko and Leach 2007; Tytler 2007; Tytler and Prain 2010; Tytler, Prain and Peterson 2007).

Representation as a discursive practice and a learning process

One of the discursive practices in the science community is its discourse. Science explanations, for example, can be represented in different modes (e.g., linguistic, numerical, graphical, tabular) as well as through an integration of modal forms (i.e., multimodal representation). In fact, once it is recognised that all science concepts are presented through representations, even the language on a page, then 'learning about new concepts cannot be separated from learning both how to represent these concepts as well as what these concepts signify'. Combining a growing appreciation of this aspect of science discourse with analyses of students' meaning making, through their language, drawings, gestures and so on, has resulted in recognition of the key role of 'representational

R

negotiation' in understanding how students learn. 'Representational negotiation' is how students negotiate their meaning of a phenomenon (e.g., evaporation) across contexts and time through various forms of representation (Tytler, Prain and Peterson 2007, pp. 311–17).

Students' understanding of a phenomenon can be communicated through conversations, gestures, drawings, movement and other forms of expression. These expressive forms are really just the students using various kinds of representation to communicate their meaning making to others, such as peers and teachers. When students use these representations, whether they are verbal/written (e.g., science journals), visual/spatial (diagrams, 3-D models, interactive computer simulations), mathematical (measurements) or kinaesthetic (role-playing), they are developing their conceptual understanding. Detailed analyses of case studies of young students learning science over several years suggest that these 'expressive' forms of representations usually include students making reference to context, perceptions, feelings, storytelling, analogy and metaphor.

This representational perspective on learning is characterised as being highly contextualised and personal. It is exemplified when primary students 'talk science' and 'draw science' in their use of individualised, personal narratives which relate whatever it is they are learning about (e.g., the process of 'dissolving') to their own life-stories (White and Gunstone 2008). The students' language and other expressive modes mediate the forms of representation (of their thinking) that they are using to show others how they are making meaning.

Learning science, from this standpoint, appreciates the role of informal methods of conceptual learning; for example, through students' personal stories and their associative (thinking) representations, such as associating 'condensation' with 'getting sweaty'. Teachers, it is argued, can complement these informal styles with more formal representations. One example could be developing a verbal scientific narrative about blood circulation by moving, metaphorically and literally, between elements in a children's book, an interactive whiteboard simulation, a model of the human body, and their own body. Students could then be encouraged to make sense of these multimodal representations to possibly transform their thinking and ideas about blood circulation, which they could 'represent' in various ways (adapted from Scott, Asoko and Leach 2007). In doing so, primary teachers would be acting on what they have always known – the use of stories, whether personal, anecdotal, historical or otherwise, can engage interest, a precursor for most learning. The difference being emphasised here is that they would now link this narrative strategy to different representational forms related to a phenomenon that science can help them understand.

Representation and literacies

Learning science through student- and teacher-generated representations is related to the development of competency in a range of literacies, both generic and science-specific. Using these literacies in science engages students and assists their conceptual learning (Prain and Waldrip 2009, 2010). Generic literacies include reading text and writing, and are associated, for example, with hands-on science tasks – they can also include imaginative writing. Science-specific literacies refer to investigation report writing, data representation (pictures, charts, tables, graphs), and diagram and model construction and interpretation – see Chapter 2 and later chapters for examples.

Participation in the discursive practices of science, in summary, emphasises that science learning:

- while a highly individualised and personal journey to meaning making, takes place within a participative setting
- can be very informal, involving, for example, story, anecdote and metaphor
- occurs through generating, interrogating and manipulating multiple and multimodal representations
- is assisted by using multiple literacies.

There are clear roles that the teacher could play to assist learning from these perspectives.

Activity 1.7 Constructivism: your initial reactions

- In what ways, from your own learning experiences, can you identify with the notion that we construct ideas rather than absorb them? Can you relate these notions to rote and meaningful learning?
- What sense do you make of the concept of a border crossing into the world of science? What might be the implications for a teacher of five- to six-year-olds?
- Why do you think we may learn more effectively if we participate in multirepresentational model building? Illustrate your thinking by representing your thoughts on how you would explain the occurrence of 'day and night'. How could you manipulate the 'model' you have generated? What teacher roles might you adopt in helping students to learn about day and night through multiple and multimodal representations?
- What interpretations do you place on 'learning through participation'? Does this relate to developing competencies associated with discursive practices in science? If so, in what ways?
- From the above descriptions of constructivism, what implications do you draw for teaching? Are the implications different for personal compared to social constructivism? What about sociocultural perspectives? Justify your answers. Do these teaching implications differ from your current views on teaching science (see Activity 1.6 on p. 11)? If so, how?

Constructivist learning: overall summary

In summary, then, there is no single constructivist position on learning science. Various perspectives on learning science, such as those using a sociocultural lens, can be related to a 'broad' rather than a narrow interpretation of what it means for students to 'sense-make' or 'meaning-make' or construct ideas (and competencies) rather than 'receive' them. These different constructivist views indicate that learning science occurs on social and individual planes and that it can be viewed along two dimensions – namely, acquiring/developing ideas and participating in a (scientific classroom) learning community. Further, by developing competencies in the discursive practices of science, students' science learning has been recognised as representational in nature. Each perspective has something to offer teachers in helping them to understand how their students learn science (Sinatra and Mason 2008).

As stated in the introduction to this chapter, we should teach science in a way that is consistent with what we know about how students learn science. The social constructivist learning perspective has long been the 'prevailing view of learning in science' (Goodrum, Hackling and Rennie 2001, p. 17; Hackling and Prain 2005), although recent studies are suggesting variations in emphasis. Table 1.1 summarises the *essential* features of this view of learning and the generally accepted teaching implications – the latter are developed in the 'Teaching science constructively' section of this chapter on p. 19.

With some qualifications, these basic social constructivist learning premises (and their teaching implications) can be a helpful summary and guide for teachers. Firstly, it is now accepted that, in most instances, conceptual change is evolutionary rather than revolutionary. The expression 'conceptual change' suggests a rapid replacement of ideas, which rarely happens, and so 'conceptual development' and 'conceptual formation' are being used more often to describe what is happening. Another qualification is that the nature of conceptual formation and change is usually domain-specific (e.g., relating to astronomy or energy concepts) and not generic (Vosniadou 2008). It is also becoming increasingly recognised, even with younger learners, that there is a two-way flow between students' abilities to carry out scientific inquiry and to grow conceptually within specific knowledge areas (such as the properties of materials). This would include, for example, the use of multimodal learning and multiple and interactive representations. Conceptual development does not proceed without scientific inquiry; that is, one does not occur without the other (Metz 1998; also refer back to the section 'Changing emphases in learning primary science' on p. 3).

If you are seeking more detailed descriptions of constructivism and its implications, you are encouraged to consult Harlen (1999, 2009), Leach and Scott (2000) and Tytler (2002a, 2002b, 2007), all of which have been written for teachers. For detailed reviews of the research underpinning the various constructivist positions, see Bell, Cowie and Jones (2009), Leach and Scott (2002, 2008), and Scott, Asoko and Leach (2007).

TABLE 1.1 Basic premises of learning constructively in science and the main teaching implications

LEARNING CONSTRUCTIVELY MEANS THAT	TO TEACH BASED ON THESE ASSUMPTIONS MEANS
• *learners must construct their own knowledge on the basis of what they already know and the sociocultural context in which they find themselves* • *learning is an active process in which learners try to make sense of their experiences to construct understandings about the world* • *learners develop knowledge and ideas in science that make sense to them by linking new information to their existing conceptual frameworks* • *new information is incorporated into learners' existing mental structures in ways that are meaningful to them (it is not written on a clean slate).*	• *selecting activities that build upon and respect learners' prior experiences and that are perceived by students to be purposeful and interesting* • *developing teaching programs that develop concepts rather than the learning of facts; the concepts selected should be embedded in issues that are meaningful to students* • *fostering the development of science language and communication skills* • *providing opportunities for students to 'learn how' rather than just 'learn that'.*

Source: Derived from Goodrum, Hackling and Rennie 2000, pp. 17–19. See website at http://www.dest.gov.au/sectors/school_education/publications_resources/profiles/status_and_quality_of_science_schools.htm

Activity 1.8 Conceptual growth: two case studies

Read a primary teacher's experience of teaching for conceptual change related to her students' understanding of heat, found in Watson and Kopnicek (1990) – available at http://www.exploratorium.edu/ifi/resources/workshops/teachingforconcept.html

- What does the story say to you about how some children learn? Would 'conceptual growth' describe what happened better than 'conceptual change'?
- Can you identify with the children in the class? Why?
- Discuss with colleagues how the teacher's experiences and the authors' reflections exemplify elements of personal and social constructivist learning.
- From what you have read up to this point, would more current thinking about how students learn (e.g., through representation) have influenced the authors' interpretations of what happened in this lesson sequence? If so, how?

It is difficult to briefly illustrate here the nature of conceptual growth from longitudinal studies of individual students over several years. Different interpretations of what conceptual growth means can result from such reports. If possible, now read the case study of Calum and the growth of his ideas about evaporation from Year 1 to Year 6 (Tytler, Prain and Peterson 2007, pp. 321–6).

- How does the interpretation of this case study compare with your responses to the above teacher's report about heat?
- What messages, as a teacher, do you derive from exploring these two descriptions and interpretations?

Constructivism: some important caveats

There are several issues that are further developed in the online Appendix 1.4. Firstly, it is acknowledged that there are other learning theories apart from constructivism – some of these are outlined and the position taken in this book is explained. Secondly, it is emphasised that constructivism also refers to a theory of knowledge as well as to a theory of learning. This can have profound implications for how we teach science. Thirdly, there are critiques of various interpretations of constructivism, as is evident in the earlier discussion of the different constructivist perspectives. These critiques all have something to offer and some have added considerably to our evolving understanding of how students learn; an example is 'hot' conceptual development and change, which refers to the influence of affective and other personal factors on conceptual learning (see the '"Hot" conceptual considerations: student engagement' section of this chapter on p. 25).

Appendix 1.4 Constructivism: some important caveats

Teaching science constructively

'Teaching science constructively' really means teaching science based on constructivist views of how students learn science. The previous section overviewed these major research-based views. In this section, the discussion revolves around the teaching implications of these perspectives and moves from general teaching principles through to specific teaching strategies and techniques.

Ausubel (1968) stated many years ago that we need to ascertain where the learner is at and teach accordingly. This dictum is as true today as it was then, except that 'teach accordingly' is not as straightforward as it may superficially seem. A teacher cannot simply be a clear explainer of ideas. This is obvious when we realise you can have teaching without learning. Teachers must recognise and value the voices/views of their students. What, then, characterises teaching that acknowledges student ideas so that students learn constructively?

From a personal constructivist perspective, teachers would focus on the knowledge of the individual learner and would see learning as an active and continuous process; they would take the view that learners have the final responsibility for their own learning and that teaching is the 'promotion of opportunities and support for learning' (Tytler 2002a, p. 18). A social constructivist perspective would place more emphasis on how classroom environments support learning. It would focus 'attention on the social processes operating in the classroom by which a teacher promotes a discourse community in which students and teachers "co-construct" knowledge' (ibid., p. 19). Sociocultural perspectives further emphasise situated or context-based learning, as well as the role of language and symbols. A focus on students building ('constructing') competencies in the practices of science within their classroom science learning community (that is, discursive practices) flows from these latter two positions. Table 1.1 (see p. 18) captures some, but not all, of these teaching emphases.

A set of teaching approaches or a referent

Constructivist teaching need not merely be seen as a set of teaching and learning strategies and models (a 'method'). It can also be viewed as a 'referent' (Tobin, Tippins and Gallard 1995, p. 47). The former reduces the theory to a limited set of teaching methods that are considered consistent with the general theory (see, for example, Taylor 1998, p. 1113). Seeing constructivism as a referent interprets it as a 'set of beliefs about knowledge and a set of reflective tools that enable the science teacher to consider the optimal ways to mediate the process of learning science' (Tobin, Tippins and Gallard 1995, p. 47). Such teachers could be said to have a 'constructivist mindset'. In this book, constructivism is considered a referent, but teaching techniques, strategies, models and schemata based upon its underlying principles are emphasised. We would agree with Hewson, Beeth and Thorley (1998, p. 200), who argue that, when certain teaching conditions are met, 'learning will occur more frequently for *more* students' than when the conditions are not met (emphasis in original). Consequently, these conditions are highlighted. Of course, teachers may adapt the suggestions made but they are considered a starting point for teachers who wish to put their pedagogy on a more constructivist footing.

Teachers with a constructivist mindset will interpret all teaching strategies and approaches differently and hence use them in particular ways. This point needs to be emphasised. When asked, some teachers say they are teaching constructively because they are, for example, integrating subjects, using groups or hands-on activities (Skamp 2000). But if they are not teaching with a constructivist mindset, this can be fallacious thinking. From a constructivist viewpoint, a hands-on activity is 'useless if "[students'] hands are on, but their heads are out"' (Resnik 1997, cited in Ben-Ari 2001, p. 51). Similarly, if students are working in groups, but not with a group focus, there may be little learning occurring. On the other hand, if the teacher has set up a cooperative learning context, then the outcomes could be quite different. There is now evidence that collaborative learning appears to be related to conceptual growth and improvements in scientific inquiry; other advantages of collaborative learning include students being exposed to other students' science views and being required to express their own science views, on which others may comment. In some instances, such discussion may even mirror a community of young science investigators, as in the science community (Metz 1998; Treagust 2007), and hence exemplify social constructivist ideals.

Constructivist teaching approaches

Although there are varying constructivist and related perspectives on how students learn science, there is sufficient research evidence to provide guidelines for teachers at the macro and micro levels. A macro implication of a social constructivist perspective would be that students engage in inquiry-oriented or investigative learning as a community of (science) learners; at the micro level, this perspective could guide the nature of the dialogue that occurs between teacher and student(s). The teaching guidelines derived from the different constructivist and related perspectives on learning can, in various ways, be complementary in facilitating learning (Harlen 2009; Hewson, Beeth and Thorley 1998; Scott, Asoko and Leach 2007; Stard 1998) and are relevant to primary learners of all ages (Metz 1998). Each chapter of this book devotes a major section to constructivist learning (and therefore teaching) approaches as they apply to the conceptual area that is the chapter's focus.

In the following discussion, there will be consideration of the following:

- *constructivist teaching models or schemata* – The most well-known models/schemata are introduced, with brief descriptions of their major phases. Most were developed from a personal constructivist perspective but several of these have been adapted to integrate teaching implications from social constructivist and related viewpoints. A broad schema for what sequencing might encompass from a social constructivist perspective is included.
- *the main components of constructivist teaching models or schemata* – Wandersee, Mintzes and Novak (1995, p. 191) have categorised the components of these models/schemata into two groupings: those that focus on 'externalising and modifying the learner's knowledge structure' (where 'structure' can be interpreted differently, from 'knowledge pieces' to 'frameworks'; see the earlier section 'Why the change in emphasis? Alternative conceptions' on p. 7), and those that 'address the need for self-monitoring and controlling the events of learning'. The latter relate to what are termed 'metacognitive strategies', that is, strategies that help students learn how to learn. Preceding these two categories is a focus on 'student engagement', an understated component of earlier thinking about constructivist teaching approaches (Tytler 2007). Other teaching implications of recent developments in how students learn science can be partially accommodated in these two groupings. Within the strategies suggested for 'modifying the learners knowledge structure' there is an emphasis on approaches teachers can take when learning is seen as 'representation'.
- *teacher and student roles* – If the components in the above sequences are implemented with a constructivist mindset, then this has implications for the roles of students and teachers. These are overviewed.
- *the question, 'Does constructivist teaching work?'* – Constructivist teaching does make a positive difference to students' science learning. The research evidence supporting this claim is summarised.

Constructivist teaching models or schemata

At the macro level, there are several constructivist models or schemata available to guide teachers in the sequencing of lessons. These mainly derive from personal constructivist perspectives but, as has been emphasised, social constructivist strategies and sociocultural influences can be integrated into these models, including learning which is facilitated by representation, multiple literacies and other strategies. Case studies or vignettes (presented throughout chapters 2 to 12) illustrate teaching implications at the micro level.

These lesson sequences usually include at least four components, which need not necessarily be seen as discrete; for example, students can reflect on their ideas and the learning process at various points in any learning sequence. The four components are:

1 initially set the context and gain interest, then (or concurrently)
2 find out students' ideas
3 select strategies that would help students develop, form and modify their ideas, and then
4 ensure that students reflect on their ideas and their learning processes.

The sequence would need to be set within the framework of your science syllabus or framework (McGuigan and Russell 1997). The following models or schemata are the most well known and can all be accommodated within this basic sequence:

- *the 5E model (engage, explore, explain, elaborate, evaluate)* – This model is based on a number of earlier instructional models, but especially the 'learning cycle' comprising exploration, concept development and expansion (or application of the concept to new situations) (Bybee et al. 1989; also see the 'Planning for conceptual development with the 5E instructional model' in Chapter 3, p. 116). More recently, the model has been used by the Australian Academy of Science's *Primary Connections* (AAS 2005) project; all chapters in this book make reference to *Primary Connections* units and the teacher and student roles in each phase (Marek 2008 also outlines these roles), while some chapters (e.g., 3, 10 and 12) further illustrate 5E sequences. *Primary Connections* also has cooperative learning as an integral component and stresses literacy–science connections, multiple literacies and multiple representations. Social constructivist emphases are readily apparent.
- *the Interactive Teaching approach* – Especially designed with primary teachers in mind, this model focuses on student questions (Biddulph and Osborne 1984) and has been adapted in various ways (see, for example, Chapters 4, 6 and 9). Many of the teacher and student roles associated with this model are consistent with a social constructivist perspective (Tytler 2002b, 2007).
- *the Children's Learning in Science Project (CLISP) model* – This conceptual-change scheme (see, for example, Driver and Oldham in Leach and Scott 2008; full details in Scott, Dyston and Gater 1987) encourages the exchange of ideas, reflection on understandings, and the modification of ideas based on evidence gained from testing them. The influence of group discussion upon individual learning is promoted. Exemplar sequences are presented in several chapters of this book (e.g., Chapters 5 and 8).
- *the Generative Teaching approach* – This approach involves cognitive conflict (Osborne and Freyberg 1985; for an example, see the Chapter 6 section 'Teaching approaches that value students' ideas and small-group work' on p. 230). Class discussion and the generation of shared ideas is emphasised; social and cultural processes are implicit (Tytler 2002b).

Other models or schemata, based mainly on personal constructivist assumptions, have been reported; for example, Harlen (1992, 2007) describes the Science Processes and Concept Exploration (SPACE) sequence; some aspects of this are referred to in chapters 8 to 12. Later research has questioned some of the emphases within the above models, as originally described, but schemata such as the 5E approach (e.g., as in *Primary Connections*) have embraced several of these more recent findings. Teachers have sometimes used these schemata as described while others have adapted them to suit their particular purposes (see, for example, chapters 3 and 8).

The more recent constructivist and related perspectives – for example, sociocultural interpretations – do not readily translate into such schemata or models. Rather, more general pedagogical principles tend to guide teaching decisions, although some of these can be incorporated into the above models. For comparison, a sequence with a social constructivist emphasis, suggested by Leach and Scott (2002), is outlined next.

Sequences from a social constructivist perspective

The engagement of students within lesson sequences with a social constructivist emphasis would probably take on a different character to that of the above sequences. Translation of social constructivist views of learning into direct guidance for teachers is more subtle but certainly not less important. Clearly, from this perspective, teachers have a key role in introducing scientific ways of thinking about our world (e.g., the social development of science ideas, including concepts and processes) and then must facilitate talk and thought by their students through, for example, engaging them in the discursive practices of science. As summarised by Tytler (2007, p. 34), such classrooms would conduct many exploratory activities and talks, as well as support group and whole-class conceptual discussions. There would be an emphasis on activities and conversations that mirror the NOS, especially an understanding and use of evidence and the role of representations in advancing thinking. As just stated, the earlier schemata (e.g., the 5E approach) can incorporate many of these features (Tytler 2007), but teachers would need a 'sociocultural constructivist mindset'.

M&E

One interpretation of a 'teaching sequence' from a social constructivist perspective (Leach and Scott 2002, pp. 122–5) is that it needs to:

- **stage a scientific story** – The teacher develops a 'scientific story' over a sequence of lessons related to familiar phenomena and the story is expressed using the ideas and conventions of 'school science social language'; for an example, see the case study 'A condensation sequence' in Chapter 9 (p. 367) relating to the 'science story' of condensation for a middle primary class.

  To assist in the development of the story, multiple representations of science concepts would be used (an example is given in the 'Changes of state: teaching about particles' section of Chapter 9; see p. 381) and attention paid to different types of teacher talk, especially collaborative dialogue, but also the presentation of ideas by the teacher (e.g., in such a way that exemplifies characteristics of the NOS, as in the examples given in chapters 8 to 10). It is still critical that students' existing understandings are elicited and other students and the teacher engage with these ideas.

- **support student internalisation** – Making the scientific story available is clearly different to students making sense of this (possibly) different way of seeing familiar phenomena, as it might be presented in one of the earlier personal constructivist schemata. These differences have been previously emphasised when discussing the attributes of students' conceptions and the complexity of students' conceptual development. To assist students to internalise aspects of the 'scientific story', teachers need to be 'continually monitoring ... and *responding* to students' understandings during the whole lesson sequence' and not just at a specific point when a science view is introduced – as might happen if some of the personal constructivist schemata were interpreted too rigidly. As previously mentioned, the 'elicitation of ideas' can be seen as a continual process, but this and the 'monitoring of ideas' need to be planned.
- **hand over responsibility to students** – This involves the need for students to apply their ideas and to make them 'their own'. Initially, this could be facilitated by scaffolding, which would be reduced over time. Many examples, and levels, of scaffolding are illustrated in this book (see the online Appendix 1.5).

Appendix 1.5
Scaffolding strategies

This description of 'sequencing' focuses more on micro-decisions, such as the talk which surrounds activities rather than the activities themselves. An example would be the style of communication that the teacher used: interactive rather than non-interactive, an example of the latter being the teacher doing the telling; dialogic, or eliciting students' points of view and taking them into account, rather than authoritative, or focusing on the science point of view (Leach and Scott 2008, pp. 666–7). Activities, therefore, are only important 'insofar as they act as reference points in the development of the scientific story' under consideration. The role of the teacher is critical.

Other attempts to translate social constructivist perspectives into schemata do not appear as inclusive as that just described. Teaching science using inquiry-oriented approaches can be consistent with social constructivist premises. Inquiries could be literature/media-based or field experience/classroom experiment-based. One inquiry-based sequence has five phases: initiation, design and planning, performance, interpretation, reporting and communicating (Hodson and Hodson 1998a). These inquiry steps could be integrated into all of the above models, either as a subsection of them (e.g., planning and carrying out a fair test; see the 'Fair test investigations' section in Chapter 2, p. 69) or across several of the suggested schemata steps. It is important that inquiry-oriented approaches do not become ends in themselves but facilitate movement towards 'the "big ideas" of science that enable understanding of our world and active participation in decision-making involving science and technology' (Harlen 2009, p. 34). Chapters 2 to 12 provide examples of such inquiry steps and how they relate to conceptual development.

Main components of constructivist models and schemata

All the schemata and models include strategies which aim to orient and engage learners, and externalise and modify the learner's current thinking, as well as strategies that address the need for self-monitoring and controlling the events of learning. These are overviewed next, together with commentary on how they are integrated into Chapters 3 to 12.

'Hot' conceptual considerations: student engagement

The teacher must initially set the context and gain student interest. This is called 'situational interest', which is momentary interest because of events that have occurred within the teaching situation. This is clearly important, but a major affective goal in science is 'personal interest' – this is when students want to learn about science topics. Personal interest develops over time and is relatively stable (Koballa and Glynn 2007).

As constructivist strategies value students' ideas and use them as a focus for many of the activities within a lesson sequence, student interest and engagement sometimes naturally follow. However, this cannot be assumed. 'Hot' factors such as classroom context and the students' goals, expectations and needs, as well as their interest and other affective attributes, must be considered (Duit, Treagust and Widodo 2008; Koballa and Glynn 2007).

Claxton (1989) commented that 'cognition does not matter if you're scared, depressed or bored'. In fact, recent neuroscience findings indicate that cognition and affect are linked brain functions; we may remember more if learning is associated with emotional arousal (McCrory 2008). Interest and engagement therefore must be very high on your agenda as you plan and teach science with a constructivist mindset. The learning environment is pivotal to sustaining engagement and later chapters provide many stimulating ideas which will assist in the development of a community of engaged science learners; even so, an appreciation of the importance of teacher–student relationships cannot be overlooked in maintaining that engagement.

Apart from the intrinsic characteristics of constructivist learning that are associated with high student satisfaction – acknowledging students' voice, student-centred learning, a focus on investigation with a strong emphasis on firsthand experience, and use of multimodal approaches to representing ideas – there are many other strategies a teacher can use to arouse and sustain interest and curiosity. It's critical to ensure that science is linked to students' lives and personal interests in order that the context is meaningful/relevant (Bell 2007; Tytler 2007). Meaningful/relevant contexts need to be seen through a wide lens: it is not just, for example, that content may relate to popular media, but that students perceive that the content would assist them in being responsible members in their local and global communities, now and in the future. Explicit references to content related to 'science as a human endeavour' (see the section of the same name later in this chapter, on p. 42), like school–community links such as 'citizen science' initiatives within lesson sequences (Zeegers, Paige and Lloyd 2010), and inclusion of socioscientific issues, would all help students see 'meaningfulness' in what they are learning. Further, it needs to be remembered that, for some students, a topic will be relevant simply because it is fascinating.

Other ideas for engaging students' interest include: create a wonder-filled environment with access to assorted materials, ICT and events; include outdoor learning and interactions with natural settings; have students raise questions; allow students to make multiple observations; encourage hypothesising; and model 'wonder' as a teacher. For an extensive list of ideas suitable for young learners, see McWilliam (1999), and for older learners see Palmer (2004).

The importance of 'hot' considerations cannot be overemphasised. Research suggests that students' personal interest in school science starts to decline towards the end of primary school, with girls' interest declining more than boys. Many factors may be responsible for these declines, but engaging students in learning science can only help capture and, over time, retain interest in science. When students are learning science they need to have their hearts, hands and heads all 'turned on'. The approaches suggested in this book strongly urge teachers to talk with students conversationally about the topics being studied, and to use various means to explicitly bring out a topic's relevance and its applications. Today's students need to see that science helps other people as well as helping them to self-actualise and express themselves creatively (Koballa and Glynn 2007; Osborne 2008; Skamp 2007).

Externalising ideas: starting with learners' preconceptions

As Ausubel (1968) implied, the elicitation of learners' ideas is a prerequisite to modifying and changing their alternative conceptions about how the world works. This book suggests strategies and techniques to elicit the conceptions that your students may hold. Also, various elicitation strategies are used to help you recognise your own ideas. In most chapters, children's and your own conceptions are then compared and contrasted with scientists' views. Three focuses are evident in these sections:

1 the *elicitation techniques* themselves – Most are procedures you could use in your classroom to elicit the existing ideas your students hold about events and phenomena. Examples you will encounter are drawing, concept maps, interviews of various types and student diaries. White and Gunstone (1992) and Naylor, Keogh and Goldsworthy (2004) catalogue a wide variety of elicitation strategies. You are encouraged to trial some of them with students.

   When using elicitation strategies, you need to be conscious of the social context. Children need to feel comfortable when sharing their ideas. A trusting and supportive environment certainly helps. Some contexts encourage idea sharing more than others: for example, the use of concept cartoons in collaborative groups has been shown to be effective (Keogh and Naylor 2004) and was even more so when puppets were used (Keogh et al. 2006). Other factors that may influence whether students wish to share their thoughts include the relationship between teacher and child/class, communication patterns in the class, and the view of learning the teacher holds. Also, it is often important to probe beyond a young student's first response (Fleer and Robbins 2003b).

2 the *alternative conceptions* you and your learners may hold – If possible, you need to ascertain how strongly these beliefs are held from an affective perspective; for example, emotionally (Sinatra and Mason 2008). Most chapters of this book have sections dealing with 'adult' (teacher) *and* (primary) 'student' conceptions related to the concepts and understandings which are the focus of the chapter. Summaries of research findings related to students' ideas on particular science concepts are included. These are linked to strategies to modify and develop particular ideas.

3 the *currently accepted scientific views* about the objects and phenomena that are the focuses of particular chapters – The scientists' views are, in most instances, explicitly stated, but are at times integrated with discussion of particular reader activities, or consideration of students' ideas. It is important you try to clarify your own ideas by referring to scientists' views, so that you can more effectively interact with your students and help to increase their acceptance of scientific ideas. To assist you, a section at the end of chapters 3 to 12 refers to concepts and understandings for primary teachers. (Also see the 'Your own science background' section of this chapter on p. 45.)

What is being emphasised here is that students' ideas must be made explicit. However, when familiarising yourself with the research about students' ideas related to particular objects and phenomena, you need to interpret any proposed classification and/or categories of students' alternative conceptions with care. This is because for individual children there may be many personal experiences which influence the type of responses they provide (Robbins 2005; Tytler and Prain 2010; Tytler, Prain and Peterson 2007).

Students' ideas need to be considered alongside any ideas that teachers and others (explicitly) introduce. To determine which ideas have the greater status, students need to apply standards of evidence rather than simply accept ideas on the basis of teacher authority. All ideas, those of teachers and students, are but hypotheses that may give rise to questions and other hypotheses. Students need to determine which ideas have the support of the strongest evidence – this can be assisted by the ideas being the subject of (fair) testing, further observations, discussion, argumentation and so on. 'Argumentation' is a strategy by which we try to convince others of the validity of our claims; teachers try to help students distinguish between 'claims' and 'evidence'. Students' conceptions, therefore, are inputs to effective lesson development which could result in scientific investigations, classroom debate and other activities (Gil-Perez et al. 2002; Harlen 2009).

The elicitation of ideas need not occur separately from other aspects of teaching. It has been separated here for the sake of emphasis. Students' existing ideas may be elicited during the development of a topic, rather than just at the commencement of it, but the latter is more common. Eliciting students' ideas, although critical, is insufficient by itself. Further decisions need to be taken as to how to address the ideas that students have; approaches have already been alluded to, and below, as well as in later chapters, many other suggestions are made.

Modifying learners' ideas

After determining students' preconceptions of an object, event and/or phenomenon, you can use a range of teaching or learning approaches to facilitate conceptual growth and change or, as suggested by Hewson, Beeth and Thorley (1998), to raise or lower the status of particular ideas within a students' thinking. This is the purpose of the models and schemata guiding the development of constructivist lesson sequences. These approaches all include, in some form, elements of scientific inquiry – or 'thinking and working scientifically', as it is described in Chapter 2. They will usually involve hands-on explorations and often more structured scientific investigations, although hands-on activities are not necessarily essential. There is increasing evidence that students' ideas can be influenced by simple conversation, collaborative talk and reflection; this is done usually, but not always, with the assistance of teacher scaffolding. One way for teachers to think of this talk is as the 'negotiation of meaning' (Appleton 2007; Miyake 2008): this talk can relate to hands-on tasks in class but also to recalled everyday experiences as well as other catalyst stimuli such as interactive media. Many of these ideas are exemplified through content-specific examples in later chapters of this book.

Some generic strategies have been suggested for modifying students' concepts. Harlen (2001, p. 16) has categorised the responses teachers could try depending on the basis of students' existing conceptions. If the students' ideas:

- indicate limited experience, then provide experiences that challenge the idea
- are based on perception rather than logic, then investigate further, leading to a different interpretation of the logic

- are focused on one experience, ignoring others, then ask students to go on thinking – anything else?
- indicate faulty reasoning, then help students to test their ideas more rigorously
- are tied to a particular context, then encourage them to apply their idea in a different context to see if it still works
- suggest a misunderstanding of words, then ask them for examples of what they mean and introduce scientific terms alongside the ones students use
- do not lead to other ideas, then arrange for the sharing of ideas or scaffold more scientific ways of explaining things.

Other examples would include making imperceptible changes perceptible; encouraging the planning of investigations to test student hypotheses and predictions, sometimes along with the accepted scientific idea; helping students to generalise (Watt, in Sherrington 1998), using predict–observe–explain (POE) procedures that may include discrepant events (Liem (1987) describes over 400); and using analogies. Many others could be listed; see, for example, Harlen (1999). Although some of these strategies directly challenge students' ideas, others use more subtle approaches. This is significant as several researchers have cautioned against an overuse of discrepant or conflicting events to modify students' ideas, believing this may be especially relevant with young primary learners (Harlen 1999).

Social constructivist principles are incorporated into strategies aimed at modifying student ideas because they encourage student talk about the ideas involved: this could be expressed as 'talking science to others' and 'talking science to ourselves' (Lemke, cited in Leach and Scott 2000, p. 51). Specific teaching strategies are now emerging that encourage the use of focused conceptual discussion, debate and argumentation in science lessons, all of which can be applied at the primary level (Leach and Scott 2000; Naylor, Keogh and Downing 2007; Osborne et al. 2001; Shakespeare 2003; Trippett 2009; Yoon et al. 2010). Having an appreciation of the degree of affective commitment a student may feel towards an alternative conception can also guide the choice of strategies; for example, using refutation text, which simply disputes students' ideas, may be sufficient for ideas held with low commitment, while argumentation strategies may be required for ideas held with high commitment (Sinatra and Mason 2008). The use of 'argumentation' will be more effective if teachers are explicit about its structure and steps, introduce the terms 'claims' and 'evidence', use groups of four to six students, and absent themselves from discussions from time to time (Martin and Hand 2009; Naylor, Keogh and Downing 2007; Trippett 2009). These strategies address social constructivist concerns about how students learn and are illustrated in several chapters (see, for example, 3, 7, 8 and 9).

'Student talk' is clearly very important in assisting conceptual development. Although there is evidence that students enjoy focused talk in science lessons, and that such talk can be productive, this is not always the case. Specific techniques may be required to encourage and teach children how to talk – for example, with the aid of puppets, using talk rules, and outlining the discourse needed for science, such as asking, 'How can you tell …' or 'What is the evidence for …?' (Brigland, Way and Buckle 2008, p. 26; Keogh and Naylor 2007). When students feel free to talk about their science, then the other guidelines outlined by Keogh and Naylor (2007) can help make the talk more scientifically productive; for example, having students pool ideas so that all are considered valuable – an individual's self-esteem is thereby not threatened because they feel their ideas are being judged as wrong. Other techniques and strategies which may lay the groundwork for further developing negotiated and conversational science talk include 'diversifying learning strategies, seating arrangements and student responsibilities; being more "present" with students by asking for details, explanations of thought processes, and elaborations; fostering conversations in which teachers and students are listened to, and thoughts are formed, modified and reshaped; and recognising and discussing power relationships inside and outside the classroom' – these and further ideas are expanded on in Patchen and Cox Peterson (2008, pp. 1010–11).

On another level, a student's preferred learning approach and their view of learning and the nature of knowledge may impact on their thinking processes as they engage in experiences aimed at their conceptual development. Students who orient towards a view of knowledge that is constructivist (rather than realist) may more readily engage in learning that leads to a change in ideas (Novak 2005). Also, students' views of learning and teaching can differ markedly and hence affect their approach to learning – Mitchell (1992, p. 80) lists a range of students' views and describes how this affects their learning. Consequently, the approach used to explicitly teach the NOS and the agency students are given in taking responsibility for their own learning can be influential.

Representational and related strategies

As stated earlier, learning science ideas can be interpreted as students using various expressive modes to *represent* their meaning making, and this can occur in multiple and multimodal ways. These *student-generated representations* become the focus of negotiation during teaching and learning activities. The development of students' generic and science-specific literacies can assist in such negotiations. Strategies focusing on students' representational learning can be readily integrated with the above suggestions for modifying learners' ideas. They have been separated here in order to emphasise them.

Constructivist sequences centred on representational learning must have a clear conceptual focus, otherwise teachers will not be able to effectively scaffold the meanings in students' representations of their thinking. Also, as highlighted using other constructivist perspectives, topic selection that 'takes account of [students'] interests, values, aesthetic preferences and personal histories' is preferred, and the selection of learning experiences with a 'strong perceptual context' (e.g., hands-on, experiential) is encouraged so that students can use 'perceptual clues to make connections' between phenomena, events and objects and their representation(s). To further develop students' thinking using this approach, teachers could:

- introduce multiple and multimodal representations of science concepts. These can be teacher- and/or student-generated. The variety of forms of representation suitable for primary students would include descriptive (verbal, graphic, tabular), experimental, mathematical, figurative (pictorial, analogous and metaphoric) and kinaesthetic (embodied; e.g., use of gesture and physical action).
- incorporate a range of representational challenges and in so doing 'open up discussion' of the student- and teacher-generated representations of phenomena. This could include eliciting students' ideas, having them 'integrate different representations meaningfully', and applying representations to new situations. Representations are to be seen as tools

R

for learning. not as end-products as, for example, a diagram of the water cycle might be treated in a traditional science lesson. Doing this will encourage students to use language and representations in a reasoning way.

- (related to the above bullet) facilitate students' active translation and manipulation of these representations. This could involve assisting them in articulating and making explicit the 'rules' (e.g., arrows, location) in their own representations, but also physically changing the variables to test in a physical representation.
- use ICT (e.g., the interactive whiteboard) as a means of encouraging the collaborative sharing of ideas about representations and their manipulation
- ensure the representations used are discussed with an explicit focus on their function, form, parts and purposes
- utilise and develop a diversified range of formal and informal writing types that are used for different purposes. These purposes could include the justification and understanding of topics; the need for students to move backwards and forwards in their developing texts; and connection with an audience that is meaningful to students.
- constantly link science-specific literacies to students' 'everyday discourses, values and representational capacities'
- incorporate writing support for using science-specific literacies – for example, see the Chapter 2 section 'Literacy in science' (see p. 84) and the Science Writing Heuristic (Hand and Keys 1999).

Based on Hubber and Haslam (2009); Murcia (2010); Prain and Waldrip (2009), pp. 77, 79; Prain and Waldrip (2010); Tytler and Prain (2010); and Tytler, Prain and Peterson (2007).

As indicated, these strategies could be integrated with or adapted for use in most of the previously suggested sequences. The 5E scheme in the *Primary Connections* project (Hackling 2008), for example, has an explicit focus on the representation of ideas and generic and science-specific literacies, and incorporates a range of techniques to develop the latter.

Monitoring and controlling learning

Implicit in many of the constructivist schemes for sequencing lessons is the use of metacognitive strategies. These strategies assume that 'students who know how to monitor and control their own learning are empowered to engage in more purposeful "meaning making" (Novak 1987) and, as a result, can be expected to recognise and attempt to correct inconsistencies in their own thinking' (Wandersee, Mintzes and Novak 1995, p. 193). What is involved in metacognition is exemplified in Hennessey's (1991) description (cited in Hewson, Beeth and Thorley (1998), p. 206):

> Tammy sat cross-legged, leaning back against the wall, with palms cupped together as if holding something in her hands, gazing down at them with a meditative expression on her face. Her teacher asked her what she was thinking about: 'Oh, I was just sitting here holding my thoughts in my hands. Not really, I was just pretending as if I could really hold what I was just thinking ... I thought that, if I could sort of hold my thoughts in my hands, I could better look at them to see why my ideas are intelligible, plausible, and fruitful to me.' With that, Tammy took her cupped hands, placed them on her head as if returning her thoughts to her mind, stood up and continued the conversation about her thoughts with her teacher.

Students need to be metacognitive to experience conceptual change because they have to recognise their own ideas, assess them and decide whether to modify or change them (Gunstone 1994; Gunstone and Northfield 1992; Vosniadou, Vamvakoussi and Skopeliti 2008) or give them a different status. In other words, we need to help students be 'intentional' in their learning – they have to play an 'active intentional role'. Factors that may influence being intentional are personal interest, perceived importance of topic, whether the topic is thought to be of any utility value, achievement goals, whether students believe they will do well on a topic (self-efficacy), and how much control students think they have in a situation. The relationship between these factors and intentionality can be complex (Sinatra and Mason 2008).

When primary students respond to questions about a topic or lesson concerning what they have learned, understood and/or believed, and how their ideas have changed, they are being metacognitive. Metacognitive processes are therefore 'inherent in the process of conceptual change'; further, they assist students to accept an idea because of its 'internal authority' rather than that imposed by the teacher or the curriculum (Duit, Treagust and Widodo 2008, p. 633; Hewson, Beeth and Thorley 1998, pp. 205, 213).

Various strategies have been identified which may improve students' self-regulation and learning in science. These include authentic project-based inquiry (Hume 2009), even for younger learners (as described in Chapter 2); using small collaborative learning groups and techniques, such as thinking aloud together (see examples in Chapter 9); and discussions involving mental modelling (as in chapters 5 and 8 to 10). ICT can support these strategies and hence improve metacognitive learning (Schraw, Crippen and Hartley 2006). One example is the use of software packages like Inspiration (see http://www.inspiration.com) that allow students to develop and continually update concept webs and maps while working in small groups.

The Project to Enhance Effective Learning (PEEL; see http://www.peelweb.org; also Mitchell 2009) provides detailed accounts of how teachers have used metacognitive strategies in science and other subjects. There is also a CD that contains 650 ideas based on metacognitive principles which can be applied to science learning (Mitchell, Mitchell and McKinnon 1999), and *PEEL in Primary Schools* (see http://peelweb.org/index.cfm?resource=primary%20peel) and *PEEL from a Primary Perspective* (Flack et al. 1998) provide further metacognitive teaching suggestions. As children as young as eight can be conscious of their own learning (Hodson 2002), teachers at all primary levels can integrate metacognitive approaches into their science-learning sequences.

Metacognitive approaches can enhance conceptual learning. When one Year 6 teacher fully integrated activities such as classroom discussion, annotated drawing, concept mapping, and keeping diary-like notes, but with a metacognitive orientation, he found that more students could apply their learned science concepts in an unfamiliar setting (Georghiades 2006). This is what the application phase of constructivist sequences focuses on.

To learn effectively takes effort, with learners being ultimately responsible for their own learning, although the teacher can facilitate acceptance of that responsibility. Gradated 'scaffolding' in the form of structures and processes which enable learners to develop a greater degree and/or higher level of understanding about their learning can encourage metacognitive thinking – see, for example, Appleton (2007, pp. 514–17), Fleer (1990),

Hodson and Hodson (1998a, 1998b), and the online Appendix 1.5. Teachers, though, need to be aware that students, especially older ones, may have 'very firm (and very limited) views as to what constitutes appropriate teacher behaviour and classroom activities'. Hence, if techniques or strategies are used that require students to think about their learning, then teachers may have to work to gain student cooperation in some instances (Baird and Mitchell 1986, p. 50). The extent to which this may be a problem will probably depend on the type of teaching used and hence on the type of learning that students have previously experienced.

Appendix 1.5
Scaffolding strategies

You may have realised that the 'reader activities' in this and other chapters are an attempt to encourage *you* to take control of your own learning. In what ways are they doing this?

Teacher and student roles

Teaching for constructivist learning implies various roles for the teacher and the student that are markedly different to those in traditional transmission classrooms (Hewson, Beeth and Thorley 1998). You have been asked to suggest some of these roles in earlier activities (e.g., Activity 1.6; see p. 11). To further guide your decisions and actions, checklists of teacher behaviours consistent with facilitating constructivist learning are offered in the online Appendix 1.6. One of these checklists (Harlen 2009) advocates three additional emphases in science teaching and learning along with the personal constructivist position. They are:

Appendix 1.6
Checklists of notes for constructivist teachers

1 discussion, dialogue and argumentation strategies that stress the importance of language in learning and the value of talking, including how it relates to evidence – this derives from a sociocultural orientation
2 inquiry
3 formative assessment.

The subsequent table of teacher roles provides a very balanced summary of the multifarious tasks that teachers need to keep in mind to facilitate effective science learning, and the other actions they need to consider. Student roles are also listed across each category – each of the emphases listed above receives attention in later chapters. It is worth noting that the effective implementation of these roles is probably dependent on your appreciation of the theoretical underpinnings of constructivist learning, as outlined earlier in this chapter (Holt-Reynolds 2000) – that is, you need to teach with a constructivist mindset.

A

Your decisions and behaviours *will* have an impact on how well students understand how their world works. As you teach, you relate to students as well as facilitate their learning – both are intertwined. Guidance as to how to structure learning environments (for example, the sequencing of strategies within, and across, lessons) and your own choice of actions (such as questioning and talking) is provided in all chapters of this book. A word of caution, though: Darby (2005) found that students' perceptions of their teachers' impact on their learning may have as much to do with how the teacher relates to their students as the teachers' approaches, strategies and techniques. You therefore need to be aware of: the passion you display towards your teaching and the content; the characteristics you exhibit, such as being friendly and non-threatening; whether you have established a comfortable learning environment; and the ways in which you support learning by being encouraging, understandable and attentive. Being conscious of these relational attributes is consistent with a sociocultural perspective on how students learn and, as discussed earlier, 'hot' conceptual change.

Does constructivist teaching work?

Teaching with a constructivist mindset will not work miracles. It is sobering to realise that there are no studies that have 'found that a particular student's conception could be completely extinguished and then replaced by the science view'; it is probably more correct to assert that 'peripheral conceptual change' occurs when students modify their thinking (Duit, Treagust and Widodo 2008, p. 630). This conclusion is consistent with the range of unintended conceptual outcomes that can result when students encounter a science idea (from any source) as described earlier in this chapter in the 'General characteristics of students' alternative conceptions' section (see p. 8). Many years ago, it was thought that four conditions needed to be satisfied to effect conceptual change:

1. Learners must be dissatisfied with their current explanations.
2. The new idea must be intelligible in that it must be able to link in some way to the learner's thinking so that the learner can see that an experience can be viewed differently.
3. The new concept must be seen to be plausible in that it can provide ways to explain an event not readily explained by existing ideas
4. The new idea must be considered fruitful in that it can be applied in other situations (Posner et al. 1982).

Although these 'conditions' still have value from a personal constructivist standpoint, it is now appreciated that conceptual development and change is a far more complex phenomenon, as shown, for example, by the earlier discussion of sociocultural and representational learning perspectives. Many students' ideas, for instance, are contextually bound. This means that students may describe or explain a phenomenon or event (e.g., evaporation) in two or more different ways, one of which may be scientifically more correct depending upon the context(s) (e.g., in which evaporation is being studied). Another example is an appreciation of how numerous personal factors (e.g., students' personal histories with an idea) may impact on the formation of concepts.

Considering the ongoing search for a clearer understanding of students' conceptual development, you might well ask if there is research evidence that constructivist approaches work. The short answer is 'yes'; there is substantive evidence that teaching with a constructivist mindset makes a positive difference to students' science learning.

How effective is constructivist teaching?

What we do know is that, for most learners, traditional science teaching does not work, as evidenced by the 'alternative conceptions' research. In contrast, 'there is convincing evidence from studies of learning in many different areas to support a constructivist view of learning' (Harlen 1999, p. 39). Advances in neuroscience may also support constructivist approaches (Gil-Perez et al. 2002). An overview of research studies into the effectiveness of personal and social/sociocultural constructivist and representational approaches to learning is presented next; the online Appendix 1.7 provides more substantive details of some of these evaluations.

Appendix 1.7
Constructivist teaching: Does it work?

Personal constructivist emphases

There is substantive research evidence, from several hundred studies, that students taught using 'conceptual change' strategies, such as the 5E schema, have shown positive conceptual growth – see reviews by Brown and Abell (2007), Guzetti et al. and Wandersee, Mintzes and Novak, cited in Duit and Treagust (1998), and Murphy and Alexander (2008). These and other reviews (Duit, Treagust and Widodo 2008) indicate that questioning and addressing students' existing conceptions is essential, but other features characteristic of a 'conceptual change supporting environment' must also be present. Interventions differ in their impact on conceptual learning: videos and text tend to have weak, sometimes negative effects, while hands-on activities and physical demonstrations coupled with explanations seem to be more effective. Cognitive conflict strategies have had mixed success, depending on how they are implemented (Clement 2008). There is some evidence supporting the careful use of analogies and teacher interactive/dialogic talk that builds on students' ideas (Duit and Treagust 1998; Leach and Scott 2000). Some researchers, though, have questioned the overall effectiveness of relying only on personal constructivist approaches (Tytler 2007; Tytler and Prain 2010).

Social constructivist emphases

Current thinking supports this approach to science learning, although the research evidence is, as yet, not as substantive as that underpinning 'conceptual change' strategies (Traianou 2006). Some studies found students whose primary teachers' practice supported a social constructivist view of learning made significant gains in the level and depth of their learning (Murphy et al. 2001, Murphy and Davidson 2002). These findings were supported by the positive results of a comparative study of 15 classes at the lower secondary level, which focused mainly on social constructivist principles, but especially the communication styles used (interactive and dialogic) as well as the 'learning demand' of the introduced ideas (refer back to the section 'General characteristics of students' alternative conceptions' on p. 8); the comparison classes did not have this focus (Leach and Scott 2008).

Representational and multiple literacy emphases

There are some early, but promising, findings emerging from approaches focusing on representational learning (e.g., Hubber and Haslam 2009). Also, the use of diversified science writing tasks and 'writing to learn' processes has improved science conceptual outcomes as well as attitudes towards, and engagement with, science (Prain and Waldrip 2010). The *Primary Connections* initiative (AAS 2005) is underpinned by a belief that representational issues and multiple literacies are central to learning science – small-scale evaluations have shown significant improvements in procedural and conceptual learning (Hackling, Peers and Prain 2007).

Effectiveness studies: a summary

It must be appreciated that it is simplistic to assess the effectiveness of constructivist approaches categorised in these ways. This is because there are often elements of each perspective in the particular approaches that have been reported in the research literature. Also, as Leach and Scott (2008, p. 670) comment, 'it is naïve to assume that any educational practice can be shown to be straightforwardly "better" than other practices'. However, there is sufficient evidence – e.g., see the summary in Duit, Treagust and Widodo (2008, p. 636) – to clearly show that the constructivist pedagogy espoused in this book will lead to more effective long-term conceptual gains than traditional rote- and fact-based approaches, as well as those that do not intellectually engage the learner, have no conceptual direction, and do not acknowledge the agency of students.

Implementing science curriculum requirements with a constructivist mindset

Learning outcomes

Effective primary science teaching involves implementing lesson sequences that aim to meet your syllabus or framework learning outcomes. These normally include knowledge, process and affective outcomes. Constructivist teachers need to identify the key conceptual learning outcome(s) for a lesson sequence. This means asking what is the science concept (or concepts) towards which you are hoping to facilitate conceptual growth. Often these are not explicitly stated in curriculums. At the end of chapters 3 to 12 are the major science concepts related to particular topics – used with the guidance provided, they should assist you in identifying the conceptual focuses for lesson sequences. You should also be guided by what you know about primary students' alternative conceptions, in relation to your sequence's conceptual focus. This background is in each of chapters 3 through 12 and will help you assess the 'learning demand' of the concept (see the earlier section 'General characteristics of students' alternative conceptions' on p. 8); that is, how different the scientific perspective is from the everyday perspective of the concept. For example, think how differently the concept 'energy' is used in conversation compared to its meaning in various science disciplines such as biology and physics. Learning demand also can be affected by whether students believe something is real or not (e.g., gas) as well as other factors (Scott, Asoko and Leach 2007).

Syllabus learning outcomes always include reference to science inquiry processes and usually the NOS. All constructivist perspectives strongly advocate lesson sequences with a focus on the discursive practices of science (e.g., inquiry skills, representational thinking and creativity, as outlined in Chapter 2). If these learning outcomes are to be met, then rarely, if ever, would science be taught as a one-off lesson. Constructivist models, schemata and strategies have been developed to address just this concern. This has pragmatic implications: it means you need to allocate several lessons to a particular idea and the lessons should not be separated by too much time if learning is to be effective (Nutall, in Appleton 2007).

Assessment for student learning

In this book, the focus is on classroom formative assessment that improves (not proves) students' science learning, rather than summative assessment. Diagnostic assessment may, in part, be aligned with the elicitation of students' ideas.

Within a constructivist framework, formative assessment is where teaching and learning meet; it is 'active assessment' that links thinking, learning and assessment (Naylor and Keogh 2007). It is where teachers and students access students' initial and developing conceptions as they try to make meaning of their experiences, so that teachers (mainly) can facilitate ongoing conceptual progress. 'Formative assessment does improve learning' – Black and William (1998, p. 61), concluded after a review of 578 articles. Primary teachers have characterised formative assessment as: 'responsive, uses written, oral and non-verbal sources of evidence, a tacit process, integral to teaching and learning, done by both teachers and students to improve teaching and learning, highly contextualised and involves managing dilemmas' – abbreviated from Bell (2007, p. 973). Formative assessment can be planned by the teacher or be interactive (on the spot); it can be initiated by students or the teacher.

A

Teachers formatively assess students on their progress towards science learning outcomes. This embedded assessment will relate to students' ability to use the various processes and skills involved in scientific inquiry, and scientific and socioscientific decision making (and hence the discursive practices of science), as well as their understanding and application of science ideas, including their expanding knowledge of the NOS. Further guidance as to what to assess is sometimes provided by 'achievement standards' in some curriculums (e.g., ACARA 2011).

Formative feedback by either teachers or students is effective when it is substantive; that is, related to the content of the work and linked to the purposes of the task and/or the students' strengths and weaknesses in the task (and not the student). Substantive feedback means learners need to think about their conceptual and other learning; it aims to encourage students to be 'intentional' about their learning or metacognitive. Self-assessment therefore has a clear role in constructivist teaching (Harlen 2009; Naylor and Keogh 2007). A challenge for teachers is that primary students' perceptions about assessment might be confined to summative tasks and not include issues such as clarifying their understanding (Bell 2007; Cowie 2009). How might you address this belief and help student learning become more intentional?

Common formative assessment approaches could embrace the elicitation strategies mentioned earlier. Consideration should be given to using teacher observations (structured or otherwise), concept maps, interviews and conversations, learning stories (e.g., journals), concept cartoons, true–false statements, card sorts, POE tasks, graphic organisers, annotated drawings, deliberate mistakes, and odd-one-out; for details, see Naylor, Keogh and Goldsworthy (2004). Assessment formats should provide opportunities for multiple and multimodal responses. Resources that may assist are the Science Education Assessment Resources or SEAR (see http://cms.curriculum.edu.au/sear) and *Primary Connections* (AAS 2005), which provide guidance about formative (and summative) assessment strategies. It must be remembered that student responses can be influenced by the formative assessment processes used (Carter and Rua, in Appleton 2007). Some helpful assessment rubrics that can assist teachers' decision making in the complex domains related to the development of scientific literacy (Murcia 2009) and participation in socioscientific issues (Sadler, Barab and Scott 2007) have been published. For the latter, four levels of practice were identified after analysing discussions with Year 6 students – they were complexity, perspectives, inquiry and scepticism.

A pragmatic assessment issue is how to assess individual students when there are, for example, 30 in the class. While having knowledge of where individual students are at is naturally valuable, it will be more realistic at times to know where your 'class of students' is at. Hence teachers can use some of the suggested assessment formats (e.g., concept cartoons) to engage students in a mixture of whole-class, group, pair and individual activities that require them to share and defend their thoughts with peers and teachers.

Formative assessment is therefore 'active', a part of ongoing learning, and can be done by teachers, peers and individual students (i.e., self-assessment) (Naylor and Keogh 2007).

Most chapters of this book refer to one or more of these aspects of assessment (e.g., chapters 3 and 10) as well as diagnostic and, on occasion, summative assessment.

Integration and learning

Integration into the primary curriculum is one of the pragmatic responses to the increasing pressure on teachers to meet expanding societal expectations related to children's learning. It has no agreed meaning (Appleton 2007) and, therefore, when it is used we need to be able to answer the following questions:

- *What* is being integrated (subject matter)?
- *How* is it being integrated? For example, is it integrated via:
  - correlating various subject matters, often involving links between subjects
  - a theme or topic
  - contributions to areas of practical thinking and living, such as living in a sustainable society
  - or the student's own inquiry?
- *Why* is one wanting to integrate? (Pring 1976, pp. 99–111)

The reasons for integration, apart from pragmatism, are what should focus teachers' attention when considered from a science learning standpoint. Selecting science contexts that relate to students' real-life experiences was considered by primary teachers in England as the best way to improve science learning. Using cross-curricula approaches was a 'highly motivating' way to meet this goal (Wellcome Trust 2005, p. 20). Other reasons include the complementary and synergistic outcomes that might evolve from combining two or more subject areas; some examples are provided below.

M&E

In this book, many examples are provided of how science can be integrated with other subjects: you need to ensure that integration of science using one of Pring's four methods still retains the integrity of science as a way of knowing (science inquiry), a body of knowledge (science understandings) and a human endeavour (ACARA 2011). This is a critical point, as although integration may lead to improved motivation, interest and higher-order thinking skills, this can occur at the cost of reduced conceptual understanding; this is a planning issue you need to keep in mind (Venville et al. 2002).

Activity 1.9 Drama and science

Drama-based teaching methods have been shown to improve students' conceptual growth and their attitudes towards school science (Francis 2007). How many drama techniques can you think of that could be used to help students understand how science and/or their world works? Try to think of a specific science topic or idea to use with each technique. Why do you think using drama could result in improved learning?

You will find it helpful to read Francis' article 'The impact of drama on pupils' learning in science' after you have brainstormed this activity. Afterwards, speculate as to whether other subjects (e.g., visual art; also see 'The role of mathematics in investigations' section in Chapter 2, on p. 73) could have similar positive effects, and why.

Literacy–science connections

Science and literacy share many cognitive processes and discursive practices, and these can be integrated across the same content. Common processes can relate to approaches to investigation, similarities in applying knowledge to new situations, and use of similar metacognitive processes (sequencing, making inferences and analysing). Both areas share discursive processes of 'requiring the making of meaning through language, text, signs and symbols' (Girod and Twyman 2009, pp. 13–14). The *Primary Connections* initiative (Hackling, Peers and Prain 2007) is built upon these premises.

Does the blending of science and literacy benefit students? Science learning was enhanced when students expressed their science understandings using multiple literacies (Tytler, Peterson and Prain 2006). Improved learning across both subjects resulted when there was an integrated focus on reading, writing, listening and talking outcomes with science conceptual and procedural outcomes (Hackling and Prain 2005; Varelas, Pappas and Rife 2006). An interesting study of Year 2 classes compared the impact of a hands-on-inquiry-oriented science lesson sequence with a blended science literacy sequence which focused on some discursive practices of science (e.g., processes of science inquiry through various text types). The topic was terrariums/habitats and lasted eight to 10 weeks. The tentative findings provide food for thought. Students experiencing the blended curricula made greater conceptual gains and understood more about the NOS. The hands-on sequence probably had a more positive influence on interest and affect towards science. These results suggest that simply acting like a scientist (e.g., doing a science investigation) may not lead to an appreciation of the NOS (also see Appleton 2007), but if the NOS were taught concurrently and explicitly then multiple science learning outcomes may be possible.

M&E

Many suggestions about how to use literacy–science connections to advantage are available. A 2006 issue of *Primary Science Review* (no. 92) focused on how children's stories can instigate effective science learning, while Mawer (2007) developed an Australian sequence around Jeannie Baker's *The Story of Rosy Dock*. Numerous examples of science literacy connections are given in later chapters (see especially Chapter 2).

Technology–science connections

In general, when technology is in the primary curriculum, its aim is to develop technologically literate people who

> understand the designed world, its artefacts, systems and the infrastructure to maintain them; have practical skills in using artefacts and fixing simple technical problems; are able to identify practical problems, design and test solutions; recognise risks and weigh costs and benefits associated with new technologies; can evaluate, select and safely use products appropriate to their needs; and contribute to decision making about the development and use of technology in environmental and social contexts.
>
> **Rennie, in Goodrum and Rennie 2007, p. 6**

There can be considerable overlap, therefore, between scientific and technological literacy, and consequently it is not surprising that most current primary science curriculums refer to 'technology', as the two areas can reinforce each other's goals. In your science curriculum, which facets of technological literacy are included or implied? What do you see as the benefits and/or limitations of the approach taken in regard to technology in your science curriculum? Also, in your curriculum, what is (are) the stated or suggested relationship(s) between science and technology? Is it one or more of the following: technology as applied science; science and technology as independent fields/communities; technology as giving rise to scientific understanding; and/or science and technology as equal and interacting fields/communities (Gardner in Jones 2009, p. 24)? In some primary science and/or technology curricula, 'design technology' is emphasised. This relates to technological problem-solving and is identified with the 'investigate, design, make/produce, and evaluate/appraise' process; in other science curricula, technology is sometimes only seen as applied science. Clearly, there can be different emphases.

Examples of some of the above types of interaction between technology and science are presented in various chapters of this book – Chapter 6 revolves around a science and design technology case study; Chapter 3 includes numerous technology examples; Chapter 8's case study poses challenging questions about science and technology outcomes; and Chapter 10 briefly refers to some additional science and technology possibilities. Although the relationship(s) between the two areas is seen as positive in this book, it is acknowledged that the role of technology in science teaching may 'still [be] hotly contested' by some (Cross 1999, p. 702), and that studies have found that the transfer of scientific knowledge to technological solutions can be problematic and hence an issue for teachers of science and technology (Jones 2009; Venville et al., 2002). For a more comprehensive treatment of primary school technology, you should consult other references; for example, Fleer and Jane (2004) and Harriman (1998).

Information and communication technology (ICT)

ICT is now integral to many countries' primary curricula. It is not to be equated with technology as described above, although it is clearly associated with it. ICT refers to a range of computer-associated tools and resources, including those used as a means of communication. These can overlap and all can assist science learning. ICT *tools* can be used:

- for collecting data and modelling – spreadsheets, databases and data loggers/sensors; the last mentioned for the detection of changes in variables such as temperature, light and sound intensity
- as a means of visualisation – digital telescopes and microscopes, flash animations, video clips, interactive whiteboards (IWB)
- for creating virtual situations – for example, virtual visits to other planets and experiments using simulations and real time webcams.

ICT *reference source* examples are CD-ROMs and the Internet; both can be also interactive. ICT as a *means of communication* would include digital cameras, word processing, IWBs, wiki spaces, iPods, virtual learning environments, PowerPoint, email and websites (which could include linking with, and communicating through, national and international communication networks). Many journals include reviews of ICT tools and reference sources for primary science, such as *Primary Science*, *Science and Children* and *Teaching Science*.

ICT

At present, the main ICT applications in primary science appear to be word processing, data logging, graphing, simulations, modelling, analysing data using databases and spreadsheets, and accessing information using the Internet, with the last mentioned a very common application of ICT. The use of many of these ICT applications in primary science appears still to be limited, but more case studies are now appearing: for example, using SMART Boards to help kindergarten children learn about living and non-living things and investigation processes (Preston and Mowbray 2008); using IWBs across K-6 classes for a range of *Primary Connections* topics (AAS 2005; Murcia 2010); and seven- to nine-year-olds using podcasts when learning about the body (Rodrigues and Connelly 2008). Research findings on the impact of ICT on primary students' conceptual learning in science are still limited. However, what is already apparent is the critical nature of the teacher's role in scaffolding the use of ICT, if effective science learning is to occur (Harlen 1999; Murcia 2010; Songer 2007). It has been found that, when ICT is used in primary science as a 'digital' information 'resource' for a topic, as it commonly is, then if it is not used with the appropriate scaffolds, science learning can be quite limited (Hoffman et al. 2003; McCrory Wallace, Kupperman and Krajcik 2000). The online Appendix 1.8 provides suggestions for the more effective use of 'digital' information 'resources' and the types of ICT 'cognitive tools' in primary science that may become more readily available.

Appendix 1.8 Turning a 'digital resource' into a 'cognitive tool'

Using ICT in science does have a strong motivational effect and can engage attention as well as fulfil other roles (e.g., assessment) (Ball 2003; Hart 2003; Murphy 2003; Oakes 2009). Several of these affective functions are clearly important for effective learning – for example, see the '"Hot" conceptual considerations: student engagement' section of this chapter on p. 25. However, ICT may have a limited impact on conceptual learning if teachers are not asking themselves basic questions about how they are using it for science learning. Teachers need to think about the difference between ICT 'for doing science' and ICT 'for learning science' (Songer 2007, p. 472) and to reflect on whether their students are using ICT as a tool 'to learn about a topic with' or 'to learn about a topic from' (O'Conner 2003). In summary, when selecting ICT to assist learning, teachers need to decide whether it is better than other available means and/or if it is the only means by which the science learning outcomes can be achieved. One acronym to guide teachers is WISE: Is the use of ICT '*worthwhile*, genuinely *interactive* and based on sound *scientific* and *educational* principles' (Hawkey 1999, p. 4; emphasis added). ICT is only one of a range of tools or aids that can be used to assist learning in science, but one with significant potential. This is the view taken in this book.

Content areas for constructivist learning

The selection of content within primary science curriculums is a contentious issue (Fensham 1998; Tytler 2007). For pragmatic reasons, this book focuses on the conceptual areas found in most primary science curriculums. As some have argued that there is more than one scientific way of knowing, Tytler has questioned what 'knowledge or knowledges' would be appropriate (Tytler 2007, p. 5). Several science curriculums now include 'multicultural science', and this is referred to on several occasions in this chapter (see the 'Science and culture: Indigenous and cultural knowledge' section on p. 42) and in later chapters.

Content strands in primary science curriculums

Many countries (e.g., the UK, USA, Australia and New Zealand) have similar strands in their primary science curricula, with only minor variations in emphasis. All include reference to science as a human endeavour, a way (or ways) of knowing and a body of evolving knowledge. The science knowledge and understandings component is usually related to the biological, physical, chemical, earth and space sciences.

In Australia, for example, the proposed national curriculum (ACARA 2011) comprises three interrelated strands: science inquiry skills (SIS), science as a human endeavour (SHE) and science understandings (SU). In the 'science understandings' strand, various topic areas are included (see Table 1.2) and these would be typical of most countries. It is also expected that technology and socioscientific issues, such as environmental and sustainability matters, will be related to, or sometimes be the context or catalyst for, the science understandings being studied. In this book, this curriculum structure has been used to guide the selection of content because teachers have to implement curriculums. Hence, Chapter 2 focuses on the NOS, including scientific investigations, and the remaining chapters refer to the physical, chemical, earth and space sciences (chapters 3 to 5, and 8 to 12) and the biological sciences (chapters 6 and 7) as in Table 1.2. Aspects of 'science as a human endeavour' (and the NOS) are illustrated in each of chapters 3 to 12.

Within the 'science as a human endeavour' strand, the following could be included: NOS, everyday science, science in the community, science and culture, influence of science, collaboration in science, and the contributions of scientists. Each is mentioned across chapters 2 to 12, and three are introduced below; namely, Indigenous knowledge, socioscientific issues, and careers in science and related fields.

In the 'science inquiry' strand, various science skills (e.g., observing, communicating, predicting, analysing, proposing explanations) and methods of investigation, including fair testing, are emphasised. Evidence and the critiquing of it are also key ideas, and the safe use of a variety of equipment is included. These are all considered in Chapter 2 and further exemplified in later chapters.

TABLE 1.2 Major emphases within the science understandings strand of curriculums

SCIENCE DISCIPLINE	MAJOR TOPICS AND EMPHASES (AND THE CHAPTERS WHICH FOCUS ON THEM)
Physical and Chemical sciences	Focus on movement, pushes and pulls, forces and motion (Chapter 3); sounds, light and electricity (chapters 4 and 5); everyday materials and their properties, forms and uses (Chapter 8); liquids and solids and changing materials (Chapter 9)
Biological science	Focus on living things and how they are grouped; their growth and change, structure and function, life cycles and interactions, and relationships between them (chapters 6 and 7)
Earth science	Focus on daily and local environment; Earth's resources and their use; change and major events on the Earth's surface (chapters 11 and 12)
Space science	Focus on daily and local environment; day and night; space and our solar system (Chapter 10)

Source: Adapted from ACARA 2011, The Australian Curriculum: Science, http://www.australiancurriculum.edu.au/Science/Curriculum/F-10.

Most primary science curriculums expect that teachers will use science activities to enhance students' general capabilities in areas such as literacy, numeracy, ICT and creativity, as well as make reference to cross-curriculum dimensions like sustainability and Indigenous knowledge. The making of links with other curriculum areas is also encouraged. Most of these issues have been overviewed in this chapter, while specific examples are given in later chapters (e.g. creativity in Chapter 2).

Of course, emphases differ across age ranges and this is reflected within curriculums. In younger years, a focus on an awareness of self and one's local world would predominate. Focuses in later primary years can vary but a recognition of what can be investigated scientifically and then the carrying out of such investigations is usually included (ACARA 2011). Later chapters acknowledge these differences.

Activity 1.10 will convince you that the content in Table 1.2 is consistent with your own curriculum content.

Activity 1.10 How is the science content organised in your primary science curriculum?

Survey your primary science curriculum. Locate the various 'science understandings' topics, as well as the 'science inquiry' and 'science as a human endeavour' components.

Skim the chapters of this book and hence 'sense' how it might help you implement your curriculum.

Science as a human endeavour

'Science as a human endeavour' has become more explicit in curriculums as questions have been asked about the relevance of school science content. Just how is science related to you and I and to humanity in general? Need it be perceived as a subject divorced from much of society or with which we have no agency? This strand addresses such questions, and three of its components are developed next. One way to engage students with the general purpose of this strand could be to have them consider their own family history. You could use yours as an example. Develop the approach using narratives and storytelling, as this usually engages students. Piggott (2008) did this and shows how he personalised science topics with stories from his own family history. Students could then contribute to such stories from their own personal spheres or research their own family background from a science topic perspective. Creating family time lines can help bring alive historical developments in science.

Science and culture: Indigenous and cultural knowledge

Science may not be viewed in the same way in different cultures. Those that hold this (contested) view refer to 'multicultural science education'– an edition of the journal *Science Education* (85(1), 2001) focuses on this topic – and argue that 'different societies might define and organise science differently' (Lui and Lederman 2007, p. 1285).

Proponents of multicultural science education holds that the 'scientific' world views of some cultures and peoples differs from the generally accepted characteristics of the NOS, which they argue are also culturally dependent (McKinley 2007) and should be referred to as

S&C

Western modern science (WMS). Examples of different world views could be different perspectives on the relationship between 'nature and humanity' or 'ways of thinking' – the natural world, for instance, may not be seen as an object for investigation in some non-Western cultures. Religious beliefs may also influence views about the nature of scientific knowledge (Lui and Lederman 2007).

Of particular interest to science education has been the relationship between Indigenous knowledge (IK), meaning the 'world views of Indigenous peoples', and WMS. IK does now have a place in many science curriculums, although how it should be included remains problematic (McKinley 2007). Having discussions with students about IK, whether it be Chinese medicine or Australian Aboriginal uses of plants, can lead to the questioning of how such knowledge was derived, and this can be compared and contrasted with accepted characteristics of the nature of WMS. In this way, IK is still valued (Lui and Lederman 2007). One interpretation of IK and WMS is that they both:

- have empirical databases using observations of the natural world accumulated and stored over time, but IK databases were accumulated over a longer period and included more qualitative information compared to WMS databases; and
- can construct theories (models) and make predictions and are subject to verification, but they test predictions and treat results differently.

Some argue that if IK has these characteristics, then WMS should not be presented in the science curriculum as the only 'scientific' way to 'see' the world. Others believe that there are only some aspects of IK that can be characterised in this way (McKinley 2007). Irrespective of these positions, you will need to determine how you are going to integrate IK with the topic you are teaching.

What is clear from the limited research in this area is that primary teachers of science need to acknowledge that their students (and maybe themselves) may hold different world views to that of WMS. This is important because if students' world views are ignored, then their motivation may be reduced and learning impeded. Accepting that IK has a place in the science curriculum is consistent with a sociocultural perspective on learning; it acknowledges that culture and its associated language(s) can influence students' ideas about their world and that teachers need to sensitively help students make 'border-crossings' (McKinley 2007; McKinley and Stewart 2009; Prain and Waldrip 2009). Several chapters of this book (e.g., 6, 10 and 12) refer to IK and other ways of knowing about our world.

Science in the community: careers in science and related areas

Upper primary students need to be exposed to science-related careers, as many may make life choices at such an early stage (Osborne 2008; Ramsay, Logan and Skamp 2005). What is important at this age is not giving 'vocational guidance' but to open students' eyes, not just to what people do *in* science (e.g., marine or forensic scientists) but also to the wide range of occupations that could be considered *from* science; that is, careers in which science could be helpful, from being a town planner to a hairdresser (Hannam 2008).

The influence of science, or science in the community: socioscientific issues

For many primary teachers, socioscientific issues (or SSI) may be new (Dolan, Nichols and Zeidler 2009). They are usually scientific topics that are controversial and engage students in dialogue, debate and discussion. Moral reasoning and an evaluation of ethical concerns are involved in attempts to reach a resolution of the issues. They require evidence-based reasoning. Several guidelines have been formulated to assist primary teachers in using SSI.

Before engaging students in a SSI, teachers need to:

- ensure students have an adequate background related to the science concepts that will be discussed in considering the SSI
- be aware of background information about the topic in order to use appropriate questioning during debates and similar discussion (e.g., How was the evidence obtained? What if you were in another's shoes? Are there moral issues to be considered?)
- appreciate that students' moral beliefs, alternative conceptions, lack of personal experiences and content knowledge, limited scientific reasoning skills and level of emotional maturity will influence their level of engagement with SSIs
- prepare tasks using scaffolded websites or other available sources of diverse perspectives about the issue that students can read, discuss and evaluate
- consider using guided discussion before advancing to debate and argumentation strategies; this implies becoming familiar with the key steps and processes involved in debate and argumentation.
- *While* engaging students in a SSI, teachers need to:
- be sensitive to students who may feel dissonance when they are required to negotiate, resolve conflicts and enhance the quality of their arguments
- expect and model tolerance, mutual respect and sensitivity
- assist students in developing critical thinking skills – analysis, inference, explanation, evaluation, interpretation and self-regulation – so that they wish to seek the 'truth' and be open-minded, analytical, systematic and judicious (adapted from Zeidler and Nichols 2009, pp. 49, 51–4).

SSI

For three descriptions of SSIs used in Year 5 classes and related to the Earth, life and physical sciences, see Dolan, Nichols and Zeidler (2009). A novel strategy used in a SSI approach with upper primary students, after they had experienced many cross-curricula activities related to the issue, was to engage in a mock court case about a threat to their local environment. These students played the roles of jury, prosecutors and the press while adults played the parts of judge and defendants (Evans, Green and Ling 2008). SSIs are an exciting approach to introduce to your classroom: they may change the way you and your students think about school science.

Conditions for effective primary science: an overview

Is it possible to bring together all the considerations related to effectively learning science? One large-scale Australian research initiative, the School Innovation in Science (SiS) project, found that the following eight components contributed strongly towards effective science learning:

- students are encouraged to actively engage with ideas and evidence
- students are challenged to develop meaningful understandings

- science is linked with students' lives and interests
- students' individual learning needs and preferences are catered for
- assessment is embedded within the science learning strategy
- the NOS (how science works) is represented in its different aspects
- the classroom is linked with the broader community
- learning technologies are exploited for their learning potentialities (Tytler 2002b, p. 35; 2002c, p. 9; 2003, p. 285).

The online Appendix 1.9 expands on each of these components and Campbell (2006) describes how one teacher integrated them into their teaching practice.

Appendix 1.9
The SiS components

All these components are embraced here. This book focuses on meaningful understandings and teaching approaches to assist students in modifying their ideas. Active engagement with ideas and the development of competencies in the discursive practices of science are illustrated through numerous examples and can be built into all the constructivist approaches and strategies. 'Evidence' is a critical concept embedded within these approaches, especially that gained through investigative (inquiry-oriented) science teaching practices – the importance of various forms of student and teacher talk in conjunction with such inquiries is stressed here. The NOS is regularly revisited in different sections of this book, as various content strands are discussed. Engaging student interest also receives a special mention, and its key role as a prerequisite to effective conceptual learning is emphasised. Reference is made to how science can be linked to students' preferences, interests and needs, especially through a focus on science as a human endeavour, which includes linking science with the community. Assessment is exemplified as integral to the constructivist teaching and learning processes advocated. How ICT can play a significant role in one or more of the above receives a mention in each chapter.

Underpinning all these SiS components, Osborne (2008) would argue, is the idea that students need a new vision for learning science. Teachers need to make it clear, as students get older, that they are not learning science because they will study it at secondary school, but rather because science offers, and continues to develop, ideas that have transformed the way we think about our world and ourselves, making life more comprehensible. Also, teachers need to make it explicit that science has offered us an understanding of how we know these ideas (i.e., how science works). This latter aspect (i.e., the NOS) is critical in helping students critique decisions based wholly or in part on science; for example, the nature of scientific evidence when discussing a socioscientific issue. Osborne's vision for science education should be essential reading for all teachers, especially as teachers are the key to quality student learning in science (Tytler 2007). You need to have a clear vision as to why you are teaching science to young students to help them to see the big picture or 'map'.

Your own science background

If your students are going to gain a more scientific understanding of science concepts, then it certainly is advantageous if you, as their teacher, have some ideas about what lies at the end of a sequence of lessons (Jarvis, Pell and McKeon 2003); that is, that you have some knowledge of the 'conceptual territory' of the lessons (Leach and Scott 2002). This is important irrespective of whether you are teaching children who are starting primary school or those in the middle and later years (Novak 2005). Several examples are provided in this book of students in years 1 to 3 engaging in significant conceptual conversations with their teachers and each other.

However, it is critical that you appreciate the distinction between *learning* and *knowing*, and to value learning more than knowing. We are always developing deeper understandings of ideas, whether scientific or otherwise. You and your students are on a lifelong learning path of conceptual development (Harlen 2009). Your own knowledge of science ideas needs to be seen in this light. As one university lecturer commented:

> As teachers we tend to hide [our doubts about certain science ideas considered inviolate; for example, a definition of a solid] because we believe we should know the answers. I certainly used to believe that most [other lecturers] already knew the 'right' answer and that such learning signified a personal, and shameful, deficit in my own ... knowledge. I am now convinced that such learning needs to be celebrated. It can, with benefit, be shared with colleagues, and the fact the teacher is still learning *should* be shared with students.
>
> **Goodwin 2002a, p. 395**

As you think about your own science knowledge, celebrate your questions, queries and advances. An inability to respond – say, to a particular science task – does not necessarily indicate your lack of knowledge. Understanding takes time and depends on context and, at a later time, you may reconstruct your ideas differently. Further, you may need access to equipment and/or need to carry out various actions such as observations and testing of ideas to assist in the way you respond to a question or task about science (Traianou 2006). Discussion with colleagues can also be affirming. This approach often reveals 'latent understanding waiting to be awakened' (Harlen 1999, p. 76); that is, you know more than you realise! Also appreciate, if you do not already do so, that when you are helping your students to learn science ideas, with a constructivist mindset, you will sometimes 'present' ideas to them in ways suggested in various sequencing models, but more often 'contend' ideas with them (Goodwin 2002b). This is when there is interactive dialogue about ideas and you can collaboratively decide which ideas are supported by more evidence and have the higher status. To 'contend' ideas is a healthy perspective.

You may also find that your knowledge about particular science ideas will develop as you think about and teach a particular topic; this has certainly been the experience of some teachers. When teachers appreciate how children learn constructively, this mindset has overcome their concerns about teaching science, which usually has related to confidence about particular knowledge areas. This may happen because such teachers listen to their students to find out their ideas and questions about a topic and respect their responses. As they believe students' learning in science is important, they tend to research and read to enhance their knowledge of a topic so they can assist students in reconstructing their understanding of events and phenomena. Teachers unaware of the significance of students' alternative conceptions and how they learn science may simply become frustrated and move on (Ackerson 2005). What is important here is that when you find yourself in such a situation, you may need access to the appropriate resources for clarifying and expanding your knowledge. This book assists in this regard – see the early sections of most chapters.

The online Appendix 1.10 also lists several resources written for primary teachers wishing to clarify their own science ideas. You can take heart from one study that found when primary teachers' reading was focused, their science content knowledge improved (Smith 1997).

Appendix 1.10 Background science books for primary and middle school teachers

Our understanding of science ideas can often be messy. Providing decontextualised statements of science concepts and understandings (as in the next section) are not necessarily the best way forward. However, if you interact with them from where you are at and see them as one set of tools in an ongoing development of your own conceptual understanding, then they have their place.

Concepts and understandings for primary teachers

Listed at the end of each of the conceptual chapters of this book are some of the key conceptual ideas and understandings related to that chapter's topic area. They are considered to be ideas with which a primary teacher should be familiar when teaching the topic. These concepts and understandings include the ideas towards which we are encouraging students to move. Some would be appropriate for learners at lower primary level, others for middle and upper primary learners, some for only the most able students, and some just as guidance for teachers as they interact with their students and lay the foundations for later learning. It is imperative that teachers use this list to audit their own understanding, not as a list of concepts to be directly taught (transmitted) to students at various levels or stages, which would contradict this book's message. Where necessary, you may have to talk with colleagues and/or consult other resource material in order to understand (meaningfully learn) rather than superficially (rote) learn the listed ideas.

Also, it needs to be appreciated that it is virtually impossible to definitively state that concept or understanding X should be understood by all students at stage A. Many of the conception studies in this book emphasise this fact. It is for this reason that various concepts and understandings are not listed for particular stages, ages or grade levels. Only signposts for teachers can be provided, and these signposts are within the introductory sections of most of the chapters. In many cases, we can make a good judgement about what is appropriate and can usually tell fairly quickly what is inappropriate if we know what we are looking and listening for as we interact with our students.

What is special about these lists of concepts and understandings is that many of them are derived from alternative conception studies and hence address the (alternative) views that many students and teachers may hold; see Driver et al. (1994a, 1994b).

Summary

Constructivism is well grounded as a theory of learning. Teaching approaches to help learners 'sense-make' or develop and (re)construct their ideas based on the variants of this theory, and related theoretical insights, are still being researched and refined. Conceptual change in science, for example, is now better described as science learning which results in 'various changes in perspectives' (Duit, Treagust and Widodo 2008, p. 641). Many useful pedagogies have been proposed, albeit from different constructivist standpoints, and considerable research evidence supports their general success in improving conceptual learning. This book describes, with examples, a range of models, schemata, strategies and principles to guide lesson sequences. Reflection on these ideas and approaches, it is hoped, will help you to develop your own coherent constructivist mindset and pedagogy.

Appendix 1.11 Two model teachers guided by constructivist mindsets

Activity 1.11 Reflecting on exemplary practice

Before reading further, you may wish to read about two exemplary teachers of primary science who were guided by constructivist mindsets. The factors that characterised each teacher's practice are described in the online Appendix 1.11 (Beeth and Hewson 1999; Tytler, Cripps Clark and Darby 2009). They are a worthy set of guidelines to aim for.

Compare their pedagogy with some of the recent thinking about learning through representation and the influence of sociocultural and affective factors.

Look back at your initial conceptions of 'learning science' and 'teaching science' (see Activities 1.5 and 1.6 on pp. 10–11). Devise a list of priorities that you want to investigate and reflect upon over the next year as you improve your practice in teaching primary science.

Search me! science education

Search me!

Explore **Search me! science education** for relevant articles on teaching primary science constructively. Search me! is an online library of world-class journals, ebooks and newspapers, including *The Australian* and the *New York Times*, and is updated daily. Log in to Search me! through http://login.cengage.com using the access code that comes with this book.

KEYWORDS

Try searching for the following terms:

- constructivism
- social constructivism
- sociocultural constructivism

Search tip: **Search me! science education** contains information from both local and international sources. To get the greatest number of search results, try using both Australian and American spellings in your searches: e.g., 'globalisation' and 'globalization'; 'organisation' and 'organization'.

Appendices

In these appendices you will find material related to teaching primary science constructively that you should refer to when reading Chapter 1. These appendices can be found on the student companion website. Log in through http://login.cengage.com using the access code that comes with this book.

Appendix 1.1 What do you think science is?

After an introduction that argues how our beliefs about the NOS influence our classroom practice and our students, there are five tasks that challenge your thinking about this question. They relate to your views of a scientist, the implications of media reports about science, a series of thought-provoking questions (e.g., about evidence and dinosaurs), determining your 'nature of science' profile, and a closed questionnaire seeking your views about science. The appendix concludes with how your epistemological view of science will influence your teaching and lists a consensus view of the NOS as well as what would be appropriate for primary students.

Appendix 1.2 Learning science, doing science and learning about science

Provided is an interpretation of these three ways of thinking about science and how they could be related to hands-on, heads-on and hearts-on science.

Appendix 1.3 Teachers' conceptions of 'learning science', 'teaching science' and 'science'

The views that teachers hold about the responses to the questions posed in Activities 1.5 and 1.6 (see pp. 10–11) are summarised.

Appendix 1.4 Constructivism: some important caveats

Further details are provided about learning theories other than constructivism, constructivism as a theory of knowledge, and critiques of constructivism.

Appendix 1.5 Scaffolding strategies

The purposes of three different levels of scaffolding and what teachers could do at each level are listed.

Appendix 1.6 Checklists of notes for constructivist teachers

Three checklists are provided. The first lists a range of teacher actions consistent with constructivist pedagogy. The second contrasts actions a teacher could take which either encourage students to access science experiences or limit their access to them. From a social constructivist perspective, you should be orienting your actions to the access-enhancing column. The third provides a summary of teacher roles across the spectrum of constructivist perspectives as well as referring to science inquiry and assessment. The source of another checklist is listed.

Appendix 1.7 Constructivist teaching: Does it work?

Further details evaluating the effectiveness of personal constructivist approaches to teaching are provided.

Appendix 1.8 Turning a 'digital resource' into a 'cognitive tool'

Guidelines for using digital resources more effectively are outlined, as well as examples of sophisticated cognitive tools that may become more widely available in future years.

Appendix 1.9 The SiS components

More complete descriptions of these eight components are outlined.

Appendix 1.10 Background science books for primary and middle school teachers

Eight titles are suggested.

Appendix 1.11 Two model teachers guided by constructivist mindsets

The characteristic actions of two exemplary teachers of primary science are detailed.

Appendix 1.12 Another view about 'alternative conceptions'

This appendix outlines Novak's (2002, p. 555) reasons for why 'students' alternative conceptions' is an inappropriate term and should be replaced by 'limited or inappropriate propositional hierarchies'.

References

Ackerson, V. 2005. How do elementary teachers compensate for incomplete science content knowledge? *Research in Science Education*, 35 (2/3), pp. 245–68.

Aikenhead, G. 1996. Science education: Border crossing into the subculture of science. *Studies in Science Education*, 26, pp. 1–52.

Appleton, K. 2007. Elementary science teaching, in S. Abell & N. Lederman (eds). *The Handbook of Research on Science Education*. Mahwah, NJ: Lawrence Erlbaum Associates, pp. 293–535.

Australian Academy of Science (AAS). 2005. *Primary Connections*. Canberra: AAS.

Australian Curriculum, Assessment and Reporting Authority (ACARA). 2011 Australian Curriculum: Science. Available at http://www.australiancurriculum.edu.au/Science/Curriculum/F-10 (accessed 12 April 2011)

Ausubel, D. 1968. *Educational Psychology*. New York: Holt, Rinehart & Winston.

_____. 2000. *The Acquisition and Retention of Knowledge*. Dordrecht: Kluwer Academic.

Baird, J. & Mitchell, I. (eds). 1986. *Improving the Quality of Teaching and Learning: An Australian Case Study – the PEEL Project*. Melbourne: Monash University Press.

Ball, S. 2003. ICT that works. *Primary Science Review*, 76, pp. 11–13.

Beeth, M. & Hewson, P. 1999. Learning goals in an exemplary science teacher's practice: Cognitive and social factors in teaching for conceptual change. *Science Education*, 83, pp. 738–60.

Bell, B. 1993. *Children's Science, Constructivism and Learning in Science*. Geelong: Deakin University Press.

_____. 2007. Classroom assessment of science learning, in S. Abell & N. Lederman (eds). *The Handbook of Research on Science Education*. Mahwah, NJ: Lawrence Erlbaum Associates, pp. 965–1006.

_____, Cowie, B. & Jones, A. 2009. Theorising learning in science, in S. Ritchie (ed.). *The World of Science Education: Handbook of Research in Australasia*. Rotterdam: Sense Publishers, pp. 85–105.

Ben-Ari, M. 2001. Constructivism in computer science education. *Journal of Computers in Mathematics and Science Teaching*, 20 (1), pp. 45–73.

Bencze, J. 2000. Democratic constructivist science education: Enabling egalitarian literacy and self-actualisation. *Journal of Curriculum Studies*, 32 (6), pp. 847–65.

Bereiter, C. 1994. Constructivism, socioculturalism and Popper's world 3. *Educational Researcher*, 23 (7), pp. 21–3.

Biddulph, F. & Osborne, R. 1984. *Making Sense of Our World: An Interactive Teaching Approach*. Hamilton: University of Waikato Centre for Science and Mathematics Education Research.

Black, P. & William, D. 1998. Assessment and classroom learning. *Assessment in Education*, 5 (1), pp. 7–74.

Brigland, A., Way, H. & Buckle, A. 2008. Tackling talk through science. *Primary Science*, 102, pp. 25–6.

Brown, P. & Abell, S. 2007. Examining the learning cycle. *Science and Children*, 46, pp. 58–9.

Bruner, J. 1962. *On Knowing: Essays for the Left Hand*. Cambridge: Harvard University Press.

Bybee, R., Buchwald, C., Crissman, S., Heil, D., Kuerbis, P., Matsumoto, C. & McInerney, J. 1989. *Science and Technology Education for the Elementary Years: Frameworks for Curriculum and Instruction*. Washington, DC: National Center for Improving Science Education.

Campbell, C. 2006. A teacher's attempt at change in a science classroom. *Teaching Science*, 52 (3), pp. 18–26.

Caravita, S. 2001. A re-framed conceptual change theory? *Learning and Instruction*, 11, pp. 421–9.

Claxton, G. 1989. Cognition doesn't matter if you're scared, depressed or bored, in P. Adey (ed.). *Adolescent Development and School Science*. London: Falmer, pp. 155–61.

Clement, D. 2008. The role of explanatory models in teaching for conceptual change, in S. Vosniadou (ed.) *International Handbook of Research on Conceptual Change*. London: Routledge, pp. 417–52.

Cobb, P. 1994. Constructivism in mathematics and science education. *Educational Researcher*, 23 (7), p. 4.

Cowie, B. 2009. The evolution of assessment purposes and practices, in S. Ritchie (ed.) *The World of Science Education: Handbook of Research in Australasia*. Rotterdam: Sense Publishers, pp. 235–48.

Cross, R. 1999. The public understanding of science. *International Journal of Science Education*, 21 (7), pp. 699–702.

Darby, L. 2005. Science students' perceptions of engaging pedagogy. *Research in Science Education*, 35 (4), pp. 425–45.

Davison, J., Kenny, J., Johnson, J. & Fielding, J. 2004. Use learning objects to bring an exciting new dimension to your classroom. *Teaching Science*, 50 (1), pp. 6–9.

diSessa, A. 2008. A bird's-eye view of the 'pieces' vs. 'coherence' controversy (from the 'pieces' side of the fence), in S. Vosniadou (ed.) *International Handbook of Research on Conceptual Change*. London: Routledge, pp. 35–60.

Dolan, T., Nichols, B. & Zeidler, D. 2009. Using socioscientific issues in primary classrooms. *Journal of Elementary Science Education*, 21 (3), pp. 1–12.

Driver, R., Squires, A., Rushworth, P. & Wood-Robinson, V. 1994a. *Making Sense of Secondary Science: Support Materials for Teachers*. London: Routledge.

_____. 1994b. *Making Sense of Secondary Science: Research into Children's Ideas*. London: Routledge.

Duit, R. 2006. Bibliography – Students' and Teachers' Conceptions and Science Education. Available at http://www.ipn.uni-kiel.de/aktuell/stcse/stcse.html (accessed 30 March 2010).

_____ & Treagust, D. 1998. Learning in science: From behaviourism towards social constructivism and beyond, in B. Fraser & K. Tobin (eds). *International Handbook of Science Education*. Great Britain: Kluwer Academic, pp. 3–25.

_____, Treagust, D. & Widido, A. 2008. Teaching science for conceptual change: Theory and practice, in S. Vosniadou (ed.) *International Handbook of Research on Conceptual Change*. London: Routledge, pp. 629–46.

Evagorou, M. & Osborne, J. 2007. Argue-WISW: Using technology to support argumentation. *School Science Review*, 89 (327), pp. 103–10.

Evans, S., Green, M. & Ling, G. 2008. Putting your elders on trial. *Primary Science*, 101, pp. 28–31.

Fensham, P. 1998. *Primary Science and Technology in Australia*. Perth: National Key Centre for School Science and Mathematics, Curtin University of Technology.

Fitzgerald, A., Dawson, V. & Hackling, M. 2009. Perceptions and pedagogy: Exploring the beliefs and practices of an effective primary science teacher. *Teaching Science*, 55 (3), pp. 19–22.
Flack, J., Mariniello, S., Osler, J., Saffin, A. & Strapp, K. 1998. *PEEL from a Primary Perspective.* Project for Enhancing Effective Learning. Melbourne: Monash University.
Fleer, M. 1990. Scaffolding conceptual change in early childhood. *Research in Science Education*, 20, pp. 114–23.
_____. 1999. Children's alternative views: Alternative to what? *International Journal of Science Education*, 21 (2), pp. 119–35.
_____. 2009. Understanding the dialectical relations between everyday concepts and scientific concepts within play-based programs. *Research in Science Education*, 39 (2), pp. 281–306.
_____ & Hardy, T. 2001. *Science for Children* (2nd edn). Sydney: Prentice Hall.
_____ & Jane, B. 2004. *Technology for Children*. Sydney: Pearson Australia.
_____, Jane, B. & Hardy, T. 2007. *Science for Children* (3rd edn). Sydney: Pearson Education.
_____ & Robbins, J. 2003a. Understanding our youngest scientific and technological thinkers: International developments in early childhood science education. *Research in Science Education*, 33 (4), pp. 399–404.
_____. 2003b. 'Hit and run research' with 'hit and miss' results in early childhood science education. *Research in Science Education*, 33 (4), pp. 405–32.
Francis, M. 2007. The impact of drama on pupils' learning in science. *School Science Review* 89 (327), pp. 91–102.
Gabel, D. 1996. The complexity of chemistry: Research for teaching in the 21st century, in W. Beasley (ed.) *Chemistry: Expanding the Boundaries Proceedings of the 14th International Conference on Chemical Education.* Brisbane: Royal Australian Chemical Institute, Chemical Education Division, pp. 43–9.
Georghiades, P. 2006. The role of metacognition activities in the contextual use of primary pupils' conceptions of science. *Research in Science Education*, 36, pp. 29–49.
Gil-Perez, D., Guisasola, J., Moreno, A., Cachapuz, A., Pessoa de Carvalho, A., Torregroas, J., Salinas, J., Valdes, P., Gonzalez, E., Duch, A., Dumas-Carre, A., Tricarico, H. & Gallego, R. 2002. Defending constructivism in science education. *Science & Education*, 11, pp. 557–71.
Gilbert, J. 2006. On the nature of 'context' in chemical education. *International Journal of Science Education*, 28 (9), pp. 957–76.
Girod, M. & Twyman, T. 2009. Comparing the added value of blended science and literacy curricula to inquiry-based science curricula in two 2nd-grade classrooms. *Journal of Elementary Science Education*, 21 (3), pp. 13–22.
_____ & Wong, D. 2002. An aesthetic (Deweyan) perspective on science learning: Case studies of three fourth grades. *The Elementary School Journal*, 102 (3), pp. 199–224.
Goodrum, D. 1987. Upper primary science, in K. Tobin & B. Fraser (eds). *Exemplary Practice in Science and Mathematics Education.* Perth: Curtin University Press, pp. 69–81.
_____, Hackling, M. & Rennie, L. 2001. *The Status and Quality of Teaching and Learning of Science in Australian Schools*. Canberra: Department of Education, Training and Youth Affairs.
_____ & Rennie, L. 2007. *Australian school science education: National action plan 2008–2010.* Canberra: Department of Education, Science and Training.
Goodwin, A. 2002a. Is salt melting when it dissolves in water? *Journal of Chemical Education*, 79 (3), pp. 393–6.
_____. 2002b. Teachers' continuing learning of chemistry: Some implications for science teaching. *Chemical Education: Research and Practice in Europe*, 3 (3), pp. 345–59.
Gunstone, R. 1994. The importance of specific science content in the enhancement of metacognition, in P. Fensham, R. Gunstone & R. White (eds). *The Content of Science Education: A Constructivist Approach to Its Teaching and Learning.* London: Falmer Press, pp. 131–46.
_____ & Northfield, J. 1992. Conceptual change in teacher education: The centrality of metacognition. Paper presented at the annual meeting of the American Educational Research Association, San Francisco.
Hackling, M. 2008. *An overview of Primary Connections Stage 3 research outcomes 2006–2008*. Canberra: Australian Academy of Science.
_____, Peers, S. and Prain, V. 2007. *Primary Connections:* Reforming science teaching in Australian primary schools. *Teaching Science*, 53 (3), pp. 12–16.
_____ & Prain, V. 2005. Primary Connections. *Stage 2 Trial Research Report*. Canberra: Australian Academy of Science.
Hand, B. & Keys, C. 1999. Inquiry investigation. *The Science Teacher*, 66, pp. 27–9.
Hannam, N. 2008. Careers from science: The importance of the careers message. *School Science Review*, 89 (328), pp. 123–7.
Hanrahan, M. 2006. Highlighting hybridity: A critical discourse analysis of teacher talk in science classrooms. *Science Education*, 90 (1), pp. 8–43.
Harlen, W. 1992. *The Teaching of Science*. London: David Fulton.
_____. 1998. Teaching for understanding in pre-secondary school science, in B. Fraser & K. Tobin (eds). *International Handbook of Science Education*. Great Britain: Kluwer Academic, pp. 183–97.
_____. 1999. *Effective Teaching of Science: A Review of the Research*. Edinburgh: Scottish Council for Research in Education.
_____. 2001. Taking children's ideas seriously. *Primary Science Review*, 67, pp. 14–17.
_____. 2007. Holding up a mirror to classroom practice. *Primary Science Review*, 100, pp. 33–4.
_____. 2009. Teaching and learning science for a better future. *School Science Review*, 90 (333), pp. 33–42.
Harriman, S. C. 1998. *Carrots, Kites, and Traffic Lights*. Melbourne: Curriculum Corporation.
Hart, G. 2003. Quality ICT resources to enhance science teaching. *Primary Science Review*, 76, pp. 6–8.
Hawkey, R. 1999. Exploring and investigating the natural world: Sc1 on line. *Primary Science Review*, 60, pp. 4–6.
Hellden, G. 2005. Exploring understandings and responses to science: A program of longitudinal studies. *Research in Science Education*, 35 (1), pp. 99–122.
Hewson, P., Beeth, M. & Thorley, N. 1998. Teaching for conceptual change, in B. Fraser & K. Tobin (eds). *International Handbook of Science Education*. Great Britain: Kluwer Academic, pp. 199–218.
Hodson, D. 2002. A new metaphor for teaching: Science teacher as anthropologist, in *Rethinking Science and Technology Education to Meet the Demands of Future Generations in a Changing World.* International Organisation for Science and Technology Education Symposium Proceedings. Foz do Iguaca, Parana, Brazil, July.
_____ & Hodson, J. 1998a. From constructivism to social constructivism: A Vygotskian perspective on teaching and learning science. *School Science Review*, 79 (289), pp. 33–41.
_____. 1998b. Science education as enculturation: Some implications for practice. *School Science Review*, 80 (290), pp. 17–23.
Hoffman, J., Hsin-Kai, W., Krajcik, J. & Solaway, E. 2003. The nature of middle school learners' science content understandings with the use of online resources. *Journal of Research in Science Teaching*, 40 (3), pp. 323–46.
Holt-Reynolds, D. 2000. What does the teacher do? Constructivist pedagogies and prospective teachers' beliefs about the role of the teacher. *Teaching and Teacher Education*, 16, pp. 21–32.
Hubber, P. & Haslam, F. 2009. The role of representation in teaching and learning astronomy. Paper presented at the Australasian Science Education Research conference, Geelong, July.

Hume, A. 2009. Authentic scientific enquiry and school science. *Teaching Science*, 55 (2), pp. 35–41.
Jarvis, T., Pell, A. & McKeon, F. 2003. Changes in primary teachers' science knowledge and understanding during a two year in-service program. *Research in Science & Technological Education*, 21 (1), pp. 17–42.
Johnson, P. 2005. The development of children's concept of substance: A longitudinal study of interaction between curriculum and learning. *Research in Science Education*, 35, pp. 41–61.
Jonassen, D. 2008. Model building for conceptual change, in S. Vosniadou (ed.). *International Handbook of Research on Conceptual Change*. London: Routledge, pp. 676–93.
Jones, A. 2009. Exploring the tension and synergies between science and technology in science education, in S. Ritchie (ed.). *The World of Science Education: Handbook of Research in Australasia*. Rotterdam: Sense Publishers, pp. 17–27.
_____, Moreland, J. & Cowie, B. 2006. Pedagogical content knowledge and classroom interactions in science and technology classrooms. Paper presented at Australasian Science Education Research Association Conference, Canberra, July.
Kelly, G. 2008. Learning science: Discursive practices, in A. de Mejia & M. Martin-Jones (eds). *Encyclopedia of Language and Education, Vol. 3: Discourse and Education*. New York: Springer, pp. 329–40.
Keogh, B., Dowling, B., Maloney, J. & Simon, S. 2006. Puppets bringing stories to life in science. *Primary Science Review*, 92, pp. 26–8. Available at http://www.puppetsproject.com (accessed 10 May 2006).
_____ & Naylor, S. 2004. Children's ideas, children's feelings. *Primary Science Review*, 82, pp. 18–21.
_____. 2007. Talking and thinking in science. *School Science Review*, 88 (324), pp. 85–90.
Khishe, R. & Abd-El-Khalick, F. 2002. Influence of explicit and reflective versus enquiry-oriented instruction on sixth graders' views of nature of science. *Journal of Research in Science Teaching*, 40 (1), 2002, p. 104.
Koballa, T. & Glynn, S. 2007. Attitudinal and motivational constructs in science learning, in S. Abell & N. Lederman (eds). *The Handbook of Research on Science Education*. Mahwah, NJ: Lawrence Erlbaum Associates, pp. 75–102.
Leach, J. & Scott, P. 2000. Children's thinking, learning, teaching and constructivism, in M. Monk & J. Osborne (eds). *Good Practice in Science Teaching*. Buckingham, UK: Open University Press, pp. 41–55.
_____. 2002. Designing and evaluating science teaching sequences: An approach drawing upon the concept of learning demand and a social constructivist perspective on learning. *Studies in Science Education*, 38, pp. 115–42.
_____. 2008. Teaching for conceptual understanding: An approach drawing on individual and sociocultural perspectives, in S. Vosniadou (ed.). *International Handbook of Research on Conceptual Change*. London: Routledge, pp. 647–75.
Liem, T. 1987. *Invitations to Science Enquiry* (2nd edn). Thornhill, Ontario: S17 Science.
Logan, M. & Skamp, K. 2008. Engaging students in science across the primary secondary interface: Listening to the students' voices. *Research in Science Education*, 38 (4), pp. 501–27.
Lui, S. & Lederman, N. 2007. Exploring prospective teachers' worldviews and conceptions of the nature of science. *International Journal of Science Education*, 29 (10), pp. 1281–1307.
Lunn, S. 2002. 'What we think we can safely say ...': Primary teachers' views of the nature of science. *British Educational Research Journal*, 28 (5), pp. 649–72.
Marek, E. 2008. Why the learning cycle? *Journal of Elementary Science Education*, 20 (23), pp. 63–9.
Martin, A. & Hand, B. 2009. Factors affecting the implementation of argument in the elementary science classroom: A longitudinal study. *Research in Science Education*, 39 (1), pp. 17–38.
Mathews, M. 1995. *Challenging New Zealand Science Education*. Palmerston North: Dunmore Press.
_____. (ed.). 1998. *Constructivism in Science Education*. Dordrecht: Kluwer Academic.
Mawer, M. 2007. Linking science and literacy: The story of Rosy Dock (S1–S3 unit). Paper presented at the Science Teachers Association of NSW Primary conference, Sydney, September.
McCrory, P. 2008. Getting them emotional about science. *Education in Science*, 229, pp. 32–3.
McCrory Wallace, R., Kupperman, J. & Krajcik, J. 2000. Science on the web: Students online in a sixth grade classroom. *Journal of Learning Sciences*, 9 (1), pp. 75–104.
McGuigan, L. & Russell, T. 1997. What constructivism tells us about managing the teaching and learning of science. *Primary Science Review*, 50, pp. 15–17.
McKinley, E. 2007. Postcolonialism, indigenous students and science education, in S. Abell & N. Lederman (eds). *The Handbook of Research on Science Education*. Mahwah, NJ: Lawrence Erlbaum Associates, pp. 199–226.
_____ & Stewart, E. 2009. Falling into place: Indigenous science education research in the Pacific, in S. Ritchie (ed.). *The World of Science Education: Handbook of Research in Australasia*. Rotterdam: Sense Publishers, pp. 49–68.
McWilliam, S. 1999. Fostering wonder in young children: Baseline study of two first grade classes. Paper presented at the annual meeting of the National Association of Research in Science Teaching, Boston, MA. Available at http://www.eric.ed.gov/PDFS/ED441833.pdf (accessed 30 March 2006).
Metz, K. 1998. Scientific enquiry within reach of young children, in B. Fraser & K. Tobin (eds). *International Handbook of Science Education*. Great Britain: Kluwer Academic, pp. 81–96.
Mitchell, I. 1992. The class level, in J. Baird & J. Northfield (eds). *Learning from the PEEL Experience*. Melbourne: Monash University.
_____. 2009. *Teaching for Effective Learning: The Complete Book of PEEL Teaching Procedures* (4th edn). Melbourne: PEEL Publishing (Monash University).
_____, Mitchell, J. & McKinnon, R. 2000. *Peel in Practice* (CD-ROM). Melbourne: PEEL Publishing (Monash University).
Miyake, N. 2008. Conceptual change through collaboration, in S. Vosniadou (ed.). *International Handbook of Research on Conceptual Change*. New York: Routledge, pp. 453–78.
Murcia, K. 2009. Re-thinking the development of scientific literacy through a rope metaphor. *Research in Science Education*, 39 (2), pp. 215–29.
_____. 2010. Multi-modal representations in primary science: What's offered by interactive whiteboard technology? *Teaching Science*, 56 (1), pp. 23–32.
Murcia, K. & Schibeci, R. 1999. Primary student teachers' conceptions of the nature of science. *International Journal of Science Education*, 21 (11), pp. 1123–40.
Murphy, C. 2003. *Literature Review in Primary Science and ICT: A Report for NESTA Futurelab*. Bristol: NESTA Futurelab.
Murphy, P. & Alexander, P. 2008. The role of knowledge, beliefs and interest in the conceptual change process: A synthesis and meta-analysis of the research, in S. Vosniadou (ed.). *International Handbook of Research on Conceptual Change*. New York: Routledge, pp. 583–616.
_____ & Davidson, M. 2002. Exploring the characteristics of effective teaching and learning: A study of primary science classrooms. Paper presented at the Annual Meeting of the NARST, New Orleans, April.
_____, Qualter, A., Simon, S. & Watt, D. 2001. *Effective Practice in Primary Science: Report of an Exploratory Study*. London: Nuffield Foundation.
Naylor, S. & Keogh, B. 2007. Active assessment: Thinking, learning and assessment in science. *School Science Review*, 88 (325), pp. 73–80.

_____, Keogh, B. & Downing, B. 2007. Argumentation and primary science. *Research in Science Education*, 37 (1), pp. 17–39.
_____, Keogh, B. & Goldsworthy, A. 2004. *Active Assessment*. London: David Fulton.
Novak, J. 2002. Meaningful learning: The essential factor for conceptual change in limited or inappropriate propositional hierarchies leading to empowerment of learners. *Science Education*, 86, pp. 548–71.
_____. 2005. Results and implications of a 12-year longitudinal study of science concept learning. *Research in Science Education*, 35, pp. 23–40.
Oakes, M. 2009. 'Hard to teach' topics in science. *Education in Science*, 231, pp. 20–1.
O'Conner, L. 2003. ICT and primary science: Learning 'with' or learning 'from'. *Primary Science Review*, 76, pp. 14–16.
Osborne, J. 1996. Beyond constructivism. *Science Education*, 80 (1), pp. 53–82.
_____. 2008. Engaging young people with science: Does science education need a new vision? *School Science Review*, 89 (238), pp. 67–71.
_____, Erduran, S., Simon, S. & Monk, M. 2001. Enhancing the quality of argument in school science. *School Science Review*, 82 (301), pp. 63–9.
Osborne, R. & Freyberg, P. 1985. *Learning in Science: The Implications of Children's Science*. Auckland: Heinemann.
_____ & Wittrock, M. 1983. Learning science: A generative process. *Science Education*, 67 (4), pp. 489–504.
_____. 1985. The generative learning model and its implications for science education. *Studies in Science Education*, 12, pp. 59–87.
Palmer, D. 2004. Theory into practice: Ideas for arousing student interest in science classes. *Teaching Science*, 50 (2), pp. 51–3.
Park, D. & Lee, Y. 2009. Different conceptions of the nature of science among preservice elementary teachers of two countries. *Journal of Elementary Science Education*, 21 (2), pp. 1–14.
Patchen, T. & Cox Peterson, A. 2008. Constructing cultural relevance in science: A case study of two elementary teachers. *Science Education*, 92 (6), pp. 994–1014.
Phillips, D. C. 1996. The good, the bad and the ugly: The many faces of constructivism. *Educational Researcher*, 24 (7), pp. 5–12.
Piggott, A. 2008. Across the generations or 100+ years of stories. *School Science Review*, 89 (329), pp. 83–9.
Porlan, R. & del Pozo, R. 2004. The conceptions of in-service and prospective primary school teachers about the teaching and learning of science. *Journal of Science Teacher Education*, 15 (1), pp. 39–62.
Posner, G. L., Strike, K. A., Hewson, P. W. & Gertzog, W. A. 1982. Accommodation of a scientific conception: Toward a theory of conceptual change. *Science Education*, 66 (2), pp. 211–27.
Prain, V. & Waldrip, B. 2009. Representation and learning in science in Australasia, in S. Ritchie (ed.). *The World of Science Education: Handbook of Research in Australasia*. Rotterdam: Sense Publishers, pp. 69–84.
_____. 2010. Representing science literacies: An introduction. *Research in Science Education*, 40 (1), pp. 1–3.
Pratt, N. 2009. Process skills and knowledge: What does it mean to learn science? *Primary Science*, 106, pp. 27–30.
Preston, C. & Mowbray, L. 2008. Use of SMART Boards for teaching, learning and assessment in kindergarten science. *Teaching Science* 54 (2), pp. 50–3.
Pring, R. 1976. *Knowledge and Schooling*. London: Open Books.
Prinsen, M. 2001. Teaching the dog to whistle: Case study exploring the professional development needs of teachers implementing a new constructivist-based science syllabus. Unpublished BEd (Hons) thesis. Rockhampton: Central Queensland University.
Ramsay, K., Logan, M. & Skamp, K. 2005. Primary students' perceptions of science and scientists and upper secondary science enrolments. *Teaching Science*, 51 (4), pp. 21–6.
Rennie, L. 2005. Science awareness and scientific literacy. *Teaching Science*, 51 (1), pp. 10–14.
Robbins, J. 2005. 'Brown paper packages'? A sociocultural perspective on young children's ideas in science. *Research in Science Education*, 35 (2), pp. 151–72.
Rodrigues, S. & Connelly, E. 2008. Podcasting in primary science: Creative science education. *Primary Science*, 105, pp. 33–5.
Russell, T. & Martin, A. 2007. Learning to teach science, in S. Abell & N. Lederman (eds). *The Handbook of Research on Science Education*. Mahwah, NJ: Lawrence Erlbaum Associates, pp. 1151–78.
Sadler, T., Barab, A. & Scott, B. 2007. What do students gain by engaging in socioscientific inquiry? *Research in Science Education*, 37 (4), pp. 371–91.
Schraw, G., Crippen, K. & Hartley, K. 2006. Promoting self-regulation in science education: Metacognition as part of a broader perspective on learning. *Research in Science Education*, 36, pp. 111–39.
Scott, P., Asoko, H. & Leach, J. 2007. Student conceptions and conceptual learning in science, in S. Abell & N. Lederman (eds). *The Handbook of Research on Science Education*. Mahwah, NJ: Lawrence Erlbaum Associates, pp. 31–56.
_____, Dyston, T. & Gater, S. 1987. *A Constructivist View of Learning and Teaching in Science*. Leeds: Centre for Studies in Science and Mathematics Education, Leeds University.
Shakespeare, D. 2003. Starting an argument in science lessons. *School Science Review*, 85 (311), pp. 103–8.
Sherrington, R. (ed.). 1998. *ASE Guide to Primary Science Education*. Hatfield, UK: ASE.
Sinatra, G. & Mason, L. 2008. Beyond knowledge: Learner characteristics influencing conceptual change, in S. Vosniadou (ed.). *International Handbook of Research on Conceptual Change*. New York: Routledge, pp. 560–81.
Skamp, K. 2000. Working constructively. *Queensland Science Teacher*, 27 (2), pp. 28–33.
_____. 2001. Student teachers' conceptions of effective primary science practice: A longitudinal study. *International Journal of Science Education*, 23 (4), pp. 331–51.
_____. 2007. Conceptual learning in the primary and middle years. The interplay of heads, hearts and hands-on science: More than just a mantra. *Teaching Science*, 53 (3), pp. 18–22.
Smith, R. 1997. 'Before teaching this I'd do a lot of reading': Preparing primary school teachers to teach science. *Research in Science Education*, 27 (1), pp. 141–54.
Songer, N. 2007. Digital resources versus cognitive tools: A discussion of learning science with technology, in S. Abell & N. Lederman (eds). *The Handbook of Research on Science Education*. Mahwah, NJ: Lawrence Erlbaum Associates, pp. 471–91.
Stard, A. 1998. On two metaphors for learning and the danger of choosing just one. *Educational Researcher*, 27 (2), pp. 4–13.
Taber, K. 2001. The mismatch between assumed prior knowledge and the learner's conceptions: A typology of learning impediments. *Educational Studies*, 27 (2), pp. 159–71.
Taylor, P. 1998. Constructivism: Value added, in B. Fraser & K. Tobin (eds). *International Handbook of Science Education*. Great Britain: Kluwer Academic, pp. 1111–23.
_____, Fraser, B. & Fisher, D. 1997. Monitoring constructivist classroom learning environments. *International Journal of Science Education*, 27 (4), pp. 293–302.
Tobin, K., Tippins, D. & Gallard, A. 1995. Research on instructional strategies for teaching science, in D. Gabel (ed.). *Handbook of Research on Science Teaching and Learning*. New York: Macmillan, pp. 45–93.
Traianou, A. 2006. Teachers' adequacy of subject knowledge in primary science: Assessing constructivist approaches from a sociocultural perspective. *International Journal of Science Education*, 28 (8), pp. 827–42.
Treagust, D. 2007. General instructional methods and strategies, in S. Abell & N. Lederman (eds). *The Handbook of Research on Science Education*. Mahwah, NJ: Lawrence Erlbaum Associates, pp. 373–91.

Trippett, C. 2009. Argumentation: The language of science. *Journal of Elementary Science Education* 21 (1), pp. 17–25.
Tsai, C. 2002. Nested epistemologies: Science teachers' beliefs of teaching, learning and science. *International Journal of Science Education*, 24 (8), pp. 771–83.
Tytler, R. 2002a. Teaching for understanding in science: Student conceptions research and changing views of learning. *Australian Science Teachers' Journal*, 48 (3), pp. 14–21.
_____. 2002b. Teaching for understanding in science: Constructivist/conceptual change teaching approaches. *Australian Science Teachers' Journal*, 48 (4), pp. 30–5.
_____. 2002c. School innovation in science (SiS): Focusing on teaching. *Investigating: Australian Primary and Junior Science Journal*, 18 (3), pp. 8–11.
_____. 2003. A window for a purpose: Developing a framework for describing effective science teaching and learning. *Research in Science Education*, 33 (3), pp. 273–98.
_____. 2007. Re-imagining science education: Engaging students in science for Australia. *Teaching Science*, 53 (4), pp. 14–17.
_____.& Cripps Clark, J. 2009. Educating the whole child through science: A portrait of an exemplary primary science teacher. *Teaching Science*, 55 (3), pp. 23–33.
_____ & Peterson, S. 2003. Tracing young children's scientific reasoning. *Research in Science Education*, 33 (4), pp. 399–404.
_____ & Peterson, S. 2005. A longitudinal study of children's developing knowledge and reasoning in science. *Research in Science Education*, 35, pp. 61–98.
_____, Peterson, S. & Prain, V. 2006. Picturing evaporation: Learning science literacy through a particle representation. *Teaching Science*, 52 (1), pp. 12–17.
_____ & Prain, V. 2010. A framework for re-thinking learning in science from recent cognitive science perspectives. *International Journal of Science Education*, 32 (15), pp. 2055–78.
_____, Prain, V. & Peterson, S. 2007. Representational issues in students' learning about evaporation. *Research in Science Education*, 35 (1), pp. 63–98 and 37(3), pp. 313–31.
Varelas, M., Pappas, C. & Rife, A. 2006. Exploring the role of intertextuality in concept construction: Urban second graders make sense of evaporation, boiling and condensation. *Journal of Research in Science Teaching*, 43 (7), pp. 637–66.
Venville, G., Wallace, J., Rennie, L. & Malone, J. 2002. Curriculum integration: Eroding the high ground of science as a school subject? *Studies in Science Education*, 37, pp. 43–83.
Vosniadou, S. 2008. Conceptual change research: An introduction, in S. Vosniadou (ed.). *International Handbook of Research on Conceptual Change*. New York: Routledge, pp. xviii–xxiii.
_____, Vamvakoussi, S. & Skopeliti, I. 2008. The framework theory approach to the problem of conceptual change, in S. Vosniadou (ed.). *International Handbook of Research on Conceptual Change*. London: Routledge, pp. 3–34.
Vygotsky, L. 1978. *Minds in society: The development of higher conceptual processes*. Cambridge, MA: Harvard University Press.
Wandersee, J., Mintzes, J. & Novak, J. 1995. Research on alternative conceptions in science, in D. Gabel (ed.). *Handbook of Research on Science Teaching and Learning*. New York: Macmillan, pp. 177–210.
Watson, B. & Kopnicek, R. 1990. Teaching for conceptual change: Confronting children's experiences. *Phi Delta Kappan*, May, pp. 680–4.
Wellcome Trust. 2005. *Primary Horizon: Starting Out in Science*. Available at http://www.wellcome.ac.uk/primaryhorizons (accessed 3 March 2006).
White, R. & Gunstone, R. 1992. *Probing Understanding*. New York: Falmer Press.
_____. 2008. The conceptual change approach and the teaching of science, in S. Vosniadou (ed.). *International Handbook of Research on Conceptual Change*. London: Routledge, pp. 619–28.
Williams, J. 2007. Do we know how science works? A brief history of the scientific method. *School Science Review*, 89 (327), pp. 119–25.
Wong, D. & Pugh, K. 2001. Learning science: A Deweyan perspective. *Journal of Research in Science Teaching*, 38 (3), pp. 317–36.
Yoon, S., Bennett, W., Mendez, C. & Hand, B. 2010. Setting up conditions for negotiation in science. *Teaching Science*, 56 (3), pp. 51–5.
Zeegers, Y., Paige, K. & Lloyd, D. 2010. 'Operation Magpie': Are we engaging teachers with science and students with the community. Paper presented at the Australasian Science Education Research conference, Port Stephens, NSW, July.
Zeidler, D. & Nichols, B. 2009. Socioscientific issues: Theory and practice. *Journal of Elementary Science Education*, 21 (2), pp. 49–58.

Endnotes

1 Novak (2002, p. 555) has argued that 'students' alternative conceptions' is an inappropriate phrase and should be replaced by 'limited or inappropriate propositional hierarchies'. His reasons are detailed in the online Appendix 1.12.

2 The word 'context' is used many times in this and later chapters. It is fundamental to a sociocultural perspective on learning. It has no simple definition but rather various interpretations. Gilbert (2006, p. 960) uses a description which sees a context as 'a focal event' embedded in its cultural setting which has four attributes: '(a) a setting ... in which mental encounters with focal events occur; (b) a behavioural component of the encounters ...; (c) the use of specific languages; and (d) a relationship to extra-situational background knowledge'. Space does not permit expansion here but as teachers you need to be aware that 'context' has multiple meanings and that it is integral to appreciating how students respond to a situation (e.g., how the use of a globe when students are learning about the shape of the Earth can influence their responses – see chapter 10). Gilbert explores the various meanings of 'context'.

2

Thinking and working scientifically

by Rosemary Feasey

> The most exciting phrase to hear in science, the one that heralds the most discoveries, is not 'Eureka!' (I found it!) but 'That's funny ...'
>
> Isaac Asimov

Introduction

Science curricula across the world have moved on in the past 40 years, from discovery learning to a focus on national standards for subject knowledge and skills development in the 1990s, to more recent moves in science education towards a more sophisticated approach which recognises the complex relationship between science, scientists and the public's understanding of science, and its applications and implications both locally and globally. Harlen (Harlen and Qualter 2009, p. 35) suggests that pupils need a basic understanding of what science is, how it works and its limitations, as well as lifelong learning, since we cannot learn everything in those few years at school. She continues by stating that these two aspects

> are summed up in the notion of developing scientific literacy, where the emphasis is not on mastering a body of knowledge but on having, and being able to use a general understanding of the main or key ideas and in making informed decisions and participating in society. I like to define it as being comfortable and competent with broad scientific ideas, with the nature and limitations of science and with the processes of science and having to use these ideas in making decisions as an informed and concerned citizen.

This move towards developing young people's understanding of the impact of science on their daily lives and their responsibility as global citizens, is an important one and has a significant effect on teaching and learning in primary science.

In this chapter, the key characteristics of thinking and working scientifically have been expanded to take into account the concept of science as a human endeavour and the impact on science in the classroom of the need for children to appreciate the applications and implications of science in everyday life, and how young people can respond to this.

Science in the school curriculum

It is important that consideration is given to the rationale behind teaching science. The following two activities offer opportunities to consider personal viewpoints on the teaching of science in primary schools.

Activity 2.1 The role of science in the curriculum

Imagine that the government has decided to abolish all science teaching and you are charged with providing its defence and offering reasons for continuing to include science teaching in the curriculum. What reasons would you give to persuade the government of the centrality of science in the curriculum and in the education of the individual?

A variety of responses could be offered to the above question, ranging from, for example, the need to maintain economic viability and service the nation's health care, to the idea that young children enjoy the practical nature of science and the view that science can be as intrinsically interesting and creative to an individual as art and music.

Scientific literacy

This second activity focuses on a concept central to this and the remaining chapters.

Activity 2.2 Why scientific literacy?

The following quotations advocate the need for scientific literacy.

> The main function of such input is not to make scientific decisions – the best people to make scientific decisions are scientists – but to inform scientists by making explicit the social context and moral environment in which their science policy decisions must work.
>
> **Gregory 2001**

> The ever-growing importance of scientific issues in our daily lives demands a populace who have sufficient knowledge and understanding to follow science and scientific debates with interest, and to engage with the issues science and technology poses – both for them individually, and for our society as a whole.
>
> **Millar and Osborne 1998, p. 2001**

- What, in your opinion, is scientific literacy?
- What reasons do the authors of the quotations give for the need for science education?
- What do the authors suggest is the basis of scientific literacy in an individual?
- How similar or different is the idea of scientific literacy as an underlying philosophy for science education to the reasons you offered in Activity 2.1?
- What do you think a person needs to be scientifically literate?
- Do you consider scientific literacy to be only a white Western viewpoint of science? What other perspectives on science could be taken? How do you think culture affects approaches to science in education?

NOS

If the aim of science education is to develop a scientifically literate population, then those involved in science education face an enormous task. Feasey (1996) suggests that most people are scientifically illiterate and often hold negative and contradictory viewpoints of science. On the one hand, the public is intrigued by science; this is manifested by the popularity of television science programs and, on the other, science is construed as unglamorous, responsible for the problems of the world and elitist – too difficult for the average person. In fact, few people, as Rassan (1993, p. vii) points out, 'feel any need to apologise for knowing very little about science; in fact many seem to take a perverse pride in their ignorance'.

Feasey and Gott (1996) suggest that there are different elements that provide the foundation for a scientifically literate individual. The first is science understanding, which relates to an understanding of key ideas and facts in science, such as those related to forces and electricity. Sound knowledge and an ability to apply such concepts in a range of contexts are essential. The second element is the development of science inquiry skills, which include an understanding of evidence that focuses on the individual's understanding of how and why scientists collect evidence, and an ability to challenge the reliability and validity of evidence in order to decide on its believability.

The individual needs to understand the centrality of evidence in science: past, present and future. For example, the scientific facts and theories we use today in science are the product of many years of rigorous testing and the eventual acceptance of evidence from those tests as reliable and valid by the scientific community. In the future, new theories and facts will be subject to similar scrutiny and become part of people's everyday bank of scientific knowledge, which identifies science as a human endeavour. Today, evidence is crucial to a whole range of people engaged in scientific activity. For example, industrial chemists must present believable evidence before placing a new drug on the market. Engineers continually test established science ideas in different contexts with a range of materials and at different locations; only when the evidence suggests it is safe to do so will construction of a building, bridge or machine go ahead.

However, there are other elements that are also important, such as the need for a questioning attitude towards science. This is not to infer a negative approach to science, but a healthy scepticism and a willingness to challenge science and the evidence it offers. In addition, communication has been placed under the curriculum spotlight. Communication is central to scientific literacy – individuals must understand the way in which knowledge, ideas and policies are communicated so that they, in turn, can communicate their questions and concerns to the scientific community, interest groups and also politicians.

In developing scientifically literate individuals, many national curricula have included a focus on developing students' critical appreciation of scientists, their theories and ways of working, and how science has impacted on daily life.

If we turn to national curricular documents from different countries around the world, we find that concepts of evidence are embedded in the different strands and curriculum statements – for example, in the three strands of the Australian Science Curriculum (ACARA 2011), namely science as a human endeavour (SHE), science inquiry skills (SIS) and science understanding (SU). The New Zealand primary science curriculum (New Zealand Ministry of Education 2010a) acknowledges that:

> Science is a way of investigating, understanding, and explaining our natural, physical world and the wider universe. It involves generating and testing ideas, gathering evidence – including by making observations, carrying out investigations and modelling, and

communicating and debating with others – in order to develop scientific knowledge, understanding, and explanations. Scientific progress comes from logical, systematic work and from creative insight, built on a foundation of respect for evidence. Different cultures and periods of history have contributed to the development of science.

Given that the underlying philosophy of science education is to develop a scientifically literate person, how then can the different elements be translated into the school curriculum and how do they relate to the component 'thinking and working scientifically'?

Perceptions of science and scientists

In Chapter 1 you were asked to think about the NOS; you might now like to refer to any ideas you had while working through that section to see if and where there are any connections with the following discussion. Thinking and working scientifically emphasises an approach that seeks to adopt some of the ways in which scientists construct and acquire knowledge. It is a particularly interesting move to include this element in the science curriculum when one considers that research into the public understanding of science indicates that scientists belong to one of the least understood professions. Yet, as teachers (and part of that general public), we demand that children emulate an approach that is characteristic of a group of people we know little about and often distrust. Therefore, a useful starting point in an attempt to understand how scientists work is to acknowledge our own and children's prior understanding of scientists. The following tasks adopt a constructivist approach to elicit personal frameworks of understanding about scientists and science.

Activity 2.3 Our views of scientists

- Consider your image of a typical scientist.
- Draw a picture of that image and list the main characteristics of the person you have drawn.
- What is your opinion of scientists and the work they do? Make a note of your views.

Activity 2.4 Children's views of scientists

A similar activity was carried out with children from School A. The response of one child, Katrina, can be seen in Appendix 2.1 (see p. 96).

- How does your drawing of a scientist compare with the picture created by Katrina, aged 10?
- Make a list of any common characteristics between your picture and Katrina's drawing.
- Carry out this activity with children from your own school. Were the results similar or different to Katrina's drawing?

Activity 2.5 Children's views of scientists

The following is an extract from a teacher's notes for a session with children in years 2 and 3 exploring children's ideas of scientists and science.

What is a scientist?

An introductory session for the children to draw and write, without collaboration, about their views and ideas on science and scientists.

1 Brief, open-ended discussion to outline the activity.
2 The children were given a worksheet and, as they worked, their verbal comments were noted on the sheets.
3 Reflection time to share ideas and opinions, with a deliberate focus on the children asking questions and giving feedback.
4 The teacher listing ideas under the following categories:
   - We think scientists are ...
   - We think scientists work on ...
   - Working scientifically means ...

Teacher evaluation

Careful close questioning revealed that most children really had no ideas or opinions. One of only two children with a parent working in the scientific community had personal opinions. Only one child drew a child 'doing science' and her illustration showed a stereotypical magic show kind of experience.

These provided valuable insights into the range of conceptual frameworks held by children about scientists. They also provided a timely reminder that it's all right to literally not know. Many children used key words such as 'test', 'laboratory', 'experiment' and 'test tubes'. When questioned closely there was a great deal of confusion about what those things were and what they meant. Many children identified television as the source of these ideas and words.

Consider

- How are the responses different to those in Activity 2.4?
- What issues arise from the children's responses in activities 2.4 and 2.5 for teaching science?

Activity 2.6 Changing stereotypical images of scientists

The stereotypical image of a scientist offered by most children and many adults usually includes some of the following characteristics:

- male
- balding
- glasses
- clever-looking
- white laboratory coat
- manic look
- scientific instruments in pockets.

Consider

- What is it that influences, and therefore helps to create, the stereotypical image of a scientist that children and the public in general hold?
- What image(s) of scientists do current TV science current affairs/update programs (e.g., *Catalyst*) present?
- Do they assist or hinder a more appropriate view of scientists?
- How close do you think the stereotypical image is to a scientist working in the community; for example, in industry or health care?
- What can a teacher do to help children develop a more balanced view of scientists and their work?

- Developing positive images of scientists and science requires a whole-school approach and the support of colleagues. Make notes on how this might be accomplished within a whole-school situation.
- How will you help children to develop an understanding that different cultures have their own contribution to make to how we view the world and what we know?
- How will you develop children's understanding of how Indigenous groups have developed their understanding of natural phenomena, through observing, exploring and testing their ideas in everyday life?

By the time children reach the age of five, many have firm ideas about what a scientist looks like and does. Equally, they have firm ideas about who can become a scientist – usually, an exclusive group of white males, bespectacled, balding, eccentric and highly intelligent. Schibeci (1989, 2006) provides examples of children's responses to 'draw a scientist' tasks.

In a research study, preservice teachers were asked to engage children in a 'draw a scientist' task. Nearly 20 years after Schibeci's research, the outcomes were still very similar. The analysis of the data collected by the preservice teachers indicated that:

> Most children's drawings showed the stereotypical 'scientist'; white lab coat, with beakers and test tubes, working in a laboratory and carrying out experiments. Some children said that scientists have useful skills and make lots of money. The majority thought scientists were mostly males, but some recognised that there are female scientists too. Most drawings showed stereotypical images of scientists as looking weird, bad and mad.
>
> **Jane, Fleer and Gipps 2007**

Wardle (2001, p. 12) found that children in his class told him 'that a scientist was invariably male and old, had wild, grey, sticking up hair, a long white (laboratory) coat, big glasses and spoke in a funny voice'.

Such stereotypical images of the mystical scientist and the eccentric boffin engaged in ritualised science are not only incorrect, but are also damaging because they suggest to children that only an exclusive group of people can engage in science. Equally, a lack of understanding of what a scientist is and does can cause difficulties, since either perception suggests to children that there is no commonality between the way a scientist works and the way children think and work in science. Perhaps more importantly, it offers a subtle but damaging suggestion that what scientists say and do are inaccessible to the public and cannot be challenged.

Changing our own and children's perceptions of science and scientists can be achieved through developing an understanding of how scientists work. Consideration of what scientists think about and do can help to challenge ideas and assumptions as well as create a framework for developing a scientifically literate individual, which encompasses the skills, concepts and attitudes that are component parts of thinking and working scientifically.

Who better to offer this insight than scientists themselves? The following section explores the key features of thinking and working scientifically in the science curriculum and draws on the ways of thinking and working that scientists value in themselves and their peers, thereby providing useful parallels with the classroom and the development of scientific literacy.

Thinking and working scientifically

Many adults and children share the misconception of the manic scientist making his strange discoveries and assume that this is what science is about. Certainly, science is about discovery, although this is a much smaller part than most people realise. The majority of science is repetitive work, the painstaking collection of evidence.

How many of us have experienced that moment when we have discovered something about our environment and, having made personal sense of it, felt enormous elation and excitement, as if we alone are the only ones to know about and see it? Most of us have watched children experiencing a moment of great intensity, such as seeing for the first time a spider making a web or looking through a hand lens at an insect, and seen the sheer rapture and excitement expressed on the young face, followed by the huge desire to share that moment. Is that how the scientist feels on making a discovery and taking a step forward in his or her own learning? In terms of the whole of humanity, some moments are more profound than others, but undoubtedly all are milestones in an individual's intellectual and emotional development, whether the person is four or 34, six or 60, and should not be underestimated.

Profound though these moments are, it would be naive to assume that this is the everyday reality of the scientist or the child. Such discoveries are not made in a vacuum; they depend on personal interest and the ability to use knowledge and experience to make sense of events. The teacher has an important role to play in structuring classroom situations and experiences to offer opportunities for discovery as one of only a number of ways to gain access to science.

Experience alone, however, is insufficient for a person to develop as scientifically literate. Crucial to understanding experience will be the development of an ability to think about and approach the world in a scientific way. What kind of thinking enables children to view and make sense of the world through a pair of science spectacles? We can find a basis to describe the components of thinking and working scientifically by considering how scientists approach questions and problems.

Activity 2.7 Scientists' thinking and working

Consider

- What do you think scientists think about and what do you think they do when they are engaged in solving a problem or answering a question? Record your responses in a table using the headings 'What scientists think about' and 'What scientists do'.
- Ask some children, across different grades if possible, the same questions and record their responses. Compare them to your own. The list in Appendix 2.2 (see p. 97) is a compilation of adult and child responses to Activity 2.7.
- Do you know a scientist? If you do, discuss your ideas and those of children with that person.

At this stage you might like to refer back to Activity 1.2 (see p. 5) and the online Appendix 1.1 (Task 1) and consider what similarities and overlap there is with your ideas.

Appendix 1.1
What do you think science is?

Science skills

The list in Appendix 2.2 (see p. 97) is divided into two – what scientists think about and what scientists do. Taking the latter first, it refers to the notion that scientists need to be able to do certain things, such as use equipment and create tables and graphs. These could be referred to as the tools of the trade, the skills necessary to being a scientist. These are the mechanical aspects of doing science that the individual needs to master and are essential to working scientifically: that is, the doing of science. In the classroom these include:

- using equipment
- measuring
- recording information, including drawing tables
- communicating information, including drawing graphs
- identifying possible risks and controlling obvious risks.

Evidence in science

These tools of the trade or skills are of use only if the individual understands what they are for and when and how to use them. In order to use them, an individual needs to have an understanding of what Gott and Duggan (1995, p. 26) describe as the 'thinking behind the doing'. The column in Appendix 2.2 recording what scientists think about represents the kind of thinking scientists engage in when faced with a problem or a question. Central to the thinking behind the doing (thinking scientifically) is the need for evidence. When scientists think and work scientifically, they spend much of their time attempting to validate their own or someone else's theory, answering questions or solving problems. Central to their thinking and ways of working is the need to create a set of evidence that is believable and therefore acceptable to others. In order to do this, they require an understanding of what Gott and Duggan (1995, p. 30) term 'concepts of evidence'. In its simplest form, this means that they have to consider the questions 'What will I have to think about doing to collect data (evidence) to help me solve my problem or answer my question?' and 'What will I have to think about doing to make sure that my data (evidence) is believable to myself or others?'.

NOS

An understanding of why it is important to produce believable data and how to produce data that can withstand scrutiny is important in developing abilities in thinking and working scientifically. Thinking scientifically is the knowing how, why and when that underpins working scientifically (the doing), which includes using a range of skills, such as handling equipment, measuring and drawing graphs, when carrying out tests.

Thinking scientifically has its parallels in the classroom, where children need to have an understanding of why, how and what to do in order to engage in practical activities in science. When children work without that understanding, they do so only at a mechanistic level, superficially going through the motions of doing and using skills that sometimes characterise primary science practical work and comprise nothing more than busy work. In order to appreciate what it means to understand evidence, Table 2.1 offers an overview of concepts related to evidence based on an analysis by Gott and Duggan (1995).

In thinking and working scientifically, scientists use their understanding of evidence to answer questions and solve problems in such a way that it produces believable evidence. Australia

TABLE 2.1 Concepts of evidence

CONCEPTS OF EVIDENCE	EXAMPLES
Ideas and decisions related to the design of a test	Understanding the idea of a fair test; that is, what to change, what to keep the same and what to measure. Knowing why the design must produce believable results
Ideas and decisions related to measurement	Knowing why measurements are important, what to measure and when and how to take measurements. Considering how frequently to take measurements and when it is appropriate to take repeat measurements
Ideas and decisions related to data handling	To understand why results need to be recorded and which methods are appropriate. To understand how graphs are used to communicate results and which graphs to use according to the type of data collected. To know how to recognise patterns in data and use data to answer the original question or problem
Ideas and decisions related to the evaluation of evidence (data)	To know how to evaluate evidence (data) and how the test was carried out in terms of the believability of the information in order to answer a question or solve a problem

Source: Adapted from Gott and Duggan 1995

has a history of scientific inventions and developments that have been subject to the same need for believable evidence. They include the flame ionisation detector, an instrument that detects minute quantities of organic gases used in pollution control; prototypes for solar-powered telecommunications and water heating systems; and a diagnostic ultrasound technique for the examination of human organs and the developing foetus. These are major scientific advances and, although none can be directly translated into practical activities for the primary classroom, there are other examples that can be used with children.

All new scientific and technological advances go through a process which demands scientific dialogue and challenge. An important element of concepts of evidence as explored in Table 2.1 is the need for constructive criticism or argumentation.

The Discussions in Primary Science (DiPS) project (see http://www.azteachscience.co.uk/resources/cpd/discussions-in-primary-science/view-online.aspx) provides many ideas for developing children's ability to enter into discourse in science, and suggests that

> Science knowledge and ideas are constructed and can be challenged and changed as new evidence is produced. Many children (and adults) think all scientists do is put on white coats and work alone in laboratories. In fact scientists collaborate and talk to others ... as much as they [do] in laboratories. **Doing** science requires **talking** science.

AstraZeneca Science Teaching Trust 2010

Concepts of evidence are embedded in many national science curricula. For example, the Australian Science Curriculum (ACARA 2011) identifies the idea of evidence as a key element of science inquiry skills, in curriculum statements such as:

- Year 1 – Compare observations with predictions and use observations as evidence to support students' ideas and to answer questions posed.

- Year 5 – Reflect on the process of the investigation to evaluate the quality of evidence and to suggest improvements to the planning of investigations.

The New Zealand curriculum (New Zealand Ministry of Education 2010b) also demands that children collect and use evidence in science, as in the following example:

- Level 5 – Understand that scientists' investigations are informed by current scientific theories and aim to collect evidence that will be interpreted through processes of logical argument.

Activity 2.8 Scientific and technological developments

The following three examples of Australian scientific and technological developments might be used to help children to appreciate that they can tackle similar problems to scientists (if you are not an Australian reader, find an example from your own country). Choose one of the developments and imagine you are the person working on the invention, but working in the present. What would you need to think about and do to test your invention and produce believable evidence, which could be presented to your peers in the scientific community and to the public?

Consider your plan of action in light of the ideas from Activity 2.7 (see p. 61) and the information about the concepts of evidence in Table 2.1 (see p. 63).

Plastic lenses

The first plastic spectacle lenses in the world were made in South Australia in 1960. The major advantage of plastic as a material for lenses is that it is only one-third the weight of glass, an important factor for people who require thick lenses in their spectacles.

Canvas bag for storing and transporting water and which also cools it

This invention by Sir Thomas Livingstone Mitchell (1792–1855) was based on Aboriginal water carriers. A bag of plain canvas, once wet enough, holds water quite well. Easy to carry, it has the added advantage of cooling its contents as the water seeps through to the outside and evaporates. Over 150 years later, the bag is still in use by travellers.

The Hill's Hoist rotary clothesline

This was invented by motor mechanic Lance Hill in 1945. It is a rotary line that fits in a small space, can be raised and lowered and, with its rotary arm, gives easy access to all areas of the clothesline.

For further information about these and other inventions, see Ingpen (1982), Encel (1988) and Kendra (1990).

S&C

In approaching the development and testing of any one of the inventions noted above, you would have been involved in thinking and working scientifically. Many of the things you had to think about and do will have had parallels with ideas from earlier activities.

Testing out ideas requires a synthesis of an understanding of key concepts, an understanding of evidence and competence in a range of skills. Figure 2.1 illustrates that combination.

The synthesis of the elements of knowledge with understanding, skills and evidence is the thinking behind the doing: it is the ability to think about the question or problem and decide what to do in order to produce believable evidence from which one can draw conclusions and a solution.

Of course, in everyday life only a small group of people are directly involved in scientific research; however, many people use one or more of these understandings and skills in a range of employment, from biochemist to local pharmacist or engineer to mechanic. Perhaps

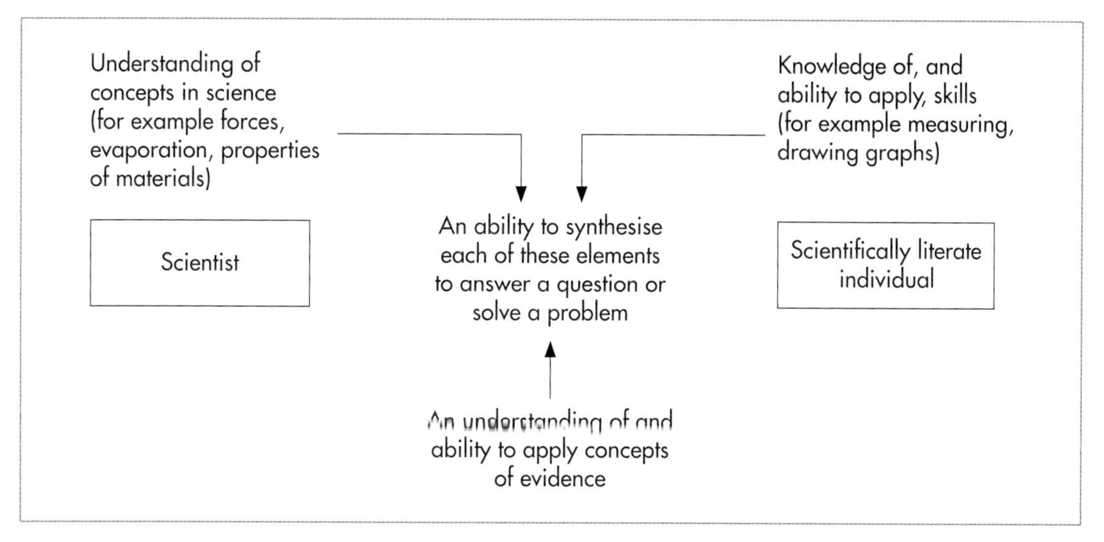

FIGURE 2.1 Synthesising understanding towards scientific literacy

more importantly, an understanding and ability to synthesise the different components is required for the individual to decode science information from a variety of sources. Where someone possesses these abilities, that person is more likely to be scientifically literate and able to take part in debate, make personal decisions or contribute to decisions in the local community and worldwide. It would be useful to revisit socioscientific issues in Chapter 1 (see under 'The influence of science, or science in the community: socioscientific issues' on p. 43) to appreciate one way of placing these ideas into the broader context of children constructing ideas about thinking and working scientifically.

While it is important to consider past scientific inventions and discoveries, it is equally important for children to develop an appreciation of how scientists in their own country are currently contributing to human understanding of the world. At the time of writing this chapter, scientists in Australia and New Zealand were researching some of the following, which have the potential to affect people on both a local and global scale:

- *Malaria* – Scientists in Australia have found that proteins produced in the human immune system could help lead to the creation of a malaria vaccine (see http://www1.voanews.com/english/news/health/Breakthrough-for-Australian-Scientists-Working-on-Malaria-Vaccine-82156332.html).
- *Spinal problems* – Scientists in Australia have now developed an understanding of the cause of ankylosing spondylitis, a type of inflammatory arthritis that targets the joints of the spine (see http://www.physorg.com/news184314943.html).
- *Breast cancer* – Australian scientists are experimenting with stem cell breast-growing techniques, which means that breast cancer patients could have a choice of treatments in addition to implants and reconstruction (see http://www.telegraph.co.uk/health/6548802/Australian-scientists-to-start-breast-regrowth-trial.html).
- *Water to electricity* – Scientists in New Zealand are exploring how tidal power can be harnessed as a renewable energy source (see http://www.voxy.co.nz/national/niwa039s-energy-projects-turning-tide/5/45053).

- *Bite or chew* – New Zealand scientists have found out that taking bigger bites and chewing less was better for people watching their weight, as the food was broken down more slowly in the stomach (see http://www.stuff.co.nz/the-press/news/christchurch/3583620/To-chew-or-not-to-chew-what-suits-you).

SSI S&C

So, how might these developments by scientists be used in primary science? Well, some could be used for debate, such as asking 'What are the positives and disadvantages of tidal power?' On the other hand, perhaps children could test the question 'Which makes you feel fuller, biting or chewing food?' Both approaches of course demand that children use different elements of thinking and working scientifically, and help to show children that they can engage in science in the same if not a similar way to professional scientists.

In everyday life, people will make many personal decisions requiring the application of scientific knowledge and skills. Some of those will be very personal – for example, relating to a health issue – while other decisions will contribute to global difficulties.

However, making decisions is not easy; who and what do we believe? People often believe everything they read in the newspapers, but the new oracle might be the Internet. Should we believe everything we access on the Net? How do we use our science to make reasoned judgements having looked at a range of evidence?

Take, for example, sustainability and global warming. How much do people understand? Are they able to make decisions using their scientific understanding to contribute to a global reduction in carbon emissions? What is your response to the following, from an Internet site? What information would you need to make a decision? What would it take to convince you to change your ways? Do you think that everything on the Internet is true? Do you think some Internet sites might be biased?

> There are many ways to reduce global warming, some very simple whilst others require a more focussed approach. At least do the easy ones! Scientists are convinced that the single biggest cause of global warming today is the release of carbon dioxide through burning of fossil fuels in our power stations, vehicles and transport systems.
>
> **Speak Out:**
> Our countries need new national legislation and laws to direct us towards solving climate problems and to reduce global warming. Better legislation and laws will help to ensure we develop cleaner cars and cleaner power plants. We also need to agitate for government rebates on installing solar power, solar hot water, or wind power in our homes.

Global Greenhouse Warming.com 2010

On what evidence will you make your choice? Good science education aims to provide individuals with the tools to be active decision makers of the future. How do we do this? This links to socioscientific issues, as outlined in Chapter 1, where exposure to a range of issues across the primary years can help children to appreciate that scientific and technological developments may affect different people in different ways. With this understanding, they can go on to develop an appreciation that people will naturally have different viewpoints which help us to construct our ideas and form our own perspectives.

Having identified the key components of thinking and working scientifically, the important question for science education becomes how the theoretical rhetoric is translated into the practicalities of teaching and learning in the classroom.

Activities to develop thinking and working scientifically

The key components, skills, concepts of evidence and knowledge and understanding of concepts in science (such as forces and electricity) can be developed through different kinds of activities that form the foundation of a menu of activities for teachers in science. The activities on the menu should be used on the basis of fitness for purpose; that is, teachers should decide what it is they want children to learn (learning outcome) and then match the most appropriate activity type to the outcome and the children. The following describes the possible range of activities as offered by Feasey in Sherrington (1993, pp. 55–8) and Aubrey (1994, pp. 83–5).

Observation and measurement activities

Observation is frequently listed under scientific skills. However, as Feasey (in Aubrey 1994, p. 76) indicates, 'its inclusion in this context is misplaced ... Observation is conceptually based'. The individual observes (using all or some of the five senses) something through his or her own personal conceptual framework and one can enhance that observation by challenging those frameworks or demanding that the individual pays greater attention to detail, notes similarities and differences, or uses certain equipment to enhance observation. However, the crucial element is that the person observing is asked to make connections between what that individual can see and the individual's science knowledge and understanding. Science relies on qualitative observations (for example, colour and texture) and quantitative observations (measured observations).

Classification activities

Classification is central to science. As one way of organising and making sense of the world, such things as materials, animals and plants are all classified into groups according to their similarities; for examples, see chapters 6 and 8. Interestingly, although we might think that classification relies on looking for differences, in fact it is recognising similarities which is the most important thing to do. We invariably group things that have the same characteristics and give that group a name.

In science education we ask children to classify (sort) many different things to help them develop their knowledge and understanding – for example, materials that are shiny to teach properties of materials; invertebrates with three pairs of legs, three parts to the body and two pairs of wings to help them understand the group called insects.

Skills activities

Skills activities are designed to teach and develop the tools of the trade, the mechanics of doing science, such as teaching children how to use a thermometer or produce a graph. Skills need to be taught. Take, for example, drawing and using tables. Some children, if asked the question 'What is a table?', will answer 'You have your dinner on it', or they might say 'Tables are like times tables in maths'.

Such statements from children only serve to reinforce that the teacher should not take anything for granted and that teaching the skill of drawing a table, as well as why and when to use it, is essential. Tables are important in science as they are the key to organising data and supporting investigations. Tables can help to:

- organise what children need to do
- organise what to use
- encourage children to observe accurately
- define what children have to measure
- define the beginning and end of an activity.

Equally, children need to be taught the skills of drawing graphs and using a range of equipment in science, and of course we must not forget that as children progress through their primary school they should become increasing more independent in relation to recognising and developing appropriate suggestions for managing risk in their science activities.

Illustrative activities

Illustrative activities are designed to teach or illustrate a particular knowledge and understanding concept. This type of activity is usually recipe-like – a set of instructions that children follow and which leads to a defined outcome; for example, a set of instructions that tells children how to find out which substances dissolve in water, designed to develop the concept of dissolving.

Research activities

There are areas of science that do not lend themselves to extensive hands-on experience; for example, finding out about the planets Jupiter and Mars, or plant reproduction. Research activities enable children to develop skills for extracting the relevant information from a variety of sources to support and develop concepts. Children could, for example, use books, CD-ROMs, the Internet, videos, posters, leaflets and television programs. Research activities help to broaden children's knowledge and understanding of concepts and facts as well as their ability to communicate science. However, it is important to encourage children to represent their research to avoid verbatim copying. Sometimes, asking children to communicate what they have found out to a specific audience can help to support children in reconstructing what they have learned from their research activity. Osborne (1991) suggests that learning experiences using these sources must require reflective, not receptive, interaction with the resource. Research activities would benefit from the use of directed activities related to text, where children reformulate their understanding from text in a variety of ways, such as discussions and creating diagrams and tables of information.

Survey and secondary data activities

Surveys are where children are engaged in activities where they have to collect information. Some information will be qualitative; for example, where in the schoolgrounds different plants grow. Other information might lead to numbers (quantitative data); for example, favourite foods, or how many hours children exercise each week as part of a healthy living project.

The collection of numerical data provides essential information which, when analysed, can help to develop children's understanding in different concept areas as well as their understanding of evidence and the development of skills, such as drawing tables and graphs. For example, a survey that contrasts habitats in the schoolgrounds could assist children in understanding the conditions required by different animals and plants to survive. Here, children develop concepts related to evidence, such as looking for relationships and patterns and challenging the believability of data, as well as concepts about living things. They can also develop skills related to using a range of equipment.

Scientists frequently use secondary data – someone else's data. The public is given secondary data whenever they are given statistics about issues such as bovine spongiform encephalopathy (BSE), severe acute respiratory syndrome (SARS) and pollution. Children need to consider secondary data and draw conclusions from it using the information they are given, since that is what a literate public needs to do in real life. Children also need to develop an understanding that they do not have to accept the data as true; they also need to be encouraged to be critical of it and ask questions about its validity and to look for anomalies and seek alternative conclusions.

Although young children do collect data, they often ignore it when developing conclusions. They tenaciously hold on to their original ideas, ignoring the data in favour of these, or decide, as very young children sometimes do, that theirs was best or the winner regardless of what really happened. Giving children data produced by others, even if they are the results from another group in the class, can be very useful, because it challenges children to engage with information that they have not generated.

Exploration activities

The importance of exploration should not be underestimated. Exploration is simply where children try something out to see what happens; for example, when young children pour water over a waterwheel and observe the effect, or when older students put two or more bulbs in a circuit to find out what will happen This hands-on experience is crucial; however, it is important that there is a balance of child- and teacher-initiated exploration. The teacher needs to provide purpose and challenge in children's exploration. Exploration can help to develop not only conceptual understanding across different areas – from forces and electricity to living things – but can also help develop skills, such as using equipment, and positive attitudes in science, such as perseverance, open-mindedness and questioning.

Fair-test investigations

The final category from the menu, investigations, is the most complex of all the activities. The essence of investigations is the synthesis of the understandings described in Figure 2.1 (see p. 65). The term 'investigation' is often used to mean any activity in which a child is engaged in practical work; it has become a catch-all term for any practical activities in science, such as observation, exploration and illustrative activities. In the context of thinking and working scientifically, investigations are defined as activities in which children use their conceptual knowledge and understanding of areas such as forces, materials and light alongside an understanding of concepts of evidence and skills to find a solution to a problem or question. In this context, investigations will invariably involve children carrying out a fair test.

Activity 2.9 Thinking and working scientifically

An example of a question that might be investigated is: Which sugar dissolves the fastest? Consider what kind of things you will have to think about and do to investigate this question?

Plan your investigation. Record your ideas in a table using the headings 'What I will have to think about' and 'What I will have to do'. Remember to consider the idea of evidence and the need to produce a set of evidence that is believable.

In your own plan, the different elements of thinking and working scientifically listed in Figure 2.1 (p. 65) can be identified. Table 2.2 provides an analysis of the component parts of thinking and working scientifically in Activity 2.9.

Activity 2.10 Children thinking and working scientifically

What do you think children need to think about when they plan an investigation? Jot down your ideas.

- How do you think children should communicate their planning of an investigation?
- What about those children for whom writing a plan is challenging and might detract from their enjoyment of science?
- What kind of alternatives are there for children to communicate their planning in ways that might best suit their ability, in particular their literacy level?

Your list might have contained some or all of the elements of thinking and working scientifically from earlier activities. No doubt your list makes interesting reading when compared to the list of what people think scientists think about and do from Activity 2.7 (see p. 61).

TABLE 2.2 Dissolving sugars – thinking and working scientifically

SCIENCE CONCEPTS	SKILLS USED	EVIDENCE	SYNTHESIS
Dissolving	Weighing sugar	Understanding the need to have a fair test so that results are believable	The ability to put the individual parts together in a whole investigation to find out which sugar dissolves fastest
Heat/temperature	Measuring time using a stopwatch	Knowing that measurements are needed as evidence	Understanding why and how to carry out an investigation so that the evidence about which sugar dissolves the fastest is believable
	Drawing a table	Understanding that a table organises data and can show patterns	To be able to use the evidence (data) from the investigation to answer the question: Which sugar dissolves the fastest?

Certainly it will indicate that, when children carry out investigations, they have much in common with scientists. Your list should also suggest that the number, range and intellectual level of decisions children have to make is high. What is evident is that investigations make a complex set of demands on children related to thinking and working scientifically, and that the range and type of decisions they make and the things they do are not available to them in the other activities on the menu; hence, the uniqueness of investigations. Perhaps most importantly, investigations demand that children are the decision makers. When teachers take control and tell children what to do, the activity changes and becomes an illustrative activity. However, a word of caution – this is not to infer that children should carry out investigations without support or, indeed, that there is no need to teach investigations.

Your list of ideas for how children might communicate their plan could include:

- telling someone orally
- creating a concept map
- making a list of ideas
- setting out their resources, photographing and annotating
- demonstrating what they are going to do

The following section suggests how the teacher can develop teaching sequences to develop children's ability to carry out investigations themselves.

Teaching investigations

Investigations demand that children pull together in one activity a basic set of requirements which include:

- basic skills of measurement
- basic skills of drawing tables and graphs
- how to carry out a fair test
- how to record data
- why evidence is important in science and how to collect valid and reliable data
- how to critically question data and other scientific information
- knowing when it is appropriate to carry out an investigation to solve a problem or answer a question
- how to synthesise skills and understandings in an investigation
- how to communicate what they did and what they found out.

There is no reason why, individually, these elements cannot be taught directly using a range of activities from the menu (that is, observation and measurement skills and so on, as discussed above – see the section 'Activities to develop thinking and working scientifically' on p. 67) and then children's ability in the individual components be brought together in an investigation. At the same time, children should also be taught the model of an investigation so they understand the nature of this type of activity in science. It is naive to assume that children learn these things via osmosis. Fleer et al. (1995, p. 7) suggest that 'teaching science, in many respects, is no different from, say, the teaching of language and mathematics'. Science, like maths and language, requires a high level of teacher input to ensure student success in this area of the curriculum.

Considerable planning and teaching is required to ensure that each of the components of thinking and working scientifically are developed in order that children can become more autonomous in their learning. One would expect to plan how to teach science concepts such as forces, energy, electricity or plant growth; thinking and working scientifically is no different in its need to be developed in a systematic and structured way.

In order to plan an investigation, children require a mental model of an investigation. Consequently, the teacher needs to consider how to help develop that framework. Figure 2.2 illustrates a 'planning house' to support young children (five to seven years old) in planning investigations in a range of contexts. The planning house is taken from the GINN Star Science Scheme (Phipps et al. 2000), which develops children's ability to plan investigations using this approach.

R

Although the basic format of an investigation does not change, the way it is planned can be presented in interesting and stimulating ways. The questions asked to support children might change according to the age and ability of the children. It is also possible that the questions could be produced by the children themselves.

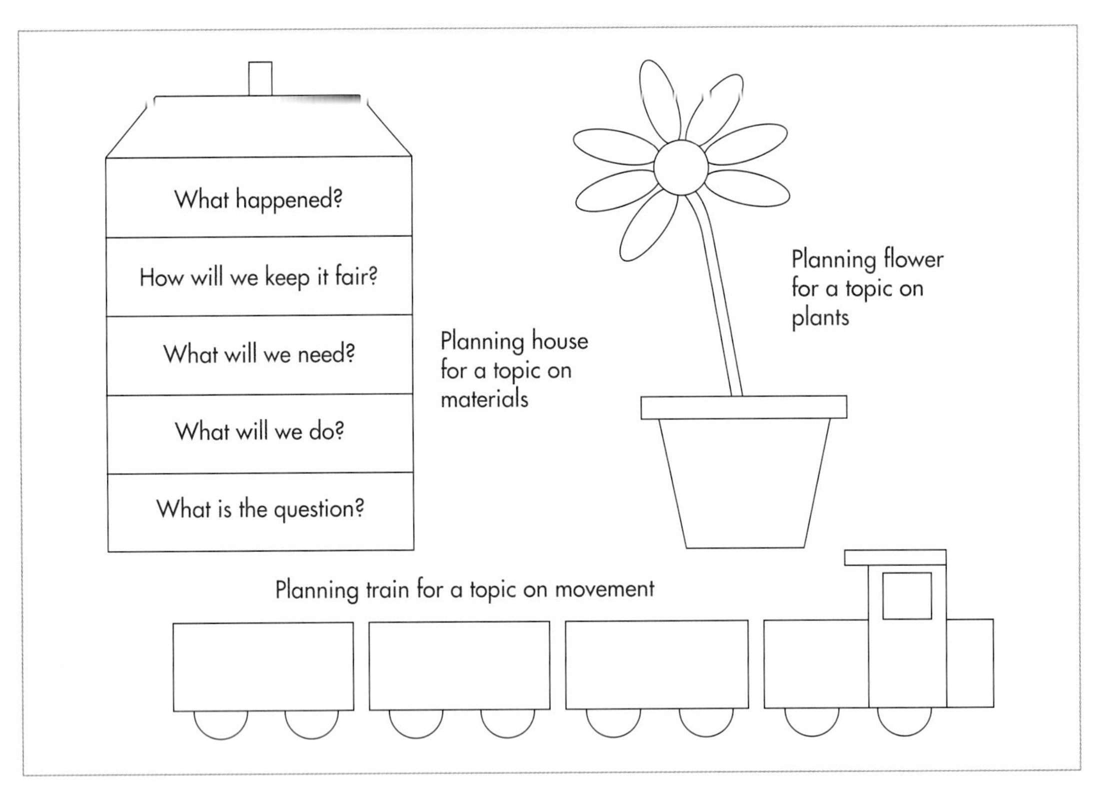

FIGURE 2.2 Approaches to planning investigations

Source: Adapted from Phipps et al. 2000

The use of a planning framework with children should change over time. Initially, it might be predominantly teacher-directed, then, gradually and with increasing confidence, children could take over the board until the time arrives when this type of crutch is no longer required and the children go solo, discarding the board altogether. The aim is to offer children a mental model of an investigation and to support them, through questions, in planning their investigation. An alternative method of planning is the poster system (see the online Appendix 2.3), which is taken from Goldsworthy and Feasey (1997) and uses Post-its to support the children through planning an investigation. The online Appendix 2.3 offers a simple method for the teacher and children to determine what kind of table or graph should be used for data generated in different kinds of investigations. For other suggestions in assisting students in planning fair tests and developing graphs, see the appropriate appendices in the *Primary Connections* units (AAS 2005–10).

The role of mathematics in investigations

Appendix 2.3
Posters to plan investigations

> At the heart of science is mathematics. Science relies on the use of mathematics to provide quantitative evidence on which scientists make comparisons, note patterns and trends, make generalisations and draw conclusions. In exactly the same way, school science relies on children's mathematical understanding as a base, upon which to develop their ability in primary science.

Feasey and Gallear 2000, p. viii

In thinking and working scientifically, children will be drawing on their knowledge and understanding of mathematics and their personal numeracy, using it as a tool to support and develop their science. When engaged in science, children use a range of mathematical tools; for example, measurement, estimation, calculations such as averages, subtraction and addition, as well as the creation and use of tables and graphs.

Mathematics is such an important part of thinking and working scientifically that it cannot be left to chance. To support this process, the teacher should help the children develop an understanding that:

- using numbers makes science easier because we can use patterns in numbers to make comparisons and draw conclusions
- by using numbers we can be more confident in what we say and think; telling someone that the temperature rose by 18°C is more authentic than saying 'It went up a lot.'
- by measuring accurately, carrying out calculations and drawing graphs correctly, our ability to do science can improve
- carrying out calculations helps us to see patterns and relationships that may not be apparent from the original numbers
- drawing graphs presents data in a different format and can help us see patterns and trends
- estimating can help us to decide what to do in an investigation.

Feasey and Gallear 2000, p. viii

Feasey and Gallear go on to suggest that children do not automatically transfer skills from mathematics into science. Therefore, an important role for the teacher is to draw children's attention to those areas where they need to apply their mathematics in science. The teacher should:

- remind children how to use measuring equipment
- challenge children to use numbers and insist on accuracy
- ask children what kind of sums, calculations and measurement they need to do in their investigations
- challenge children to use tables and graphs to present data.

The teacher should make sure that, when planning science and investigative work, there are opportunities for children to generate numerical data at the appropriate level. The type of investigative question provides the framework for the investigation and the level of mathematics required. It is important that the questions children investigate make appropriate demands of them; for example:

1 Which sugar dissolves the fastest?
2 How does the amount of water affect the rate at which sugar dissolves?

Both of these questions are related to dissolving sugars and each makes different mathematical demands on the children. Question one demands that children time how long it takes the sugars to dissolve and create a graph that shows the results, whereas the second question requires children to make links between two sets of data, the measurement of water and time, and to produce a line graph.

Of course, with all data we need to make sure that children develop the ability to critique the information and decide whether they consider it to be reliable and valid.

Appendix 2.4 Choosing the type of graph

Graphs in science

Chapter 1 discusses the idea of science-specific literacies (see the section 'Representational and related strategies' on p. 29) and refers to data representation such as pictures, charts, tables and graphs, the last an area of mathematics used in science that children find particularly challenging. The ability to read, interrogate and draw conclusions from a graph is central to thinking and working scientifically. The ASE and King's College Science Investigations in Schools Project (the AKSIS Project) carried out research and curriculum development that explored children's ability to produce and use graphs in investigations. The project found that:

> Over 75% of their graphs were incorrectly constructed and most pupils regarded graphs as an end in themselves. Only a few pupils referred to their graphs when considering their evidence. We found that many teachers we interviewed recognised the difficulty pupils had with graphs, but few had made a point of teaching pupils about the construction and use of graphs.
>
> **Goldsworthy, Watson and Wood-Robinson 1999, p. 2**

Activity 2.11 Graphs in science

Consider

Think about your own understanding and ability when producing and using graphs and, if applicable, how children you have taught manage graphs.

Which aspects of producing and using a graph do you think children might find difficult? Make a list. For each item on your list, explain why children might find this aspect difficult.

The AKSIS Project indicates that there are many aspects of creating and using graphs that cause children problems. Compare the list below with your own list and consider the similarities and difficulties.

Children find it difficult to:

- physically draw a graph
- use the language of graphs; for example, 'axis', 'horizontal', 'vertical', 'scale', 'bar', 'line'
- decide which way the axis should go
- work out which data go on which axis
- understand the relationship between a table and a graph
- read and tell the story of a graph
- appreciate that the way a graph is drawn and the scale used can sometimes give an inappropriate impression of the data
- appreciate the difference between a line graph and a bar chart
- choose the right scale for the axis
- understand that the wrong scale can affect the meaning of a graph
- transfer information from a table to a graph
- plot a line of best fit.

Activity 2.12 Supporting strategies for graphs in science

Refer to the list you created for Activity 2.11.

- For each item on your list, explain why children might find this aspect difficult.
- Create a two-column table similar to the table below and for each of the difficulties that children experience with graphs, consider a teaching strategy to support children in overcoming their difficulty.

DIFFICULTY	SUGGESTED STRATEGY
Physically draw a graph.	Give children squared paper or partly prepared graphs so they fill in the appropriate parts.
Use the language of graphs; for example, axis, horizontal, vertical, scale, bar, line.	

The AKSIS Project concluded that teachers must teach children directly how to construct and use graphs and not assume that children will transfer learning from mathematics to science or learn through osmosis. The AKSIS Project offered a range of strategies that might help children when constructing and using graphs, suggesting that teachers should:

- teach children the terminology of graphs (for example, 'scale', 'axis')
- use paper with large squares to support the drawing of graphs (for example, 1 cm or 2 cm squared paper)

- give children partly prepared graphs (for example, with axis drawn, blank boxes for labelling)
- give children examples of graphs in which there are deliberate errors for them to spot, then challenge them to explain what is wrong and how they could improve the graph
- create with the children a crib list of things to remember when producing a graph, to remind children what to do and what to include
- ask children to describe the graph using visualisation; that is, get the children to imagine that they are the graph and tell the story of what is happening in the graph.

The final suggestion is an important one since it can help to scaffold children in terms of making sense of a graph. Children can be asked to visualise the story of the graph or, alternatively, walk their fingers along a bar chart or line graph and try to explain what was happening in the investigation; for example, the temperature went up, then it went down 10°C.

The AKSIS Project found that many children did not refer to their graphs once they had been created, meaning that they did not use their graphs to:

- look for trends and patterns
- search for anomalies
- find the line of best fit
- interpolate and extrapolate
- ask questions from the data that could lead to further investigative work.

Goldsworthy and Feasey (1997) suggest that teachers use effective questioning to focus children's attention on different aspects of the graph and to help them interpret the data presented. For example, the teacher could ask:

- What is the story of your graph?
- Are there any results that do not seem to fit the pattern? Which ones?
- Can you explain why this is?
- Do you think your results give an accurate picture of the way things worked in your test? Why?
- What kind of pattern can you see in your results? Describe the pattern.
- Are there any surprises in your graph? What are they? Why have they occurred?
- How does the pattern in your graph match what you thought would happen?
- How many statements can you make about the graph? Make a list.
- Look at this graph. How do you think the children who made this graph carried out their investigation?
- If we continued the line of the graph, what do you think the pattern would be? Why do you think that?

Initially, it will be the teacher who models questions about tables and graphs, but gradually, as children become more expert in creating and interpreting graphs, there should be a gradual shift in responsibility for questioning and interrogating data from the teacher to the children.

Graphs are an important part of thinking and working scientifically. We should take seriously the need to plan for teaching aspects of graphing because if we do not do so, then we are at risk of failing children of all abilities and failing to develop scientifically literate individuals.

Activity 2.13 Developing aspects of thinking and working scientifically

- Consider your own science teaching. Which elements of thinking and working scientifically do you teach?
- How do you or could you ensure that the elements of thinking and working scientifically are developed?
- How would you teach children to draw tables and develop an understanding of why and when tables should be used?
- How will you ensure that children's understanding and use of graphs (usually taught in mathematics) is developed in the context of science?
- How would you support children in planning an investigation? Which kind of teaching and learning strategies do you or could you use?

You might find Hackling (2005) and the appendices in the *Primary Connections* units (AAS 2005–2010) especially useful to support your thinking in this activity.

High teacher input in science should not be taken to infer a move away from a child-centred approach to a didactic approach to teaching, or one that does not acknowledge and, indeed, develop the ability of the individual to become autonomous and a decision maker in his or her own learning. In terms of investigations, structured development of the component parts of investigations and the development of a model for investigations offer children the tools and understandings that allow them to be more autonomous.

What becomes evident is that, to succeed, the teacher needs to consider a range of strategies and carefully plan the route for the children. Fensham (1988, p. 58) acknowledges that 'successful teaching is a complex activity which requires the teacher to select from the diversity of possible strategies and actions the ones most appropriate for his/her existing classroom conditions'.

NOS

Referring back to the menu of activities (Aubrey 1994; Sherrington 1993), the menu can be viewed as a toolkit for teachers, where they choose from the range of activities on the basis of fitness for purpose.

Activity 2.14 Activity audits

Using plans for a sequence of lessons or a topic you have just taught, or one you are about to teach, carry out an activity audit. Look at your planning and carry out the following suggestions:

- Using the activity types from the menu (that is, observation, illustrative skills, survey, research, exploration, investigation, classification), count how many times children were engaged in a particular type of activity.
- Analyse your results. What do you think your results suggest about the teaching and learning opportunities you offer? What changes could you make?
- Carry out a similar audit to find out how frequently you teach and develop children's ability in the component parts of science: for example, planning, conducting investigations, processing data, evaluating findings, using science and acting responsibly (Hackling 2005).

You should also consider your science syllabus.

- Select a *Primary Connections* unit (AAS 2005–10) and complete a similar analysis to the above. What activity types were used and for what purposes? Has your selected unit provided an adequate range of teaching and learning opportunities? Justify your response.

Schemes of work

At this point, it would be useful to refer back to Chapter 1 and consider how the ideas relating to adopting a constructivist model/schema for planning can support the development of a scheme of work. In particular, consider how the following four components could be developed when planning:

- setting the context and gaining interest, then (or concurrently)
- finding out students' ideas
- selecting strategies that would help students develop, form and modify their ideas, and then
- ensuring that students reflect on their ideas and their learning processes.

Table 2.3 provides an example of activities from a scheme of work for the topic 'forces'. A scheme of work is a plan of action that indicates the learning intentions and associated activities for a topic. The two examples illustrate how the menu of activities can be used to teach the topic 'forces' and elements of thinking and working scientifically alongside each other at lower and upper primary levels.

While reading this section, you should consider how the teaching of the NOS (see the section 'The nature of science: how it works' in Chapter 1, p. 5) is threaded through the schemes of work. The schemes of work illustrate that science should not be taught in one-off lessons, but as a sequence of carefully planned sessions that link to each other. Some activities will take a short amount of time, while others (for example, investigations) might span several lessons. There will, of course, be exceptions – for example, when a rainbow is seen over the school and children's interest is so high that exploring making rainbows using prisms flows naturally from the incident. The schemes of work on forces illustrate how the teachers began by using the constructivist approach to elicit children's ideas on forces and friction, and then, through a set of related activities, developed children's conceptual understanding of forces alongside aspects of thinking and working scientifically. The upper primary scheme in particular illustrates how teacher planning identifies skills that need to be taught in order that children can use them in an investigation later in the scheme of work. For example, it can be seen that the teacher identifies the need to develop children's skill in using a Newton meter. The intention here is to teach this skill in the most economical way, first by a teacher demonstration of Newton meters followed by a skills activity in which children practise using a Newton meter around the classroom. Once this skill is established, the teacher can offer an investigation later in the topic in which children have to solve a problem using knowledge and understanding of friction relating to escape roads on hills, which demands that they are able to know why, when and how to use a Newton meter in a fair-test investigation.

One of the key advantages of this approach to planning is that the columns relating to outcomes for knowledge and understanding of concepts and thinking and working scientifically are also the expected outcomes for any assessment. For example, if children are expected to learn how to use a Newton meter, then the assessment is 'Can children use a Newton meter?'

Choosing from the full range of activities enables the different aspects of thinking and working scientifically to be taught and developed. In planning science, the teacher should make appropriate choices, aware that by not using the full range, the potential for the balanced development of children in science will be restricted.

In order to ensure progression, it is necessary to plan when to focus on each of the component parts over a series of science topics. For example, decisions will need to be made about when to introduce the construction of tables, bar charts and line graphs, and where and how the use of thermometers, Newton meters and light sensors will be taught. Such decisions need to be made at the whole-school level as well as in relation to specific classes and individual children. Crucially, the teacher will also need to provide opportunities in which children can develop their understanding of evidence and bring together the various understandings to tackle a testable question or problem.

As you read through Table 2.3, consider how the 'learning outcomes' will be aligned with 'learning outcomes' in your syllabus (e.g., see ACARA 2011).

Activity 2.15 Progression and continuity in thinking and working scientifically

Use your science syllabus to respond to the following questions. If your syllabus does not refer to thinking and working scientifically or related objectives or outcomes, then you may find it useful to refer to Hackling (2005) and/or *Primary Connections* units (AAS 2005–10) across their various stages.

Consider

- How would you develop the elements of thinking and working scientifically with children in your class in relation to other classes in the school?
- How do you think that a school could ensure progression and continuity in these elements across the year groups?
- Discuss with colleagues the idea of creating a development plan to ensure that the elements of thinking and working scientifically are developed across the school.

Effective questioning in science

In terms of a scientifically literate individual, the ability to question and challenge science and the evidence it presents is crucial if the individual is to be able to participate in a democracy. A scientifically literate person needs to be an effective questioner, someone who can use his or her knowledge and understanding alongside the ability and confidence to ask the right question at the right time.

TABLE 2.3 Examples of activities from a scheme of work on forces

FORCES: LOWER PRIMARY				
LEARNING OUTCOMES – FORCES KNOWLEDGE AND UNDERSTANDING	**ACTIVITY**	**LEARNING OUTCOMES – THINKING AND WORKING SCIENTIFICALLY**	**CLASSROOM MANAGEMENT**	**COMMENTS**
Children can explain their understanding of the word 'force'.	Children drew a picture about the word 'force'.	Making sense of everyday words in terms of science understanding.	Whole-class discussion followed by individual pictures and a sentence.	Interesting results; one child drew his dog having puppies and wrote, 'Mitsi forced her puppies out of her'.
Children know that a push or a pull can make things move.	Illustrative activity. Children put stickers on objects that they can push and make move.		Small-group activity. Children rotating through this activity. When all groups through, whole-class discussion.	Feedback and questions help to clarify and reinforce understanding.
Children are able to complete a table on pushing and pulling.	Illustrative activity. Clothes box offered for children to select garments from, put on the garments and decide whether they pushed or pulled into them.	Children recorded their answers in a table created by the teacher.	Small-group activity. Children rotating through this activity. When through, whole-class discussion.	Teacher discussed the table and how to use it with the children. Children completed table with confidence.
Children can use their knowledge about forces in the context of an investigation.	Investigation: Which car goes the furthest?	Children are introduced to using a planning house with teacher assistance. Children record results on a table.	Small-group activity. Children rotate through this activity over a week.	Children found it easier to use the planning house since the questions supported them. Idea of fair test required a lot of discussion. Help was needed to decide how to measure and fill in the table.
FORCES: UPPER PRIMARY				
Children can explain what they know about the word 'friction'.	Brainstorm the word 'friction' – teacher scribes children's responses.	Children consider scientific meanings of words.	Whole-class discussion.	Interesting answers included, 'Friction is something you read, it's not a fact book'.
Children know what friction means.	Children asked to carry out a series of illustrative activities, which include rubbing		Whole-class carried out activities in small groups. At the end of the session, whole-	Children have experience of friction in a range of contexts; for example, slipping

	different materials together, walking up PE benches with different shoes to feel friction.		class discussion for feedback.	on shiny floors or polished surfaces. The activities served to reinforce their ideas.
Children know that a Newton meter measures a pull force and can use a Newton meter to measure a pull force/friction force.	Skills/exploration. Introduce children to a Newton meter. Children use Newton meters to measure the force it takes to move objects; for example, to open a door, pull a shoe.	Using measuring equipment.	Whole-class teacher demonstration followed by small groups – limited equipment, therefore groups rotated through this activity at different times during the week.	The teacher introduced Newton meters and explained what they were and how they worked, after which children explored using the equipment.
Children know that a Newton meter measures a pull force, that there are different Newton meters, and that we need to choose the right one for the job.	Skills activity. Children given a selection of Newton meters from which they have to decide which is the best one for a range of tasks; for example, pulling a brick, opening a drawer, moving an empty box.	Draw and complete a table of results on their own. Choose the appropriate Newton meter from a range.	Small groups – limited equipment, therefore groups rotated through this activity at different times during the week. Whole-class discussion of results from different groups.	Children soon understand that there were different Newton meters with different scales.
Children are able to apply their knowledge of friction in an investigation.	Investigation. Plan and carry out a fair test to find out which is the best shoe for wearing in PE.	Plan and carry out a fair test. Use appropriate Newton meter. Produce good evidence.	Small groups – limited equipment, therefore groups rotated through this activity at different times during the week. Whole-class discussion of results from different groups.	Children were confident with planning and carrying out an investigation. They were able to choose and use the correct Newton meter. Children made connections between results and the material and pattern of shoe soles.
Children apply their knowledge and understanding of friction in a problem-solving situation.	Investigation. Children offered the scenario of which surface to use for escape roads for vehicles which have problems on steep hills. Create a fax report on the investigation.	Plan and carry out a fair test. Choose and use appropriate Newton meter. Create believable evidence. Use evidence to help answer problem.	Small groups – limited equipment, therefore groups rotated through this activity at different times during the week. Whole-class discussion of results from different groups.	Children created a fax report to a fictional company. The children had to decide what essential information on the investigation and results the company would require.

As with investigations, the teacher needs to be a good model for children since questioning is a linguistic form which, like all language, requires modelling. In science, the teacher is the key to the development of the child as an effective questioner. It is the teacher's ability to synthesise language and science that is important, as is offering a range of questions that have scientific outcomes.

Teachers need to refine the art of asking questions that are scientifically focused. Simplistic though this statement may appear, the reality is that, all too frequently, teacher questions in science lack this crucial component. The ability to ask the right question at the right time is probably one of the most difficult aspects of teaching. The essence of effective questioning in science is in the ability to link the linguistic form of the question to a scientific outcome. It requires that:

- the teacher employs a range of question stems; for example: Why? What? What if? Where? How? What do you think? Which?
- questions are carefully linked to a scientific outcome (scientific outcome + appropriate question stem = effective question)
- the teacher offers a range of questions in a variety of contexts
- the teacher encourages children to ask a range of questions in a variety of contexts
- children have access to questions in a written as well as an oral form
- the classroom houses questions (for example, question mobiles, questions on display boards) that encourage dialogue between children as well as between adults and children.

As teachers, we must ensure that our questions to students do not remain at a lower cognitive level; it is not that these are inappropriate, simply that they do not challenge higher order thinking skills. Teaching and learning in science should include both lower and higher cognitive level questions, thus challenging students to engage in higher order thinking where appropriate. It is useful to remind ourselves of Bloom's Taxonomy, which, in a revised adaptation (Wilson 2006), suggests that questions should relate to:

- creating
- evaluating
- analysing
- applying
- understanding
- remembering.

NOS

Where students apply knowledge to different contexts in science, they are engaging in higher order thinking. For example, students who are asked to analyse data, or a phenomenon, are using higher order thinking skills. Where teachers challenge students to synthesise experiences or information or evaluate data and investigations, students are having to engage in a different kind of thinking; similarly, where students are asked to engage in persuasive argument, they are being challenged to use higher order skills.

Questioning is pivotal to the development of thinking and working scientifically and has a role in all aspects of science activity.

The success of children as effective questioners in science depends largely on the classroom environment. Asking the right question at the right time is not the easiest task for the teacher and invariably requires the teacher to apply the fitness-for-purpose rule.

The investigation planning board is a strategy involving the use of a selection of questions to move children from one stage of an investigation to the next. All questions should demand that children put on their 'science spectacles' and construct their responses in terms of their scientific knowledge and their emerging scientific literacy.

Activity 2.16 Effective questioning in science

Teacher questions have been categorised in a range of ways; see, especially, Harlen (2003, Chapter 5). Record your own questions while working with a group or arrange for a colleague to log the questions you ask children.

- Does an analysis of your questions show whether you use the full range of question types offered in Harlen (2003, Chapter 5)?
- Consider your own practice. How could you encourage children to become more effective questioners?
- How do you react to and manage children's questions? How could you improve your practice?
- How could you use questions to encourage children to put on their science spectacles and make connections using their science knowledge and understanding?
- What strategies could you adopt to encourage a questioning approach in your classroom? Make a list.

Did you notice how many different question stems you used? Question stems include the following:

who	what	could	how	if	should	would	will	do	did
what	if	can	where	have	when				

Students need to be challenged to use different question stems since they lead to different ways of answering them. For example, question stems such as '*How can we* ...' lead to activity; '*What if* ...' leads students to speculating what might happen; '*Can* ...' offers students the opportunity to take a creative leap into suggesting alternative scenarios or solutions to problems.

You might have made a number of suggestions in relation to strategies you could use in the classroom, including:

- big books with questions inside
- question thought-showers (brainstorms)
- postboxes into which children place questions written on slips of paper
- mobiles which can be used to group questions on a similar topic
- games such as 'What am I?'
- the activity 'Hot Seating' where students ask questions to another student on a specific topic
- asking students to prepare questions for a visit or visitor.

Although encouraging children to raise questions is important, of equal importance is providing children with opportunities to answer their own questions and then communicate their findings. When students are asked questions, they should be given thinking time to prepare their answers; they should also be allowed to articulate their answers using, for example, 'science buddies'. Working with their science buddy, they can talk through their response, thereby rehearsing and possibly revising what they want to say before committing their answer to paper or to the public audience of their peers and the teacher. The need for children to be supported in developing their ability to ask questions should not be underestimated, and teacher scaffolding and appropriate activities are important.

Literacy in science

While reading this section, it will be useful to keep in mind the ideas relating to blending science and literacy and its benefits to students which Chapter 1 outlines (see the section 'Literacy–science connections' on p. 38). There, it's suggested that science and literacy share many cognitive processes and discursive practices and that these common processes are activated in scientific investigations and where children apply knowledge to new situations. These premises are important aspects of children developing as scientifically literate adults and they are major components of the *Primary Connections* initiative (AAS 2005–10).

The development of a scientifically literate person relies on the individual's understanding of and ability to communicate science; the quality of this is reliant, in turn, on the individual's literacy skills.

Feasey (1999, p. iv) suggests that:

> Science and literacy are inextricably linked. Without personal literacy individual children will find it more difficult to engage with science and certainly doors to a range of literature, both fiction and non-fiction, will be closed. In society where the written and oral word explodes on to screens, airwaves and paper, one might be as bold as to suggest that to be a literate individual is one's birthright as much as food, water, shelter, love and safety.
>
> The partnership between science and literacy is two way: science offers natural contexts for the use and development of literacy skills and understanding whilst literacy helps to offer the individual access to the exciting and challenging world of science.

Thinking and working scientifically demands a level of literacy that ensures that children have the tools to organise the ways in which they think, work and communicate. While science might place an emphasis on the written word, the importance of the spoken word should not be underestimated. Discussion allows children to articulate their thinking, rehearse ideas, rephrase and reorganise their thinking, and share ideas, as well as challenge other people. This kind of dialogue is central to thinking and working in a scientific way; if we are to develop confident, articulate children, then science lessons should be a healthy mix of the spoken and written word.

Whenever children are engaged in thinking and working scientifically, they use and develop a range of literacy skills. In the next set of activities, we will consider this range in the context of an investigation. The starting point for this investigation is the use of a concept cartoon by Keogh and Naylor. If you have not come across these before, then you should read Keogh and Naylor (2000a, 2000b; see also http://www.conceptcartoons.com).

Activity 2.17 Using concept cartoons as a starting point for investigations and multiple literacies in science

Concept cartoons can be used as a starting point for fair-test-type investigations. The teacher shows the concept cartoon to the children and asks them to think about the alternative viewpoints of the children in the cartoon and then allows them to discuss their ideas as a group.

- Look at the snowman concept cartoon (see Figure 2.3).
- If the teacher used this as a starting point, what generic and science-specific literacy skills would children be using and developing?
- Make a list.

A

Using the snowman concept cartoon as a starting point for discussion, the teacher and the children will read and think about the different options offered by the cartoon characters. If children are allowed to discuss the different options in the cartoon, they may be able to articulate their thinking behind their choice; they will also listen to their peers and might reconsider their original ideas. Some children will argue and offer reasons why they think that they are correct and someone else is wrong. Many children will make links to scientific ideas

FIGURE 2.3 The snowman concept cartoon

Activity 2.18 Literacy links in carrying out a fair-test-type investigation

Once children have planned their investigation, then the next stage is, of course, carrying out their fair test.

- What range of generic and science-specific literacy skills do you think children might employ when working as a group to carry out a fair-test investigation to find out which will make the ice pop melt faster: material or no material?
- Make a list of your ideas.

and to everyday personal experiences. An experienced teacher will engage children in responding to questions that might be designed to challenge children, and will encourage them to make links to other scientific ideas or their own personal experiences. All of this helps to develop children's literacy skills, albeit in the context of science; the teaching and learning sequence also makes use of children's literacy.

Keogh and Naylor (1997, p. 28) suggest that the scientific issue related to this cartoon is 'whether the coat is an insulator or whether it actually generates heat'. Some children might suggest that the coat will keep the snowman warm, so he will melt. It is this thinking that can be tested and children can then be asked to think about what they will do to test their prediction.

Once children begin to articulate thinking through discussion, they can be encouraged to talk about testing their idea. Where the teacher suggests that they could use ice pops, children invariably suggest that they could wrap the ice pop in fabric and compare it with an ice pop that is not wrapped.

A

In planning how to test their ideas, children could be engaged in using and developing any of the following generic literacy skills:

- discussion
- listening to others
- asking questions
- engaging in persuasive argument
- sequencing ideas
- making notes
- explaining the plan to the teacher.

Keogh and Naylor (1997, p. 12) rightly suggest that 'concept cartoons can be a powerful means of providing starting points for investigations, particularly for children who have poor literacy skills, who lack confidence in science or who are reluctant learners, as well as the more able'.

The list of literacy skills employed by children is impressive. You might have some of the following in your list:

- listening to others
- negotiating ideas and ways of working
- following oral instructions
- reading plans and measurements
- recording results
- reading, analysing and drawing conclusions from results.

Of course, carrying out the investigation, analysing results and drawing conclusions are not the end of children's work when thinking and working scientifically. The next step is to share their work with others, to engage in communicating what they have planned, done and found out to an audience.

Feasey and Siraj-Blatchford (1998, p. 4) suggest that in primary science, children are, to an extent, mirroring how the scientific community works. In the classroom, the importance of children communicating their science should not be underestimated. This is an example of one interpretation of social constructivism (see the section 'Social constructivism' in Chapter 1 on p. 13) in which students move from one culture (everyday) to another (science), and this of course requires careful scaffolding by the teacher.

Sharing their investigation and results with others (their audience) is not just a matter of telling someone what they did, what happened and what they found out. It should be a process in which children are confident enough to be challenged by others regarding their predictions, ways of working, data, conclusions and the believability of their results. This is the scientific process – it mirrors how the scientific community works and it involves using and developing a wide range of literacy skills.

Feasey and Siraj-Blatchford (ibid.) suggest that children need to be exposed to a range of audiences other than themselves and the teacher. When children are given an authentic audience, they are faced with the necessity of having to keep the needs of that audience in mind. The audience might, for example, not know anything about the investigation; on the other hand, it might be very familiar with the activity – two different audiences requiring different information. Communicating to an audience demands that children consider the following questions:

- Who is my audience?
- What do they need to know?
- What do I want to tell them?
- How can I make it interesting and keep their attention?
- Which parts should I tell them about?
- How should I present the information?

The audience might be any of the following:

- other children in the school
- parents
- the local community
- children in a local school or one in another part of the country or world
- local industry
- any combination of these.

Communication is a tool that enables children to learn from others and extend their own personal boundaries. The ability to communicate and receive communication, like all other aspects of thinking and working scientifically, needs to be taught and the teacher should make explicit how he or she plans to develop this aspect in a topic or series of science activities.

Children should be exposed to a wide range of recording and communicating to help them develop an understanding of how information can be communicated in a variety of ways and the need to think carefully about what they want to tell their audience.

Of course, we must remember that communication in science does not only refer to written communication. Equally important is verbal communication, and that is a precursor to writing. Children should be given opportunities to talk through ideas first before writing about them; in this way they can verbalise ideas, share and practise them, and also refine them. In the section 'Evidence in science' (see p. 62), the idea of argumentation in science was introduced, as was the need for discourse. Referring again to the DiPS project, we must make sure that the teacher talks less in science. The DiPs project reminds us that: 'Talking together improves critical thinking and helps children to think about their ideas and compare them to the ideas of others including what scientists say. Talk rather than writing allows children to rehearse their thinking in a collaborative and safe learning environment. Ideas are easier to change' (AstraZeneca Science Teaching Trust 2010).

The AstraZeneca Science Teaching Trust (AZSTT) website (http://www.azteachscience.co.uk) contains a number of continuing professional development units relating to children talking in science, which you will find useful in developing your understanding of this area; it also provides access to teachers who talk about different aspects of discourse, and the activities they used with primary children.

Activity 2.19 Literacy links and communicating investigations

Before completing this activity, return to Chapter 1 (see the section 'Teaching science constructively' on p. 19), which refers to multimodal ways of representing ideas in science. Most children are used to communicating their science using more-traditional forms such as prose, pictures, diagrams, tables and graphs. However, there is a wide variety of different ways of communicating, thinking and working scientifically.

Make a list of alternative ways of representing ideas in science. You should be able to come up with at least 15 suggestions. Think laterally and include approaches such as role-playing, newspaper articles and emails.

In order to be able to use the variety of approaches to communication that you have in your list, children need to develop a range of skills related to literacy, such as using:

- carefully constructed sentences
- scientific and everyday language
- correct grammar
- salient points
- bullet points
- questions
- connectives
- the language of persuasive argument.

So, not only should they be able to use different methods of communication, but they should also have a sound underlying understanding of why communication is important and what forms are appropriate in different contexts.

In your list of different ways of communicating science, you might have included newspaper articles. This is a rich seam in terms of:

- using and developing literacy skills
- communicating to an audience

- needing to communicate scientific thinking and ways of working
- using ICT to communicate their work; for example, PowerPoint, animation.

This symbiotic relationship between literacy and science can have a profound effect on the quality of science education. If children were asked to write a story in a language lesson, one would expect them to use correct punctuation, grammar and spelling, to write in an appropriate way that matches the genre and to make the story interesting. Why, then, shouldn't the same thing happen in science? Why is it that some teachers are willing to sacrifice spelling and language conventions to make sure that the children's ideas flow in science? How many times has a teacher suggested that what is being taught is science, not language; that it is the science that is important, so spelling and grammar don't matter as much? Of course they matter. How can individuals develop as scientifically literate adults if they are unable to communicate their science in an appropriate way?

Thinking and working scientifically: digital technologies

ICT

In today's education environment, people refer to different kinds of literacy. Already in this chapter we have referred to scientific literacy, and allied to this is the idea of 'digital literacy'. Hague and Payton (2010, p. 1) suggest that:

> To be digitally literate is to have access to a broad range of practices and cultural resources that you are able to apply to digital tools. It is the ability to make and share meaning in different modes and formats; to create, collaborate and communicate effectively and to understand how and when digital technologies can best be used to support these processes.

In relation to thinking and working scientifically, digital literacy is an important concept, given that we would expect that children should become increasingly competent in the use and development of a range of digital technologies in science activities.

As Hague and Payton (ibid.) remind us, digital technologies are embedded in the daily life of most children who have mobile phones, use YouTube, have MP3 players and communicate with friends and family using the Internet. This means that most children are already manipulating data in their daily lives, and the application of these competences to activities in school will be a natural extension of their home lives. In primary science, we can take advantage of their increasing understanding of different applications in a range of ways; for example, prompting children to:

- take photographs of how they are working in science
- create video sequences to communicate to others what they have done
- enter into email conversations with scientific experts
- create podcasts of science-related interviews, or use them to report science investigations on the school website
- use computer sensors (data loggers) to monitor sound, light, pH, heart rate, movement over time and so on
- create PowerPoint presentations to communicate plant investigations underpinned by knowledge and understanding of plants

- participate in video conferences to share ideas, data and environments with children from other schools, and replicate investigations
- use digital microscopes to view and capture photos of a range of objects; for example, different materials, to understand their properties
- speak into digital microphones to collect, store and download observations or data from investigations
- explore sound amplifiers.

Teachers need to consider the difference between ICT 'for doing science' and ICT 'for learning science' (Songer 2007, p. 472). Crucially, what is most important is that children should, by the end of their primary years, have had experience using a range of these technologies in a wide variety of contexts in science, to both do science and learn science, and eventually should become confident in choosing and using the most appropriate technology for the job.

Thinking and working scientifically: creativity

Have you ever thought about what kind of people and attributes you associate with creativity and, indeed, how this links to thinking and working scientifically? Take a few moments to reflect on creativity and your views.

Activity 2.20 What is creativity?

Make notes of your thoughts on the following questions:

- What words do you associate with the term 'creativity'?
- When you think about creativity, who, in society, do you think of as being creative?
- What do you think are the characteristics of a creative person?
- Do you consider yourself to be a creative person? Why?

As part of a small-scale research project, 100 teachers completed a research questionnaire on various aspects of creativity. When asked which words were associated with creativity, common responses included 'artists', 'musicians', 'clever' and 'intelligent'. The majority of respondents linked creativity to the arts, yet creativity is at the heart of science: think of the incredible imaginative leaps taken by scientists such as Einstein, Curie, Galileo, Franklin, Lewis H. Latimer, Charles Drew and Barry J. Marshall.

Thinking and working scientifically needs creativity, and just as we want students to understand what it means to think and work like a scientist, we should also encourage them to think of science as being a creative endeavour.

Feasey (2006, p. 1) suggests that:

> Constructivism advocates working from where the child is in their understanding, therefore acknowledging children's personal perceptions of the world. So, one of the main sources of thoughts on creativity should come from children themselves who offer remarkable insights into creativity and creative people.

So what do students think a creative person is in relation to science? Working with children aged eight to 11 years, Feasey (2006, pp. 10–11) asked them to describe a creative person in science. Their responses included the following:

- let loose imagination
- show thoughts
- encouraging and joyful
- reach new heights
- express imagination
- let your mind take over
- think the unthinkable.

The students' suggestions are intuitive and pertinent; perhaps they should have been asked to join a committee to report on creativity in education by the government in England! The National Advisory Committee on Creativity and Cultural Education (NACCCE), in its report *All Our Futures: Creativity, Culture and Education*, offered a definition of creativity that would be useful to explore in relation to teaching and learning in science education. The NACCCE report suggested four features of creativity:

M&E

1. **Imagination** – Imaginative activity is generating something original that is alternative to what might be expected.
2. **Pursuing purposes** – Creativity is aimed at meeting an overall objective or problem solving.
3. **Being original** – This could be original in relation to the individual and previous work, or original in relation to their peer group, or historically original; that is, unique in terms of human endeavour.
4. **Judging value** – Value could be in terms of, for example, effectiveness, usefulness or whether the output is enjoyable.

Activity 2.21 Thinking about creativity

The rhetoric of creativity is eloquent and many have written on the subject, including Sternberg (1999). However, for young children the rhetoric requires translation into the practicalities of science in the classroom; it is the teacher's role to mediate this and to offer suitable opportunities that support the development of children's creative potential.

Consider

- What do you think are the characteristics of the creative child in science?
- What would you be looking for in how children were thinking and working?
- Make notes about your ideas and then compare them with the research data in Table 2.4.
- Is there anything that surprises you? If so, what and why?

The data in Table 2.4 comprise some interesting responses by teachers to the first question posed in Activity 2.21. Given that creativity is about creating something original, the score for this aspect is very low. Equally, perseverance and determination also have low scores, but on reflection, some of the greatest advances in science have been the result of

TABLE 2.4 Characteristics of the creative child in science

CHARACTERISTICS OF THE CREATIVE CHILD IN SCIENCE	PERCENTAGE RESPONSE BY TEACHERS
Questioning/inquisitive/curious	21.8
Uses previous knowledge	18.2
Independent/lateral thinker	15.5
Interested/enthusiastic	12.7
Participates/gets involved	10.9
Verbal/communicates ideas	10.9
Originality	4.5
Perseverance	3.6
Determination	1.8

determined people who persevered in difficult circumstances or were willing to work long hours for many years to reach their goal.

Science does not always provide easy solutions, whether you work as a research scientist or a nine-year-old child in the classroom. Originality, perseverance and determination need to be encouraged because the creative child might:

- have ideas that are different to those of other children
- take intuitive leaps of the imagination that are not appreciated by others
- think laterally to find a solution that might not be obvious to peers
- pay attention to detail, working more slowly than others
- have a higher level of interest than other children.

One element that does not appear in the teachers' list is 'risk taking', an aspect of creativity that is rated highly by the authors of *All Our Futures* (NACCCE 1999), who considered risk taking to be central to creativity. Creative people are those who are prepared to take risks in their thinking and ways of working. The concept of risk is not in terms of physical danger, but in stepping outside the norm, thinking the unthinkable or not agreeing with peers just to conform. Risk taking in thinking and working scientifically is crucial if we want children to challenge their own and other people's ideas and ways of working, to think the unthinkable or to try something in a different way. However, Gunstone (2002) offers a salutary note in relation to risk taking in science. He suggests that risk taking can be uncomfortable and may require people to do things differently. He goes on to explain that in risk taking there is the likelihood that this may lead initially to being less confident and less expert. This is true for children exploring new ideas and approaches to working and for teachers trying to be creative and changing their philosophy and approach to science teaching.

Activity 2.22 Developing creativity

If creativity is an important element of thinking and working scientifically, then can creativity be increased by conscious effort?

In the classroom, what should the teacher consider in order to offer an appropriate learning environment, one that will encourage and develop children's creative potential in science?

Make a note of your ideas on this, and then compare them with the data in Table 2.5. The table shows research data relating to responses by teachers to the question: Can creativity be increased by conscious effort?

Finally, reflect on where you think opportunities for 'creativity' are offered in your own syllabus or in the key resource that you use, such as *Primary Connections* (AAS 2005–10). Be critical in your reflection. Do you think that the syllabus or resource is creative in both its approach and in supporting the development of children's creativity? Will you need to reconsider any elements?

In regards to the data in Table 2.5, few people would argue with the top four responses – they seem logical and quite straightforward. However, a number of the other responses are worth some discussion. For example, the idea that structured and organised activities should be central to creativity might be an anathema to some people, suggesting the opposite to encouraging creativity. After all, some people might think that creativity requires freedom for children to express themselves and be engaged in activities without constraints. In reality, that freedom and encouragement to take risks, to try new ideas and ways of working, require carefully thought-through approaches to ensure appropriate teaching and learning opportunities that will support children in realising their creative potential.

TABLE 2.5 Can creativity be increased by conscious effort?

TEACHER RESPONSES ON HOW CREATIVITY CAN BE INCREASED BY CONSCIOUS EFFORT	PERCENTAGE RESPONSE BY TEACHERS
Dialogue/open-ended	19.8
Practical/problem solving	19.8
Variety of teaching methods	17.4
Activities pitched at achievable/correct level	9.3
Structured/organised activities	9.3
Interesting activities	9.3
Encouraging pupils	2.3
Being inventive	1.2

In developing children's ability to think and work scientifically, there are a number of things that the teacher could do to encourage the development of creativity. Feasey (2003) suggests that the teacher could, for example, ask will science lessons:

- be different – scintillating?
- challenge?

- remove barriers to risk taking and failing?
- offer resources and learning spaces that stimulate creative contributions?
- allow problem solving?
- use up-to-date ICT applications?
- provide opportunities for children to learn in environments beyond the school that excite and engage interest?
- allow children to meet and work with different people?
- offer collaborative ways of working?
- stretch the imagination?

Duffy (1998, p. 1) quotes Chukovsky, who stated that the 'future belongs to those who do not rein in their imagination'. If there is to be a creative future for science, then teachers must not rein in their creative approach to teaching and learning and children must be allowed to explore their creative potential in science.

Science as a human endeavour

'Science as a human endeavour' is an important part of children making sense of how science relates to and impacts on everyday life. When developing activities relating to thinking and working scientifically, the teacher should aim to locate examples of where people from different cultures and careers use those skills and concepts related to being a scientist. Indigenous knowledge, for example, is clearly based on observation, exploration of the natural world, and communicating from generational ways of working and a knowledge base that has accumulated over these generations, whether it is managing an environment or using constellations for navigation. Students need to be involved from an early age in debate about science and its impact on themselves and their local community, as well as make sense and form ideas about global issues. They should be challenged to think about how science impacts upon their own lives and who uses science in everyday life. For example, the school cook uses science since he or she needs to understand hygiene, food groups, health and safety, and microorganisms. A local builder requires knowledge of forces, materials, thermal and electrical conductors, as well as understanding the environmental implications of a building project.

NOS

S&C

Children need to appreciate their emerging role in relation to scientific and technological developments that affect their lives, whether it is a wind farm being located in their community or the conservation of an animal species. They need to know about the issues and make sense of scientific information and opinions before drawing their own conclusions. At the same time as developing students' ability as scientifically literate persons, where they are able to be critical of science and its applications and implications, teachers have a duty to ensure that students understand who in the world uses science. Children need to develop an understanding of the importance of thinking and working scientifically in relation to the workplace and potential careers. Examples of the links between careers and science in everyday life are provided in later chapters. In so many occupations, people with the ability to think and work scientifically are valued because they are able to think logically, problem-solve, manage data, and draw conclusions. Science education also develops a range of personal capabilities

(Bianchi 2003; also see http://www.personalcapabilities.co.uk/about_personal_capabilities) that employers value, such as:

- teamwork
- self-management
- creativity
- communication
- problem solving
- tenacity
- a positive self-image.

We must remember that thinking and working scientifically is not just applicable to the science lesson, but is a cognitive framework that will serve children well in their adult and working lives.

Summary

This chapter began by considering the idea of scientific literacy. It has been argued that thinking and working scientifically provides the foundation for the individual to be able to question science and challenge personal assumptions and those offered by society. The role of the teacher is to plan to assist the development of those elements that a person needs to be scientifically literate; thinking and working scientifically with their pupils should be emphasised. The key is to do it in a manner that is interesting, exciting and challenging and enables children to experience the awe and wonder of their world as it unfolds for them. If teachers can successfully combine the elements discussed in this and other chapters, then perhaps children might respond in equally exciting and challenging ways. Of course, we would hope that they would, as suggested by the quote at the beginning of this chapter, not think 'Eureka!' but 'That's funny ...'

Two poems have been included as examples of children working in science and exploring how to communicate their work in science in a creative way – see the online Appendix 2.5. The children who wrote the poems engaged in creative thinking and working scientifically. Their responses are testament to exciting and motivating learning experiences.

Appendix 2.5
Students' poems

Acknowledgements

Thanks to Elizabeth May, teacher at Aberfeldie Primary School, for sharing her material, which appears in Activity 2.5.

Search me! science education

Search me!

Explore **Search me! science education** for relevant articles on thinking and working scientifically. Search me! is an online library of world-class journals, ebooks and newspapers, including *The Australian* and the *New York Times*, and is updated daily. Log in to Search me! through http://login.cengage.com using the access code that comes with this book.

KEYWORDS

Try searching for the following terms:

- Investigations
- Scientific literacy

Search tip: **Search me! science education** contains information from both local and international sources. To get the greatest number of search results, try using both Australian and American spellings in your searches: e.g., 'globalisation' and 'globalization'; 'organisation' and 'organization'.

Appendices

In these appendices you will find material related to thinking and working scientifically that you should refer to when reading Chapter 2. These appendices can be found on the student companion website. Log in through http://login.cengage.com using the access code that comes with this book. Appendices 2.1 and 2.2 are included in full below.

Appendix 2.1 Illustration of a scientist by Katrina, age 10

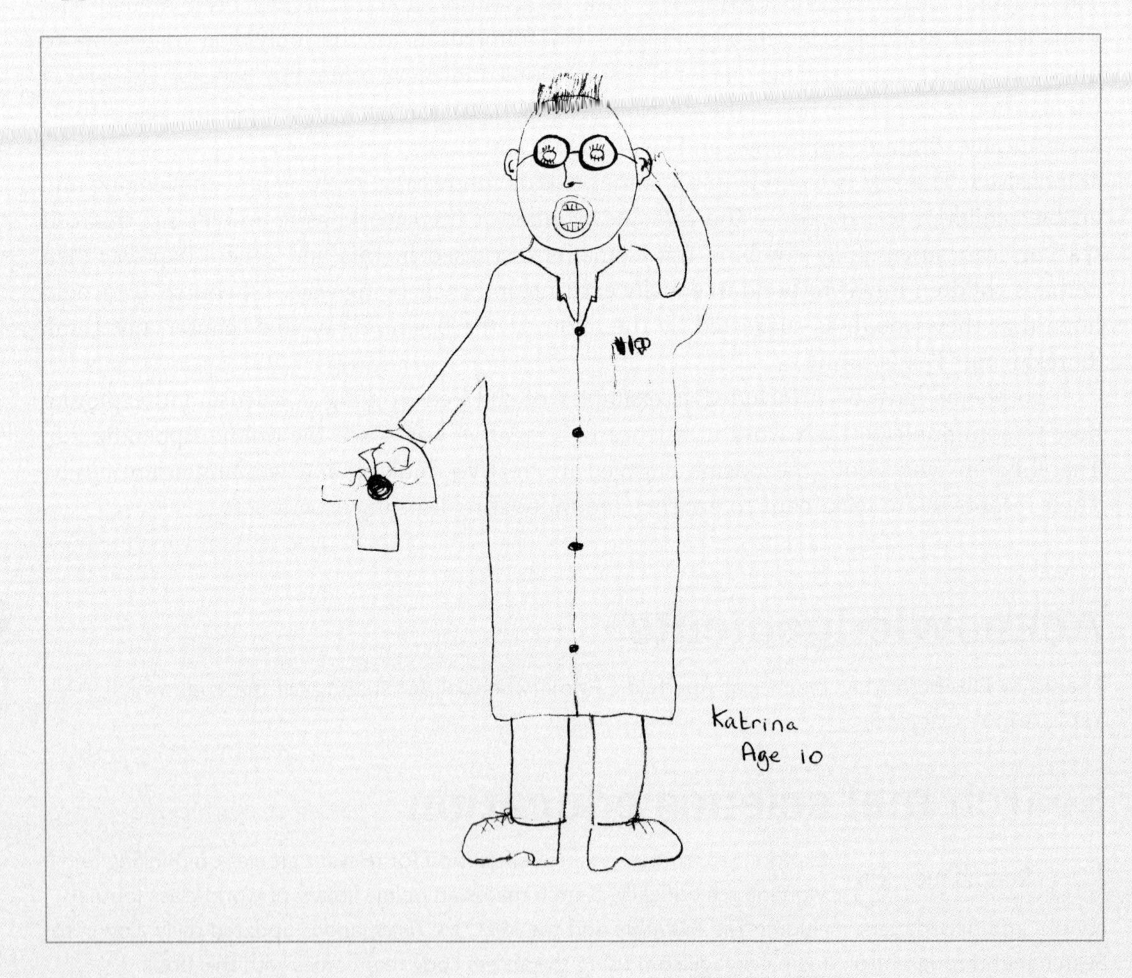

Appendix 2.2 What scientists think about and do

WHAT SCIENTISTS THINK ABOUT	WHAT SCIENTISTS DO
How to do things	Use equipment
What to measure, what to use	Measure things
How many times to do something	Make things fair
Where to find out about things	Try things out
Whether they are right	Investigate
How to prove they are right	Experiment
How to record, what to record	Record things
What information is needed, how and where to find it, who to talk to, what to read	Draw graphs
What the information says	Use a computer
What they need to know	Look at results
What is the problem, how to solve it	Write reports
Whether data is valid and reliable	Ask questions
Whether they believe the information	Make things
Look at and think about results	Test things
Draw conclusions	

Appendix 2.3 Posters to plan investigations

Various scaffolds for younger and older students that would assist them to plan investigations are illustrated in this online appendix.

Appendix 2.4 Choosing the type of graph

This online appendix assists in choosing the most appropriate type of graph to use.

Appendix 2.5 Students' poems

Two children's poems with a science focus are shown in this online appendix.

References

AstraZeneca Science Teaching Trust. 2010. Discussions in Primary Science (DiPS). Available at http://www.azteachscience.co.uk/ext/cpd/dips/index.htm (accessed 15 September 2010).

Aubrey, C. (ed.). 1994. *The Challenge of Science: The Role of Subject Knowledge in the Early Years of Schooling*. London: Falmer Press.

Australian Academy of Science (AAS). 2005–10. *Primary Connections*. Canberra: AAS.

Australian Curriculum Assessment and Reporting Authority (ACARA). 2011. Australian Curriculum: Science K-10. Available at http://www.australiancurriculum.edu.au/Documents/Science/Curriculum/F-10 (accessed 12 April 2011).

Bianchi, L. 2003. Better Learners. *Primary Science Review*, 80, pp. 22–4.

Duffy, B. 1998. *Supporting Creativity and Imagination in the Early Years*. Buckingham, UK: Open University Press.

Encel, V. 1988. *Australian Genius: 50 Great Ideas*. Sydney: Atrand.

Feasey, R. 1996. Initial teacher training: Developing the right kind of primary science teacher. *Paper presented at Culturas de Aprendizagem, Castelo Branco*, Portugal, September.

_____ 1999. *Primary Science and Literacy*. Hatfield: ASE.

_____ (ed.). 2001. *Science Is Like a Tub of Ice Cream: Cool and Fun*. Hatfield, UK: ASE.

_____ 2003. Primary science creative futures. *Primary Science Review*, 78, pp. 21–3.

_____ 2006. *Creative Science Achieving the WOW Factor with 5–11 Year Olds*. London: David Fulton Publishers.

_____ & Gallear, R. 2000. *Primary Science and Numeracy*. Hatfield: ASE.
_____ & Gott, R. 1996. A proposal for a UK National Centre for Initial Teacher Training in Primary Science. *Paper presented at the meeting of the Society of Chemical Industry*, London, May.
_____ & Siraj-Blatchford, J. 1998. *Key Skills: Communication in Science*. Durham: Golden Offset.
Fensham, P. (ed.). 1988. *Development and Dilemmas in Science Education*. London: Falmer Press.
Fleer, M., Hardy, T., Baron, K. & Malcolm, C. 1995. *They Don't Tell the Truth about the Wind*. Melbourne: Curriculum Corporation.
Global Greenhouse Warming.com. Strategies to reduce global warming. Available at http://www.global-greenhouse-warming.com/reduce-global-warming.html (accessed 9 October 2010).
Goldsworthy, A. & Feasey, R. 1997. *Making Sense of Primary Science Investigations*. Hatfield, UK: ASE.
_____, Watson, R. & Wood-Robinson, V. 1999. *Getting to Grips with Graphs – Investigations*. Hatfield, UK: ASE.
Gott, R. & Duggan, S. 1995. *Developing Science and Technology Education Investigative Work in the Science Curriculum*. Buckingham, UK: Open University Press.
Gregory, J. 2001. Public understanding of science: Lessons from the UK experience. Available at http://www.scidev.net/en/features/public-understanding-of-science-lessons-from-the.html (accessed 20 April 2010).
Gunstone, R. F. 2002. Curriculum change, in J. Wallace & W. Louden (eds). *Dilemmas about Teaching and Learning*. London: Routledge Falmer, pp. 231–44.
Hackling, M. 2005. *Working Scientifically*. Perth: Education Department of Western Australia.
Hague, S. & Payton, S. 2010. *Digital Literacy across the Curriculum – a Futurelab Handbook*. Bristol: Futurelab.
Harlen, W. 2003. *The Teaching of Science in Primary Schools* (3rd edn). London: David Fulton.
_____, Macro, C., Reed, K. & Schilling, M. 2003. *Making Progress in Primary Science – A Handbook for Inservice and Preservice Course Leaders* (2nd edn).London: Routledge.
_____ & Qualter, A. 2009. *The Teaching of Science in Primary Schools* (4th edn). London: David Fulton Publishers.
Ingpen, R. 1982. *Australian Inventions and Innovations*. Adelaide: Rigby.
Jane, B., Fleer, M. & Gipps, J. 2007. Changing children's views of science and scientists through school-based teaching. *Asia-Pacific Forum on Science Learning and Teaching*, 8 (1). Available at http://www.ied.edu.hk/apfslt/v8_issue1/janefleer/janefleer4.htm (accessed 12 June 2010).
Kendra, J. 1990. *Made for Australia*. Sydney: Harcourt Brace Jovanovich.
Keogh, B. & Naylor, S. 1997. *Starting Points for Science*. Crewe: Mill House Publishers.
_____ 2000a. *Concept Cartoons in Science Education*. Cheshire: Millgate House Publishers (see also *Investigating*, 17 (1), pp. 47–8).
_____ 2000b. Teaching and learning in science using concept cartoons. *Investigating*, 16 (3), pp. 10–14.
Millar, R. & Osborne, J. 1998. *Beyond 2000: Science education for the future*. London: Nuffield Foundation National Research Council.
National Advisory Committee on Creativity and Cultural Education (NACCCE). 1999. *All Our Futures: Creativity, Culture and Education. Report to the Secretary of State for Education and Employment and the Secretary of State for Culture, Media and Sport*. London: DfEE.
New Zealand Ministry of Education. 2010a. The New Zealand Curriculum Online: Learning areas – Science. Available at http://nzcurriculum.tki.org.nz/Curriculum-documents/The-New-Zealand-Curriculum/Learning-areas/Science (accessed 18 October 2010).
_____ 2010b. The New Zealand Curriculum Online: Learning areas – Science: Science curriculum achievement aims and objectives. Available at http://nzcurriculum.tki.org.nz/Curriculum-documents/The-New-Zealand-Curriculum/Learning-areas/Science/Science-curriculum-achievement-aims-and-objectives#4 (accessed 18 October 2010).
Osborne, J. 1991. Approaches to the teaching of AT16 – the Earth in space: Issues, problems and resources. *School Science Review*, 72 (260), pp. 7–15.
Phipps, R., Feasey, R., Gott, R. & Stringer, J. 2000. *New Star Science Teachers' Resource Book*. Aylesbury, UK: Ginn.
Rassan, C. C. 1993. *The Second Culture: British Science in Crisis – the Scientists Speak Out*. London: Aurum Press.
Schibeci, R. 1989. Images of scientists 1989. *Investigating: Australian Primary Science Journal*, 5 (3), pp. 24–7.
_____ 2006. Student images of scientists: What are they? Do they matter? *Teaching Science*, 52 (2). pp. 12–16.
Sherrington, R. (ed.). 1993. *ASE Primary Science Teachers' Handbook*. London: Simon & Schuster.
Songer, N. 2007. Digital resources versus cognitive tools: A discussion of learning science with technology, in S. Abell & N. Lederman (eds). *The Handbook of Research on Science Education*. Mahwah, NJ: Lawrence Erlbaum Associates, pp. 471–91.
Sternberg, R. J. (ed.). 1999. *Handbook of Creativity*. Cambridge: Cambridge University Press.
Wardle, C. 2001. Literacy links to science and scientists. *Primary Science Review*, 69, pp. 12–14.
Wilson, I. 2006. Beyond Bloom – A new version of the cognitive taxonomy. Available at http://www.uwsp.edu/education/lwilson/curric/newtaxonomy.htm (accessed 18 October 2010).

3

Movement and force

by Russell Tytler, Linda Darby and Suzanne Peterson

Introduction

How – and why – do things move? How do we describe how they move? This chapter looks at ideas and activities concerning movement and force. It deals with two major issues: firstly, ideas children have about motion and the strategies for teaching about motion in the primary school program. This will include some discussion of the different contexts in which movement and force can be studied. Secondly, it looks at the wider context of studying movement and force, linking it with technology and science as a human endeavour.

Background to the chapter

Two of the authors (Russell and Suzanne) were involved in a longitudinal study of children's science learning and, as part of that, have explored, through activities and interviews, ideas about movement and force and air and flight. Some of the material in this chapter relates to the insights generated from this exploration. Another specific input into the chapter comes from work that Linda has been undertaking with her science teacher education students around literacy and unit design based on the *Primary Connections* 5E framework. Other activities, in particular the unit sequence, derive from work that Suzanne conducted with her own years 3 to 4 class. Some of the wheels and language activities are based on the ideas of Tom Radford, who was an earlier contributor to this chapter.

Thinking about movement and force

What makes things move? How do they slow down and stop? How do forces come into this? Below is an activity sequence that might be used to introduce ideas about movement and force. Think about what your own response to the activities might be – even try the activity out and think about variations!

An activity sequence to elicit and challenge children's ideas

Push and pull

Start with a toy that moves. What do you do with it? Which toys do you push and which toys do you pull?

Using butcher's paper, create a list of things that you push, things that you pull.

Acting on playdough

Children are given playdough and asked to say what they might do with it. In small groups, they explore with the playdough.

Gather them back together to talk about the different things they did: twist it, squeeze it (squeeze is two pushes – one from each side), press it. Introduce diagrams, with arrows as a visual symbol to represent pushing and pulling. Ask children to draw what they did to the playdough.

Establish, through discussion, the effectiveness of diagrams and arrows to represent/communicate what is happening – a lower secondary school teaching sequence based on this idea is described in Hubber, Tytler and Haslam (2010).

Challenge the children to think about what is happening when they squeeze an aluminium can, sit on a sponge cushion or squeeze a toothpaste tube.

Pushing and pulling a table

Up-end a table onto the floor. Challenge children to pull it. What happens if someone stands on it? What happens if two people stand on it? What difference does it make if two people both pull? What happens if one person pulls and the other pushes from the other end?

Loop an elastic length (such as stockings) around the table and investigate how much force is needed by one person, two people etc. to pull the table.

Activity 3.1 **Representing force**

- Discuss with friends what view of force is being encouraged by the sequence above.
- Note that in each case the force is represented as coming from an outside agency – a hand, a rope. Is this always true of force?
- Discuss the way the sequence builds multimodal representations of forces acting on objects, including body–kinaesthetic, verbal, and visual/symbolic.
- Think about and sketch, using arrows to represent forces, the forces acting on a

tennis ball that finds itself in the circumstances listed below:
- resting on the tennis court
- floating in a pond
- in contact with a racquet
- bouncing
- flying through the air
- moving up in a vertical throw.

- As this chapter unfolds, collect a set of ideas and questions for further consideration to use as a science journal to help clarify your changing ideas.
- After the discussion on scientists' ideas (below), you should return to review your responses and questions.

Scientists' ideas about movement and force

The theory of movement we accept as the official scientific view was largely developed by Isaac Newton in the 17th century. Newton argued that we should think of force as causing changes in motion rather than motion itself. Forces are the way we describe the effect of external influences on an object. An internal force, such as gripping a steering wheel or pushing on the brake, will not directly slow your car down or otherwise affect its motion, but the friction on the car tyres from the ground will. Thus, a ball or a box, for instance, will move along a flat surface forever, unless an external force (such as friction or a kick from a boot) acts to stop it, speed it up or deflect it sideways. The box, if it is sitting on the ground, will experience a gravitational force down, balanced by a reaction force up of equal size. Thus, the total force on the box is zero. The reaction force is similar to the force acting on you if you stand on a trampoline as it is stretched. In fact, the ground will be ever so slightly stretched in the same way, much as a plank of wood or a wooden floor might be, so that it exerts a force upwards on you.

Forces do not always result from direct contact. Take the case of gravity. Newton showed that the weight force acting on an object is due to gravitational attraction from the Earth and that this force, which causes apples to fall, is also the force that keeps the moon orbiting the Earth. Let's sum up a short list of scientific ideas about force before moving on to challenge and refine our views:

- Forces cause changes in motion and are not, unlike momentum and energy, associated with the motion itself.
- Forces are our way of describing the way external effects (pushes, pulls, gravity, support) can influence the motion of things.
- A force is an effect on an object, not a property of the object or its motion. If you want to explain a change in motion (speeding up, slowing down, swerving), then you must look for an external effect and not at the object or something inside it.
- Forces occur in action–reaction pairs. Thus, if your standing body pushes down on the ground, the ground will push back up on you.
- Pairs or sets of forces will add together to affect motion. But, as the addition must take into account direction, opposing forces can cancel each other out.
- Common forces include contact forces (physical pushes, support or traction from the ground, friction, air or water resistance opposing motion, force from wind) and field forces (gravity, magnetic forces, electric field forces).

Activity 3.2 Further questions, measuring force

- Discuss each of the scientists' ideas above, sketching some examples to illustrate each idea.
- Following this, add to your list of ideas and questions to be further investigated in your science journal.
- Try out the following friction activity sequence, which is designed to provide a way of measuring force.
- Friction is not a straightforward idea. Discuss whether there are situations in which friction is necessary for movement to occur (think of a world without friction). What forces are acting to cause movement in these cases?
- Further refine your science journal before considering the section below, which discusses informal ideas that people often have about movement and force.

Moving a brick – investigating friction

The challenge is to find a way of moving a brick using as small a pulling force as possible. You could use a plastic bottle filled with water instead of a brick. An elastic band (or several joined together) can be used to pull. The amount of stretch is a measure of the elastic band force. You can attach the elastic band to the brick by tying and taping a piece of cloth or cord around it.

- Check the fact that the rubber band will stretch more as the pull gets greater.
- Discuss some possible ways of making the brick move with as little pull as possible. Some possible things to use when investigating include surfaces with different degrees of smoothness (cloth, sandpaper, aluminium foil, plastic), pencils, skewers or other rolling things.
- You can record the amount of stretch needed to just overcome the friction force to move the brick by placing a streamer alongside the band and cutting it to the stretch length. Work out a way of displaying the results to best bring out the trend.
- Predict what will happen to the friction force when one, then two, then three bricks on top of each other are pulled on the same surface. Test your prediction using streamers.
- Discuss how you might draw a diagram, in which you use arrows for force, to represent what is happening.
- A sideline challenge: use your elastic band to measure the weight of a solid object, then see what seems to happen to its weight when it is dunked in water. Discuss how you might represent your findings by using a force diagram.

Alternative conceptions about movement and force

Many studies of informal ideas about movement and force (for example, Gilbert and Watts 1985; Gunstone 1987; Ioannides and Vosniadou 2001) have shown the difficulty that children and adults have with the idea of force and the causes of movement. In fact, children's ideas about force have been shown to have a lot in common with the ideas of earlier scientists, such as Aristotle and the medieval impetus theorists who thought of force, or 'impetus', as residing in moving objects (McCloskey 1983).

Gilbert and Watts (1985) identified a list of intuitive (and scientifically unacceptable) rules that children have been found to use in explaining motion. These are:

- Forces are to do with living things (things, such as gravity, friction or jet propulsion, are not forces, but people can apply force).
- Constant motion requires a constant force (rather than constant motion resulting from no force and a net force causing speeding up or slowing down or deflection).
- The amount of motion is proportional to the amount of force (faster-moving objects are thought to have or need a greater force, whereas, according to the scientific notion of force, an object, such as with a spacecraft, can be moving very fast even with no force on it).
- If an object is not moving there is no force acting on it, and if a body is moving there is a force acting on it in the direction of motion. (This is not true. For instance, there is no forward force on a rolling or sliding object – friction will act in a direction opposite to the motion. A stationary person standing in a room is subject to two forces, as discussed above.)

Further, there is a lot of evidence that children are of the view that force is a property of a body, a sort of power or energy within the body that stays with it but gradually diminishes as the body slows down. This idea has more in common with old ideas about impetus or the concept of momentum than with the scientific view of force. These ideas have been repeatedly identified in a range of studies. It has been argued (Stein, Larrabee and Barman 2008) that the identification of the extent of such beliefs through elicitation tasks can help teachers plan effective instruction.

A Victorian study (Adams, Doig and Rosier 1991) has examined students' ideas in a range of science areas, including movement and force. A brief look at some of their findings for Year 5 students is instructive. You might like to think about how you would answer their questions (see Figure 3.1) yourself before reading the interpretations.

The skateboard rider stops kicking (see Figure 3.1). Why does the skateboard stop? Forty per cent of Year 5 students' responses were uninterpretable and 48 per cent gave responses

Look at the rider in the picture. If the rider stops kicking, the skateboard will stop. Can you explain why?

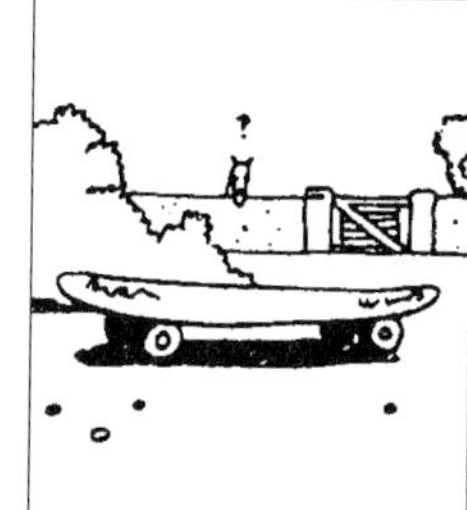

The skateboard in the picture is not moving. The ground is flat. What forces, if any, might there be on the skateboard?

The skateboard in the picture is rolling down a slope. Why does the slope make the wheels turn and the skateboard move? Can you explain why?

This skateboard rider is trying to pull three friends up the hill but can't make them move. What forces, if any, might there be on the skateboard?

FIGURE 3.1 Probes of understanding about force

implying that, without a kicking force, there is no motion. A further 4 per cent thought the rider supplied a force, energy or power to the skateboard, which stops when this is used up. In fact, the skateboard experiences a gravitation force down, a reaction force up from the ground and a frictional force that acts against the motion and slows the skateboard down. These cartoon probes can be used in class as elicitation techniques to promote discussion and rethinking of ideas about force.

Adams, Doig and Rosier (1991, p. 23) analysed the responses on all items and were able to divide Year 5 students' views about motion into five levels along a continuum. All but 4 per cent of students had views substantially at odds with the scientific notion. Most had few or very confused ideas. At the upper end, the bulk of students were at a transitional stage in which they have some awareness of the scientific notion of force but retain elements of intuitive conceptions, such as the necessity of force for movement and the idea of force residing in an object. Students at this level may explain that a thrown ball is given a force by the thrower and that force now resides in the ball and keeps it moving.

Alonzo and Steedle (2009) used multiple-choice tests to develop and refine a Force and Motion Learning Progression that describes a sequence of conceptions of increasing sophistication, identifying common errors at each stage. Thus, at level 1, students understand force as a push or a pull, but may think that a force is caused by living things or is an internal property of a moving object. At level 2, students understand the links between force and motion, but may think that an initial force is carried with the object but may dissipate as it slows, or that if there is no motion there is no force. At level 3, students understand that an object will be at rest if there is no force or if there is no net force acting on it, but may think that objects will slow down naturally even without force, or that speed rather than acceleration is proportional to force such that an object will come to rest if forces are in balance. Such learning progressions are receiving increasing attention as ways of thinking about and planning for student learning.

It has been shown that children, even if they have learnt to use scientific ideas in school, often revert to their life–world beliefs when dealing with situations outside school. It is as if they think in two domains, one relating to the classroom and one to their out-of-school lives (Solomon 1983). The scientific concept of force is neither intuitive nor easy to grasp, but is important for interpreting many situations. Adults and children need support to clarify their understandings about gravity, friction and what we mean by force. Much of the teaching at primary school could concentrate on developing a more consistent language in which force and energy are distinguished.

A difficulty in building up a consistent, scientifically acceptable view of motion seems to be the influence of friction, which acts to slow things down and which is associated with the decrease of energy of motion and its conversion to heat energy. It is our common experience that things do stop when we stop pushing them. It takes a leap of imagination to be convinced that a brick would keep going steadily across a floor or field if it wasn't for frictional effects. Compare, though, what would happen if the brick were slid across a skating rink, or think about how carefully you need to walk across an icy surface. Wheels have the effect of reducing the amount of friction because, with a wheel, the point of contact with the ground is not sliding, but is momentarily stationary. There is some friction from the ground, however. The amount will depend on the surface (think of riding a bicycle in sand) and on factors such as how freely the wheel spins.

A British study of primary school teachers' ideas about force (Kruger, Summers and Palacio 1990) showed similar conceptions to be held by adults. The difference in this case was that while the teachers showed considerable confusion, they were much more reflective about the state of their knowledge and were concerned about achieving a consistent view of motion. Some of their responses in 'interviews about instances' based on prompt cards are given below (Kruger and Summers 1988, pp. 6–10). They illustrate a range of informal views about motion.

> **Instance 1: A person is riding a bike. There are no brakes, he is not pedalling but is slowing down. Is there a force on the bike?**
>
> Yes, momentum [The force he exerted when he was pedalling] ... that's gone ... almost stored in the system. Stored energy a stored force ...
>
> [the force that was there when he was pedalling] hasn't got really expended yet ... It's running down.
>
> **Instance 2: A golf ball has been hit and is now on its way down, falling freely to land on the green. Is there a force on the golf ball?**
>
> now it's losing the force from him and its own weight's going to make it start dropping and gravity ... and air resistance this way.
>
> Well, as it's still going forward, I would have thought yes [the force of the hit is still with it] because if there were no force it would be dropping straight down ...
>
> I'm not sure whether once the thing has been set in motion, you discount ... the impetus ... there is in one sense a force that has been but whether you still talk about [it] once the actual work is finished [and it's] set in motion, I don't know ... the weight ... is vertically downwards but because the ball is already in motion the ball doesn't go vertically, but that may not be anything to do with the force because the force is actually vertically downwards.
>
> **Instance 3: A box is pushed and then is sliding down a slope. Is there a force on the box?**
>
> Does friction come in? ... I don't know if you call it a force ... because it stops ... if you think of skiers, they want minimum friction ... So it's stopping it moving, it's not a force.

In the student conception literature, the concepts associated with force and motion have received a lot of attention. Planinic et al. (2006) showed these naive conceptions to be particularly strong, not only in that students answered according to them, but also that the students had high levels of confidence in these naive beliefs. This was particularly true of the following conceptions:

- constant force produces constant velocity
- heavier bodies fall faster
- a body can be at rest while an unbalanced force (gravity) acts on it.

They hypothesised that the strength of these naive conceptions about force and motion comes from the fact that these are topics with which we have everyday and longstanding familiarity. Chi (2005), on the other hand, argues that conceptions such as 'force is the property of a moving body' are difficult to shift because they require a category change for force, from a property to an effect. Other naive conceptions, such as that an insect is not an animal, do not involve such a shift and therefore are not so robust.

Activity 3.3 Exploring the concept of force

- Think about the situations shown in Adams, Doig and Rosier's (1991) paper. How do your own ideas about movement and force compare with students' ideas identified by their study?
- What are the key difficulties you have with the scientists' notion of force?
- Consider each of the intuitive rules identified by Gilbert and Watts (1985). Build up a list of counterexamples for each rule that is clearly untrue (for example, constant motion can't require a constant force; an ice puck will slide without a force).
- Reflect on the teachers' interview statements (Kruger and Summers 1988). Can you identify the nature of the misuse of the word 'force' in each case? Discuss your views with a colleague.
- Review your ideas and questions in your science journal. What has been clarified and what uncertainties remain?
- Discuss and jot down some activities that might be used in a primary school classroom to get children talking about their ideas concerning movement and force.

Effective teaching about movement and force

The constructivist principles involved in teaching about movement and force are no different to those that might be used for any topic, but there are some special issues that have been addressed in the literature. Parker and Heywood (2000) showed that learners rarely think about floating and sinking in terms of forces. In a later study involving primary teachers and secondary trainee teachers, Heywood and Parker (2001) showed that

- by providing tactile experiences of forces acting on floating objects (the upthrust from the water, the weight force down) and the opportunity to explicitly identify the forces and the direction in which they act
- by investigating what happens when weight is kept constant while size varies
- and through encouragement of personal reflection on their learning,

teachers could not only identify the balancing of forces involved in floating and sinking, but could also extend the idea of balanced and imbalanced forces to a range of other situations. These features are, of course, classical conceptual change strategies. Our own research (Tytler and Peterson 2005) has shown that children improve markedly in their ability to investigate the flight of parachutes and whirlybirds (see the case study 'Children engaged in air activities' on p. 110) when they are able to conceptualise the weight force and upthrust from the air as separate and capable of independent manipulation. Hart (2002), using a variety of examples, shows the importance of clarifying with students the meaning they attach to the ideas surrounding force.

What are the factors that make such activities productive for young learners? Hadzigeorgiou (2002) argues that with young children it is more important to build experiential foundations than to explicitly teach concepts, and that these foundations include attitudes such as curiosity and experience in working in situations that provide a rich environment for exploration and feedback. Hadzigeorgiou worked with preschool children aged four and a half to six years on a challenge activity that involved them constructing as tall a tower as they could on a sloping surface using cans of varying diameter and weight. He was able to show that children whose teacher provided a structured experience (clarifying questions, asking for predictions, pointing out significant results and providing further challenges) learnt much more successfully than those who were given the task without further support. Clearly, the teacher's role in providing such scaffolding is critical. The study found that these young children could gain an understanding of what affected the stability of such structures without recourse to advanced ideas such as centre of gravity.

Vosniadou and Ioannides (2001) investigated the nature of a learning environment that successfully challenged Year 5 children's intuitive ideas about force and resulted in improved learning. They characterised this as consisting of:

- taking into consideration and making explicit students' prior knowledge
- the use of measurements (as with Activity 3.2; see p. 102), representations (such as force arrows) and models
- creating cognitive conflict through challenge activities (for example, by showing that the force needed to move an object is less than its weight, leading to a discussion of friction)
- paying attention to the order in which concepts are introduced
- talking through language problems with the class (for example, what energy or force might mean to scientists that is different to everyday meanings).

Yuruk, Beeth and Andersen (2009) worked with a teacher on a secondary level force-and-motion unit emphasising engagement in metaconceptual processes – that is, helping students become aware of their understandings and the way these are changing, monitoring their learning processes, and evaluating competing conceptions for their ability to explain real phenomena. The approach involved instructional activities such as poster drawing, group debate, journal writing and class and group discussion. They found that the class developed a higher level of conceptual understanding than a matched class taught using traditional instruction. At the primary level, Carruthers and de Berg (2010) worked with Year 6 students on a small-group inquiry and argumentation magnetic force sequence (e.g., Zembal-Saul 2009). They found that students spontaneously developed push–pull notions of force and were capable of working with rudimentary elements of argumentation (making claims supported by evidence). An understanding of forces interacting in pairs, however, was more difficult to achieve, in that students tended to think of magnets causing forces on nails but not vice versa.

Technology challenges involving the investigation of structures, or machines, offer a rich context in which to explore force ideas. Bennett (2009) investigated the use of practical activities with construction kits, involving geared machines. He concluded that students did not readily transfer their knowledge of gears to new design situations, and that simply giving children a kit with a set of instructions was not likely to encourage learning. The teacher's intervention in guiding activity was crucial in focusing attention and developing thinking. The team developed a four-level approach to intervention:

1. What appears to be the problem? Can you explain it to me?
2. How have you tried to solve the problem?
3. Have you thought of ...? (providing strong direction)
4. Here, let me show you.

Such an approach could be used for any design or investigation activity. It provides a natural way to assess students' learning through determining the level of support a group needs to successfully solve a task.

Literacy and learning about force

Recent research has emphasised that learning involves developing an increasing capability to participate in the discursive practices (ways of talking and of doing things) of a subject area. In this view, science is acknowledged as a mixture of languages involving representations in a variety of modes, including such things as diagrams, text-based explanations and reports, tables and graphs, two- or three-dimensional models and even gestures (Lemke 2004). Tytler, Peterson and Prain (2006) explore the use of multiple representations in developing explanations of evaporation (see Chapter 9). Hubber, Tytler and Haslam (2010) explore the teaching and learning of force from a representational perspective. Russell and McGuigan (2001) argue that learning involves being increasingly able to use different representations to explain ideas, and their work provides further exploration of some of the ideas about force discussed above. In a study of teachers working with this principle, they showed that effective learning about gravity occurred as students were challenged to represent their explanations in different ways (force and other diagrams, written and verbal explanations, models) and think about how these differed from their own ideas. This approach is sometimes referred to as *representational redescription*.

R

The national curriculum project *Primary Connections* (Australian Academy of Science 2005) links science with literacy learning. As part of this focus, the units explore how science can be used to support students' general literacy skills, but also articulates the particular literacies associated with science. In a stage 1 unit, *Push-Pull*, students are asked to generate labelled diagrams identifying motion, and pushes and pulls, and also descriptive observations of movement they had investigated. If we were to think about developing with students a vocabulary of force and motion, then that vocabulary in the wider sense must encompass not only words, but conventions to do with arrows and labels, time-lapse drawings, charts and graphs, and possibly even gestures. Thus, a discussion of the features of scientific representations is an important part of teaching and learning. Hubber, Tytler and Haslam (2010) argue for a representation-focused pedagogy that comprises the structuring of challenges involving the generation, negotiation and evaluation of representations. Understanding, from this perspective, involves the capacity to coordinate multiple representations and to appreciate that each brings a specific aspect of a phenomenon into focus. They showed that the challenge for teachers with this approach was, firstly, managing discussion of a variety of student inputs, and secondly, adopting the perspective that there are many ways in which science ideas can be represented, and not a single, given, 'correct' way, as is often assumed in texts.

Primary Connections has two more units dealing with force and motion besides 'Push–Pull'. These are 'On the Move' and the stage 2 unit 'Smooth Moves', which introduces gravity and arrow representations of force.

Activity 3.4 Representing force and motion

- Revisit the representations of force and motion you generated in Activity 3.1 (see p. 100). Were the representations of you and your group similar? Could these be used to generate some useful conventions for representing force and motion?
- Discuss, in your group, your views about what is happening in each of the skateboard scenarios from Adams, Doig and Rosier's (1991) paper. In coming to an agreement, note the different representations you draw upon to convince your colleagues, perhaps including verbal descriptions, diagrams with arrows or other devices, gestures or models.
- Discuss whether there is a particular set of representations you feel should be taught for this topic, and whether variation in students' representations should be negotiated when teaching force and motion.
- Discuss how the idea of representational redescription fits with Vosniadou and Ioannides' (2001) principles, and what implications there might be for your own planning.

Learning about air and flight

Flight, which relates directly to movement and force, is a rich area for science investigation in the primary school. Often, however, units on flight focus on the technology of such things as kite or paper plane construction, rather than on the science underlying flight.

The science of flight involves consideration of the forces due to air: air resistance, the uplift of air on wings or the buoyancy of air in balloons. Thus, a discussion of air should precede any discussion of a flight sequence, particularly for younger children, who may not have a clear conception of the existence or properties of air (see Chapter 8).

Where does air exist? Children can be readily taught to say 'air is everywhere', but when asked if air is in a closed jar or cupboard, inside a room, under a table or in an open box, their responses can be surprising. Young children tend to associate air with wind, which is perceptible. If you wave a piece of paper in front of a child's face and ask what is happening, the explanation comes in two quite distinct forms: one from children who think that air is created by the moving paper and one from those who think it is present but simply caused to move by the paper.

Some activities to establish the presence of air and the idea that air takes up space that you might like to discuss are as follows:

- Children attempt to collect air in plastic bags from various places, including cupboards, to explore their ideas.
- A tissue is squashed into an empty glass which is then upturned and plunged under water. Will the tissue be soaked?

You can find activities and children's ideas on air and flight in Tytler (2002) and in the *Primary Connections* units 'Push–Pull', 'On the Move' and 'Smooth Moves' (AAS 2005).

CASE STUDIES

Children engaged in air activities

What follows is a series of flight activities and children's responses to them that illustrate ideas about the forces involved in flight. The transcripts have been taken from a longitudinal study (Tytler and Peterson 2005) that follow a number of children over their primary school years to gain insights into their developing understandings. Some of the transcripts will show the responses of the same children interviewed some years apart to illustrate the nature of this development.

Paper drop

This is a predict–observe–explain (POE) sequence focusing on air resistance and flight. Try it yourself. In each case you should predict what will happen before trying it. The results will be unexpected.

In pairs, drop:

- two sheets of A4 paper, one held horizontally and one vertically
- an uncrumpled sheet of A4 paper and a crumpled sheet of A4 paper
- an A4 sheet of paper and an A4-size book
- an A4 sheet resting on top of an A4-sized book.

In the paper drop activity, most students will arrive at a conclusion that the air resists the A4 paper but that dropping it when held vertically minimises the surface that is pushing through the air and hence it will drop quickly (at least initially, until it skews off course). Younger students will often say the crumpled paper drops more quickly than the A4 sheet because it is heavier. Density, or compactness, is often confused with weight, so this is a good opportunity to have that discussion. The book falls more quickly because it has greater weight to overcome the action of air on its surface.

The real surprise and challenge offered by this activity is the fact that the paper, in the final drop, falls along with the book. The reason is that the book is pushing the air that would be resisting the paper on its own. The paper is not needing to force through air and effectively falls as it would in a vacuum. This may remind you of the experiment conducted by Neil Armstrong on the moon. He dropped a hammer and a feather to find that, in the absence of an atmosphere, they fall at identical rates – as argued by Galileo. Some people argue that the paper is in the book's slipstream, which is, in fact, the same explanation. Some argue that air comes around the back of the book because of turbulence and holds the paper on. This is substantially incorrect, although there is turbulence and there is some complicated science associated with turbulence.

One five-year-old child explained this counterintuitive result very quickly and convincingly by pointing out that the paper acted just like another page in the book, and so would be expected to fall with it.

There is a lot of fun to be had with this activity by dropping the book and paper from different heights or varying the position and extent of overlap of the paper.

Parachute

The following flight activities, featuring parachutes and whirlybirds, are common as part of primary school science sequences. These activities were constructed in the longitudinal study as explorations of the effect of canopy size and weight in the case of the parachute, and wing size and weight for the whirlybirds. These activities work well as probes for eliciting children's ideas and also as investigations that help establish ideas of variable control. The transcripts are taken from parts of interview sequences with the same children over their first four years of school and are reported in Tytler and Peterson (2005).

In grade prep (kindergarten year), Anna experimented with parachutes and described how they worked in terms of uplift from the parachute and linked to the idea of hot air:

Interviewer	How did the parachute make it a better landing?
Anna	Because it was lighter ... because it's got a gas inside it because it's like a hot balloon and it's a bit fatter.

Eighteen months later in Year 1, children were shown four parachutes, two with large and two with small plastic bags as canopies; each had either a small or a large plastic model of a parachutist in it. Children were challenged to separate out the effect of canopy size and weight. It is a common finding with children's conceptions that, within the change, there are ideas they keep returning to. Anna's idea of parachutes making people lighter seems to persist. Anna seems to regard the model and canopy as one object, differing in weight. Thus, when challenged to explore the effect of canopy size, she compares two with the same size canopy:

Anna	This one will fall a bit faster than this one.
Interviewer	Why is that? They've got the same size canopy.
Anna	No, but this one is a bit lighter than this one – this one has nothing inside it and this one does.

Whirlybird

Whirlybirds, or spinners, are intriguing flight devices that can be manipulated to vary in the way they spin and drop. The instructions are set out in Figure 3.2.

Whirlybirds provide the opportunity for the development of students' knowledge of investigations: hypothesising, fair testing, measuring, recording and reporting. Timing the fall is difficult; comparing different designs two at a time is probably the most productive thing to do if you don't have access to a stopwatch and a balcony from which to drop them. To keep track of what is happening, children should modify one aspect at a time and, preferably, retain each modified design.

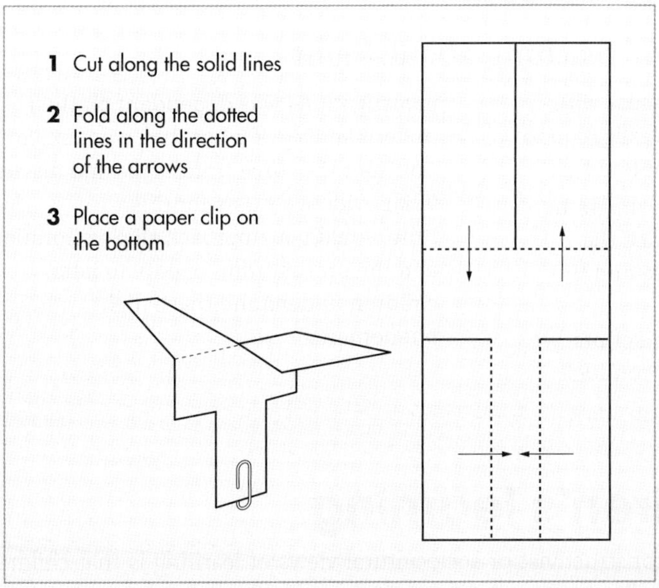

1 Cut along the solid lines

2 Fold along the dotted lines in the direction of the arrows

3 Place a paper clip on the bottom

FIGURE 3.2 Whirlybird design

Children were shown six paper whirlybirds that spin as they fall. There were three pairs of whirlybirds, each with different wingspans – short, medium and long; one of each pair had a paperclip attached. Students were asked to work out which whirlybird fell the slowest. They were challenged to give their reasoning and, where appropriate, to provide evidence to support their assertions. Children performed this task at the end of grade prep and again at the end of Year 2.

This activity is a variable control activity, but there were interesting instances of a range of approaches.

Karen, Year 2

In grade prep, Karen, without really considering the science, had organised a play-off between pairs of whirlybirds, with winners playing winners to find the slowest. In Year 2, when she is posed the task, she immediately drops one whirlybird and starts entering a number in a table on a sheet of paper. This confuses the interviewer until he realises she is counting under her breath to measure the time of fall:

Interviewer	How did you know what to write in it? Oh ... the time ... oh. OK, OK.
Karen	Medium with clip ... small with clip ... 10 seconds ...

By this time Karen is unstoppable. She had launched straight into a planned method for measuring and recording. She proceeds to fill in the table showing times for each whirlybird, identified appropriately by wing length and clip/no clip before reviewing results:

Karen	So probably the one without clip takes the longest to go down. And the small one doesn't with the clip. It takes the ... smallest to go down and it, um, it wasn't ... [?] The big one with clip took 10 seconds and so the big ones are probably first, and the mediums could be, like, second, but the ones with the clips ... and so it's like ... 'cause, like, the ones without folded up are better than the ones with the clips 'cause the clip's heavier and it pulls them down.

Karen's investigation does not conform to the interviewer's expectation of a variable control experiment, but it is nevertheless very effective in identifying what is happening. However, she finds it hard to construct a clear test to demonstrate the effect of the wings.

Calum, Year 2

When Calum was in grade prep he had not come up with a clear view of why whirlybirds fly, mentioning neither weight nor air. His exploration was therefore unfocused. In Year 2, however:

Interviewer	What do you think causes that spin?
Calum	Probably … the paperclip to pull it down and, um, because of the air pushing up … instead of them going up it would have to push, the air would have to push it away so it goes round.
Interviewer	What I'd like you to do is to do an experiment with this just to find out which one actually drops slowest to the ground and why that one is slowest.

Calum chooses the two long-winged whirlybirds, one with and one without a paperclip.

Interviewer	OK. So, what do you think?
Calum	Probably it's because the paperclip pulls it down. And you can see the one without the paperclip … spinning …

In the intervening year, the class had been exposed to whirlybird and paper plane flight as classroom activities, but it is not clear how much explicit teaching about the science occurred and the outcome was clearly different for the different children.

The science of whirlybirds

The spinning effect is due to the action of air on the wings as it rushes past the dropping whirlybird. You can check this by holding the whirlybird and pushing up on one wing with your finger. The body moves back as the wing is forced up. Pushing up on the other wing has the opposite effect. You can see that the net result is a spinning set of forces. Flipping the wings causes them to spin in the opposite direction.

The longer the wings, the slower the drop because of the uplift on the greater wing area. The more paperclips, the faster the drop and spin because of the greater weight.

Activity 3.5 **Strategies for teaching about flight**

Try the above flight activities and discuss with colleagues how best to set up the investigation to explore the science ideas and present findings.

- Construct some annotated force diagrams to clarify the factors that influence the flight of the paper, the parachute and the whirlybirds.
- In the paper featuring these transcripts, we argued that children's ability to conduct investigations was dependent on their understanding of what they were exploring. Consider the three transcripts and discuss what evidence exists for this.
- How would you approach this exploration and the presentation of data to help children understand these flight phenomena?

Assessing children's learning

One of the major implications of constructivist or sociocultural views of learning is that children's understandings need to be continually monitored as part of the ongoing discourse of the class. If the conceptual conversation is to be rich, the teacher should be encouraging all children to participate in the generation and evaluation of ideas. Particularly from a representation-focused pedagogical perspective, these ideas can be diverse. A child can best be supported to embrace richer and more flexible understandings if their current beliefs and understandings are known. A discussion of the purposes and forms of assessment can be found in Chapter 1 (see the section 'Assessment for student learning' on p. 35). Formative assessment is used to shape and plan the teaching program and support children's learning. It is important that a teacher takes children's ideas seriously and encourages and supports them to monitor their own understandings. Summative assessment involving judgements about children's understanding at the end of a teaching and learning sequence has more to do with credentialling than supporting learning. Nevertheless, being clear about what learning is intended and having a view about the different levels of understanding that can be held by children are powerful aids to planning and responding. The way assessment can be embedded in unit planning is further discussed in the section 'Learning about simple machines' later in this chapter (see p. 120).

A

Levels of understanding of movement and force

Ultimately, at the end of a classroom sequence on movement and force, we should be attempting to arrive at some overview of the level of each child's understandings. If a sequence has a coherent, conceptual agenda, it is not adequate to simply add marks given on many different items. We need a clear assessment framework.

An assessment rubric for movement and force

One of the methods primary teachers have used to help plan and assess children's understandings is the use of assessment rubrics. These are descriptors of different levels of, for instance, understanding of movement and force, cast quite broadly so they can be used across activities and probes. What follows is a possible example based on the research into alternative conceptions described above.

- *Level 1* – Can describe simple situations involving movement and force using appropriate terminology, such as 'push', 'pull' and 'speeding up'. Can make sensible observations of a variety of movements using appropriate, simple language.
- *Level 2* – Can make observations about movement and force situations and generate interpretations based on patterns, such as 'heavy things need more force to move' or 'whirlybirds with longer wings fall slower'. Uses the language of push and pull, but may harbour varied notions of forces residing in moving objects and has confused ideas about gravity and friction.
- *Level 3* – Can describe a variety of motions in detailed terms and can attempt reasonable explanations of different motions using representations of the actions of single forces of different types (gravity, friction, forces due to air, pushes and pulls). Has a basic understanding of gravity and upthrust. May still harbour a variety of alternative conceptions.
- *Level 4* – Can describe more complex motions in specific terms, such as 'vibrating' or 'orbiting'. Can coordinate representations of combinations of forces to explain an object's motion in situations where they may oppose or where an object is balancing under the action of different forces. Can attempt interpretations of complex motions (such as whirlybird flight) in terms of detailed consideration of different forces.

Activity 3.6 Applying assessment rubrics

- One of the problems with any broad judgements about levels of understanding is that some contexts are more difficult to interpret than others and may trigger different ideas. Discuss with colleagues the feasibility of making judgements across a variety of activities to arrive at a level.
- Choose one of the levels described and generate examples of what this level of understanding might entail for a range of the activities described in this chapter.
- Children can often perform at a higher level with the encouragement and support of an adult. It is possible, within each level, to make judgements about the extent of support the child needs to perform at that level. 'Beginning level 2' means the child can operate at the level only with strong and specific cues. 'Consolidating' indicates the child can achieve the level independently in some circumstances, but needs scaffolding for others. 'Established' involves the child operating consistently and independently at the level for many situations. Discuss whether you think these distinctions are sensible.
- Review the transcripts in this chapter (see the earlier case studies on p. 110) of children's responses. Discuss whether you could assign a level to each instance and what issues are involved in attempting to do so.
- Look back over your science journal, at the growth in your understandings. Could you describe your growing understanding in terms of the achievement of levels?

Assessing children's approach to exploration

This chapter has argued that children learn best if their ideas are challenged and supported using investigative activities in which they test ideas against evidence. We have argued that the ability to approach, design and carry through investigations is dependent on children's level of knowledge. Thus, we need to develop ways of assessing investigations as well as understandings. In analysing children's performance on a range of tasks, such as the whirlybird or parachute tasks or open explorations of animal behaviour, we developed descriptions of different levels at which children approached exploration. These levels are shown, with examples, in Table 3.1. They represent qualitatively different ways of reasoning about ideas using evidence. We believe they represent a mixture of children's level of curiosity, their intuition and their understanding of the way ideas are used in science. For example, science explanations are a specific genre of reasoning with language that needs to be modelled for young children. For further discussion of the importance of understanding how ideas and evidence are related in science, of argumentation, and particularly the importance of understandings of the NOS, refer to Chapter 1.

A

In Table 3.1, illustrations of the whirlybird exploration are given, together with examples of the types of conceptual understanding that tend to be associated with each level.

Reasoning in science, which is closely related to children's coordination of ideas and evidence in Table 3.1, is often also related to the idea of 'argumentation', where students are encouraged to make judgements about claims (ideas) based on evidence, and engage with

TABLE 3.1 Children's approach to exploration: a reasoning rubric

LEVEL	WHIRLYBIRD EXAMPLE
1 Ad hoc exploration No systematic observations or comparisons are made nor use of a guiding explanatory purpose given. Exploration at this level is restricted to low-level interpretation that lies close to what is observed.	Children explore one whirlybird or parachute at a time without explicitly comparing characteristics. They do not explicitly seek patterns, but focus on the flight of individual whirlybirds.
2 Inference searching The inference could be about patterns in what is observed or about explanatory ideas. Children explore on a try-it and-see basis without a noticeable sequence, but leading to some hypotheses or inferences. They notice things, comment and infer underlying patterns or causes. Exploration at this level is data-led, but with some conceptual interpretation.	Children compare pairs of whirlybirds based on some factor of interest and without a plan based on an idea. They may be able to interpret the outcomes in terms of whether long or short wings fall faster or, if asked, even provide some explanation.
3 Hypothesis checking The hypothesis could be about relations between variables or about theoretical ideas. Children carry out focused observations or interventions that involve trying out an idea or following up a prediction with some conceptual basis. Explorations have a recognisable hypothesis driving them. Exploration at this level is theory-led, but does not necessarily separate variables.	Children take the lead in developing a strategy to check ideas (for example, longer wings slow down the whirlybird) about what affects whirlybird flight. Generally, at this level they are working on an explanatory hypothesis, such as that long wings catch the air more.
4 Hypothesis exploring Strategic search for evidence to refine or distinguish between hypotheses or rule out other possibilities. Setting up checks of ideas generated and dealing explicitly with the possibility of confounding variables or other limitations on experimental design. Exploration at this level acknowledges the interdependence of data and theory.	For the whirlybird flight, children spontaneously try out competing ideas; for example, by altering conditions in an ordered way, controlling for variables of weight and wingspan, and exploring the interaction between stability and weight and wingspan. The ideas and explanations at this level are often complex and speculative.

competing claims and evidence. Simon, Erduran and Osborne (2006) have worked with teachers to develop a model for introducing argumentation activities into science classrooms, aimed at modelling the way in which knowledge is supported with evidence in science. The UK work on argumentation has produced the curriculum materials *Ideas, Evidence and Argument in Science Education* (IDEAS), which are being widely used. These involve activities that challenge students and encourage them to hypothesise and resolve claims and counterclaims on the basis of evidence. Argumentation is also discussed in Chapter 1 (see under 'Main components of constructivist models and schemata' on p. 24).

Activity 3.7 Assessing to support exploration

- Discuss how each of the transcripts presented in this chapter might relate to this approach to an exploration rubric.
- Discuss one of the exploratory activities you carried out recently with colleagues. Can you identify how different features of your exploration might relate to the different levels?
- If a major part of the purpose of science is to encourage and support children to explore ideas, how could you devise an appropriate assessment, feedback and reporting regime to effectively encourage this?

Planning for conceptual development with the 5E instructional model

In technological applications, forces are utilised and manipulated in a variety of ways to perform useful functions. The next two sections of this chapter will model how the 5E scheme can be used flexibly to plan coherent learning sequences on the manipulation of forces using simple machines. This section describes how the 5E framework discussed in Chapter 1 (see the section 'Constructivist teaching models or schemata' on p. 22) has been applied by teachers and curriculum writers in lesson and unit planning. Then, the principles of simple machines are explored in three contrasting lesson sequences planned according to the 5E framework.

As discussed in Chapter 1, the 5E scheme is gaining currency as a framework for thinking about effective teaching and learning. The model has so far had applications in program development in schools (for example, for lesson and unit planning), commercial curriculum resources (for example, *Primary Connections*) and policy development (for example, a modified version of the 5E scheme, the 'e^5', is central to a school improvement model initiated by the Victorian Department of Education and Early Childhood Development). Bybee and his colleagues developed the 5E instructional model in 1989, influenced by historical instructional models – for example, John Dewey emphasised the importance of experience and reflection, and a commitment to supporting deeper learning through inquiry, and Johann Friedrich Herbart emphasised that effective pedagogy allows students to discover the relationship among experiences – and the contemporary three-phase learning cycle of Atkin and Karplus (1962) that promoted *guided discovery* (see Bybee et al. 2006).

To recap, the 5E instructional model consists of five phases: engagement, exploration, explanation, elaboration and evaluation. Each phase has a specific function and contributes to coherent instruction by the teacher and the formulation of better scientific and technological knowledge, attitudes and skills in the learner. The scheme is based on the premise that students learn best when allowed to work out explanations for themselves over time through a variety of learning experiences structured by the teacher (Hackling 2006). Social construction of meaning is promoted through collaborative work and the joint construction of explanations based on common experiences – social constructivism is discussed further in

Chapter 1; see the section 'Social constructivism' on p. 13. The focus on experience to build conceptual understanding means that the 5E framework supports an inquiry approach to science teaching.

The *Primary Connections* (AAS 2005) resource is an example of how the 5E scheme can act as the framework for planning inquiry-based primary science units. As described in Chapter 1, the resource integrates science and literacy. Table 3.2 describes the purpose of each phase based on Bybee et al.'s (2006) original model, and outlines how the framework as applied in *Primary Connections* incorporates an emphasis on multiple representations to develop the literacies of science. Each phase advocates particular science literacy strategies that allow students to engage with different textual forms and represent data and ideas in multiple ways.

TABLE 3.2 The 5E model with a science literacy focus and assessment framework

PHASE	PURPOSE	ROLE OF TEACHING AND LEARNING ACTIVITY, EMPHASISING LITERACY PROCESS AND PRODUCT	ASSESSMENT FRAMEWORK
Engage	• *An object, event, problem or question is used to engage students with the topic and to elicit students' current knowledge and experiences.* • *Activities make connections between past and present learning experiences, expose prior conceptions, and point student thinking towards the learning outcomes of current activities.*	• *Activity or multimodal text set context and establish topicality and relevance.* • *Motivating/discrepant experience creates interest and raises questions.* • *Open questions, individual student writing, drawing, acting out understanding, and discussion to reveal students' existing ideas and beliefs so that teachers are aware of current conceptions and can plan to extend and challenge as appropriate.*	Diagnostic
Explore	• *Objects and phenomena are explored through teacher-guided hands-on activities.* • *Current conceptions, processes and skills are identified and conceptual change is facilitated.* • *Activities help learners use prior knowledge to generate new ideas, explore questions and possibilities, and design and conduct preliminary investigations.*	• *Investigations to experience the phenomenon, collect evidence through observation and measurement, test ideas and try to answer questions.* • *Investigation of text-based materials (for example, newspaper articles, Web-based articles) with consideration given to aspects of critical literacy, including making judgements about the reliability of the sources or the scientific claims made in the texts.*	Formative
Explain	• *Students' attention is focused on particular aspects of the prior learning experiences.* • *Students explain their understanding of concepts and processes by drawing on their experiences from the engagement and exploration phases.* • *Conceptual clarity and cohesion are sought as the teacher introduces new concepts and skills.*	• *Student reading or teacher explanation to access concepts and terms that will be useful in interpreting evidence and explaining the phenomenon.* • *Small-group discussion to generate explanations, compare ideas and relate evidence to explanations.* • *Individual writing, drawing and mapping to clarify ideas and explanations.*	Formative (sometimes Summative)

PHASE	PURPOSE	ROLE OF TEACHING AND LEARNING ACTIVITY, EMPHASISING LITERACY PROCESS AND PRODUCT	ASSESSMENT FRAMEWORK
		• *Small-group writing/design to generate a communication product (for example, a poster, oral report, formal written report or PowerPoint presentation, cartoon strip, drama presentation, letter) with attention to form of argumentation, genre form/function and audience, and with integration of different modes for representing science ideas and findings.*	
Elaborate	• *Activities allow students to apply concepts in contexts, and build on or extend understanding and skills.* • *Through new and challenging experiences, students gain a deeper and broader understanding, more information and adequate skills.*	• *Student-planned investigations, exercises, problems or design tasks to provide an opportunity to apply, clarify, extend and consolidate new conceptual understanding and skills.* • *Further reading, individual and group writing may be used to introduce additional concepts and clarify meanings through writing.* • *A communication product may be produced to re-represent ideas using and integrating diverse representational modes and genres consolidating and extending science understanding and literacy practices.*	Formative Summative
Evaluate	• *Students assess their knowledge, skills and abilities.* • *Activities permit evaluation of student development and effectiveness of the teaching program*	• *Discussion of open questions or writing and diagrammatic responses to open questions – may use same/similar questions to those used in 'Engage' phase to generate additional evidence of the extent to which the learning outcomes have been achieved.* • *Reflections on changes to explanations generated in 'Engage' and 'Evaluate' phases to help students be more metacognitively aware of their learning.*	Summative

Source: Adapted from Bybee et al. (2006) and AAS (2005).

The *Primary Connections* units provide a model for how a holistic approach to assessment can be promoted in 5E lesson sequences using the diagnostic, formative and summative assessment framework (see Table 3.2). (Chapter 1 has more detail on assessment in the context of discussions about externalising and modifying students' ideas, metacognition and formative assessment; see the section 'Assessment for student learning' on p. 35.) Diagnostic assessment typically occurs during the 'engage' phase, and formative assessment during the 'explore', 'explain' and 'elaborate' phases, while summative assessment commonly occurs in the 'elaborate' and 'evaluate' phases. Appendix 3.1 (see p. 140) gives examples of assessment strategies that teachers can use within this assessment framework.

Activity 3.8 Classifying assessment activities

The three forms of assessment used by *Primary Connections* are diagnostic, formative and summative assessment. A holistic assessment framework should incorporate each of these forms, recognising that each form is based on assumptions about what and how learning can be achieved, and the role of the teacher and learner in the learning process.

The assumption behind *diagnostic assessment* is that students come to the classroom with understandings about the world and experiences that may be useful to the science topic. Diagnostic assessment strategies include gathering information about what students already know and are able to do, allowing for identification of gaps or alternative conceptions in prior learning, eliciting students' questions or wondering about a phenomenon or topic, and using students' prior knowledge and experiences to inform the teaching and learning process.

The assumption behind *formative assessment* is that there is a gap between the current level of knowledge and what the learner can potentially do. Formative assessment strategies include focusing on the student during the learning process, guiding and informing the teacher so that the next steps for the students' learning can be planned, monitoring whether students' questions are being responded to, and helping the students to learn and have positive learning outcomes.

The assumption behind *summative assessment* is that students' learning can be measured against learning outcomes or standards. Summative assessment tasks include evaluating the achievement of learning outcomes, providing a judgement of what has been learnt and what change has occurred in the students' understanding or performance, generating grades for assessment, and providing comparative information about what has been learned.

- Look at one of the *Primary Connections* units that deal with force and motion: 'On the Move', 'Push–Pull' (both level 1) or 'Smooth Moves' (level 2). Describe the assessment framework used in the unit by identifying and classifying assessment activities as diagnostic, formative or summative. For each activity, describe what a teacher can learn about the student and their progress.
- When planning a learning sequence on the action of forces, the teacher ensures there are opportunities for students to display their understandings at many points during the sequence. Classify each of the following activities as non-assessment, diagnostic, formative or summative activities – you do not know the context in which these activities are being used; that is, when the activities will occur and the teacher's purpose. Also discuss the role that context plays in determining the purpose of an assessment activity. The activities are as follows:
  - Individual and class discussion takes place about students' responses to a circus of activities designed to challenge preconceptions.
  - Once a week, the teacher asks students to complete a journal that answers teacher-specified questions aimed at student reflection on concepts, concerns, feelings and responses to activities. This is collected and the teacher responds to each reflection in an affirmational (positive) way.
  - The teacher circulates as the groups brainstorm ideas.
  - Students develop a PowerPoint presentation of a machine they have designed.
  - Students complete a sequence of questions from a textbook.
  - A 10-question quiz is conducted at the beginning of every second lesson that relates to previous lessons and introduces the concepts of the next two lessons.
  - The teacher uses targeted questioning as they enact a role-play of gears.

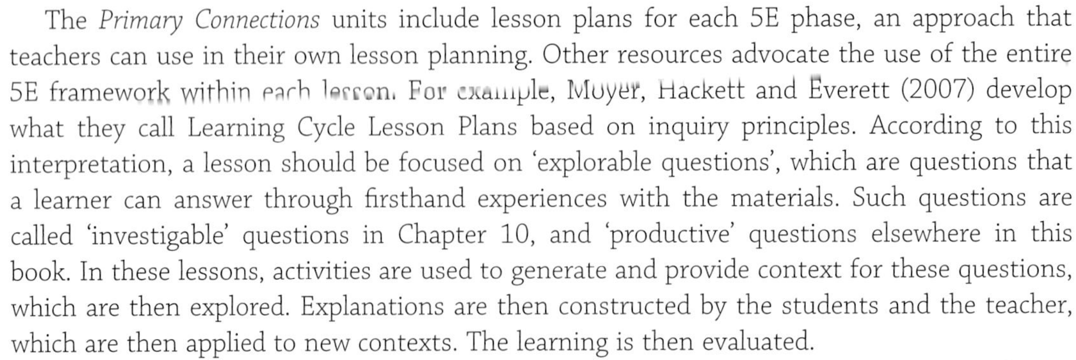

The *Primary Connections* units include lesson plans for each 5E phase, an approach that teachers can use in their own lesson planning. Other resources advocate the use of the entire 5E framework within each lesson. For example, Moyer, Hackett and Everett (2007) develop what they call Learning Cycle Lesson Plans based on inquiry principles. According to this interpretation, a lesson should be focused on 'explorable questions', which are questions that a learner can answer through firsthand experiences with the materials. Such questions are called 'investigable' questions in Chapter 10, and 'productive' questions elsewhere in this book. In these lessons, activities are used to generate and provide context for these questions, which are then explored. Explanations are then constructed by the students and the teacher, which are then applied to new contexts. The learning is then evaluated.

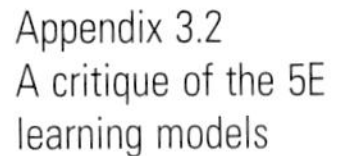

Appendix 3.2
A critique of the 5E learning models

As demonstrated above, the 5E model has a wide range of applications. As with any schema, there are benefits and issues associated with adopting the 5E model for planning instruction. A critique of the 5E model can be found in the online Appendix 3.2.

Learning about simple machines

In this section, we will explore different types of simple machines and how they work to modify force and motion. Following that, we will use the 5E model to plan learning sequences for simple machines.

The science of simple machines

Forces have the capacity to affect the motion or change the shape of objects. The history of the invention of technologies designed to move or modify objects (such as piles of dirt, doors or bicycles, or even ourselves) is the history of how we have learnt to modify the action of forces to our advantage. We are talking here of the principles of simple machines, which can:

- magnify a force applied
- reduce a force opposing motion
- change the direction of a force
- speed things up or slow things down.

The compelling aspect of simple machines that makes them a worthwhile topic in primary schools is the familiarity children will have with examples of such machines in their lives: in the kitchen, children use scissors, can-openers and egg-beaters; in the shed, they see wheelbarrows, shovels, screwdrivers and pliers; and for leisure activities, they use bicycles, scooters and mechanical toys. Once learnt, the principles of simple machines offer a new way of looking at how these everyday devices work. Students may be prompted to use the term 'mechanical advantage' to describe how the simple machine makes life easier – mechanical advantage refers to the fact that you can use a machine to produce a large load force (e.g., to lift a large weight) using a small effort force.

Figure 3.3 shows an interpretation of the simple machines embedded in a can opener. The relative size of the load force (from the blades onto the can) and the effort force are in inverse ratio to the distance of each from the fulcrum. Similarly, the length of the turning handle gives a mechanical advantage in proportion to this length compared to the radius of the blades that do the cutting. The gears are of equal size and rotate in opposite directions at the same rate. The blade is also a machine, acting as an inclined plane magnifying the sideways tearing force on the can that is larger than the pressure applied by the point of the blade.

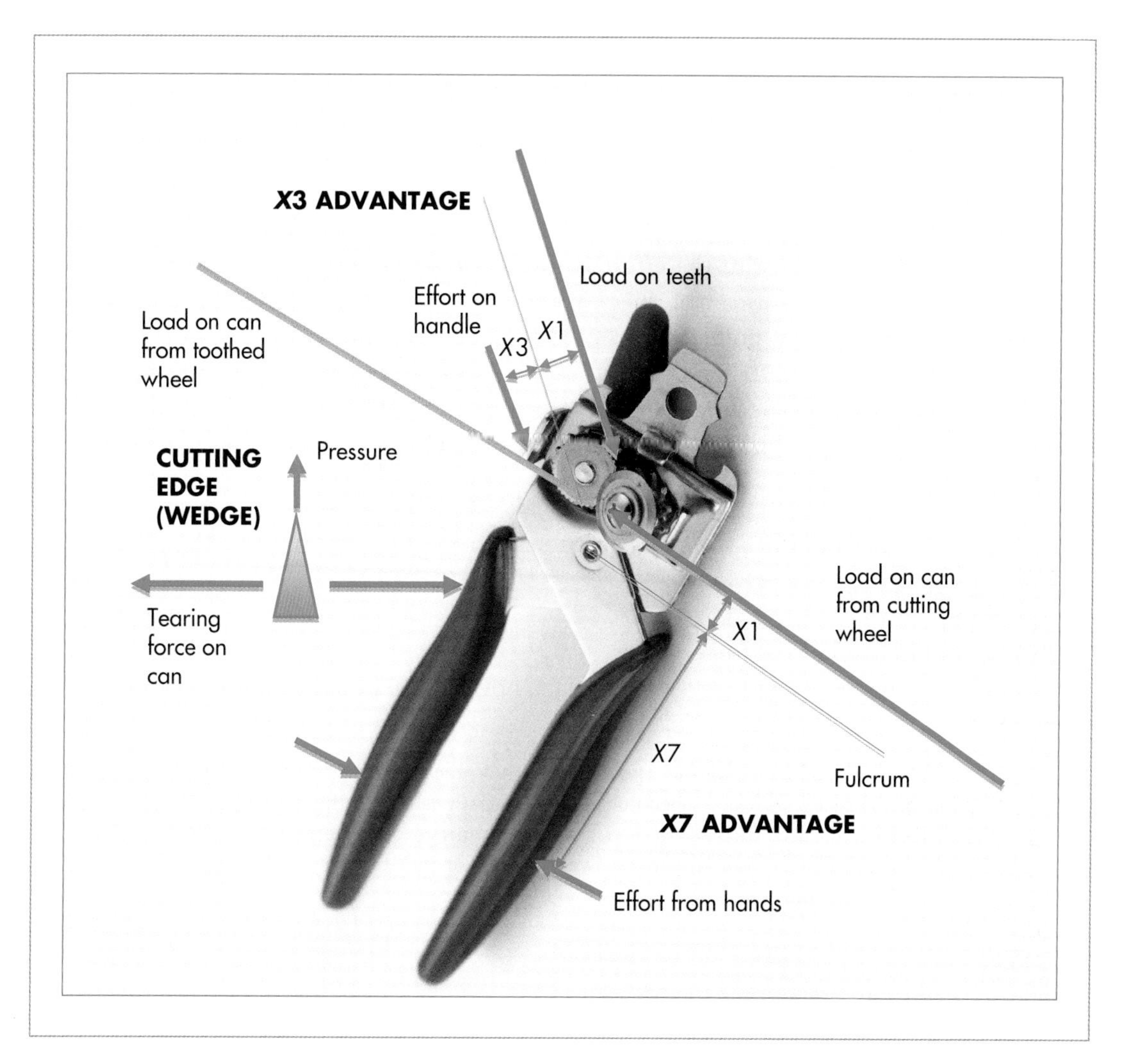

FIGURE 3.3 The can-opener as a simple machine

Everyday examples of simple machines can be examined to show the different ways in which they manipulate force. The principles of a short list of simple machines and the mechanical advantages that they provide are briefly described below.

Levers

Using a lever involves rotating a lever arm around a pivot (also called a fulcrum) to create a 'load' force. If the distance from the fulcrum to where the effort force is applied is further than that from the fulcrum to the load force, then the load force is larger than that applied; the force is magnified. An example would be using a screwdriver to open the lid of a paint tin, where a small force applied at the screwdriver handle creates a large force at the screwdriver tip to prise open the lid. You can find details of simple machine principles on many websites, such as that of the Commonwealth Scientific and Industrial Research Organisation (CSIRO; see, for example, http://www.csiro.au/scope/episodes/e73.htm).

There are three types of levers:

- 1st-class levers – the fulcrum is between the effort force and the load force (e.g., scissors a crowbar)
- 2nd-class levers – the load force is between the fulcrum and the effort force (e.g., a wheelbarrow, nutcracker)
- 3rd-class levers – the effort force is between the fulcrum and the load force (e.g., a fishing rod).

The ratio of the load force to the effort is in inverse proportion to the distance of these from the fulcrum. Thus, for a pair of pliers, if the force applied by the hand is five times as far from the fulcrum as the point of grip of the pliers, then the load (grip) force which is applied to the object in the pliers will be five times the effort force from the hand. This represents a five-fold mechanical advantage. The way this works, including for the turning handle on a can-opener, is shown in Figure 3.3 (see p. 121).

Gears

Gears are used to transfer force from one gear to another. Gears can do three things:

1. speed things up
2. slow things down
3. change the direction of the force.

Lehrer and Schauble (1998, pp. 4–5) explain how gear trains (sequence of interlocking gears) work:

> Each tooth on the driving gear must push one tooth on the gear that it drives. The turning speeds of the two gears must depend on the number of pushing teeth and the number of teeth that get pushed. Pairs of meshed gears must turn in opposite directions, or that small gears turn faster than large ones. Every other gear in a train of meshed gears will move in the same direction.

Pulleys

A pulley changes the direction of the force or effort. A pulley with one wheel does not reduce the force required to lift the load; however, it is easier to pull down than it is to pull up. A pulley with two wheels makes it easier to lift a load by halving the effort required (Oxlade and Hawken 1998).

Inclined planes, screws and wedges

Inclined planes increase the distance covered (e.g., in pushing a wheelbarrow up a ramp rather than lifting it) but decrease the effort needed to move that distance. This principle also applies to screws, which are essentially inclined planes threaded around a central shaft, and to wedges and blades, which are like portable compound inclined planes. In each case, a small force translates into a large load force at right angles to the effort, but over a smaller distance.

Wheels and axles

Wheels reduce the friction that opposes motion. This happens because for a wheel, the point of contact with the ground does not scrape along the ground. The key idea with the wheel is the relationship of the axle and vehicle to the wheel.

As an example of the way in which simple machines have transformed our lives, let's consider the role of wheels in humankind's attempt to manipulate and improve its surroundings. Wheels are often described as humanity's greatest invention. So much of human social history has to do with transport, with getting things from one place to another with a minimum of fuss and energy expenditure. Some of the great mysteries of ancient civilisations involve questions of transport. For example, how did the Egyptians get the stones in place on the pyramids? How were the Easter Island statues transported to their sites and erected?

The most plausible theories concerning the pyramids involve the use of rollers. The roller, in fact, is the forerunner to the wheel. The disadvantage of rollers is that they are not attached to a carriage. They have to be repeatedly taken from the back of whatever is being transported and brought to the front again. Having an axle solves that problem, though attaching the axle and arranging for the wheel to be able to turn freely is a problem in itself. Some axles are fixed, with the spinning of the wheel made relatively frictionless by using ball bearings or lots of grease. Other axles spin, such as the driving axle on a car that causes the wheel to turn. If you look carefully at a bicycle wheel, you will see that the axle is fixed and the wheel spins around using a sleeve arrangement. When they make model carts in a classroom, children can have a lot of difficulty in solving this wheel–axle problem.

Because of the need for strength to carry loads over rough terrain, early wheels were solid and bulky items. Once road-building technology improved, it was possible to use a lighter design of wheel, which made carts easier to pull. Early spoked wheels were constructed from wood, but once metal technology developed, the use of taut wire on bicycle wheels and metal spokes on trains and buggies became possible. Early coach wheels used metal bands that were heated and placed on the wooden rims, then hosed down with cold water. The resulting contraction caused the band to grip tightly on the rim. The metal gave protection against the wear and tear caused by constant contact with the road surface. Rubber technology allowed for a lighter tyre that improved grip on the road and gave a gentler ride because of the natural stretch of the material. Inflatable tyres are lighter again and make use of the compressibility of air to give a natural springiness. The history of transport is thus a good vehicle for a discussion of materials and technology.

Children's ideas about simple machines

In the research on simple machines, interest has focused on the extent to which children can develop generalised understandings of machine principles through direct observation and exploration, and at what point they need guidance in developing these understandings.

In researching children exploring rollers and ramps, Liu (2000) found that physical actions and reasoning/conceptual understandings developed together, as students were able to conceive various relations between characteristics such as slope and speed of rolling, and elaborations on the different characteristics of rollers, as they experimented.

Lehrer and Schauble (1998) found that primary school children perceive gear trains in a variety of ways, focusing on different aspects such as direction, plane of turning, and motion. But while children can develop superficial understandings quite readily through direct observation, deep understanding does not readily emerge without considerable reflection.

Children's reasoning became more general, formal and mathematical as problem complexity increased, and Lehrer and Schauble argue that formal mathematical reasoning about gears may develop when this provides a clear advantage over simple causal generalisations, such as 'If I turn a gear this way, the adjoining gear always turns the other way'. Lehrer and Schauble recommended the use of context-based technology and design problems to focus children's attention on developing and revising hypotheses to develop explanations that would account for observed regularities.

When examining gears and mechanical advantage in the context of Lego robotic design, the understandings of nine- to 10-year-old children of direction of turning, relative speed and number of revolutions were enhanced, but they had difficulty providing the rationale for choosing gear arrangements that made the robot faster or slower (Chambers, Carbonaro and Murray 2008). Because research has shown that it is difficult for students to understand mechanical advantage through direct observation of gear functioning (Chambers, Carbonaro and Murray 2008; Lehrer and Schauble 1998), guided-inquiry instructional approaches are advocated to support conceptual development during construction design activities.

There is increasing interest in embodied cognition as the idea that we understand the world through bodily relations; for example, in perceptions of touch or spatial relations (as in mental imagery and spatial metaphors). We can see how this may be a critical factor in children's understanding of forces, and how machines work. We can perceive, for instance, the effect of a lever by imagining ourselves pushing. Thus, an advocated strategy is to have children work with large-scale levers or pulleys or inclined planes to provide sensory experiences around which perceptually based understandings can develop. Having children explore simple seesaws, for instance, can provide powerful insights into how balance is affected by the relationship between distance from the fulcrum and relative weight. In an important sense, this notion of embodied cognition shifts the focus of attention away from ideas being situated purely in the mind, to a realisation that our understandings are distributed in our local environment (distributed cognition) as well as being highly perceptual in nature.

Designing lesson sequences for simple machines using the 5E model

In this section, we demonstrate how the 5E framework can be applied in different ways to the planning of units about simple machines for early and middle years students. Three units of six lessons each have been constructed to illustrate how activities can be used differently across a 5E sequence.

Exemplar 5E lesson sequences

Table 3.3 outlines three units focusing on simple machines. Units 1 and 2 include activities informed by preservice teachers planning for teaching years 3 and 4, and unit 3 is based on an early years sequence developed by practising teachers. Unit 1 explores levers, inclined planes, gears and pulleys in a variety of contexts, leading to an elaboration of simple machine principles through a design challenge. Unit 2 uses the context of the playground to explore levers and inclined planes, with elaboration where the students design playground or game equipment that incorporate levers and inclined planes. Unit 3 focuses solely on wheels, exploring the nature of wheels in detail. All units culminate in a design challenge relevant to the contexts explored through the unit. The key activities in each lesson are listed; the numbered and italicised activities are described in more detail in the next section. Note that some activities are incorporated in more than one unit sequence to illustrate that there is more than one way to use activities within the 5E framework. Activities may be applied at different points in the unit and serve different purposes in the learning sequence.

Activities for use in the simple machine lesson sequences

The italicised activities numbered 1 to 13 in Table 3.3 are elaborated on in the following sections. These descriptions include different approaches to the activities, links to the 5E framework and assessment, and links to technology.

1 Mystery boxes

Children are asked to rotate around four tables, each with a mystery box, butcher's paper and pens. Children take the items out of the box one by one, name them, and compare them to each other. On a piece of butcher's paper, children write responses to questions such as: How

TABLE 3.3 Simple machines units using the 5E instructional model

LESSON	UNIT 1	UNIT 2	UNIT 3
1	**Engage** Challenge: How can I separate this piece of paper into two pieces with straight edges? *Mystery boxes* (1) KWL chart – complete K and W Discuss and add key words raised during the lesson to a word wall	**Engage** *Analysis of simple machines in pictures* of different playgrounds and Honda Accord commercial *video* (7) KWL chart – complete K and W	**Engage** Looking at *wheels on tricycles and pushers* (10)
2	**Explore** *Exploring levers* (2) *Exploring gears* (3) *Exploring pulley systems* (4) *Exploring inclined planes* (5)	**Explore** *Exploring levers*(2) *Exploring inclined planes* (5)	**Explore** *Role-modelling wheels* (11) Exploring the science of rolling down slopes: *ramproll* (12)
3	**Explain** *Representing force in simple machines* (6)	**Explain** *Classification and differentiation activities* (9)	**Explain** *Excursion to a science resource centre,* such as a museum (13)
4	**Elaborate** *Analysis of simple machines in video and pictures* (e.g., Rube Goldberg video) (7) *Design challenge*: design a Rube Goldberg machine (8)	**Elaborate** *Design challenge*: design and construct playground equipment or game equipment (8)	**Elaborate** *Design challenge*: cart construction (8)
5	**Evaluate** Complete design of machine in groups. Students present design and as a class decide how the group designs will be put together as one machine. KWL chart – complete L	**Evaluate** Students present their equipment and are peer-, teacher- and self-assessed KWL chart – complete L	**Evaluate** Hold a cart derby in the schoolgrounds Teachers and students evaluate their designs

do you think the tool works? Draw it and label what you know. What is it used for? What would we do if it did not exist? What makes them all different?

Mystery boxes may contain objects or photos such as:

- gears of different sizes (e.g., can-opener, egg-beater)
- levers: first-class levers (e.g., seesaw, claw hammer, crowbar, scissors, pliers); second-class levers (e.g., wheelbarrow, paper cutter, door, nutcracker, garlic press, bellows, bottle opener); third-class levers (e.g., fishing rod, hammer, cricket bat, hockey stick, golf club, tennis racket, shovel, pitchfork, hoe, broom, tweezers, ice tongs, children's arms and legs)
- pulleys: pulleys and weights, photos of pulleys for lifting objects (e.g., flag and flagpole, pulley clothesline, blocks and tackles, large cranes, chain hoists, hydraulic systems)
- inclined planes and ramps (e.g., screws, wedges, nails, toys that have ramps).

2 Exploring levers

For each of the following activities, children are asked to explore the materials and consider the problems posed. As they carry out the activity, children can discuss what will happen if the weight is changed, the fulcrum moved, or the load or effort increased. Children can log predictions prior to each change and record observations in a science journal. The language of fulcrum, effort, load and force can be introduced to students, and the teacher can model how to use this language when making predictions. Students may also be encouraged to identify some 'explorable' questions that arise out of the activities.

By exploration of the following, children should be able to:

- recognise and name the fulcrum, effort, load and force (output)
- name the three types of levers
- investigate changes that occur in force and load when a fulcrum is used.

Lifting the load

In this activity, children are asked to use a 30 cm rule as a first-class lever: tape a pencil as the fulcrum at the 5 cm mark and connect weights to one end of ruler as the load. Students should now find the easiest way to lift the load; that is, how to lift it using minimal effort. Begin by asking students how difficult/easy it will be for them to lift the weight, and why they think that? Is it possible to lift the load using another weight less than that of the load?

Catapult

Children are asked to use a 30 cm rule as a first-class lever: tape a pencil as the fulcrum at the 5 cm mark and place an eraser as the load at the short end. Students drop beanbags onto the longer end, trying to get the eraser to hit a piece of paper on the wall. Discuss with students what they think will happen when they drop the weight from a certain height, and why? Mark where the eraser hit on the paper and note in the journal the weight dropped (force) and the position of the pencil (fulcrum) in centimetres.

Making a wheelbarrow

In this activity, children are asked to use a 30 cm rule as a second-class lever: place a weight at 10 cm on the rule as the load, rest the long end on a table, and connect a force meter to the shorter end. Students measure how much force is required to lift the load when it is placed at different points along the rule. It is worthwhile discussing with students how the force meter acts like a spring balance that tells them the amount of force applied to the lever.

Going fishing

In this activity, children are asked to use a 30 cm rule as a third-class lever: tape string onto the 30 cm end of the rule and tie a magnet onto the end of the string, then hold onto the 0 cm end of the rule with one hand, attach a force meter at 15 cm, and place metal objects ('fish') of different weights on the table. Pull upwards on the force meter to pull up the fishing rod. Students are using the rule as a fishing rod and measuring the amount of upwards force required to 'catch' or lift fish of different weights. They can also explore the effect of placing the force meter (effort) at different points along the rule. Discuss with students what they think will happen to the amount of force required to lift a weight if it is placed closer to the supporting hand (fulcrum) and closer to the string (load), and when different fish (load) are caught.

Extension

Students can brainstorm, draw and build other examples of when they might use first-, second- and third-class levers.

3 Exploring gears

Gear chains

In this activity, children are asked to create different gear trains and record changes in the direction and speed of the gears. Try gear trains with different numbers of gears. Have students predict the direction of the last gear, or how many revolutions (turns) the different gears will do in relation to other gears. Students can document the number of revolutions against the number of teeth (e.g., gear 1 has 10 teeth and turns once when placed with a smaller gear with five teeth that turns twice).

Students should be able to:

- identify the driver, the follower and the teeth
- recognise that, within a gear train, the more teeth a gear has, the slower it will rotate
- recognise that gears in a gear train will alternate direction.

Extension

Students can be introduced to the idea that gear trains transfer force in other directions (e.g., right angles).

4 Exploring pulley systems

Broom pulleys

Students are introduced to pulley concepts by building a pulley system using a broom and rope (view this activity on video at http://www.csiro.au/scope/clips/e71c01.htm). This activity involves children holding brooms (act as the pulleys) opposite each other at 2 metres apart. Tie a rope around one broom (first broom), now pull the rope hard while the person with the broom provides resistance. How much force is required? Wrap the rope around the second broom so that the rope comes back towards the first broom (the second broom should try to stay stationary), now pull and feel whether this changes the amount of effort needed to move the fixed broom. How much rope is pulled through to get the fixed broom to move? Children can experiment by comparing how often the rope is wrapped around the brooms (number of pulleys) with the effort required and the length of rope pulled through.

Students should be able to:

- recognise that pulleys make things easier to lift or move
- recognise that more than one pulley will decrease the effort required by spreading the load over a greater distance. Two pulleys reduce the load by half, three pulleys by one-third.

Pulley challenge

Students can use a variety of materials to find the easiest way to lift heavy objects, or to transfer an object from one place to another. Students can be given a challenge, such as to move a heavy object from one side of a canyon to another (between tables), or to lift the object from the floor onto a table. Children can be given a variety of objects, such as tables that can be upturned to use the legs as anchorage points, a load (such as a small bucket of sand or weights in a rack), pulleys (single and double), rope or strong twine, and string and sticky tape for attachment purposes. Ask students to examine the materials first and discuss how they could be used as fixed or moving structures. As an added challenge, students may be required to shift a large weight with a smaller weight.

Extension

Students can begin to explore the direct relationship between increases in the number of pulleys and the length of string being pulled and the effort required. 'Explorable' questions can be developed to guide this.

5 Exploring inclined planes

Measuring the force required to shift a load up a ramp

In this activity, children explore the usefulness of ramps in lifting loads. Ramps can be constructed out of any flat, stiff object, such as wood, a table or a book raised at one end. Children pull objects up a ramp and measure the amount of force required using either a force meter or by measuring the stretch of an elastic band attached to some sort of weight. By adjusting the height and length of the ramp, students can see the change in force required to lift a load.

Students should be able to:

- recognise that less effort is required to push or pull a load up a ramp than to lift it straight up
- identify that the longer the ramp (and therefore a lower gradient slope), the less effort is required to move an object up to a particular height.

Looking at screws

This activity allows students to construct screws, reinforcing the idea that screws are inclined planes wrapped around a shaft. This activity would come after students have an understanding of the mechanical advantage of inclined planes when pushing or pulling an object upwards, but that greater distance is required. Students can be encouraged to think about how we can arrange a ramp so that it takes up less space but still has the necessary length; for example, how ramps would be used to get to the top of a three-storey building, alluding to spiral staircases. How are screws the same as spiral staircases? In order for students to see the inclined plane in a screw, encourage them to change the way a two-dimensional paper ramp (a right-angled triangular piece of paper) looks by twisting it around a pencil, starting with the shorter vertical edge against the pencil. Have students do this with 2-D paper ramps at different angles in order to see how screws are formed, comparing the number of turns and distance travelled, and drawing conclusions about the effect of each option. If possible, have them experiment with real screws of different sizes, screwing them into a block of wood: Does the distance between the grooves make a difference to the amount of effort required?

Extension

Students can be introduced to the idea that a decrease in the length of the ramp also means an increase in 'incline' or 'angle'. Students may investigate the effect that friction has on the effort required to move an object, using 'explorable' questions such as: What does changing the surface of the ramp do to the amount of force required to lift an object along it?

6 Representing force in simple machines

As an 'explain' activity, this would involve classroom discussion about what is actually happening in the simple machines that students have experienced during the 'explore' lessons. The emphasis here is on encouraging students to think about where force is being applied, where the load is, the motion that results from the force, and how the machine has modified the force; that is, the mechanical advantage. Students can be encouraged to think about how they would 'represent' force (usually a push or pull, or a twist in the case of handles), load, motion and mechanical advantage on a diagram. It is better to begin with students using their own representations before the scientific convention of arrows is introduced. Students can then use arrows to represent the direction and size of the force applied, and the direction and speed of the resultant motion. (Refer to the earlier section on representing forces in this chapter, 'Acting on playdough' on p. 100, for a possible sequence for this.) Remembering the activities from previous lessons, use two common simple machines to draw annotated pictures, using arrows to represent force and movement.

7 Analysis of simple machines in videos and pictures

In this activity, students view and analyse DVDs/videos and pictures of simple machines. This activity can be constructed in many ways depending on its purpose and placement in the learning sequence. If used as an 'engage' activity, the focus is on introducing everyday examples of simple machines, or setting the scene for a problem-based sequence (as in unit 2), as well as for motivational purposes. If used during the 'explain' phase, students can be encouraged to apply their new knowledge to the events in the video; for example, students or the teacher could choose sequences in a Honda Accord commercial (see http://www.youtube.com/watch?v=uyN9y0BEMqc) where simple machines are used, demonstrating students' ability to identify and classify different machines. Students could also describe the mechanical advantage of each. Used at the beginning of an 'elaborate' lesson, as in unit 1, a 'Rube Goldberg machine' video (see the next section) or the *Wallace & Grommit: the Curse of the Were-Rabbit* trailer (see http://trailers.apple.com/trailers/dreamworks/wallace_and_gromit) can provide stimulus for student designs. There are many examples of these on YouTube.

8 Design challenges

This activity integrates with the design–make–appraise strand associated with the various design technology curriculum frameworks. All three units culminate in a design challenge, mainly because the topic, simple machines, lends itself to integrating technology as an application of scientific principles. Technology activities can work well as summative assessment tasks, so long as the assessment focuses on how students apply the scientific principles of simple machines. Do students understand how the design features of the machines change the direction of the force, increase or reduce the speed of the force, or magnify the strength of the force?

For any design challenge where construction is involved, students engage with three processes: designing, making and appraising, the order of which may change depending on the purpose of the activity (Fleer and Jane 2004). Students may be encouraged to: design first, make the item according to the design and write about it, then appraise with the teacher in terms of improvements or changes that could be made and add this to the written explanation (DMA); make and write about the item, appraise with the teacher, and develop the design (MAD); or play with and appraise the materials, design the item, and make and evaluate and write about it (ADM).

Designing a Rube Goldberg (or chain reaction) machine

A Rube Goldberg machine is an overly complex and confusing machine that is designed to perform a very simple task. Rube Goldberg was an American cartoonist who was well known for depicting complex devices that accomplished something simple through complex means, often through a chain reaction sequence beginning with a stimulus and ending in the desired action. The design only, or design and construction, of such a machine could be done individually, in groups or as a class. Computer animation software can also be used to create or manipulate machines, then observe the effects; for example, Pivot Stickfigure Animator (see http://pivot-stickfigure-animator.en.softonic.com) and Crayon Physics Deluxe (see http://www.crayonphysics.com). Potential contexts for the construction are:

- a class project, where small groups design and construct separate sections of a large machine that utilises the whole classroom
- as a board game, similar to the Mouse Trap board game; see Barlow, Kramer and Glass (1963)
- for a particular action, such as opening the door or cleaning the whiteboard.

Designing playground or game equipment

Students can design playground or game equipment that incorporates levers and inclined planes (as in unit 2). Students develop annotated diagrams within a storyboard to show how the equipment is used, where the levers and inclined planes are, and how they modify forces.

Designing a cart

Children can design carts (as in unit 3) for a particular purpose, such as for running quickly down a slope or travelling as far as possible in a straight line across the floor. They are asked to solve the problem of how the wheel and axle are attached so that the wheels go around. This can be an individual or paired activity, according to children's needs. Different versions of the activity are discussed in the online Appendix 3.3.

Appendix 3.3
Cart construction and thinking technologically

Provide assorted materials – boxes, containers, card, tape, string, scissors, straws, wooden skewers, container lids, pieces of polystyrene, Blu-Tack and plasticine.

Older children can be asked to make a drawing of their design first. This can then be modified after the construction and a report written on why modifications had to be made. The task can be focused on how well the vehicle rolls down a ramp or along the ground.

Activity 3.9 Cart construction activities

- Read the online Appendix 3.3 and discuss the following issues in light of the vignettes and discussion.
  - The materials children are given have a big impact on the style of cart they construct. A limited range of materials limits the range of possible solutions. Materials that are difficult to work with can not only frustrate children, but can also encourage significant problem-solving behaviours. What range of materials would be appropriate for cart construction? Why?
  - A teacher should encourage a range of solutions to design problems, but should make sure each child constructs a successful cart. This may involve suggesting specific solutions or pointing out problems before they happen. If children cannot make a cart, they cannot move forward to considering aspects of motion. How appropriate would it be to give children worksheets showing how a cart can be made?
  - Design drawings should be a part of any construction activity. Usually, though, the design is worked out as the child grapples with the materials. Drawings can often be more useful as part of reports written after the construction phase. Do you agree with this view? What should such a drawing show?
  - In any sort of learning activity, and particularly with individual investigative or construction work, the purpose of the child and that of the teacher can be very different. What aspects of children's purposes in constructing carts might interfere with learning about wheels and motion?
  - An English study identified measurement as being an undeveloped science skill in primary school, despite emphasis on processes in the English national curriculum. Children do not tend to undertake controlled measurements or understand their purpose. They need to be encouraged to do so. How is this best done for cart construction?
- Construct a land yacht that will move in a controlled manner – driven by the breeze from a fan – across the floor. Compare different models of land yachts and draw up a list of statements about the effect of features such as the length of the axles, the materials used, the size and position of the sail and the wheel–axle arrangement. Plan and conduct a controlled investigation to verify one of these statements.
- Locate the statements in your State/Territory or the Australian science curriculum related to measurement and investigation. Discuss the sorts of expectations and responses to activities involving carts that might illustrate each of the outcome statements at the different levels.

Activity 3.10 Developing science concepts through construction activities

Cart or machine construction is essentially a technology activity that can take place without any reference to science concepts. At what point does technological knowledge become scientific knowledge? A class of children has successfully built a variety of carts. What activities or questioning strategies might be used during and after the construction to extend children's understandings about force, energy and motion?

From your science syllabus/standards document, locate the science outcome statements relevant to motion and construction activities. What sort of things might children do, say or write that would illustrate these outcomes? Use activities from this chapter to generate examples of children's responses that would illustrate the range of outcomes.

9 Classification and differentiation activities

When used in the 'explain' phase, these activities enable students to recall, reorder, consolidate, build on and transform the knowledge and experiences gained through the 'explore' activities. The following activities could be completed in sequence or developed further individually. Usually, a discussion about what has occurred up until now (e.g., types of simple machines, the examples already seen, what the machines did) would precede the activities.

For a classification activity, as a class, use texts that describe the key ideas of levers and ramps (either teacher-prepared or generated in class) to classify a selection of common household simple machines.

For a differentiation activity, take two items and develop a Venn diagram of the similarities and differences in their uses and how they make life easier. Focus on how the machine in question is designed to change the direction of force applied, increase or reduce the speed of the force applied, or magnify the strength of the force applied.

Activity 3.11 'Explaining' force principles through representation

There are two approaches to the 'explain' phase in units 1 and 2. Unit 1 focuses on how to represent force principles in diagrams, while unit 2 uses classification activities and Venn diagrams to differentiate between the different principles. In this activity, you are asked to think about, firstly, how you would represent your own understandings of simple machines, and secondly, how explanations might be generated flowing from the previous exploratory activities.

- For each type of machine, develop a representation of how the machine works to secure mechanical advantage. To do this, you might use a combination of: drawings with force representations, annotated comments, mathematical expressions, role-plays, time sequence drawings or annotated photographs.
- Work out a way of representing (such as with graphs, annotated diagrams, tables, Venn diagrams or concept maps) the different sorts of simple machines and examples of each.
- Represent, perhaps using a table, Venn diagram or other classification device, the key principles of force alteration involved in a range of simple machines. You might like to think about how an interactive whiteboard might provide a flexible way of grouping a variety of machines under different principles.
- Discuss these representations with colleagues. What are the various ways in which knowledge of simple machines can be effectively communicated?
- Develop, in your group, a strategy for scaffolding children's ideas that emerge through the exploratory activities, to generate representations that constitute productive explanations of simple machine principles.

10 The language of wheels: tricycles and pushers

This activity is intended to introduce children to the vocabulary and basic concepts associated with wheels and motion and to give them experience in observing similarities and differences. This can be extended into a substantial language activity.

- Bring a pusher (or pram) and a tricycle into the classroom and seat the children around them so that they can look at the items from different angles. You may like to put the vehicles on a table so that the children can look up at them from underneath.
- Ask the children to look at them close up and from a distance.
- Make a class list of the parts that children notice: one list for the pusher and one for the tricycle.
- Look for things that the pusher has in common with the tricycle.
- Discuss the differences between them.
- Ask the children for their ideas on what specific parts are used for.
- Help them to decide how the pusher and tricycle would be made to move or stop.

For older children, a discussion of where the energy comes from and what energy is will be appropriate (see Chapter 5).

11 Role-modelling wheels

This activity helps children to get a feel for how the roundness of wheels helps with motion.

- Use the classroom, corridor or mats outside for rolling activities.
- Ask individual children to roll along the ground.
- Talk about how they will get going.
- Discuss how they will stop, what keeps them rolling and which shape is best for keeping going. Observe these efforts closely.
- Invite suggestions from children about different shapes for rolling.
- Ask the rollers to talk about the differences they felt in one complete roll.
- Group the children into threes or fours.
- Experiment with rolling: ask one group member to start off the roller; have another person stop the roller in three ways:
  - by remaining still so that the roller experiences the force of a stationary object
  - by interrupting the roll by some movement so that the roller experiences the force of a brake
  - by rolling over a cushion.
- Change around so that everyone has a turn in each role.
- If you have access to a slope, have the children compare rolling up and down and stopping on a slope so that they experience the difference that a bit of falling makes to the rolling.
- Compare the different need for pushing and pulling.
- Discuss the differences in their experiences and link to the pusher and the trike in terms of shapes, brakes, getting going and so on.

Prep children could be asked to make wheels from playdough. Their wheels could be spherical, flat-like pancakes or long and thin. What is a wheel? Are all round things wheels? Could an apple or orange be made into a wheel?

12 Exploring the science of rolling down slopes: ramproll

This activity is very open-ended and exploratory and can be used to work with children on measurement principles, experimental design and analysis using tables or other means. It is a classic engagement with ideas and evidence activity.

Jam jar POE

Take two identical jars, one full of jam (or honey or even plasticine will do) and the other empty. Predict which, when rolled down a slope, will go furthest along the carpet (see Figure 3.4 for the arrangement). Discuss your prediction in a group. Try this so that you are confident you have a clear result. Discuss what you think is happening. Is the difference due to the speed they roll down the slope?

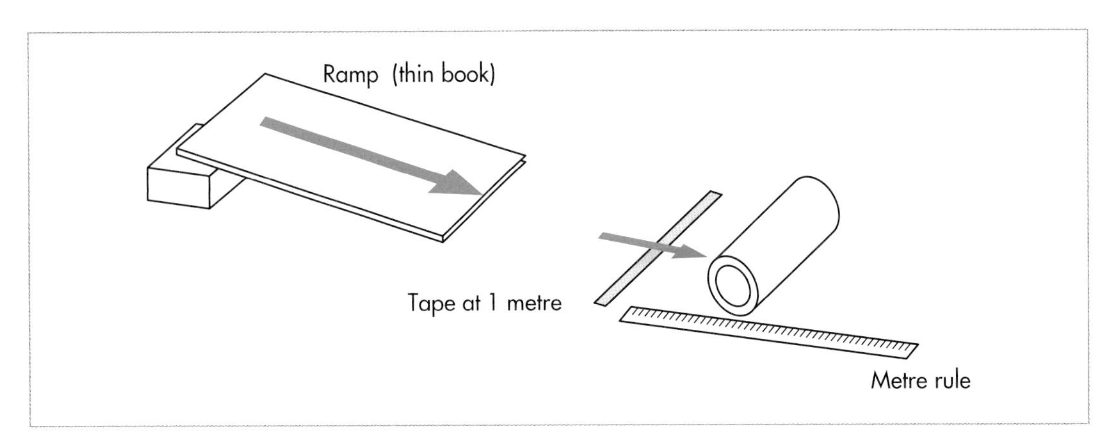

FIGURE 3.4 Ramproll

How far does it roll?

Work with a number of cylindrical rollers (e.g., various-sized batteries, plastic pipes, cardboard tubes, steel rods, pens and wooden rods). Your task is to work out what makes a difference to how far they can roll along the floor.

Try each roller, rolling it down the slope (a book will do) and writing in the table how far it goes. Measure each distance twice to check.

How fast does it roll?

Compare a number of rollers, two at a time, to determine the order of speed with which they roll down a ramp. Try a series of cylinders and balls, hollow and solid, light and heavy. Line them up in order of speed. What patterns can you see?

Activity 3.12 Exploring rolling

The science underlying the ramproll activity

Hollow things tend to roll further than solid things because they have less weight for the same diameter; this reduces friction. If the friction was the same, they would all go the same distance as the energy gained rolling down the slope is used up. Paradoxically, hollow things tend to roll slower down the slope because more energy goes into the rotation due to the distribution of matter further from the centre of the object. Less energy, gained as they drop in height, goes into speed. Nevertheless, they roll the furthest along the carpet. Despite the slower starting speed, they have greater energy of rotation, which keeps them moving, rather like a flywheel that has a lot of rotational energy and momentum compared to the weight being closer to the axle. For a similar reason, spheres roll faster down a slope than cylinders; carts with good spinning wheels should roll faster again. Larger-diameter rollers go further because they have less friction and can ride over the bumps in the carpet. Carts with large wheels, for example, are much better in rough terrain because small wheels tend to get stuck in the ruts.

The question of weight is not straightforward, but less weight seems to go further, probably because of having less friction from the carpet, which it squashes down. In the cart construction activity, weight does tend to help carts go further, but the optimum amount of weight depends on the surface because of friction effects.

Children will tend to associate speed with weight. This is similar to the (mostly mistaken) idea that heavy things fall quicker. Children will often interpret their results to match this idea. However, there are plenty of opportunities to challenge this notion.

Children and adults will sometimes identify hollowness as a factor but tend to attribute the motion to the effect of air circulation.

- Try the constructing carts and ramproll activities. Discuss with colleagues the different ideas about movement and force that are relevant to these activities.
- Sketch in your science journal annotated diagrams of the objects rolling down a ramp and along the floor to show the forces acting in each case.
- Construct a list of possible factors that affect the motion.
- What alternative ideas do you think children would bring to these activities that could be worked with and challenged to advance their thinking?

13 Excursion to a science resource centre

Science resource centres include zoos, aquariums, museums, science activity centres (such as Questacon in Canberra or Scienceworks in Melbourne), and environmental centres or field sites of various types. Taking children to a science resource centre can be highly motivating and can prompt significant learning through interaction with quality exhibits and activities and expert interpreters. Such excursions can be used at various points in a unit sequence: at the beginning or early in the sequence to promote interest in the topic and to provide a reference point for later experiences; at the 'explain' stage to allow students to build explanations based on prior experiences in collaboration with information from the centre exhibits; or as a finale to the children's study.

In one implementation of unit 3, the children attended Scienceworks, which had a display on 'wheels' at the time that focused on the technology of wheel design as it had developed over time. Children explored a range of wheeled vehicles and played with wheeled toys, role-modelled wheels, and were exposed to the language of wheels. They looked at old stagecoaches, penny-farthing bicycles and a range of historic and modern vehicles. A sheet of prompt questions was used by accompanying parents to elicit ideas and explore such things as wheel size, axle arrangements, brakes and gearing. Questioning approaches were modelled by the teacher at the first exhibit, a Cobb & Co. coach. This use of the centre was consistent with research into the role of informal settings in science learning.

Studies of informal learning explore different sources of public knowledge about science. These influence the ideas children bring with them to the classroom. Television, newspapers and magazines, the Internet and other electronic media, and experiences within the family, all form part of the science learning environment. Rennie (2007) points out that, while education tends to be narrowly equated with what goes on in schools, museums and resource centres provide an alterative informal setting for science learning.

Griffin (1994) argues that visits to informal science education settings should reflect what we know about the way in which students learn science, about exemplary teaching practices and family group learning behaviours in informal settings, rather than principally be the worksheet-dominated visits that tend to be the norm for schools. It is quite common in museums to see schoolchildren sitting in the middle of a display hall collaboratively filling in worksheets using observations made by only a few children. Griffin argues for inclusive, learner-centred approaches to museum visits and offers a set of guidelines:

- Embed the museum visit firmly in the classroom-based learning unit.
- Use a learner-centred approach in which the students are finding the answers to their own questions rather than their teachers' or the museum's questions.
- Encourage students to gather questions while at the museum as well as finding answers.
- Apply learning methods used by informal groups, such as an orientation period and decreasing detailed examination of exhibits over the visit.
- Develop strategies and approaches to learning that recognise and complement the particular learning environment.
- Recognise that students and teachers need to adapt to and learn to use this different type of learning setting.
- Create a close link between museum and school learning, but recognise the different roles of each.

Activity 3.13 Planning for a visit to a science centre or museum

- Carry out a Web search of a museum or science resource centre to identify exhibits that relate to simple machines (for instance, Melbourne's Scienceworks has machines as part of its Nitty Gritty Super City exhibit (see http://museumvictoria.com.au/scienceworks/education/education-kits/nitty-gritty). Evaluate any information kits or teacher or student resources associated with the exhibit.
- How might you embed a visit to the centre as part of a learning sequence focused on force and motion? Devise some pre- and post-visit classroom activities that would maximise children's learning from a visit to the centre.
- Generate a range of questions that children could explore during the visit or that might arise out of the visit.

Activity 3.14 Designing learning sequences

- Use activities relating to force to plan five consecutive lessons according to the 5E framework for a year 3 or 4 class. Select activities from earlier in this chapter, from Appendix 3.1 for assessment strategies (see p. 140), or from others that you know. You might like to base your sequence on a

theme (e.g., three little pigs, dad's shed, your favourite sport).
  - Provide a rationale for each part of the sequence and for the assessment strategy used.
  - Share your sequence with another group – How are they different and similar in how the activities are sequenced and in the rationales used?
- Design a worksheet that could be completed before, during and/or after one of the 'explore' phase activities in the above sequence. Make sure the worksheet is structured in such a way that directions for students' observations and thinking are linked to the intended learning outcomes of the activity. Considering that this is an 'explore' phase activity, how much direction should the worksheet give students in their 'explorations'? As an added challenge, embed in the worksheet opportunities for student inquiry where students construct and investigate 'explorable' questions that might arise out of the activity.

Activity 3.15 Planning assessment for learning about simple machines

- As explained earlier, three types of assessment are diagnostic, formative and summative. Each is appropriate at different stages of a unit of work. For one of the three units of work described in Table 3.3 (see p. 125), identify which tasks may be used for diagnostic, formative and summative purposes. Discuss and report on other ways you could explicitly build assessment into the sequence.
- Looking at the earlier section on constructing assessment rubrics (see the section 'An assessment rubric for movement and force' on p. 113), develop an assessment rubric for one of the assessment tasks you proposed. Be sure to carefully consider what the task allows you to see about students' science knowledge, skills and attitudes. Are there other criteria that you regard as worth assessing?
- Look at your State/Territory's science curriculum or the Australian curriculum. Map your criteria with the curriculum standards or outcomes.

Science as a human endeavour

Several sections of this chapter have dealt with the human dimensions of force and machines. This was particularly explicit in the simple machines section, which is all about how machines are designed to alter forces to allow us to do things we could not otherwise do, such as crack nuts, move large weights or even lift our arms. The section on the history of wheels and transport shows clearly the wider social impact of force and machines. Even with only a basic understanding of force and its application to motion, the human dimensions are clear. In teaching about force, we can appeal to a wealth of everyday experiences to give a sense of the embodied nature of forces – the way we can 'feel' forces in a diagram or description of how hard we have to push a trolley, how gravity acts, or how air takes hold of a kite. We can explore our bodies to find examples of the application of forces – the lever action of the forearm or the structural properties of the skeleton.

Thus, the science of force and motion can be related to personal experience to bring relevance. It can also be related to the story of human use and manipulation of force through machines. In relation to the world of work, engineers design machines to make our lives easier, and for every machine there will be a host of professions that use them.

There are many activities we can include in a unit on force that would link the topic with human use and human interest:

- Children could research and report on machines throughout history, such as waterwheels, catapults, steam trains or chariots.
- Children could write letters to manufacturers or engineering organisations asking questions about machine design, thus encouraging questioning skills and supporting literacy practices.
- Children could create a time line of machines throughout history, or of the development of the bicycle or gears.
- The human dimension of the topic could underpin further investigations based on the conceptual principles of a machines unit. The question 'What machines are seen in ...?' could be the underlying question for surveys and transport studies, from simple ways of categorising machines (tricycles, bobcats, gymnasium equipment, kitchen implements) according to their uses, to more complex issues of engineering works and community life.
- Studying means of transport where wheels would not be useful could enlarge the children's view of the world.
- Technological design activities where machines were invented for particular human use would open up discussion of science and human needs, and potentially of ethics and values associated with science and technology.
- There could be interesting connections made between the topic of machines and design technology, with children investigating and building the wheels used in entertainment – for example, unicycles, Ferris wheels, merry-go-rounds – or simple machines used in hunting and gathering societies, with the study of materials and systems involved developing into a connected topic.
- An understanding of community life and those members who use simple machines as part of their employment or their means of movement could be a rich source of new information and insight. This might include interviewing people who use wheelchairs, or potters, gardeners with shovels and wheelbarrows, and tyre repairers. The rights and responsibilities of being a community member could be explored through services for other people and the different occupations of people in the neighbourhood.

Activity 3.16 Science as a human endeavour

- Review the discussion about forces and simple machines in this chapter and list the variety of ways these topics link with a 'science as a human endeavour' strand, as in the Australian science curriculum (ACARA 2011). These could refer to human use, the impact of forces and machines on human culture, science professions that involve their development or use, and values and ethical issues associated with the topic.
- Revisit the curriculum sequences you planned in Activity 3.14 (see p. 136). Which activities add to students' understandings of science as a human endeavour?
- Create a list of ways in which this strand could be more explicitly represented in the sequence, including assessment.

Summary

In this chapter, the authors have drawn on various aspects of their teaching and research to raise some issues about the teaching and learning of movement and force and issues to do with teaching and learning, assessment and schooling more generally. These include:

- the way children's conceptions of movement and force relate to scientific conceptions
- the role of exploratory activities in conceptual learning
- approaches to assessment of learning
- the role of literacy practices in learning science
- scientists' and children's conceptions of simple machines
- planning simple machine sequences using the 5E framework
- the management of, and the learning that arises out of, simple machine activities
- science as a human endeavour in the context of force and motion.

Concepts and understandings for primary teachers

Movement and force are areas in which many informal, non-scientific ideas are held by children and adults. The major scientific understandings related to movement and force are discussed below as a summary of much of the discussion in this chapter.

On forces and motion

- A force can be thought of as a push or pull, but essentially, it is an effect on an object that causes a change in its motion: speeding up, slowing down or changing direction.
- Forces are an external influence on an object, such as a hit from a bat or a pull from the Earth's gravity, and are not something the object possesses within it. Thus, when a sliding object slows down, or when a golf ball falls to Earth, it is not because of an initial force being used up, but because of friction acting in the first case, and gravity in the second.
- Kinetic energy (the energy associated with motion) and momentum, on the other hand, are associated with a moving object and will reduce as the object slows.
- Friction is a pervasive force that causes things to slow down, but it is also the force that helps us accelerate when we are running or riding a bicycle.
- Forces come in pairs, so if an object such as your hand exerts a force to hold up a brick, the brick will exert a force back on your hand that is equal but opposite in direction. Action–reaction force pairs always act on different objects.
- Common forces include contact forces (physical pushes, support or traction from the ground, friction, air or water resistance opposing motion, force from wind) and field forces (gravity, magnetic forces, electric field forces).
- Forces will add together to affect objects. Forces can oppose each other, such as the upward force from the ground opposing gravity when we are standing. They can also add, such as when two people pull a cart along together.
- Situations of balance (such as for mobiles or seesaws) and twisting can be understood as due to forces acting off-centre, not through the fulcrum or rotation point.
- An object rolling down a slope will be accelerated by the force of gravity (its motion is somewhat akin to a slow-motion freefall).

On flight and falling

- Falling things are subject to a gravitational (weight) force down and an opposing force of air resistance up.
- Except for very spread-out objects, such as pieces of paper or feathers, for which the air resistance can be as large as the weight, heavy things do not fall faster (that is, they do not accelerate more) than lighter things. Thus, a tennis ball and a cricket ball will fall with essentially the same acceleration (at least over short distances).
- Nor do heavy things roll faster down slopes, but there are effects on speed of rolling that are due to shape (for example, hollowness, spheres versus cylinders).

- Planes stay in the air because of the greater force of air on the underside of their wings compared to the top side. This is due to the shape of the wings and airflow patterns.

On simple machines

- Simple machines are devices for altering the nature and effectiveness of a force to better get jobs done.
- Levers are usually classified as first-, second- or third-class according to the relationship between the effort, load and fulcrum. For a lever, there is a simple relationship between the output load force and the effort, linked to the relative distance of these from the fulcrum.
- Screws and wedges are specialised examples of the inclined plane, which magnifies the size of a force and changes the direction.
- Gears and gear chains change the speed and direction of motion as well as the size of forces.
- Pulleys change the direction of a force and in cases of multiple pulleys are useful in increasing the magnitude of a force.
- Wheels are devices for reducing the force of friction. They do this since, at the wheel's contact point with the ground, it does not slide. Large-diameter wheels (or rollers) tend to reduce the friction associated with rough surfaces.

Search me! science education

Search me!

Explore **Search me! science education** for relevant articles on movement and force. Search me! is an online library of world-class journals, ebooks and newspapers, including *The Australian* and the *New York Times*, and is updated daily. Log in to Search me! through http://login.cengage.com using the access code that comes with this book.

KEYWORDS

Try searching for the following terms:

- Learning about force
- Simple machines
- Flight

Search tip: **Search me! science education** contains information from both local and international sources. To get the greatest number of search results, try using both Australian and American spellings in your searches, e.g. 'globalisation' and 'globalization'; 'organisation' and 'organization'.

Appendices

In these appendices you will find material related to movement and force that you should refer to when reading Chapter 3. These appendices can be found on the student companion website. Log in through http://login.cengage.com using the access code that comes with this book. Appendix 3.1 is included in full below.

Appendix 3.1 Assessment strategies

DIAGNOSTIC	FORMATIVE	SUMMATIVE
• *Graphic organisers – concept maps, mind maps* • *Journal entries* • *K and W of KWL charts* • *Pre-tests*	• *Concept maps – the idea is to have an assessment piece that incorporates the possibility of reviewing and revising and resubmitting*	• *End-of-topic tests* • *Repeated probing activity from beginning* • *Practical reports* • *Posters*

DIAGNOSTIC	FORMATIVE	SUMMATIVE
• *Questioning at the beginning of a unit* • *Concept cartoons* • *Brainstorming as a class, in groups, or individually* • *Probing activities* • *Targeted observations of groups or individuals* • *Analysis of stories based on description, application or evaluation* • *Question wall, graffiti wall or word wall where items are added or removed at points throughout the unit*	• *Feedback – the idea is to listen to the students by providing opportunities for students to share their thoughts* • *Attention to individual learning needs of students – the idea is to be flexible, to provide tasks that cater for different levels of learning* • *Conferences with students* • *Observations* • *Question-and-answer sessions* • *First drafts* • *Quizzes* • *Science journals, where students make a record of observations, new knowledge and reflections (each entry may be structured by the teacher to be specific to the lesson, or have a generic format for the entire sequence)* • *A class flip-book that is added to at the end of each lesson to build up a compilation of socially constructed ideas*	• *Dioramas* • *Stories* • *Texts of different genres (e.g., song)* • *Models with explanation, such as explanatory text or presentations* • *L of KWL charts*

Appendix 3.2 A critique of the 5E learning models

This online appendix provides a critical appraisal of the 5E instructional strategy, identifying its strengths and limitations when applied to lesson and unit planning.

Appendix 3.3 Cart construction and thinking technologically

In following the sequence described as unit 3, a number of cart-construction classes were observed and these observations form the basis of the case study in this online appendix.

References

Adams, J., Doig, B. & Rosier, M. 1991. Science Learning in Victorian Schools: 1990. *ACER Research Monograph no. 41*. Melbourne: Australian Council for Educational Research.

Alonzo, A. & Steedle, J. 2009. Developing and assessing a force and motion learning progression. *Science Education*, 93 (3), pp. 389–421.

Atkin, J. M. & Karplus, R. 1962. Discovery or invention. *The Science Teacher*, 29 (2), pp. 121–43.

Australian Academy of Science (AAS). 2005. *Primary connections*. Canberra: AAS. Available at http://www.science.org.au/primaryconnections/index.htm (accessed 5 September 2006).

Australian Curriculum, Assessment and Reporting Authority (ACARA). 2011. K-10 Australian Curriculum: Science. Available at http://www.australiancurriculum.edu.au/Science/Curriculum/F-10 (accessed April 2011)

Barlow, G. A., Kramer, H. & Glass, M. 1963. Mouse Trap (board game). Springfield, MA: Milton Bradley.

Bennett, R. 2009. An investigation into some key stage 2 children's learning of foundation concepts associated with geared mechanisms. *Journal of Design and Technology Education*, 1 (3), pp. 218–29.

Bybee, R. W. 1989. *Science and Technology Education for the Elementary Years: Frameworks for Curriculum and Instruction*. Washington, DC: The National Center for Improving Instruction.

_____, Taylor, A., Van Scotter, P., Carlson Powell, J., Westbrook, A. & Landes, N. 2006. *The BSCS 5E Instructional Model: Origins and Effectiveness*. Colorado Springs, CO: BSCS.

Carruthers, R. & de Berg, K. 2010. The use of magnets for introducing primary school students to some properties of forces through small-group pedagogy. *Teaching Science*, 56 (2), pp. 13–17.

Chambers, J. M., Carbonaro, M. & Murray, H. 2008. Developing conceptual understanding of mechanical advantage through the use of Lego robotic technology. *Australasian Journal of Educational Technology*, 24 (4), pp. 387–401.

Chi, M. 2005. Commonsense conceptions of emergent processes: Why some misconceptions are robust. *Journal of the Learning Sciences*, 14 (2), pp. 161–99.
Fleer, M. & Jane, B. 2004. *Technology for children: A research approach*. Frenchs Forest, NSW: Pearson Education.
Gilbert, J. & Watts, M. 1985. Force and motion, in R. Driver, E. Guesne and A. Tiberghien (eds). *Children's Ideas in Science*. Milton Keynes: Open University Press, pp. 85–104.
Griffin, J. 1994. Learning to learn in informal science settings. *Paper presented at the conference of the Australasian Science Education Research Association*, Hobart, July.
Gunstone, R. 1987. Student understanding in mechanics: A large population survey. *American Journal of Physics*, 55 (8), pp. 691–6.
Hackling, M. 2006. Primary Connections: A new approach to primary science and to teacher professional learning, *in ACER, Proceedings of the Annual ACER Research Conference, Boosting Science Learning – What will it take?* Canberra, ACT: ACER, pp. 74–9.
Hadzigeorgiou, Y. 2002. A study of the development of the concept of mechanical stability in preschool children. *Research in Science Education*, 32, pp. 373–91.
Hart, C. 2002. If the sun burns you is that a force? Some definitional prerequisites for understanding Newton's laws. *Physics Education*, 37 (3), pp. 234–8.
Heywood, D. & Parker, J. 2001. Describing the cognitive landscape in learning and teaching about forces. *International Journal of Science Education*, 23 (11), pp. 1177–99.
Hubber, P, Tytler, R. &.Haslam, F. 2010. Teaching and learning about force with a representational focus: Pedagogy and teacher change. *Research in Science Education*, 40 (1), pp. 5–28.
Ioannides, C. & Vosniadou, S. 2001. The changing meanings of force: A developmental study, in D. Psillos, P. Kariotogloux, V. Tselfes, G. Bisdikian, G. Fassoulopoulos, E. Hatzikraniotis and M. Kallery (eds). *Proceedings of the Third International Conference on Science Education Research in the Knowledge Based Society, Vol. 1*. Thessaloniki: Aristotle University of Thessaloniki, pp. 96–8.
Kruger, C. & Summers, M. 1988. Some primary school teachers' understanding of the concepts force and gravity. *Primary School Teachers and Science (PSTS) Project, Working Paper No. 2.* Oxford: Oxford University Department of Educational Studies.
_____, Summers, M. & Palacio, D. 1990. An investigation of some English primary school teachers' understanding of the concepts force and gravity. *British Educational Research Journal*, 16 (4), pp. 383–97.
Lehrer, R. & Schauble, L. 1998. Reasoning about structure and function: Children's conceptions of gears. *Journal of Research In Science Teaching*, 35 (1), pp. 3–25.
Lemke, J. 2004. The literacies of science, in E. W. Saul (ed.). *Crossing Borders in Literacy and Science Instruction: Perspectives on Theory and Practice.* Newark, DE: International Reading Association and National Science Teachers Association, pp. 33–47.
Liu, X. 2000. Elementary school students' logical reasoning on rolling. *International Journal of Technology and Design Education*, 10 (1), pp. 3–20.
McCloskey, M. 1983. Intuitive physics. *Scientific American*, 248, pp. 114–22.
Moyer, R. H., Hackett, J. K. & Everett, S. A. 2007. *Teaching science as investigations: Modeling inquiry through learning cycle lessons*. Upper Saddle River, NJ: Pearson Education.
Oxlade, C. & Hawken, N. 1998. *Machines*. New York: Gareth Stevens Publishing.
Parker, J. & Heywood, D. 2000. Exploring the relationship between subject knowledge and pedagogic content knowledge in primary teachers' understanding of forces. *International Journal of Science Education*, 22, pp. 89–111.
Planinic, M., Boone, W., Krsnik, R. & Beilfuss, M. 2006. Exploring alternative conceptions from Newtonian dynamics and simple DC circuits: Links between item difficulty and item confidence. *Journal of Research in Science Teaching*, 43 (2), pp. 150–71.
Rennie, L. 2007. Learning science outside of school, in S. K. Abell and N. G. Lederman (eds). *Handbook of Research on Science Education*. Mahwah, NJ: Lawrence Erlbaum Associates, pp. 125–67.
Russell, T. & McGuigan, L. 2001. Promoting understanding through representational redescription: An illustration referring to young pupils' ideas about gravity, in D. Psillos, P. Kariotoglou, V. Tselfes, G. Bisdikian, G. Fassoulopoulos, E. Hatzikraniotis and M. Kallery (eds). *Proceedings of the Third International Conference on Science Education Research in the Knowledge Based Society, Vol. 2*. Thessaloniki: Aristotle University of Thessaloniki, pp. 600–2.
Solomon, J. 1983. Learning about energy: How pupils think in two domains. *European Journal of Science Education*, 5 (1), pp. 49–59.
Stein, M., Larrabee, T. & Barman, C. 2008. A study of common beliefs and misconceptions in physical science. *Journal of Elementary Science Education*, 20 (2), pp. 1–11.
Tytler, R. 2002. Using toys and surprise events to teach about air and flight in the primary school. *Asia–Pacific Forum on Science Learning and Teaching*. Available at http://www.ied.edu.hk/apfslt/v3_issue2 (accessed 15 September 2010).
_____ & Peterson, S. 2005. A longitudinal study of children's developing knowledge and reasoning in science. *Research in Science Education*, 35 (1), pp. 63–98.
_____, Peterson, S. & Prain, V. 2006. Picturing evaporation: Learning science literacy through a particle representation. *Teaching Science, the Journal of the Australian Science Teachers Association*, 52 (1), pp. 12–17.
Vosniadou, S. & Ioannides, C. 2001. Designing a learning environment for teaching mechanics, in D. Psillos, P. Kariotoglou, V. Tselfes, G. Bisdikian, G. Fassoulopoulos, E. Hatzikraniotis and M. Kallery (eds). *Proceedings of the Third International Conference on Science Education Research in the Knowledge Based Society, Vol. 1*. Thessaloniki: Aristotle University of Thessaloniki, pp. 105–11.
Yuruk, N., Beeth, M. & Andersen, C. 2009. Analyzing the effect of metaconceptual teaching practices on students' understanding of force and motion concepts. *Research in Science Education*, 39 (4), pp. 449–75.
Zembal-Saul, C. 2009. Learning to teach elementary school science as argument. *Science Education*, 93 (4), pp. 687–719.

Children's books about machines and wheels

Brinley, B. 2001. *The Mad Scientists Club*. Cynthiana, KY: Purple House Press.
Cristaldi, K. & Adams, L. 2003. *No More Training Wheels*. Sydney: Cartwheel Books at Scholastic.
Gaff, J. 1990. *Tell Me about Wings, Wheels and Sails*. London: Kingfisher.
Graham, B. 1986. *Pig's Wild Cart Ride*. Melbourne: The Five Mile Press.
Hicks, C. 2006. *The marvellous Inventions of Alvin Fernald*. Cynthiana, KY: Purple House Press.
Kubler, A. (illustrator). 2002. *The Wheels of the Bus Go Round and Round.* Sydney: Child's Play (International)
Low, W. 2009. *Machines Go to Work*. New York: Henry Holt & Company.
Lund, D. & Neubecker, R. 2008. *Monsters on Machines.* London: Harcourt.
McMahon, P. & Godt, J. 2000. *Dancing Wheels*. New York: Houghton Mifflin.
Mitton, T. & Parker, A. 2002. *Amazing Machines Truckload of Fun*. Boston: Kingfisher.
Newton, D. & Newton, L. 1990. *Bright Ideas: Design and Technology*. London: Scholastic Publications.
Paterson, B., Niland, K. & Niland, D. 1988. *Mulga Bill's Bicycle*. Sydney: Collins.
Pollard, M. 1996. *Making Things Move*. London: Macmillan Educational.
Prince, A. & Laroche, G. 2006. *What Do Wheels Do All Day?* Boston: Houghton Mifflin.
Prior, T. 1985. *Grug and His Bicycle*. Sydney: Hodder and Stoughton.
Simon, S., Erduran, S. & Osborne, J. 2006. Learning to teach argumentation: Research and development in the science classroom. *International Journal of Science Education*, 28 (2/3), pp. 235–60.

4

Electricity

by Peter Hubber and Valda Kirkwood

Introduction

This chapter deals with a topic that is very familiar to teachers and students alike. Electricity is very much part of our daily lives and yet very few people assert a confidence in having a scientific understanding of what happens within the electrical devices that are used on a regular basis. From this perspective, 'the conceptual area of electricity is one which presents considerable challenge to both prospective and practising teachers' (Heywood and Parker 1997, p. 869), given that most primary science curricula include content related to electricity. The first part of this chapter discusses the scientific meanings of terms associated with electricity within the context of explaining a simple electric circuit as can be found in torches. An outline of students' understandings of electric circuits follows.

This chapter also includes the story of how Catherine, who felt she had insufficient background knowledge, taught electricity to her years 3 and 4 composite class (eight- to nine-year-old girls and boys) using the context of torches, an everyday piece of technology with which the children were familiar. Particular consideration is given in this chapter to the main components of the approach – interactive teaching – used by Catherine.

Finally, this chapter describes ways in which the teaching and learning of the topic of electricity reflect 'science as a human endeavour'. Following the summary, this chapter then lists concepts and understandings for primary teachers, to be used in framing the teaching and learning experiences students will have in the topic of electricity.

Teachers' and scientists' understandings of electricity

The study of electricity in schools is usually in two main areas that are described as static electricity and current electricity. Static electricity deals with the separation of electric charges that occur in phenomena such as lightning and the shocks one may receive from touching a doorknob, or another person, after walking on carpet. Static electricity is not as relevant to our daily lives as current electricity. Current electricity relates to the motion of electric charges and deals with many aspects of our lives, such as the operation of electrical devices in the home (e.g., lighting, toasters, refrigerators and lawnmowers). Current electricity plays an important part in communication through the telephone, television and computer, and manufacturing goods (most industries rely on electric motors). The focus of this section will be on current electricity and the development of an understanding of simple electric circuits as can be found in torches.

Electricity is very much a part of our daily lives, a fact that is brought home to us in times when power blackouts occur. Due to our daily experiences with electricity, scientific terms such as 'electricity', 'power', 'current' and 'energy' are part of our everyday language. However, the meaning given to such terms by people are often quite different to those accepted by the scientific community. This may lead to confusion in the minds of learners as electrical terms are introduced and discussed in the classroom.

High on the list of terms with several everyday meanings is the term 'electricity' itself. For example, a dictionary definition of electricity gives two different scientific views of the term:

1 a fundamental form of kinetic or potential energy created by the free or controlled movement of charged particles such as electrons, positrons, and ions.
2 electric current, especially when used as a source of power.

World English Dictionary 1999

The first definition equates electricity with a form of energy, while the second definition equates it with current. When three separate classes of primary school children were asked their ideas about electricity, there was a significant number of students who associated 'electricity' with 'power', as the following quotes indicate:

Electricity is power from water high in the mountains (Year 2)
Electricity is the power that is in the lights (Year 3)
Electricity is a power that makes things work like lights, stoves, microwaves, telephones and heaters (Year 4)

Is 'electricity' current, power or energy? From a scientific perspective, it is not any of these terms. The term 'electricity' only refers to a field of science or class of phenomena in the same way the words 'physics' or 'optics' are used. Therefore, in teaching about simple electric circuits it is best to refrain from using the term 'electricity' in your explanations as it can be interpreted in a number of ways by the students. However, it is important to discuss with students the different everyday uses of the term 'electricity.' When students use it in explanations of electrical phenomena, or when it is used in textbooks, videos and other resources, the teacher needs to ask, 'What meaning of electricity is being used here?'

The discussion of other terms relating to electricity will be undertaken within the context of the operation of a torch in the following sections. But before having this discussion, it is important to clarify your own understanding of electricity by completing activities 4.1 and 4.2.

Activity 4.1 Meanings given to electrical terms

- Write down what you understand the following terms to mean: electricity, voltage, current, energy and power. Compare your understandings with others.
- Look at different dictionary and science textbook definitions of electricity, voltage, current, energy and power. What understandings of these terms do you gain from reading these? Do any of them connect with your understandings? Do any of the definitions give conflicting views? Discuss and compare your findings in this activity with others.
- In the Australian national curriculum, the topic of electricity is to be taught in upper primary years with a particular science understanding for students, stated as 'Electrical circuits provide a means of transferring and transforming electricity'; and 'Energy from a variety of sources can be used to generate electricity' (ACARA, 2011). What meaning is being ascribed to the term 'electricity' in this statement?

Activity 4.2 Torch-drawing and making

- Draw a torch, label it and indicate on the drawing how you think the torch works.
- Using the equipment list in Appendix 4.1 (see p. 171), design and make a torch that works.
- Now, if necessary, draw another picture of the torch, showing how it works.
- Can you identify the differences between your first and second drawing of the torch? What are they? Do the changes indicate you have learnt something? If so, what?
- Has the experience raised some questions for you? What are they?

How a torch works

This section explains the operation of a torch using the concepts and understandings for primary teachers listed at the end of this chapter (see p. 169). The main components of a torch include a battery or series of batteries, connecting wires, a switch and a light bulb inside a parabolic mirror. Figure 4.1 shows two representations of a torch. One is a schematic view of the torch and the other is a symbolic representation, called a 'circuit diagram', which uses standard symbols for each component – refer to Appendix 4.2 on p. 171 for a list of symbols that represent other circuit components. The circuit diagram is the conventional representational form that is used by experts to communicate the layout of electrical systems. Besides being used for communication purposes, circuit diagrams are also used as a problem-solving and reasoning tool.

The torch will operate when the switch is closed. If you refer to the schematic view of the torch in Figure 4.1, you should notice that when the switch is closed, a complete (unbroken)

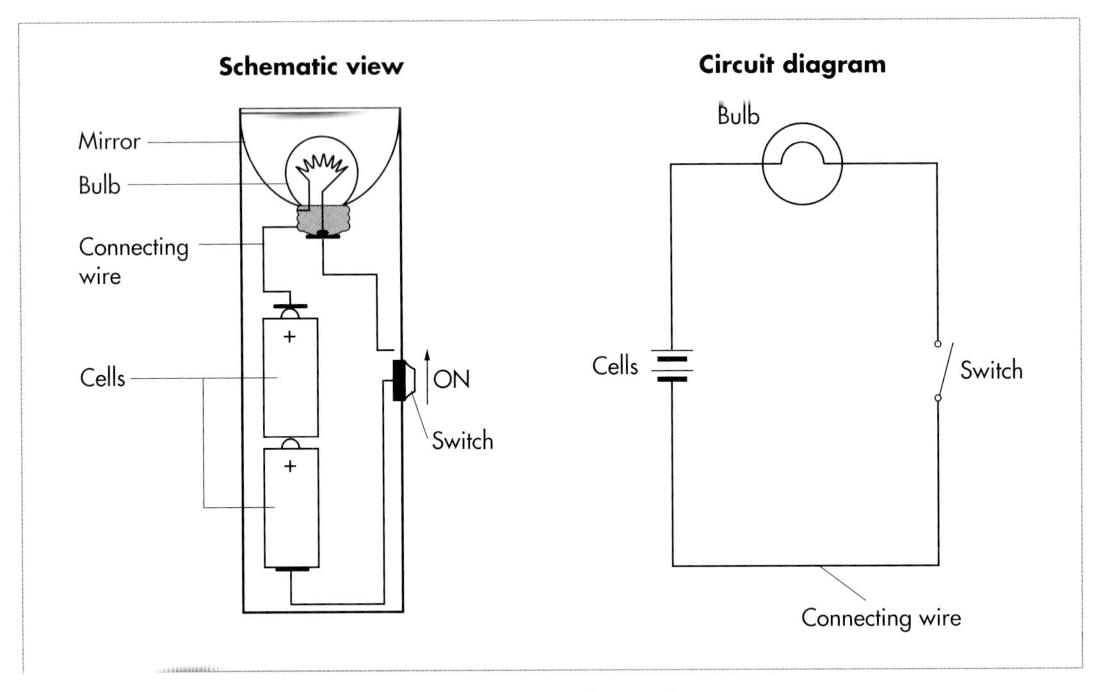

FIGURE 4.1 Schematic view and circuit diagram of a torch

pathway, or one complete loop, is created that includes the wire, batteries (also called cells), switch and the filament in the bulb (see how the pathway leads up the centre of the bulb, through the filament and out the side of the bulb). Such a pathway is called an 'electric circuit', or more precisely, a 'closed circuit'. It is this pathway that is re-represented in the symbolic representation of the circuit diagram.

The complete loop can have a different order for the positions of the switch, bulb and wire and the torch will still work. The batteries can be placed in various positions as well but must be connected so that the positive ends or the negative ends of the batteries are not in direct contact. You may have noticed this yourself when placing batteries into an electrical device. If they are not placed in the correct order the device will not operate. If the loop is disconnected at any point, then the arrangement is called an 'open circuit'. The bulb will not glow in an open circuit.

Activity 4.3 Exploring a torch

This activity involves you investigating the operation of a torch. For this exploration activity, you will require:

- a working torch and an old torch that can be pulled apart
- an old torch bulb or a household bulb
- a new battery and a 'dead' battery.

Dismantle the working torch. Determine for yourself the following. (If the observations are difficult to make then you may need to destroy an old torch.)

- A wire or metallic strip connects the battery(ies), bulb and switch in a complete loop.

- If there is more than one battery, how are they arranged? What voltage is written on the side of each battery?
- What is the mechanism of the switch for opening and closing the circuit?
- Look at the bulb. What labelling is present on the side? What does it mean?
- What is the placement of the bulb in relation to the curved mirror? Is this important? Why have a curved mirror?

Destroy an old torch bulb (a normal household bulb may be easier) to determine for yourself that:

- there is a conducting path from the base of the bulb through the filament to the side of the bulb
- the metallic base of the bulb is separated from the side of the bulb by insulating material.

Cut open a new and a dead battery lengthways (be careful undertaking this task as the substances inside are corrosive). What similarities and differences do you observe between the new and dead battery?

Electricity is electrons

Torches light up their surroundings – they transform the energy stored in the battery into light, which travels outwards from the torch and is reflected from objects back to our eyes. Energy is generated in the battery by a chemical reaction and is referred to as 'chemical energy'. This energy ultimately gets transformed into light and heat energy at the bulb. However, the battery and the bulb are usually some distance apart and so we need to consider how that energy gets transferred from the battery to the bulb. To understand how these energy changes can take place, we need to consider the requirement of having a closed circuit which contains a pathway made of metal.

The energy is carried by very tiny particles called electrons. These particles are part of all atoms that make up all substances, including the wires in the closed circuit. While electrons are part of all substances, they are generally not free to move away from the atoms they are attached to, except in some substances such as metals. Metals are part of a group of substances known as 'conductors' that contain electrons that are free to move; these electrons are referred to as 'free electrons'. Substances that contain electrons that are not free to move are called 'insulators'. A good example of an insulator is a material made out of plastic.

The function of the battery in the torch

To understand how the free electrons in the wire get energy from the battery, we need to consider what happens in the battery when a closed circuit is formed. The battery contains different chemicals that will undergo a chemical reaction when the closed circuit is formed. During the chemical reaction, electrons move away from one of the substances to the other via the closed circuit. Therefore, the chemical reaction in the battery creates movement of the free electrons in the closed circuit. The battery may be considered to provide the push to move the electrons. However, if the closed circuit gets broken, the battery no longer provides a push to the electrons as the chemical reaction ceases.

The chemical reaction provides a push on the free electrons in the wire by creating an electric field. We can think about how an electron behaves in an electric field by likening it to

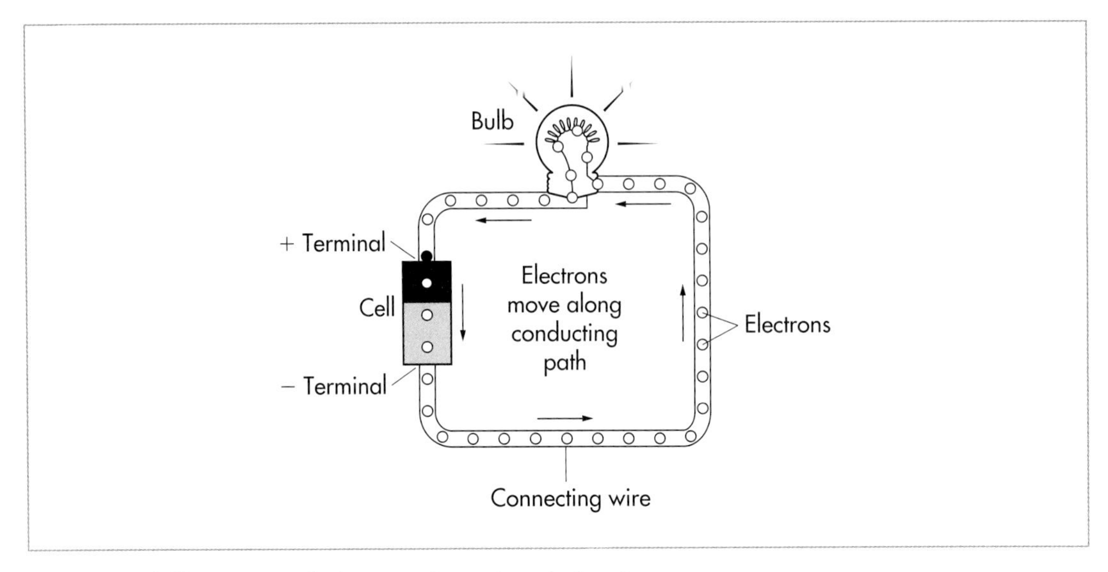

FIGURE 4.2 Movement of electrons in a closed circuit

how objects behave in a gravitational field, or gravity. The Earth creates a gravitational field that will attract objects to its surface, even though the objects may be kilometres away. Imagine if you could turn off gravity by using a switch. This would result in objects being suspended in the air. Switching on gravity would result in all the suspended objects moving at once. While we cannot turn off gravity, we can turn off electric fields. Once the closed circuit is formed, the chemical reaction begins, which creates an electric field all along the circuit resulting in all the free electrons moving simultaneously. Breaking the closed circuit stops the chemical reaction, ceasing the electric field and the free electrons.

The electric field is created in only one direction. This direction is from the negative end, or terminal, of the battery (flat end) around the closed circuit, towards the positive terminal (end with a bump). The electrons are pushed in the direction of the electric field (see Figure 4.2).

The idea that a battery creates an electric field in one direction in the wire explains the need to place two batteries into a circuit (see Figure 4.3) in a particular way. The batteries should be connected so that the positive terminal of one battery connects to the negative terminal of the other. In this way, the electric fields that are created by each battery result in the electrons moving in the one direction. In the arrangement of the batteries shown on the right-hand side of Figure 4.3, each battery will create an electric field in opposite directions resulting in no movement of the free electrons.

The voltage of the battery gives a measure of the strength of the electric field that pushes the free electrons around the closed circuit. Therefore, a 6-volt battery will give four times the push on the free electrons than a 1.5-volt battery. This has the effect of not only increasing the flow of electrons in the closed circuit, but also the amount of motion energy each free electron receives. The flow of electrons is called an 'electric current'. Therefore, the greater the voltage of a battery in a closed circuit, the greater will be the electric current.

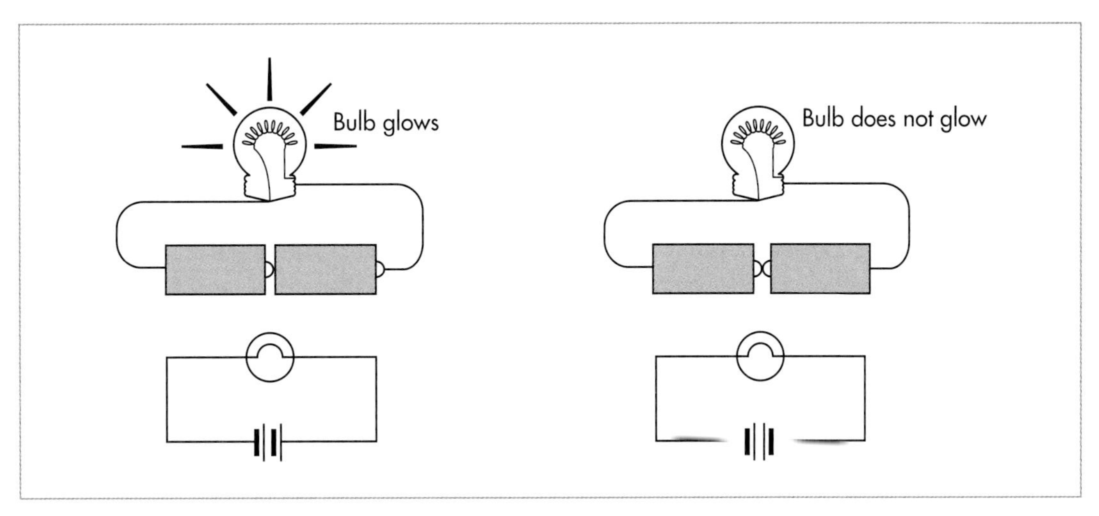

FIGURE 4.3 Arrangements for two batteries in a torch circuit

Energy transformations and transfers in the torch

The section above explained how the battery is responsible for the movement of free electrons in a closed circuit. Using this information, we can explain the energy transformations that occur in the torch. The energy in the chemicals of the battery (chemical energy) is transformed into motion energy of the electrons. As they speed around the closed circuit, they collide with the atoms in the filament in the bulb, thus transferring their motion energy to the atoms. This causes the atoms to vibrate. The vibrating atoms transform their motion energy into heat and light energy.

It should be noted that all the energy the free electrons receive is not transferred to the atoms in the bulb. The electrons also collide with atoms in the wire and the battery, giving these atoms motion energy. This results in the wire and the battery heating up slightly. It has been mentioned above that the voltage of the battery gives a measure of the push, and the amount of motion energy, given to the electrons. This means that if the batteries in the torch circuit were to be replaced by batteries of larger voltage, then the bulb will be brighter, as more energy is being transformed. However, in practice, this may result in the bulb blowing, as bulbs are designed for specific voltages to operate correctly.

Electric current, resistance and power in the torch

The voltage of the battery, being related to the electric field, affects the size of the current; the greater the battery voltage, the greater the resultant current in the circuit. The size of the current is also affected by the ease with which the free electrons are able to move in the circuit. In the torch circuit, the filament does not provide an easy path for the free electrons as the wire of which it is made is much narrower than the other wires in the circuit. Because of the narrow path, the free electrons have most of their collisions in the filament, which means most of the energy coming from the battery is transformed in the filament.

The property of a substance that restricts the path of free electrons moving through it is called the 'resistance'. A closed circuit that contains a high resistance will have fewer moving free electrons than a circuit with a low resistance. In other words, a closed circuit with a high resistance will have a small current, while a closed circuit with a low resistance will have a large current. In general, the size of the current in the circuit depends on the resistance in the circuit.

A quantity that is used when discussing electric circuits, and which is used in everyday speech, is 'power'. Power relates to the rate at which energy is transformed in the electric circuit. The unit of power is the watt, where 1 watt represents 1 joule of energy being transformed per second. For example, a household light bulb rated at 100 watts will transform 100 joules of electrical energy into light and heat energy for every second that the light bulb is operating.

Energy, voltage, current, resistance and power: the connections

The torch circuit explanations given above have referred to such quantities as energy, voltage, current, resistance and power. All these quantities are related. These relationships can be summarised as:

- The battery voltage determines the amount of energy given to each electron; the higher the voltage, the greater the energy.
- The battery voltage and resistance in the circuit determine the size of the electric current; greater currents occur with larger battery voltages and/or less circuit resistances.
- The amount of energy transformed at the bulb depends on the amount of resistance the bulb has compared to the rest of the circuit. In a torch circuit, most of the free electrons' energy gets transformed in the bulb as it has most of the circuit resistance.
- The rate at which energy is transformed in the electric circuit is the power.

Sending out the light beam in the torch

The bulb in the torch emits light in all directions. However, a directed beam is needed for an effective torch. The mirror produces this directed beam by reflecting all light that reaches it from the filament in much the same direction. To achieve this, the filament is placed at the point known as the focal point of the mirror. Figure 4.4 illustrates what happens – notice that at each point at which the light hits the mirror, it is reflected in a symmetrical way so that the labelled angles are equal. (These mirrors can also be used in reverse, concentrating parallel beams of light onto one point; some telescopes incorporate this idea.)

Different types of circuits: series and parallel

The torch circuit is only one type of simple circuit that can be made with wires, switches, batteries and bulbs. There are two types of circuits, described as series circuits and parallel circuits. The torch circuit is an example of a series circuit in that the conducting path that includes the batteries, bulb, wires and switch makes a single loop. One can have any number of batteries, bulbs, wires and switches. In a series circuit, the only condition is that there is only one conducting loop. The parallel circuit is one that contains more than one conducting loop. Consider the two circuits shown in Figure 4.5. Both circuits contain a battery, wire and two

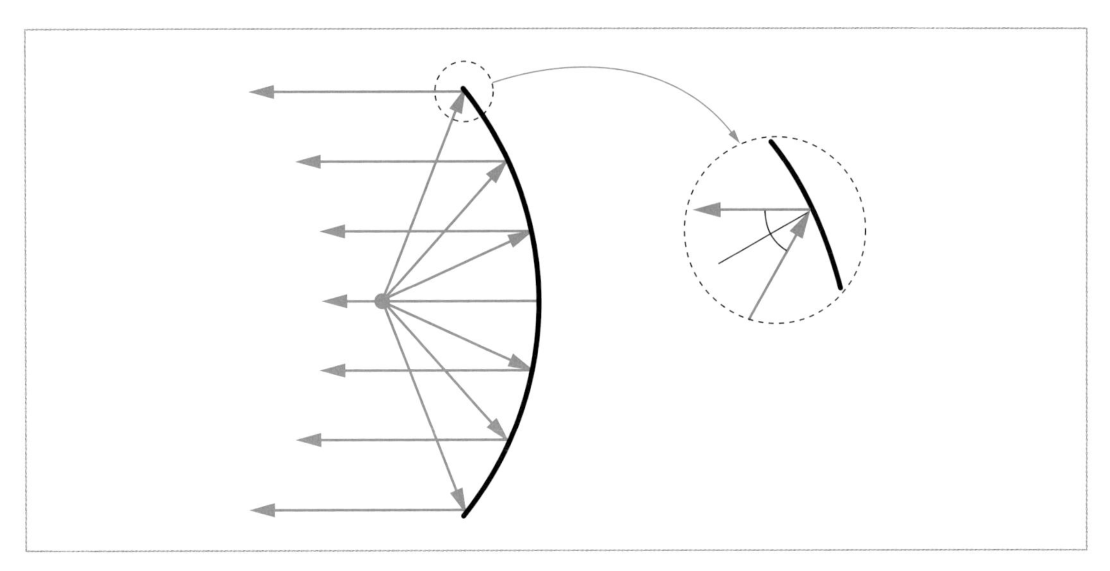

FIGURE 4.4 Reflected light from a torch mirror

bulbs. However, the circuit on the left contains one conducting loop, and so is a series circuit, while the circuit on the right contains two conducting loops, and is therefore a parallel circuit.

Even though the two circuits shown in Figure 4.5 contain the same components, there will be different outcomes as to the brightness of the bulbs. The following observations are made for these circuits:

- The bulbs in the series circuit are less bright than the bulb in the torch circuit; the battery will last longer in the series circuit than the one in the torch circuit.
- The bulbs in the parallel circuit are the same brightness as the bulb in the torch circuit; the battery will not last as long as the one in the torch circuit.

The two-bulb series circuit observations can be explained by the existence of one continuous conducting loop. The battery supplies the same amount of energy, as determined by the battery voltage, to each of the electrons in this loop. In the torch circuit, most of this energy is transformed in the single bulb (some is transformed in the wires), but in the

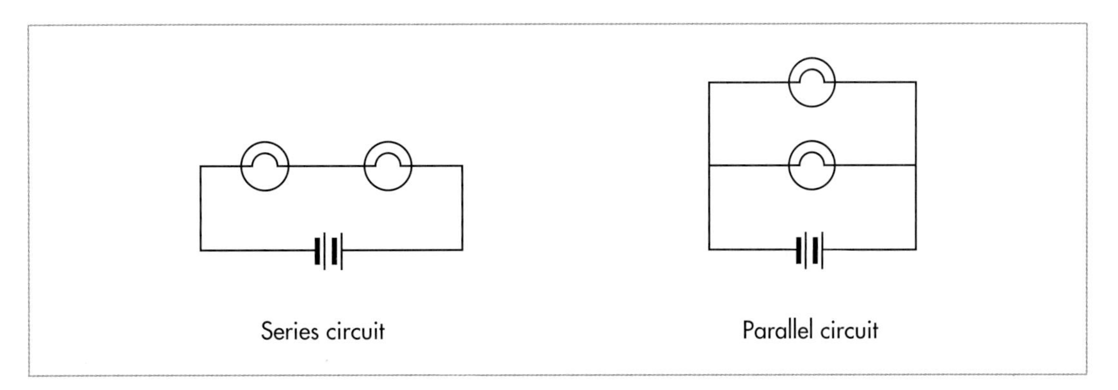

FIGURE 4.5 Series and parallel circuits

two-bulb series circuit, most of the energy will be shared among both bulbs (in any loop, the energy supplied by the battery is transformed in all resistances in proportion to their size). In addition, the two bulbs nearly double the resistance in the loop and so the current will be halved (half as many electrons flowing per second). Therefore, the dual effects of less current and sharing of energy will result in each bulb in the two-bulb series circuit receiving about one-quarter the energy received by the torch bulb. As a result, each bulb in the two-bulb series circuit will be less bright than the torch bulb. The smaller current means less electrons flowing per second to the battery and so the battery's chemical reaction will run slower (and for longer) in the two-bulb series circuit than in the torch circuit. This means that the two-bulb series circuit battery will last longer.

The parallel circuit contains two battery loops in which each loop contains one battery and one bulb. The battery sets up electric fields in both loops. If one considers that each loop operates independently of the other, then the parallel circuit is just two torch circuits. For each loop we apply the rule that in any loop, the energy supplied by the battery is transformed in all resistances in proportion to their size. As the loop contains only one battery and one bulb, then the current and bulb voltage will be the same as in the torch loop. Therefore, the bulbs in the parallel circuit will be just as bright as the bulb in the torch circuit. However, the difference between the parallel circuit and the torch circuit is that the current flowing into and out of the battery of the parallel circuit will be twice that of the torch circuit. The one wire from the battery feeds into two branches, each containing a bulb, and then feeds back into the one wire back to the battery. The free electrons from the battery separate into both branches of the parallel circuit. This is similar to how water flows into branched pipes of an irrigation system. The greater current in the parallel circuit will mean the battery in this circuit will die more quickly than the battery in the torch circuit.

Examples of household appliances that employ series and/or parallel circuits are Christmas tree lights and electric blankets. Christmas tree lights are manufactured as one of two types of circuits. In one type all the bulbs are in series, and in the other type all the bulbs are in parallel. Can you think of any disadvantages or advantages of these types? Electric blankets are made with two resistance wires (one wire is longer than the other) sewn into the blanket. A switch, with different heat settings, is connected to the wires in such a way that one setting connects one of the resistance wires in a complete circuit, another setting connects the other resistance wire to make a complete circuit, and another setting connects both wires to make a parallel circuit. Which circuit arrangement would match each heat setting? Can you draw a circuit diagram that shows the two resistance wires and the switching mechanism?

A

Activity 4.4 Concept map

You are now in a position to discuss the terms listed in Activity 4.1 ('electricity', 'current', 'voltage', 'energy' and 'power'; see p. 145). How does your understanding of these terms differ now?

Construct a concept map using each of the terms 'battery', 'wires', 'switch', 'circuit', 'open circuit', 'closed circuit', 'series circuit', 'parallel circuit', 'electron', 'voltage', 'current', 'ampere', 'voltage', 'power', 'resistance', 'energy', 'chemical energy' and 'light energy'.

Activity 4.5 Testing your understanding of electric circuits

For this activity you will need to access the journal article by Lee and Law (2001). It contains a test (p. 130) designed to elicit students' alternative conceptions of electric circuits. It consists of nine problems drawn from published studies that were known to elicit alternative conceptions.

Use the ideas discussed in this section to answer the first five questions. Keep in mind that if there are any changes to a circuit in terms of resistance, then you need to consider what changes occur in current and voltage over the whole circuit.

The answers to these questions are in Appendix 4.3 (see p. 171).

Students' understandings of electricity

There has been considerable research into children's understanding of electricity – for example, Georghiades (2004), Jaakkola and Nurmi (2008), Keinonen (2007), Osborne et al. (1990), Osborne and Freyberg (1985), Pilatou and Stavridou (2004) and Shipstone (1985). There have also been studies that have explored preservice and practising teachers' understandings of electricity – for example, Buck et al. (2007), Mahapatra (2006), Mulhall, McKittrick and Gunstone (2001) and Pardham and Bano (2001). This research indicates that people have a view that there is a source, such as a battery, and a consumer, such as a bulb or a motor. From this perspective, Shipstone (1985) suggests that people believe that

> Electricity, current, power, volts, energy, 'juice', or whatever, is stored in the source and flows to the load where it is consumed. The battery is usually seen as the active agent or 'giver' in this process, with the load being the 'receiver', but it is also common to find the load regarded as the active agent, as a 'taker', drawing what it 'needs' from the battery.
>
> Maixhle 1981, cited in Shipstone 1985, p. 35

For many who have received instruction in electricity, the thing that gets used up is often associated with electric current. For a circuit that contains a battery and a bulb, the following models were found (see Figure 4.6).

1 *Clashing currents model* – The current from each end of the cell clashes in the bulb to provide the light (see A in Figure 4.6).
2 *Consumption model* – Some of the current from one end of the cell is lost as it passes through the bulb (see B in Figure 4.6).
3 *Source-sink model* – The second wire is unnecessary (see C in Figure 4.6) as current from one end of the battery is used up in the bulb.
4 *Same current model* – The current remains uniformly the same at all points in the circuit (see D in Figure 4.6).

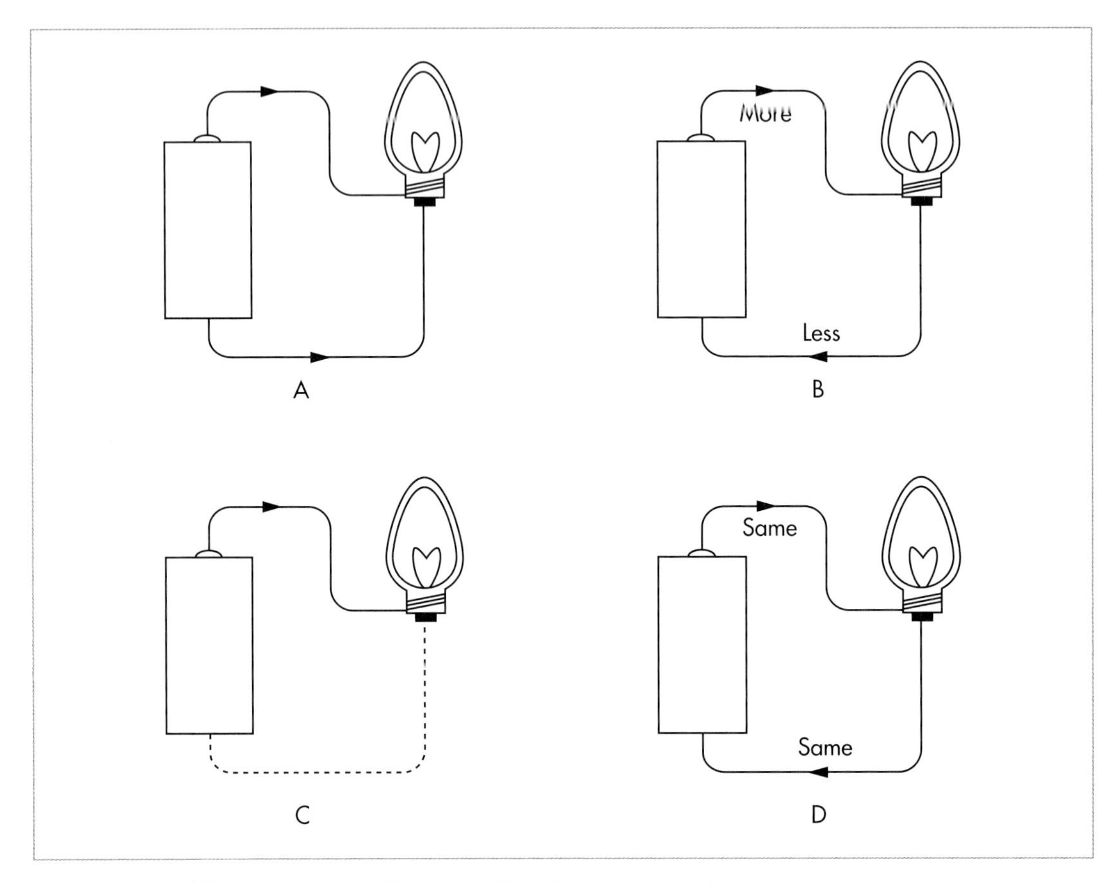

FIGURE 4.6 Different models of how an electric current works

Activity 4.6 Models of electricity transfer and electric current

This activity involves an evaluation of different models of electricity transfer and electric current.

- Which model best represents the way you think about how an electrical current works?
- Devise tests to check each of the ideas listed.
- Check them out.
- Which model do you think is best now?
- Ask other people what they understand by the term 'electrical current' and how they think it works.
- What similarities or differences can you find in the way other people think about electrical current and the way you think about it?

CASE STUDY

Applying an interactive teaching approach

This section details the different stages in an interactive teaching approach adopted by a teacher named Catherine with her years 3 and 4 composite class (eight- to nine-year-old girls and boys). In teaching electricity, Catherine used the context of torches, an everyday piece of technology with which the children were familiar.

Preparation: choosing the approach

Catherine decided she wanted to use an interactive teaching approach (Biddulph and Osborne 1984, see the online Appendix 4.4; Fleer, Jane and Hardy 2007). She was aware it was an approach that could incorporate other teaching approaches, such as process, discovery and transmission (Biddulph and Osborne 1984; Fleer, Jane and Hardy 2007; Kirkwood 1991) within its framework.

Appendix 4.4 Interactive teaching approach

An interactive teaching approach is an approach that uses focus activities to elicit children's questions, which are then investigated to find answers. The approach is called interactive to convey 'the sense of an interchange of talk among people who respect each other's ideas. From a teacher's point of view, this begins with a genuine desire to know what a child thinks (and why)' (Biddulph and Osborne 1984, p. 13). Teachers are expected to

- identify children's present ideas
- provide children with stimulating experiences either to confront and explore those ideas or as a basis for developing ideas
- help children develop, clarify, modify and extend their ideas through seeking answers to questions or through checking proposed answers
- encourage children to reflect critically on how they came by an idea and whether it is a sensible and useful one
- assist children to communicate their developing ideas
- help children realise their explanations of why things behave the way they do and that they are frequently neither 'right' nor 'wrong'; rather, they are consistent with the evidence or inconsistent, useful or less useful, plausible or implausible, intelligible or unintelligible
- convey to children awareness that their genuine ideas are valued.

Biddulph and Osbourne 1984, p. 13

The various elements of the interactive teaching approach described are consistent with the changes in emphases in conceptual change processes outlined in Chapter 1 (see the section 'How effective is constructivist teaching?' on p. 33). The types of student engagement with science embedded in this approach map very well with the discursive practices of the science community.

Preparation: choosing the topic

Catherine had taught environmental and biological topics, but had avoided teaching physical science topics. She decided the time had come for her to overcome her fear of teaching in areas where she felt she had insufficient background knowledge. She was conscious that the lack of opportunities for learning in science had a negative impact on children's growth and development and on their life choices. The professional development course in which she had been introduced to the interactive teaching approach had focused on the topic of electricity. With the support of one of the authors, Valda, she decided to teach electricity in the context of torches, since they were part of the lives of the children she was teaching.

Exploration: beginning the topic

Before actually teaching the topic, Catherine did a lot of reading to further develop her own understandings. She planned the topic and sought resources to support her plan: for example, lots of books and posters from libraries and resource centres, which she put up on the walls around the classroom. Table 4.1 provides an overview of the topic as a whole.

M&E

At the start of the topic, Catherine called the class together on a mat and began signalling in Morse code using a torch. She asked the children if they knew what she was doing. It was not a phenomenon any of them had experienced, so Catherine talked with them about how people used to communicate across distances before the invention of the telephone. She put up a poster of Morse code (see the online Appendix 4.5). The children had to

TABLE 4.1 Overview of Catherine's torch unit

LESSON	CONTENTS
1	Introduction to electricity unit with a torch flashing Morse code
2	Pretorch drawing. Designing and making a torch. Post-torch drawing. Questions arising
3	Investigating questions; for example, completing or redesigning torches, circuit making
4	Electrician's or scientist's circuit language and symbols. More circuit making, including switches
5	Incorporating ammeters into circuits
6	Investigating conductors and insulators
7	Visit from an electrician parent
8	Science party: circuit game, spelling, science game, practical assessment, magic show, food

Appendix 4.5
Morse code

work out what it was she was communicating. They were very excited when they discovered it was one of their names being transmitted by the light energy. This initial activity was designed to set the context and gain students interest (refer to the '"Hot" conceptual considerations: student engagement' section of Chapter 1, on p. 25).

Exploration: focusing activity

In order to establish the children's prior ideas, Catherine asked them how they thought a torch worked and suggested they each draw a picture of a torch, labelling all the parts and incorporating the torch's workings, a strategy she had used before with the class. Catherine reiterated aspects of drawing and labelling diagrams that she had gone through with them on previous occasions.

She stressed the importance of clarity of drawing and labelling to communicate their understanding to another person. Annotated drawing has been found to be very useful in many facets of science learning (Georghiades 2004), as well as in other curriculum areas. It is one way of eliciting children's preconceptions (Glynn 1997). Georghiades (2004), in working with Year 5 students in the topic of electricity, found that annotated drawings were useful in enabling students to represent their understanding. Children visually representing their ideas form an important aspect of a multimodal approach to teaching and learning science (Kress et al. 2001; Lemke 2004).

Following that introduction, Catherine provided materials (see Appendix 4.1 on p. 171) for the children to design and make their own torches, either on their own or in groups. A systematic description and analysis of classroom grouping practices can be found in Baines, Blatchford and Kutnick (2003). These researchers are part of a major project called SPR*in*G (Social Pedagogic Research into Group-work; see http://www.spring-project.org.uk), the aim of which is to enhance the learning potential of students working in classrooms. Strategies to support group work in primary classrooms, especially cooperative learning, are also explored in Galton and Williamson (1992). In addition, each of the units within *Primary Connections* (AAS 2005) explicitly refers to group work within their introductions and appendices.

The children were allowed to develop and make their own designs. Catherine went around the groups and helped where necessary. During these interactions with the children, Catherine focused on the children's ideas. This included getting children to articulate their own ideas, communicating other children's ideas, challenging children's ideas, getting children to challenge each other's ideas in non-threatening ways, and seeing if children could link what they were doing to what they already knew and to what they had previously experienced. She also acted as a naive investigator herself, asking questions like 'I wonder why ...?' and 'I wonder what would happen if ...?'. During these interchanges, a lot of valuable information surfaced regarding children's considerable experiences with electricity, such as the girl who had her own mobile phone, with its battery, and the boys who had considerable experience with putting circuits together during their playing with, for example, Lego.

By the end of the torch making, some interesting designs had developed. Catherine was very surprised to find that one boy in particular, who appeared to have little aptitude for learning in any other curriculum area, designed in a very short time an elegant torch that worked. Without this experience of seeing that child's capability in the torch making, Catherine was sure she would have finished the year of teaching still unaware of his abilities and knowledge in this area. For Catherine, it was a convincing argument for continuing to teach physical science in her primary school classroom.

After this challenge, children were asked once again to draw how they thought a torch worked (see Figure 4.7 for

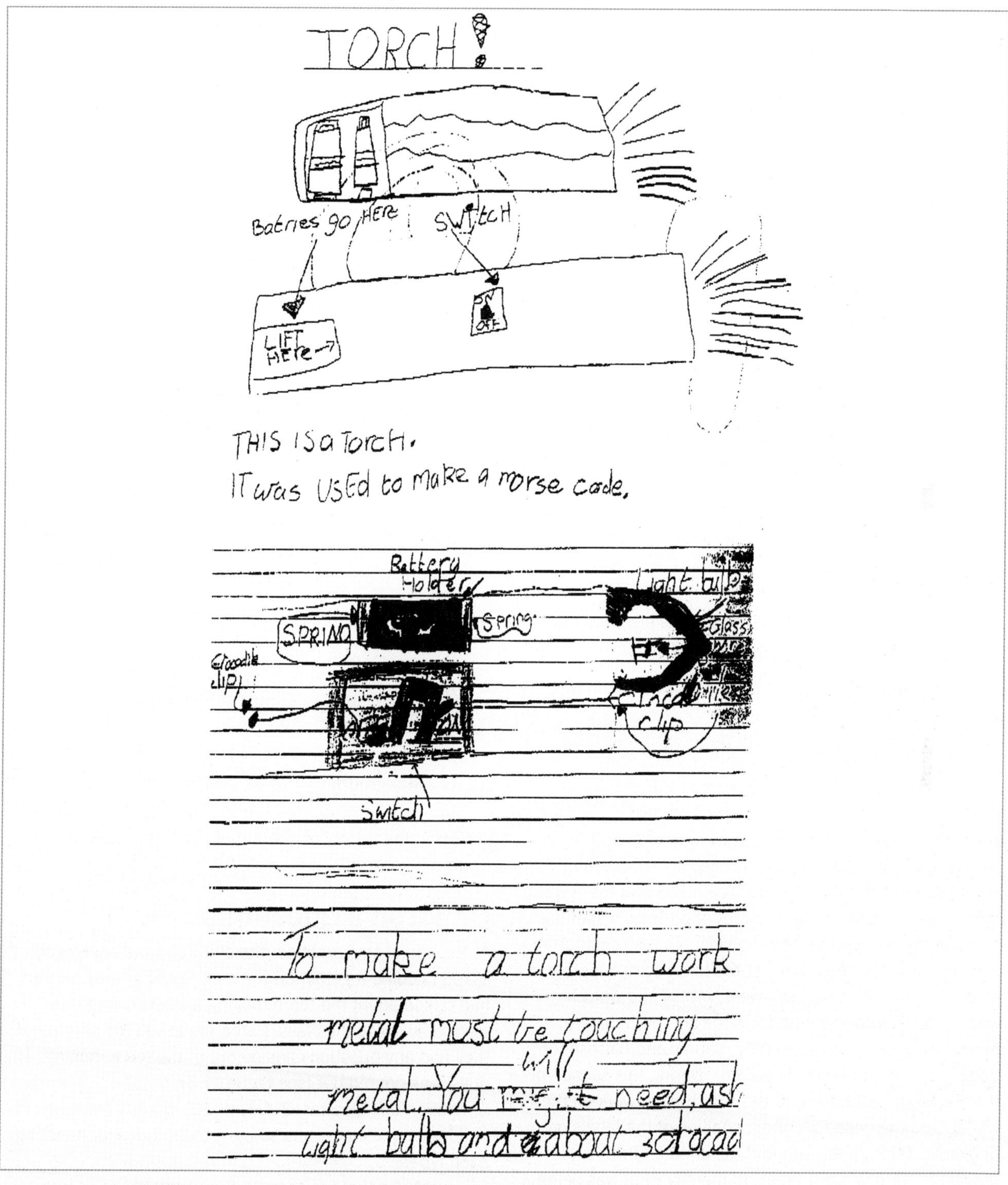

FIGURE 4.7 One child's pre- and post-torch drawings

What happens if you have something covered with plastic and the power point?
Solar power – what is it?
Why don't birds standing on power lines get electrocuted?
Why can't people go under trees when it's lightning?
What makes things blow up?
How do batteries work?
What's inside a battery?
How does a wire work?
How does electricity get through things?
What things will electricity go through?
What things won't it go through?
What is inside a globe that makes it light up?
Can I make three globes light up at once?
How does lightning/electricity work?
What is electricity?
Is electricity and lightning the same?
Who made electricity?
Was it the person who made the first light globe?
Is electricity from a battery the same as electricity from the switch?
How much power is there in one volt?
Why do we have to get out of the swimming pool when it's stormy?
What is a volt?

FIGURE 4.8 Torch questions from the years 3 and 4 composite class (eight- to nine-year-old girls and boys)

one child's pre- and post-torch drawings). Pre- and post-task or topic drawing can be used very successfully as an assessment strategy. It should be noted that there was a significant increase in the accuracy of children's spelling between the first and second drawings. In their first torch drawing, children spelt many of the torch components in a variety of ways (for example, wiers, wiyess and woes for wires; glob, bowlb and bolb for globe and bulb; batere, bateries, batters, battries, batery, batires and battrise for battery; swich and swith for switch). During this section of the teaching, Catherine emphasised through work on the whiteboard the need for correct spelling of the labels on diagrams. Throughout, she told the children that readers needed accurately spelt labels to identify what was being communicated through the drawings. In their final drawing, when most of the children spelt most of the components correctly, the variety of alternative spellings was not evident.

Children's questions

At the end of this exercise, the children were once again asked to come together on the mat. Some of the children had talked about the torch making and explained their torch drawing to the class. Catherine asked the children if they had any questions arising out of the torch making. The questions poured out (see Figure 4.8).

Harlen and Qualter (2004, p. 30) identify five types of children's questions, and suggest techniques for handling each type:

1 questions that are comments or expressions of interest arising from some stimulus – teachers need to acknowledge the stimulus

2 questions of a philosophical kind – teachers may request that the child explain their question
3 questions asking for factual information – teachers may supply the answer directly or suggest a source where the child might find it, or undertake to find it at a later time
4 questions that are investigable by the children – the time devoted to enabling children to investigate their question rather than supplying them with a direct answer not only means they will learn more than the simple answer, but also means they learn to develop the skills and attributes to think and work scientifically
5 questions that require a complex answer – these may be philosophical questions that need to be clarified with the children and which can often be turned around so that they become questions that are investigable.

Specific investigations: electricity sources and receivers

In the next session, the children began investigating some of these questions. Initially, Catherine decided to focus on providing an experience with wires, cells and light bulbs to develop the children's expertise in constructing circuits. While they were on the mat, she introduced them to electricians' and scientists' shorthand ways of representing switches, light bulbs and so on (see Appendix 4.2 on p. 171).

Some of the questions referred to receivers of energy, such as a light bulb. Others related to the source of electrical energy, for example, the cell, or battery, as it is often called in everyday language. Catherine decided to answer these questions by getting Valda to carry out a demonstration in front of the whole class. A dead cell, a hacksaw, safety glasses, a pair of rubber gloves and plenty

Activity 4.7 Drawing as an assessment strategy

- The use of pre- and post-drawings is a valuable assessment strategy. Devise a set of criteria for assessing a set of torch drawings.
- Compare your scheme with one drawn up by some student teachers (see Appendix 4.6 on p. 172). The set of criteria could be used as a formative tool to measure the extent of individual children's knowledge and represent assessment for learning (refer to the 'Assessment for student learning' section of Chapter 1 on p. 35). Alternatively, the change in children's drawings between pre- and post-drawings could be used as a summative assessment strategy at the end of a topic and represent assessment of learning.

A

Activity 4.8 Questions: their categorisation and links to curriculum

- In what ways do the children's questions differ from yours (see your responses to Activity 4.1)?
- Can you categorise these questions according to Harlen and Qualter's five categories?
- Which questions require reference to books, videos or other sources of information?
- Which questions are investigable?
- Which questions are difficult to answer but can be turned around and formed into questions that are more readily investigable?
- Can you think of other ways to categorise the questions?
- What understandings might develop from the investigation of these questions?
- Where do these understandings fit into your science curriculum?

of paper provided the materials for her to carefully cut open a cell for children to see what was inside. What children saw was quite different from what many had anticipated. Catherine told them the name and the purpose of each component and wrote them on the whiteboard. During this process, Catherine explained how some things in everyday life, such as a battery, had different names in science, such as cell. This highlights a 'learning demand' for students as they negotiate the differences between the everyday and scientific meanings of terms (Scott, Asoko and Leach 2007). Catherine explained that, to scientists and electricians, a battery was two or more cells connected together.

Specific investigations: conductors and insulators

Some materials allow electric charges (electrons) to be transferred through them (conductors); some do not (insulators). After the torch making, the putting together of circuits and the cutting open of a cell, the children in Catherine's class enjoyed taking objects from their classroom environment and testing them to see whether they conducted electricity. At this point, Catherine decided it was an appropriate time for children to learn to connect an ammeter into the circuits they now felt confident in making. Incorporating the ammeter into the circuits enabled children to see more easily if there was current flowing through or not; a light bulb had proved to be a less readily recognisable source of evidence. The children had great fun trying not only the group of objects Catherine had brought together for them to test, but also almost any moveable thing within the classroom. This experience provided the children with lots of opportunities to predict, test, classify and then generalise. Many decided that things containing metal conducted, whereas things made of plastic did not.

Providing explanations for electric circuits

A particle-based approach to explaining electric circuits to primary school students is advocated by Summers, Kruger and Mant (1998), in which electricity is associated with particles or, more specifically, electrons. Other ideas associated with this concept include:

> Electric current consists of electrons moving in one direction.
> The battery provides the push to make the electrons move.
> The battery voltage is a measure of the push.
> The electrons are already in the wire and the components of the circuit (the battery is not the source of electricity).
>
> Summers, Kruger and Mant 1998, p. 168

R

These authors recognise that such an explicit particulate approach, where current is identified as moving electrons and the battery as supplying the push, is not common in published materials for primary science. However, they have found evidence in their research that 'supports the view that primary school children readily acquire these ideas and use them confidently when talking about electricity and circuits' (Summers, Kruger and Mant 1998, p. 169). More details about this approach can be found in a teaching resource developed by these researchers (Summers, Kruger and Mant 1997).

The initial activity of making a torch develops in the students an understanding that a complete circuit is required for the torch to operate effectively. The necessity for a complete circuit brings into conflict a source-sink model (see Figure 4.6 on p. 154) in which there needs to be only a single connection between the battery and the bulb for the bulb to light up. Another activity that brings into conflict a source-sink model involves providing each student with a battery, bulb, bulb holder and one piece of wire. Students are then given the task of finding a way to make the bulb light up. (Try this out for yourself. How many ways can the bulb be lit?)

A teaching sequence described in Summers, Kruger and Mant (1997) began with an initial activity involving the building of complete circuits to make the bulb light up. Discussions followed as to the children's explanations of the necessity for a closed circuit. The teacher then presented an alternative view to the students' explanations, namely the scientist's view that electricity consists of electrons (already present in the wire) and a bulb, and that a current occurs when the electrons move in one direction. These ideas were shown diagrammatically in chart form (see Figure 4.9).

The term 'electricity' is associated specifically with electrons in this teaching sequence. Having defined electricity in one particular way, importance should be placed by the teacher on maintaining this definition and not changing to other everyday uses of the word, such as 'energy' or 'power'.

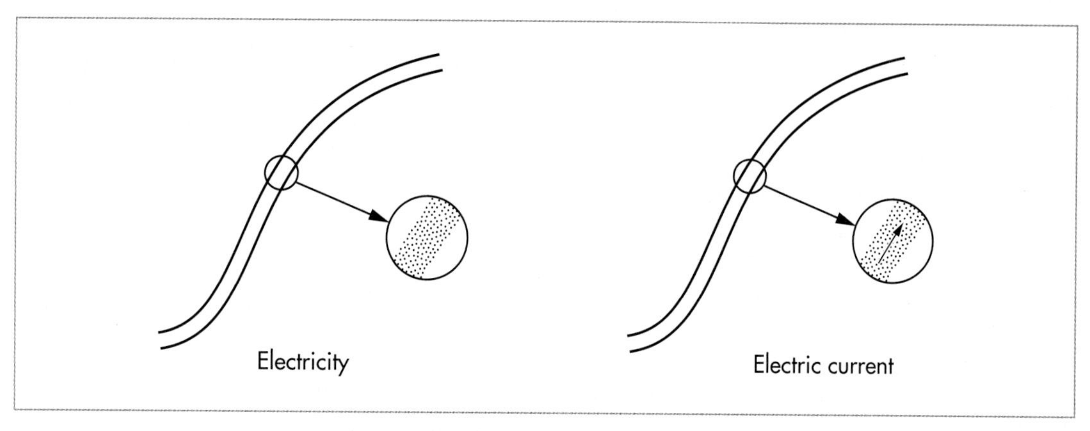

FIGURE 4.9 A view of electrons moving in a wire

Source: Summers, Kruger and Mant 1998, p. 158

Different-sized batteries were then given to the children for them to examine. The voltage of the battery was explained to the children as a measure of the push the battery applies to the electrons in the wire. A larger voltage produces a greater push. To bring into conflict the consumption model (see Figure 4.6 on p. 154), the teacher introduced an ammeter to test the size of the current in different parts of the circuit. Apart from tests with the ammeter, the teacher also introduced devices (for example, buzzers) that only allow electrons to pass through them in one direction. These demonstrations reinforce the view that electrons only travel in one direction from the push applied by the battery. Another reason for using other devices (not necessarily ones that allow current in only one direction) is to provide a wider range of experience so that children do not view the light bulb as a unique object. Small motors are examples of such a device.

To reinforce and develop the ideas already introduced, the teacher then presented the students with a bicycle chain analogy. In this analogy, as with all analogies, links are made between the analogue and the target phenomenon. The analogues of the links in the chain, the pedals and the wheel are compared, respectively, to the target electrons, battery and the bulb (see Figure 4.10). From this analogy, the moving links represent the electric current and the push on the links by the pedals (from the cyclist) represents the push on the electrons by the battery. This analogy is useful in helping to convince a child that

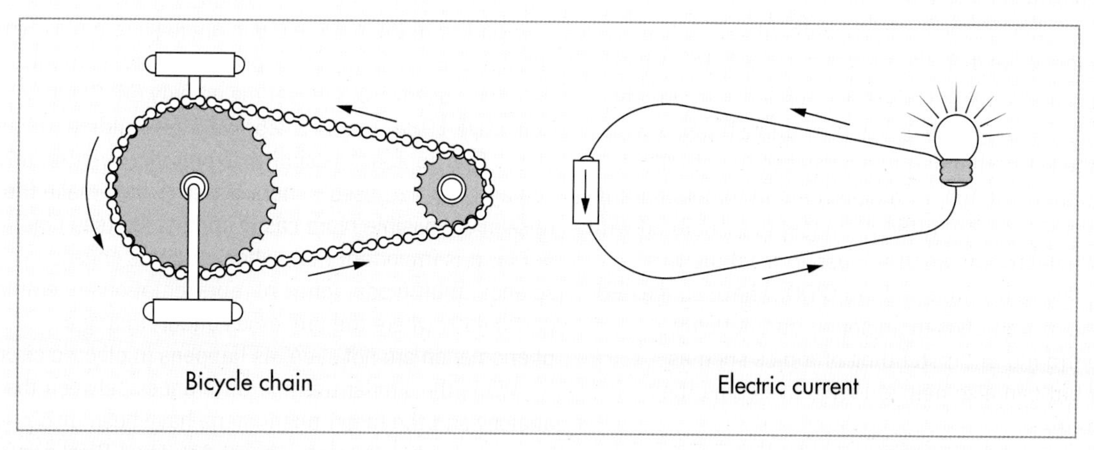

FIGURE 4.10 Bicycle chain analogy

Source: Summers, Kruger and Mant 1998, p. 161

current is not used up (consumed) by the bulb in a simple circuit.

While this analogy is successful in presenting the view that current is conserved in the circuit, children still cling strongly to the idea that something is used up in the circuit (Summers, Kruger and Mant 1997). After all, why does the battery run flat and why do we have to pay for electricity? It is energy transformation, not current, that accounts for the something that is consumed. Therefore, energy transformation and current conservation need to be made explicit in the introductory teaching of electricity. The bicycle chain analogy should then be extended to include the cyclist. This 'human' energy goes into turning the back wheel (motion energy). As the cyclist pushes on the pedals, he or she becomes fatigued and loses energy.

The bicycle chain analogy accounts for the following:

- The bulb does not glow if there is a break anywhere in the circuit – the cogwheel does not spin if there is a break anywhere in the chain.
- The current is conserved (same everywhere in the circuit) – no links get lost and the number of links moving past a particular part of the chain is uniformly the same all along the chain.
- The bulb lights almost instantaneously when the battery is connected – the wheel turns at the same time there is a push on the pedals.
- Energy travels from one location to another for the electric circuit and the bicycle. However, in the electric circuit, energy is transformed whereas in the bicycle, energy is transferred.
- A switch turns the bulb off – the brake stops the wheel. However, whereas the bulb turns off almost instantaneously when the switch is opened, the wheel may take some time to stop when the brake is applied.

The use of analogies helps in providing students with a view of the world that can't be accessed; electricity is an unseen phenomenon. Analogies help in understanding the relationships of the different concepts associated with the topic of electricity, and their use in teaching this topic is widespread. Chiu and Lin (2005) have suggested that the use of well-chosen analogies can not only promote profound understanding of topics such as electricity but can also help students overcome their misconceptions.

It is important to note that the use of analogies may not be an effective pedagogic tool for students' understanding if attention is not given by the teacher to use of an analogue that students have experienced, generally, in their daily lives. From this perspective, the bicycle chain analogy is an appropriate one as most children have had experience riding bicycles.

Confusion can surface for children when they inadvertently connect unlike aspects that an analogue has with a target concept. It is recommended that teachers explore with children the similarities and differences between the concept being explored and the analogue it is being linked to. In the bicycle chain analogy given above, some of the differences between the analogue and the target are:

- For most children's bicycles, one doesn't need to always keep pedalling for the back wheel to turn around. This would suggest that when you disconnect the battery, the bulb should keep glowing for a little while. In reality, the bulb stops glowing immediately the battery gets disconnected.
- There isn't an effective bicycle chain mechanism to explain what happens in series and parallel circuits.

A number of analogies are used to explain electrical phenomena, many of which are applicable for secondary school students or for teachers to assist their own understanding – examples of various analogies are described in Cosgrove (1995), Dupin and Joshua (1989), Heywood and Parker (1997) and Jabin and Smith (1994). One such analogy, which is often used in secondary schools, uses a water reticulation system to model series and parallel circuits. Students often have difficulty understanding this analogy because of their limited experiences with water reticulation systems.

Apart from the bicycle chain analogy, role-play-type analogies are also useful for primary school students. Catherine used a role-modelling activity (McClintock Collective 1988, pp. 218–25) with the children in her class. This approach is similar to one that Brigitte, a New Zealand teacher, used in her classes to help make the unseen or invisible more easily understood (see Figure 4.11). It is important that, when students engage in science, multimodal forms of representation are employed, particularly where the processes underpinning a phenomenon are not seen, as happens in electric circuits.

While the similarities and differences between the analogy and the target phenomena have been discussed for the bicycle chain analogy above, Activity 4.9 asks you to undertake this task.

This is a role play activity used to explain some of the potentially confusing concepts involved in understanding more about electricity. It can be used for students of all ages and the discussion extended or simplified to suit the age and understanding of the students. It involves going outside and setting up a circular running track with several obstacles in the way. The obstacles may be benches to jump or climb over, tyres to step in and out of or playground bars to swing along. You explain the track to the students. Then you position them one after the other along the track in starting positions, emphasising that they are not to pass each other and must try hard to keep the group together and not straggle. The final incentive is to tell them before they start moving around the track that you have a bag of sweets and that they will get a sweet each time they pass you. So, say 'Go' and let them run around the track several times.

After everyone has completed the circuit several times a discussion on the process can be started. The question is: How was what we just did similar to electricity making our light globes glow? The children will come up with many ideas; for example:

The track was the electrical circuit.
The teacher was the battery or energy source giving each student new energy to run the circuit again.
The students were the charges running around and around the circuit. Note that they flowed around the circuit together as a group. Individually, they are electrons making up the current.
The obstacles represent light globes or other appliances in a circuit.
The obstacles required a lot more energy to pass through or over than just running on the track. The energy used in passing is the voltage. Globes use a lot of voltage; the wires themselves require little voltage.
At the end of the circuit, the runners were rewarded with a sweet. This represents the current getting another boost of energy or voltage to send them around again.

track = circuit
sweets from teacher = battery or energy source
students = charges
energy = voltage

This role play can be extended.

One student can sit on the sidelines and count the number of children who pass in a minute (or a few minutes). This represents the ammeter.

You could add a short circuit potential to your track by giving the students the option of missing one of the obstacles. Observe their choices and discuss what happens in a real circuit with most of the current taking the easiest and least energy-intensive route, but with some still choosing to complete the obstacle course.

The energy is all used up when the current returns to the battery. It must be given a boost before flowing around again. This can be shown in class using a voltmeter in a circuit connected around the battery.

battery or energy source = energy giver
current = flow of electrons
voltage = energy used

Hopefully, this clarifies some key words and concepts about electricity and is a lot of fun as well. Good luck.

FIGURE 4.11 Brigitte's role-modelling of an electric circuit

Source: (Acknowledgement to Brigitte Glasson, Christchurch, NZ)

Activity 4.9 Role-modelling of electric circuits

Try Brigitte's role-modelling approach in your whole group.

- How does it help to clarify your understandings of electricity?
- What analogy is being used?
- How is this analogy similar to and different from electricity?
- How does this analogy compare with the bicycle chain analogy described above?
- What is your preferred analogy for electricity or an electric current?
- How is your analogy similar to and different from electricity?

R

Teaching and learning about electric circuits with animated simulations

In recent times there has been a growing trend to use highly illustrated 'static' visual representations rather than relying on largely text based representations of information (Lowe 2003). The introduction of newer technologies to instruction has led to the use of dynamic visual representations such as animations. These are often used to show learners something that can't be seen in the real world, such as the movement of molecules (Ainsworth 2008). In the case of electric circuits, animations can assist learners in visualising the motion of electrons and can highlight key ideas, such as that electrons are already contained in the wire and that an electric current is the movement of electrons in one direction. While it has been recognised that animations may play a role in engaging students, this may not be the case in supporting the learning of science ideas shown in the animation (Lowe 2003). Non-interactive animations can contain too much graphical information that is constantly changing, leading to the learner missing the key features of the animation that explicate the science idea. On the other hand, interactive animations that allow the teacher and/or student to control the rate of the animation may provide more opportunities for the learner to appreciate the science idea being shown. As with analogies, it is important for the teacher to discuss with their students those features of the animation that match the target phenomenon, as well as pointing out those features of the animation that don't. This can be more easily achieved if the teacher has some control over the running of the animation.

Hennessey, Deaney and Ruthven (2006) have suggested that interactive simulations may be used by students as a reasoning tool whereby they can explore and visualise the consequences of their reasoning. Interactive simulations give instant feedback in the form of dynamic graphics or numerical representations of how variables are interrelated. The University of Colorado has produced Web-based interactive simulations that are freely available for teachers to use (see http://phet.colorado.edu/index.php). One such interactive simulation is a circuit-construction kit (see Figure 4.12). Learners are able to construct their own circuits that may include any number of batteries, bulbs, switches and resistors. The construction kit also allows learners to take quantitative measurements using a virtual ammeter or voltmeter. The students are able to test the functionality of their constructions, where a closed circuit is represented by the bulb that glows and the movement of electrons is represented as spheres within the wires moving around the circuit (see Figure 4.12). Students might use the circuit-construction kit as a reasoning tool to explore various features of a functioning electric circuit; for example, the electric circuit will not function unless there is a continuous conducting loop for the electrons to flow. Students might also examine the effect of placing a switch into various parts of the circuit.

The students might be given problems that require the construction of complex circuits, such as 'Build a circuit

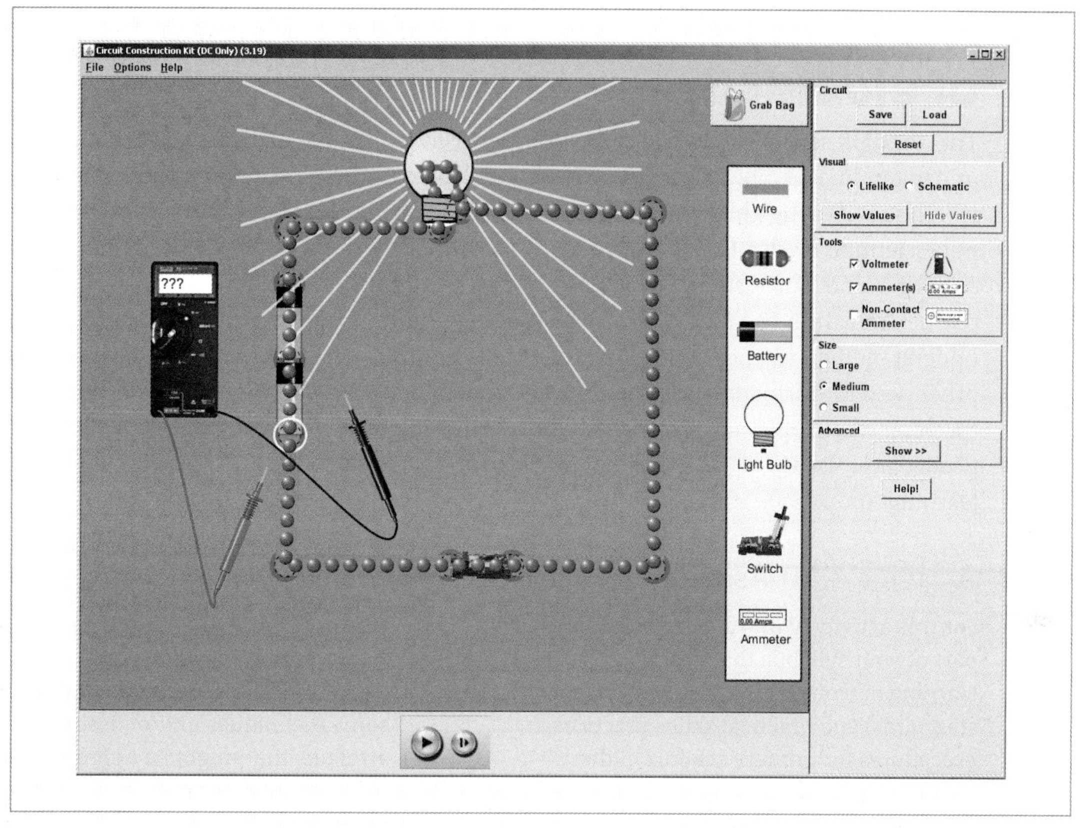

FIGURE 4.12 Circuit-construction kit (DC only)

Source: University of Colorado 2009

ICT

that uses two batteries, two bulbs and three switches, where one switch operates all bulbs and each of the other switches only operates a single bulb'. The students might then use the construction kit to test out their ideas in constructing a virtual circuit, and then build the actual circuit using real electric components. While there are benefits for students in building and testing electric circuits using computer simulations, such activities should not be seen as substitutes for practical activities with real electrical components, but as complementary to one another (Jaakkola and Nurmi 2008).

Unanswered questions

A child in Catherine's classroom has a father who is an electrician. Catherine invited him to the school for the class to meet and to ask him questions. The most important aspect of this encounter was the emphasis on safety that the father reinforced. His story of a workmate being flung across a room when he unwittingly touched a live wire left strong impressions on the children. While the issue of safety had been incorporated throughout the unit, the real-life story from a parent emphasising safety captured the issue's importance in a way that was impossible for the teachers involved to match.

Assessment of students' understandings of electricity

The omnipresence of electricity in children's lives results in them constructing their own understanding of how electricity works, well before they are taught the topic of electricity in schools. During this time, a lot of the terminology of electricity is acquired and used in non-scientific ways. In addition, many of the children's ideas about electricity are non-scientific and surprisingly resistant to change (Shipstone 1985). It is important for the teacher to assess the students' prior understandings of electricity and then to monitor their understandings as they engage in the scientific ideas of electricity within the classroom. In other words, formative assessment or assessment for learning becomes important when teaching the topic of electricity. Refer to Chapter 1 for a discussion related to the pedagogical issue of assessment for learning (see the section 'Assessment for student learning' on p. 35).

A

Activity 4.10 Assessing Catherine's class

For this activity, you will need to access the part of your syllabus that describes the learning outcomes (or levels of attainment or standards) and practical skills, practices and procedures for primary school children.

- Read through the activities that Catherine's class undertook as part of the interactive teaching approach. Using your syllabus, determine which, if any, learning outcomes, levels of attainment or standards, and practical skills, practices and procedures match with the activities that were undertaken. Does your syllabus suggest that Catherine's topic on electricity was appropriate to the year or age level of her composite years 3 and 4 class (eight- to nine-year-old girls and boys)?
- Which concepts underpinned the topic of electricity taught by Catherine? You may like to choose from the list of concepts in the later section 'Concepts and understandings for primary teachers' (see p. 169).
- List all the assessment techniques or artefacts that may have been assessed in Catherine's unit on electricity. Include some techniques and/or assessable artefacts that might be appropriate given the description of the activities that were undertaken.
- From your list, determine those that represent formative assessment and those that represent summative assessment. Do any of the techniques qualify for both formative and summative?
- Refer to the *Primary Connections* unit of work entitled 'It's Electrifying' (AAS 2007). This document describes how diagnostic, formative and summative assessment is embedded in the 5E teaching and learning model applied to the topic of electricity. Compare and contrast the assessment opportunities described in this document with those that were adopted by Catherine.

Some final comments about Catherine's class

The children in Catherine's class decided they had learnt a lot from this unit on torches, though, needless to say, the students still had unanswered questions at the end. In reflecting on the key teaching and learning features of Catherine's class, refer to Activity 4.11.

Activity 4.11 Teaching strategies, teacher roles and thinking and working scientifically

This activity asks you to reflect on the torch unit that the children in Catherine's class experienced.

- What teaching strategies did Catherine use?
- Compare your list with a list student teachers produced after experiencing the same process (see the online Appendix 4.7).
- Reflect and comment on the similarities and differences.
- What teaching roles has Catherine exemplified?
- Again, compare these teaching roles with the list student teachers produced after experiencing the same process (see the online Appendix 4.7).
- Compare them with the suggested roles for a science teacher in Harlen (2009) – see the online Appendix 1.6 – and reflect and comment on the similarities and differences.
- In what ways were children enabled to learn about thinking and working scientifically (see Chapter 2) and to develop scientific skills and attitudes during the torch unit?
- Where do these skills fit into your science curriculum?
- Catherine adopted an interactive teaching approach to her topic of electricity. Compare Catherine's teaching approach with the 5E model advocated by The Australian Academy of Science's *Primary Connections* unit 'It's Electrifying' (AAS 2007). What similarities and differences do you see between these two teaching approaches?

Appendix 1.6
Checklist of notes for constructivist teachers

Appendix 4.7
Teaching strategies perceived by primary teachers-to-be

Science as a human endeavour

There are characteristics of science as a human endeavour that can be readily incorporated into lessons focusing on electricity. These include the NOS (e.g., the characteristics of asking questions and solving problems) and the influence of science (the effects of science and technology on our lives, as well as how science is used).

The nature of science

The Australian national science curriculum (ACARA 2011) describes scientists as people who explore the world around them and share information about what they find. They work by asking questions and solving problems, creating knowledge that can be used to make predictions. These descriptions of scientists' roles are reflected in the activities undertaken by the students in

Catherine's class. The students began by exploring an everyday electrical device in the form of a torch, which led them to asking questions which were subsequently solved by them.

A key activity of scientists is modelling, which has been advocated above in the use of analogical models to explain the behaviour of electrical circuits. In using the animated modelling circuit-construction kit software, students get to solve problems, which is another key activity undertaken by scientists.

Activity 4.12 **Using models to solve problems**

In this activity you are asked to use modelling software to solve some electrical problems. Access the University of Colorado's circuit-construction kit (see http://phet.colorado.edu/en/simulation/circuit-construction-kit-dc) and use it to model the electrical circuit for a house lighting system that includes the following features:

- All lights in the house can be turned off at a main switch.
- Some lights have their own on/off switch.
- There are pairs or trios of lights on the same switch (such as multiple downlights in the one room).
- There is at least one light that can be operated from two switches (such as lights in a hallway, with switches at both ends of the hallway).

The influence of science: the effect of science and technology on our lives

There is little doubt that electricity plays a significant part in most people's daily lives. This becomes apparent in times of power stoppages. Apart from taking into account the use of electrical items, consideration also needs to be given to the role played by electricity in the production of non-electrical items. For example, a daily newspaper is not only produced by presses that run on electricity, but often contains news items obtained from news sources through electronic communication. Students could investigate the way electricity and electrical appliances have changed the way we live. This could be undertaken with respect to various periods in history; for example, the period of time prior to the introduction of mains electricity or the period before the introduction of computers. Students could also explore electricity use in isolated communities that don't have mains electricity, or in communities that don't use electricity, such as Amish or tribal cultures.

SSI

In Catherine's teaching sequence, she invited an electrician into her class to answer students' questions and focus on electrical safety. In this way, she showed how people in the local community can use science in a range of ways, and use scientific knowledge to act responsibly. Biddulph and Osborne (1984) recommend inviting experts to the school. Teachers who have used this interactive approach have brought in many and varied people. For example, a local bike shop owner was brought in to answer questions for a unit on bikes, a local clock maker for a unit on clocks, a heart specialist for a unit on hearts, a podiatrist for a unit on feet, a wildlife specialist for a unit on lizards, an expert in the Indigenous use of natural materials for a unit on colour, and a weaver for a unit on dyeing materials. It is not always necessary, however, for the expert to come to the classroom. Mary took her class of five-year-olds to the local university to have their questions on smell answered, and for them to experience aspects of smell not accessible in their school. Local secondary schools may also be able to provide access to equipment and knowledge.

A recent federal government initiative, the *Scientists in Schools* program (see http://www.scientistsinschools.edu.au/index.htm) was designed to engage and motivate students in their learning of science, and to broaden awareness of the variety of exciting careers available in the sciences. The program is open to research scientists and engineers, postgraduate science and engineering students, and people involved in applied sciences, such as doctors, vets, park rangers and so on, who go on to work with teachers and their classes in schools.

When developing a unit, always think about the role your community – colleagues at school, parents, and staff at local secondary schools and tertiary institutions, to mention a few – may be able to play.

Summary

This chapter has explored a complex topic for students and teachers alike. The concepts and understandings for primary teachers listed in the next section have been applied to explain the processes involved in the operation of an everyday electrical device such as a torch. Some of the difficulties students experience in learning about electrical circuits have been outlined. A case study about an interactive teaching approach has been given, which focused on a teacher who felt she had insufficient background knowledge. This chapter has also given some attention to teaching about electrical circuits with the use of analogies, and has provided a discussion of the formative and summative aspects of assessment. Finally, this chapter has discussed electricity in terms of 'science as a human endeavour' using the characteristic themes of the NOS (asking questions and solving electrical problems) and the influence of science (the effect that electricity has on our lives).

Concepts and understandings for primary teachers

The following list of concepts and understandings came out of research that identified 'a set of electricity concepts which can be acquired readily by primary school teachers and taught effectively to children' (Summers, Kruger and Mant 1997, Preface). These concepts and understandings underpin a scientific understanding of the operation of simple electric circuits, such as those found in torches, Christmas tree lights and electric blankets.

- An electric circuit is a complete (unbroken) pathway.
- Electrons are very, very tiny particles.
- An electric current consists of a flow of electrons.
- Electrons are part of all atoms that make up all substances.
- The electrons are in the wires all the time.

- Conductors have free electrons, which can move.
- The battery provides the push to move the electrons.
- The battery voltage is a measure of the push.
- A chemical reaction in the battery creates an electric field, which produces the push.
- All the electrons move instantaneously.
- The size of the current in a circuit depends on the resistance.
- A series circuit has all the components in a line. There is only one pathway.
- The current is the same all around a series circuit.
- In a series circuit adding more bulbs increases the resistance and decreases the current. The bulbs are dimmer and equally dim.
- A parallel circuit has branches. There is more than one pathway.
- Identical bulbs in parallel are as bright as one bulb alone. The current in each branch is the same.
- The current in the battery leads is the sum of the currents in the separate branches.
- In a bulb, moving electrons collide with fixed atoms in the filament causing them to vibrate.
- The vibrating atoms emit light and heat.

Summers, Kruger and Mant 1997, p. 15

Acknowledgements

The authors of this chapter would like to acknowledge Dr Fred Biddulph, University of Waikato, and the late Roger Osborne for their work in the Learning in Science Project (LISP) (primary) supported by the then New Zealand Department of Education, which developed an interactive teaching approach; Margaret Bearlin and Dr Tim Hardy, directors of the Primary and Early Childhood Science and Technology Education Project (PECSTEP), University of Canberra, and the teachers involved in that professional development program by the ACT Education and Training Council and the ACT Schools Authority; the University of Melbourne for grants to support research associated with the Science and Technology Education Project (STEP), within which the contents of this chapter were developed; and Fairfield Primary School and the children of Catherine's class who provided the basis for this chapter.

Search me! science education

Search me!

Explore **Search me! science education** for relevant articles on electricity. Search me! is an online library of world-class journals, ebooks and newspapers, including *The Australian* and the *New York Times*, and is updated daily. Log in to Search me! through http://login.cengage.com using the access code that comes with this book.

KEYWORDS

Try searching for the following terms:

- Electricity
- Teaching energy

Search tip: **Search me! science education** contains information from both local and international sources. To get the greatest number of search results, try using both Australian and American spellings in your searches: e.g., 'globalisation' and 'globalization'; 'organisation' and 'organization'.

Appendices

In these appendices you will find material related to electricity that you should refer to when reading Chapter 4. These appendices can be found on the student companion website. Log in through http://login.cengage.com using the access code that comes with this book. Appendices 4.1, 4.2, 4.3 and 4.6 are included in full below.

Appendix 4.1 Materials for torch making

Allow for one electrical kit between two children.

A kit contains:

- one switch
- two cells
- two battery holders
- five crocodile leads
- two bulbs in bulb holders.

This kit is required for most of the activities. A plastic lunchbox is an ideal container and readily available. A card of contents attached to the lid makes it easy for the children to check that the kit is intact after use. It is useful to order extra bulbs and bulb holders as these pieces of equipment are easily broken.

Appendix 4.2 Electricians' and scientists' shorthand for drawing electrical circuits

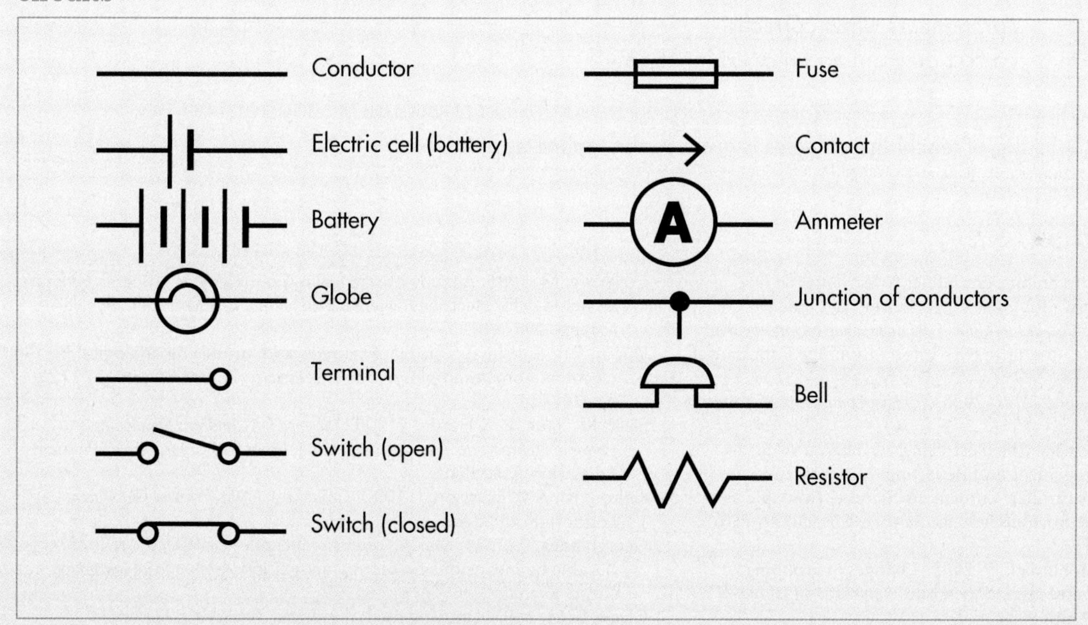

Appendix 4.3 Answers to electricity questions

The following are brief answers to the first five questions in a set of nine (Lee and Law 2001, p. 130).

1 Arrangement (a) only. This is the only arrangement that has a complete circuit.

2 a Model 4. The current remains the same at all points in a series circuit.

  b Place an ammeter into different parts of the circuit to verify that the current remains the same at all points in the series circuit.

3 a A and B have the same brightness as both bulb voltages are the same (assuming the bulbs are identical).

  b The current is the same at all points along the conducting path as it is a series circuit.

4 There is more current in circuit 1 than in circuit 2 as the resistance in the circuit has increased with the added bulb. The brightness of Y increases (voltage increases), whereas the brightness of X decreases (voltage decreases).

5 a The motor's speed will decrease as it now has less voltage and less current.

b Moving the position of the motor relative to the lamp in the series circuit will not affect the motor's speed.

c The lamp will remain at the same brightness, as the lamp's voltage and current will remain the same.

Appendix 4.4 Interactive teaching approach

This online appendix gives an outline of the interactive teaching approach.

Appendix 4.5 Morse code

The morse code for each letter of the alphabet can be found in this online appendix.

Appendix 4.6 Criteria for assessment of pre- and post-torch drawings

0 some components (switch [s] or battery [b] or bulb [g])

1 all components (s, b, g) – no connections

1½ wires and components shown

2 components wired together (for example, one-way)

3 some idea of circuit (for example, b or g or s [switch hardest])

4 components wired correctly

5 correct diagram and nice explanation

Appendix 4.7 Teaching strategies perceived by primary teachers-to-be

A listing of teaching strategies is given in this online appendix.

References

Ainsworth, S. 2008. How do animations influence learning?, in D. Robinson and G. Schraw (eds). *Current Perspectives on Cognition, Learning, and Instruction: Recent Innovations in Educational Technology that Facilitate Student Learning*. Charlotte, NC: Information Age Publishing, pp. 37–67.

Australian Academy of Science (AAS) 2005. *Primary Connections*. Canberra: AAS.

_____ 2007. *Primary Connections*. 'It's Electrifying'. Canberra: AAS.

Australian Curriculum, Assessment and Reporting Authority (ACARA) 2011. F-10 Australian Curriculum: Science. Available at http://www.australiancurriculum.edu.au/Science/Curriculum/F-10 (accessed March 2011)

Baines, E., Blatchford, P. & Kutnick, P. 2003. Changes in grouping practices over primary and secondary school. *International Journal of Educational Research*, 39, pp. 9–34.

Biddulph, F. & Osborne, R. 1984. *Making Sense of Our World: An Interactive Teaching Approach*. Hamilton: University of Waikato, Centre for Science and Mathematics Education Research.

Buck, G. A., Margaret, A., Macintyre Latta, M. A. & Leslie-Pelecky, D. L. 2007. Learning how to make inquiry into electricity and magnetism discernible to middle level teachers. *Journal of Science Teacher Education*, 18, pp. 377–97.

Chiu, M-H. & Lin, J-W. 2005. Promoting fourth graders' conceptual change of their understanding of electric current via multiple analogies. *Journal of Research in Science Teaching*, 42 (4), pp. 429–64.

Cosgrove, M. 1995. A study of science-in-the-making as students generate an analogy for electricity. *International Journal of Science Education*, 17 (3), pp. 295–310.

Dupin, F. J. & Joshua, S. 1989. Analogies and 'modelling analogies' in teaching: Some examples in basic electricity. *Science Education*, 73 (2), pp. 207–24.

Fleer, M., Jane, B. & Hardy, T. 2007. *Science for Children: Developing a Personal Approach to Teaching* (3rd edn). Frenchs Forest, NSW: Pearson Education Australia.

Galton, M. & Williamson, J. 1992. *Groupwork in the Primary Classroom*. London: Routledge.

Georghiades, P. 2004. Making pupils' conceptions of electricity more durable by means of situated metacognition. *International Journal of Science Education*, 26 (1), pp. 85–99.

Glynn, S. M. 1997. Drawing mental models. *The Science Teacher*, 64 (1), pp. 30–2.

Harlen, W. 2009. Teaching and learning science for a better future. *School Science Review*, 90 (333), pp. 33–42.

_____ & Qualter, A. 2004. *The Teaching of Science in Primary Schools* (4th edn). London: David Fulton Publishers.

Hennessy S., Deaney R. & Ruthven K. 2006. Situated expertise in integrating use of multimedia simulation into secondary science teaching. *International Journal of Science Education*, 28, pp. 701–32.

Heywood, D. & Parker, F. 1997. Confronting the analogy: Primary teachers exploring the usefulness of analogies in the teaching and learning of electricity. *International Journal of Science Education*, 19 (8), pp. 869–85.

Jaakkola, T. & Nurmi, S. 2008. Fostering elementary school students' understanding of simple electricity by combining simulation and laboratory activities. *Journal of Computer Assisted Learning*, 24, pp. 271–83.

Jabin, Z. & Smith, R. 1994. Using analogies of electricity flow in circuits to improve understanding. *Primary Science Review*, 35, pp. 23–6.

Keinonen, T. 2007. Electricity: Sixth graders' thoughts on it. *International Journal of Learning*, 14 (3), pp. 127–34.

Kirkwood, V. M. 1991. Approaches to teaching and learning in early childhood science. *Australian Journal of Early Childhood*, 16 (3), pp. 9–13.

Kress, G., Jewitt, C., Obborn, J. & Tsatsarelis, C. 2001. *Multimodal Teaching and Learning*. London: Continuum.

Lee, Y. & Law, N. 2001. Explorations in promoting conceptual change in electrical concepts via ontological category shift. *International Journal of Science Education*, 23 (2), pp. 111–49.

Lemke, J. L. 2004. The literacies of science, in E. W. Saul (ed.). *Crossing Borders in Literacy and Science Instruction*. Arlington, VA: NSTA Press, pp. 33–47.

Lowe, R. K. 2003. Animation and learning: Selective processing of information in dynamic graphics. *Learning and Instruction*, 13, pp. 157–76.

Mahapatra, M. 2006. Exploring physics teachers' concepts of simple DC circuits. *Paper given at GIREP*, Amsterdam, Netherlands, August.

McClintock Collective. 1988. *Getting into Gear*. Canberra: Curriculum Development Centre.

Microsoft Corporation 1999. *World English Dictionary*. Seattle: Microsoft Corporation.

Mulhall, P., McKittrick, B. & Gunstone, R. 2001. A perspective of confusions in the teaching of electricity. *Research in Science Education*, 31, pp. 575–87.

Osborne, J., Black, P., Smith, M. & Meadows, J. 1990. *Electricity*. Liverpool: Liverpool University Press.

Osborne, R. & Freyberg, P. 1985. *Learning in Science: The Implications of Children's Science*. Auckland: Heinemann.

Pardham, H. & Bano, Y. 2001. Science teachers' alternative conceptions about direct currents. *International Journal of Science Education*, 23 (3), pp. 301–18.

Pilatou, V. & Stavridou, H. 2004. How primary school students understand mains electricity and its distribution, *International Journal of Science Education*, 26 (6), pp. 697–715.

Scott, P., Asoko, H. & Leach, J. 2007. Student conceptions and conceptual learning in science, in S. Abell & N. Lederman (eds). *The Handbook of Research on Science Education*. Mahwah, NJ: Lawrence Erlbaum Associates, pp. 31–56.

Shipstone, D. 1985. Electricity in simple circuits, in R. Driver, E. Guesne and A. Tiberghien (eds). *Children's Ideas in Science*. Milton Keynes: Open University Press, pp. 291–316.

Summers, M., Kruger, C. & Mant, J. 1997. *Teaching Electricity Effectively: A Research-Based Guide for Primary Science*. Hatfield, UK: Association for Science Education.

_____ 1998. Teaching electricity in the primary school: A case study. *International Journal of Science Education*, 20 (2), pp. 153–72.

University of Colorado. 2009. Physics Education Technology (phET) Interactive Science Simulations: Circuit Construction Kit (DC only). Boulder: University of Colorado. Available at http://phet.colorado.edu/index.php (accessed 25 May 2010).

5

Energy

by Peter Hubber and Valda Kirkwood

Introduction

'Harvesting energy from all sources', 'Energy for sale – going cheap', 'Go solar – there's plenty of energy wasting', 'Nuclear energy is the key to our future', 'Switch to green energy', 'Red Bull is the most popular energy drink in the world!', 'This product has a 5-star energy rating', 'Conserve energy – switch off your lights'. These are just some of the headlines and advertisements mentioning energy that we encounter at home, in the workplace and during our leisure time.

Chapters 3 and 4 have covered aspects of two forms of energy – motion and electricity. This chapter looks at the invented and abstract nature of the concept of energy, and the problems involved in teaching the concept in schools. There is much debate as to whether this topic should be taught in primary schools (Duit 1987; Warren 1986). Nevertheless, energy is a significant concept in the science curriculum. This is evident in current Australasian P-10 curriculum documents, where energy has status as a major knowledge strand. Within the Australian national curriculum (ACARA, 2011), the term 'energy' is only mentioned once in the content descriptions. However, different forms of energy are mentioned in several parts of the curriculum. The content descriptions and year levels are: 'Light and sound are produced by a range of sources and can be sensed' (Year 1); 'Heat can be produced in many ways and can move from one object to another' (Year 3), and; 'Energy from a variety of sources can be used to generate electricity' (Year 6).

A valid reason to address the scientific basis of the concept of energy in primary schools is that the word 'energy' is part of each learner's familiar world. Students are interested in aspects of their world related to energy and the experiences they bring to the classroom provide a lively basis for teaching and learning. This is most evident in relation to perhaps the most significant environmental issues of our time: global warming and climate change. Most children understand that the earth is getting hotter, that global weather patterns will change and that the polar ice caps will melt if the greenhouse effect increases. They appreciate this while at the same time holding alternative conceptions as to the processes that lead to an enhanced greenhouse effect (Ekborg and Areskoug 2006). Teachers can assist students with a progressive clarification of understanding about energy throughout schooling by clarifying and exploring their own understanding.

The omnipresence of the concept of energy in all the major areas of science has resulted in scientists' inability to come to a consensus view on a useful definition of the term that would suit all purposes. In addition, researchers have found that preservice and practising teachers have difficulties in understanding the scientific aspects of energy. Given the complexities of energy, the intention of the first part of this chapter is to develop a definition of energy that will be useful to primary teaching. The rest of the chapter explores ways in which children can learn about energy. The final sections of the chapter discuss another form of energy, light, and explore science as a human endeavour within the context of global warming and the greenhouse effect.

Scientists' and teachers' understandings of energy

Energy is one of the most important concepts in science. It has a central role to play in the understanding of phenomena within each of the main branches of science; namely, biology, chemistry and physics. A clear definition of energy can be found in physics textbooks and most dictionaries. They define energy as 'the capacity to do work'. The term 'work' has been defined as 'the quantity of energy transferred when an object moves in the direction of a force applied to it' (Lofts et al. 1997, p. 318). While this definition comes from the area of physics, it would only satisfy a physicist in some contexts, and it certainly would not find resonance with a chemist or biologist or, indeed, a primary school teacher.

This may be due to the fact that the energy concept is considered within many and varied contexts and is used by scientists and teachers in a variety of ways and meanings. For example, a biologist thinking about the energy flow from the sun through an ecosystem, or a science teacher teaching students about energy content in foods, do not consider the underlying concepts of work, force and displacement. Neither does the chemist who thinks about energy as being associated with the behaviour of chemicals and their reactions. One of the most famous equations of modern times is attributed to Albert Einstein. It relates energy and mass through the equation $E = mc^2$ (E – energy, m – mass, c – speed of light). Apart from its application to nuclear physics, this equation has little relevance to other areas of science. Millar (2005, p. 3) states that 'energy is an abstract, mathematical idea. It is hard to define "energy" or even to explain clearly what we mean by the word'. Millar also notes that 'the word "energy" is widely used in everyday contexts, including many which appear "scientific" – but with a meaning which is less precise than its scientific meaning, and differs from it in certain respects'. Sefton (2004) points out that there is no universal definition of energy.

Clearly, the definition of energy given above is not useful within a primary school science setting. But before there is further discussion of scientists' and teachers' views of the energy concept, it is important for teachers to clarify their own understanding of the energy concept in order to help students develop a better understanding.

Activity 5.1 Your own understanding of energy

- Write three sentences using the word 'energy' as you would in your everyday life.
- What is the meaning of the word 'energy' in these sentences?

- Ask other teachers and adults to provide you with sentences that include the word 'energy'.
- Several researchers (Kruger 1990; Kruger, Palacio and Summers 1992; Pinto, Couso and Gutierrez 2005; Trumper 1997) have explored preservice and practising primary teachers' views of energy (see Appendix 5.1 on p. 213). From this research it was determined that some of the main categories of understanding energy are as follows:
  - Energy is associated with living things.
  - Energy is associated with moving objects.
  - Energy is a life force.
  - There is no distinction between energy and force.
  - Energy is a hidden force present all the time in a substance, waiting to be released.
  - Energy comes from the sun.
  - Energy is not conserved.

  Label the sentences you have written yourself and collected from other teachers and adults that relate to each of these categories. You may find a sentence that fits into more than one category or you may find you need to construct a new category.
- You will find mention of the following terms in various curriculum documents: 'energy form(s)', 'energy source', 'energy receiver', 'electricity', 'thermal energy', 'light', 'movement', 'sound', 'kinetic', 'potential', 'chemical', 'elastic', 'gravitational and nuclear energy', 'energy transfer', 'energy transformation', 'energy efficiency', 'energy dissipation', 'energy resource', 'energy degradation', 'conservation of energy' and 'change'. Some of these terms you will recall from schooling or from other prior experiences, while others will be new terms. Before these terms are introduced in this chapter, it is useful to clarify the ones with which you are familiar or vaguely recall, to identify those that are completely new, and begin to explore the relationships or links that you perceive between them. Create a concept map that makes sense of all the terms in the list you are familiar with. State the way they are linked on the connecting line if you are able to. Examples from your day-to-day life may help illustrate aspects if you are unable to find a descriptive word.
- Compare your concept map with the concept map in the online Appendix 5.2. What differences can you detect?

Appendix 5.2
Concept map of energy

In many non-physics science textbooks, the authors do not introduce energy via the formalised physics definition. This may be due to the authors' reluctance to provide an extensive discussion of the underlying concepts of work, force and displacement, or maybe they have a wider view of energy than this. As an example of an introductory statement, Watson (1996, p. 182) states: 'It is hard to say exactly what energy is. It is much easier to describe what energy does. Energy makes things work, it produces changes and makes things happen'. Another example is given by Morgan et al. (1967, p. 343), who state: 'Energy is a concept – an idea – a way of describing the cause of certain events; we find it useful to think of it in terms of the effects it can produce, rather than trying to think of what it is'.

In providing useful pathways for understanding energy, it is better to focus on how energy is associated with systems undergoing change rather than on what energy actually is. Solomon (1992, p. 119) referred to Richard Feynman, whom she described as 'the greatest modern teacher of university physics ... who simply refused to define the term energy'. To Feynman, energy was an invented construct that was useful in explaining situations undergoing change. From this perspective, energy can be better understood as a 'job-doing' capability (Kruger et al. 1998) – it is something that can do useful jobs for us. A study of energy then involves considerations of the types of jobs and tasks performed, the source(s) of energy and what is happening to the energy as the job or task is being undertaken.

Activity 5.2 Dictionary definitions and scientists' understanding of energy

- Look at several dictionary definitions of energy. What understandings of energy do you gain from reading them? Do any of them connect with your understanding of energy?
- Look at several different science books containing ideas about energy, including biology, chemistry and physics texts, as well as those you would find in a primary school classroom or library. What similarities and differences do you find in these texts with respect to the
  - definition, if given, of the energy concept
  - context in which energy is introduced
  - topic areas in which energy is discussed
  - understanding of energy that is conveyed?
- Make a list of the different science areas with which energy is used within the texts.
- Discuss with others and compare your findings in these activities.

What should be made clear is that energy is not a concrete entity but an attribute. Kruger et al. (1998) draw on the analogy of the age of a fossil. The age of a fossil is certainly not a physical quantity but an attribute of the fossil. Just as one can apply certain techniques and thinking to determine the age of the fossil, one can also apply certain ways of thinking to calculate amounts of energy.

Drawing on the view that energy is an invented construct that refers to a job-doing capability, we can explore the various characteristics of energy. These are discussed in the following sections.

Energy storage

Examples of energy storage

Energy can be stored and accumulated in various ways. Consider the following examples:

- A moving object has movement or motion energy. This type of energy is called kinetic energy.
- Spring and spring-like materials (objects which, when squashed or stretched, return at least some way towards their natural shape) contain elastic potential energy.
- A raised object contains gravitational potential energy.
- Hot and cold objects contain internal energy.
- Fuel and oxygen systems contain chemical potential energy.
- Food systems contain chemical potential energy.
- Batteries contain chemical potential energy.

- Electricity power stations contain either chemical potential energy (stored in the coal and oxygen system – coal-fired power stations) or gravitational potential energy (stored in the raised water – hydroelectric power stations).

While a number of terms are used when referring to energy storage, as described above, there are just two basic ways of storing energy – kinetic energy and potential energy.

Kinetic energy

Kinetic energy is the energy stored in a moving object, whereas potential energy is energy stored where there is no movement. Both kinetic energy and potential energy relate to the macroscopic world (e.g., consideration of moving cars or raising objects into the air) and the microscopic world (e.g., considering the temperature and stability of objects from an understanding that objects are made up of tiny vibrating particles that are connected to each other). It should also be noted in these discussions of stored energy that the amount of energy we think of as 'stored' depends on the process that will take place. The amount of stored chemical energy in coal, for example, might depend on the conditions under which coal is burnt.

Kinetic energy is related to the speed as well as the mass of an object. The greater the speed and/or mass of a moving object, the greater its kinetic energy. For example, while a car and a truck may both be moving at 100 km/h, the truck will have more kinetic energy due to its greater mass. It takes more effort to get it to that speed and more effort to stop it.

A stationary object, such as a car, is seen from a macroscopic perspective to have zero kinetic energy, as the car is considered to be one large particle. However, from a microscopic perspective, the car is composed of tiny particles in constant motion. Therefore, each of these particles, called atoms, has kinetic energy. It is this microscopic view of moving tiny particles with kinetic energy that is central to understanding the scientists' view of the concept of temperature of an object. The average kinetic energy of the particles within a substance gives a measure of the temperature of the substance. If the particles are stationary, they have no kinetic energy and the temperature is zero. The point at which particles become stationary is a theoretical limit and refers to a temperature of approximately −273°C, also known as absolute zero. Scientists have achieved temperatures that come within a fraction of a degree of absolute zero.

Potential energy

Potential energy is related to an object's position in relation to other 'connected' objects. Forces of attraction or repulsion connect objects to other objects. In the example of a raised object, there is a connection between the raised object and the Earth – this is gravity or the gravitational force of attraction. One can imagine gravity as a spring (Kruger et al. 1998) connecting the centre of the Earth with the raised object (see Figure 5.1). As the object is raised, the spring is further stretched, thereby storing more potential energy. Using this analogy, an object on the surface of the Earth also contains potential energy. This analogy also gives the scientific idea that gravitational potential energy is stored within the object and Earth system.

This view of connectedness also occurs on a microscopic level between the particles that make up objects. Due to electrostatic forces of attraction and repulsion, atoms are

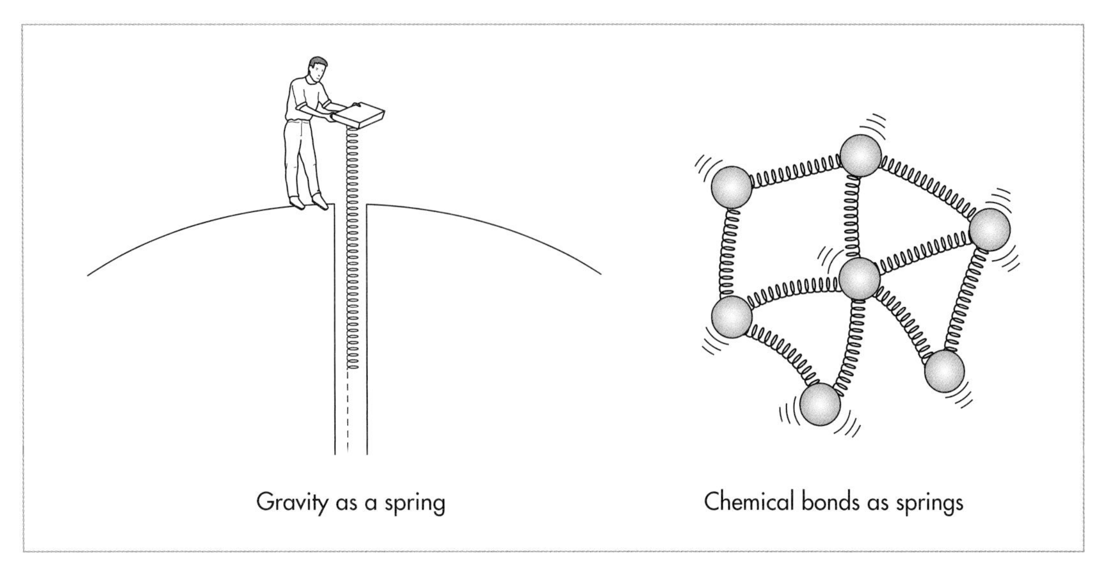

FIGURE 5.1 Pictorial representations of potential energy

connected to other atoms to form molecules, which are themselves connected to other molecules. These connections can be imagined to be springs that are stretched and compressed as particles vibrate with kinetic energy (remember that at all temperatures above absolute zero, the particles have kinetic energy). If the distance between neighbouring particles (atoms/molecules) increases, so does the potential energy of the particles (see Figure 5.1). The connections between atoms/molecules are called 'chemical bonds'. As with the raised object–Earth system, the potential energy is stored within the atoms/molecules systems.

Reinterpreting examples of energy storage

We can now reinterpret the examples of energy storage given above in terms of the ideas about kinetic and potential energy.

- In the example of a stretched rubber band, potential energy is stored as the particles within the rubber band move further apart, stretching the chemical bonds between the particles.
- Any substance at a temperature above absolute zero will contain particles with kinetic energy and potential energy. The sum of these energies is called 'internal energy'. A substance that is heated will increase in internal energy as its particles will vibrate faster (increase in kinetic energy) and stretch the chemical bonds (increase in potential energy). It should be pointed out that the internal energy of a substance is often mistakenly referred to as 'heat'. Some textbooks define heat as the transfer of internal energy from one substance to another in the process of heating or cooling.
- When a fuel such as wood, oil, petrol or gas is burnt, a chemical reaction occurs between the fuel and oxygen in the air. This results in new substances being formed. On a microscopic level, the atoms within the molecules in the fuel and oxygen get rearranged to form different molecules associated with different substances; for example, when petroleum gas is burnt, the main products of the chemical reaction are water and carbon dioxide. The release of energy in the form of heating the space around the reaction can be accounted for in

terms of the internal energies of the substances involved in the reaction. This energy comes from the fact that the total internal energy of the reactants (petroleum gas and oxygen) is more than the total internal energy in the products (water and carbon dioxide). A number of textbooks give the view that energy is only stored within the fuel but this example suggests otherwise. The energy has been stored within the fuel and oxygen system. Also, it is the change in energy associated with a process that is of interest, not some absolute value.

- In considering foods, the digestion process involves a chemical reaction between the food molecules and other substances so that there is a rearrangement of atoms to form new molecules. As with fuels, the total internal energy of the reactants is more than that of the products and so the energy that is 'left over' goes into warming the body, in addition to allowing organs and muscles to function. Just like fuels, an important carbohydrate – glucose – will react with oxygen to create water and carbon dioxide. This process is called 'oxidation' or 'combustion'.
- In batteries, chemical reactions occur when a complete circuit is set up. Again, the energy is stored within the reactants that are involved in the chemical reactions.

Energy as a job-doing capability

From the discussion of the storage of energy and the concept of energy as a job-doing capability, we need to consider the energy changes that take place when the job is being done. Kruger et al. (1998) draw an analogy between energy and money. They suggest that energy resembles money in that:

- it can be stored and accumulated
- it does not have an effect until it is transferred from one place to another
- when energy gets transferred, jobs get done.

As with any analogy, it is important to recognise where the analogue doesn't match up with the target phenomenon. For example, in this analogy, consider the analogue of monetary payment linking with the target phenomenon of energy transfer. Monetary payment for a job can occur before, during or after the job has been completed. However, in the target phenomenon, the transfer of energy occurs at the same time the job is being completed.

Analysis of a job

To draw out the main characteristics of energy consider the following job: A carpenter wishes to drive a nail into a block of wood.

In analysing the various aspects of this job, we first of all need to consider the system, which defines the boundaries in which the job is being done. Let's consider that the system involves the hammer, carpenter, nail, block of wood and surrounding environment (which includes the air and table or ground supporting the wood). Assume that the carpenter is holding the hammer.

The first situation that undergoes change is the lifting of the hammer. This is a job in itself and so it requires energy, which comes from the carpenter. She gives the hammer kinetic energy (in moving the hammer) and gravitational potential energy (in raising the hammer). Just before reaching the highest point of its motion, the hammer's kinetic energy transforms into gravitational potential energy (the hammer slows down to a stop). This results in the hammer stopping at the highest point on its path. At this point, the hammer only has gravitational potential energy (we can imagine the gravity 'spring' getting further stretched).

Upon dropping the hammer, the gravitational potential energy gets transformed into kinetic energy. There will also be some contribution to the kinetic energy of the hammer from the carpenter. Upon contacting the nail, a number of changes occur:

- Sound is generated.
- The nail moves.
- The wood surrounding the nail moves, separating to allow entry of the nail.
- There is a slight increase in the temperature of the nail and hammer (the temperature increase is more noticeable after several blows of the hammer).

Each of these changes involves an energy change. From the kinetic energy of the falling hammer, some of the energy goes into production of sound. Sound, as will be described later, is a travelling vibration generated from the source. This travelling vibration is the transfer of kinetic energy. The sound travels in all directions in the air and through the block of wood. The nail is pushed into the block of wood, which separates. Some of the kinetic energy of the hammer is also transferred to the nail, which in turn is transferred to the wood. Some of the kinetic energy in the hammer also transfers, on a microscopic scale, to kinetic energy of particles within the hammer, nail and wood, resulting in an increased temperature of these objects (increasing the kinetic energy of the particles within a substance increases its temperature).

Within the system described for this situation, the total energy remains constant – scientists refer to this concept as 'the law of conservation of energy' or the First Law of Thermodynamics. For most jobs, this is not readily observed. In the case of the falling hammer, after striking the nail into the block of wood, the hammer, nail and block of wood come to rest. From a macroscopic perspective, each of these objects has lost its kinetic energy. However, from a microscopic perspective, the energy can be accounted for if one considers that it has been transferred into the kinetic energy of the particles within the nail, block of wood and hammer, and particles within the surrounding environment. The transfer of kinetic energy results in sound being heard and an increase in the temperature of all objects within the system. It is important to note here that, unless we consider the environment surrounding the system under study, then the law of conservation of energy breaks down. For example, if the system in the nailing example had boundaries that only included the carpenter, hammer and nail, then energy would be viewed as not being conserved. The specification of the system with appropriate boundaries is central to a comprehension of energy conservation.

Another important point to note is that the transfer of energy results in it spreading out. It is shared among many more particles than it was in the initial situation of the falling hammer. This spreading out of energy is termed 'dissipation of energy'.

In considering the job of driving a nail into a block of wood, the transfer of kinetic energy from the hammer to the nail is considered 'useful', whereas the transformation of energy into sound and heating the environment is not – it is considered 'wasteful'. Scientists refer to the production of wasteful energy within jobs as 'degradation of energy'.

Key characteristics of energy

From the carpenter hammering a nail example, we can draw out a number of key characteristics of energy that are pertinent in situations in which change occurs. They are as follows:

- Energy can transform from one type into another.
- Energy can transfer from one location to another.
- Energy degrades and dissipates but remains conserved.
- Energy is associated with systems undergoing change.

The following sections discuss each of these characteristics.

Energy can transform from one type into another

Energy can change from one form into one or more other forms: for example, when a light bulb is switched on, electrical energy is changed into light energy and thermal energy. This is called 'transformation of energy'. Often in transformations, forms of energy are produced that are not thought of as useful: for example, light bulbs produce not only useful light energy, but also wasteful thermal energy. Machinery that is designed to produce movement often also produces thermal energy and sound energy, which do not contribute to the efficiency of the machine. With each transformation, the amount of useful energy becomes less and less. This is called 'the degradation of energy' (see also above).

Energy can transfer from one location to another

Forms of energy are seen as being able to move: for example, electrical energy moves in wires from the power station (the electricity source) to your washing machine, refrigerator, computer or hair dryer (all electricity receivers). Sound energy moves through air and other media (as those living near sports stadiums and outdoor concert venues will know). Light energy reaches our light receivers (our eyes) when it is emitted from primary sources (for example, stars) or scattered from objects (that is, secondary sources, such as a mirror). Thermal energy travelling through space, from electric heaters in winter and from the sun in summer, enables us to keep warm. This movement of energy from one location to another, called the 'transfer of energy', can be experienced by students; for example, with a string telephone; a pot on a stove, especially if it has a metal handle; an electric fence; an image on a cinema screen; or a bat hitting a ball.

The transfer of energy between a source and a receiver occurs in three basic ways:

1. as a collision effect in which particles at the source collide with neighbouring particles; each particle influences its neighbours and the movement is transferred
2. as a collision effect where particles in the source physically travel to the location of the receiver and collide with particles at this location
3. as a radiation effect where particles in the source can affect particles in the receiver across space.

The transmission of sound from the source to the receiver is a collision effect of the first listed type. As the sound source vibrates, it collides with surrounding particles, setting them vibrating. These neighbouring particles collide with surrounding particles, also setting them vibrating. The travelling vibration through collisions with particles is known as a sound wave.

The kinetic energy of the vibrating source is dissipated through being transferred to neighbouring particles. As sound is a collision effect, it explains why sound:

- requires a medium to travel; sound cannot travel through space
- travels much faster in solids and liquids than in gases as the particles are closer together
- lessens in intensity with increasing distance from the source.

The transfer of thermal energy through materials, called 'conduction', is also a collision effect. For example, if one places one end of an iron bar into a fire, the whole bar soon heats up. This can be explained as the particles in the end of the bar gaining kinetic energy from the gas particles in the fire through colliding with them. The increased movement of the particles in the bar creates internal collisions within the bar. The collisions raise the kinetic energy of the particles within the bar, thus raising its temperature. The transfer of thermal energy can also occur when particles at the source travel to the receiver to then transfer kinetic energy in a collision effect. Hot air currents are an example of this effect. This type of energy transfer is called 'convection' and only occurs in liquids and gases.

The transmission of light through space occurs without a medium and is thus a 'radiation' effect. From a scientific perspective, light is a general term for many types of radiation, all of which travel at the speed of light. Scientists refer to this general notion of light as 'electromagnetic radiation'. Some forms of electromagnetic radiation are visible light, infra-red radiation, ultraviolet radiation, radio waves, microwave radiation, X-rays and gamma radiation. Each type of radiation is distinguished by the manner in which it is produced and the strength of the energy (X-rays have a greater strength than infra-red radiation).

In the production of electromagnetic radiation, there is a change in the energy source – this occurs on a microscopic scale. For example, a burning log sends off thermal radiation or infra-red radiation due to the agitated state of the particles (or, more specifically, the electrons) within the burning materials. In fact, all materials with a temperature greater than absolute zero will contain particles in an agitated state and so will emit thermal radiation. The higher the temperature, the more radiation is emitted. The radiation travels through space and can be absorbed by objects some distance away. Energy from the sun comes to the Earth in this way. The absorbed energy increases the internal energy of the particles within the objects, which in turn can produce a number of changes. For example, the electromagnetic radiation from the sun can result in the following changes to the human body if absorbed:

- an increase in temperature; the absorbed energy increases the kinetic energy of the skin particles
- the vision cells at the back of the eye are agitated, initiating the vision process
- chemical reactions are initiated, causing tanning or burning; these can affect the normal cell reproduction of the body and lead to cancer.

While our eyes cannot detect thermal radiation, night goggles, as used by soldiers in the field, can. The goggles detect varying levels of infra-red radiation and produce a thermal image of the surrounding area.

Another example of a radiation effect is the transfer of electrical energy within a wire in an electric circuit. A chemical reaction in the battery creates electromagnetic radiation that travels along the wire at the speed of light. The electrons within the wire absorb this radiation, creating an electric current. A more detailed description of this process can be found in Chapter 4.

Energy degrades and dissipates but remains conserved

The interconnections of different forms of energy led to the principle of the conservation of energy (refer back to the description of the First Law of Thermodynamics in the 'Analysis of a job' section; see p. 180). In any change, the amount of energy in the system was seen to remain the same; the forms of energy may change, but the overall amount of energy remains the same. Mathematical expressions were developed to describe this constancy. As Davies (1984, p. 68) writes:

> The law of energy conservation embodies a wide range of physical experiences that, in the absence of the energy concept, would all have to be discussed separately. Energy lets us connect many ideas together, and as such it could be deemed beautiful.

Feynman (1963, section 4, p. 1) states:

> There is a certain quantity, which we call energy, which does not change in the manifold changes that nature undergoes. That is a most abstract idea, because it is a mathematical principle; it says there is a numerical quantity that does not change when something happens. It is not a description of a mechanism or anything concrete; it is just a strange fact that we can calculate some number and when we finish watching nature go through her tricks and calculate the number again, it is the same.

Potential energy in its many guises was invented to make the sums add up when there appeared to be missing energy with the use of only the manifest or obvious forms of electricity, thermal energy, light, movement and sound. Potential energy should then be considered as an abstract construct invented by scientists to allow them to link phenomena that would otherwise be thought of as separate entities.

The scientific concept of the principle of the conservation of energy is in sharp contrast to the everyday discussion of the need for energy to be conserved. Individuals are familiar with such things as stickers above light switches saying 'Switch off the light – conserve energy', posters saying 'Conserve energy – travel by public transport, not by car', or newspaper headlines that state 'The world's energy supplies are running out'. Such statements refer to the conservation of energy resources and not the scientific law of conservation of energy. These statements present a view that energy gets used up, which means that society must use it wisely or find ways to renew it. This view is compelling given that once an energy resource, such as petrol, is used for some task (driving an engine), it is perceived to be no longer there.

While the conservation of energy law represents constancy amid change, there is a decline in the useful energy available to do jobs. Ross (1988) used the terms 'fuel value' or 'available energy' to describe the useful energy stored in energy resources, which is used up to drive systems. Ross points out that these terms have a meaning closely related to scientists' understanding of free energy, which in turn is related to the Second Law of Thermodynamics. According to Ross (1988, p. 439), 'The Second Law of Thermodynamics says that although energy can be converted from one form to another the process is never 100 per cent efficient', as some energy is always 'lost as waste heat'. Free energy gets used up in the sense that it gets degraded to thermal energy and spreads out. Solomon (1982, p. 419) reinterpreted this law for a student's understanding in terms of renaming it 'The Running

Down Principle', which states that 'In all energy changes there is a running down towards sameness in which some of the energy becomes useless'. In the example of the hammering of a nail, the running down towards sameness means that the dissipation of energy results in a slight rise in temperature of each of the objects within the system. As energy transfers often result in very tiny rises in temperature, conservation and dissipation of energy can't be easily demonstrated.

The important point to note here is that, while the conservation of energy principle means that energy does not get lost, it should not be interpreted without also considering that in any change there is also energy degradation and energy dissipation. This is the essence of the scientists' first and second laws of thermodynamics.

The conservation of energy implies that it can be measured. Energy is measured in joules (calories is a disused imperial unit for energy – there are just over 4 joules for every calorie). Power is another often-used quantity related to energy. It is a measure of the rate at which energy is transferred and is measured in units of watts. For instance, by turning on a 100-watt light bulb, 100 joules of energy is being transformed from electrical energy to light and thermal energy every second that the bulb is operating. Thus, in a minute, it transforms 60 000 joules (60 kJ) of energy.

Energy is associated with systems undergoing change

A major characteristic of energy is that it is always associated with situations undergoing change; that is, whenever a change occurs, energy, in some form, is associated with it. We should not perceive energy as a substance, but rather as something associated with change.

It has to be remembered that scientists view energy in different systems in different ways. Examples of different environmental systems are biological (e.g., the human body or a pond system), chemical (e.g., a chemical reaction between vinegar and baking soda or the rusting of a bike left out in the rain), geological (e.g., an earthquake, the erosion of a landscape by water) and physical (e.g., a person rollerblading along a path or a torch lighting up the path ahead).

The concepts of energy conservation, dissipation and degradation need to be understood within isolated systems. These are systems that include all objects in which the transformation and transfer occur. In the example of the carpenter driving a nail into a block of wood, the closed system needs to include the surrounding environment in addition to the carpenter, hammer, nail and block of wood, otherwise the energy conservation law breaks down. Therefore, the system and environment are key ideas in understanding energy conservation.

An important point to note about analysing the energy considerations within particular changes in a system is that change begets change; that is, for every change there is a preceding change that initiated that change. Therefore, the consideration of a particular change should be viewed as part of a continuous sequence of changes. Consider the example of the hammering of a nail into a block of wood. The first change that occurs within this situation was the lifting of the hammer by the carpenter. However, this change was initiated by a previous change. Energy was expended within the carpenter's arm muscles (the arm gained gravitational potential energy in being raised, elastic potential energy in stretching the muscles and kinetic energy in being moved). Preceding this change there are still further changes such as:

- chemical reactions associated with the digestion of food and respiration of oxygen
- chemical reactions associated with the process of photosynthesis in food production
- nuclear reactions within the sun that result in the emission of electromagnetic radiation
- the astronomical processes involved in the formation of our solar system.

Dare we extend the changes back further to the big bang with the creation of the universe? And what about before then?

An understanding of energy as associated with change must be recognised within the context of an endless sequence of changes. From this perspective, the concept of conservation of energy as one where energy is neither created nor destroyed can be well appreciated. What is required when analysing the energy considerations of situations is to be specific about a beginning and an end point to the situation and to have boundaries to the system strictly defined.

Activity 5.3 **Concept map of understanding energy**

A

Activity 5.1 (see p. 175) asked you to construct a concept map on the basis of your previous knowledge of energy terms. Now, having read the section on scientists' and teachers' understandings of energy, revisit your concept map and make appropriate changes.

In addition, construct a new concept map that encapsulates the key concepts about energy listed at the end of this chapter in the section 'Concepts and understandings for primary teachers' (see p. 212).

Children's understandings of energy

A large number of studies have explored children's understandings of the concept of energy (Boylan 2008; Duit and Haeussler 1994; Forde 2003; Kaper and Goedhart 2002; Kruger et al. 1998; Lee and Liu 2010; Lijnse 1990; Liu and McKeough 2005; Solomon 1983, 1988; Stylianidou 1997; Trumper 1993; Watts 1983). Trumper (1997) has also explored preservice primary teachers' understandings of energy. Recent reviews by Millar (2005) and Papadouris, Constantinou and Kyratsi (2008) of conception studies confirm earlier findings that some of the most popular and persistent alternative conceptions are as follows:

- Energy is related only to living things – an anthropocentric view. Living things get tired and less active without energy. Human energy is rechargeable through food or by resting. Some children have contradictory uses for human energy. For example, there may be the view that energy is built up as a result of sports training, but children can be told to run outside in the playground to use up energy.
- Only objects in motion have energy.
- Stationary objects relate to energy only from the perspective of them as an energy store; for example, batteries, power stations, oil and coal store energy. Energy is therefore a causal agent, a source of activity based or stored within certain objects.
- Energy is a concrete entity. In some circumstances, such as in the heating of an object, energy is a fluid-type material that flows from hot objects to cooler ones.

- Energy is not an entity that is conserved because one has to pay for it. The energy required for devices such as light bulbs and engines gets consumed.
- Conservation of energy is associated with everyday care not to waste energy, which differs from scientists' conservation of energy principle.
- Energy is found in foods; it only gets harnessed when the food is eaten. Some children believe that 'high energy' foods convert directly into energy or somehow crack open to release their 'store of energy' when consumed.
- Energy is considered a fuel. Fuel is energy, rather than fuel contains or is a source of energy.
- There is confusion between 'energy' and 'force'. Other words, such as 'motion' and 'power', are also frequently used interchangeably with energy.

Watts (1983) categorised students' views of energy into seven ways of thinking about energy. These ways of thinking, none of which represents scientists' views, were substantiated by Forde (2003) and Trumper (1993) and are described as

1. anthropocentric – energy is associated with living things
2. depository – some objects have energy and expend it
3. ingredient – energy is a dormant ingredient within objects, released by a trigger
4. activity – energy is an obvious activity
5. product – energy is a by-product of a situation
6. functional – energy is seen as a very general kind of fuel associated with making life comfortable
7. flow transfer – energy is seen as a type of fluid transferred in certain processes

Activity 5.4 Eliciting people's understandings of energy

In addition to writing sentences containing the word 'energy' (see Activity 5.1 on p. 175), the LISP (Energy) team, like many other researchers, used interviews about instances or events (Osborne and Gilbert 1980) to survey students' understanding of energy in the case-study classrooms (LISP (Energy) 1989). They used four scenarios representing different biological systems (a growing plant – seedling to bush) and a growing animal (kitten to cat), a chemical system (a burning candle when first lit and when well burnt down) and a physical system (a child at the top of a slide and the same child at the bottom of the slide a moment later). The following questions were asked:

1. If you used all your senses, what changes would you observe? (Use sight, smell, touch and hearing.)
2. During these changes, is any energy involved? Yes/No/Don't know.
3. Explain your answer to question 2.

Answer the questions for each scenario yourself. Draw eight pictures, two for each scenario, and pose the same questions for each scenario to a small group of colleagues and a small group of children.

- Develop another task that you think might also explore people's understanding of energy.
- Test out this task with a small group of colleagues and a small group of children at different ages (for example, 5, 8 and twelve).
- What did you find from carrying out the above tasks? Were there any differences between the adults and the children? Were there any differences between the children

A

of different ages? What have you learnt from the experience about:
- people's understandings of energy
- the tasks
- your own understandings of energy
- gaining access to people's understandings
- your skills at gaining access to people's understandings?

- Discuss and compare your findings to these activities with others.

Children's questions about energy

Three primary school teachers asked their classes (a Year 2, Years 3 and 4 composite and a Year 5 class) to write down something the students would like to find out about energy. What was clear from the children's responses was that the children readily identified with the word 'energy', but at the same time had an enormous range of questions about the energy concept. This is illustrated in the sample of responses given below. Many of the questions imply the alternative ideas described above. The questions have been separated into different categories. There did not appear to be any significant differences between the year levels, although the Year 2 students concentrated on energy in living things.

Some of the children's questions about energy related to people and other living things were:

- How does energy run through our body? (Year 2)
- Can people's energy be measured? If so, how? (Year 3)
- How do people play sport and get energy? (Year 3)
- Could we survive without energy? (Year 4)
- I would like to learn how much energy a human would use each day. (Year 5)

Some questions about energy related to non-living objects were:

- Why do some toys need energy? (Year 2)
- Does the moon have any energy? (Year 3)
- Do 'unalive' things have energy? (Year 4)

Some questions about energy related to food were:

- What types of food give you energy? (Year 3)
- How do you get energy from food? (Year 4)
- What's the best food that would make you have energy? (Year 5)
- If I do not eat and don't have energy, how can I get energy without eating? (Year 5)

Some questions about energy sources were:

- Where might you find energy instead of your car, yourself, your food or your drink? (Year 3)
- How much energy does a battery need? (Year 4)

Some questions about the genesis of energy were:

- Who invented the word 'energy'? (Year 4)
- How is energy produced? (Year 5)
- Can science create energy? (Year 5)

Some questions about what energy is were:

- Is energy dangerous? (Year 3)

- Can you see energy? (Year 3)
- Is energy like air? (Year 3)
- What can you make energy out of; for example, sticks, rocks? (Year 4)
- What is inside energy? (Year 5)

Some questions about energy types were:

- What is magnet energy? (Year 3)
- How many different energies are there? (Year 4)
- Is static electricity energy? (Year 4)
- How do batteries work inside and how does petrol give cars energy? (Year 5)

Some questions about energy storage and transfer were:

- Why is light energy so fast? (Year 3)
- How do you put energy in food? (Year 3)
- Can you take energy and put it somewhere else? (Year 3)
- How is energy transferred from a battery to something else? (Year 5)

Changing students' ideas about energy

There are various opinions about how energy should be taught. Some are informed by research – see Millar (2005) for one review of this area – while some are not informed by research at all. Some are people's ideas about what they think would be improvements, based on action research of their interventions in classroom experiences. This section outlines some teaching and learning strategies that are informed by constructivist principles.

Elicitation of children's ideas and questions

The previous section 'Children's understandings of energy' (see p. 186) illustrated well how children have a wide range of preinstructional ideas, many of which are different to currently accepted scientific ideas. In addition, the section 'Children's questions about energy' (see p. 188) indicates that children readily identify with the concept of energy and have an interest in finding out about many aspects relating to it. Therefore, the dual approaches of eliciting students' preconceptions and questions about energy would prove fruitful in terms of planning activities to satisfy the interests of the students, in addition to addressing areas of non-scientific thinking.

The methods used by researchers to elicit people's ideas about energy are applicable within a classroom setting to elicit students' ideas. For example, the following two examples employ the technique of interview about instances or events (Osborne and Gilbert 1980). In the first example, Trumper (1990) and Forde (2003) asked students to write down three things they associate with the word 'energy' and then to write sentences that link their associations with the word. The students were also shown eight pictures that consisted of a

- chemical reaction
- power plant
- box being pushed up an incline
- train
- growing plant
- radiator
- lighted lamp
- football player.

From these pictures, the students were to select three they considered had the involvement of energy and explain their choice in one or two sentences using the word 'energy'.

In the second example, Trumper (1997) describes a questionnaire that was developed and validated by Kruger, Palacio and Summers (1992). The questionnaire comprised statements and also drawings of different situations. For each statement, respondents were to indicate one of the following: true, false, don't understand, not sure. For example, two statements accompanying a picture of an electric bar heater with the bars glowing were: 'The energy from the power station that supplies this heater did not exist before it was generated at the station' and 'Only some of the energy from the heater goes into heating up the room'.

The section 'Children's questions about energy' (see p. 188) produced a wealth of questions through the teacher asking the students to write down a question they would like to know about energy. Small-group and whole-class discussions should also be fruitful in extracting questions from the students.

Activity 5.5 Interpreting children's understandings of energy

The interview-about-instances technique described above was employed by two Year 5/6 teachers who implemented a teaching sequence about energy. A written questionnaire designed to elicit the students' ideas about energy was administered to the students at the beginning of the sequence (see the online Appendix 5.3). Many of the ideas about energy that the teachers used to construct their questionnaire and teach the students in their lesson sequence can be found in the section 'Concepts and understandings for primary teachers' at the end of this chapter (see p. 212). Student responses to question 1 (a) and 4 of this questionnaire are given below.

Student responses to question 1 (a) (see the online Appendix 5.3):

There are all types of energy like wind energy, solar energy (power).

Solar energy we can get from the sun by solar panels; wind energy we can get that from wind turbines.

A power or force that makes lights light up and fans turn.

Energy helps power up lights and other appliances.

If you're full of energy you are able to do lots of exercise.

Energy powers up your body.

Energy is non-visible unless you look through microscope.

One type of energy is conetic [sic] energy the energy of movement.

Another type of energy is stored energy which is in batteries.

Energy makes all things happen.

When I kick a football it has potential energy.

People, animals and plants can all get energy from food and water.

Almost everything that happens is caused or has something to do with energy.

There are lots of types of energy, sound energy, heat energy, potential energy, etc.

Potential energy is when something wants to move but it can't and when released it has energy.

Energy can be absorbed by any object.

Energy can be found anywhere.

Electricity is a form of energy.

Energy is power, balls have air inside them and energy is air, the air helps it bounce.

A

ONLINE

Appendix 5.3
A pre-test questionnaire about energy for Year 5/6 students

Student responses to question 4 (see the online Appendix 5.3):

	TRUE	FALSE	DON'T UNDERSTAND	NOT SURE
a	65.7%	22.8%	2.9%	8.6%
b	80%	17.1%	2.9%	0.0%
c	55.9%	11.8%	23.5%	8.8%
d	27.8%	58.3%	2.8%	11.1%
e	38.2%	32.4%	14.7%	14.7%
f	30.0%	53.3%	10.0%	6.7%
g	62.1%	20.7%	0.0%	17.2%
h	20.7%	58.6%	0.0%	20.7%
i	86.7%	6.7%	3.3%	3.3%

Study the students' responses to question 1 (a):

- Which of the statements indicate a scientific understanding of an energy concept?
- Which of the statements indicate an alternative conception?
- As this information represents data from a formative assessment task, in what way(s) can this data inform the planning of activities and teaching approaches for the topic of energy?

The student responses for question 4 are given as percentages for the Year 5/6 classes.

- For each statement, determine the scientifically acceptable response.
- For each statement, determine the key energy ideas that are being elicited by the question (refer to the online Appendix 5.3).
- Based on these percentages, what is your interpretation of the level of scientific understanding of the students at the beginning of the science topic?
- As this information represents data from a formative assessment task, in what way(s) can this data inform the planning of activities and teaching approaches for the topic of energy?

Refer back to the 'Children's questions about energy' section (see p. 188). Select some questions that could form part of a teaching sequence. In what way(s) might they be used?

Addressing alternative conceptions about energy

Getting started

It should be clear from the discussions within the section 'Scientists' and teachers' understandings of energy' (see p. 175) that the concept of energy is best understood as a job-doing capability and considered only in situations where change occurs. The LISP (Energy) team considered it was most helpful if energy was always associated with situations undergoing change; that is, whenever a change occurs, energy, in some form, is associated with it. This avoids the perception of energy as a substance, but rather allows it to be perceived as something associated with change. However, Millar (2005) points out that teachers will likely find themselves talking about energy sources which imply that energy can be stored in certain places and transferred to other places. This treatment of energy as a 'quasi-material' substance is not strictly correct, but Millar (2005, p. 12) argues that 'this is not a serious hindrance to the later development of a more precise scientific understanding of energy', a conclusion others agree with (Duit 1987; Kaper and Goedhart 2002).

In undertaking a constructivist approach to the teaching and learning of energy, the initial discussions with students should centre on what they already know. Given that many students understand energy from a human-centred perspective, then the concept of energy as a job-doing capability should be applied to changing situations involving humans.

The initial teaching of energy considerations from a human-centred perspective should be extended to a more general perspective to address a common alternative conception that energy is a property of, or an idea restricted to, living things (Trumper 1993). Kruger et al. (1998) suggest that useful tasks that can be undertaken by both humans and machines be discussed; for example, a person who collects rubbish lifts a rubbish bin into the back of a rubbish truck by hand – this same task can be analysed in the circumstance where the hydraulic arm of the garbage truck picks up the rubbish bin. If students believe that possession of energy implies the capability of doing a job, and that the same (or very similar) jobs can be performed by human beings and machines, then they should come to realise that it is reasonable to suggest that energy is not only a property of, or an idea restricted to, living things, but can also be applied to non-living things.

Emphasis on change

In exploring energy changes where jobs or tasks are undertaken, students need to be directed to evidence of change. The effects associated with different forms of kinetic energy (thermal energy, sound, electricity, movement) and electromagnetic radiation (light) are readily observed by children. However, this is not the case with the different forms of potential energy (elastic, chemical, gravitational, nuclear). To decide whether any of these forms of energy are associated with a changing situation, learners need to look for clues, such as changes in height (gravitational), changes in elasticity or springiness (elastic) or changes in state (solid, liquid or gas), colour, smell and appearance (chemical).

Another problem with considerations of potential energy may arise when students use their everyday understanding of the word 'potential', meaning that something has the ability to be or do something. This everyday understanding can lead to confusion when confronted with scientists' use of the word 'potential' in this context. For example, students believe there is potential thermal energy in an unplugged electric heater, potential light in an unlit candle and potential sound in a switched-off radio, whereas scientists make use of the term 'potential' in relation to the position of an object in respect to other objects. The LISP (Energy) group suggests that a discussion of these different meanings is helpful at secondary school, but that introducing these very abstract notions of energy to primary-school-age students is very problematic. However, Kruger et al. (1998) suggest the use of analogical models to assist students, including primary school students, in understanding potential energy. For example, in an analogy likening gravity to a spring that connects an object to the centre of the Earth, just as raising the object increases the stretch of the spring and spring energy, then raising the object increases the gravitational potential energy.

In addition to the use of analogies, teachers may refer to the various potential energies without using the term 'potential'. For example, gravitational potential energy may be referred to as height energy and chemical potential energy as just chemical energy or fuel energy.

In the section 'Scientists' and teachers' understandings of energy' (see p. 175), there was a discussion of how the different forms of energy can be reduced to just two types – kinetic and potential. While Ellse (1988) argues that one should eliminate all other forms of energy, Trumper (1993, 1997) suggests that too early a separation of different forms of energy may not be helpful. Therefore, in the initial teaching of energy, students should get acquainted with as many different forms of energy as possible.

Teaching conservation, degradation and dissipation of energy

Students experience transformations of energy in their lives all the time; for example, when playing musical instruments, playing with toys, and using household appliances and simple machines or implements (kettles, pegs, ballpoint pens, light bulbs, toasters, torches, heaters) and more complicated artefacts such as telephones. Transformations of energy often produce forms of energy that are not considered useful, such as when light bulbs produce 'wasteful' thermal energy. It is worthwhile, then, discussing with students the topic of wasteful versus useful forms of energy. This contributes to an understanding of the degradation of energy.

The scientific concept of the principle of the conservation of energy sharply contrasts with everyday discussion of the need for energy to be conserved. Confusions can arise later in schooling when students interpret the conservation of energy with the conservation of energy resources. Teaching needs to take into account students' everyday understanding and, while in later schooling a discussion might be helpful, in primary schools the correct usage of the phrase 'conservation of energy resources' would perhaps be more so.

The concept of the conservation of energy is a very sophisticated one that is challenging even for adults. The LISP (Energy) team thought it inappropriate for primary school students. However, several researchers (Ogborn 1990; Solomon 1982) point out that the concept of the conservation of energy, relating to the First Law of Thermodynamics, can be more readily understood if energy is taught via the concepts of 'useful' energy and a simplified version of the Second Law of Thermodynamics.

The Second Law says that, although energy can be converted from one form to another, the process is never 100 per cent efficient as some energy is always lost as waste thermal energy. Ross (1988) points out that students' intuitive views are similar to the Second Law and so what should be taught is that useful energy is what gets lost as it converts to wasteful energy, generally in the form of thermal energy. Teachers should admit that energy gets used up and then gradually show that 'used up' means degraded to thermal energy; thus, the energy is still there. The process of degradation of energy should also be taught in tandem with the process of dissipation of energy. In this way, students may be able to account for the apparent loss of wasteful energy in terms of it being shared with the environment. In another teaching approach to understanding the conservation of energy principle, Brook and Wells (1988) suggest that students in the first instance need experiences with situations in which changes are perceptually obvious, to give plausibility to the principle.

Liu and Collard (2005) and others (Dawson 2006; Liu and McKeough 2005) suggest that the progression of students' conception of energy is characterised by the following distinct stages:

1 perception of energy as activities or abilities to do work
2 identification of different energy sources and forms
3 understanding the nature and processes of energy transfer
4 recognition of energy degradation
5 realisation of energy conservation.

This progression of increasingly complex energy ideas needs to be considered when planning to teach the topic of energy.

Energy changes within defined systems

Systems are situations in which energy changes occur. They require boundaries to be explicitly articulated: Do they or do they not include human action to set the system in motion? (For example, is the person switching on the torch an important part of the system?) Energy changes should be seen as part of a continuum of changes, so it is necessary to indicate a start and finish to any particular energy change within a system. A closed system is one in which the energy is conserved within the system. If, for example, the system that involves hammering a nail into a block of wood does not include the environment, then, after the nail has been hammered into the wood, one would not be able to account for the energy transferred from the hammer. When discussing systems undergoing change, it is important that teachers also discuss and perhaps negotiate the limits or boundaries of the system under investigation.

Students can readily experience the transfer of energy. The transfer of energy is not to be confused with the transformation of energy. Unfortunately, the words 'transfer' and 'transformation' are sometimes used interchangeably, without apparent recognition that they each have a specific meaning. Confusing the two terms is not helpful for learners. Although discussion about these different terms and how confusions can arise if they are used incorrectly is considered helpful in later schooling, at primary levels the correct usage of the terminology by teachers may prevent future learning difficulties.

Symbolic representations of energy changes

Several authors (Ametller and Pinto 2002; van Huis and van den Berg 1993; Brook and Wells 1988; Kruger et al. 1998) advocate the use of symbolic representations to assist students in understanding the main ideas relating to energy. Ametller and Pinto (2002, p. 308) suggest that 'Symbolic representations have shown their power to transmit the idea of system, energy transfer, energy degradation or energy conservation'. What is important when employing symbolic representations is for the teacher to point out the features of the representation that match the concept as well as those features that do not match. Given the abstract nature of the concept of energy, multiple and multimodel representations of various processes involving systems undergoing change need to be exploited by teachers as they make the ideas concrete for students (Kress et al. 2001) – for a full discussion, refer to the Chapter 1 section 'Representation as a discursive practice and a learning process' (see p. 15).

R

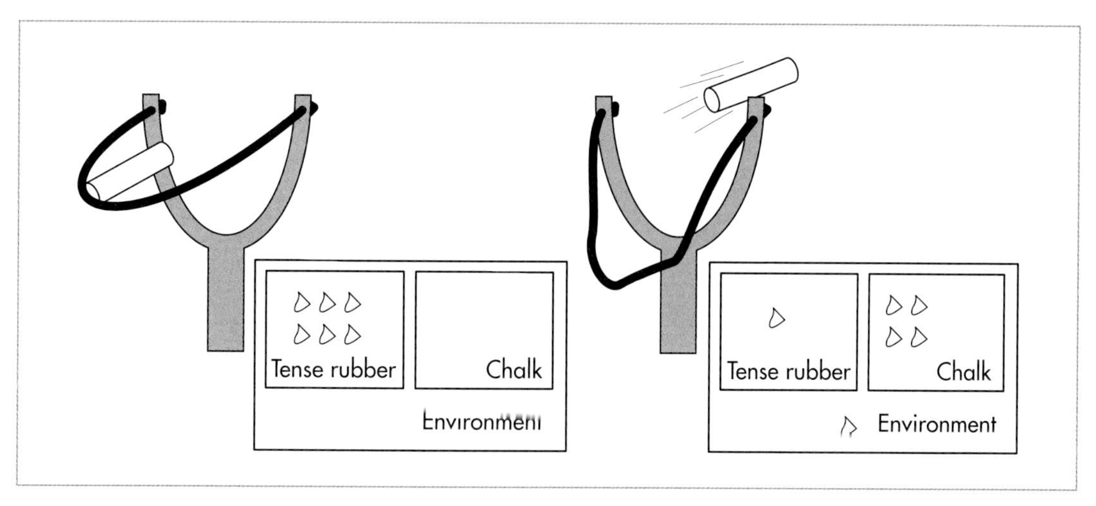

FIGURE 5.2 Energy changes in a slingshot

Source: Ametller and Pinto 2002, p. 292

One example of a symbolic multimodal representation of energy transfer, taken from Ametller and Pinto (2002), is shown in Figure 5.2. The features of this representation include:

1 a figurative representation showing cartoon images of the initial and final situations
2 an abstract representation showing a scientific–technical drawing that centres on energy transfer and energy conservation
3 a written statement of the changing situation.

Small triangles are the symbols of energy. By considering the number and location of the icons in the abstract drawing, one can visualise the transfer and conservation of energy. This representation can be extended to include cartoons and drawings of the intermediate phase of the changing situation.

Another example of symbolic representation of energy transfer and/or energy transformation, taken from Brook and Wells (1988) and advocated by Millar (2005), uses an arrow representation called a Sankey diagram, and is shown in Figure 5.3.

The system undergoing change is represented by arrows of varying widths. The width of each arrow is a measure of the amount of a particular form of energy present. Whatever the changes that take place within the system, the total width of the arrows remains constant, indicating in a diagrammatic representation the conservation of energy principle. To draw such representations, teachers need to discuss with students a well-defined system with a beginning and an end to the changing situation under study. The students then consider the location of the energy at the beginning, the end and, possibly, the intermediate phases. Millar (2005) points out that it is better to ignore the intermediate phases as they often involve quite complex processes, and to consider events and processes where the beginning and end points can be easily described and demonstrated.

Wright (2007) advocates the use of bar graphs as a way of representing energy for the middle years of schooling. It elicits higher-order thinking about the concepts and results in

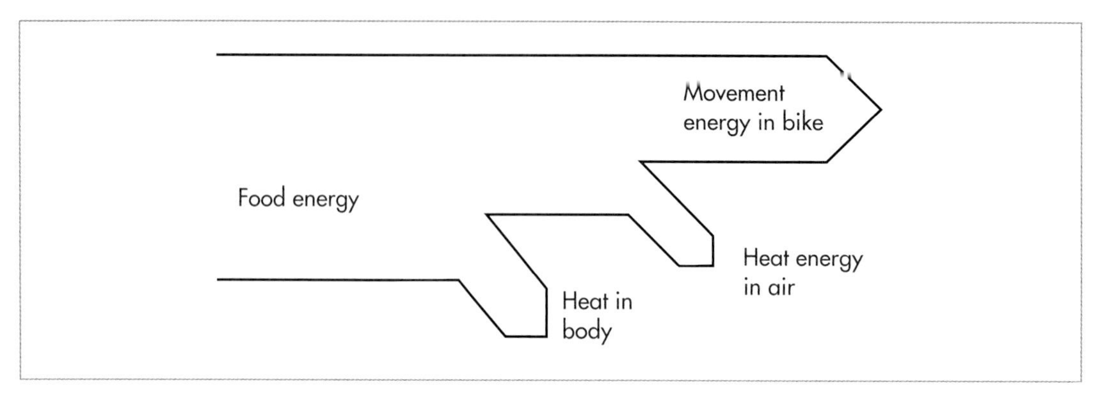

FIGURE 5.3 Energy changes in riding a bicycle

Source: Brook and Wells 1988, p. 84

useful explanations for real-life events. Bar graphs are forms of representation that are familiar to primary level students and can assist them to represent different forms of energy as well as clearly show the application of the law of conservation of energy.

Lawrence (2007) points out that a description of the world in terms of energy needs to be quantitative, but while a full quantitative description is not possible for younger students, the different representational modes, such as triangular symbols, bar graphs and Sankey diagrams, do allow for some quantification. As evidence of what can be constructed by primary school students in explaining systems undergoing change from an energy perspective, the two teachers who developed the energy pre-test questionnaire (see Activity 5.4 on p. 187) provided the following group task to their Year 5/6 students. The students were to work in groups of three to investigate the energy changes that take place in a household device. They were to present their findings in a poster that described the energy changes in their device using two or more representations. Figure 5.4 gives the poster presentations for three groups which investigated an umbrella, a stamp and an egg beater.

Chapter 4 discussed the benefits for the student learning of science concepts of the use of interactive animations (see the case study 'Applying an interactive teaching approach' on p. 155). A useful interactive animation that explains the conservation of energy using kinetic, gravitational and thermal energy within the context of skateboarding can be found at http://phet.colorado.edu/en/simulation/energy-skate-park. The animation involves the simulation of a skateboarder travelling the length of a track. Various graphs showing how the energy forms change during the time the skateboarder traverses the track can be shown. The software that runs the animation allows the user to stop or change the speed of the animation at any time. It also allows the user to change the shape of the track, the amount of track friction and the amount of gravity.

Addressing confusions in terminology

Much confusion arises in students about the notions of energy forms and energy resources as they are not understood to be the same thing in science. Electricity is a form of energy, whereas coal, wood (trees), oil and so on are energy resources that can be used to produce

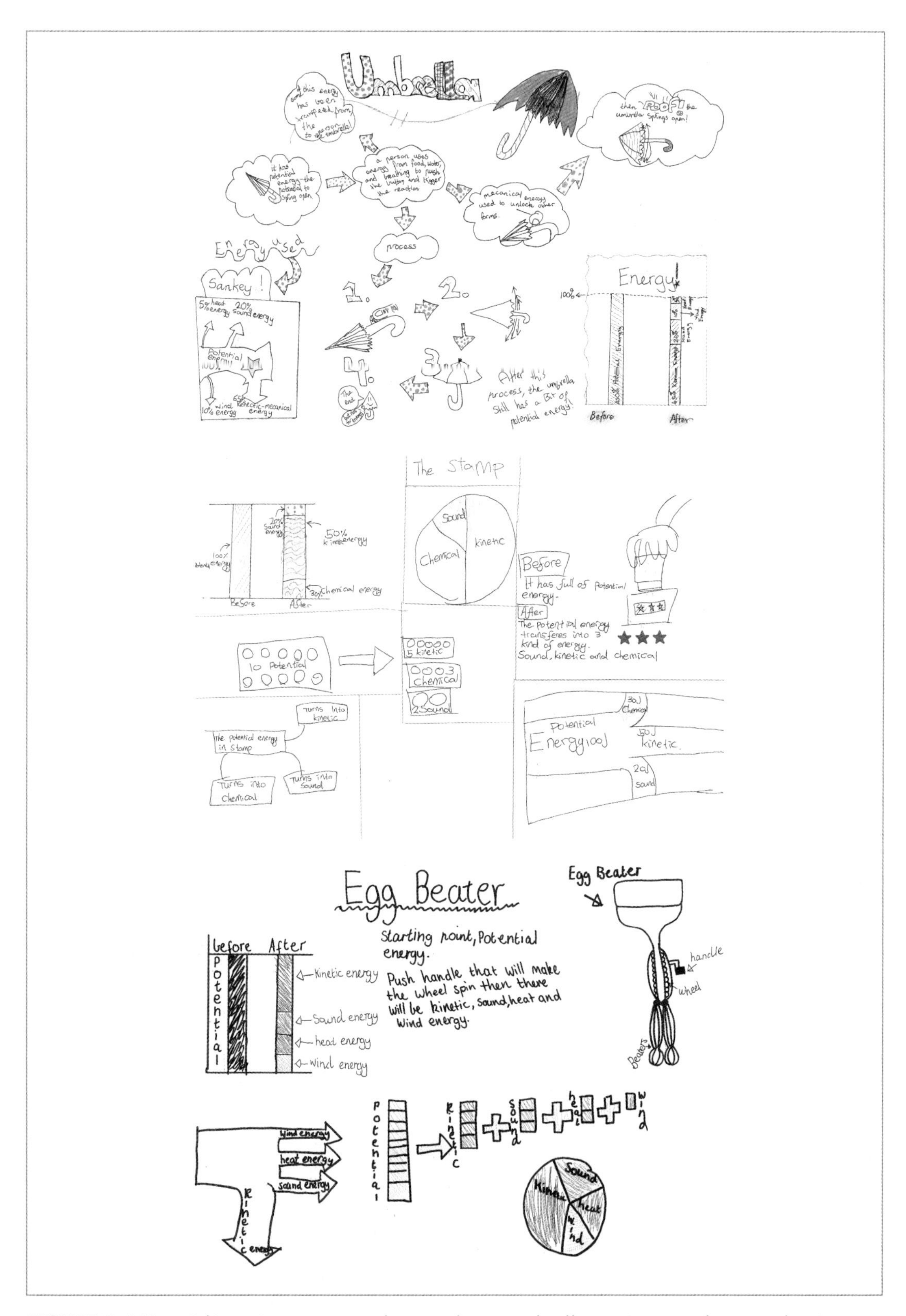

FIGURE 5.4 Year 5/6 posters: energy changes in an umbrella, a stamp and an egg beater

electricity. Solar energy is a form of radiant energy, whereas the sun is the energy resource. Often associated with energy resources are terms such as 'renewable/non-renewable energy', 'direct/indirect energy' and 'primary/secondary energy'. These terms can cause even more confusion for learners. They are best used clearly and unambiguously, with the meanings being clarified or otherwise avoided. Millar (2005) points out that students often associate energy with fuels or food and so there may be some benefit in focusing on the word 'fuel' in relation to 'fuel use' and 'fuel consumption' rather than 'energy use' and 'energy consumption'. This focus offsets problems for students' understanding of conservation of energy.

Another area of confusion for students is distinguishing between energy and other concepts used in science. Energy and force are not the same thing in science understandings. A force in science can be thought of as a push or a pull that can result in different outcomes: a change in shape or acceleration or a reaction. Magnetism is a force, not an energy form. Energy and power are not the same thing in science understandings either. Power is the amount of work done in a particular time. Food is not energy. People get the energy they need to maintain their body system (for example, heating, movement, electric nerve impulses and more) from the chemical reactions between the digested food particles and the oxygen breathed in, in a process called respiration. The digestion and respiration of food involves the transformation of chemical energy to other forms, such as thermal energy and movement. Kruger et al. (1998) suggest that, in addressing the misuse of scientific terms, teachers need to be aware of their own language use as well as the language used by the students.

For further strategies for teaching energy, refer to the section 'Teaching about energy using analogies' which appears later in this chapter (see p. 205). In this section, examples of analogies are given to assist students in understanding aspects of the abstract concept of energy. The next section discusses an alternative approach to teaching of energy, which focuses on studying topics of different forms of energy rather than topics on energy.

Activity 5.6 Tasks or jobs for addressing alternative conceptions

The teaching strategies to address the alternative conceptions about energy outlined above can – and should – be applied to different contexts in which different tasks or jobs related to energy are thought about and discussed. The following exercises require you to compile lists of tasks or jobs, or activities or demonstrations, that a teacher can use in the classroom to show:

1. that both living things and machines are capable of completing jobs
2. evidence of obvious energy change
3. evidence of transfer, or transformation, of energy
4. energy changes in everyday events
5. the distinction between transfer and transformation of energy.

- Construct a list of tasks that are performed by humans or other living things, as well as by machines. For example, walking up the stairs or taking a lift can accomplish the task of getting to the second floor of a building.
- Construct a list of jobs or tasks in which there is evidence of obvious change. For example, a burning candle has several obvious changes, such as smoke emission, light and thermal energy emission, the wick burning and the wax melting.
- What demonstrations or student activities can be undertaken in the classroom that show energy transfer or transformation? Which toys are

appropriate? For example, a toy bow and arrow shows energy transformation from potential energy into kinetic energy. A game of skittles shows how kinetic energy can be transferred from the bowling ball to the skittles.

- Construct scenarios where energy changes occur. Be explicit about what constitutes the system in which the changes occur and about the boundaries, in terms of specifying a start and a finish. If energy changes are seen as part of a continuum of changes, extend each scenario forwards or backwards to see where they lead. For each scenario draw a symbolic representation in one of the forms described above (see Figure 5.2 on p. 195 and Figure 5.3 on p. 196). For example, consider the scenario of turning on the switch that sets in motion a wind-up toy car. One particular start would be after the spring inside the toy has been wound up and an end a few moments after the switch has been turned on. From an energy continuum perspective, one may consider the energy changes prior to releasing the spring, such as kinetic energy from the hand of the person twisting the handle, which came from potential energy in the muscles in the arm, and so on.
- What demonstrations or student activities can be undertaken to clearly differentiate between transfer of energy and transformation of energy? For example, let a marble roll from rest down an inclined track that levels out. On the level part of the track, the rolling marble collides with a row of marbles. The student will observe that the rolling marble stops and a single marble at the end begins to roll. If two marbles are let go at the top of the incline there will be two marbles at the end of the row that will move after the collision. This activity demonstrates a transformation of gravitational potential energy into kinetic energy in the first marble, and a transfer of kinetic energy from the first marble to other marbles in the row.

Teaching and learning about light

Students' understandings of light

There has been an enormous amount of research into students' understandings of light (Hubber 2002). While a significant amount of this research has focused on secondary and tertiary students, there have been researchers who have included students of primary school age as part of their studies (Adams, Doig and Rossier 1991; Feher and Rice 1986, 1988, 1992; Guesne 1985; Rice and Feher 1987; Stead and Osborne 1980; Wang and Xie 2002), and other researchers whose studies focused entirely on primary school students (Fleer 1996; Osborne et al. 1993; Selley 1996a, 1996b; Shapiro 1988, 1994). There has also been some research into preservice primary teachers' understandings of light (Bendall, Goldberg and Galili 1993; Heywood 2005). In terms of the science ideas that underpin the teaching of and learning about light, you are directed to the list of concepts contained in the section 'Concepts and understandings for primary teachers' at the end of this chapter (see p. 212).

Activity 5.7 Exploring understandings about light

The following stimulus ideas (drawn in pictures in the book) and associated questions come from two resources. One is titled *Concept Cartoons in Science Education* (Naylor and Keogh 2000, pp. 131–48), and the other is ACER research (Adams, Doig and Rossier 1991, pp. 111–20); both concern a small group of students talking about light.

- In one picture (Naylor and Keogh 2000, p. 137), designed to explore students' ideas about seeing in the dark, a student says, 'We will be able to see when our eyes adjust to the dark', while another says, 'We won't see anything if there is no light in the cave'. What do you think?
- In another picture (ibid., p. 145), designed to explore students' ideas about how far light travels, one student says, 'The big torch shines further than the little torch', while another says, 'They both shine the same distance'. A third student says, 'The big torch lights up a bigger area', and a fourth student says, 'They both shine the same distance, but the big torch shines more brightly'. What do you think?
- In stimulus pictures exploring how students regard the relationship between light and vision (Adams, Doig and Rossier 1991, pp. 116–19), one student says a candle makes light, another that a book makes light, another that the moon makes light and a fourth that a television makes light. Explore each statement. Which is correct? Discuss how the candle, book, the moon and the television can be perceived.
- Use these ideas to explore the concepts of a small group of colleagues and a small group of children at different ages (e.g., five, eight and 12 years old). What did you find? Were there any differences between the adults and children? Were there any differences between the children of different ages? What have you learnt from the experience about:
  - people's understandings of light
  - the task you used
  - your own understandings of light
  - gaining access to people's understandings
  - your skills at gaining access to people's understandings?
- Compare your experiences, findings and learning with others.

A

The research cited above has found common views held by primary school children about light. These views cover a variety of areas and include the following:

- Light is often equated only with its source (e.g., light is in the light bulb), with its effect (e.g., the sun lights up the landscape) or with a property (e.g., light is shiny).
- Light is not seen as an entity.
- Light from dim sources travels a small distance or does not travel at all, while light from bright sources travels a great distance. There is also a view that daylight affects the production or propagation of light from a source.
- The formation of shadows is viewed as a trigger effect in which light incident on an object triggers the production of a shadow that travels from the obstacle to the screen. Shadows are sometimes thought of as reflections of dark light.
- The vision process is understood in a number of ways. These include:
  - Light only needs to illuminate an object for it to be seen.
  - The object and observer need to be 'bathed' in light for the object to be seen.

  - Light illuminates the object and something emanates from the eye in the act of seeing the object.
- Magnifying glasses increase the amount of light.
- Children rarely associate colour with light. Colour is seen as an intrinsic property of an object.

CASE STUDY

Changing students' ideas about light

A conceptual change model undertaken in a case study conducted by Hubber (2002) was successful in changing secondary school students' concepts of light. A few of the teaching and learning approaches that were employed in this case study are applicable for use in primary schools, albeit at senior levels, and are discussed in this section. As a comparison teaching sequence, the *Primary Connections* unit 'Light Fantastic' (see http://www.science.org.au/primaryconnections/curriculum-resources/light-fantastic.html) employs a 5E teaching and learning model.

Conceptual change instructional model

A key feature of various conceptual change instructional models is the elicitation of students' preconceptions and the use of cognitive conflict to resolve situations where non-scientific views are exposed (Davis 2001). The model adopted in this case study comprises a sequence of five phases: orientation, elicitation of ideas, restructuring of ideas, application of ideas and reviewing the change in ideas. The phases are briefly described as:

- *orientation* – an activity designed to give students the opportunity to develop a sense of purpose and motivation for learning the topic
- *elicitation* – students make their ideas explicit, bringing them to conscious awareness
- *restructuring* – occurs if alternative conceptions were elicited in the previous phase. Once the ideas are out in the open, the clarification and exchange of ideas occurs through discussion. The different views, including those of the scientists, are then evaluated, either through discussion or exposure to conflict situations. If any of the students don't offer the scientists' view, then this needs to be given by the teacher alongside the other views before evaluation occurs
- *application* – students are given the opportunity to use their developed ideas in a variety of situations
- *review* – the students are invited to review how their ideas have changed between their thinking at the start of the topic and now.

Orientation and elicitation of ideas

The initial strategy used was called a 'postbox' strategy (Bell 1993). It undergoes three phases. In the first phase, the students provided anonymous written responses to a set of questions, such as that shown in Figure 5.5.

The questions were designed to elicit the students' alternative conceptions of light. For primary school use, the questions may need to be reworded and more pictures used to assist the students in understanding what is being asked. The concepts that underpin the questions are appropriate for primary school students.

In the second phase of the strategy, the responses to each question were separated and then collected – posted – into separate boxes (e.g., shoeboxes). The class was divided into six groups, each of which had the task of analysing the responses from one of the boxes.

The final phase of the strategy had each group reporting their findings to the rest of the class. Most of the groups gave oral reports, although the group responsible for question 1 drew diagrams on the blackboard. Another approach to reporting would be for each group to produce a poster. The group places the students' responses into different categories, which are then pasted onto

FIGURE 5.5 Questions about light

Q1	(Picture of a student looking at a tree is shown with the sun in the background.) Draw arrows to show how light from the sun helps the student see the tree.
Q2	(a) If you were placed in a room where there was no light would you be able to see? (b) What about spending some time in this room? Would you be able to see? (c) Can a cat or an owl see a mouse in a room where there was no light? Why?
Q3	If you were directly behind someone can you attract their attention by just staring at them? Why?
Q4	Which of the following make light? [Pictures could be used.] Glowing coals of a fire at night, glowing coals of a fire during the day, moon, mirror, television operating during the day, television operating during the night, glow-in-the-dark sticker at night, glow-in-the-dark sticker during the day.
Q5	[Picture of a firefly, a glow-in-the-dark toy.] How far does light travel from a firefly (a) during the night? (b) during the day?

butcher's paper with appropriate labels. The groups can still give oral reports, but displaying the posters provides a permanent reminder of the different views held by the students. This is important during the restructuring and review phases.

In completing the postbox activity, the students were very animated and motivated. This may have been due to the fact that they were discussing and analysing their own views and at no stage was there any judgement from the teacher as to who was right and who was wrong. The students' views were also elicited in an anonymous and thus non-threatening manner. The analysis and reporting of different views allowed for student-generated discussions and an appreciation by all the students that there can be a range of different concepts to explain or describe the same phenomenon. The teacher's role through this process was one of questioner and someone who ensured that different views were clarified and understood by everyone. For each of the questions, there was at least one of the students who expressed the scientific view and so the teacher did not need to present this view to the class. What follows are some representative responses made by the students.

Question 1

A Arrows were drawn going to the tree and then to the student.

B Arrows were drawn just going to the tree.

Question 2

A (a) No
(b) No
(c) No, because there is no light reflecting off the things, so the mouse would be hidden

B (a) No
(b) Yes, because your eyes adjust to the dark
(c) Yes, because they have better eyesight

Question 3

A No, because you can't feel vision

B Yes, because you can sense them

Question 4

The class as a whole only agreed on glowing coals at night, television operating at night and glow-in-the-dark sticker at night.

Question 5

A (i) Just a little bit around it
(ii) It doesn't give off any light.

In the clarification and exchange of ideas, more information was gained from the students in relation to their ideas. For example, the following responses came from the students' diagrams for question 1 – A: 'Light just hits the tree and the sun just reflects off to make the object appear … her eyes pick up the light … it travels to her eyes'; and B: 'Light has to be on the tree' and, when asked if the light needed to go anywhere else, the student said, 'No, just on the tree'. The responses to the view that cats and owls can see in the absence of light were 'Cats can see in the dark because their eyes glow' and 'because they are night hunters'.

Restructuring of ideas

One of the greatest issues for teachers in using a constructivist approach is what to do when you have elicited the views from the students and there is a range of alternative conceptions present. In addressing the alternative conceptions in this case study, the teacher first of all focused on question 3 (see Figure 5.5 on p. 202) by asking the students to generate and carry out an experiment to test if one could feel a stare (incidentally, this question is meant to expose the view that something emanates from the eye in the vision process). The students came up with the following experiment. Three students were blindfolded and placed at various points in the classroom. At a designated moment, the rest of the class stared at one particular student. If any of the blindfolded students could feel a stare, they were to raise their hands. Following the experiment, the class came to a unanimous view that one could not feel a stare as nothing emanates from the eye in the staring process. The experiment raised a number of discussions and experimentation related to a person's awareness of things about them, such as the effectiveness of a person's senses other than vision and the extent of one's peripheral vision.

In testing out more of the students' ideas, the teacher set up exposure to a conflict situation in which students could experience zero light conditions. The school had a photography darkroom to enable this to happen. While few primary schools would have such a room, a storeroom would work just as well, particularly if it had no windows, leaving only the edges of the door to mask out.

It soon became apparent that very few students had experienced zero light conditions. After they had, they then found the scientific view – that you need light to enter your eyes to see something – quite plausible. While in the darkroom, the class experienced the dilation effect on one's pupils in dim conditions. This was also quite surprising to the students. Discussions then ensued in relation to some animals, such as cats, which have the ability to dilate their pupils very well in dim conditions.

To provide another plausible argument in support of the scientific concept of vision over other views, the teacher drew on an analogy linking a burglar alarm system with the senses. The main feature of a burglar alarm is a sensor, such as a touchpad, which, when touched, sends an electrical signal along some wires to an alarm. In the analogy, the following connections are made between the senses and the burglar alarm:

- sense receptor – sensor pad
- nerves connecting the sense receptor to the brain – wires
- brain – alarm.

In running the analogy, each of the senses were discussed. For example, to experience taste, something physical must be placed on your tongue – once your tongue is touched, a message gets sent to the brain via the nerves in the same way an electrical signal is sent along the wires to the alarm once it is touched. The same analogy was given for touch, smell and hearing. It now seems plausible to the students that, if the analogy worked for hearing, touch, taste and smell, then it should also work for sight – something physical, such as light, must enter the eye to activate the vision process.

Application of newly constructed ideas

The darkroom activity and analogy discussion proved successful in changing the students' ideas of vision, making them more scientific. The vision concept was seen as a key concept as it proved fruitful when the class explored other concepts of light. For example, to address the students' claims that light does not travel far from dim luminous objects (glow-in-the-dark toys, for example), discussions centred on the perspective that, if one can see a luminous object, no matter how dim it may be, then light from it must be at least reaching your eyes. The students then reasoned that their eyes could act as light detectors, so if one could see a luminous object, whether it be during the day or night, then light is at least reaching one's eyes. It then became plausible for the students to believe the scientific concept that light from a luminous object keeps travelling until it interacts with matter. The vision concept was also fruitful in addressing the alternative conception that light incident on non-mirrored objects just remains on the object or reflects in a regular way.

Prior to the class discussion using the vision concept, an 'interview about instances/events' technique (Osborne and Gilbert 1980) was used to further explore the students' ideas about how far light travels from luminous objects during night and daylight conditions. Twelve cards showing different events were distributed around the classroom. Each card showed a picture of a luminous or non-luminous object looked at by a person. An appropriate caption accompanied each picture. Some of the cards matched, in that the same situation was shown in daylight and during the night. One caption read, 'The person is looking at the television during the day'. The students were each given a survey sheet and asked to move around the room looking at each card. For each card they were to decide if the object gave off light and, if so, how far from the object the light travelled (the students were to select from four alternatives – it stays on the object, it travels halfway to the person, it travels just to the person, or it keeps travelling until it hits something). When the students completed their survey sheets, their responses were debated in a general

Activity 5.8 Analysis of a conceptual change model

The case study described above used a conceptual change instructional model.

- Make a list of the main concepts about light that framed the teaching sequence.
- Describe how aspects of the case study match up with the different phases of the model.
- Discuss with others the merits, or otherwise, of this conceptual change model for other topics in science.

class discussion that included the vision concept as mentioned above.

Review of changed ideas

The teaching sequence continued for a couple more weeks, exploring several concepts in a number of areas related to light, including shadow formation, reflection off mirrored and non-mirrored surfaces, refraction effects, images in mirrors and lenses, and the colours of light and objects. There were several occasions when discussions revolved around previously learnt concepts and reflections on how the students' ideas had changed.

A POE activity

Another successful activity used in the teaching sequence related to the area of shadow formation is applicable for use in primary school settings. The activity employed a predict–observe–explain (POE) (White and Gunstone 2008) technique that required students to carry out three tasks. In the first task, a physical situation was demonstrated to the students. This involved an obstacle being placed between an unlit candle and a screen. The students were to predict the size and shape of the shadow formed by the obstacle when the candle was lit. The students were also asked to explain their predictions. In the second task, the candle was lit, and the students were to describe what they saw. In the final task, discussions ensued to reconcile any conflict between the students' predictions and observations. During these discussions, diagrams with arrows and lines were introduced to indicate that light travels in straight lines and, as such, can explain the size and shape of the shadow formed. The drawing of lines and/or arrows to represent the direction and propagation of light should not be difficult to understand for primary school students, although they may need to have a view that light is an entity that travels away from luminous objects. Shapiro (1994) gave several examples of primary school students drawing pictures in which light was represented as straight lines and/or arrows.

Following discussions on diagrams representing light propagation as arrows or lines, the students were given an activity sheet indicating different situations that created shadows or illuminated shapes. They were required to make predictions of the size and shape of the shadows or illuminated shapes and then to test their predictions by setting up the situations. The situations included:

- a small candle, two obstacles and a screen
- two small candles, an obstacle and a screen
- a small candle, a card with a circular hole in it and a screen.

For this POE activity, candles could be replaced with torches that act as sources of light and screens can be butcher's paper attached to the wall.

Teaching resources beyond the activities described in this case study and the *Primary Connections* unit 'Light Fantastic' are contained in the topics 'Night is a big shadow' and 'My special torch' (Fleer et al. 1995), and *There's an Emu in the Sky* (Malcolm 1995). Langley, Ronen and Eylon (1997) advocate the use of computer simulation tasks. For example, The Learning Federation Project (Curriculum Corporation 2010) has produced a number of such tasks, called Learning Objects, titled 'mixing colours (Years P–2)', 'optics and images (Years 5–6)', 'light and reflection (Years 5–6)', 'additive colour (Years 5–6)' and 'subtractive colour (Years 5–6)'.

Throughout this chapter, a number of analogies have been drawn, not only for use within the classroom but also to assist the reader in understanding the abstract concepts related to energy.

Teaching about energy using analogies

A significant pedagogical issue for teachers is the determination of ways to assist students to construct meaning for scientific concepts, particularly if the concepts are abstract in nature. This issue has particular relevance in this chapter as it deals with understanding aspects of the abstract concept of energy. A pedagogical approach that can be used by teachers is to use analogies, connecting real-world experiences with the abstract scientific concepts.

Treagust (1995, p. 46) describes an analogy as a 'process of identifying similarities between two concepts. One concept is familiar, referred to as the analogue, and the other concept, the unfamiliar, is called the target. Usually the target relates to the scientific concept'. For example, in the previous section 'Scientists' and teachers' understandings of energy' (see p. 175), a gravity spring analogy was used to explain the concept that a raised object stores energy in the form of gravitational potential energy.

In this analogy, the target is gravity or, more precisely, gravitational force, and the analogue is a spring. Just as one can readily appreciate that energy is stored in a stretched spring, then, in a similar way, a raised object stores energy if one can imagine a spring joining the object and the centre of the Earth (see Figure 5.1 on p. 179). Raising an object stretches the gravity spring, so the object stores more energy. The concept of stored energy is similar to both the analogue and the target. Another concept that is characteristic of the target and the analogy is that the energy is stored in the system containing the Earth and object rather than the object itself. This is illustrated in the analogue by the spring connecting the object and Earth. The use of the spring analogy has value in addressing a common alternative conception that objects on the surface of the Earth have no gravitational potential energy. The gravity spring is still stretched for objects on the surface of the Earth.

The use of a good analogy in the classroom allows students to conceptually link new ideas with things that are already familiar to them and is therefore consistent with a constructivist perspective. It is important for the teacher to be cognisant of the students' prior knowledge, particularly as it relates to the analogue. Students need to be familiar with the characteristics of the analogue; the choice of analogue should be made with this in mind.

The features – attributes – that are shared by the target and the analogue should be pointed out in what is termed 'mapping' (Treagust 1995). Just as it is important to map shared attributes, it is equally important to point out to the students those attributes that are not shared. There may be features of the analogue that are unlike those of the target, which may lead to the generation of alternative conceptions. Such features are called 'negative attributes'. For example, a particular feature of the spring analogue is that the tension force within the spring increases as it stretches. This directly opposes the target feature, in which the gravitational force decreases with distance. The analogue and the target may also contain features that cannot be mapped – such features are called 'neutral attributes'. For example, a neutral attribute of the spring is the composition of the spring itself.

A teaching approach in the use of analogies developed by Glynn (1991) and used by Treagust (1995, p. 52) consists of the following six steps:

1 Introduce the target concepts to be learnt.
2 Cue the students' memory of the analogous situation.
3 Identify the relevant features of the target concept and the analogue.

4 Map out the similarities between the target and the analogue.
5 Draw conclusions about the target concept.
6 Indicate where the analogy breaks down.

Treagust refined this approach to three phases that begin with a 'focus' phase, followed by an 'action' phase and finishing with a 'reflection' phase. In the focus phase, the teacher needs to determine if the concept is difficult, unfamiliar or abstract for the students. If any of these is the case, then the use of an analogy is warranted. The focus phase also elicits the students' preinstructional ideas of the concepts and determines the extent of familiarity the students have with the analogue. The action phase discusses the shared and unshared attributes of the target and analogue. The reflection phase analyses the analogy in terms of clarity and usefulness, which may result in a need to refocus if the students see the analogy as not useful or confusing.

Analogies that are useful for the classroom are those that contain analogues that are much simpler than the target and have features that are well understood by the students. Analogies that are pictorially presented or deal directly with the students' real world, such as money, people, food and relationships, are also beneficial to use. Research indicates that personal analogies are not only viewed by students as being enjoyable, but also lead to enhanced understanding of the science concepts (Chiu and Lin 2005; Treagust 1995).

A personal analogy (see Figure 5.6) that students find enjoyable is described as a line-of-students analogy, which explains the concept of sound energy transmission as a wave phenomenon. In this analogy, the medium in which sound travels (target) is compared to students standing in a single line (analogue). In the mapping of attributes:

- the sound source is compared to a student at one end of the line
- the ear of listener is compared to a student at other end of line
- particles in the medium are compared to students between the sound-source student and the ear-of-listener student.

A role-play is established in which the sound-source student begins to vibrate, causing a collision with a neighbouring student. Through a series of collisions, a travelling vibration is observed to travel from the sound-source student and ear-of-listener student. If one follows Treagust's approach to the application of analogies in this instance, then teachers need to be

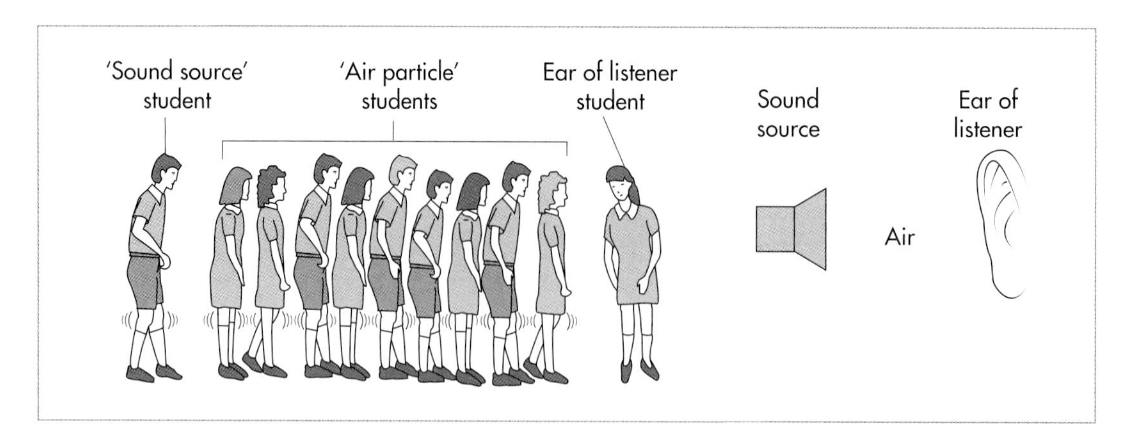

FIGURE 5.6 Line-of-students analogy for sound transmission

aware that in the focus phase they need to be looking for a common alternative conception held by students that sound can travel through space. In mapping the analogy with respect to the medium and the students, the students need to realise that they are representing particles in the medium and so they should imagine that there is space between neighbouring student particles. It should be noted that most primary students do not have a sub-microscopic view of matter consisting of particles with spaces between them (see Chapter 9).

While the use of personal analogies can be successful, the teacher needs to be aware of students not attaching human characteristics, such as feelings or behaviour, to inanimate objects and concepts. In the example of the line-of-students analogy, a boisterous group of students may result in the energy of the vibrating source increasing with distances as successive students push harder after being knocked.

Another problem with the use of analogies is that the students may understand the analogue as reality itself (Grosslight et al. 1991). This may be overcome with an emphasis on where the analogue does not link with the concept, or the application of multiple analogies to explain the same concept or different aspects of the same concept. Another analogy that explains sound energy transmission as a wave phenomenon makes use of slinkies (helical springs). The mapping of this analogy relates the medium with a stretched slinky. Vibrations that are initiated at one end of the slinky are seen to travel the length of the slinky. The following activities explore other analogies, particularly those that relate to energy.

The personal analogy of sound energy transmission described above can be modified slightly to explain the transfer of thermal energy through materials. The transfer of thermal energy in this way is known as 'conduction'. For example, if one holds one end of a metallic rod and places the other end in an open fire, one soon finds that the entire rod heats up as thermal energy is conducted along the entire length of the rod. In this analogy, the metallic rod (target) in which the thermal energy travels is compared to a line of students (analogue). The students represent the metallic particles of the rod. A student at one end of the line of the 'metallic particle' students represents the 'heat source' student, while another student at the other end represents the 'hand of the rod holder' student.

A key concept in this analogy is the idea that the temperature of an object is related to the motion energy of its constituent particles. As an object gets heated, its particles move faster; they vibrate at a faster rate. Therefore, in the analogy, the 'heat source' student vibrates, causing a collision with a neighbouring 'metallic particle' student, which makes this student vibrate. Through a series of collisions, vibrations are set up along the line of 'metallic particle' students to finally reach the 'hand of the rod holder' student. The increase in vibration of the students represents an increase in the temperature of the rod and the hand of the person holding the rod.

This analogy can also be used to explain why some objects expand when they get heated (the particles in the object vibrate faster, thus creating more space), but it can't explain why some objects, such as those made out of plastic or ceramics, don't conduct thermal energy as well as metallic objects. As a teacher, you could discuss with the students how the analogy could be modified to account for such effects.

Activity 5.9 **Analysing classroom analogies relating to energy**

The following list of analogies relate to explanations of concepts of energy. Discuss with other teachers each of the following analogies' applicability to the primary school classroom. Keep in mind not only where the analogy fits, but also where it breaks down.

- The sound energy transmission analogy has been described above. How can the analogy be used to explain the observation that sound:
  - cannot travel through space
  - travels faster and further in solids, such as a metal rail, than in the air
  - reflects off solid barriers?
- The dissipation of energy is like the spreading out of money after you receive your pay packet. The amount of money you receive for your pay remains the same, but it gets spread around: for example, pocket money is given to your children and money is spent to buy groceries or pay bills. This money, in turn, gets spread further as children spend their money and shopkeepers pay workers and suppliers of raw materials. How might this analogy be undertaken as a role-play in the classroom?
- The degradation of energy is like the decrease in the value of money as it gets converted from one type of currency to another. For example, if one pays $100 to a money exchange, some is taken for a fee, say $3, while the rest, $97, is converted into another currency. Repeated conversions progressively decrease the amount. How does this analogy show that energy:
  - comes in different forms
  - is conserved
  - dissipates?
- The use of marbles can be used in analogies to explain a number of energy concepts. For example, an open tray containing several marbles is shaken to show how the tiny particles that make up an object behave when heated. Shaking the tray faster will result in some marbles being expelled from the tray. How does this action of the analogue relate to the target in terms of heating an object?
- In another example, a line of marbles is set up between two rails. Striking one end of the line of marbles with a rolling marble results in a marble at the other end of the line moving. How can this analogue explain conduction of thermal energy?
- In another example, a tray with holes in it is positioned above another tray. The top tray is held at an incline and marbles are allowed to roll from the top of the tray. Some marbles will fall through the holes while others will continue to roll to the bottom of the top tray. How can the analogue explain energy degradation? What do the marbles represent in the analogy?

Science as a human endeavour

NOS

There are characteristics of science as a human endeavour that can be readily incorporated into lessons focusing on energy. These include the NOS (e.g., science ideas and understandings change as new evidence becomes available) and the influence of science (e.g., the effects of science and technology on our lives, as well as how science is used).

The nature of science

In recent times, scientists have seen evidence of global warming and climate change which has led them to investigate possible causes. This then led to debate among scientists, which has been played out in public forums, involving differing views of whether or not these phenomena exist, and what the underlying reasons for this might be. The current thinking is that most scientists have come to the view that the increase in global warming is due to an enhanced greenhouse effect resulting from increased levels of greenhouse gases released into the atmosphere through human activity (Boyes, Stanisstreet and Yongling 2008). This new scientific understanding came about through debates where scientists with differing views put forward claims based on evidence gathered through scientific means, and where an ultimate consensus view was reached. This process of construction of new knowledge is a key feature of the processes that characterise the NOS (see the Chapter 1 section 'The nature of science: how it works' on p. 5).

A large number of studies have explored students' understandings of the concepts related to the greenhouse effect, ozone depletion, global warming and climate change (Anderson and Wallin 2000; Boon 2009; Boyes and Stanisstreet 1993; Boyes, Stanisstreet and Yongling 2008; Boylan 2008; Daniel, Stanisstreet and Boyes 2004; Lee et al. 2007; Rye, Rubba and Wiesenmayer 1997). Preservice primary school teachers' views have also been researched (Ekborg and Areskoug 2006; Papadimitriou 2004). Some of the most popular and persistent alternative conceptions found in students' and preservice teachers' thinking are detailed in Chapter 12, where there is also further discussion of students' ideas and the consequent pedagogical implications related to this topic.

A useful resource about climate change has been produced by the Intergovernmental Panel on Climate Change (IPCC), which is an organisation instituted in 1988 by the United Nations Environment Programme and the World Meteorological Organization to engage the world's leading experts in reviewing the most up-to-date, peer-reviewed literature on the scientific and technical aspects of climate change. The resource, called the Climate Change Information Kit (see http://unfccc.int/resource/docs/publications/infokit_2002_en.pdf) summarises in simple language the most up-to-date findings on climate change. A listing of the key concepts associated with the greenhouse effect, ozone depletion, global warming, climate change and renewable energies can be found in the 'Concepts and understandings for primary teachers' section below (see p. 212). These concepts are expanded on in Chapter 12.

A key activity of scientists is modelling, which involves the creation of models that not only explain phenomena but also make predictions. This is pertinent for environmental scientists, who use sophisticated mathematical models to predict such things as future increases in global temperature and sea levels according to increases in atmospheric levels of greenhouse gases. Activity 5.10 gives students some insight into the modelling process, where they are able to predict global mean temperatures through changes in concentrations of greenhouse gases.

Activity 5.10 Using models to make predictions

In this activity, you are asked to use modelling software to make some predictions. Access the simulation called The Greenhouse Effect, which can be downloaded from http://phet.colorado.edu/en/simulation/greenhouse.

Open the 'Glass Layers' section of the animation. You will notice that sunlight is represented by yellow particles of light called photons. The sunlight particles are absorbed by the Earth, heating it. The Earth then

radiates energy in the form of infra-red radiation, which is represented by red photons.

- Where there are no glass layers shown, this simulates an Earth without any atmosphere. What global mean temperature does this animation software predict?
- Insert one glass layer. What is the effect on the sunlight and infra-red photons? What is the effect on the temperature.
- Insert a second and then a third glass layer. What is the effect on the temperature?
- Open the 'Greenhouse Effect' section of the animation. Begin the animation using today's atmospheric setting. What is the effect on the sunlight and infra-red photons? What is the effect on the temperature? Contrast these effects with the simulation using a glass layer.
- Adjust the greenhouse gas concentration. What is the effect on the sunlight and infra-red photons? What is the effect on the temperature?

A lot of public debate about climate change relates to the level of confidence in scientists' ability to predict future global temperature rises. This is not surprising as weather forecasters are sometimes unable to correctly predict the weather only a few days into the future. In teaching about global warming, teachers need to point out to students that a key role for scientists is modelling, which doesn't always produce predictions that are 100 per cent accurate. However, the models that scientists construct are constantly being revised as more information is gathered.

The influence of science: the effect of science and technology on our lives

The significant use of fossil fuels for humankind's energy needs has created a major environmental problem, with an enhanced greenhouse effect leading to global warming and climate change. On a personal level, people need to reduce their energy consumption from fossil fuels, and on a societal level, people need to explore the use of non-renewable energies. Students can get some insight into what they personally contribute to the greenhouse effect, and what they can do to reduce energy consumption from fossil fuels, by undertaking Activity 5.11.

Activity 5.11 Calculating my personal contribution to the greenhouse effect

In this activity, you get to estimate what effect your home life has on greenhouse gas emissions. You will need to access the Australian Greenhouse Calculator (see http://www.epa.vic.gov.au/GreenhouseCalculator/calculator/default.asp), a program that will calculate your yearly greenhouse gas emissions based on the information you provide. The calculator will also compare your greenhouse emissions with that of a 'typical' house and also a 'green' house.

Access the calculator and input information related to your home.

- What is your annual CO_2 contribution to the greenhouse effect?
- How does this contribution compare to other homes in your area? How does this contribution compare to a 'green' home?
- What aspect of your personal living contributes the most CO_2 to the greenhouse effect?
- In what realistic ways can you limit your annual CO_2 contribution to the greenhouse effect?
- In what ways are your CO_2 contributions related to your energy costs?

Two useful curriculum resources for teachers include the Australian Sustainable Schools Initiative (AuSSI) (see http://www.environment.gov.au/education/aussi) and Energy Savers (see http://www.forteachersforstudents.com.au/Origin/teachers.html). The AuSSI website contains curriculum resources that can assist teachers in broadening their students' understandings of energy, guide students through an energy audit process, and plan relevant activities to allow students to make inquiries about energy and sustainable energy options and practices in a locally relevant way. The AuSSI website also refers to case studies of Australian schools that have developed a range of sustainable practices on their grounds and in their communities. Energy Savers is a new teachers' resource for upper primary students that helps bring issues on climate change and energy efficiency 'to life'; the website also contains a useful online energy efficiency calculator.

SSI

By undertaking audits like that described in Activity 5.11, teachers and students can be prompted to realise that, on a personal and school level, they can make a positive difference in limiting the phenomenon of global warming.

Summary

This chapter has explored one of science's overarching and complex concepts, that of energy. Everyone has their own understanding of energy and will use the word differently in a variety of contexts. For the primary school teacher, the different contexts may involve daily life as distinct from teaching energy within a biological, chemical or physical system.

It is helpful to realise that developing our understanding of energy is a slow, laborious and accretional process. At primary school we need to be aware that, as teachers, we have learnt science at secondary school and maybe in a tertiary establishment. We take into the classroom an understanding of energy that has developed partly as a result of those experiences. We need to acknowledge that each of the children we teach also will come to class with an understanding that may involve several meanings of the word 'energy', depending on the context. When we teach an energy topic, we need to consider the curriculum we teach. The key concepts of energy and light (listed in the next section) act as powerful scrutineers for this. Clarifying our own understanding of scientists' understanding of the concept is a useful way forward, ensuring that we provide experiences that will help learners understand the meanings of energy that are appropriate for scientific contexts. This may make energy appear to be a complicated and difficult concept. Well, it is, but energy is not alone in this – it shares the commonality of being complex with many other scientific concepts. This chapter described how energy related to global warming and the greenhouse effect can be incorporated into primary school science lessons as evidence of how science ideas and understandings change as new evidence becomes available, and how science and technology affect our lives.

If we can begin to convey to learners the complexity and invented nature of the concept, provide them with lots of experiences of different forms of energy, and develop their understandings of these forms and a knowledge of the properties of various energy forms, then primary school learners will have a sound basis for future learning.

Best wishes for your energetic journeying with learners.

Concepts and understandings for primary teachers

Key concepts associated with energy

- Energy is an invented idea that helps us make sense of processes of change.
- Energy may be considered as a job-doing capability.
- Energy is associated with systems undergoing change.
- Energy is manifest in many forms. Different forms of energy include thermal, chemical, light, gravitational, sound, elastic, movement/motion/kinetic, nuclear and electrical.
- Energy is stored in two basic forms – kinetic and potential. Kinetic energy relates to the speed and mass of an object. Potential energy relates to the position of objects with respect to other connected objects.
- Energy forms transfer from one place to another.
- Methods of energy transfer include conduction, convection and radiation.
- Energy forms have various energy sources.
- A source of energy may produce more than one form of energy.
- Energy can be transformed from one form to another.
- Changes occur when energy is transformed and these changes may be observed and measured. Energy is measured in joules. The rate of energy change is power and is measured in watts.
- Energy is neither created nor destroyed. When energy changes in form, the total amount of energy remains constant.
- In systems undergoing change, energy spreads out from the source. This is called 'dissipation of energy'.
- In transforming energy for our use, some energy is used for the chosen task, while the rest appears in non-useful forms and is lost.
- Energy eventually 'runs down' by being transformed into low-level heat. This is called 'degradation of energy'.

Key concepts associated with light

- Light is an entity that travels very fast.
- Light travels in straight lines. Ray diagrams show light travelling in straight lines as arrows.
- Shadows are the result of an object blocking the passage of light.
- Luminous objects emit light; non-luminous objects reflect light.
- Light from luminous objects keeps travelling until it interacts with matter. In interacting with matter, light reflects, transmits and/or gets absorbed.
- Light from each point on a luminous object travels in all directions.
- Transparent objects enable light to be transmitted, whereas opaque objects do not.
- Mirrored surfaces will reflect a beam of light in one direction, whereas non-mirrored surfaces will reflect a beam of light in all directions.
- The vision process is enacted when light enters the eye directly from a luminous object or through reflection from a non-luminous object.
- Images of objects in mirrors, lenses and other transparent objects are caused by light from the object being redirected in some way.
- Light from the sun and other similar luminous objects is a mixture of all the colours of the rainbow.
- The colour of an object is determined by what colour of light it reflects.

Key concepts associated with the greenhouse effect, ozone depletion, global warming, climate change and renewable energies

The following are expanded upon in Chapter 12.

- The natural greenhouse effect is responsible for keeping Earth's average temperature at 14°C. Without the natural greenhouse effect, the Earth's average temperature would be −18°C.
- An enhanced greenhouse effect is due to increased concentrations of greenhouse gases in the Earth's atmosphere, produced by human activity, and is responsible for global warming and climate change.
- Greenhouse gases in order of abundance in the Earth's atmosphere are water vapour, carbon dioxide, methane, nitrous oxide, ozone and chlorofluorocarbons (CFCs).
- There is a difference between incoming (solar) radiation from the sun and outgoing (infra-red) radiation by the Earth.
- When averaged over a year, the incoming energy in both the Earth and its atmosphere equals the outgoing energy.
- The absorption by greenhouse gases of infra-red radiation rather than solar radiation underlies the greenhouse effect.

- An enhanced greenhouse effect results in changing weather patterns and sea levels rising.
- The enhanced greenhouse effect is a separate effect to the depletion of the ozone layer in the Earth's atmosphere.
- Ozone is a substance that contains molecules containing three oxygen atoms (O_3) and absorbs ultraviolet (UV) radiation from the sun.
- The reduction of the ozone layer will cause an increase in UV radiation on Earth. An excess of UV rays has been linked to skin burns, skin cancer, cataracts and harm to certain crops and marine organisms. However, excess UV radiation is not linked to global warming.
- Ozone-depleting substances (ODSs) are widely used in refrigerators, air conditioners, fire extinguishers, dry-cleaning (as solvents for cleaning), electronic equipment, and as agricultural fumigants. ODSs include CFCs, halon, carbon tetrachloride, methyl chloroform, hydrobromofluorocarbons (HBFCs), hydrochlorofluorocarbons (HCFCs), methyl bromide and bromochloromethane (BCM).
- CFCs and ozone are also greenhouse gases but carbon dioxide does not deplete the ozone layer.
- A renewable energy source (e.g., biomass from plants) is one that can be replenished in a short amount of time. A non-renewable energy source (e.g., coal) cannot be replenished in a short period of time.

Search me! science education

Search me!

Explore **Search me! science education** for relevant articles on energy. Search me! is an online library of world-class journals, ebooks and newspapers, including *The Australian* and the *New York Times*, and is updated daily. Log in to Search me! through http://login.cengage.com using the access code that comes with this book.

KEYWORDS

Try searching for the following terms:

- Renewable energies
- Energy
- Teaching energy

Search tip: **Search me! science education** contains information from both local and international sources. To get the greatest number of search results, try using both Australian and American spellings in your searches: e.g., 'globalisation' and 'globalization'; 'organisation' and 'organization'.

Appendices

In these appendices you will find material related to energy that you should refer to when reading Chapter 5. These appendices can be found on the student companion website. Log in through http://login.cengage.com using the access code that comes with this book. Appendix 5.1 is included in full below.

Appendix 5.1 Difficulties in understanding energy by some preservice and practising primary teachers

Various researchers (Kruger 1990; Kruger, Palacio and Summers 1992; Pinto, Couso and Gutierrez 2005; Trumper 1997) have explored preservice and practising primary teachers' understanding of energy. The following list indicates some of the difficulties some teachers have with the concept of energy:

- There is a lack of ability to differentiate between the concepts of force and energy.
- Energy is not associated with motion.
- Energy is only associated with motion.
- Energy is not an attribute of stationary objects.
- Kinetic energy doesn't depend on speed.
- Energy is only found in living things.

- Energy is seen as human liveliness.
- Energy is a life force that is within all living things.
- Energy is a hidden force present all the time in a substance waiting to be used.
- Energy comes from the sun, the original source of all energy.
- Energy is not conserved – it can be created and/or destroyed.
- There is non-acceptance of the idea of energy degradation.
- Gravitational energy is either unrecognised or misunderstood. Other forms of potential energy are unrecognised or unclear.
- Energy is considered to be a concrete entity rather than an abstract one.

Appendix 5.2 Concept map of energy

The concept map in this online appendix shows links between many of the key ideas associated with energy.

Appendix 5.3 A pre-test questionnaire about energy for Year 5/6 students

A pre-test questionnaire suitable for Year 5/6 students is given in this online appendix.

References

Adams, R., Doig, B. & Rosier, M. 1991. *Science Learning in Victorian Schools: 1990*. Melbourne: Australian Council for Educational Research (ACER).

Ametller, J. & Pinto, R. 2002. Students' reading of innovative images of energy at secondary level. *International Journal of Science Education*, 24 (3), pp. 285–312.

Anderson, B. & Wallin, A. 2000. Students' understanding of the greenhouse effect, the societal consequences of reducing CO_2 emissions and the problem of ozone layer depletion. *Journal of Research in Science Teaching*, 37 (10), pp.1096–111.

Australian Curriculum, Assessment and Reporting Authority (ACARA). 2011. F-10 Australian Curriculum: Science. Available at http://www.australiancurriculum.edu.au/Science/Curriculum/F-10 (accessed March 2011).

Bell, B. 1993. *Children's Science, Constructivism and Learning in Science*. Geelong: Deakin University Press.

Bendall, S., Goldberg, F. & Galili, I. 1993. Prospective elementary teachers' prior knowledge about light. *Journal of Research in Science Education*, 30 (9), pp. 1169–87.

Boon, H. 2009. Climate change? When? Where? *Australian Education Researcher*, 36 (3), pp. 43–65.

Boyes, E. & Stanisstreet, M. 1993. The 'greenhouse effect': Children's perceptions of causes, consequences and cures. *International Journal of Science Education*, 15 (5), pp. 531–52.

_____, Stanisstreet, M. & Yongling, Z. 2008. Combating global warming: The ideas of high school students in the growing economy of South East Asia. *International Journal of Environmental Studies*, 65 (2), pp. 233–45.

Boylan, C. 2008. Exploring elementary students' understanding of energy and climate change. *International Electronic Journal of Elementary Education*, 1 (1), pp. 1–15.

Brook, A. & Wells, P. 1988. Conserving the circus? *Physics Education*, 23, pp. 80–5.

Chiu, M-H. & Lin, J-W. 2005. Promoting fourth graders' conceptual change of their understanding of electric current via multiple analogies. *Journal of Research in Science Teaching*, 42 (4), pp. 429–64.

Curriculum Corporation 2010. The Le@rning Federation Schools Online Curriculum Content Initiative. Available at http://www.thelearningfederation.edu.au/default.asp (accessed 20 June 2010).

Daniel, B., Stanisstreet, M. & Boyes, E. 2004. How can we best reduce global warming? School students' ideas and misconceptions. *International Journal of Environmental Studies*, 61 (2), pp. 211–22.

Davies, P. 1984. *Superforce: The Search for a Grand Unified Theory of Nature*. London: Heinemann.

Davis, J. 2001. Conceptual Change, in M. Orey (ed.). *Emerging Perspectives on Learning, Teaching, and Technology*. Athens, GA: Department of Educational Psychology and Instructional Technology, University of Georgia. Available at http://projects.coe.uga.edu/epltt (accessed 28 September 2010).

Dawson, T. L. 2006. Stage-like patterns in the development of conceptions of energy, in X. Liu and W. Boone (eds). *Applications of Rasch Measurement in Science Education*. Maple Grove, MN: JAM Press, pp. 111–36.

Duit, R. 1987. Should energy be illustrated as something quasi-material? *International Journal of Science Education*, 9 (2), pp. 139–45.

_____ & Haeussler, P. 1994. Learning and teaching energy, in P. Fensham, R. Gunstone and R. White (eds). *The Content of Science*. London: Falmer Press, pp. 185–200.

Ekborg, M. & Areskoug, M. 2006. How student teachers' understandings of the greenhouse effect develops during a teacher education programme. *Nordic Journal of Science Education (NorDiNa)*, 5, pp. 17–20.

Ellse, M. 1988. Transferring not transforming energy. *School Science Review*, 69, pp. 427–37.

Feher, E. & Rice, K. 1986. Shadow/shapes. *Science and Children*, 24 (2), pp. 6–9.

_____ 1988. Shadows and anti-images: Children's conceptions of light and vision, II. *Science Education*, 72 (5), pp. 637–49.

_____ 1992. Children's conceptions of colour. *Journal of Research in Science Teaching*, 29 (5), pp. 505–20.

Feynman, R. 1963. *Lectures on Physics*. San Francisco: California Institute of Technology.

Fleer, M. 1996. Early learning about light: Mapping preschool children's thinking about light before, during and after involvement in a two week teaching program. *International Journal of Science Education*, 18 (7), pp. 819–36.

_____, Hardy, T., Barron, K. & Malcolm, C. 1995. *They Don't Tell the Truth about the Wind: A K–3 Science Program*. Melbourne: Curriculum Corporation.

Forde, T. 2003. 'When I am watching television I am not using any energy': An empirical study of junior science students' intuitive concepts of energy. *Irish Educational Studies*, (22) 3, pp. 71–88.

Glynn, S. M. 1991. Explaining science concepts: A teaching-with-analogies model, in S. M. Glynn, R. H. Yeany and B. K. Britton (eds). *The Psychology of Learning Science*. Hillsdale, NJ: Lawrence Erlbaum, pp. 219–40.

Grosslight, L., Unger, C., Jay, E. & Smith, C. 1991. Understanding models and their use in science: Conceptions of middle and high school students and experts. *Journal of Research in Science Teaching*, 28 (9), pp. 799–822.

Guesne, E. 1985. Light, in R. Driver, E. Guesne and A. Tiberghian (eds). *Children's Ideas in Science*. Milton Keynes, UK: Open University Press.

Heywood, D. S. 2005. Primary trainee teachers' learning and teaching about light: Some pedagogic implications for initial teacher training. *International Journal of Science Education*, 27 (12), pp. 1447–75.

Hubber, P. 2002. A Three Year Longitudinal Investigation into Six Secondary School Students' Understandings of Optical Phenomena. Unpublished PhD thesis, Bendigo: La Trobe University.

Kaper, W. H. & Goedhart, M. J. 2002. 'Forms of energy', an intermediary language in the road to thermodynamics? Part II. *International Journal of Science Education*, 24 (2), pp. 119–37.

Kress, G., Jewitt, C., Ogborn, J. & Tsatsarelis, C. 2001. *Multimodal Teaching and Learning: the Rhetorics of the Science Classroom (Advances in Applied Linguistics)*. London: Continuum International Publishing Group.

Kruger, C. 1990. Some primary teachers' ideas about energy. *Physics Education*, 25, pp. 86–91.

_____, Palacio, D. & Summers, M. 1992. Surveys of English primary teachers' conceptions of force, energy, and materials. *Science Education*, 76 (4), pp. 339–51.

_____, Summers, M., Mant, J., Childs, A. & McNicholl, J. 1998. *Teaching Energy and Energy Efficiently Effectively*. Hatfield, UK: Association for Science Education.

Langley, D., Ronen, M. & Eylon, B-S. 1997. Light propagation and visual patterns: Preinstructional learners' conceptions. *Journal of Research in Science Teaching*, 34 (4), pp. 399–424.

Lawrence, I. 2007. Teaching energy: Thoughts from the SPT11–14 project. *Physics Education*, 42 (4), pp. 402–9.

Learning in Science Project (LISP) (Energy). 1989. *Energy for a Change*. Hamilton: University of Waikato, Centre for Science and Mathematics Education Research.

Lee, H. & Liu, O. 2010. Assessing learning progression of energy concepts across middle school grades: The knowledge integration perspective. *Science Education*, 94, pp. 665–88.

Lee, O., Lester, B. T., Ma, L., Lambert, J. & Jean-Baptiste, M. 2007. Concepts of the greenhouse effect and global warming among elementary students from diverse languages and cultures. *Journal of Geoscience Education*, 55 (2), pp. 117–25.

Lijnse, P. 1990. Energy between the life-world of pupils and the world of physics. *Science Education*, 74 (5), pp. 571–83.

Liu, X. & Collard, S. 2005. Using Rasch model to validate stages of understanding the energy concept. *Journal of Applied Measurement*, 6, pp. 224–41.

_____ & McKeough, A. 2005. Developmental growth in students' concept of energy: Analysis of selected items from the TIMMS database. *Journal of Research in Science Teaching*, 42 (5), pp. 493–517.

Lofts, G., O'Keefe, D., Robertson, P., Pentland, P., Hill, B. & Pearce, J. 1997. *Jacaranda Physics 1*. Brisbane: Jacaranda Wiley.

Malcolm, C. (ed.). 1995. *There's an Emu in the Sky*. Melbourne: Curriculum Corporation.

Millar, R. 2005. Teaching about Energy. Research Paper 2005/11. York: Department of Educational Studies, University of York.

Morgan, D., Best, E., Lee, A., Nicholas, J. & Pitman, M. (eds). 1967. *Biological Science: The Web of Life*. Canberra: Australian Academy of Science.

Naylor, S. & Keogh, B. 2000. *Concept Cartoons in Science Education*. Cheshire: Millgate House Publishing and Consultancy Ltd.

Ogborn, J. 1990. Energy, change, difference and danger. *School Science Review*, 72 (259), pp. 81–5.

Osborne, J., Black, P., Meadows, J. & Smith, M. 1993. Young children's (7–11) ideas about light and their development. *International Journal of Science Education*, 15 (1), pp. 83–93.

Osborne, R. & Gilbert, J. 1980. A method for investigating children's concept understanding in science. *European Journal of Science Education*, 2 (3), pp. 311–21.

Papadimitriou, V. 2004. Prospective primary teachers' understanding of climate change, greenhouse effect, and ozone layer depletion. *Journal of Science Education and Technology*, 13 (2), pp. 209–307.

Papadouris, N., Constantinou, C. P. & Kyratsi, T. 2008. Students' use of the energy model to account for changes in physical systems. *Journal of Research in Science Teaching*, 45 (4), pp. 444–69.

Pinto, R., Couso, D. & Gutierrez, R. 2005. Using research on teachers' transformations of innovations to inform teacher education: The case of energy degradation. *Science Education*, 89, pp. 38–55.

Rice, K. & Feher, E. 1987. Pinholes and images: Children's conceptions of light and vision. I. *Science Education*, 71 (4), pp. 629–39.

Ross, K. A. 1988. Matter scatter and energy anarchy. *School Science Review*, 69 (248), pp. 438–45.

Rye, A. J., Rubba, J. A. & Wiesenmayer, R. L. 1997. An investigation of middle school students' alternative conceptions of global warming. *International Journal of Science Education*, 19 (5), pp. 527–51.

Sefton, I. 2004. Understanding energy, in Proceedings of 11th Biennial Science Teachers' Workshop, 17–18 June 2004, University of Sydney.

Selley, N. J. 1996a. Children's ideas on light and vision. *International Journal of Science Education*, 18 (6), pp. 713–23.

_____ 1996b. Towards a phenomenography of light and vision. *International Journal of Science Education*, 18 (7), pp. 837–46.

Shapiro, B. 1988. What children bring to light, in P. Fensham (ed.). *Developments and Dilemmas in Science Education*. London: Falmer Press, pp. 73–95.

_____ 1994. *What Children Bring to Light*. New York: Teachers College Press.

Solomon, J. 1982. How children learn about energy or does the first law come first? *School Science Review*, 63 (224), pp. 415–22.

_____ 1983. Learning about energy: How students think in two domains. *European Journal of Science Education*, 5, pp. 49–59.

_____ 1988. Messy, contradictory and obstinately persistent: A study of children's out-of-school ideas about energy. *School Science Review*, 65, pp. 225–9.

_____ 1992. *Getting to Know about Energy in School and Society*. London: Falmer Press.

Stead, B. & Osborne, R. 1980. Exploring science students' concepts of light. *Australian Science Teachers Journal*, 26 (3), pp. 84–90.

Stylianidou, F. 1997. Children's learning about energy and processes of change. *School Science Review*, 79 (286), pp. 91–7.

Treagust, D. F. 1995. Enhancing students' understanding of science using analogies, in B. Hand and V. Prain (eds). *Teaching and Learning in Science – the Constructivist Classroom*. Sydney: Harcourt Brace.

Trumper, R. 1990. Energy and a constructivist way of teaching. *Physics Education*, 25, pp. 208–12.

_____ 1993. Children's energy concepts: A cross-age study. *International Journal of Science Education*, 15 (2), pp. 139–48.

_____ 1997. The need for change in elementary school teacher training: The case of the energy concept as an example. *Educational Research*, 9 (2), pp. 157–73.

United Nations Environment Program (UNEP). 2002. The Climate Change Information Kit. Available at http://unfccc.int/resource/docs/publications/infokit_2002_en.pdf (accessed 20 June 2010).

University of Colorado. 2009. Physics Education Technology (phET) Interactive Science Simulations: Energy Skate Park & the Greenhouse Effect. Boulder: University of Colorado. Available at http://phet.colorado.edu/index.php (accessed 25 May 2010).

van Huis, C. & van den Berg, E. 1993. Teaching energy: A systems approach. *Physics Education*, 28, pp. 146–53.

Wang, T. & Xie, Y. December 2002. Where is light? A survey on the alternative conception of light and shadow. *Asia–Pacific Forum on Science Learning and Teaching*, 3 (2), article 11.

Warren, J. W. 1986. At what stage should energy be taught? *Physics Education*, 21, pp. 154–6.

Watson, G. 1996. *Science Works. Book 3*. Melbourne: Oxford University Press.

Watts, D. 1983. Some alternative views of energy. *Physics Education*, 18, pp. 213–17.

White, R. & Gunstone, R. 2008. The conceptual change approach and the teaching of science, in S. Vosniadou (ed.). *Handbook of Research on Conceptual Change*. Mahwah, NJ: Lawrence Erlbaum, pp. 619–28.

Wright, A. 2007. Capturing the essence of energy: A graphical representation. *Teaching Science*, 53 (2), pp. 24–8.

6

Minibeasts: linking science and design technology

by Wendy Jobling and Beverley Jane

Introduction

This chapter is a narrative, based on a case study of a class of Year 5 students who studied minibeasts (small animals – invertebrates) they found in their school environment. In this particular curriculum unit, science and design technology are linked in a symbiotic way; that is, they are dependent on each other – other interpretations of science and technology connections are presented in Chapter 1 (see the section 'Technology–science connections' on p. 38). Before the students could conduct their scientific explorations they had to design and make two technological products – a minibeast catcher and enclosure – for the specific purpose of collecting and housing the animals to be studied. Activities in this unit were student-centred, consistent with an inquiry-based teaching and learning approach espoused by the national *Primary Connections* resource (AAS 2005), and based on constructivist learning principles as described in Chapter 1.

Embedded in this narrative you will find ways of addressing the following issues of concern to primary teachers:

- putting constructivist teaching and learning theories into practice
- planning learning experiences within a meaningful context
- engagement through focusing on student-generated questions
- involving students in learning experiences that require the collection and analysis of data to support or refute a hypothesis, illustrating the nature of science
- relating content to science as a human endeavour
- linking several learning areas
- making use of limited resources
- selecting assessment strategies.

The students shaped the story because they had ownership of their work.

> My aims for this unit of work were to have the children looking at their school environment and becoming aware that there are a lot of elements to that school environment. You've got your trees and shrubs, but on and under and perhaps even within the bark of those trees and shrubs you've got animals and they make up a whole ecosystem. In order to study those, the children were needing to find some way of

> capturing the small animal they wanted to study further, and then, through asking and posing their own questions, finding out more about that animal and having a very much child-centred unit of work so the children feel that they had ownership of what they were doing. The children come up with the design for the enclosure they're going to put the animal into, they design how they are going to catch that animal, they pose the questions that they're interested in finding out about that particular animal. (Wendy, teacher)

Participation in the unit encouraged the students to value their natural environment, take responsibility for aspects of their learning, work cooperatively and to communicate orally as well as in written form. The time frame for the unit was eight weeks, with sessions of one-and-a-half hours per week.

Following on from this unit, the school concerned developed a greater focus on environmental issues, including sustainability both within the school and the broader community. Actions have included:

- a recycling program for paper, plastics and food waste (food waste is recycled through composting, a worm farm and free-range hens)
- an organic vegetable garden that uses some of the recycled material above
- a greenhouse where indigenous plants are grown for use in local parks
- water conservation and harvesting via rainwater collection tanks
- student participation in programs such as the Model Solar Vehicle Challenge.

These actions have been incorporated into the curriculum at all levels. Many urban and rural schools also have programs where student learning in science is linked to the community.

Taking account of prior knowledge

As highlighted in other chapters, the ideas students bring to the classroom affect how and what they learn. Here, students' and scientists' ideas are either integrated into the narrative of the learning unit or referred to in related sections following the narrative. When planning the unit, Wendy took the students' prior knowledge into account by finding out what the students had done in earlier years at school and drawing on her own past experiences with these students. She began the unit with the students revisiting trees and shrubs that they had studied earlier in the year.

> Before the children went and designed their bug catcher and their bug enclosure we revisited some work that we did earlier in the year when we took a broad look at the school environment, and in the first session we went out and looked at the trees and shrubs within the school, having one tree or shrub for each group. The children selected it and they were really observing textures, leaf shapes, posing questions as to what advantage to the tree it might be to have that type of bark, whether it was going to give protection from insect attack or even the climatic conditions. You are really building up that spirit of inquiry and questioning and that leads naturally into looking at the creatures found in and around it, and the sorts of questions you could generate for yourself from observing those and wondering what they are and how they live, even how they reproduce. (Wendy, teacher)

A

This introductory activity that Wendy described was important because it encouraged students to link new knowledge with their existing knowledge (Harlen 2009; Harlen and Qualter 2009). These students' prior knowledge was about snails. In previous years they had found out the snail's preferred habitat, food and other survival needs. The students had also done activities that involved looking at drawings of snails that showed no two snail shells were the same.

Other prior knowledge would have been gained during the children's informal learning experiences at home with their families (Heliden and Heliden Onnestad 2006). Research by Jane and Robbins (2004, 2006) has shown the importance of intergenerational learning involving grandparents. Their study of everyday interactions between grandparents and their grandchildren has documented a range of science and technology activities that foster young children's understandings of these subjects (Robbins and Jane 2005, 2006). Teachers who hold a constructivist view of learning value these everyday concepts and plan lessons to link with and build on these understandings.

Students observing and recording

In order to develop their observation skills, Wendy required the students to re-examine their trees and shrubs and search for animals living on or around these plants. The activity was introduced using focus questions:

> Are different-shaped leaves on the plant?
> Are they adult or juvenile leaves?
> Can you suggest reasons for any differences? For example, how could the leaves help the plant?
> Closely examine a leaf from your plant. What is its shape, size and texture? What are the advantages for the plant of having these particular leaves?
> Look at the colour and texture of the bark. What is its function?
> Find out the name of your plant. What are its origins and the ideal growing conditions for your plant? You could think up an experiment.
> Consider the trees as part of an ecosystem. How might you present the information about the animals on and under your plant? (Wendy, teacher)

NOS

The latter point encourages students to explore ways of representing and communicating their science understandings. The value of multiple student representations and multimodal representations has been highlighted (Carolan, Prain and Waldrip 2008; Tytler et al. 2009) (see the Chapter 1 section 'Representational and related strategies' on p. 29).

Magnifying glasses were available for students to help them examine the leaves more closely. After the class had observed their trees and surrounds, students shared the observations they had recorded. For example, Lisa reported that she had noticed different coloured leaves on her tree (see Figure 6.1). The leaves in the sun were green, whereas those in the shade were brown. Lisa also observed that the juvenile leaves were shaped differently from the more mature leaves.

Most schools have access to equipment such as digital cameras and microscopes (e.g., Dino-lite). Students can take digital photographs to record features like those noted by Lisa.

FIGURE 6.1 Mature and juvenile leaves on Lisa's tree

These photographs can then be inserted into students' written work. The computer digital microscope can be used in the same way. A feature of both devices is the capacity to make short video recordings that can highlight features such as the ways in which some animals move; for example, the movements of some young giant stick insects are similar to leaves moving in the wind.

Activity 6.1 Teacher posing questions

As facilitators, teachers can aim to capitalise on reporting back to the class sessions by initiating questions designed to extend the students' ideas.

- Think of some questions related to Lisa's observations that would help develop students' understanding of what is required for plants to make their own food.
- Would you introduce the process of photosynthesis here? Why?

Making links with the English literacy area

The blending of science and literacy (see the 'Science and literacy' section of Chapter 1 on p. 38) can be developed more effectively through teaching skills in relevant contexts. Science and design technology can provide the opportunity for students to learn a range of generic (e.g., writing) and science-specific (e.g., report genres) skills through activities such as reporting the results of an experiment or study. This incorporates reading, vocabulary development, speaking and listening. The *Primary Connections* resource has been developed with such a focus, and includes specific techniques such as how to use a science journal, vocabulary development through the use of a word wall, and ways of sharing information using a class science journal (AAS 2005). *Science Essentials* (Brenner 2005) is another resource incorporating literacy.

Kelly used the word 'camouflage' in her report to the class. When she was asked to explain the meaning of this word, she said, 'to blend in'. Kelly's comment illustrates the overlap that inevitably occurs with other learning areas.

Students find it empowering to use the correct terminology. For example, when Neil described a paper model of his group's technological product, he confidently used the word 'prototype' when talking to his peers. When new words are introduced in a meaningful context, language becomes more relevant to students.

All aspects of language are important in the learning of science and technology. No matter what subject area or topic, all teachers at all levels need to see themselves as teachers of language. Wellington and Osborne (2001) discuss the importance of language and offer suggestions on how to address this issue. They refer to the research that, over many years, has highlighted the language of science as being one of the main difficulties in the learning of science. As demonstrated above, this chapter gives examples of children's language and its development in the science and technology sociocultural context. A further example is the students' research questions and how they sought to answer these; their developing awareness of the need to base conclusions on evidence is apparent.

In conjunction with their plant study, students were encouraged to make careful observations of evidence of any small animals living on the leaves or bark of their tree. Students could take digital photographs, including digital microscope video clips, and incorporate these into a PowerPoint presentation or poster to share with the class. Students sought and gained additional information through reading texts. Some useful teaching and learning strategies related to the reading and writing of non-fiction are given by Wray and Lewis (1997).

ICT

Activity 6.2 Linking with English literacy

How do you think the plant study described above could be linked with the English literacy area? Relate your ideas to learning outcomes as described in your English syllabus, considering the different modes of communication: writing reports, procedural texts detailing the construction of each product, speaking and so on. Chapters 1 and 2 of this book provide many suggestions.

Authentic technological activity

The observation and recording task described in the previous section led to the technological activity of designing, making and appraising or evaluating a small animal catcher and enclosure in order to capture and house the animals (such as a slater, which is also known as a wood louse) for closer study. As mentioned previously, in this unit there was a symbiotic interaction between science and design technology.

Wendy's view of technology was influenced by one of her State Department of Education's curriculum publications and her training as a facilitator of in-service professional development courses in technology education. We define technology education as 'a process, a way of thinking and doing which satisfies needs in society and the environment' (Jane and Jobling 1995, p. 193; cf with Chapter 1). As students engage in a technological process, they draw on technological knowledge and, frequently, scientific knowledge in order to generate their products.

Students were required to work in small groups. Wendy framed the technological activity using the following open-ended design brief:

> **Design brief**
> Design and make:
> 1 A device for safely catching a small animal, such as a slater.
> 2 A suitable container to house the animal until you finish your study (the animal will then be returned to its original habitat).

Some students recognised that the technological task set by the teacher was authentic in that it had a real purpose.

> The part I liked best was making the bug house and actually using it.
>
> (Kelly, student)

Activity 6.3 Authentic (design technology) tasks

Consider how you would define an 'authentic' task in 'design technology' and in 'scientific inquiry' – for similarities and differences in the ways in which the latter has been interpreted, see Hume (2009). What do you consider to be the key features of each? You may want to look at your education system's information about design technology or speak to someone who works in an industry requiring the application of design and technology skills and knowledge. Subject associations such as the Design and Technology Teachers Association (DATTA; http://www.datta-australia.asn.au/index.html) are also invaluable sources of information; an Internet search will provide you with contact details for your local association.

Designing technological products

To focus the students' attention on the aspects of a useful design, Wendy asked the class what they needed to think about before deciding on their design. The resultant class brainstorm revealed that the students thought it important to consider safety aspects and the animals' welfare. At this point, a discussion concerning ethics ensued. Students were encouraged to empathise with the animals they were to study and to realise the importance of treating them in an ethical way. It was this discussion that prompted several student groups to place vegetation from the animals' natural environments in the designed enclosure. When baby slaters emerged in one enclosure, the students firmly believed that reproduction occurred as a result of the high-quality artificial environment that they had created.

Also during the initial brainstorm, students suggested that the design drawings should be to scale and be clearly labelled to show measurements. The type and cost of materials were also important factors to be considered. Wendy provided mainly recycled materials, such as paper, cardboard, plastic soft-drink bottles and pantyhose. People in the local community can often provide schools with materials considered by local businesses to be waste. Teachers can request these materials through school newsletters or by directly approaching local businesses, such as timber yards, or retailers who often have large quantities of surplus cardboard or packaging. Some areas have organisations for recycling a whole range of materials. Subject associations for science and technology can usually provide information on such organisations.

Classroom observations showed that the Year 5 students developed technological capability or know-how using basic materials.

> I can see that the children have developed enormously over the year, and again just using very simple materials, so we haven't spent a great deal on resources to get them to the stage where they can be given quite a variety of problems to solve and a selection of materials to use and then go about going through that process of designing, then making and then evaluating. (Wendy, teacher)

Evaluating technological products

Early on in the unit, Wendy emphasised to the students the importance of evaluating their technological products.

> When reporting back to the class you need to report on
> - how well your catcher worked
> - how well your bug house worked
> - the animals you found and where you found them, such as on the tree or shrub in the leaf litter or bark around the plant. (Wendy, teacher)

A

Throughout the unit, students shared their ideas and experiences by giving oral presentations to the class. Many students evaluated their products by talking about how well their catchers and enclosures worked and suggested improvements (see Figure 6.2).

> With our catcher it worked well but the bugs could climb up the top and get out, and the slaters could also roll off. So next time I think we had better put something over the top and a flap to close the front. (Liz, student)

After carrying out their investigations, the students answered specific questions about their technological products; these written reports were compiled into a class book called 'Year 5 Bug Catchers and Enclosures'. This book reveals that these students were capable of evaluating their products and redrawing their designs to incorporate modifications as improvements. In evaluating their animal-catching devices and enclosures, most students thought their products worked well. Most groups indicated that during the process they experienced some problems, but they were able to recommend changes to their designs. Although the majority of students thought the available materials were suitable, they also suggested different materials that would be better. In Kelly's written evaluation, she recommends using plastic instead of paper for the bug catcher and house combination (see Figure 6.3).

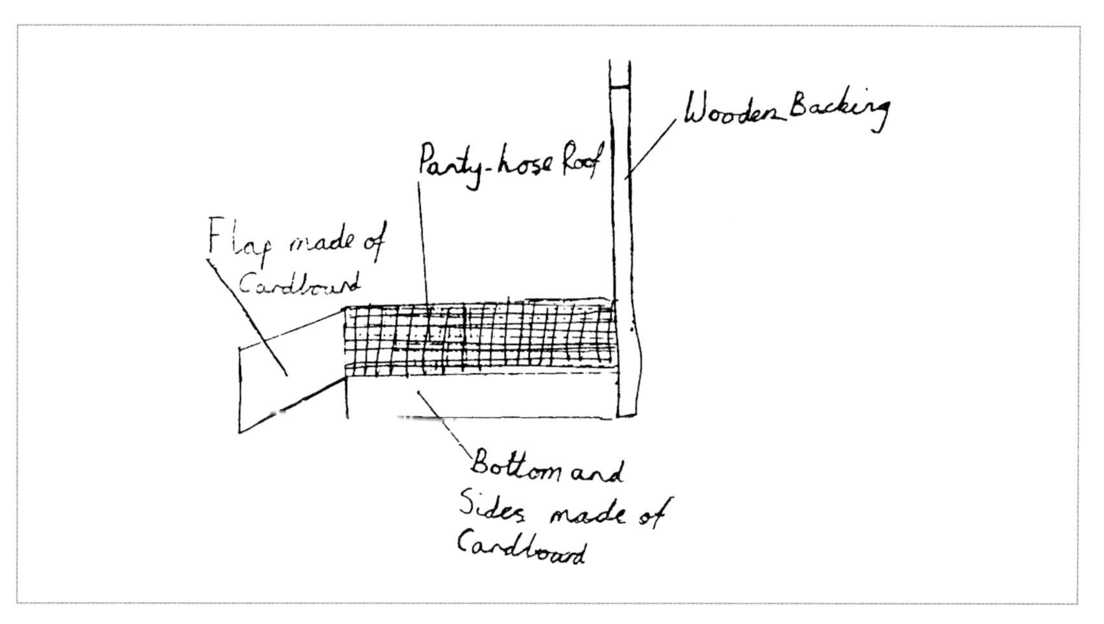

FIGURE 6.2 Liz's bug catcher design, with improvements

I thought ours was good. The bug catcher, it didn't work, but I think it would have worked better if we had plastic to make the house out of. If it was airtight, it would have worked better. (Kelly, student)

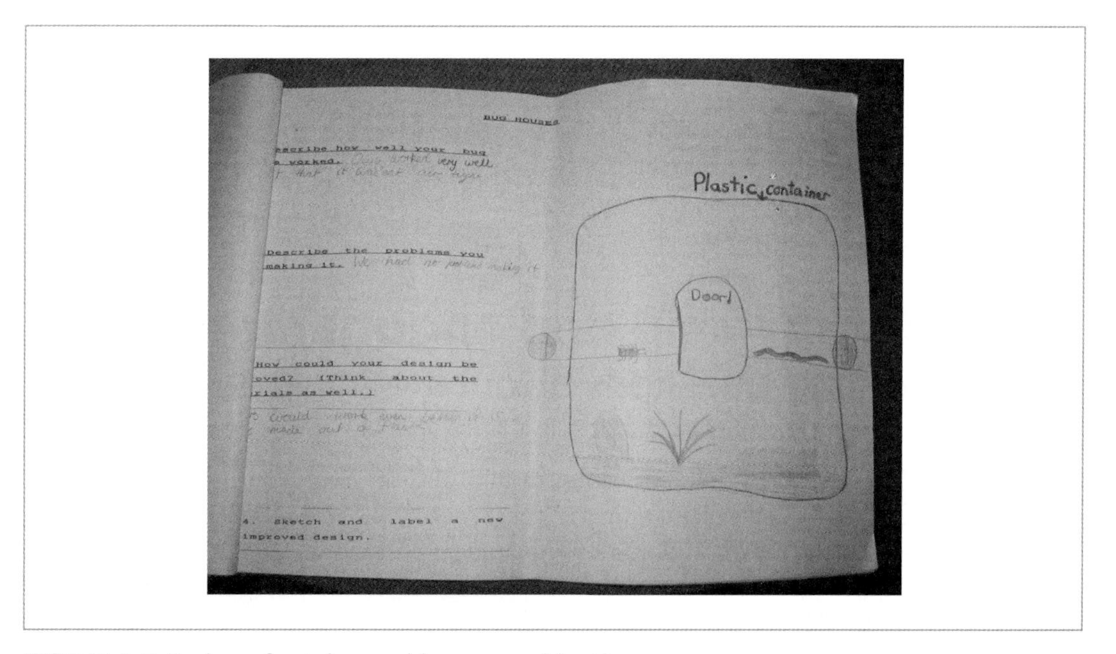

FIGURE 6.3 Design of catcher and house combination

FIGURE 6.4a Dorsal view of slater taken with a computer microscope

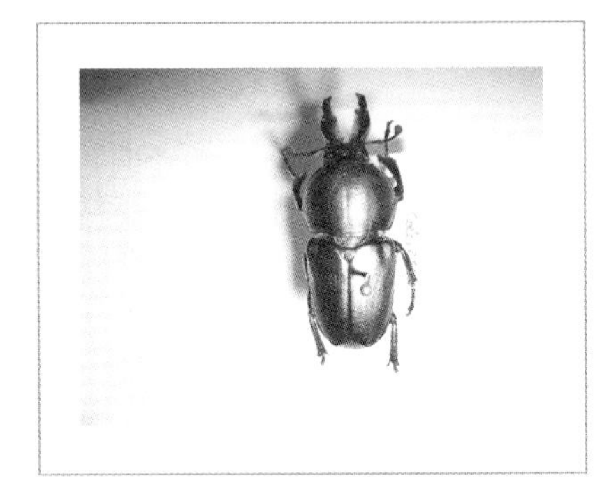

FIGURE 6.4b Golden green stag beetle coastal habitat

Source: Collection: Cuong van Huynh

ICT allows students' photographs (see figures 6.4a and 6.4b) to be inserted into word-processed documents or electronic slide shows. Short audio recordings can be made and added to slide shows.

Students were required to present their information orally. They were keen to communicate this in recorded form via video technology. The video titled *Children Linking Science with Technology* (Jane 1994; see http://www.deakin.edu.au/arts-ed/education/sci-enviro-ed/video/video_files/video3.php) shows the unit from the teacher's and the students' perspectives. The video makes a case for linking science and technology in the primary curriculum using a project approach.

Student ownership and engagement

Students asking questions about the minibeasts they found

After observing the animals in their enclosures or minilabs, the Year 5 students' curiosity led them to ask questions about their animals. These questions would fit into each of the categories suggested in Chapter 2. Some of the students' questions could be answered by referring to reference books or the Internet. Wendy, as facilitator, guided students to resources such as the library reference section and the insect poster on the display board. Barry enjoyed seeking information in this way:

> I liked catching the bugs and making all the things. I liked learning about all the slaters and looking up books to find out information. (Barry, student)

Emma and Liz asked several questions about the slug they found in the leaf litter under their tree. Their questions necessitated that they carefully observe their slug as well as refer to biology books. Scientists classify slugs in the Phylum Mollusca and the Class Gastropoda. The dark areas under the slug's skin fascinated both girls. During their literature search they identified this area as being the slug's intestines. They also discovered that their slug was called the Leopard slug. Figure 6.5 shows that Liz was capable of making detailed observations.

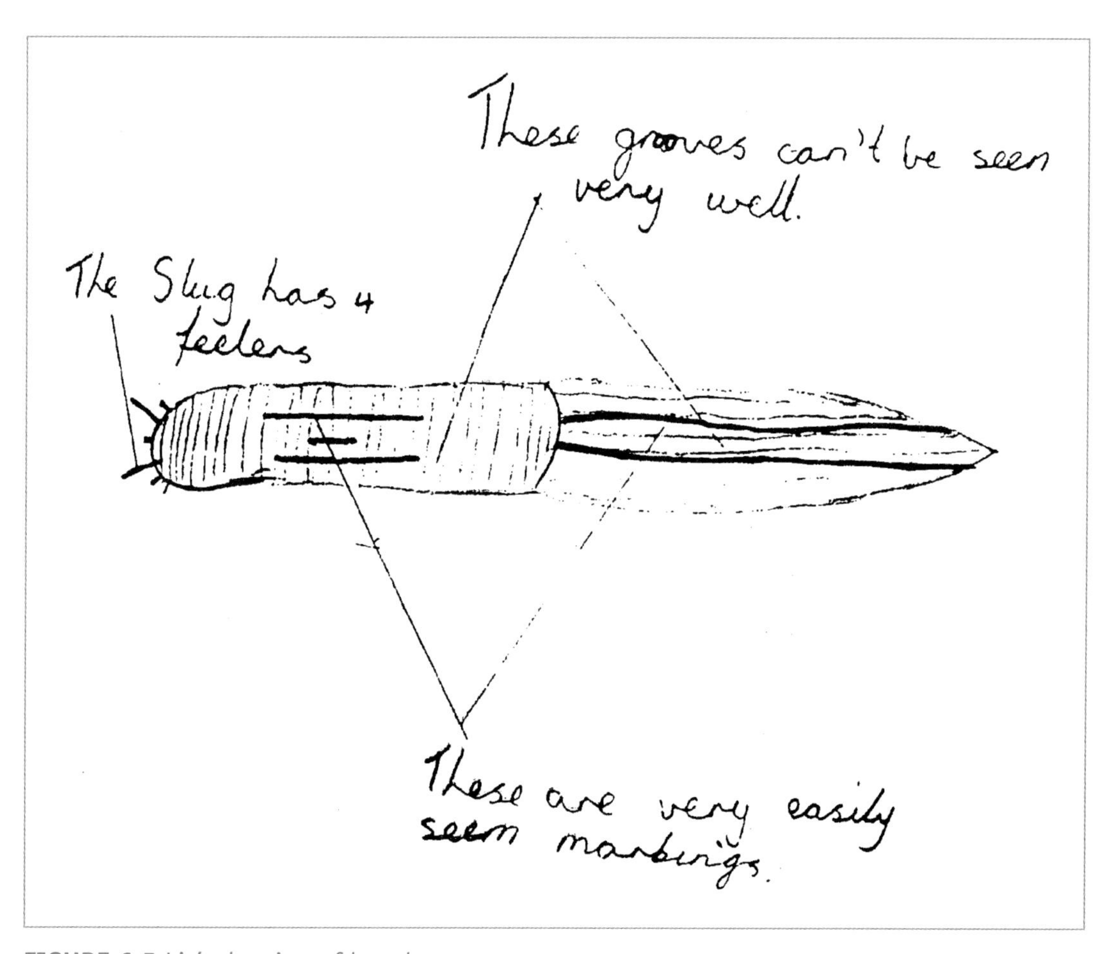

FIGURE 6.5 Liz's drawing of her slug

Below is Liz and Emma's report showing their questions and written answers. They used reference books to answer some of these questions.

Our research questions

1 Do slugs prefer light or dark places? They prefer dark, damp places.
2 What do slugs eat? Slugs eat juicy green leaves.
3 What is the average size of the Great Grey Slug? The average size is 10 inches, or 25 cm.
4 Do slugs come in different colours? Yes! Slugs come in different colours.

5 What is the name of the slug we caught? The name of the slug we caught is the Leopard Slug.
6 What do the intestines of a slug look like? (Liz and Emma, students)

Emma, who worked with Liz, preferred to draw the small animals they found from different perspectives. Her comprehensive written report clearly identified how the technological products were designed to achieve their purpose. She was aware of the problems encountered by her group, evaluated the effectiveness of the products in terms of their function, and suggested modifications to the designs. Emma drew the animals in the enclosure (made from a recycled plastic soft-drink bottle) and labelled the living and non-living components of the animals' habitat.

Neil was very interested in how slaters move, as was Morris, whose questions and answers are shown below as 'Slater park facts'. Mary's questions follow on and these reveal that she wanted to find out the position her slater would occupy in the schoolground ecosystem.

Slater park facts

1 How many legs do slaters have?
14
2 How does a slater move?
A slater moves its back legs first.
3 What's its scientific name?
'Isopod' is the scientific name.
4 How do you tell a male from a female?
The female has a yellow stomach. (Morris, student)

Mary's questions

1 How can you tell a male slater from a female slater?
2 Statistics:
- How many legs?
- What is the average length?
- What is its habitat?
3 What foods does it prefer?
4 How does it move?
5 Babies: colour, size?
6 What are its enemies?
7 Do they live in groups?
8 How does it protect itself?
9 Does it have a big role in the ecosystem?
10 What is its scientific name? (Mary, student)

The students' questions indicate what they wanted to know about the animals they found. Many were interested in the animals' body parts and the function of each part. Observations of groups of slaters led students to discover structural differences for themselves. Some students were also curious about the animals' behaviour. They asked, 'Why do slaters curl up into balls?' By the end of the unit, Katy understood that the slaters' behaviour was a means of protection from predators.

> When we were digging around to find insects on our tree we couldn't really find any at first, because we didn't know that when slaters try to protect themselves they curled up into little balls. We didn't know what they were. We thought they were little things that fall off trees, little seeds, and in the end we found some slaters and a little worm. (Katy, student)

Other students also generated questions and explanations for the behaviour of their small animals. Initially, Liz did not know that the animal she found was a slug.

WENDY: Why do you think the animal is so still?
EMMA: I think it is dead.
WENDY: What other reasons might there be?
LIZ: It might be scared.

This example shows the importance of the teacher's questioning skills as a scaffolding technique that extended the girls' thinking. The response also shows the tendency for young students to interpret animal behaviour in human psychological terms – see Deakin University's online science and environmental education resources for a discussion of students' alternative conceptions (http://www.deakin.edu.au/arts-ed/education/sci-enviro-ed/early-years/animals.php). Kallery and Psillos (2004) also discuss anthropomorphism and animism in early years' science, including teachers' conscious and unconscious use. They found that the majority of the teachers in their small-scale study who chose to use anthropomorphism and animism did so because of their lack of knowledge about the science topic or issue being taught. This extended to their pedagogical knowledge, where the teachers reported that they did not have appropriate ways to teach science concepts or explain phenomena. Unconscious use was attributed to three factors:

- It had been one way they had acquired their own knowledge.
- It was the way they had learned to present science to young children.
- They were influenced by the use of metaphors in everyday language; a concern about the use of such metaphors is their potential for misleading learners.

Science concepts related to ecosystems

The Year 5 students were capable of generating explanations for their observations. For example, Moira accounted for the lack of animals near Cameron's tree in terms of the interdependence between plants and animals. Other students found several kinds of animals living harmoniously in the same habitat. By identifying the food requirements of each animal, students were able to locate the place each animal occupied within the ecosystem. They drew food chains, beginning with plants as producers and progressing the chain through to small animals as consumers. They learnt that the earthworm is a special kind of consumer called a 'decomposer'. Snails and slugs eat plants, making them 'herbivores', or first-order consumers. These animals are prey for second-order consumers, such as birds. A food web is more complex and consists of many food chains. The book *Life in a Rotten Log* (Atkinson 1993) is a fascinating story of the ecosystem of a tree after it fell to the forest floor. A more recent publication of the same name (Penny 2004) also describes the process. *Monsters in Your Garden* (Cushing 2003) is an interesting book for children that relates to backyard ecosystems, while *Ask Dr K Fisher about: Minibeasts* (Llewellyn 2008) covers information about survival, habitats and food sources.

NOS

Activity 6.4 Food web in the schoolground ecosystem

Use the above description to show two simple food chains. Then, represent part of the schoolground ecosystem as a food web, showing the interrelationships between slugs, slaters, worms, insects, leaf litter, shrubs, trees and birds.

- One group of students found a spider and a centipede. Slaters are food for spiders and centipedes, so where would these animals fit in your food web?
- Consider the ways in which such information can be represented, such as your students role-playing a food web using knitting wool to make the connections.
- Discuss the strengths and weaknesses of each representation in terms of what is and is not communicated.

Refer to other chapters of this book (e.g., 1, 3 and 8) for further ideas about learning with a representational focus.

Children's conceptions of animals

Moira talked about insects. Research has shown that following a study of minibeasts, some students no longer view insects as animals. For example, Byrne (1993, p. 4) reports that: 'Recently I asked a group of Grade 5 children whether they regarded insects and spiders as animals. Less than a handful in the whole class were generous enough to award these creatures this status'.

The differences in students' definitions of animals and those of scientists has been a research focus for many years. Bell and Barker (1982) investigated how children's views of what constitutes an animal differ from those of scientists. A total of 15 five-year-olds and 23 nine- to 10-year-olds responded to the question 'Is ——— an animal?' The majority of both groups of children responded correctly that a dolphin and a horse are animals. Forty per cent of the five-year-olds responded correctly that a person and a spider are animals. However, of the nine- to 10-year-olds, 74 per cent gave a correct response that a person is an animal, but only 17 per cent gave a correct response that a spider is an animal.

Further to this, there have been studies dealing with students' ideas about vertebrates and invertebrates. The range of definitions of insects held by students is discussed by Shepardson (2002) in his study of students from kindergarten to Year 5. He found that primary students tended to view the interactions between humans and insects as detrimental to the former. Students such as Moira showed through their studies during this unit that they had gained a broader understanding of the role of insects and other invertebrates within an ecosystem (consistent with recommendations for student explorations of insects in their natural settings). The need for firsthand experiences is also discussed by Braund (1997) in his earlier study of students' ideas about animals with and without backbones (see also the Chapter 7 section 'Children's conception of animals'; p. 249).

Activity 6.5 Children's views of animals

- Consider the data presented above and briefly describe how students' views of what an animal is changed with age.

Suggest possible reasons why these changes may have occurred. In trying to bring students' understanding closer to the

scientific view of an animal, what criteria would you use to characterise an animal?
- Select an age level and describe a sequence of activities that you would use to:
  - identify the students' prior conceptions of an animal
  - challenge their conceptions in order to move them closer to those of scientists.

Scientists continually update classification systems

Students need to understand that the process of classification is something that everyone engages in and that classification systems are created for a purpose. In the case of the classification system used by scientists, the purpose is to communicate insights about living things, and it is like a common language. In the past, scientists classified living things into two major groups or kingdoms, these being the kingdom of plants and the kingdom of animals. Following the technological development of high-powered microscopes, scientists discovered microscopic organisms that did not seem to fit neatly into the two-kingdom categories. Accordingly, to better explain our world, the scientific community devised a new classification consisting of five kingdoms: Kingdom Animalia, Kingdom Plantae, Kingdom Fungi, Kingdom Protoctista and Kingdom Monera (see the 'Concepts and understandings for primary teachers' section at the end of this chapter, on p. 242, for details).

It is important to note that the classification of living things is not static but rather reflects the NOS as an ongoing process and is at times a cause for debate. An example of this is the Kingdom Protoctista, where it is common for the term 'protista' to be used. Rothschild (1989) discusses the terms protozoa, protista and protoctista, and relates some of the history involved, including issues such as the classification of some organisms as animal or plant and unicellular or multicellular – this is also mentioned briefly by Haselton (see http://www.bio.umass.edu/biology/conn.river/protoc.html).

Until recently there were 30 orders of insects in the Class Insecta. In 2001, this number was increased to 31 with the discovery of a new order. How was this discovery made? Zompro received some amber (solidified tree sap) that contained insect larvae that were completely different from any he had seen before. The distinct body shape and diet of this particular insect meant that it did not fit into any of the existing orders. Zompro and his team of collaborators (Adis, Moombolah-Goagoses and Marais) gave the mystery insects with armoured covering the common name 'gladiators'. Because the insects looked like a cross between a walking stick and a mantis, the team created the scientific name *Mantophasmatodea* from *manto* the mantids and *phasmatodea* the phasmids leaf or stick insects (Adis et al. 2002). Insects from this order are also known as Heelwalkers.

Another recent example of scientists updating knowledge came about when X-ray video technology revealed how beetles breathe: 'Even the most up-to-date biology textbooks, if they address insect respiration, now need revision. With the help of a high-energy particle accelerator, researchers have documented bugs breathing in a manner never before thought possible: like mammals' (Wright 2003).

The implication for teachers is that, as classification systems are human constructions, they are being updated as new discoveries are being made. The human construction of such systems also needs to be considered in relation to culture, including Indigenous ways of knowing. A 1998 study (Chen and Ku) involving aboriginal students from Taiwan (grades 2, 4 and 6) found, for example, that over half of the students classified earthworms and spiders as insects, and 'nearly 10% in each grade classified a turtle, a snake, or a frog as an insect' (p. 61). The students lived in a mostly natural environment where one of their food sources is wildlife. Thus, students' experiences and knowledge of animals differed from those experienced by their non-aboriginal peers even though both received 'a standard Chinese education' (ibid., p. 56). Grade 2 students and nearly half of the Grade 6 students did not consider humans as animals, although the researchers could not be certain that this was a cultural influence. They noted the use of daily language in mainstream culture (accessible to the students through television and film) where common expressions tend to separate humans from animals.

Teaching approaches that value students' ideas and small-group work

Although Wendy's teaching was based on constructivism (for example, she found out the children's prior views about the topic and planned the unit to build on these), in the science and design technology unit outlined here, she did not apply any specific constructivist teaching schemata as described in Chapter 1. Three examples of lesson sequence schema that have been applied to topics similar to the case study in this chapter are described next. They could be used to guide some lesson sequences that you could implement on this theme.

Barker (1991) describes the generative teaching model (Cosgrove and Osborne 1985; Osborne and Wittrock 1985) in terms of four phases: preparation, focus, challenge and application. In this model, teacher activity in the preparation phase involves seeking the scientists' and historical views. The focus phase establishes a context and, for the topic 'What is an animal?', the interview-about-instances technique (Osborne and Freyberg 1985) was used to probe the children's ideas. Picture flash cards of a girl, an earthworm, a spider, a butterfly, a bird, a fire, a fish, a tree, a dolphin and a horse formed the basis of discussion as the students worked in groups or the teacher interacted with the whole class. Student responses showed that the majority held a restricted view of animals as large, four-legged, furry creatures found in zoos and on farms.

The challenge phase required the students to take specimen bottles outside to look for six animals they could see and six they could catch. After returning with their captured animals, the students classified them according to the following schema: the set of animals and the subsets mammals, birds, insects, worms and other groups. The captured animals (generally worms, ants, centipedes, slaters and so on) challenged the students' prior views. In this phase, the scientists' views are introduced along with supporting evidence.

In the application phase, students experienced activities to accommodate the new concept, such as a true-or-false quiz, games and puzzles. These activities assisted the students to make conceptual links with other conceptions they hold.

Although the strengths of this approach are its student-centred focus and the way in which links are made to prior views, there are some weaknesses. The impression is given that students' conceptions will be developed quickly and without embedding them within a scientific inquiry. Metz (2008) discusses the importance of students 'doing science' – also see the online Appendix 1.2 (Skamp 2007). Metz describes how even Year 1 students are able to participate in the knowledge-building practices of scientists.

Appendix 1.2 Learning science, doing science and learning about science

Another constructivist model used to guide lesson sequences is the interactive teaching approach (Biddulph and Osborne 1984; see also Chapter 4). Faire and Cosgrove (1990) provide an example of this model for the topic of mini-animals, where two animals were being investigated. Firstly, in a study of snails, students were challenged to find out how snails cope with life, and secondly, worms were investigated to show the important role they play in agriculture. Faire and Cosgrove suggest the following steps:

1 *Before views* – Ask students to write down or say what they know about the animal.
2 *Exploratory activities* – Develop several task cards that encourage the students to focus on the animal parts or behaviour.
3 *Students' questions* – Record any questions students ask about the animal.
4 *Investigations* – Students have ideas about how to investigate their questions. For the worm, these include investigations about eating, habitats, insides and senses. For the snail, these include investigations about moving, seeing, reproduction, the sex (female, male or hermaphrodite), eating, breathing, the insides and the shell.
5 *After views and reflections* – Students (individually or collectively) indicate what they now know about the animal.

Although Metz showed how students were able to participate in authentic science practices, Kanter (2009) suggests that not all project-based curricula result in an improved understanding of science content. It is important to consider the nature of such projects. Kanter (p. 526) identifies two major types: those focused on an investigation seeking to answer a 'puzzling question', and those based on performance, such as designing and implementing ways to improve the flight of a model rocket. Although there are studies showing the benefits of both, he cites examples where the latter resulted in students retaining unscientific understandings, or not learning the skills needed to allow them to design their research so that they could evaluate their data effectively.

The Faire and Cosgrove (1990) approach is in line with the 5E constructivist learning schema as described, for example, in the *Primary Connections* resource (AAS 2005): the phases are engage, explore, explain, elaborate and evaluate. Look at each of these phases in the *Primary Connections* unit 'Schoolyard Safari' (AAS 2008), which focuses on students in their second and third years of school. How do the activities in this unit compare to those in the study above and in those described in this case study of Year 5 children studying minibeasts?

Teachers can link science activities with films that the children may have seen on DVD or at the cinema. One excellent example is the film *A Bug's Life*. Following the success of the first two *Toy Story* instalments (Walt Disney/Pixar 1995, 1999), Disney and Pixar decided to create a feature film about small creatures. Director John Lasseter believes that all kids love bugs:

> We made a very tiny camera which we called a Bugcam and we dragged it through the grass and it showed up what the world looked like from a bug's point of view. We had entomologists [people who study insects], including an expert on bug movement. They gave talks to the designers and animators. We even brought live bugs into the studio.
>
> Walt Disney/Pixar 1998, p. 71

In the classroom, teachers can make meaningful literacy links by reading books such as *A Bug's Life* (also available in Spanish). Big books by Robinson and Drew (1993a, 1993b, 1993c) contain information about stick insects (phasmids), cicadas (Order Homoptera) and ants (Order Hymenoptera), respectively. Using film, fiction and non-fiction books as models, children can develop their own pages for a class book or short animated film focusing on invertebrates found in their own school environment. Websites such as that of Museum Victoria (http://museumvictoria.com.au/bugs/exhibition/index.aspx) provide information in various forms that are suitable for students.

Benefits of working in small groups

To minimise the amount of equipment and consumables required for technological activities, primary school teachers often organise the students to work in small groups. Group work was an important aspect of the unit described here. The Year 5 students in Wendy's class decided that, rather than working in friendship groups as they had done in the previous unit, groups would be formed by drawing names out of a hat.

On completion of the unit, several students commented positively about working in groups with students they did not know very well previously, as the following comments reveal:

> SAM: If I did this again I'd like to do a heaps different design for the catcher and bughouse. I'd like to work with a different group because I reckon I got to know Katy and Lee a little bit better.
>
> MORRIS: I would also like to be in another group because now we know these people. If you were in another group you would get to know other people, too.

In this unit, as well as expanding their knowledge base, students also developed their cooperative skills:

> There were some excellent examples of children who perhaps in the past have found it very difficult to accept someone else's idea and have wanted to dominate the situation of learning, and being able to realise that other people can put valuable input into whatever project is being worked on. So the real cooperative skills have been one major benefit. (Wendy, teacher)

In order to experience group work for yourself and to develop your ideas about a small animal you can easily keep in the classroom, conduct this activity with a group of friends. The appendices in each of the *Primary Connections* units (AAS 2005) describe in detail the benefits of cooperative learning teams as well as how to organise them. One strategy is to assign roles to each student, ensuring that these are rotated in order to provide each student with leadership opportunities. The website also provides this information in electronic form.

Activity 6.6 Cooperative technological activity: designing a mealworm enclosure

Mealworms are the larvae of small black beetles. You can get them at pet shops and feed them on rolled oats and sliced apple or potato. Mealworms are easily cared for and you can watch the complete insect life cycle of larva, pupa and adult. Suzuki (1989) shows the three stages and suggests keeping them in a rectangular plastic food storage box with holes in the lid. Loosely woven cloth can be cut to size and placed in-between layers of food and mealworms. In two weeks, adult beetles appear and lay eggs between the layers of cloth.

- Where do you think the beetles come from?
- What will come out of the eggs?

Although a plastic food storage box, such as an ice-cream container, is suitable for housing mealworms, you could use your creativity by working with a group of friends to design and make your own mealworm enclosure. Start by each drawing a plan of your own ideas for a mealworm house. Then look at each person's plan and pick out the good parts of the design. Take at least one idea from each person's plan to place into a final design drawing for the product. The advantage of this process is that each person's plan is being valued while pressure is put on each person to participate in the group. Consider how you could use this activity with your primary students.

Students learning about skeletal systems

The following activity suggests how you as a teacher might encourage students to work cooperatively. This activity builds on the knowledge students have gained from studying invertebrates, such as slaters, which makes it an appropriate follow-up activity to this unit.

Students' understanding of animals could be further developed if they were given opportunities to compare and contrast the structure and function of the skeletal systems of a range of animals, from small ones, such as slaters and butterflies, through to those of pets and finally to humans.

You could organise the students into groups of three or four. This can be through self-selection ('Find two friends to work with for this activity'), drawing group members' names out of a hat, or through teacher selection. One of the goals of this activity is to have each group of students work independently in order to meet the project brief. You could begin the activity with each group constructing a concept map showing what they know about skeletons. An example of such a concept map is given in Figure 6.6.

Students could share the knowledge they have about skeletons. Then, each group can pose some questions to help them compare and contrast the skeletal systems of a range of animals, including humans. Questions might include the following:

- Why do slaters have their skeletons on the outside of their bodies?
- Why are the bones in a human leg so large?
- What makes a cat's skeleton so flexible?

A variety of graphic or cognitive organisers other than concept maps can be used to show information. For example, students can use a Venn diagram to compare and contrast information about skeletons (see Figure 6.7).

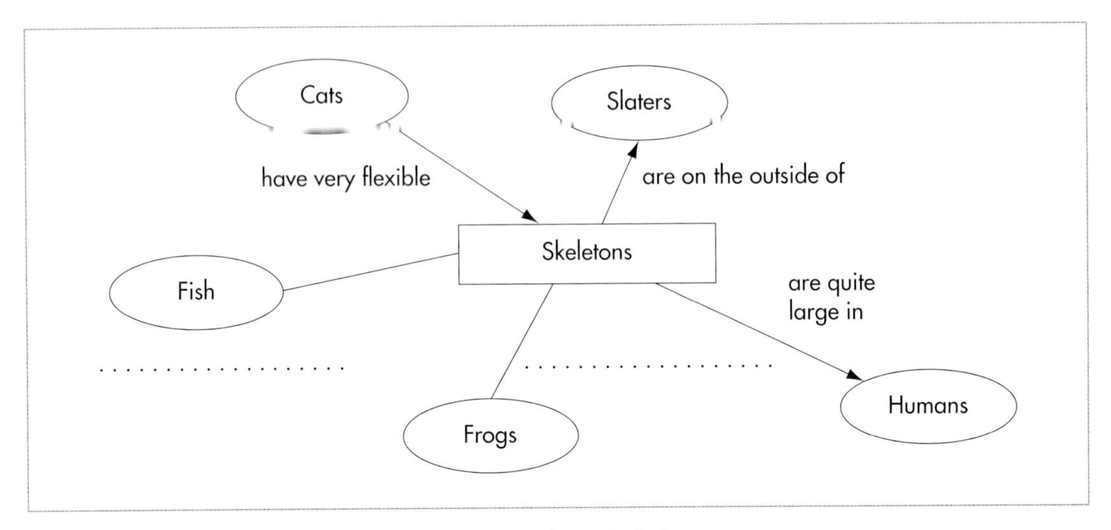

FIGURE 6.6 Concept map showing understanding of skeletons

To ensure that all students contribute to and feel that they are integral members of the group, each group member is allocated or chooses a question to investigate. On completing the task, each member can report back to the small group. The group as a whole then decides on how the information is to be presented to the class using ICT (for example, diagrams, a short recorded presentation incorporating models, a collage or a presentation, such as a PowerPoint or web pages). Preparing for the presentation will involve further sharing of roles and responsibilities, perhaps taking into consideration the particular skills of the group members. Some may be accomplished writers while others may have well-developed graphics,

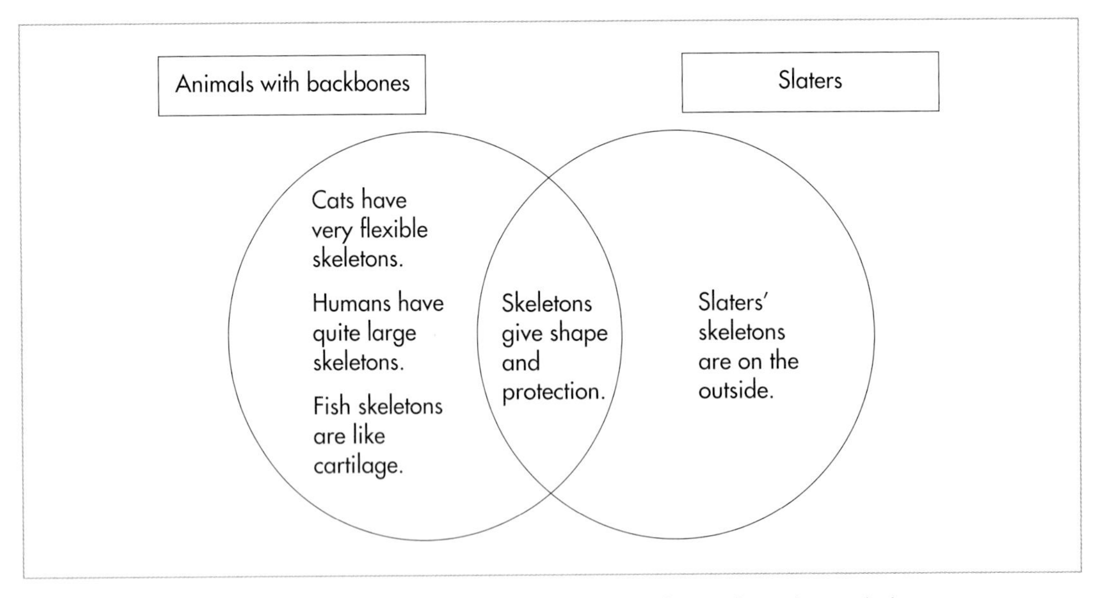

FIGURE 6.7 Venn diagram comparing and contrasting information about skeletons

ICT or other artistic skills. Group work is valuable in that it allows each individual's strength to be drawn upon and allows for the sharing of peer views.

When you teach units about the human body, the following references relating to background conceptual information for primary teachers are useful. Stephenson and Warwick (2001) focus on digestion and healthy eating, teeth and bones, as well as the benefits of exercise on the blood circulatory system and respiration. Terry (2000) refers to ideas about the human body by finding out what the children in her Year 2 class knew about the following questions.

- What does my skeleton do?
- How does my heart work?
- What foods help to keep us strong and healthy?
- What is my skin for?
- Why are our ears such a funny shape?

Activity 6.7 Finding out alternative conceptions about the human body

The students in this unit studied invertebrates, some of which had an exoskeleton (e.g., slaters). Questions about the function of such structures have been suggested earlier. Building on these previous activities, in groups, select one member to lie very still on a large piece of butcher's paper. Another member draws around the body using a felt-tipped pen to form a body outline. The group brainstorms the names of any organs they think are inside the body. Group members then have a go at drawing where they think each organ is located. Meanwhile, one member takes on the role of recorder and writes down all the questions asked during the activity. If adapting this activity for your primary students, you may want to begin with a whole-class brainstorm of organ names, with these placed on a word wall.

Human body systems and food technology

The activity described above may be extended to a unit of work on the human body and nutrition. Students can work in small groups to investigate and research a body system. They are given the opportunity to design experiments, gather information from texts and websites and, when available, interview parents who work in the health field. Experiments may be along the lines of testing the effect of exercise on the cardiovascular system. Students design the test and record and analyse the data. ICT can be used effectively at all stages of this activity. Digital photographs can be used to clarify written descriptions of the experiment. Data can be gathered using a data logger, entered onto a spreadsheet and graphs generated to assist students to see their results.

A grid of activities relating to this unit can be drawn up to allow for a range of children's learning styles and abilities (e.g., 'gifted') within the classroom. Students can have a direct say in the activities included on the grid.

DVDs/videos related to the human body are readily available and modern medical technology allows an intimate viewing of some body systems (such as the digestive system) –

the updated DVD version of *The Magic School Bus – Human Body* (Degen and Cole 1994/2005) and *Human Body* (Winston 2010) are two examples. Some museums have extensive exhibitions relating to all aspects of body systems.

The nutritional aspect of the unit can provide students with a meaningful context for food technology. Although the context is set by the teacher, ownership is then given to the students to initially gain their interest and then to promote ongoing engagement (see the Chapter 1 section '"Hot" conceptual considerations: student engagement' on p. 25). Students can be challenged by a design brief to produce a healthy school lunch (Jobling 1999).

Students work in small groups of up to four to investigate what constitutes good nutrition using the healthy food pyramid. The investigative phase of the technological process also incorporates the practicalities of what foods are suitable for inclusion in a healthy school lunch. Student menu planning has to include hygienic food storage, preparation and presentation, as well as working within time constraints. Students then produce their group's lunch at school. Spreading the groups' activities out over a week helps overcome issues of access to some facilities. Each group evaluates their lunch using criteria they helped to negotiate. In this way, the sequencing is guided by constructivist principles.

Appropriate assessment strategies

In this section we consider appropriate assessment strategies that are consistent with constructivist principles. In science, students' drawings can be an effective assessment tool. The drawings produced in Activity 6.7 (see p. 235) indicate what your group members knew about what is inside their bodies. In the primary classroom, this activity could take place near the start of a topic on the human body and later, after discussion, be repeated to show what children know by the end of the topic. The comparison of the children's before and after drawings would indicate the changes that occurred in their thinking and knowledge application. Typically, there would be more:

- body parts drawn
- organs positioned correctly
- organs correctly labelled
- details within the wrist, arms and legs
- structures present, such as the spine, breastbone, eye sockets, ears.

A

One advantage of using children's drawings as a comparative assessment tool is that the assessment is easy to organise. All children can do the drawing at the same time. For those who have difficulty writing the labels, the teacher could ask them what the body parts are called and then write the names down for the child (Frost 1997).

Children can also use the KWLH strategy for this activity. Children can draw what they know (K), ask questions indicating what they want to know (W), then communicate what they have learnt from doing the activity (L) and finally describe how they learned it (H).

Written forms of assessment are used frequently in primary science and technology. However, this type of assessment suits those students who are competent in this form of communication, but does not give students with strengths in other areas the opportunity to demonstrate their skills. All students should develop their written skills, but over a course or several units students should be encouraged to present information in more than one form. The *Primary Connections* units (AAS 2005), in each of the 5E phases, provide descriptions of assessment tasks incorporating multiple literacies and multimodal learning; these include student drawings and diagrams.

The following assessment ideas are designed to give students opportunities to communicate their ideas in a variety of ways:

- *poster presentation*, with students visually representing their ideas
- *audiovisual*, in which students present their information using videotapes or DVDs
- *making models* with appropriate labelling
- *giving a talk*
- *ICT presentations*, which may incorporate both written and pictorial forms of communication.

Assessment in the unit that is the focus of this chapter mainly involved teacher observation with anecdotal notes of student presentations. Students self-assessed their work when they reported to their peers as they made judgements about their products. During student reporting sessions, teachers can make anecdotal records of:

- students' ability to work cooperatively on a design
- technological skills
- science questions asked
- science understandings
- research skills
- individual learning styles (for example, Barry liked looking up science books, whereas Alison relied on direct observation).

Black (2008, p. 21) describes design and technology as particularly rich in terms of the ways in which they can be used to develop independence in students in terms of their learning: 'they can be required to make and take responsibility for their own decisions about their designs and subsequent products'. Effective formative assessment of technological skills is an essential component of an effective learning environment. Black provides a number of strategies, one of which is the use of teacher questioning to focus students' attention on specific aspects of their task. Such questioning and associated dialogue took place throughout this unit, not only providing students with feedback but also informing teaching.

Cumulative assessment checklists – written and oral – which contain records of students' progress in skills such as observation, recording, hypothesising, concept mapping, diagrams, evaluation, self-assessment and reporting, could also be kept. An Internet search can locate many examples of assessment rubrics. Education department websites can also provide examples and guidance about rubrics (see, for example, http://www.education.vic.gov.au/studentlearning/assessment/preptoyear10/tools/rubrics.htm).

Science as a human endeavour

Studying invertebrates within the school's grounds enabled the students in this study to gain firsthand experiences with socioscientific issues such as biodiversity; they sampled the small animals within their school community. The health of their school environment and the ecosystems within it enabled them to gain a greater understanding of science as a human

SSI

endeavour (e.g., see ACARA 2011), including the associated ethical issues. Within this strand it is important to also take into account other perspectives such as those held by Indigenous communities.

Incorporating Indigenous peoples' perspectives

S&C

In Chapter 1 (see the section 'Science and culture: Indigenous and cultural knowledge' on p. 42), the need for teachers to acknowledge that different world views may be held by their students and themselves is discussed. At the commencement of this unit, links were made to students' previous experiences and existing understandings, but other cultural or Indigenous perspectives were not identified.

The need to develop culturally responsive curricula has become an increasingly important issue in many countries, including Australia, New Zealand and Canada – in Australia, the *Primary Connections* resource (AAS 2005) has published material on Indigenous perspectives (see http://www.science.org.au/primaryconnections/indigenous). In order to design culturally responsive science curricula, we should consider Indigenous perspectives. Indigenous peoples have an intimate knowledge of the environment and base their whole education on such understanding.

In the Australian context, traditional learning was based on observation and repetition; it was also guided by stories. Knowledge belongs to appropriate individuals (elders) in the community and is shared when the youth reach the age of understanding. Children learn from older children and are not encouraged to ask questions. To this day, certain knowledge is not shared; for example, women's business is not sought by men and vice versa.

All teachers must become aware of the cultural protocols and respect them. If appropriate, it may be helpful to consider the way Indigenous peoples communicate information about the habits of animals from generation to generation, and the way dances illustrate characteristics of the animals so well, particularly the way they move. Elders could be invited to school to talk about the traditions relating to totem animals. Oodgeroo Noonuccal (1988) talks about protecting the land and her totem, Kabul, the carpet snake or Rainbow Serpent, who is seen as the giver and taker of life. Some Indigenous groups are not permitted to eat their totem animal, or at least not at certain times of the year.

Published stories can be useful starting points when designing culturally responsive curricula. One easily accessible book is *Dunbi the Owl* (Lofts 1983). This story explains why it is forbidden to harm owls.

In attempting to break down cultural barriers and plan inclusive curricula, teachers can incorporate into their integrated studies examples of bush tucker and medicines, as well as early technological implements designed by Indigenous people. People who live off the land know the multiple properties of many plants and their uses. The fruit of a particular plant may be eaten, the roots made into dye and the bark or sap used as medicines (Isaacs 1994). Fortunately, Indigenous expertise in medicinal botany is now being recognised and valued. For some illnesses, any one of 20 plants could be used, depending on the locality. In contrast to Western practice, most bush medicines are not taken orally but used as inhalants, antiseptics, rubs or liniments. Elders could be invited to the school to share their knowledge of plants and show the students some of the tools they use to gather food; see the spear example in Fleer and Jane (2004). Traditionally, Indigenous women took their children with them into the bush to gather plants such as berries, fruit, flower cones, roots and tubers; the type of food collected depended on the time of year and the locality. Gathering was a social activity.

Animals as food

Traditionally, in Australian Indigenous cultures, meat preparation and distribution were guided by important rules; for examples provided by members of an Indigenous community, read *Ngurra-kurlu: A Way of Working with Warlpiri People* (Pawu-Kurlpurlurnu, Holmes and Box 2008). Animal foods, such as lizards, snakes and tortoises, were roasted in open fires or pits or steamed in stone ovens. Insects were an important source of protein in the diet – they were usually eaten raw. Grasshoppers, termites and swarming insects, such as the Bogong moth (*Agrotis infusa*), found in the Bogong high plains in New South Wales and Victoria, were eaten. Birds and reptiles (and their eggs) were also a large component of the diet.

S&C

The most important insect food of the desert is the witchetty grub, which is eaten raw or lightly cooked in ashes. In central Australia, women collect these grubs in the roots of witchetty bushes (*Acacia kempeana*); some adult grubs grow to 10 centimetres long and 2 centimetres wide. The digging stick is an implement that women use to collect many kinds of food, including insects and grubs.

Honey ants are another important food source. In the desert near Papunya, in the MacDonnell Ranges in the Northern Territory, it is strenuous work for women to dig down more than a metre to locate honey ants. These women know that when ants become bloated with nectar they live in underground galleries where they are safe from drought. The bloated ants regurgitate nectar to feed other ants. The behaviour and characteristics of the honey ants could be contrasted with the ants urban students would find in their schoolground and home gardens. The different habitats could also be compared.

In Victoria, the traditional Koori diet includes honey from the hive of the stingless native bee. Once the hive is found, a stick is placed in a hole cut in the bottom of the hive, allowing the honey to run down the stick and be collected. Edible galls on acacias are another source of food that results from insect activity. On hunting trips, Kooris rely on these galls to supply them with juicy grubs and refreshing liquid. Crusty patches on gum leaves (called 'sugar bread') are another food source on bush food-gathering trips. In a similar way, as part of an environmental trail, students in local primary schools could examine indigenous trees in their schoolground for the presence of galls and sugar bread.

Learning outcomes are more likely to be achieved when the teacher takes into account the students' existing knowledge, skills, values, interests and cultural background. These factors should be considered when planning scientific and technological activities and providing reference materials. Indigenous students and non-English-speaking-background (NESB) students may find technological terms and concepts unfamiliar. By selecting activities and materials from other cultures, the science and technology curricula are made more inclusive. Group work, diagrams and annotated drawings, as well as instructions, all help students who have limited English vocabulary to develop technological language. Other useful strategies include explaining key terms visually and in context, modelling written and oral tasks, and making lists of equipment, concepts and processes.

Temporary care of animals

Science as a human endeavour includes 'responding to social and ethical issues', and when 'science research is influenced by societal challenges or social priorities' (ACARA 2011, p. 4). The students in this unit were involved in a study that was set in their school environment. Ethical issues concerning the treatment of the animals studied featured in both the design and use of the technological products, as well as the nature of the students' science inquiries. (Further links are also made when addressing health and nutrition issues, as described earlier when looking at human body systems and food technology.) Opportunities for students to engage in authentic experiences of science as a human endeavour can arise through incidents such as those described in Activity 6.8.

SSI

Activity 6.8 Learning from experts

Opportunities for children of all backgrounds to develop and apply science knowledge can arise through incidents where young animals are separated from their parents.

A tawny frogmouth had nested in a tree in a Victorian schoolground. One weekend, students discovered that the baby tawny frogmouth had fallen out of its nest. One student in Wendy's class and her family contacted Healesville Sanctuary (the appropriate expert) to find out what should be done. Your task, as adults, in this activity is to:

- find out from experts the feeding and housing needs of a bird found in your area (make sure you have a list of questions written out beforehand)
- design and produce an information poster or pamphlet informing others of the food and shelter requirements of your chosen bird.

A related activity suitable for primary school students would be for them to observe and record native birds and other animals in their local area. They could choose one animal and develop a disaster plan for its temporary care in case of mishap.

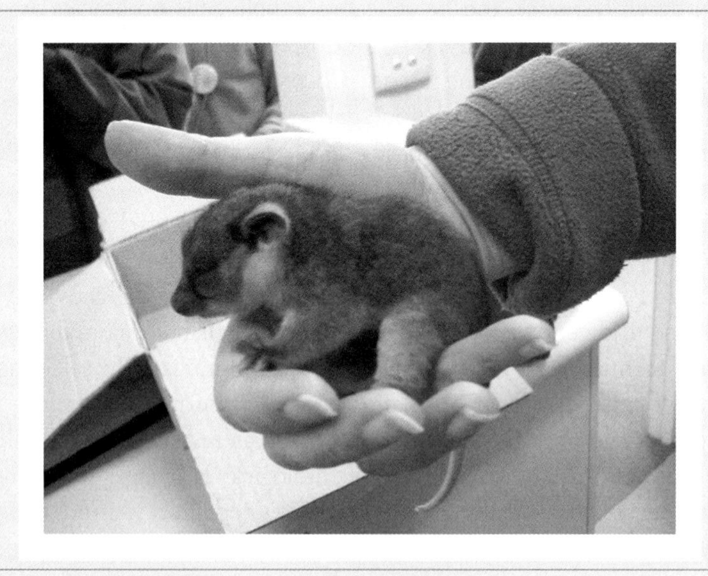

FIGURE 6.8 The possum that was saved and released

A good example (also from Wendy's school) is when a prep student found a young ringtail possum. It had apparently fallen from a tree during windy weather. A group of students became actively involved in deciding on the temporary needs of the possum until it could be delivered to a wildlife carer – coincidentally, a teacher at a neighbouring school. (The possum pictured was successfully cared for and released.)

Please note that there are particular regulations in some States and Territories that apply to the use/care of animals in schools and that these should be consulted by teachers to avoid any difficulties. If indigenous animals are found by students, teachers should refer to their appropriate State or Territory wildlife authority.

SSI

Summary

In this chapter, the focus has been on a unit of work in which a project approach encouraged students to work in small groups on their technological products and scientific investigations. Technology was not viewed as applied science. Rather, a symbiotic relationship existed between science and technology. Three science content strands are described in some curricula. For example, in ACARA (2011) there are science inquiry skills, science as a human endeavour and science understandings. In ACARA (2011, p. 14), the relationships between technology (and design) and science are outlined. An excerpt from the teachers' notes is shown in the online Appendix 6.1, which relates to technological learning outcomes. You might like to construct a similar table (based on this chapter) that relates to the science emphasis within the unit identifying where and how student activities link to the three science content strands.

Appendix 6.1
The teacher's schematic plan of the small animal catcher and house unit: upper primary design technology component

This unit was successful due to the following constructivist teaching strategies and pedagogical principles being employed. The unit commenced from where the students were at, building and extending on what they already knew. It enabled students to use a familiar environment: the schoolground. Their levels of physical skills were taken into account and they were not expected to work beyond their capabilities. The task was appropriate for the students' conceptual knowledge and allowed for their different preferred learning styles. It was also student-centred rather than teacher-directed. Students posed questions and had ownership of the work, which allowed them to work in the way they felt most comfortable (cf. science inquiry skills in ACARA (2011)).

The unit incorporated group work, building on the strengths of individuals. It was also gender-inclusive, with discussions enabling girls to try out ideas in a safe, supportive environment with their friends. The ethical approach to the needs of the animals studied by the children (science as a human endeavour) helped to ensure that the unit was culturally sensitive. The students were given opportunities to reflect on what they had learnt and appropriate strategies were used to assess this learning. To conclude this unit required students to consider not only the ethical treatment of the animals studied (as described above) but also their place within ecosystems in their school environment.

Concepts and understandings for primary teachers

Students in this unit studied invertebrates (minibeasts) within ecosystems found in their schoolground. The school in which this unit was taken was and still is concerned with its students developing an understanding of environmental issues such as the importance of biodiversity and sustainability. It has a well-developed recycling system, including composting, worm farms, hens and an organic garden. The following information is provided in the form presented to also convey a sense of the connections between living things. Although there is perhaps not the same clear listing of concepts as in other chapters, it is important to keep in mind those that apply. The challenge of providing learning contexts and activities that will shift deeply held alternative conceptions also needs to be kept in mind (Arnaudin and Mintzes 1985).

Scientists now classify all living things into five kingdoms. (As discussed earlier, it is important to note that this is an ongoing process, as in the example of the Kingdom Protoctista or protista.) In Kingdom Animalia are the animals – organisms that eat other organisms – including the meiofauna (very small animals that can only be seen through a microscope). In Kingdom Plantae are the plants – organisms that make their own food using sunlight in the process of photosynthesis – such as trees, ferns, mosses and grasses, but excluding seaweeds. Seaweeds are in the Phylum Protoctista classification along with the diatoms and dinoflagellates and other microscopic organisms that are plant-like because they can produce their own food, though some are animal-like because they eat other organisms (Breidahl 2001). Kingdom Fungi includes toadstools, mushrooms and other fungi that obtain energy by breaking down other organisms. Previously, these decomposers were thought of as plants. In Kingdom Monera are the microbes or bacteria. Some of these microscopic organisms can make their own food while others break down organic matter.

Slaters are classified in the Phylum Arthropoda, Class Crustacea. Crustaceans have a crust (hard case) enclosing their body and usually live in sea water. However, the slater or wood louse (*Isopoda*, meaning equal foot) is a crustacean that lives on land. In the Phylum Arthropoda, the Classes include Crustaceans (animals with 10 legs: crabs, crayfish, shrimp; and animals with seven pairs of legs: slaters), Arachnids (animals with eight legs: spiders, mites, scorpions) and Insects (six legs). There are more than one million known species of Arthropods and, although these invertebrates vary considerably in appearance and size, they all have jointed bodies protected by a tough, waterproof body case (exoskeleton). This exoskeleton moults (is shed) several times during the arthropod's life so that it can grow to its adult size. Despite the hard outer covering, the animal can move easily because the exoskeleton consists of plates that are separated by flexible joints.

Snails and slugs eat plants, making them herbivores, and first-order consumers. Earthworms, snails and slugs are hermaphrodites, which means that although they reproduce sexually, each organism has both female and male gonads.

Earthworms (*Oligochaeta*) are decomposers that feed on rotting organic matter, breaking it down, thereby assisting the recycling of nutrients. Earthworms burrow into the soil, loosening it and in this way creating air spaces within the soil. They also excrete worm casts (dung). As earthworms are fast breeders, they are easily kept in the classroom in a commercially produced wormery that is available from plant nurseries and hardware warehouses. You need to purchase special compost worms, such as red wrigglers or tigers. (The bushworm is much larger than the introduced species and feeds on native plant waste such as leaves.) Canadian environmentalist David Suzuki describes earthworms as 'amazing gardeners' and gives details about how to make a wormery in *Eco-fun* (Suzuki and Vanderlinden 2001); see also Helen Cushing's (2002) *No-garden Gardening*.

Search me! science education

Search me! Explore **Search me! science education** for relevant articles on linking science and design technology. Search me! is an online library of world-class journals, ebooks and newspapers, including *The Australian* and the *New York Times*, and is updated daily. Log in to Search me! through http://login.cengage.com using the access code that comes with this book.

KEYWORDS

Try searching for the following terms:

- Biodiversity: elementary school
- Integrated curriculum: elementary school
- ICT and science

Search tip: **Search me! science education** contains information from both local and international sources. To get the greatest number of search results, try using both Australian and American spellings in your searches: e.g., 'globalisation' and 'globalization'; 'organisation' and 'organization'.

Appendices

In this appendix you will find material related to 'minibeasts' that you should refer to when reading Chapter 6. This appendix can be found on the student companion website. Log in through http://login.cengage.com using the access code that comes with this book.

Appendix 6.1 The teacher's schematic plan of the small animal catcher and house unit: upper primary design technology component

This plan describes activities, classroom management ideas and anecdotal comments about what happened during the unit's implementation.

References

Adis, J., Zompro, O., Moombolah-Goagoses, E. & Marais, E. 2002. Gladiators: A new order of insect. *Scientific American*, 287 (5), pp. 42–5.

Arnaudin, M. W. & Mintzes, J. J. 1985. Students' alternative conceptions of the human circulatory system: A cross-age study. *Science Education*, 69 (5), pp. 721–33.

Atkinson, K. 1993. *Life in a Rotten Log*, Sydney: Allen & Unwin.

Australian Academy of Science (AAS). 2005. *Primary Connections*. 'Linking science with literacy'. Canberra: AAS. Available at http://www.science.org.au/primaryconnections/index.html (accessed April 2010).

_____ 2008. *Primary Connections*. 'Schoolyard safari'. Canberra: AAS.

Australian Curriculum, Assessment and Reporting Authority (ACARA). 2011. K-10 Australian Curriculum: Science. Available at http://www.australiancurriculum.edu.au/Science/Curriculum/F-10 (accessed April, 2011).

_____ 2010b. Shape of the Australian Curriculum: Science, May 2009. Available at http://www.acara.edu.au/verve/_resources/Australian_Curriculum_-_Science.pdf (accessed October 2010).

Barker, M. 1991. *Science in the Classroom*. Hamilton: Waikato Education Centre.

Bell, B. & Barker, M. 1982. Toward a scientific concept of animal. *Journal of Biological Education*, 16 (3), pp. 197–200.

Biddulph, F. & Osborne, R. 1984. *Making Sense of Our World: An Interactive Teaching Approach*. Hamilton: University of Waikato, Science Education Research Unit.

Black, P. 2008. Formative assessment in the learning and teaching of design and technology. *Design and Technology: An International Journal*, 13 (3), pp. 19–26.

Braund, M. 1997. Primary children's ideas about animals with and without backbones. *Education*, 3-13 (25:2), pp. 19–24.

Breidahl, H. 2001. *Itty Gritty Critters: Life between Grains of Sand*. Melbourne: Macmillan.

Brenner, L. 2005. *Science Essentials: Integrating Science and Literacy* (Lower, Middle and Upper Primary edns). Melbourne: Curriculum Corporation.

Byrne, T. 1993. Invertebrates as cool animals. *Let's Find Out*, Winter (2), pp. 4–6.

Carolan, J., Prain, V. & Waldrip, B. 2008. Using representations for teaching and learning in science. *Teaching Science*, 54 (1), pp. 18–23.

Chen, S. H. & Ku, C. H. 1998. Aboriginal children's alternative conceptions of animals and animal classification. *Proceedings of the National Science Council*, 8 (2), pp. 55–67.

Cosgrove, M. & Osborne, R. 1985. Lesson frameworks for changing children's ideas, in R. Osborne and P. Freyberg (eds). *Learning in Science: The Implications of Children's Science*. Auckland: Heinemann.

Cushing, H. 2002. *No-garden Gardening*. Sydney: ABC Books.

_____ 2003. *Monsters in Your Garden*. Sydney: ABC Books.

Degen, D. & Cole, J. 1994. *The Magic School Bus – Human Body* (2005 DVD). Scholastic Entertainment: Warner Brothers' Entertainment.
Faire, J. & Cosgrove, M. 1990. *Teaching Primary Science*. Hamilton: Waikato Education Centre.
Fleer, M. & Jane, B. 2004. *Technology for Children: Research-based Approaches* (2nd edn). Sydney: Pearson Education Australia.
Frost, J. 1997. *Creativity in Primary Science*. Buckingham, UK: Open University Press.
Harlen, W. 2009. Teaching and learning science for a better future. *School Science Review*, 90 (333), pp. 33–42.
_____ & Qualter, A. 2009 *The Teaching of Science in Primary Schools* (5th edn). London: Routledge.
Haselton, A. n.d. Protoctista. Available at http://www.bio.umass.edu/biology/conn.river/protoc.html (accessed 14 October 2010).
Heliden, G. & Heliden Onnestad, S. 2006. *Students Early Experiences of Ecological Phenomena and Education for a Sustainable Future*. Proceedings of the International Organisation of Science and Technology Education Symposium, Penang Malaysia, July, pp. 165–70.
Hume, A. 2009. Authentic scientific inquiry and school science. *Teaching Science*, 55 (2), pp. 35–41.
Isaacs, J. 1994. *Bush Food: Aboriginal Food and Herbal Medicine*. Sydney: Lansdowne.
Jane, B. L. 1994. *Children Linking Science with Technology*. Video produced by Course Development Centre Video Production Unit, Deakin University Burwood.
_____ & Jobling, W. M. 1995. Children linking science with technology in the primary classroom. *Research in Science Education*, 25 (2), pp. 191–202.
_____ & Robbins, J. 2004. Intergenerational science: Grandparents encouraging children's curiosity. *Every Child*, 10 (4), pp. 8–9.
_____ 2006. *Grandparents Teaching Science and Technology in Informal Contexts*. Proceedings of the International Organisation of Science and Technology Education Symposium, Penang Malaysia, July, pp. 538–44.
Jobling, W. 1999. A healthy school lunch. *Primary Educator*, 4 (5), p. 23.
Kallery, M. & Psillos, D. 2004. Anthropomorphism and animism in early years science: Why teachers use them, how they conceptualise them and what are their views on their use. *Research in Science Education*, 34, pp. 291–311.
Kanter, D. E. 2009. Doing the project and learning the content: Designing project-based science curricula for meaningful understanding. *Science Education*, D01 10.1002/sci.20381. Available at http://onlinelibrary.wiley.com/doi/10.1002/sce.20381/abstract (accessed May 2010).
Llewellyn, C. 2008. *Ask Dr K Fisher about: Minibeasts*. London: Kingfisher Books.
Lofts, P. 1983. *Dunbi the Owl*. Sydney: Ashton Scholastic.
Metz, K. E. 2008. Narrowing the gulf between the practices of science and the elementary school science classroom. *The Elementary School Journal*, 109 (2), pp. 138–61.
Noonuccal, O. & Noonuccal, K. 1988. *The Rainbow Serpent*. Canberra: Australian Government Publishing Service (AGPS).
Osborne, R. & Freyberg, P. 1985. *Learning in Science: The Implications of Children's Science*. Auckland: Heinemann.
_____ & Wittrock, M. 1985. The generative learning model and its implications for science education. *Studies in Science Education*, 12, pp. 59–87.
Pawu-Kurlpurlurnu, W. J., Holmes, M. & Box, L. 2008. *Ngurra-kurlu: A Way of Working with Warlpiri People*, DKRC report 41. Alice Springs: Desert Knowledge CRC.
Penny, M. 2004. *Life in a Rotten Log*. Chicago: Heinemann Raintree.
Robbins, J. & Jane, B. 2005. Talking about the special things: Grandparents supporting children's learning. *Education Review, Australian College of Educators*.
_____ 2006. Intergenerational learning: Grandparents supporting young children's learning in science. Paper presented at the thirty-seventh annual conference of the Australasian Science Education Research Association, Canberra ACT, July.
Robinson, M. & Drew, D. 1993a. *Stick Insects*. Melbourne: Ashton Scholastic.
_____ 1993b. *Cicadas*. Melbourne: Ashton Scholastic.
_____ 1993c. *Ants*. Melbourne: Ashton Scholastic.
Rothschild, L. J. 1989. Protozoa, Protista, Protoctista: What's in a name? *Journal of the History of Biology*, 22 (2), pp. 277–305.
Shepardson, D. P. 2002. Bugs, butterflies, and spiders: Children's understandings about insects. *International Journal of Science Education*, 24 (6), pp. 627–43.
Skamp, K. 2007. Conceptual learning in the primary and middle years. The interplay of heads, hearts and hands-on science: More than just a mantra. *Teaching Science*, 53 (3), pp. 18–22.
Stephenson, P. & Warwick, P. 2001. Understanding the science of health: Teacher knowledge and understanding in science. *Investigating*, 17 (2), pp. 27–30.
Suzuki, D. 1989. *Looking at Insects*. Sydney: Allen & Unwin.
_____ & Vanderlinden, K. 2001. *Eco-fun: Great Projects, Experiments and Games for a Greener Earth*. Sydney: Allen & Unwin.
Terry, L. 2000. A science & technology excursion-based unit of work: The human body. *Investigating*, 16 (4), pp. 25–9.
Tytler, R., Haslam, F., Prain, V. & Hubber, P. 2009. An explicit representational focus for teaching and learning about animals in the environment. *Teaching Science*, 55, pp. 21–7.
Walt Disney Pictures and Pixar Records. 1995. *Toy Story 1*. Video produced by Pixar Records. California: Walt Disney Pictures and Pixar Records.
_____ 1998. *Adventures* n.d. California: Walt Disney Pictures and Pixar Records (other details unknown), pp. 70–4.
_____ 1999. *Toy Story 2*. Video produced by Pixar Records. California: Walt Disney Pictures and Pixar Records.
Wellington, J. & Osborne, J. 2001. *Language and Literacy in Science Education*. Buckingham, UK: Open University Press.
Winston, R. 2010. *Human Body* (repackaged). DVD produced by BBC television.
Wray, D. & Lewis, M. 1997. *Extending Literacy: Children Reading and Writing Non-fiction*. London: Routledge.
Wright, L. 2003. Ultrapowerful X-rays reveal how beetles really breathe. Scientific American (online). Available at http://www.scientificamerican.com/article.cfm?id=ultrapowerful-x-rays-reve (accessed 3 January 2011).

7

Living things and environments

by Russell Tytler, Filocha De Melo and Suzanne Peterson

Introduction

There are many possible ways to study biology, including accessing texts and websites and watching television shows. Numerous newspaper articles assume a knowledge of biology in discussing environmental issues; for example, loss of habitat, endangered species or agricultural and food issues. This chapter takes the view that a powerful way of studying living things is through investigating them in natural or constructed environments. The three sections of the chapter – invertebrates in the schoolground, studying the adaptive behaviour of small animals in the classroom, and plants in the school environment – are therefore bound by this vision. Many schoolgrounds or local parks are functioning ecosystems with a range of animals, plants and animal–plant interactions available for study. Schoolground habitats generally harbour a variety of life forms that students are normally oblivious to. The study of animal behaviour provides the opportunity to look at the adaptive purposes of animal structure and function and to develop knowledge of experimental processes used in biology.

These phenomena provide valuable opportunities for children to investigate three 'big ideas' in biological science: the interdependence of organisms in the environment, the adaptive function of structures and behaviour of living things, and the biodiversity, change and continuity that characterise any ecosystem.

This chapter draws on the authors' experience in using the local environment in research, and in teaching biological concepts to undergraduate teacher education students and to primary school children as part of school-based programs run by Deakin University. The first section on Invertebrates in the schoolground was developed as part of a research project: The Role of Representation in Learning Science (RiLS). The second section on animal behaviour was developed as part of a longitudinal research program and refined in Suzanne's classroom. The chapter covers a range of aspects of living things, from animals to plants to ecosystems, and research perspectives from the classic 1990s findings about children's conceptions, as well as more recent work on sociocultural and representational perspectives, and values. The chapter is organised as follows:

- children's ideas about living things (alive, animals)
- representations and learning about animals – illustrated through a classroom case study
- learning about life cycles, food and energy, and planning and assessing habitat activities

- investigating small animals – illustrated again through a case study and emphasising questioning
- plants in the schoolground, including children's ideas – there is also a case study of a terrarium and the cycling of matter
- science as a human endeavour – scientists' ways of working, values and living things, and links to the world of work.

How do children think about living things?

Children's conception of what it means to be alive

In teaching about life and living in the primary school, the idea of what we mean by a 'living thing' is often taken as unproblematic and obvious. However, researchers, in their attempts to investigate what children understand about what it means to be alive, have uncovered a range of interesting conceptions. For instance, Carey (1985) and Stepans (1985) found that movement – action – is the main criterion that children use to determine if an object is alive. Thus, when asked about the various life-cycle stages of a butterfly, they will often indicate that the eggs and immobile pupa are not alive, while agreeing that the caterpillar and butterfly are. In Stepans' study, for instance, most of the 24 out of 30 Year 5 students who said that lightning is alive gave striking and moving as criteria.

Both researchers found that many children give an affirmative response to the question 'Is the sun alive?' This was also the case with Year 5/6 students in the RiLS project (detailed in this chapter). Such animistic conceptions are undoubtedly influenced by everyday, metaphorical use of language that attributes purposeful movement to non-living bodies, such as 'The sun is hiding behind the clouds'.

Carey further found that children have little trouble in deciding that mobile animals are alive, but experienced greater difficulty deciding that plants, which are markedly less mobile, are also alive.. Life is often attributed to plants when they do something that children associate with movement or growth. For instance, they will state that a tree is alive when it is growing fruit. On the other hand, Inagaki and Hatano (1996) found that children as young as five group animals and plants together and differentiate them from non-living things, mainly on the basis of growth and the intake of food or water for maintaining vitality. They claim, therefore, that five-year-old children recognise an integrated category of living things.

We can help children build up a concept of living organisms by asking them to focus on the similarities shared by disparate classes of living things – for example, a gum tree and a yabby – rather than the differences that they can readily identify. This identification of shared similarities is a general strategy for helping develop most inclusive concepts, such as *living*, *animal* or *plant*. Concentration on differences alone, which is often the focus of teaching about the diversity of organisms, tends to produce less-inclusive concepts and leads to more superficial views of the living world.

A class of primary school children's conceptions of what it means to be alive can be explored using a set of cards representing a range of animals and plants, non-living but moving things such as the sun, fire, lightning and a car, and once-living things such as a plank of wood or a

Activity 7.1 Is it alive?

Build up a set of criteria you would use for deciding whether something is alive in a biological sense. For each criterion, you should decide whether it is a necessary or an optional characteristic. As we have seen, for instance, movement is a criterion that can be misleading, but you might like to consider criteria like 'breathes' (or respires) or 'responds to stimuli'.

When you have developed your criteria, think about how you would respond to a child's suggestion that fire is alive. How many of your criteria could reasonably be attributed to fire?

Is the question 'Is it alive?' the same as 'Is it a living thing?' Talk about this with some fellow students and friends. (You might decide there are important distinctions, but the example given below – concerning the freshly picked tomato – shows how fruitful a discussion of what we mean by 'alive' can be.)

If you have a chance, talk with some children about their views of what it means to be alive. Give them some instances, such as the sun or fire, to explore the boundaries of their concept.

bone. Such a probe is very productive when used for formative assessment as part of the ongoing teaching process.

Pre-tests, as an elicitation task, can be quite powerful indicators of what needs to be addressed in a lesson sequence. For instance, during the RiLS project, a pre-test used to initiate a conversation in a Year 5/6 class about the concept of 'living things', led to teachers encouraging their students to construct PowerPoint questionnaires to explore the ideas and beliefs of fellow classmates.

Activity 7.2 What can flow from a pre-test?

The teachers in the RiLS project targeted their class discussion so that it addressed students' alternative conceptions, explored via a pre-test. Shown below are the responses of Sean and Naomi, two Year 5 students, to the following question: 'Which of these do you think is a living thing?' Write beside each the main reason/s why you think so or not.

OBJECT	SEAN'S RESPONSE	NAOMI'S RESPONSE
Earth	Living – it moves, grows, lives	Living, because it changes all the time.
Fire	Non-living – fire moves but doesn't live	Living – it runs on fuel and it is full of energy
The sun	Living – like the earth	Living, because it changes all the time
A virus	Living – a virus eats, moves, grows and dies	Living – it attacks your body
A daffodil that has just been picked	It still lives but after a while it dies	Living – it drinks

If you had Sean and Naomi in your class, how would you go about helping them to address their views about the concept of living things? Discuss your response with a fellow student or friend.

Discussion arising out of a probe on what it means to be 'alive' can expose a range of subtleties concerning the concept 'alive'. For example, the classroom experience of an intermediate primary grade teacher provides an opportunity to consider your own concept of living. The class investigated the question 'Is a freshly picked tomato still alive?' The children could not resolve the question and wrote to experts in universities and agricultural institutions for advice.

Activity 7.3 Is a freshly picked tomato alive?

- In terms of your own concept of living, is a freshly picked tomato alive? Discuss this question with some fellow students or friends. You should apply the criteria you developed above to help you decide.
- What are the key questions you asked yourself in deciding this issue? Which criteria are most important?

Using the notion of similarities shared by living things, we might investigate this problem in the following way. Consider the similarities over time between two tomatoes from the same bush, one picked and one left on the bush:

- If both were green at the time of picking, one placed on a sunny shelf would change colour from green to red at a similar pace to the tomato still attached to the bush.
- At different rates, both would eventually show signs of water loss and decay. However, even when mouldy spots appear, the whole tomato tissue would not be rotten.
- Parts of the flesh of the picked and unpicked tomatoes are able to maintain themselves in a condition not dissimilar to that of a young tomato ripening on the bush.

Thus, the observation and recording of the similarities that exist between a picked tomato presumed not alive and the attached tomato presumed to be alive can assist children to explore the different processes that living things undergo.

Such an investigation is unlikely to be totally convincing. Of the 19 experts who responded to the children's letters, 17 took the view that the picked tomato was not alive. Nevertheless, there is a good case for arguing that the freshly picked tomato is alive because it is able to maintain its biological integrity for a considerable period of time after the picking.

Activity 7.4 Further thoughts on the tomato and constructivist approaches

- Consider whether this discussion has altered your view about whether the tomato is alive. Do you agree or disagree with our contention that a freshly picked tomato is alive?
- Refer to the discussions on constructivist views of learning and teaching in Chapter 1. In what ways do you think the exercise above and the discussion of what it means to be alive relate to a constructivist view? What would you hope would be the outcome of such a discussion in a classroom context?
- Think about how this relates to the strategies exemplified in this section. To what extent have these encouraged you to reflect on your own concept of living things? To what extent have your understandings about living organisms been modified?
- Probing children's prior conceptions is often a matter of finding a productive question or challenge. You might like to review activities in other chapters that perform the same function as the strategy exemplified by the 'Is it alive?' exercises.

The children's responses detailed in Activity 7.2 show that even older children have difficulty with the notion of what it means to be alive. However, children do move considerably towards a biologist's view of the living world over their primary school years. Carey (1985) suggests that this is due to the accumulation of experience and specific knowledge that all children inevitably gain over these years – knowledge, for instance, that a leopard's spots help camouflage it, or that all the animals you find on a farm have livers, or that the fish you find in a mountain stream are different to those you find in the sea.

Children's conception of animals

Young children's conception of animals tends to be largely restricted to mammals. Bell (1993), in a classic study, investigated which organisms children thought of as animals, and why. She found that, while mammals such as a cow presented little difficulty, whales were more problematic; organisms such as spiders and worms were regarded as animals by between only 20 to 50 per cent of primary school children. In fact, young children aged five to six years are more likely to think of spiders and worms as animals (the biologist's view) than are children aged nine to 10. This is thought to be because, as they gain knowledge of biological groups such as reptiles and insects and various types of worm, they lose the inclusive animal concept. This is another example of the difficulty associated with focusing on difference rather than similarity. Perhaps it also means that a little knowledge is a dangerous thing. Bell also found that there was an increasing preparedness over the primary school years (from 20 to 57 per cent) to regard humans as a class of animal. Yen, Yao and Mintzes (2007) found that Taiwanese students have similar patterns of difficulty in achieving a scientific conception of 'animal', linking the term mainly with vertebrates, and where they make distinctions it is mainly on the basis of external appearance, habitat and movement.

Does having a different conception of animals to that of scientists matter, in practical terms? Bell and Freyberg (1985) show that a class that was explicitly taught the concept of an 'animal' before a teaching sequence on consumers and producers did much better than a class without this prior teaching. They argue that the results 'clearly show that, if we can get the simple underlying words understood, then sound learning can occur. In this instance, the problem seems to be not with the more complex term consumer, but with the simpler, more common ones such as animal and living (ibid., p. 39).

Does having informal experience with living things make a difference to children's conceptions? Prokop, Prokop and Tunnicliffe (2008) found that children with pets had a better understanding of their internal organs, especially if they kept two or more, but keeping pets made no difference to their misclassification of invertebrates. They therefore advocate a greater focus in science activities on rearing invertebrates and improving children's attitudes towards and knowledge of them. Prokop and Tunnicliffe (2008) were interested in attitudes to animals as a key outcome of environmental education. They found that pet owners had more positive attitudes towards wild animals but negative attitudes towards less popular animals such as insects, spiders and rats, compared to non-pet-owners. They also found a correlation between attitudes and alternative conceptions, in that more alternative conceptions were associated with more negative attitudes.

Myers, Saunders, and Garrett (2004), in a study of children's developmental understandings of the needs of animals, found that even four- to five-year-old children recognised animals' basic physiological needs, but that their ecological and conservation needs showed strong but different developmental trends. They argue that we should not underestimate even young children's ability to understand the ecology surrounding animals (their habitats, interactions with other living things), and that drawing connections with animals children experience in their own lives can be a powerful strategy for building a conservation ethic. With a similar environmental focus, Lindemann-Matthies (2005) reported on a Swiss program, Nature on the Way to School, which supported teachers in using their local environment, including children's walk to school, to identify and study local animals and plants. The highlight of the program was a 'Nature Gallery' where students framed a plant that they especially valued. Lindemann-Matthies found that the more wild plants and animals children noticed in their local environment and could name, the more they appreciated them.

The key message that comes from these studies is that children's familiarity with and experience of animals and plants in natural settings is critically important for their conceptual learning and also their growth in positive attitudes. Knowledge and values, it seems, go hand in hand. In this chapter we will explore ways of studying animals and plants that are consistent with these messages.

Representations and learning about animals

The following case study is of a Year 5/6 sequence on animals in the school environment which included a rich range of teacher- and student-generated representations, investigative activities and discussion. It exemplifies what an explicit representational approach entails and the role of representations in:

1 supporting learning and reasoning in science
2 framing and developing science explorations.

Research findings from a study of this sequence are reported in Tytler et al. (2009). As a comparison sequence, the *Primary Connections* unit 'Schoolyard Safari' (see http://www.science.org.au/primaryconnections/curriculum-resources/schoolyard-safari.html) includes studies of individual animals (snails, worms, ants) as well as exploration of habitats, and also has a literacy focus.

CASE STUDY

R

Invertebrates in the schoolground: a teaching and learning sequence that explicitly uses inquiry with a representational approach

The case study reported here aimed to investigate the relationship between representation and learning in the context of an inquiry-based unit on animals in the schoolground at primary school level. The research team worked with two primary teachers, experienced in Years 5/6, to develop a teaching sequence 'Invertebrates in the schoolground'. The teachers combined their classes and co-taught the unit, which focused explicitly on representations to promote students' understandings of key concepts related to animals in the school environment.

Schoolgrounds provide an ideal environment where animals can be studied closely over a period of time – life-cycle changes, the interrelationships between animals and details of structure and behaviour are more accessible for observation. When investigating the schoolground, it is important to focus much of our observation on the similarities shared by all animal life. Children need to investigate a wide range of features that various animals exhibit so that these similarities become more obvious. The chance observation of a beetle might lead to a range of questions and observations.

A body of recent Australian research on the use of representations in learning science (Hubber, Tytler and Haslam 2010; Carolan, Prain and Waldrip 2008; Tytler et al. 2009; Tytler, Prain and Peterson 2007; also see Chapter 1) has identified various key principles to guide the effective planning, implementation, and evaluation of student learning. Consistent with an effective focus on conceptual learning in science generally, the teacher needs to be clear at the topic's planning stage about the key concepts or big ideas it's intended that students learn. This conceptual focus provides the basis for the teacher to consider which sequence and range of representations, including both teacher- and student-generated ones, will engage learners, develop understanding, and count as evidence of learning at the topic's end. These principles are based on previous research (Cox 1999; Greeno and Hall 1997) that argues the need to have students generate non-standard representations in reasoning and solving problems.

Teaching sequence: key characteristics of lessons and student activities

The teaching sequence 'Invertebrates in the schoolground' was designed according to the following representational principles. The various sessions of the sequence were flexible, and specifically designed to provide a rich range of student-generated representations aimed at introducing students to the ideas and methods of field biology, including sampling and mapping techniques. They were also designed to help develop in students the concepts of ecosystem, habitat, diversity of animal populations, interactions between plants and animals in an ecosystem, animal structure and function, and the adaptive purposes of animal behaviour.

The key characteristics of the sequence were:

- a focus on the methods used in science to study animals
- an inquiry approach – students asking questions, exploring and investigating
- explicit focus on representations
- students generating their own representations and using multimodal representations.

Each session consisted of between 45 and 90 minutes of class time. An overview of the sequence is presented in Table 7.1, with brief notes on the teacher actions, purpose of each session and student activities.

TABLE 7.1 The invertebrates in the schoolground activity sequence

TEACHER ACTIONS	STUDENT ACTIVITIES
1 Administration of a pre-test – teachers identified students' existing ideas about living things, the way scientists work in exploring environments, and the relationship between the structure and function of small invertebrates. *2 Class discussion to unpack students' alternative views and explanation of the ways in which scientists collect and represent evidence when studying a terrestrial habitat.*	Students completed the pre-test, debated the characteristics of living things, and worked on computers to construct interactive questionnaires about the attributes of 'living things'. They discussed what it would be like to work as a team of scientists finding out what living things exist in their schoolgrounds, and why.

TEACHER ACTIONS	STUDENT ACTIVITIES
3 Introduction to the schoolground – teachers identified different habitats in the schoolgrounds and allocated them to groups of students: G1 – The school pond; G2 – The compost heap; G3 – Under the gum tree ...	Students predicted what they might find in their habitat. They then visited the habitat to document it in their workbooks. They reported their observations back to the class and discussed inferences as to what they think happens in the habitats.
4 Methods for habitat studies – discussion about sampling, distributions and the need for developing and representing quantitative data. Emphasis on: the inquiry approach and the use of multimodal representations – what to document and how, and the need to use appropriate representations (e.g., graphs) to represent populations.	Students conducted habitat explorations with a view to complete a poster to present responses to: • what is in the habitat • what exactly happens in the habitat, or what life is like for the invertebrates living there • how the various living things in this particular habitat interact with and depend on each other.
5 Research questions – teachers helped the class generate questions for group investigations, based on questions generated by individual students according to their intrinsic interests.	Students further explored habitats to answer research questions. They recorded environmental conditions, noted the number of each invertebrate found, and drew and annotated drawings to describe body structures and behaviours.
6 Animal diversity, classification and safety issues were discussed – teachers detailed the use of branching keys, Venn diagrams and drawings focusing on similarities between the distinguishing characteristics of invertebrates: jointed bodies, number of legs, presence of wings and so on.	Students captured selected invertebrates, observed them and recorded body structures using annotated drawings and digital photographs. They used digital microscopes and the Internet to identify and classify them. They then released the invertebrates where they had been found.
7 The difference between observation and inference was established – student-generated representations were used to discuss form and function of representations to generate meaning: What does this graph represent? What doesn't it represent?	Students developed and presented posters giving an account of living things found in their habitat and discussed their preliminary ideas of how the animals and plants might interact. They then exhibited their posters around the classroom.
8 Ways of observing and describing body structures of invertebrates using different modes of representation were discussed (see the Chapter 1 *section 'Representational and related strategies' on p. 29).*	Students conducted group-generated activities using mealworms and stick insects to establish ways of describing body structures, movement and feeding, and for testing behaviour, and they used Venn diagrams to note similarities and differences.
9 Discussion about concepts of biological evolution – how biological evolution accounts for the diversity of species, how it is developed over many generations, and how adaptations include changes in structures, behaviours and physiology that enhance survival.	Students collected one invertebrate and observed it with a view to constructing a model that would represent the movement or feeding of that invertebrate. They used digital microscopes, made annotated drawings and time-sequence drawings, and described how the body structures and behaviours of the invertebrate are adaptive to the habitat.
10 3-D modelling to represent movement of invertebrates – teachers dedicated two sessions to encourage students' creativity in constructing a model that would specifically represent the movement of invertebrates.	Students used scrap materials, building blocks, modelling clay and plastic straws to build a 3-D model of the invertebrate that they then used to represent as accurately as possible how it moves or feeds.

Post-test – To explore student knowledge about concepts related to animals in the environment, the nature of science and science investigation.

The following sections discuss a range of ways in which representations are key elements in children's development of science understandings and investigative approaches and skills. These interactions will be discussed by drawing on the experience from research on the sequence described in the 'Invertebrates in the schoolground' case study, as shown in Table 7.1.

Activity 7.5 Prior knowledge about invertebrates in the schoolground

Before considering questions concerning invertebrates in the schoolground, it would be a useful exercise to examine your own expectations of what sort of animals one might find in a schoolground or local park environment.

- Think about your own experience of schoolgrounds and construct a list of animals and plants you might expect to find there. Compare your list with that of a fellow student or friend.
- Can you arrive at a list you would agree on? For each animal listed, think about where exactly you might find it. Why do you think that?

Building science concepts through representational challenges

R

Habitat-exploration activities (see Table 7.1 on p. 251) are designed to enable students to understand the concept of *habitat*: the place where the animal lives and that contains everything it needs to survive: water, food, shelter and suitable conditions for reproduction. Students need to understand the concept of animal adaptation as linked to the imperative of the species' survival. Hence, teachers need to explicitly ask students to observe and note down the characteristics of animals that are found in the environment under investigation. Scientific observation is an important capability (Haslam 1998; Haslam and Gunstone 1998) and students need support to focus on and record the structures, functions and behaviours that enable animals to interact with:

1. the physical features of the particular habitat, such as shelter, moisture and temperature
2. plants within the habitat, for shelter, feeding and as a source of food for their prey
3. other animals within the environment, such as predator–prey relations and mutualism.

These three issues form the focus for student representations for the sequence in the 'Invertebrates in the schoolground' case study (see p. 250).

The representational activities involve students in acting as scientists in observing, reasoning and communicating ideas. Table 7.2 shows how each of a number of research questions involved students in active representational work in pursuing answers.

It is important to teach children how to sample and draw up representations using tables, graphs, diagrams and cross-sections. In session 4 of the sequence outlined in Table 7.1 (see p. 251), the teacher focused strongly on the representations students might use to study the invertebrates living in their assigned habitat.

TEACHER What sorts of things you might want to record? Temperature, what was the weather like, humidity. ... Can we find out at what depth these creatures are living? Whether they interact with each other? Are there any predators? What else?

TABLE 7.2 Research questions and representations generated to answer them

RESEARCH QUESTION	REPRESENTATIONAL ACTIVITIES
How are animals in this environment suited to particular physical aspects of it?	Students show aspects of the environment: annotated drawings from different perspectives (e.g., plan and side view; sketches of plants and leaf litter; percentage of plant cover; drawings to show temperature at different points, or depths, and moisture; and descriptions of the environment and its variations.
What animals live in this particular environment and where exactly do they live?	Students show animal diversity and animal populations. Draw tables of animal numbers, present annotated drawings showing where they were found, and draw graphs with scales, labels and an animal key.
How do animal structures and behaviours support their survival?	Students draw diagrams with labels and annotations. They draw body parts from different perspectives, annotated with observations of behaviour, and functions of different parts. They then build a 3-D model to represent a particular feature, movement, of one invertebrate of their choice and present that model to the class.
How do animals in this environment interact with each other?	Students draw diagrams showing feeding relationships: food chains and food webs.

Activity 7.6 Generating questions

- Think about animals in the schoolgrounds and generate a number of questions concerning these to which you would like to know the answers.
- Which of these questions could be answered or at least inferred by observations of the animals in the schoolgrounds?
- What observations or investigations would you carry out to answer each of these questions?
- What might you record in a research notebook as a result of these observations and investigations?
- Which questions would be best answered by Internet searches and looking in books?

This discussion was aimed at helping students understand that each animal lives in a particular ecological niche, which not only involves interdependence with other living things but also dependence on non-biotic factors.

Using student-generated questions

The framing of a teaching sequence based on children's investigation of their own questions is the essence of the interactive approach (see, for example, chapters 1 and 4). The schoolground offers an ideal basis for such an approach since there are many questions to raise and investigations to carry out in the immediate school environment.

Some of the questions generated by a group of Year 5/6 students, following the pre-test and discussion, and after a preliminary visit to their allocated habitats, included: What is the population of beetles in this area? What [if any] are this beetle's unique features? Why is it living in this environment? What are the sources of nutrition for this beetle? Are there signs of damage to the environment and what impact has it had/will it have on the environment? Is this beetle endangered?

The schoolground can provide immediate access to a rich range of experiences. Investigations of animals and plants will encourage the development of science processes, such as scientific observation, measurement and experimentation, alongside the growth of conceptual understandings about animal and plant structure and function, and their interactions within a schoolground community.

This is also an opportunity for teachers to develop, in students, a conservation ethic with respect to the animals and plants in their surroundings. The understanding of ecological concepts about living things in the immediate environment, in this case the schoolground, can thus contribute to our own survival on this planet.

Within the general framework of study of various habitats in the schoolground, many opportunities will arise for you as a teacher to support children in their observations and investigations by providing suggestions about what they might look for, how they might better focus their questions, or how they might set up an investigation, as well as creating opportunities to challenge and extend their ideas. There is quite a lot of suitable, locally produced reference material, on the Internet and in books and DVDs/videos, that can be used to provide children with background knowledge to orient them to a schoolground study, or to use in exploring their questions and raising more questions.

The role of representations in reasoning and learning

Science, more than most disciplines, is communicated through multimodal means. Researchers in classroom studies where students were guided to construct their own representations of scientific ideas (Carolan, Prain and Waldrip 2008; diSessa 2004; Greeno and Hall 1997; Hackling and Prain 2005; Tytler, Peterson and Prain 2006; also see Chapter 1) claimed that students benefited from multiple opportunities to explore, engage, elaborate and re-represent ongoing understandings in the same and different representations, and noted the importance of teacher and student negotiation of the meanings evident in representations in these multiple modes, which include:

- textual – might be written but also spoken
- visual – diagrams, 3-D models, animations and simulations, photographs
- mathematical – formulae, graphs and tables
- gestural or embodied – gestures, role-plays.

Achieving an understanding in science involves being able to coordinate these representational modes and re-representing information appropriately to construct an explanation.

R

Activity 7.7 **Representing diversity in a chosen quadrat**

- Work with a partner to select and explore a habitat of your choice, and complete a PowerPoint presentation using at least five different representations that document your observations about:
  - the invertebrates found: slaters, worms, spiders or whatever you can discover (compost heaps, damp soil covered with rotting leaves and twigs, and oil from

under a rock are good places to collect invertebrates)
- biodiversity and change: the difference between animals, animal growth and their life cycles
- how you think animals interrelate: food chains, parenting functions
- animal structure and function: adaptations, physical or behavioural, that help animals survive.

- How did the habitat exploration help you develop a more inclusive concept of living things and animals? What skills do you think you are developing while completing this activity?

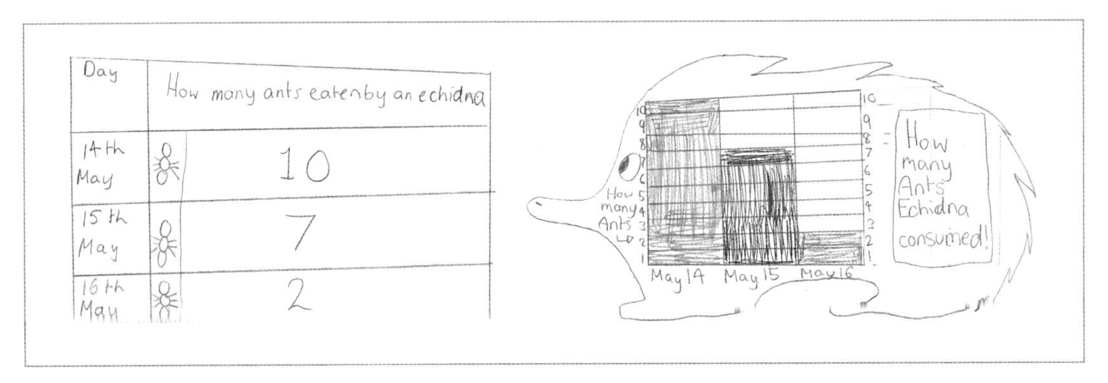

FIGURE 7.1 Student response to pre-test (left) and post-test (right)

Greeno and Hall (1997) argued that different forms of representation supported contrasting understanding of topics, and that students needed to explore the advantages and limitations of particular representations. Students need to know the function and form of the representations they use. Note the difference in the representations made by a Year 5 student (see Figure 7.1) in response to the same question in a pre-test and post-test: 'Use the most appropriate representation to show that ten ants are eaten by an echidna on the 14th of May, seven are eaten on the 15th of May and two are eaten on the 16th of May'.

Cox (1999) noted that representations can be used as tools for many different forms of reasoning, such as for initial, speculative thinking, as in constructing a diagram or model to imagine how a process might work, or to find a possible explanation, or see if a verbal explanation makes sense when re-represented in 2-D or 3-D. Students need to learn how to select appropriate representations to address particular needs, and be able to judge their effectiveness in achieving particular purposes.

Activity 7.8 Advantages and limitations of representations

Refer to your PowerPoint presentation (see Activity 7.7 on p. 255). For each of the representations in the presentation, such as a graph, a diagram and so on, ask yourself the question: What does this representation show? What doesn't it show? Then complete a table to document the effectiveness and limitations of each of the representations you presented.

REPRESENTATION	WHAT IT REPRESENTS	WHAT IT DOES NOT REPRESENT
Graph of ...		

Representation tasks supporting purposeful observation

In the RiLS study, teachers built on one student-generated question, 'Why do they move the way they do?', to challenge the entire cohort to conduct scientific observations; that is, to observe with a specific purpose in order to understand the movement of a chosen invertebrate. At this stage, the teachers' intention was to give students practice in observing and representing just one aspect of animal behaviour – that of the movement of the invertebrate.

> TEACHER Now, spend a couple of more minutes observing it move. Make it move. Observe its legs. Is there a pattern in the way it moves? See if you can show it to us in a diagram. [pauses, repeats] Is there a pattern to the way it moves?

In the RiLS study, students worked in pairs to select an invertebrate of their choice and recorded their scientific observation using representations such as drawings, annotations, scales and keys, with the sole purpose of completing a 3-D model to explain the invertebrate's movement. The class had access to magnifiers and digital microscopes. Most students included details of the invertebrate's body parts related to movement, as shown in Figure 7.2 and 7.3.

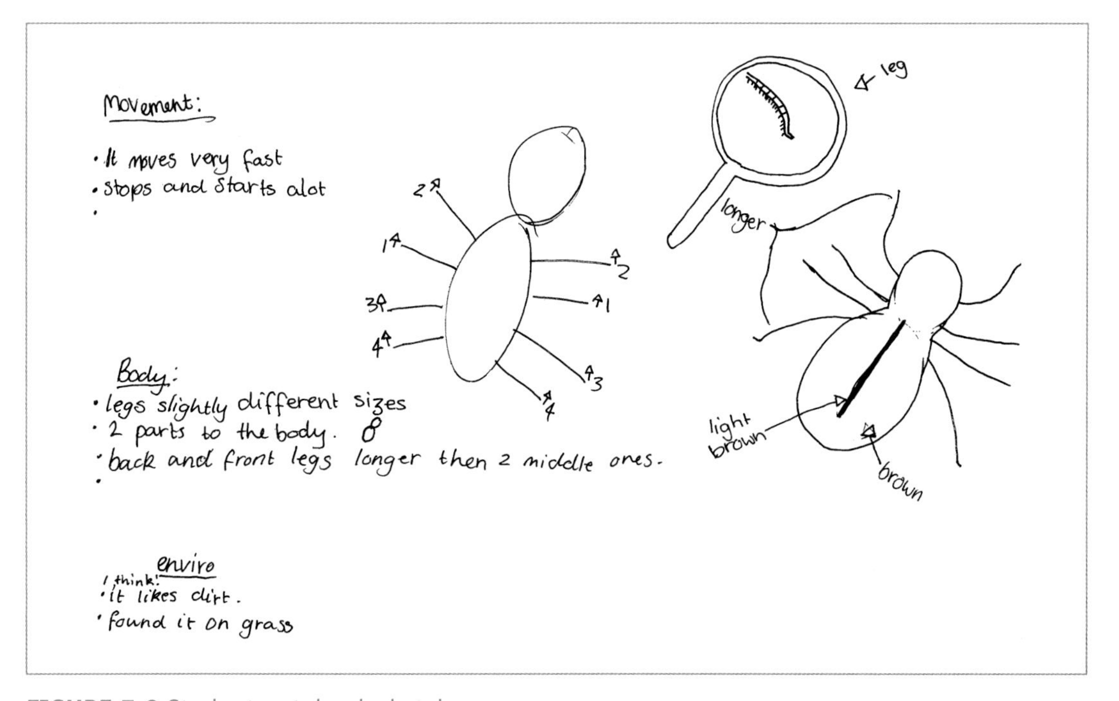

FIGURE 7.2 Student notebook sketches

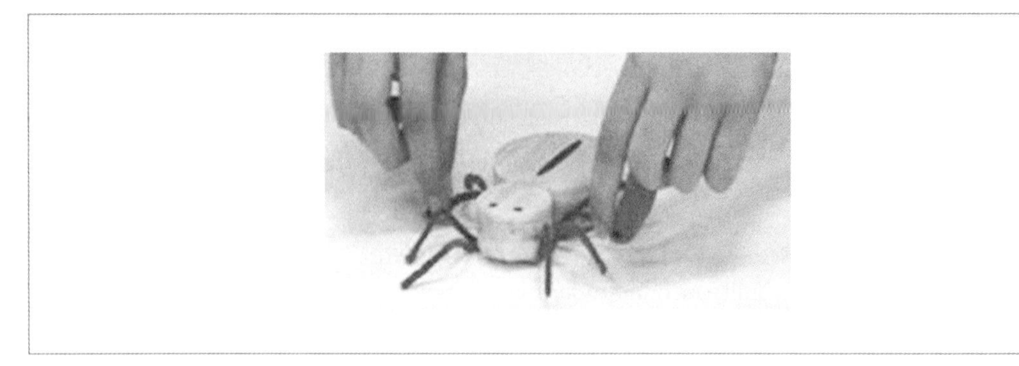

FIGURE 7.3 Melanie demonstrating movement of the spider

The representations of Melanie (Year 5) show clear evidence of observation of the spider made with the intention to record movement. In Figure 7.2, the sketch on the left has numbered lines accompanied by small arrows pointing forward. The intention here is to show the direction of movement only, whereas the place of the attachment of the spider's legs is shown in the sketch on the right. Following is an excerpt from the class presentation Melanie did with Karen, who worked with her:

KAREN We did a brown spider and it has four legs on each side.
MELANIE And we noticed that the legs were actually, were attached to the first part.
KAREN More attached to the smaller part of the body.
MELANIE The head part [raises her arms upwards, touches them to her ears, close to her torso] like this [points to her own abdomen] not down here [quickly moves her hand to the larger piece of wood that represents the spider's abdomen].
KAREN [picking up the pipe cleaners that represent the second set of legs as represented in Figure 7.2] These two move first, then that one [picks up the first set] then that one [moves the third set] and then that [moves fourth set].

Representations in science classrooms can serve many different purposes. In the sequence above, Melanie and Karen were engaged in close observation of their spider, and focused on making sense of its leg movements and the nature of the leg attachments. In other examples, children used elastic bands to reason about the sequential stretching and contracting, thinning and thickening of a worm as it moved, while another group modelled and demonstrated with gestures the undulating movement of the segmented body and legs of a centipede.

Activity 7.9 Representing and re-representing for a better understanding

- Select an invertebrate of your choice from the habitat that you explored in Activity 7.7 (see p. 255). Capture the invertebrate in a container. Conduct purposeful observations and write a paragraph to represent either how it moves or how it feeds.
- Draw it in as much detail as possible. If you have access to a magnifier, use that for fine detail. Draw with the intention to use the drawing/s or a sequence of drawings to help you to build a model that will

specifically represent its movement or feeding.
- Use the Internet and reference books to identify the chosen invertebrate.
- Build a 3-D model (use paper, wire or pipe cleaners, elastic, plasticine, containers) of the invertebrate to specifically represent the way in which it moves or feeds, and use the model to demonstrate this movement or feeding to a fellow student or friend. You should focus on the mechanism rather than on constructing a replica.

Data from the RiLS study showed that in an active classroom environment where students have agency in exploring and interpreting phenomena, the representations of science (graphs, tables, drawings, reports, photographs and models) serve as reasoning tools, central to the process of coordinating ideas and evidence in scientific explorations, and to knowledge generation and learning.

Activity 7.10 **Learning from a representational task**

- How did Activity 7.9 help you to focus your observation of the invertebrate?
- In what ways did the model serve as a tool to provide you with a better understanding of how the invertebrate moves/feeds? What specific concepts do you think you have better understood?
- Think about what you knew about the structure and lifestyle of this invertebrate before you started the drawing and subsequently built the 3-D model. Compare that knowledge with what you know now after building and presenting the model to a fellow student or friend. How has your knowledge of the invertebrate changed?
- You may find that your increased knowledge is not confined to factual information about the invertebrate's structure. Did you find yourself asking questions and drawing inferences about the use and purpose of some of the features of the invertebrate? To be able to do this is a valuable investigative skill.

Representing animal diversity

In the case study discussed above, the use of tables and graphs were central to the task of conceptualising the distribution of animals in an environment; the coordination of annotations and sketches was essential for building a picture of the variation in environments, and their different elements, Internet images and classification keys were essential for building a picture of the diversity of animals and their commonalities and differences; and the various tasks of sketching posters and modelling focused attention on and supported the development of the exploration of animal behaviour. At each stage in the generation of ideas there was the active production and refinement of representations and the coordination of these into a coherent communication.

Tytler et al. (2009) noted that the understanding of the Year 5/6 students of the diversity of invertebrates, through quadrat sampling, involved the generation of tallies of invertebrates, annotated drawings, bar graphs, formal naming systems identified on the Internet, Internet search protocols and Venn diagrams. They argue that each of these

R

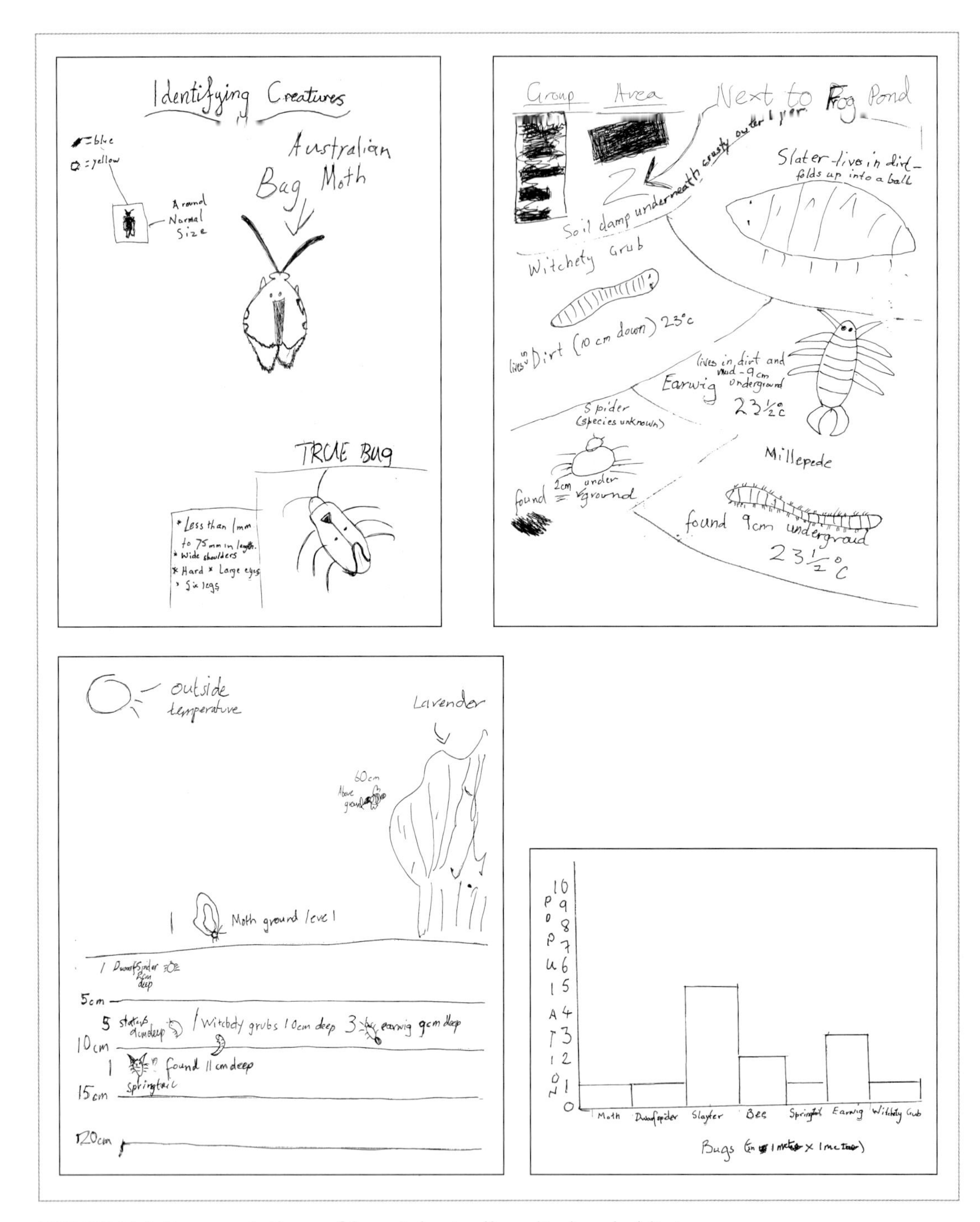

FIGURE 7.4 Representations of invertebrate diversity in a habitat

representations was part of their experience of the entire concept, and that this concept cannot be thought of as separate from these representations. This view of understanding of a concept, as the capacity to coordinate a range of representations to respond to a question or problem, is very different from the traditional view of a concept as something that can be verbally defined.

Simon is in Year 5 and worked in a group of four to observe the habitat 'Next to frog pond' as part of the RiLS case study sequence. Figure 7.4 shows four consecutive pages of Simon's notebook, in which he represents a variety of aspects of his exploration of the invertebrates in that habitat: their distribution, number, different characteristics, and details of their structures and size.

Activity 7.11 Multiple representations of diversity

This activity addresses two aspects of the role of representation in learning. Firstly, each representation of a phenomenon will be selective in drawing out a specific aspect, and will ignore other aspects. Secondly, a concept is built from a range of representations, each selective in its focus.

Look at Simon's drawings in Figure 7.4.

- In what ways do the representations on each page demonstrate Simon's growing conception of animal diversity?
- For each representation, identify:
  - the specific aspect(s) of diversity it shows
  - the aspects of diversity it does not show.

  (For example, referring to the spider drawings of Figure 7.2 (see p. 257), these show the number of legs of the spider and the sequence of movement, some detail of the structure of the legs, and, broadly, their attachment arrangements. They do not show body structures such as jointing and cover in any detail. They are selective in their focus on legs and movement.)
- Referring to Simon's drawings, discuss the claim of Tytler et al. (2010) above, that understanding a concept should be thought of as the capacity to coordinate a range of representations.
  - What is the evidence for Simon understanding 'diversity'?
  - Are there other important aspects of diversity not covered in his notes?

Helping children to classify: pictorial identification keys

Schoolground studies provide children with opportunities to closely observe a variety of animals, grouping them according to shared, observed similarities. Children could group together all those animals that have in common the presence of jointed appendages. They could then subdivide these into smaller groups using the number of appendages as the criterion. Such groupings provide accurate information for the construction of new names for the various animals. This information also assists children in approaching the Internet and books for identification purposes.

Classification maintains a prominent position in school science curricula, yet students find it difficult. The formal classification scheme in biology is based around similarities and differences. In classroom investigations, children are more often directed to emphasise specific concepts based on differences (e.g., colour, size, speed of movement) rather than the more inclusive concepts based on similarities (e.g., presence of a backbone, feathers, life-cycle stages). Schoolground invertebrate studies allow children to modify such personalised classification systems by grouping animals according to shared similarities, before consulting resources for confirmation of their reasoning.

Pictorial keys of invertebrates may be found in a number of books and on websites, and can be very useful in helping children identify animals. However, they do have a number of problems, described below. An activity that circumvents these problems and provides a common purpose for studying the range of animals in the schoolground is the production of a pictorial identification key.

Designers of pictorial keys tend to make many assumptions about the skills and understanding users have when using their keys. For example, animals depicted in pond keys are very often not drawn to scale and the enlargement or reduction is indicated by the use of a cryptic symbol, such as '(3)', indicating that the representation is three times the size of the real animal, or by a bar labelled '1 mm', showing the size of the animal relative to a millimetre. Yet scale is a sophisticated concept that many primary school children will have difficulty with, and will need to be explicitly supported. At the other extreme, in many pictorial keys published for primary school children, the details of the scale are omitted entirely, which creates a number of difficulties. Children cannot use the drawn animal as a means of identification unless its actual size is understood. The production of classroom keys gives a real purpose to accurate observation and recording. It leads to a shared resource that can be used and refined over the whole term of a schoolground habitat investigation.

Activity 7.12 Pictorial keys for assessment

- Think about a strategy for using a combination of published and class-produced keys. For instance, you might consider ways in which children could take a published key and modify it for their purposes or successively refine the classroom key. How could that be encouraged and managed? You might think of ways in which a class-produced key could be developed for display, but with published material being available to support more detailed investigations.
- Discuss how you could use the class-produced key for assessment purposes.
- Thinking about formative assessment in particular, how could you use such a key to ascertain the level of children's understandings or their observation skills? How could you use the key as a trigger for discussions about behaviour of the animals?

The nature and purpose of biological drawings

Scientific drawings represent a discrete genre of science discourse that has a long and important history, within biological science in particular. Children's drawings can be a powerful means of expressing their understandings. Biological drawings can also be used to develop children's observation skills as well as their conceptual understandings by requiring the paying of close attention to the characteristics of organisms (Hayes, Symington and Martin 1994). Children's drawings of live animals display a degree of accuracy and detail that often surprises adults.

Biological drawings are intended to represent a selective and accurate description of details of organisms and their behaviour. In the RiLS animal study, we found that when large activity books with blank pages are provided to students, they tend to use the entire page, which advantages their drawing of detail and the inclusion of multiple perspectives. Examples of strategies that could assist students in developing their drawing skills are:

- the measurement of the animal or plant so as to preserve in the drawing the length–width relationship
- the control of scale – the application of a common multiple (e.g., 3) to their measurements
- the use of vertical and horizontal axes to assist plotting of such measurements
- the use of geometric shapes to help them draw body structures
- the need to count, as with the number of appendages, body parts and so on
- the use of a black fine-pointed pencil rather than crayons or softer coloured pencils
- the use of multiple drawings to represent an animal (e.g., front view, side view, detail)
- the use of a sequence of drawings to represent motion or feeding
- the use of annotations to convey information difficult to present in a drawing.

The drawings made by children in the RiLS study exhibit some of these features. An examination of Melanie's and Karen's (see Figure 7.2 on p. 257) and Simon's (see Figure 7.4 on p. 260) drawings shows the use of scale, geometric shapes, the counting of appendages, concise pencil use, multiple perspectives, and annotations.

Representations and assessment

Children's sketches and other representations (annotations, sequence diagrams, graphs, scales) can be used as an assessment tool. The detail in the representations – what the children notice, how they choose to represent different features of the animal – can all be used as indicators of children's conceptual understandings as well as their ability to observe and represent. Annotations in particular, if they refer to the function of body parts in contributing to the animal's lifestyle and survival, can provide valuable insights into children's thinking and understanding. Such assessment can be used in a formative sense at the beginning of or during a unit to help plan strategies for challenging and developing every student's understandings.

Teachers could use a variety of art activities to support and assess science understandings in the primary school. McGrath and Ingham (1992) explored this notion using various activities involving painting, drawing, modelling and collage. They describe, for instance, the use of 2-D moveable models of different birds using lever systems to explore variations between species. This idea could be readily adapted to explorations of animals in the schoolground. Danish and Enyedy (2007) explored the way kindergarten and Year 1 students' creation of models of pollination supported their learning, and the importance of teacher scaffolding and the negotiation of meaning associated with their models.

Activity 7.13 The use of representations for assessment

- Discuss with a classmate how you could use the representations from Figure 7.4 (see p. 260) for assessment purposes. Thinking about formative assessment in particular, how could you use it to ascertain Simon's level of understanding or his observation skills?
- How could you use the images of animals for discussions about the behaviour of invertebrates in their quadrat?

Learning about life cycles, adaptation and ecology

Life cycles commonly observed in classrooms include those of silkworms, mealworms, snails, butterflies, crickets and stick insects, all of which can be purchased over the Internet. The discussion here applies to all these animals. In studying any of their life cycles, it is important to look beyond a simple description of the life cycle to its adaptive functions.

Activity 7.14 Life cycles and adaptation

- Consider these questions:
  - Why do cicadas emerge all at the one time?
  - When do they emerge in your area?
  - What factors influence the number of eggs a cicada produces?
  - Why can you find exoskeletons of cicadas in grassy areas?
- Construct a set of possible answers and discuss them with some colleagues.

There are many questions that teachers might ask to probe the concept of life cycles. As biologists, we answer questions about life cycles in terms of their adaptive advantages for the organism in its environment. This would include the reproductive advantage of cicadas/grasshoppers/crickets/dragonflies that reach sexual maturity together to maximise mating opportunities, and when there is an abundance of food and mating occurring at an advantageous time for egg deposition.

The easily observed cricket is a well-known noisy insect, well adapted to the schoolground environment, and adept at feeding, seeking a mate and laying eggs to ensure the arrival of the next generation. Crickets are an excellent example with which to teach incomplete metamorphosis. The nymph is similar to adults, without fully developed wings. Crickets grow in length each time they moult. The female cricket can be identified by the presence of a long tube-like structure called the 'ovipositor'. However, the adaptive features of life cycles are poorly understood and inadequately represented in classical life-cycle drawings such as that which appears in Figure 7.5. The representations in Figure 7.5 do not address the concepts of competition, predation, death, sexual reproduction and natural selection.

For these concepts to be well understood, life-cycle studies and models need to focus on a number of crickets rather than the traditional single animal. Life-cycle understandings central to their adaptive purpose would include the need for two adults to produce eggs, the different life histories of males and females, the proportion of eggs that result in a mature adult (not succumbing to disease, accident or predation) and the need for all stages of a life cycle to be well adapted. Part of the understanding of the adaptive purpose of metamorphosis is to broaden the ecological habitat and niche by reducing competition for space and food between the adult and offspring, thus enhancing the survival rate of the offspring.

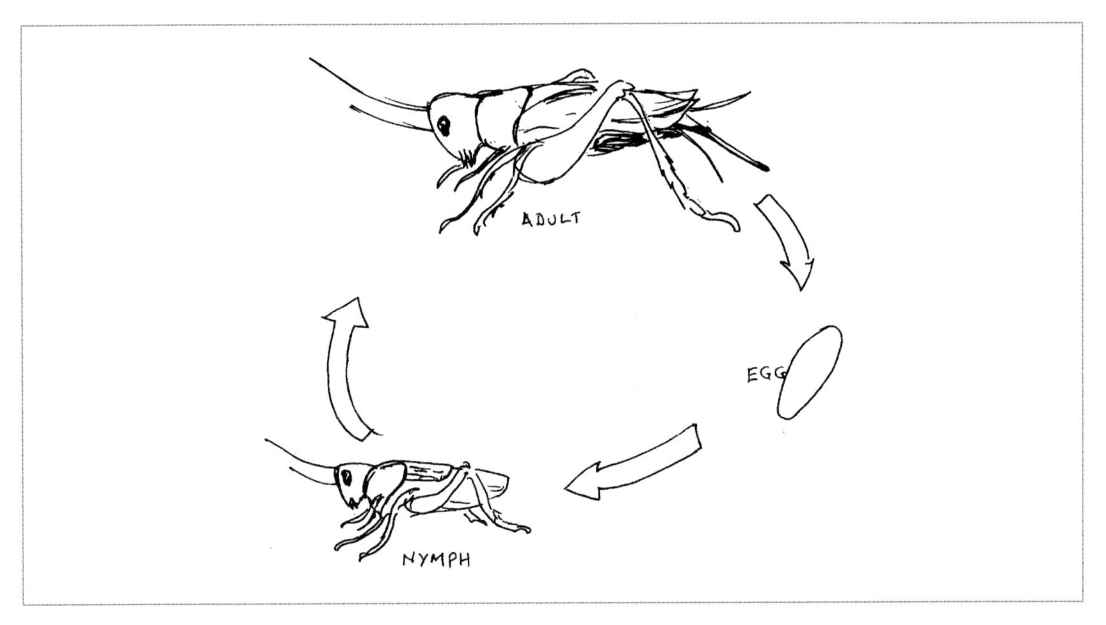

FIGURE 7.5 Life cycle of the cricket

Activity 7.15 Representations of life cycles/reproductive cycles

- Critique the classic life-cycle model represented in Figure 7.5 in terms of how well it represents the concept of adaptation and how well the complexities mentioned above can be catered for within it.
- Construct a model of the life cycle of a frog that takes into account the fact that an adult frog may go through a number of reproductive cycles in its lifetime, as well as the fact that most tadpoles do not survive to maturity.
- Discuss the advisability of referring to a reproductive cycle rather than a life cycle. What do you understand the difference between these two to be?

Children's ideas and learning about insect life cycles and adaptation

Children over the primary school years develop a sharper conception of insects (Shepardson 2002). Young children will classify spiders, worms and slaters as insects, but through to age 11 they increasingly learn to differentiate, although there is a lingering tendency to classify arthropods, particularly spiders, as insects. Part of young children's conception of insects is related to size and shape. They tend to emphasise the negative features of insects, such as biting or stinging, and do not recognise insect roles in, for instance, pollination, scavenging litter, providing a food source for other animals (birds, reptiles) and making products for human consumption (honey, wax, silk). After conducting a study of children's developing conceptions,

Shepardson recommended that children should be provided with the opportunity to observe a variety of insects and non-insects, explore insects in their natural settings as well as a variety of different insect life cycles, study a variety of insect survival mechanisms (adaptation) such as colouration, odour, mimicry and behaviour, and study the social nature of insects.

In a prior study on insect life cycles, Shepardson (1997) studied a class of Year 1 children during a teaching unit in which they kept a journal as they studied a variety of insects. Initially, the children tended to have one-, two- and mostly three-stage models of insect life cycles (ignoring the egg stage), but over the course of the unit they developed more scientific views of some insect cycles. However, their everyday experience of and ways of thinking about insects and the way the language associated with the new knowledge (pupa, chrysalis, nymph) interacted with this experience acted as a constraint to adopting a more general model of metamorphosis. Shepardson recommends the study of a number of insect life cycles, including some with fewer stages (for example, crickets and silverfish), the studying of some insects in their natural habitats, and an emphasis on the adaptive purposes of life cycles.

Children's ideas about adaptation of organisms

Young children have difficulty in seeing the features of organisms in terms of their adaptive advantages. Carey (1985) argues that young children interpret animal structures and behaviour in terms of wishes or wants in the same way as they think of humans as deciding what to do. This is related to the common finding that children commonly hold anthropomorphic views of animals (or of inanimate objects such as trains or the sun), a circumstance no doubt encouraged by children's storybooks. Through the primary school years, they gradually learn to interpret structures or behaviour in terms of the functional needs of organisms. Thus, the change in young Annaliese's thinking over four weeks, from 'that one has three things jutting out of its bottom ... and the other one, he has only two' to understanding the role of the ovipositor of the female cricket – 'the one in the middle is to lay its eggs' – represents a major shift in her view of animal structure and behaviour.

Naive views about adaptive purposes, however, can be very persistent, as Symington and White (1983) found in studying children's responses to the question 'Why do trees have bark?' Few children thought of the reason in terms of the trees' needs. The equivalent question, 'Why do plants have flowers?', was explored with students in two age groups. Their responses are reported in the plant section of this chapter (see Table 7.4 on p. 287).

Clough and Wood-Robinson (1985) refer to a confusion that exists, even for university students, between several different meanings of the word 'adaptation'. It can refer to immediate physiological changes in an individual (e.g., sun-tanning) to the characteristics of an organism that fit it for particular environments (e.g., a yabby's powerful tail), and also the process by which a population is modified towards greater fitness for its environment. This last meaning refers to the evolutionary process or natural selection, but primary school students tend to describe adaptation in the second sense, as a feature that develops in response to environmental conditions. Focusing on this sense of the meaning of adaptation makes sense in a study of animals in the schoolgrounds. This would include studying the function of the various features of the different animals in terms of the way they confer a selective advantage on individual species. It would also include focusing on the adaptive features of the life cycle of animals.

One of the difficulties in discussing adaptation with primary school children concerns their restricted sense of time. The study of animals in the schoolground over a period of weeks can give some sense of the flux in populations as they interact, or in changes of form as animals go through their life cycles. Studying the continuity of the schoolground ecosystem over the school year can give a sense of adaptations related to the seasonal cycle. Knowledge of these seasonal cycles is important for Aboriginal hunter–gatherer communities, enabling them to utilise resources, on an annual basis, over the span of a human lifetime.

Food and energy relationships within the schoolground

Children almost invariably come to schoolground study with simplistic views about the networks of food and energy relationships between organisms. Leach et al. (1996b) found that children aged between five and 11 tended to talk about predator or prey relationships as existing between one predator and one prey organism. They also found that children experience difficulty in interpreting food chains in which the energy exchange relationship is depicted by arrows (see Figure 7.6). Arrows are used to indicate many everyday concepts, and thus confusion may arise when they are used to indicate a specific relationship, such as that between predator and prey.

FIGURE 7.6 The flow of energy along a food chain

When teaching about food chains, we need to explain that there is a flow of energy from the sun through one organism to another and that some energy is always lost at each trophic level – 'trophic level' is the ecological term used to describe the position of organisms along a food chain. In a food chain, the pathway of energy begins with the sun and is one-way, whereas matter is recycled.

Children in the middle primary school years, in talking about the effect of introduced species such as cane toads, which usually feed on insects, think of their impact in terms of direct effects on local species (such as the cane toad eating crickets), rather than indirect effects such as competition for food or depletion of the environment. The notion of a food web, describing multiple energy pathways within an ecosystem (see Figure 7.7), is a much more difficult concept than a food chain, in which only one energy pathway is depicted.

Sometimes not all possible connections are shown in food webs drawn in books or those you find on the Internet, so as to not make the diagrams too complex. This creates an opportunity for you to ask students about all the possible connections.

The schoolground provides the opportunity for children to gauge the answers to a range of questions concerning nutrient relationships, particularly through the use of an interactive teaching approach. Access to animals and plants in schoolgrounds means that children are able to establish and selectively observe ecosystems to answer questions such as 'What will happen to a population of herbivores if we put them in with plants but without any carnivores?' or 'What happens in a system if the very small animals become depleted?'.

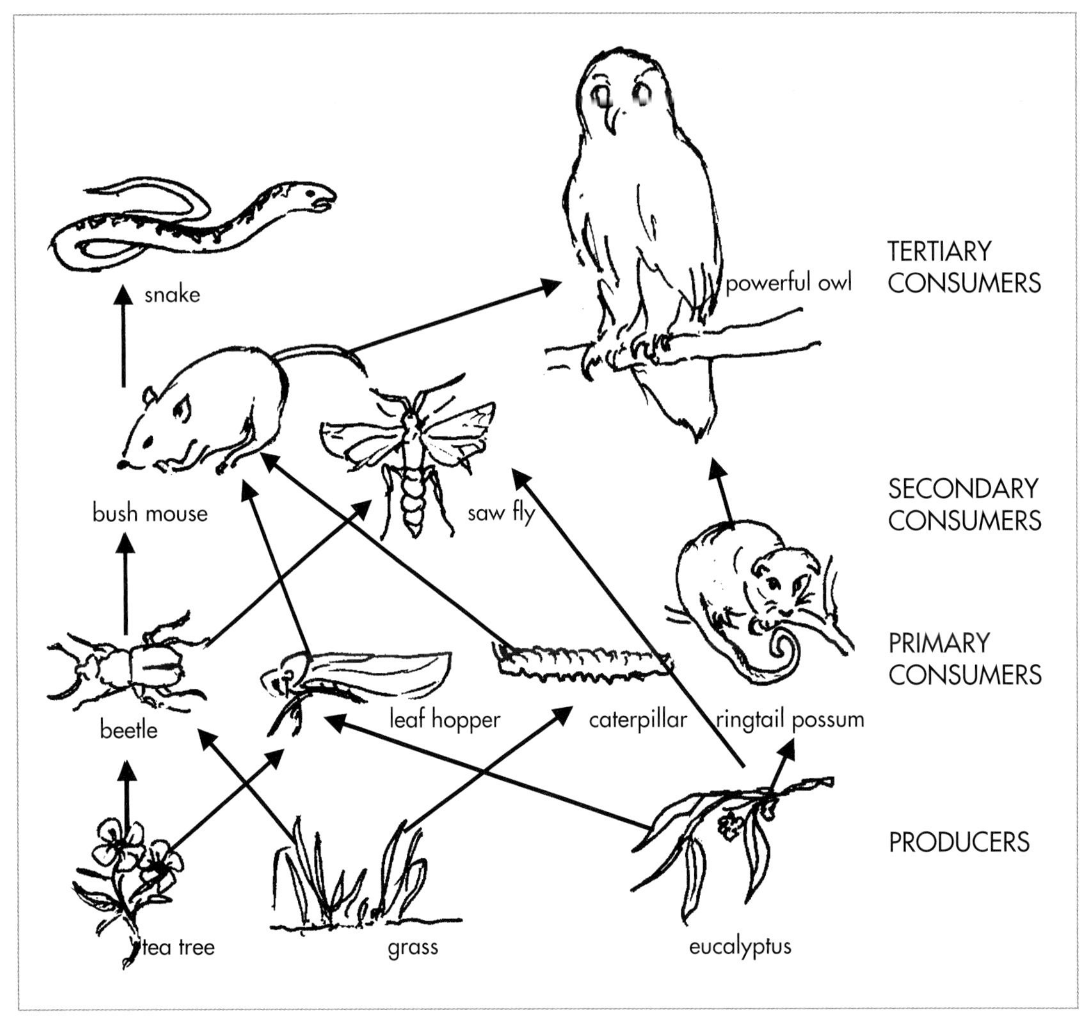

FIGURE 7.7 A forest food web

Activity 7.16 Constructing a food web

- Locate a food web diagram in a textbook or on the Internet. Use this as an example to construct a food web for the schoolground or park habitat you studied. This exercise is best done by pooling the knowledge of all your colleagues.
- Having constructed the food web, use it to generate a number of hypotheses about the effect of selected interruptions on the ecosystem. Use the form of question 'What would happen if …?' (all the tea trees were burnt, all the leaf hoppers were removed, the population of caterpillars was quadrupled and so on).
- Do you have any evidence to support your hypotheses?

- Written responses to questions such as these could provide valuable information for evaluating the understandings children have gained from working in the schoolground environment.
- Examine the outcome statements in the biological sciences topics in your science syllabus/standards/outcomes document. Discuss which level such questions are aimed at, and what sort of response you would accept as evidence of children working at this level.

The energy relationships within the schoolground between the different levels of producers and lower- and higher-order consumers is sometimes represented by a pyramid in which the numbers of the higher carnivores (for example, the kookaburras in the food chain example in Figure 7.6; see p. 267) are less than those of the lower carnivores, with the producers (grass, in this case) being present in the greatest amount. Figure 7.8 illustrates such a pyramid. The pyramid emphasises the relatively small numbers of higher-order carnivores in any ecosystem compared to the comparatively large numbers of herbivores. It takes many grasshoppers, for instance, to feed one kookaburra, and a number of kookaburras to feed one powerful owl (What would happen if the number of owls became equal to the number of kookaburras?). The reason has not only to do with the relative size of the animals, but also with the fact that a continual source of energy is needed by, for instance, the kookaburra to respire, keep its organs functioning and to move. At every stage up through the trophic levels, energy is lost to the system as heat, associated with animals' energy use in keeping alive. This implies that a large number of organisms is needed at each level to maintain a much smaller number of organisms at the level above. The pyramid is a natural consequence of the way energy flows through an ecosystem.

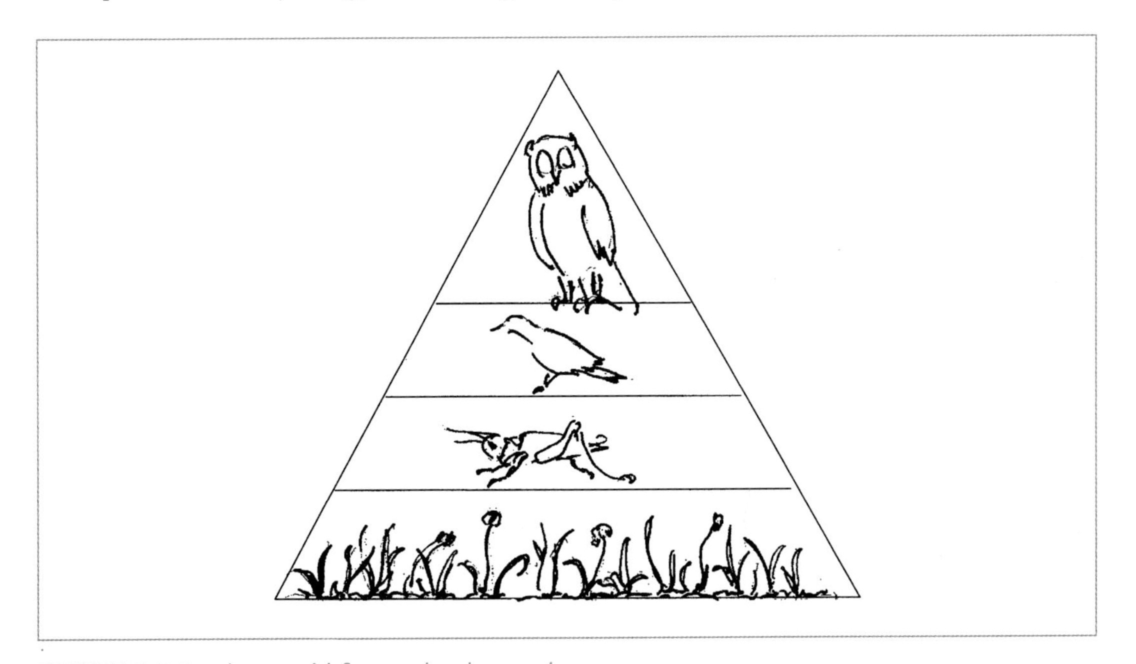

FIGURE 7.8 Food pyramid for a schoolground

Activity 7.17 The food pyramid as a representation

- Examine the food pyramid shown in Figure 7.8. What does this representation of the food pyramid show? What doesn't it show? Refer back to the discussion in Activity 7.11 (see p. 261) as an example of the specificity of representations.
- Whenever a representation is shown to the students, it is important to ask the above questions. Discuss why this might be so.
- Consider the same question in relation to:
  - the food chain of Figure 7.6 (see p. 267)
  - the food web of Figure 7.7 (see p. 268)

The concept of energy has been discussed at length in Chapter 5. In the case of animals in an ecosystem, the main energy transformation involves chemical energy (stored in organisms as chemicals such as starches, fats and protein) being transformed, through the chemical reactions between food and oxygen that constitute the process of respiration to heat energy (involved in maintaining body temperature) and movement. Some of the energy is stored as chemical energy, with a corresponding increase in body mass, but most is lost to the atmosphere as heat through radiation or convection. It is important to remember, in this context, that food itself is not energy, but is the source of energy associated with changes in the ecosystem. The energy in an ecosystem derives ultimately from the sun's radiant energy, which is transformed by plants into chemical energy by photosynthesis.

Abell and Roth (1995) found that Year 5 students who were taught this pyramid diagram tended to misunderstand it, thinking, for instance, that it represented the amount of space available for each type of organism. This very literal interpretation of scientific diagrams is not uncommon, so teachers must be careful when using scientific models such as this that students are not misinterpreting the representation. Abell and Roth recommended that children be encouraged to build up their own models representing the relationship between these population numbers, as only then will they be ready to accept the purpose of this more abstract model.

An environmental modelling game

The study of organisms found in the schoolgrounds related to the cyclic seasonal patterns in various habitats, and the interrelationships between them, can provide a platform for a range of discussions concerning the changing urban or rural environment and pressures on organisms' survival. Discussions of the availability of food could centre on the effects on the ecosystem arising from changes to one element within it.

An interesting activity that focuses attention on the *consequences of change* in an environment can be played with primary school children. This 'survival consequences game' is an example of a modelling activity that can be effective in allowing children to explore the interactions between the different elements in a system, in this case an ecosystem involving layered feeding arrangements.

The *purpose* of the 'game' is to help children understand the relationship between parts of a food chain and interference to some item in that chain. As *preparation*, the children should draw the following set of items on small pieces of card: 20 grass cards, eight insect cards, three small birds' cards and one card that depicts a raptor. The teacher can make several other cards of these items. The relationship 20:8:3:1 is only a suggestion but is effective at the conceptual level. Make a simple spinner with an equal chance for each food chain item. To *begin*: each child has one card as a label. The group arranges itself with grass sitting in a line on the floor, insects kneeling behind them, and small birds standing behind the insects. The raptor stands on a chair at the back and apex of the group. The children will see that the opening configuration could be described as a triangle or pyramid. Discuss the predominance of the grass and the single raptor. *Play* the game. The spinner is turned and the designated card must be handed in and the holder asked to return to their classroom seat. Ask the children what *two* consequences might arise from the removal of the card. For example, if a grass card is removed, then one insect card and one bird card would be removed as they are dependent on one another. As another example, if the spinner indicated a small bird was destroyed, then this would mean that more insects would survive. Hand out another insect card, but then another grass card would have to be discarded as the greater insect population would consume more of the grass inhabitants that they rely on. Turn the spinner 10 times and make the appropriate moves. Ask some children to record each move. Discuss the possible reasons for change – wilful destruction (killing birds), intentional destruction (inappropriate use of pesticides) and destruction caused by natural events (bushfires).

The game is based on the concept of an ecological web but it is representative only and the teacher would need to emphasise that it represents a closed field in that nothing except the organisms on the cards can influence the result, and there is no new grass growing to replenish the pasture. The implications of these two qualifications are important.

Teachers may wish to discuss or even model the effect of an 'open' field. A further possibility is to *enlarge* the universe of the game by including, for instance, logs with the grass cards, fungi cards, some rabbits with the small birds, and two snakes between the raptor and the bird line. Alter the spinner and play 20 spins again.

Planning and assessing habitat activities

Exploration of schoolgrounds and investigations of its inhabitants can lead to knowledge and understanding of concepts about the living world, and to the development of skills and concepts of evidence and the synthesis of these in carrying out investigations (see Chapter 2 for a detailed discussion of these aspects of thinking and working scientifically). Schoolground investigations can also lead to attitudinal development in regard to animals and their habitat, such as a greater valuing and sense of responsibility towards animals and their environment, as well as a development of interest in the detail of animal structure and behaviour and interactions between animals and plants. The activities described above can be as much concerned with these aspects of science as with conceptual knowledge.

Table 7.3 sequences some of these schoolground activities in order of increasing conceptual sophistication and describes a range of conceptual and attitudinal outcomes that could be associated with these activities. The activities are closely related to the sequence described in Table 7.1 (see p. 251) and to the discussion above. The columns in the table represent an attempt to define a progression in children's science conceptual understandings, science inquiry skills and values, and attitudes (the last mentioned having a strong association with 'science as a human endeavour') that children can be expected to move through in association with these activities. This text, by its nature, focuses mainly on the conceptual elements of learning science, including the design and interpretation of science investigations. Learning science also has a strong affective/aesthetic component that must be recognised and taken into account in teaching and learning (see the '"Hot" conceptual considerations: student engagement' section of Chapter 1, on p. 25), and Table 7.3 reinforces this aspect.

A

The table has a number of purposes; namely, to:

- provide guidance as to the appropriate habitat exploration activity for moving each student forward in terms of their concepts, processes and attitudes
- help conceptualise children's progression in each of these three aspects of science as they interact with the habitats
- show that the development of processes and value positions go hand in hand with the development of conceptual knowledge
- provide descriptors that could be used to assess children's positions along these stages of progression.

Activity 7.18 Planning and assessment

From Table 7.3 choose a number of the position descriptors related to conceptual understandings, inquiry skills and values and attitudes to formulate learning outcomes (describing more specific understandings or attitudes in a form that specifies how the outcome would be demonstrated). For each of these:

- relate the outcome to the activity described in the left-hand column. Decide what teaching strategies you would use to encourage the development of the outcome, using the appropriate activity from the left-hand column.
- Think about what evidence you might collect to judge whether the outcome applies. If possible, you should tackle this task as a group, with overlapping responsibilities so you can compare different interpretations.

It should be understood, of course, that the descriptors are not such that a child is uniquely identified by that position. Children are capable of infinite variation in their understandings, processes or attitudes, depending on the context.

- Compare the science conceptual understandings and science inquiry skills descriptors with outcomes in the corresponding strands of your science syllabus/standards/outcomes document. Can you find a matching outcome for each of the descriptors? Discuss whether you think the sequencing in Table 7.3 is appropriate.
- Discuss with colleagues the year level at which each of the activities would be most appropriate.

TABLE 7.3 Development of conceptual understandings, inquiry skills and dispositions using investigations of schoolgrounds

ACTIVITY	SCIENCE CONCEPTUAL UNDERSTANDINGS	SCIENCE INQUIRY SKILLS	VALUES AND ATTITUDES (SCIENCE AS A HUMAN ENDEAVOUR)
Random hunting	Recognises that many plants and animals coexist in the schoolground (but does not think they interact in any significant way).	Carries out simple observation sequences to answer simple questions. Sorts animals according to simple criteria. Reports on activities using simple statements.	Sees animals as interesting play objects (but sees plants as inhibiting hunting activity and shows minimal regard for animals).
Directed hunting/collecting of particular animals or types of animal.	Explains that the invertebrate supports a diversity of animals with different characteristics.	Recognises patterns in animal type and behaviour. Groups data concerning diversity of animals in simple block-and-column format. Makes a record of observations using simple drawings.	Shows interest in the variety of invertebrates. Shows concern for the welfare of larger animals. Asks the names of invertebrates rather than how or why questions.
Directed hunting/collecting in particular habitats. Collecting plants.	Can explain that: • plants and animals interact within the habitat • animals use different areas of habitat • plants have a variety of features.	Develops drawings appropriate for use in a key. Poses questions and suggests observations that might lead to an answer to a question.	Plants seen as interesting and important entities in their own right. Adopts a responsible attitude to the maintenance of the schoolground environment. Asks questions about the diversity of features of animals and plants.
Selection of animals and plants for close observation of structures and behaviour. Detailed representations such as annotated drawings of animals and plants. Modelling aspects of animal behaviour and structure and report on outcomes of observations. Design of an organism to live in a particular habitat.	Understands that plants and animals exhibit a variety of structures and behaviours that are adaptive and help them survive.	Uses conceptual knowledge to make relevant observations of structure and behaviour. Participates in planning simple investigations about animal behaviour or the purpose of plant structures. Makes detailed drawings with annotations and identifies key features when reporting on observations.	Habitats are seen as interesting discovery places. Looks for detail in features of organisms, and asks questions about purpose. Enjoys comparing and classifying animals.

ACTIVITY	SCIENCE CONCEPTUAL UNDERSTANDINGS	SCIENCE INQUIRY SKILLS	VALUES AND ATTITUDES (SCIENCE AS A HUMAN ENDEAVOUR)
Setting up of mini terrestrial habitats to study details of behaviour of animals and interactions. Feeding relationships are studied and mapped. Resources are used to extend knowledge and answer questions.	Can explain that: • organisms have structural or behavioural features (such as mandibles or life-cycle details) that work together to enable them to compete successfully in their habitat • plants and animals interact in a variety of ways • food and energy relationships can be mapped to show interrelationships.	Recognises some relevant variables and suggests ways to control these in the design of an experiment. Uses scientific knowledge to explain investigations. Finds and processes information using a range of resources. Draws conclusions with reference to data. Reports on the results of investigations in a variety of formats.	Exhibits curiosity in asking a variety of questions that link observation with conceptual knowledge. Enjoys developing methods of testing ideas. Recognises a balance between demands of study and concern for the welfare of the ecosystem. Enjoys extending knowledge through the use of a variety of resources.

Children investigating small animal behaviour

In this section, Suzanne describes, as a case study, a unit she developed and ran for her Year 3/4 children who were learning to focus on scientific investigations through working with different small animals. The unit is based on the work of Kathleen Metz, whose research (1995, 1997) indicated that young children are able to think and work at more sophisticated levels than is normally accepted. Metz's work refers to the 'scaffolding of independent inquiry', which she proposes to support her argument that Swiss psychologist Jean Piaget's work has been wrongly interpreted to justify an 'age appropriate' curriculum that restricts science education for younger children to observation, classification and manipulation of concrete materials. Metz's work in classrooms shows that children can carry out experiments that involve hypotheses, the control of variables and inferential thinking. As you read the case study, consider what elements are consistent with constructivist precepts and whether there are features of the sequence that move outside these boundaries.

CASE STUDY

Thinking about and working scientifically with small animals

Convinced that primary age children can apply their thinking and activity to rigorous and independent inquiry in a much more energetic and productive way than generally allowed, I have followed Metz's approach to investigation in units of work, one of which is described below. The approach explicitly casts children in the role of neophyte scientists, encouraging and supporting investigation and introducing learners to a range of scientific techniques and procedures. This enhances the children's view of themselves as scientists and enriches their view of the NOS (see Chapters 1 and 2).

The unit of work aims to allow the practice of science in a purposeful context that engages children's imagination (and hands) in observation, questioning, investigation, design and the management of data. It accepts that young children try to succeed and to understand what is occurring and that their inability to reason about abstract ideas is 'simply because children tend to be novices in most domains, and knowledge of the novice is limited to surface features' (Metz 1995, p. 105).

This approach requires scientific thinking more than description and classification and focuses the inquiry on goals that can be evaluated in terms of the quality of investigations and the collection and analysis of authentic data. It also provides topics that are highly motivating and that allow for diversity of interest and a broad range of metacognitive ability.

The first stage introduces an animal that is interesting to observe. I have used crickets, fish, birds and mice for this activity. The central skill is to observe behaviour and to discriminate between what is seen and the anthropomorphic conclusions that children tend to make. When observations are shared and noted on the board, the children can be asked to consider whether a fish that is looking around at everybody through the glass is an observation or an assumption. Changing animals during the observation stage helps children practise the skill and refine their notes. 'Keeping close to the glass' or 'opening and shutting mouth' become more usual observations as it is the behaviour that predominates rather than sentiment about an animal's supposed wishes.

The next stage is more demanding and children often revert to the fish-trying-to-talk and the-mice-trying-to-escape level. The process practised at this stage of the unit is the drawing of inference. From a set of personally observed behaviours, the child makes suggestions about the behaviour. You can see in Maddy's work (see Figure 7.9) that she has some knowledge about fish behaviour, but the inferences she draws still pertain to a childish identification with a fish as an equal.

When certain behaviours are established, then tallying of frequency, as non-prescribed observation and then as a time sampling, can proceed as the third stage.

Jenny's tallying of Panda the mouse is produced in Figure 7.10. She had drawn up a list of possible behaviours from a previous observation and was satisfied with the characteristics she noted as 'hiding' and 'sniffing'. She was then required to present her data in some form of visual representation and chose the bar graph (see Figure 7.11).

The time sampling was undertaken a number of times. We put the two mice, Pumpkin and Panda, in a large shallow box around which all the children could sit, at various heights, in order to map the behaviour. As well as the template of the enclosure with food, crushed paper, paper roll and water dish indicated, each child had adhesive spots numbered one to 10. Each child had to watch only one mouse and observe what the mouse was doing when I called 'Now', which I did at 15-second intervals. David's map is shown in Figure 7.12. David also had spots marked A–J, which he added to his mapping after we had removed the food tray. Each child then wrote about this event.

Here is Marianna's record of her work:

> We watched Panda, the black and white mouse, and every time Suzanne said 'Now' we put a spot on the map where Panda was. We did this ten times and Suzanne said 'Now' every fifteen seconds. Panda went to the food on time one, time three and time six. At time seven and nine and time ten she was going around the edge of the box. She was looking around the whole box just about the whole time. Then we took the food bowl away. Dots with ABCD show this and Panda stayed in the corner the whole time. We counted the same as before and we didn't have more dots but Panda stayed in the corner still.

It is important for children to consider the data and to discuss what the time sampling and observations do not tell. This was more difficult for the children and was discussed as a whole class. Some of the suggestions the children offered were:

- How far would the mouse run out from the corner to get food?
- Does being in the corner mean fear?

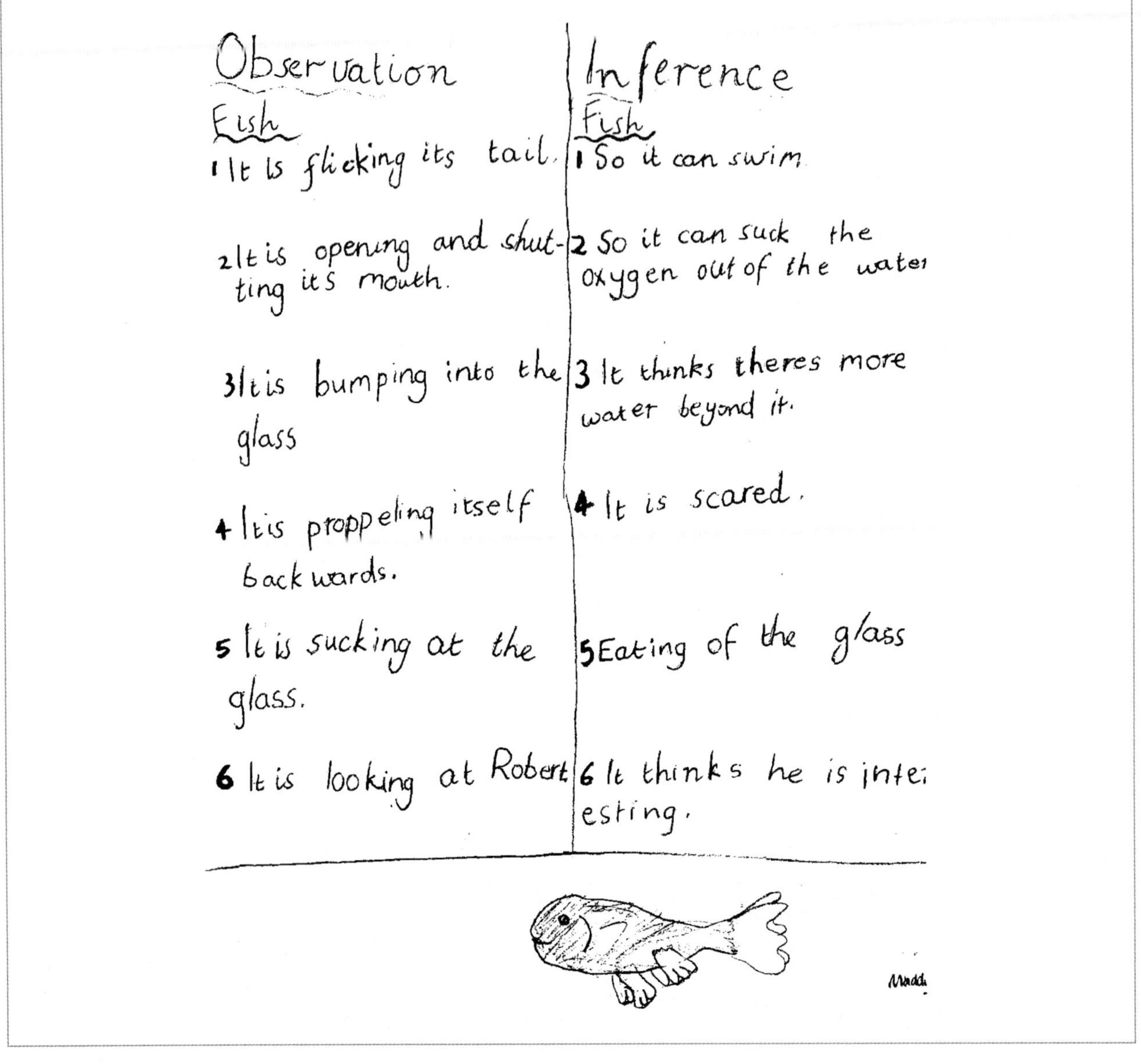

FIGURE 7.9 Maddy's list of observations and inferences

- When would a mouse hide in the cardboard roll?
- Does a mouse know it's hidden?
- Over a day, would a mouse go to dry food as much as to water?
- Would a mouse go into the wheel more than the corner if we put the play wheel in the enclosure?

We then decided which of these could be investigated in the classroom and which could not. This led to planning for different investigations, determining a research question around one of the suggestions, preparing materials and deciding whether conclusions could be realised.

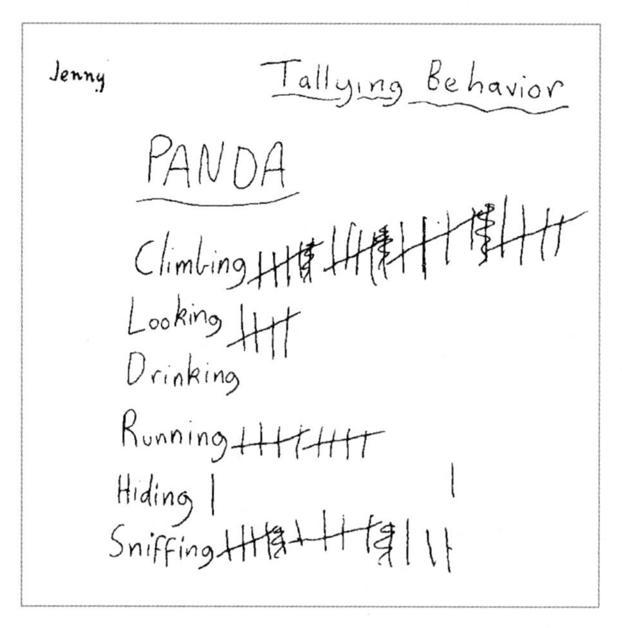

FIGURE 7.10 Jenny's tally for Panda the mouse

An extension of this stage used snails as the small animal under investigation and we observed snail behaviour employing similar stages to those outlined above.

Figure 7.13 is an early attempt by Michael to frame a question and to devise an investigation. The notion of there being an answer had never been raised, but children often feel compelled to know what the conclusive position is. After some reflection, other children decided that they should investigate the same question and devised different ways of going about it. The differing methods were analysed by other class members. An atmosphere of friendly critiquing was established as they set up a community of practice that questioned the nature of investigation in a surprisingly mature way.

The issues that were raised included:

- You would need pairs of colours with different surfaces.
- Where the snail started off would be important.

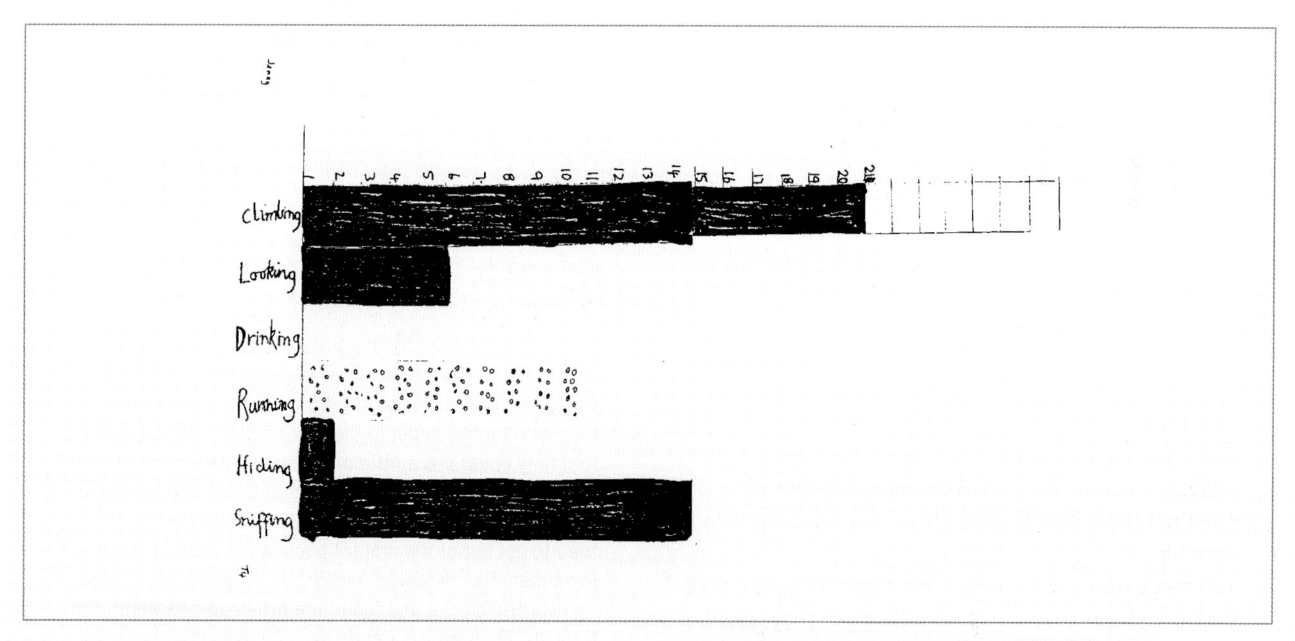

FIGURE 7.11 Jenny's bar graph

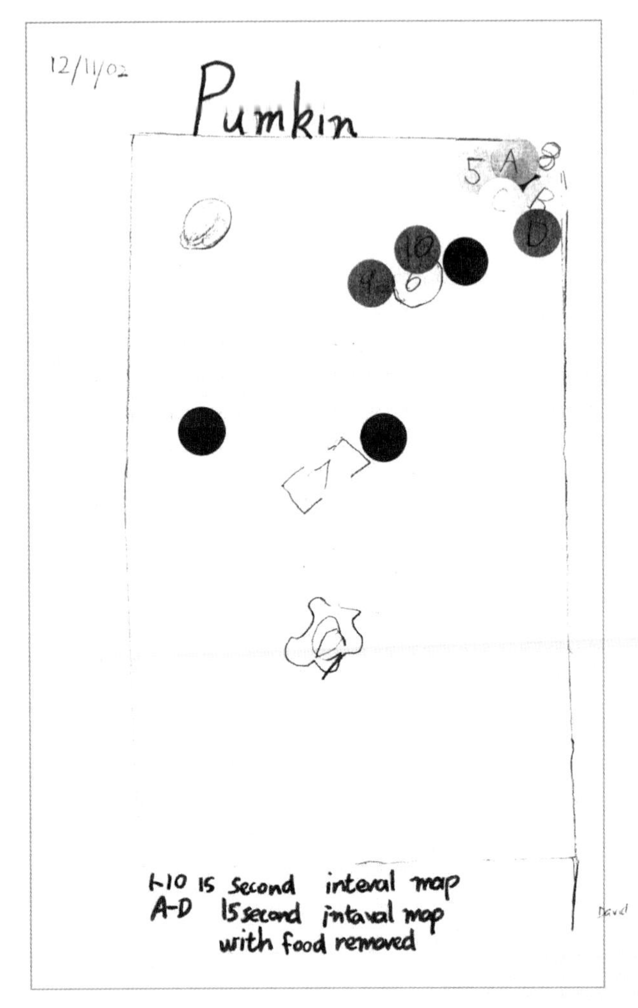

FIGURE 7.12 David's time-sampling map for Pumpkin the mouse

- Smoothness and roughness as characteristics had to be the same in all cases: 'There are different sorts of roughness'.
- Just counting over a short time wasn't enough watching.
- The closeness of each pair was important and when one child put them next to each other, this was judged to be sensible.
- You had to think where snails would go because colour might not have anything to do with where they would move. They might always move onto green.

These reflections gave rise to increased curiosity and subsequent research designs, which resulted in additional plans for watching snails in the natural habitat.

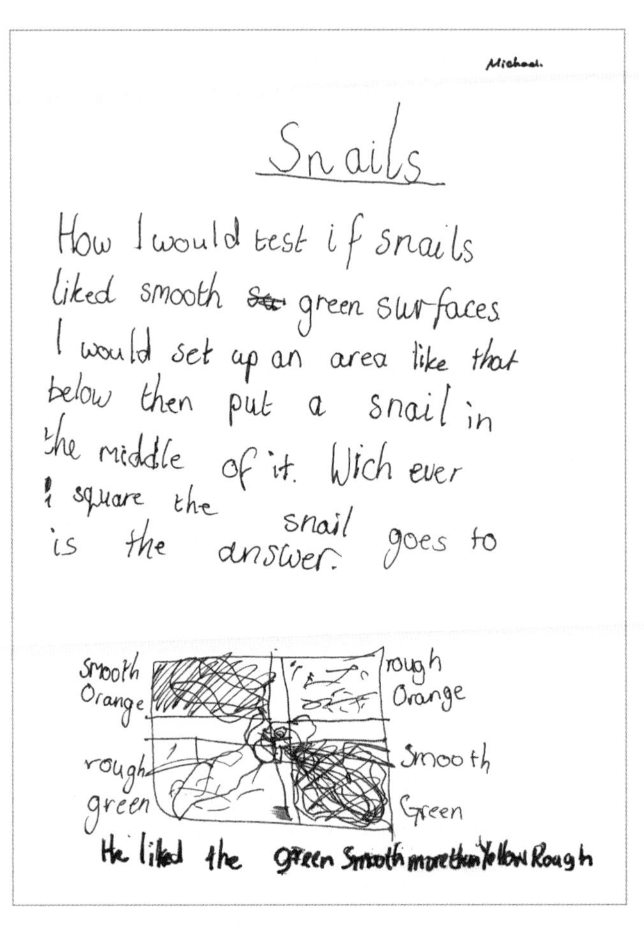

FIGURE 7.13 Michael's snail investigation

Roddy wrote a report of his experiment with the mice and discussed what his experiment did not reveal:

> We had two mice in the big bread crate with the sides covered with newspaper. Their names were Panda and Pumpkin. We wanted to see if they could cross a row of thousand blocks to get to the other mouse. We made it so that they could see a little bit through a gap in the blocks. When I put them in Panda was on one side and Pumpkin was on the other side. Panda stayed still and Pumpkin tried to get out of the crate. I pushed her back in about five times. Panda stayed mostly still. Then Panda walked up near the blocks and found the little gap and went through. Then we put Panda back by herself and put the gap in another place. Pumpkin did not look at the gap. Then Panda saw the gap and went through it. We did the same again and put the gap in other places and Panda went nearly right to it. I think Panda wanted to explore

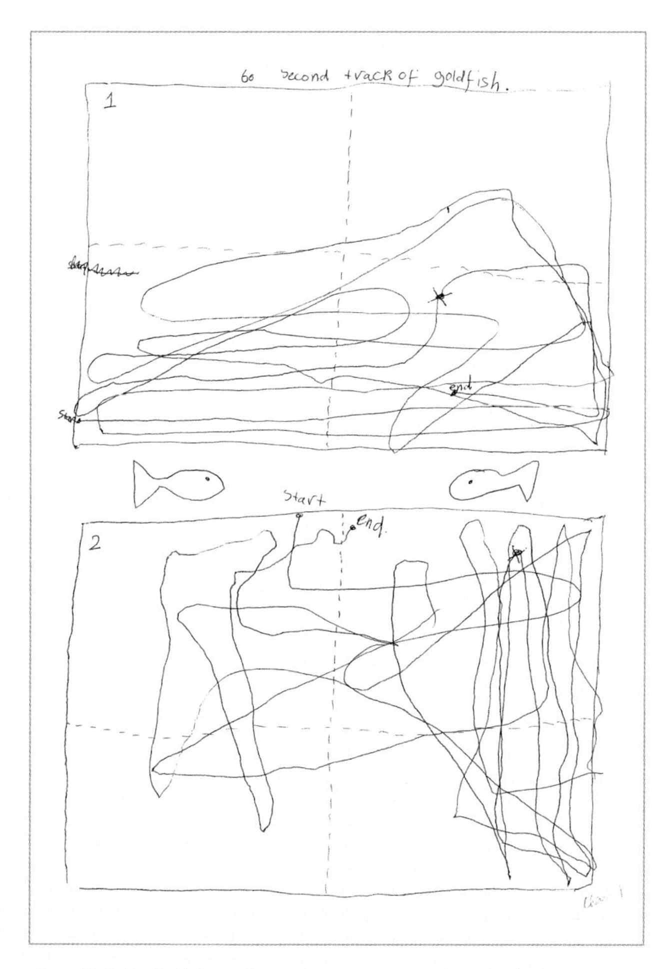

FIGURE 7.14 Tracking the movement of a fish

> and wanted to be with Pumpkin but Pumpkin just wanted to get out of the crate. When we made no gaps Panda started to walk around the crate. I thought she would try to jump over but she just walked around. Then we took all the blocks away and I thought Panda would run over to Pumpkin but she did stay still [sic].
>
> The question was to see if the mouse can cross a barrier to get to the other mouse. I don't think I know if Panda will always look for a gap and I don't know why Pumpkin didn't look for a gap. Food might be better. If they were hungry can they look for a gap? They can jump over if they are starving maybe.

Over the term, the class followed the same set of exploratory stages but different animals were used, posters were made and presented to other classes, and boxes of new pets adopted. Without direction or prompting, one child carried out multiple 60-second trackings of a goldfish movement in a tank (see Figure 7.14 for two examples) and wrote about patterns she could observe; for example, the fish only stopped once in a minute and this is shown by the star shape. There was no pattern to where it started and ended and it did not seem to return regularly to any place in the tank. Her next project was to research the effect on the goldfish of a new fish being introduced.

Another child made drawings of budgies, noting body parts and differences, including the characteristics of one type crossing its wings at the back (see Figure 7.15). Interviewing an expert revealed that this was specific to the breed known as the English budgie, which was perplexing because the child thought all budgies had to be Australian. This set her on a discussion of animal exporting and breeding practices.

It is evident that the children's ability to collect and analyse data, to plan and present investigations and to evaluate their own and others' work enriched their specific knowledge and, more importantly, engaged their interest and sense of purpose in pursuing science questions in practical ways. Not only did the unit provide a constructivist sequence in the conceptual change mould, but it also established a community of inquiry with the children learning to participate in a science discourse, communicating, affirming and challenging different ideas.

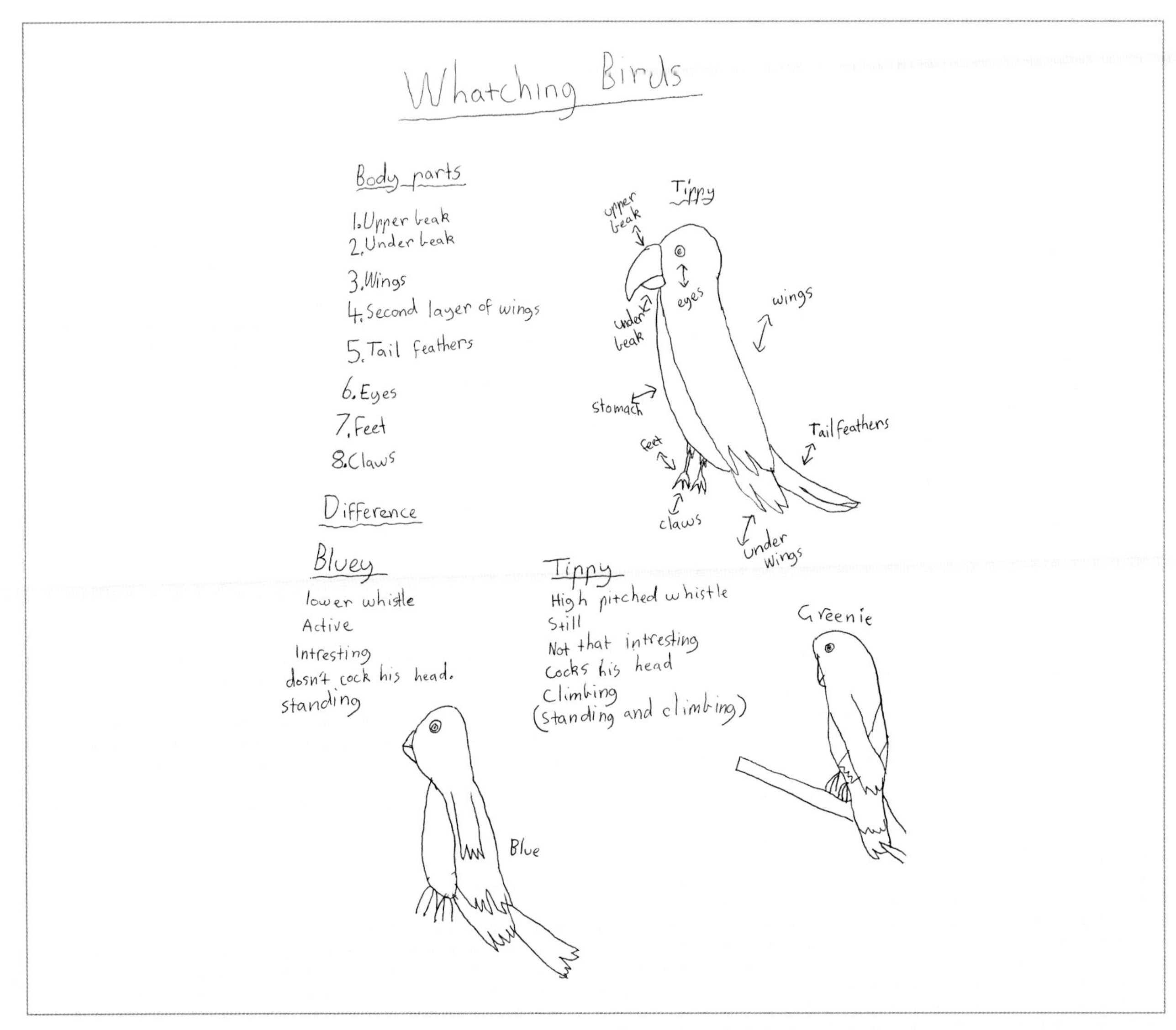

FIGURE 7.15 Drawings of budgies

Questioning and assessment

Working with animals stimulates curiosity in children and learning to ask questions stimulates metacognitive thinking. A '*Questions*' sheet pinned up for class use encourages contributions and comments, and a *community of inquiry* environment, such as that represented in the above case study, with its focus on the investigation of children's questions, encourages discussion on philosophical questions that arise. Such questions can provide insight into children's level of understanding and sophistication in inquiry, which can feed into assessment of their science capabilities.

An informal 'hierarchy' based on Elstgeest's (2001) categories can be useful in determining the sophistication of questions. Watching birds may give rise to a 'level one' style of thinking which is only attention focusing – such as 'What is its beak like?' or 'What colour is it?' At 'level two' are the *quantitative* questions which refer to counting and comparisons – for example, 'How far did the fish swim in one minute?' or 'Which mouse has the longest tail?' In science inquiry, the 'level three' *qualitative* questions are powerful, looking at similarities and differences. For example, in the animal studies, children looked for the things that snails and fish had in common as well as the differences between them. At the next level of thinking are the *action* questions, which require some implementation and refer to 'What would happen if …?' proposals, such as 'What would happen if a slater's pathway was blocked by water?' A 'level five' question style usually *poses a problem.* These are generally context-bound and dependent on prior knowledge; for example, 'How can we find out whether a mealworm goes underneath the bran to avoid light or to eat?' *Cognitive and affective linking* questions also give rise to more complex thinking as they require memories for instances and feelings, and serve to link the learner with earlier experiences and provide language for expression. This 'level six' style may include questions such as 'Have you ever seen a rabbit digging a burrow?' or 'What sort of action do they use?' 'Level seven' involves *reasoning* questions which may be metaphysical or philosophical, having no agreed definitive answer. They give rise to spirited discussion and reflection, handling assumption and examples as part of the thinking process. Children have discussed such questions as 'Are snails good mothers?' and 'How do silkworms know how and when to spin cocoons?'

The nature of science arouses curiosity and wonder which benefits all learners and appeals particularly to the talented child. One question that arose with a group of eight- and nine-year-olds was about silkworm behaviour: 'How does a silkworm know to eat a certain amount and then make a cocoon when no other silkworms are there?' Children with a special bent for philosophical speculation seek clarification and make connections that can stimulate and assist the whole group.

Science is an area of the curriculum that, at its most interesting and investigative, arouses curiosity and determination in children with particular talents who might otherwise not have been conspicuous in academic work. It can provide challenges and tasks that can satisfy children who find the classroom a problematic work space. For children with a particular talent for language, for instance, science promotes enriched writing and research skills that extend these abilities as far as the individual wishes to exercise them. It is open-ended. Science fosters critical thinking, questioning and reflection.

For the child with a physical, intellectual or emotional disability, science offers enjoyable and purposeful activity. The consistent interaction with environment, the emphasis on tactile experience and the less regimented classroom allows endeavour and some success regardless of the aspect of disability or its severity. Watching crickets in their enclosure was very stimulating for one special-needs child, for instance, and led to the development of enriched vocabulary about movement. For the NESB student, science gives a consistent opportunity and need to describe and question specific occurrences and outcomes.

Activity 7.19 Planning for thinking

- It could be argued that the case study 'Thinking about and working scientifically with small animals' focuses on quality of thinking. Discuss with colleagues just what view is represented of quality thinking in science. What are its characteristics?
- Select a science investigative activity you have previously discussed or had experience of. Generate, with colleagues, a list of questions that could arise in the activity to illustrate the question categories described above.
- Discuss how children's questions could be used to assess their level of thinking. Generate some questions relating to the case you would regard as representing a high level of thinking.

Interpreting the case study: sociocultural perspectives

In this case study, the children have been active participants in a process of inquiry, collaborating in deciding on their questions and discussing and critiquing their investigative designs. An important part of Suzanne's task as a teacher was to establish an environment in which the language of exploration and the collecting and evaluating of evidence were central elements. This case can, therefore, be viewed through a sociocultural or situated cognition lens in which the children do not so much learn to master concepts and processes as individuals (although this happens), but rather move from a position of relatively peripheral to more expert participants in the sociocultural activity that is science, represented within Suzanne's classroom. Suzanne's use of the term 'community of practice' can be related to this. These ideas are broadly based on the work of Vygotsky (1986), a Russian psychologist who emphasised the way language and culture are fundamental to how we come to learn about the world, gradually taking on established discourses as the basis for our own thinking. Note how Suzanne, following Metz, spent some energy in teaching the children the thinking tools and processes of the 'discursive' or language elements of science – the language of evidence, inference and observation, as well as particular measurement and analysis tools. The conceptual end point of the study was not strictly determined, but the quality of thinking and discussion was a prime focus. Sociocultural theorists view learning as essentially constituted within participation, and understanding as fluid and contextual rather than as stable conceptual frameworks located in people's heads. Rogoff (1998, p. 691) argued that, 'What is key is transformation in the process of participation in community activities, not acquisition of competences defined independently of the sociocultural activities in which people participate'. These ideas are discussed in some detail in Chapter 1.

This focus on 'discursive elements' (the languages and processes we use, including mathematical language and diagrams) represents a growing recognition of the importance of

R

language practices in framing thinking and understanding. There is growing recognition that science is a mix of languages involving multimodal forms of representation, and that learning science involves students being able to interpret and integrate science texts such as tables, graphs, diagrams and science reports (Lemke 2004). The representational resources that children need to develop are the literacies of science, and teaching should focus attention on introducing and negotiating productive representational resources through which students can explore and interpret phenomena (Tytler, Peterson and Prain 2006). This explicit focus on literacy underpins *Primary Connections* (AAS 2005). Pedagogies appropriate to a representational perspective have been described in some detail in the schoolground invertebrate sequence.

It is interesting to view Suzanne's practice in the case study above, through this representation lens, as she introduces students to different ways of representing animal location, movement and features, and asks them to coordinate these ideas in explanatory text. The case study represents a very rich literacy environment, which arguably is the key to its strength as a learning environment.

Activity 7.20 **Developing a view of the case study**

Discuss with colleagues the case study 'Thinking about and working scientifically with small animals' in order to come to a view about how best to consider its features.

- Write down the main elements of the sequence and identify Suzanne's role as teacher in each of these.
- Identify those characteristics of the case that provide a good illustration of constructivist principles.
- Discuss the different representational modes (e.g., written, diagrammatic, graphical) that children are asked to use in the case study to explore animals. Identify examples in the sequence where Suzanne asks students to re-represent ideas in different modes.
- Do other case studies in this text illustrate the notion that a supportive learning environment involves students in negotiating representations, and representing ideas in different modes?
- Develop a position on the extent to which you feel the teacher needs to have in mind a specific conceptual outcome or inquiry skill for a sequence such as this, as against taking a more flexible view of what individual learning might take place.
- Review some other case studies from the different chapters in this text. Do they differ in the specificity of the end conceptual position? Can sociocultural perspectives help us make sense of these other cases?

Plants in the schoolground environment

Most schools have grounds containing a variety of plant species, including a variety of native trees and shrubs as well as introduced species planted for their colourful flowers. As a result, the schoolground can be a rich source of material for studying plant structure and function. In this section, we will discuss children's and scientists' ideas about plants and introduce activities designed to clarify ideas about plant growth, plant reproduction and adaptation.

Children's ideas about plants

Children's conception of plants is much less developed than their conception of animals. Several researchers report that children think plants are only alive when they have flowers or are producing fruit. Plant movement is much more subtle than that of animals and as such it makes little impact on children's thinking.

Biddulph (1984), in a study of seven- to 11-year-old children, found that their understanding of plant reproduction was such that, when asked why they thought various plants had fruit, not one child out of 80 understood the concept that fruit functions as a seed-dispersal mechanism. Again, ambiguity between the everyday and the scientific meaning of fruit could account for some of the confusion. When asked the supplementary question 'Would the fruit be of any use to the things they grew on?', 56 per cent of nine-year-olds answered that the plant grew fruit for us to eat.

A number of studies (e.g., Leach et al. 1995, 1996a) have found that, for students even up to the age of 16, the scientific notion that the body matter of all organisms is derived from chemically transformed food poses significant problems. In the case of plant nutrition, pupils find it difficult to conceptualise plant body mass coming from invisible atmospheric gases and water (in photosynthesis plants use the sun's energy to transform atmospheric carbon dioxide and water, drawn in usually through roots, to simple sugars and starches), rather than the more solid substances of soil and water, which are usually identified as the source of food. Children's understanding of trees' need for light, rather than being framed in terms of specific energy needs, is more often explained in the non-functional statement that plants need light to grow.

Biologists define plants as organisms that produce their own food through the process of photosynthesis. There are a number of broad phyla in the realm of plants, the main one of which is angiosperms or flowering plants, but also others which include plants such as conifers, ferns and mosses and so on. You can find details of these in biology texts or most secondary school science texts. You can also find details of the life cycle of flowering plants.

Ideas about plant structure and function

In everyday language, the word 'plant' has a rather narrower meaning when compared to its use in science. Young children may mainly associate plants with things in pots in gardens and do not necessarily view a tree, a bush or grass as a plant. When asked to draw a plant, they tend to revert to stereotypical plants with flowers and single stems. A productive elicitation and exploration activity would be to focus attention on the details of a variety of plants and plant parts and identify their commonality of purpose. Tunnicliffe and Reiss (2000) found that children tend to recognise and identify plants mainly in terms of their large-scale anatomy and where they live (for example, mosses live in damp places). They recommend that children be encouraged to observe the finer details of plant structure and to link plant features with their environmental adaptive functions. They argue (p. 177) that 'we don't want pupils to have a model of the environment simply as a background against which individual organisms stand. Rather, we want pupils to understand the ways in which plants and other species affect and are affected by their environments'.

Activity 7.21 Plant parts

- Collect or work with a variety of parts from native plants from the local environment, including leaves, gumnuts or banksia fruit, tea tree seed and a variety of seed pods, cones, flowers, bark and buds.
- Identify each plant part's function. What is its role in the plant's survival?
- Draw carefully the details of selected plant parts.
- Generate a set of questions that arise from the activity.

In a similar activity with children in their first year of school, drawing plant parts from real life during a plant unit (Figure 7.16) seemed to jolt children out of their stereotypical representations of plants (Figure 7.17), encouraging them to make some detailed and focused observations of plant parts. Note how the use of pencils in Figure 7.16 allowed much more detailed representations.

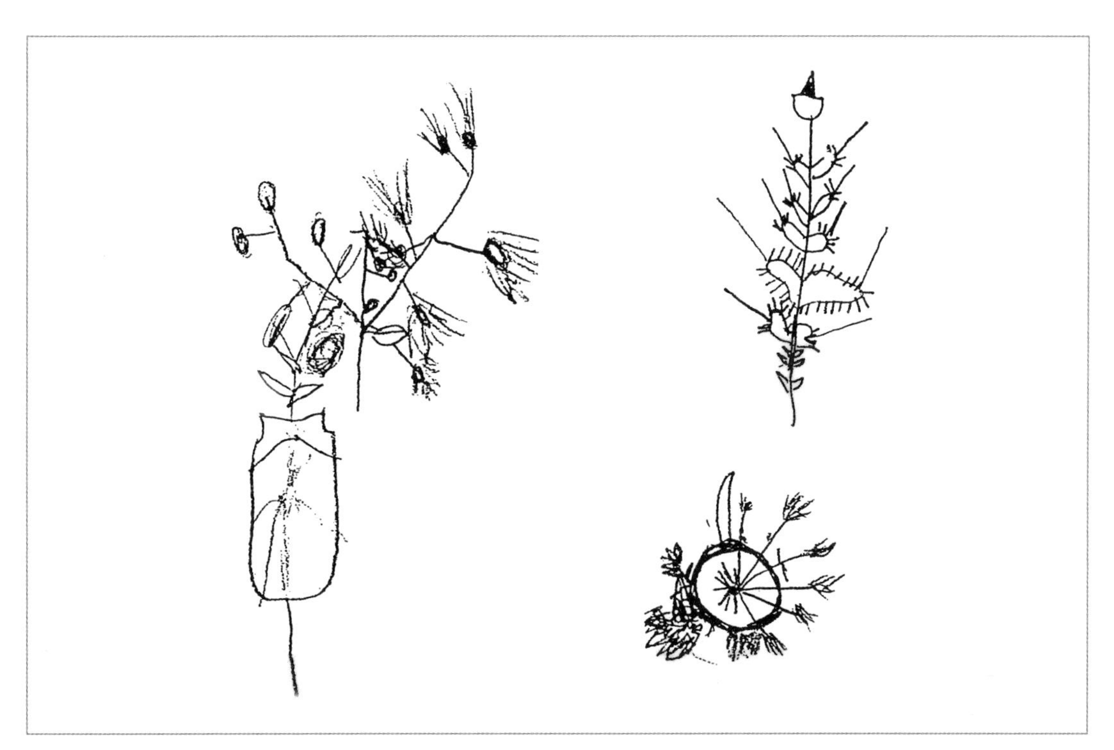

FIGURE 7.16 Children's drawings of plant parts from real life

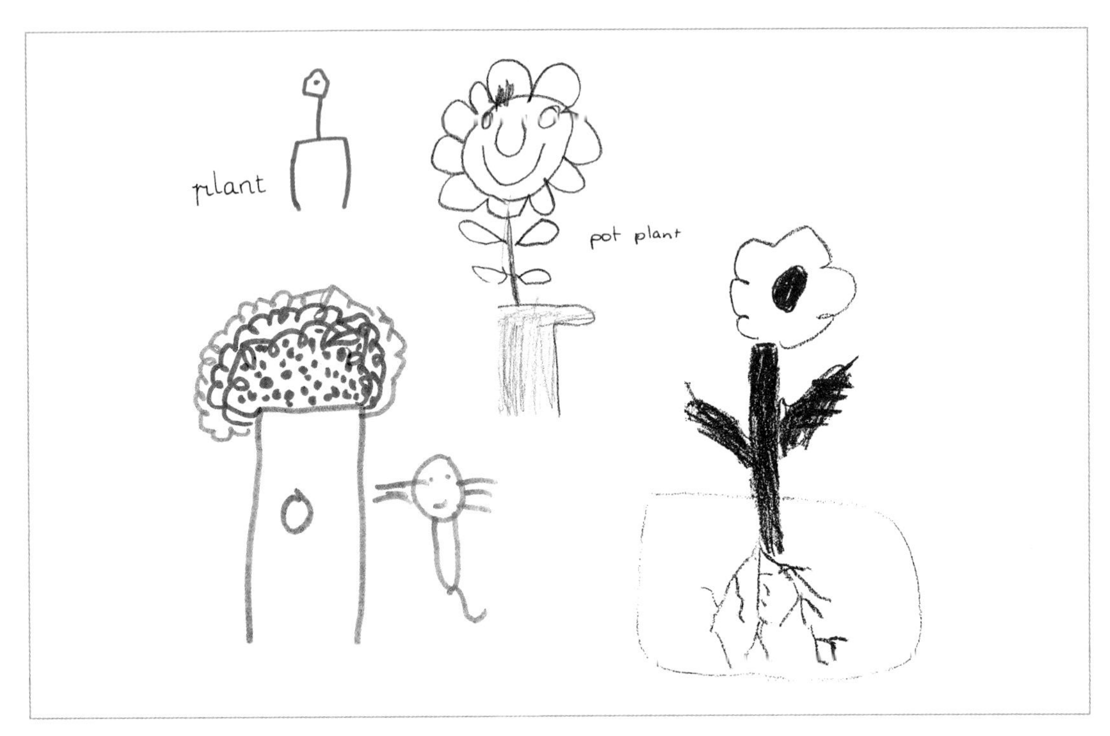

FIGURE 7.17 Children's stereotypical representations of plants

Ideas about plant reproduction

The important aspect of flowering plants is that the details of the reproductive cycle (seed–germination–seedling–plant–bud–flower–pollination/fertilisation–fruit–seed) represent an adaptation by which the plant is fitted for survival. Flowers are the mechanism for sexual reproduction in angiosperms and each flower has male and/or female parts. Pollen, the equivalent to sperm, is distributed by various means, such as insects or birds, to pollinate the female part of the flower. Pollination (transfer of pollen) occurs before fertilisation (fusion of male and female gametes), which takes place in the ovules in the ovary within the receptacle of the flower.

Table 7.4 shows the responses – from Symington and White's (1983) study – of children from two age groups to the question 'Why do plants have flowers?' As can be seen from the table, young children see the purpose of flowers mainly in terms of the needs or desires of humans, or of other animals, such as bees, rather than in terms of the adaptive needs of the plant itself.

Large, brightly coloured, perfumed petals and sweet-smelling nectar are adaptations that attract birds and insects, which assist pollen transfer from one flower to another. Flowering plants that rely on wind for pollen transfer are less conspicuous. The position of the stigma varies in different plant species. Most commonly, the stigma is found above the stamens (e.g., hibiscus) so that it is less likely for pollen from the same flower to adhere to the stigma. Advantages of this arrangement for the plant include facilitating cross-pollination rather than self-pollination. In turn, cross-pollination assures variation in the seeds (produced after

TABLE 7.4 Seven- to 12-year-olds' views about the function of flowers

THE MAIN REASON A PLANT HAS FLOWERS IS	PERCENTAGE OF PUPILS CHOOSING EACH REASON	
	7–8 YR OLD N = 83	11–12 YR OLD N = 115
***1** Because it just grows that way*	1	1
***2** To grow seeds*	5	30
***3** To make it look nice*	13	4
***4** To grow fruit*	4	13
***5** Because bees need the pollen and nectar*	45	28
***6** To show where the fruit will be formed*	6	6
***7** Because it is spring time*	10	3
***8** I don't really know why it has flowers*	6	4
***9** I've never thought about it*	11	11

fertilisation), which enhances the species' survival in a changing environment. Following fertilisation, the ovary of the flower becomes the fruit that contains, protects and nurtures the seeds. Different flowering plants have different means of seed dispersal (Why would plants be advantaged by dispersing their seeds?), such as by wind (in the case of a dandelion, for instance, which produces many seeds, each of which would have only a small chance of germinating successfully), water or by animal transport (for example, animals might eat the fruit and scatter the seeds in their faeces).

The diagrams of flowers in texts tend to show an idealised structure of a flower, with the male and female structures arranged in an ordered way within the petals. Flowers can vary a lot within this basic structure and children should experience, preferably firsthand, a variety of these so they can begin to see the broad patterns within the variety. Some flowers have a pistil with multiple heads and, in some flowers, such as the grevillea, the stamens are wrapped inside the petal and are difficult to find if you are not aware of this possibility. Some plants, such as dandelions or echium, have many tiny flowers, each with its own reproductive structure, clustered together in what most people would think of as a single flowering head.

Activity 7.22 Flower dissection

- Cut up a variety of flowers, including some Australian natives such as grevillea and eucalyptus, and some complex flowers like a daisy or dandelion. You should use a scalpel (note the need to consider safety issues) and/or scissors, and tweezers. A magnifier can be useful for identifying parts in a small flower. Identify the ovules in the ovary. Where is the stigma found and where are the stamens? Decide which flowers have a structure to assist cross-pollination.
- Identify and sketch the main parts. If running this task in a classroom, discuss the following principles with colleagues

regarding the appropriateness of praising the child who:
- makes the most observations
- makes the most counts
- formulates the most questions
- makes the most-accurate drawings
- attempts to indicate scale
- cuts up the most types of flowers
- attempts generalisations based on findings
- modifies generalisations in the light of new findings.

- Think about things that are sold by a greengrocer. For each, identify which part of the plant it represents – fruit, stem, leaf, root, flower.

One of the difficulties for children in studying the nature of flowering plants is that they have a rather narrow view of which plants have flowers. Many people think of a flowering gum as one of only a few eucalypts that have flowers, yet all eucalypts are flowering plants. Another difficulty involves the identification of fruit as a category of food, which does not correspond exactly to the biological view. For instance, a gumnut is a fruit, as is the cluster of hard woody balls on a tea tree, since they contain the seeds. A peapod is a fruit and the pea is a seed. Biologically speaking, then, any part of a plant that contains seeds is a fruit and this includes avocadoes, tomatoes, olives and pumpkin. The broccoli we eat is the flower of the broccoli plant.

Using plant trails

Given the abundance of plants in most local school environments, it seems a pity to focus classroom sequences on plants around textual representations only. Plant trails have been described in numerous publications (e.g., see the Gould League website at http://www.gould.org.au/index.asp) and come in a variety of forms. The essential nature of the trail is that it takes students through a structured observation experience: it asks them to notice significant features of plants and to become more sensitive to plants in the environment. Trails can be quite structured (e.g., asking children to collect or draw particular parts of particular species) or quite loose (e.g., asking children to search for examples of particular features of plants in general). In planning trails, it is a good idea to prepare children in advance concerning what they will be looking for by providing examples of different plant parts beforehand, for instance, or workshopping questions they might have about plants that could be explored on the trail. Botanic gardens often have sections offering insights into Indigenous people's use of plants, and information can be found, for instance, on the Australian National Botanic Gardens website (see http://www.anbg.gov.au/gardens/education/resources/index.html).

Activity 7.23 **Plant trail**

- Construct a means of collecting specimens of plants in an orderly array. This might consist of a manila folder with small plastic bags stapled into it. Each manila folder could be dedicated to a particular plant.
- Walk around the local environment with the aim of setting a trail for children. Identify plants with interesting features and select one feature to place in the folder. The features might include new and old leaves, bark, fruit, seed and flowers in different stages.
- Swap your folders with a colleague. The task for each of you now is to collect a specimen from each plant, of other designated features.

Plant growth and the cycling of matter

There are many experiments one can do concerning the growth of plants, including growing bean or lentil seeds in soil and cotton wool and charting the way shoots and roots develop, or monitoring their growth in conditions of light, or low light or differing amounts of water, soil and/or cotton wool. There are significant understandings associated with plant growth and the case study and associated discussion below is meant to show how plant growth relates to a variety of cycles of material in natural systems.

CASE STUDY

The terrarium

The first part of the story is drawn from an episode relating to a longitudinal study of children's ideas (Peterson and Tytler 2001). We had been involved in running a unit on plant growth. This included experiments with seed trays grown under different conditions, studying plants in pots and growing beans in jars. In that unit, and in interviews with selected children afterwards, Suzanne had explored their ideas about life (What does it mean to be 'living'?), the conditions for plant growth, and reproduction (Where do seeds come from, and what are they for?). Following that unit, the children had little trouble in telling us that plants need sun, soil and water to grow. Most of them knew that water came in through the plant's roots, but understandably had no real idea of how or what happened next.

The terrarium activity

There were two things we wanted to focus on in using the terrarium. Firstly, because it represents a closed system, it provides a challenge to children to think in terms of the cycling of material to nurture the plant. Our focus was on the cycling of water. The idea came from a Swedish colleague who has been using terraria to explore some very interesting ideas that older students have about the recycling of matter in plants (Hellden 1997). Secondly, we set it up as an investigation to support and explore the children's ideas about the design of experimental investigations, since we had previously worked with them on this.

We provided the following equipment:

- small fish tanks, each lined with very slightly damp potting mix
- clear plastic sheets and tape (cling film will suffice)
- water mist spray
- a variety of small seedlings.

We set the tanks up to explore the effect of the closed system and the amount of water. The children planted three small commercial seedlings in the fish tanks. These were arranged to compare what happened with and without sealing the tank and the effect of varying the amount of water.

We had an initial discussion about what the children thought would happen over the week to the plants. In general, they were convinced that the plants in the closed tanks would die, either because they would get no air to breathe or because they would grow and bump into the plastic ceiling. Interestingly, while we were discussing their expectations, condensation appeared on the sides of the closed tanks, which excited their interest. One child noticed there was more fog on the tank in the sun and speculated that this was related. We had a very animated discussion, with children explaining the fog using associations that related it to breathing or to heat. These children were thinking of the fog as an effect, not necessarily as a material on the inside of the tank.

The tanks were all left in the same place – positioned on a high sunny shelf. One week later we took them down and discussed the conditions of the plants in the various tanks. The children were absolutely amazed that the plants in the closed tanks were doing better – no-one had predicted that – and we had a very animated discussion. There were lots of ideas about the presence of water droplets on the sides and lid of the closed tank. Most of the initial ideas had to do with the lack of air or associations with closed spaces and sweating. Over the next few weeks we monitored the plants regularly, as it became clear that those in the closed tanks were thriving and the open-tank plants were wilting. At this stage we interviewed a number of the children about what they thought and found a range of levels of understanding of the cycling of water in the tank. Here are some examples:

> It got so hot water went into the plastic.
> I think it gets all hot and melts.
> Like we get all sweaty.
> [The water comes from] the hotness ... 'cause it's so hot in there ... It starts to, like, turn into water.
> ... because the moist (sic) it comes down the bottom of the soil and it gets really hot and it starts to get wet and so it grows up and up and then it goes into here and finally gets up here and spreads around. The plant got too heavy and a bit dropped off it and might flick the water up there.

One child, Hugh, made a startling leap, likening the tank to a mini globe:

> It's just like the world. Just pretend the world is a container and the clouds are fog and when the clouds get really big some of the clouds turn into water or they come down and some stay up there and when it gets really big it breaks.

Hugh thought the difference was that there is air in the world, but none in the tank.

Hellden (1997) has used terraria to explore the development of children's ideas about the cycling of matter over a number of years. He asks questions such as:

- Where does the green matter in the growing plants come from?
- If the leaf matter in the terrarium is in the forest and is replenished each year, what happens to those leaves over time?
- Why do trees lose their leaves?

Activity 7.24 Cycles in the terrarium

- Discuss with colleagues what happens with water in the terrarium. Trace its pathway and discuss its role in the plants' survival.
- Tie a plastic bag around the leaves of a bush or a tree and leave it in the sun for an hour or so. Predict, observe and discuss what happens.
- Discuss as a group Hellden's questions. Construct a list of ideas and your own questions that arise from this. Trace, if you can, the recycling of matter within the terrarium as the plant grows.
- What would happen to the weight of the terrarium as the plants grow?

Children's and scientists' understandings of plant growth and the cycling of matter

Over a 10-year study of the ideas of 25 students, Hellden found that their ideas and the way they approached thinking about plants and matter developed considerably (Hellden 2005; Hellden and Tytler 2008). He found that, gradually, their ideas became more sophisticated and their concepts more differentiated. For many students, their views seemed to be strongly linked with episodes in their early childhood and, over time, their ways of looking at phenomena retained some coherence. For some students, naive ideas persisted for many years. Examples of Hellden's findings are:

- Students do not easily understand that plant material comes from carbon dioxide and water vapour in the air, driven by light in the photosynthetic process. The idea that plants produce their own food (in ecological terms they are producers and animals are consumers) is fundamental in biology, but students tend to assume they ingest food through their roots.

- Students refer to leaves as being broken down or eaten by animals over time. They do not understand that the material in the leaves eventually is transformed into gases, including carbon dioxide and nitrogen, which are recycled through the atmosphere. Some students were convinced that the leaf litter would build up over time and the world would gradually expand.
- Many students held anthropomorphic views, maintaining that flowers were for human benefit and that leaves fell because their muscles atrophied in autumn. This leaf-centred view eventually gave way to a tree-centred view (the tree drops them), and then to a more adaptation-oriented view involving seasonal triggers within the tree shutting off the flow of nutrients.

Science as a human endeavour

There are many activities presented and discussed in this chapter that demonstrate human interactions with living things in the environment, and many more activities that could flow from these that would further support learning in this strand.

Scientists studying living things

The case study 'Invertebrates in the schoolground' involved children quite explicitly discussing how scientists might approach studying animals in the environment. The approaches to data collection and analysis, and the generation of representations such as graphs, tables, drawings and models, were couched as representing good scientific practice. The forming of children into groups around investigation of animals generated a high level of engagement in exploration, which we have described elsewhere (Tytler, Haslam, Prain and Hubber 2009) as a community of inquiry modelling the knowledge-building practices of scientists. The model-based reasoning encouraged by the animal-movement modelling task, and the challenge for students to generate representations to explore questions and construct explanations, we have argued is a more authentic representation of scientific knowledge-building practices than following a predetermined sequence of steps people might imagine represent a 'scientific method'.

In that study, a pre- and post-test question was posed, asking of two scientists researching a small beetle living in leaf litter:

1. what questions they might ask
2. what methods they would use to answer these questions
3. what their journals would look like.

In the pre-test, children had little idea of how to respond, suggesting questions like 'What are they called' and 'What do they eat', and writing very little about methods beyond 'Looking carefully'. Figure 7.18 shows a typical good-quality response in the post-test to the same question, with the student asking sensible questions and providing a range of representations that scientists might use to study the beetle and its habitat.

Similarly, Suzanne's classroom animal studies (see the case study 'Thinking about and working scientifically with small animals' on p. 274) focused explicitly on the way scientists go about framing questions and then generating data to answer them. We are thus arguing that students generating and negotiating non-standard representations leads to powerful

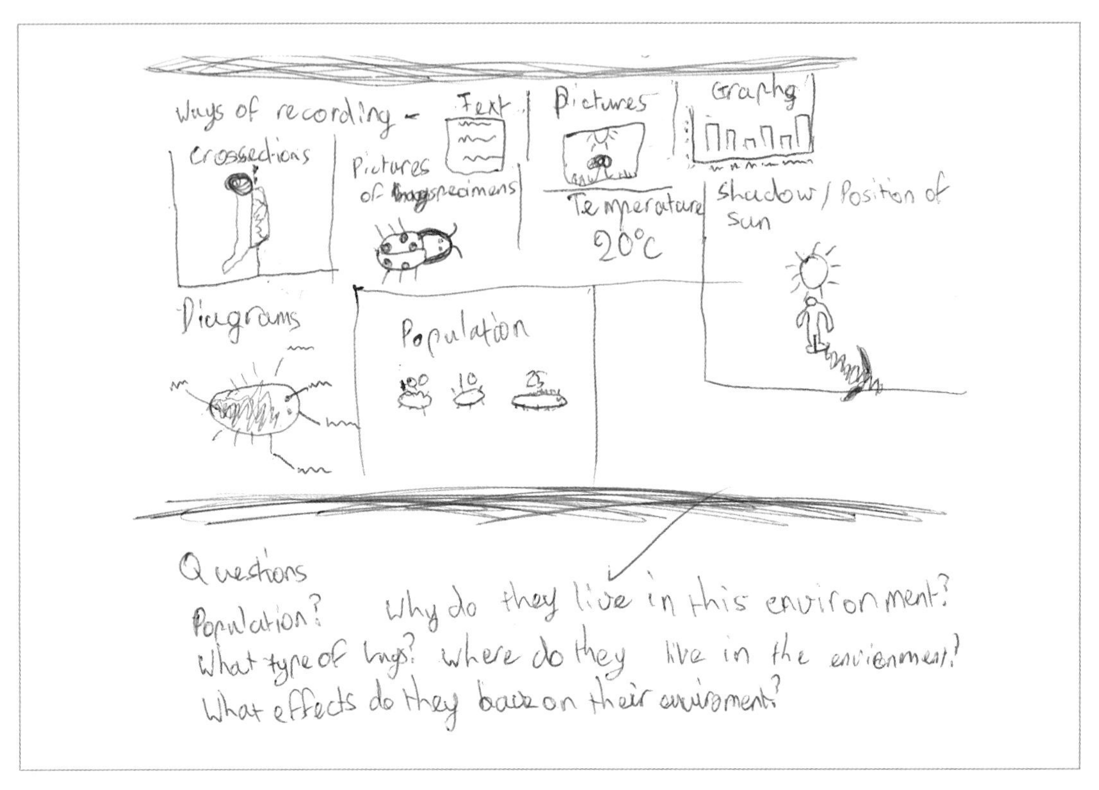

FIGURE 7.18 Response to post-test question on scientists' questions and journal entries

conceptual learning and also to better understandings of the NOS – how scientists work to build knowledge.

In terms of understandings of how other cultures have built knowledge, there are rich opportunities in a study of living things to explore how Indigenous peoples manage their environment in a sustainable fashion. Aboriginal perspectives on animals and their use of plants in healing and for sustenance in different seasons provide a different but powerful way of understanding the environment and how we need to care for it. Information on Indigenous Australians' use of plants for a variety of technology purposes, and for food or medicine, can be found on a number of websites (e.g., see http://sydney.edu.au/science/uniserve_science/school/curric/stage4_5/nativeplants/gallery).

Generating a personal ethos

M&E

Sustainability is an important theme in the science strand, and care for the environment flows naturally out of and is enriched by studies of the natural environment. Whether it concerns habitat preservation or duty of care towards classroom animals, putting children in direct contact with animals and plants directly raises these ethical questions about our human role as custodians of that which we explore and explain. While this chapter has mainly focused on conceptual knowledge, there are many points in the case studies in which issues of value and aesthetics emerge.

Carolina Castano (2010) has explored with some success the potential of studies of animals and their needs to ameliorate antisocial behaviours in schools of low socioeconomic levels in Colombia. Her results suggest that science education can have a positive impact on the attitudes of these children, promoting compassion towards animals and reducing their aggressiveness towards each other and generally. There is growing interest in the role of values in science education and the way pedagogy links with values in supporting quality learning and also positive identity responses to science (Schreiner and Sjøberg 2007; Tytler, Barraza and Paige 2010). Interest in exploring socioscientific issues, of which there are many related to animals and genetically modified foods, has recently expanded to include consideration of value positions (Zeidler and Sadler 2008; see also Chapter 1).

Bloom (1992), in a study of children exploring worm structures and behaviour, found children's thinking to be extremely fluid, progressing through a rich mix of conceptual and emotional, ethical and aesthetic commitments. In a similar vein, Per-Olof Wickman (2006) draws on the work of Dewey to argue that the traditional dichotomy between aesthetic and value positions and conceptual knowledge building is false. He argues that, for scientists, 'Aesthetic experience is everywhere evident in their daily life as scientists, in the creative moments, in finding new connections and results, and in communicating science with others, but also in the intimate relationship scientists often have with nature' (pp. 17–19).

Exploring links between aesthetics (matters of taste and judgement) and science learning in primary schools, Jakobson and Wickman (2008) pointed to very clear instances of how primary teachers of science blended aesthetic and conceptual talk in challenging and motivating students in a science class. In a study of buds on a plant, for instance, a teacher reinforced the surprise that students experienced in examining buds under a magnifying glass, and one student, Bosse, discovered a new relation:

EVA I can see!
TEACHER You may borrow this so you can also look at the buds.
EVA *Wow!*
TEACHER There you go! [inaudible] Also look at the buds ... Yes, *just super*! Did you see?
BOSSE *Wow!*
TEACHER It's *super*. Can you take it?
BOSSE Mine have red in it too.

Jakobson and Wickman 2008, p. 56

Jakobson and Wickman (2008) also quote passages where students have varying responses (delight or disgust) to worms, in which case the task of the teacher is to work with students to convince them of the intrinsic interest, if not 'cuteness', of the worms, in order for them to learn productively.

Thus, there is a close and complex relation between, and an entwining of, values and aesthetics, and conceptual learning. Thus, learning about and investigating invertebrates, and appreciating them, must proceed together. This is as true for scientists as it is for students. The developmental progressions in Table 7.3 (see p. 273) illustrate the way these strands interact. In the traditional view of science education, which denied the importance of dispositions and values, aesthetic and attitudinal dispositions were seen only in relation to their 'motivational' value in supporting knowledge building. With dispositions now an important part of the 'science as a human endeavour' dimension of science education, we can see them in a clearer perspective as an outcome in their own right and an essential aspect of deep learning.

On the social plane, the study of living things in the environment and the learning of an appreciation of the living world feed directly into the possibilities of environmental studies, and having students explore habitat change and threat, including endangered species.

Activity 7.25 Dispositions, aesthetics and learning about living things

The previous section illustrated some strategies teachers use naturally to enlist interest in living things as part of children's learning.

- Choose one of the case studies in this chapter. Discuss the strategies you might use to support children to develop a positive disposition as part of their learning in the sequence described. Are there any overt strategies used that you feel are successful in promoting aesthetic responses?
- Look at the interaction between the columns of Table 7.3 (see p. 273). Choose one level and generate a list of examples of ways in which the outcomes in each of the relevant columns could influence each other.
- For the case study you chose for the first question, generate a further sequence of activities that would support children's learning of the way science knowledge is used in the workplace.

Living things and environments and the world of work

There are many occupations that deal directly with living things and the environment. Studying aspects of horticulture, for instance, including maintaining school gardens, is a natural extension of, or context for, plant studies. Interacting with gardeners, garden suppliers or florists can put children in contact with practical knowledge about plant care and reproduction. A trip to the greengrocers can sharpen their understanding of plant structures and the difference between the scientific naming of plant parts and culturally based distinctions between 'vegetables' and 'fruit'. In rural areas, there are opportunities to interact with the agriculture or livestock industries and draw on children's experiences of this.

There are also opportunities to extend a unit on living things to include excursions to science resource centres such as a museum, aquarium or zoo. Principles for planning for such visits are discussed in Chapter 3 (see the section 'Excursion to a science resource centre' on p. 135). Local parks, rangers and environmental resource centres are also sources of insights into personal and social interactions with the living environment. The role of rangers in the preservation of environments and the management of public interactions with parks could be a fruitful extension topic. There are many occupations, and many informal interactions, that people have with living things that could form the basis of a productive context for studying living things in the environment.

Summary

In this chapter we have explored scientists' and children's ideas about a variety of life forms, and strategies for teaching and learning that focus on the study of organisms in constructed environments. Teaching and learning about animals has been discussed in the context of schoolground invertebrates, which provides the opportunity for a rich mix of activities focusing on animal structure and function, adaptation, reproduction and life cycles, and ecological concepts such as food webs and energy flow. A case study of animal behaviour introduced investigative principles and an example of a powerful way to engage children in learning about science and scientific practice. The link between the ideas and the discursive representational elements of science was discussed. In the last part of the chapter, the biology of plants was explored in the context of studies of plants in the schoolground, with an emphasis on plant structure and function, reproduction and the cycling of matter. Finally, a range of approaches to establishing understandings of science as a human endeavour were discussed in the context of studies of schoolground environments.

Concepts and understandings for primary teachers

Below are some key conceptual ideas and understandings related to biology with which a primary teacher should be familiar. You should read and interpret this list with an appreciation of its limitations as described in the Chapter 1 section 'Concepts and understandings for primary teachers' (see p. 47).

The following list of biological understandings attempts to lay out the major principles, not as things to be learnt, but as organising ideas that make sense of myriad observations and the detailed knowledge about organisms that is the basis for appreciating and making sense of the living world.

Broad concepts of living things

- There is no absolute definition of what we count as living organisms; this has changed over time. Living organisms mostly have a cellular structure but, currently, viruses are counted among living things and they do not. Living things have a variety of characteristics that are displayed to different degrees: they respire, move, respond to stimuli, reproduce and grow, and are adapted within a complex of living things within an environment.
- When a plant is picked or cut or when an animal dies, some basic life processes will occur. There is no universally agreed-upon answer to the question 'Is it alive?'
- The way in which living things are classified has also changed over time. Animals and plants are the main realms but fungi, mosses and viruses have their own separate realms.
- Animals are consumers in that they ingest food to survive. Food provides the energy for growth, movement and other life processes.
- Plants are producers and grow through the photosynthetic process by which carbon dioxide and water are used to produce starches of which the plant material is made. Sunlight drives this process.
- Animals and plants are further divided. A major category of plants is flowering plants or angiosperms. These differ from conifers or ferns, for instance, in the way they reproduce. Animals include a multitude of organisms, from microscopic creatures through insects, reptiles, mammals and so on. These different organisms are all interconnected through an evolutionary history.

Structure, function, adaptation

- Living things have various structures that enable them to survive: transport structures in plants by which water and trace elements move, digestive structures and respiratory structures in animals, reproductive structures. Each organism has particular forms of these that are essentially solutions to the business of survival.

- Biologists look at organism structures and behaviours in terms of their survival purposes. Children should be supported to do likewise.
- Each organism is adapted to a particular ecological niche, which involves interdependence with other living organisms as well as dependence on non-biotic factors.

Ecosystems

- In an ecosystem, animals and plants are interrelated through a multitude of interdependent survival needs.
- These are interconnected by the flow of energy through the ecosystem, which is the reason an ecosystem can be analysed through the notion of trophic levels, with producers at the bottom through to carnivores who prey on other carnivores at the top.
- As well as the flow of energy, there is also a cycling of nutrients and elements, such as nitrogen and water. Thus, decaying plants are consumed by scavengers and broken down, with the products becoming an organic part of soil, and then gases, such as nitrogen, carbon dioxide and water vapour. Chemical change is an integral part of life processes.

Flowering plants

- The majority of plants on Earth are flowering plants and this includes trees, grasses, cacti and other small plants, as well as the more obvious examples. All eucalypts are flowering plants, as are most deciduous and rainforest trees.
- The reproductive cycle is an important adaptation and the formation and dispersal mechanisms for seeds (contained in fruit, pods and nuts, which are the outcomes of fertilised flowers) are multifarious. Conifers do not have flowers but seeds produced in cones.
- All flowering plants have a similar reproductive cycle.

Life cycles

- Life cycles should really be called 'reproductive cycles'. In animals and plants they have unique details that are adaptive to the particular environment, including the number of offspring (or seeds) and the cycle timing and frequency, as well as mechanisms.
- In animals, the reproductive cycle can coincide with the life cycle of an organism if the adult dies after fertilisation (as with butterflies, and also effectively within some mammals, such as the male Antechinus). However, most animals will go through many reproductive cycles in a lifetime.

Animal behaviour

- Animal behaviour must be understood in terms of its adaptive function. Animals behave in ways that maximise their survival chances.
- Each species has unique behavioural characteristics, which can be studied using a range of techniques.
- For schools, it is most fruitful to study the behaviour of simpler life forms since their behaviour is not so complex and there is less tendency for children to anthropomorphise.

Search me! science education

Search me!

Explore **Search me! science education** for relevant articles on linking science and design technology. Search me! is an online library of world-class journals, ebooks and newspapers, including *The Australian* and the *New York Times*, and is updated daily. Log in to Search me! through http://login.cengage.com using the access code that comes with this book.

KEYWORDS

Try searching for the following terms:

- Living things
- Animal behaviour
- Plant growth
- Children's ideas

Search tip: **Search me! science education** contains information from both local and international sources. To get the greatest number of search results, try using both Australian and American spellings in your searches: e.g., 'globalisation' and 'globalization'; 'organisation' and 'organization'.

References

Abell, S. & Roth, M. 1995. Reflection on a fifth-grade science lesson: Making sense of children's understanding of scientific models. *International Journal of Science Education*, 17 (1), pp. 59–74.

Australian Academy of Science (AAS). 2005. *Primary connections*. Canberra: AAS. Available at http://www.science.org.au/primaryconnections/index.htm (accessed 5 September 2006).

Bell, B. 1993. *Children's Science, Constructivism and Learning in Science*. Geelong: Deakin University Press.

_____ & Freyberg, P. 1985. Language in the science classroom, in R. Osborne and P. Freyberg (eds). *Learning in Science: The Implications of Children's Science*. Auckland: Heinemann.

Biddulph, F. 1984. Pupils' Ideas about Flowering Plants. *Working paper no. 125 of the Learning in Science Project (LISP) (Primary)*. Hamilton: University of Waikato, Science Education Research Unit.

Bloom, J. 1992. The development of scientific knowledge in elementary school children: A context of meaning perspective, *Science Education*, 76 (4), pp. 399–413.

Carey, S. 1985. *Conceptual Change in Childhood*. Cambridge: MIT Press.

Carolan, J., Prain, V. & Waldrip, B. 2008. Using representations for teaching and learning in science. *Teaching Science*, 54 (1), pp. 18–23.

Castano, C. 2010. The role of science education in reducing violence towards others. *Paper delivered at the annual conference of the Australasian Science Education Research Association*, Port Stephens, NSW, July.

Clough, E. & Wood-Robinson, C. 1985. How secondary students interpret instances of biological adaptation. *Journal of Biological Education*, 19 (2), pp. 125–30.

Cox, R. 1999. Representation construction, externalized cognition and individual differences. *Learning and Instruction*, 9, pp. 343–63.

Danish, J. & Enyedy, N. 2007. Negotiated representational mediators: How young children decide what to include in their science representations. *Science Education*, 91, pp. 1–35.

diSessa, A. 2004. Metarepresentation: Native competence and targets for instruction. *Cognition and Instruction*, 22 (3), pp. 293–331.

Elstgeest, J. 2001. The Right Question at the Right Time, in W. Harlen (ed.). *Primary Science: Taking the Plunge* (2nd edn). Portsmouth: Heinemann, pp. 36–46.

Greeno, J. & Hall, R. 1997. Practicing Representation: Learning with and about representational forms. *Phi Delta Kappan*, 78 (5), pp. 361–8.

Hackling, M. & Prain, V. 2005. *Primary Connections Stage 2 Trial: Research Report*. Canberra: AAS. Available at http://www.science.org.au/reports/documents/pcreport1.pdf (accessed 10 January 2011).

Haslam, 1998. Observation in science classes: Students' beliefs about its nature and purpose. Unpublished thesis. Monash University.

_____ & Gunstone, R. F. 1998. The influence of teachers on student observation in science classes. *Paper given at the Annual Meeting of the National Association for Research in Science Teaching*, San Diego, April.

Hayes, D., Symington, D. & Martin, M. 1994. Drawing during science activity in the primary school. *International Journal of Science Education*, 16 (3), pp. 265–77.

Hellden, G. 1997. To develop an understanding of the natural world in the early ages, in K. Harnqvist and A. Burgen (eds). *Growing Up with Science: Developing Early Understanding of Science*. London: Jessica Kingsley Publishers.

_____ 2005. Exploring understandings and responses to science: A program of longitudinal studies. *Research in Science Education*, 35 (1), pp. 99–122.

_____ & Tytler, R. (2008). Insights into student learning in science: The contribution of longitudinal studies, in B. Ralle and I. Eilks (eds). *Promoting Successful Science Education: The Worth of Science Education Research*. Aachen, Germany: Shaker Verlag, pp. 19–30.

Hubber, P., Tytler, R. & Haslam, F. 2010. Teaching and learning about force with a representational focus: Pedagogy and teacher change. *Research in Science Education*, 40 (1), pp. 5–28.

Inagaki, K. & Hatano, G. 1996. Young children's recognition of commonalities between animals and plants. *Child Development*, 67, pp. 2823–2.

Jakobson, B. & Wickman, P.-O. 2008. The roles of aesthetic experience in elementary school science. *Research in Science Education*, 38 (1), pp. 45–66.

Leach, J., Driver, R., Scott, P. & Wood-Robinson, C. 1995. Children's ideas about ecology 1: Theoretical background, design and methodology. *International Journal of Science Education*, 17 (6), pp. 721–32.

_____ 1996a. Children's ideas about ecology 2: Ideas found in children aged 5–16 about the cycling of matter. *International Journal of Science Education*, 18 (1), pp. 19–34.

_____ 1996b. Children's ideas about ecology 3: Ideas found in children aged 5–16 about the interdependency of organisms. *International Journal of Science Education*, 18 (2), pp. 129–41.

Lemke, J. 2004. The literacies of science, in E. W. Saul (ed.). *Crossing Borders in Literacy and Science Instruction: Perspectives on Theory and Practice*. Newark, DE: International Reading Association and National Science Teachers Association, pp. 33–47.

Lindemann-Matthies, P. 2005. 'Loveable' mammals and 'lifeless' plants: How children's interest in common local organisms can be enhanced through observation of nature. *International Journal of Science Education*, 27 (6), pp. 655–77.

McGrath, M. & Ingham, A. 1992. Using art activities in the assessment of science in the primary school. *School Science Review*, 73 (264), pp. 33–46.

Metz, K. 1995. Reassessment of developmental constraints on children's science instruction. *Review of Educational Research*, 65 (2), pp. 93–127.

_____ 1997. On the complex relation between cognitive developmental research and children's science curricula. *Review of Educational Research*, 67 (1), pp. 151–63.

Myers, O., Saunders, C. & Garrett, E. 2004. What do children think animals need? Developmental trends. *Environmental Education Research*, 10 (4), pp. 545–62.

Peterson, S. & Tytler, R. 2001. Children and Terraria. *Investigating*, 17 (1), pp. 28–33.

Prokop, P. & Tunnicliffe, S. 2008. 'Disgusting' animals: Primary school children's attitudes and myths of bats and spiders. *Eurasian Journal of Mathematics, Science & Technology Education*, 4 (2), pp. 87–97.

_____, Prokop, M. & Tunnicliffe, S. D. 2008. Effects of keeping animals as pets on children's concepts of vertebrates and invertebrates. *International Journal of Science Education*, 30 (4), pp. 431–49.

Rogoff, B. 1998. Cognition as a collaborative process, in W. Damon (ed.). *Cognition, Perceptions and Language, Handbook of Child Psychology* (5th edn). New York: John Wiley & Sons.

Schreiner, C. & Sjøberg, S. (2007). Science education and youth's identity construction: Two incompatible projects?, in D. Corrigan, J. Dillon and R. Gunstone (eds). *The Re-emergence of Values in the Science Curriculum*. Rotterdam: Sense Publishers.

Shepardson, D. 1997. Of butterflies and beetles: First graders' ways of seeing and talking about insect lifecycles. *Journal of Research in Science Teaching*, 34 (9), pp. 873–89.

_____ 2002. Bugs, butterflies, and spiders: Children's understanding about insects. *International Journal of Science Education*, 24 (6), pp. 627–44.

Stepans, J. 1985. Biology in elementary schools: Children's conceptions of 'life'. *American Biology Teacher*, 47 (4), pp. 222–5.

Symington, D. & White, R. 1983. Children's explanations of natural phenomena. *Research in Science Education*, 13, pp. 73–81.

Tunnicliffe, S. D. & Reiss, M. J. 2000. Building a model of the environment: How do children see plants? *Journal of Biological Education*, 34 (4), pp. 172–7.

Tytler, R., Barraza, L. & Paige, K. 2010. Values in science and environmental education and teacher education, in R. Toomey, T. Lovat, N. Clement and K. Dally (eds). *Teacher Education and Values Pedagogy: A Student Wellbeing Approach*. Terrigal, NSW: David Barlow Publishing, pp. 156–78.

_____, Haslam, F., Prain, V. & Hubber, P. 2009. An explicit representational focus for teaching and learning about animals in the environment. *Teaching Science*, 55 (4), pp. 21–7.

_____, Peterson, S. & Prain, V. 2006. Picturing evaporation: Learning science literacy through a particle representation. *Teaching Science, the Journal of the Australian Science Teachers Association*, 52 (1), pp. 12–17.

_____, Prain, V. & Peterson, S. 2007. Representational issues in students learning about evaporation. *Research in Science Education*, 37 (3), pp. 313–31.

Vygotsky, L. S. 1986, *Thought and Language*. Cambridge: MIT Press.

Wickman, P.-O. 2006. *Aesthetic experience in science education: Learning and meaning-making as situated talk and action*. London and Mahwah, NJ: Lawrence Erlbaum Associates.

Yen, C.-F., Tsung-Wei Yao, T.-W. & Mintzes, J. 2007. Taiwanese students' alternative conceptions of animal biodiversity. *International Journal of Science Education*, 29 (4), pp. 535–53.

Zeidler, D. & Sadler, T. 2008. Social and ethical issues in science education: A prelude to action. *Science and Education*, 17 (8–9), pp. 799–803.

8

Materials

by Keith Skamp

Introduction

Everything is made of four 'elements': earth, air, fire and water. Each of these has certain 'powers', is related to the gods, and can be transformed into the others.

This theory, and variations on it, formed the predominant thinking about the nature of matter for centuries from about 600 BCE. This was despite some Greek philosophers proposing at about the same time that things were made of 'atoms'. The idea of atoms as we think about them today did not surface until about 1800. This is interesting when it's considered that humanity was working copper, making alloys and using the wheel from about 4000 BCE (Lievesley 2007). Clearly, the properties and uses of materials were the focus well before people wondered what materials were made of. As most, if not all, primary science curricula include a content strand related to 'materials', then you might like to reflect on this historical snapshot as you initially focus on the question 'What subject matter should comprise this content strand?'

This chapter, then, has four main sections. Firstly, teachers' and students' conceptions about materials are overviewed. The relationships between objects, materials, substances and particles are discussed. Students' conceptions about the nature, composition, properties, uses and origins of materials are identified – solids, liquids and gases receive special attention. To conclude this section, two interlocking issues are raised: Should primary students be introduced to the concepts of 'substance' and 'particle', and is there a pattern to students' development of ideas about materials? In simple terms, this first section is about *what* to teach about materials.

Secondly, there is a focus on *how* to teach about materials from a constructivist perspective. After emphasising the importance of gaining student interest in learning about materials, specific teaching strategies are suggested. The role of classroom talk in conceptual development and teacher intervention, especially in scaffolding student generation of mental models of matter, are raised. The effectiveness of these strategies is summarised. It is emphasised that helping students learn about materials and their properties is not as straightforward as it may seem.

Thirdly, models or schemata to guide the sequencing of lessons are introduced through two separate classroom-based case studies. The first describes how students explored whether the shape of an object affects its strength and is based on a conceptual-change schema. The second shows how a teacher involved her class in investigating the absorbency of fabrics. 'Scientific investigations' is a phrase often used within constructivist lesson sequences; hence, this case study could be considered a sequence within a sequence. In both sequences, you will be required to analyse the steps involved and the roles the teachers and the students played. Reference will be made to issues such as the nature of science and formative assessment. Other constructivist schemata, including the *Primary Connections'* 5E model (AAS 2005), are overviewed. The importance of planning with a conceptual focus is reiterated.

The chapter concludes with some examples of how this content area could exemplify science as a human endeavour.

Primary curricula and the study of materials

What key ideas and concepts do you think would be included in the 'materials' section of a primary science curriculum? Activity 8.1 asks you to consider this question.

Activity 8.1 What do you think primary students should know about materials?

What are the key ideas or concepts that you think would be included in the 'materials' section of a primary science curriculum? Draw a mind map around the word 'materials'. Show connections to topics, key ideas and concepts associated with the notion of 'materials'. Examples might be 'properties' and 'uses'. Properties might then branch out to aspects such as 'hardness'. Some of your connections might join 'properties' and 'uses'. Can you think of specific materials you might use with young learners? Add these to your mind map in some way. Which ideas or materials might be more suitable for lower primary teaching? Upper primary? Why? You might use different colours to show your thoughts about this aspect.

Look at your primary and related junior secondary science syllabuses. Compare your mind map with the syllabus content related to materials. Add to or amend your map (using a different colour or format). As a primary teacher, you need to have an understanding of materials at least one level beyond the primary years; the junior secondary science syllabus will help you appreciate the next level.

Survey a major resource that schools use for teaching primary science. In Australia you could look at the following *Primary Connections* units: 'Material World' (AAS 2008a), 'What's It Made Of?' (AAS 2008b), 'Package It Better' (2008c), 'Change Detectives' (AAS 2009a) and 'Spot the Difference' (AAS 2009b). What content does the resource suggest should be included in a strand related to materials?

Did your mind map and/or syllabus and/or resource distinguish between objects and materials? Materials and substances? Substances and particles? Are these distinctions important for primary teachers? For primary students? Think about these questions – they will be discussed later in this chapter.

The ideas underlying the 'materials' content strand in most primary science syllabuses include:

- objects and materials (names and features)
- form, composition (e.g., structure and parts – fibres, crystals, layers, grains) and properties of materials – how they behave (strength, transparency and so on)

- uses of materials and how these relate to their (observable) properties
- origins of materials, including their production
- physical changes in everyday materials – links with solids, liquids and gases, and processes such as evaporation, melting and condensation
- how materials can be changed, which also relates to their properties and uses (solubility, shape, heating, cooling and combining) and whether the changes are reversible or permanent. (Adapted from ACARA 2011; Russell, Longden and McGuigan 1991, p. 11.)

Did your mind map include some of these links?

This chapter focuses on materials, their properties, uses and origins, and whether they are solids, liquids and gases. How materials change (often referred to as either 'physical' or 'chemical' change) is mentioned, but this is the main focus of Chapter 9.

Teachers' ideas about materials (substances and particles)

Objects and materials

The nature of materials is one of the three big ideas of science used to explain our physical world. The other two, forces and energy, were the focuses of Chapters 3 to 5. In those chapters, the importance of being aware of the ideas primary students have about forces and energy was stressed because this background helps you to appreciate where your learners are at. Subsequent teaching decisions should depend upon such knowledge. The same applies to materials. Teachers also need to feel confident about their own understanding of concepts related to materials. This is important as 'how can teachers [you] help children to change their ideas if they [you] don't see anything wrong with the ideas they [you] have?' (Harlen 2007, pp. 14–15). Several of the following activities encourage you to try exercises that have been used with young learners, to elicit their ideas about the characteristics of materials, or with preservice or practising teachers. These activities will determine whether you need to clarify or further expand your own conceptual understanding. The exercises also provide examples of the range of elicitation techniques available to teachers for determining students' preconceptions.

Activity 8.2 What are objects made of?

Try this exercise. Obtain some steel wire, a plastic bag, sand (in the plastic bag), an apple, a potato, a plastic drinking cup, a large nail, a brick, a piece of wood, knitting wool, a milk-bottle top, a piece of cotton cloth and a newspaper. Identify each object, give it a name, say what it is made of and elaborate; group together objects made of the same materials and give each group a name.

You could try the same task with students (asking appropriate questions). Can you make sense of their responses? What do they suggest about teaching approaches to use which may help primary students find out about the nature of objects and materials?

Adapted from Sorsby 1991

How did your responses distinguish between objects and the 'materials' from which they are made? That objects are made of 'stuff' (a term that could be used with students), which scientists call 'matter' (which has weight and occupies space), is a distinction that many young learners, at least to some extent, can make. Children as young as three realise that in breaking an object, the matter (material/s) of which it is made remains the same, while not necessarily clearly distinguishing between the properties of the object and its constituent matter (material/s). This distinction between objects and materials does become more differentiated with age, but many 10-year-olds may still not readily make the distinction (Krnel, Watson and Glazar 1998).

Initially, it is important for you to appreciate that all things in the universe consist of matter and that matter can change in form and composition, but the total mass remains the same. Also, knowing how matter may be categorised will assist you in deciding on the conceptual purposes underpinning many of the activities associated with materials that you will select for your class. In this and the next chapter, the following conceptualisation will be used:

- An object is made of material(s).
- A material is composed of substance(s).
- A substance is composed of (contains) one or more element(s) (in combination).

Substances, here, are considered to be single (and pure) entities. All objects, materials, substances and elements are made of stuff – they are forms of matter (Driver et al. 1994a). See if you can identify examples for each category. Discuss your examples with a colleague to clarify your understanding. This use of terminology is not universal (for example, sometimes you will find the words 'materials' and 'substances' used interchangeably), but it is accepted in most of the literature about materials.

It is useful for you to appreciate these object, material and substance distinctions because, with the language you use and emphasise, you are laying the foundations for later learning. Further, children will readily talk about what an object (e.g., a toy, a piece of clothing, a chair) is made of – at some point, teachers will need a term to generalise across various objects: for example, all these objects are made of wood or cotton or metal or plastic. These can be referred to as different types of material. The properties of objects can then be distinguished from the properties of materials (see below), and hence their uses; other differences can also be noted, such as the origins and production of objects and materials. As will be discussed later in this chapter (and also in Chapter 9), primary students in general do not appreciate the materials/substance distinction except in clearly observable cases; however, teachers need to understand the differences so that they do not inadvertently hinder their students' conceptual development.

Some primary teachers were asked to explain the meaning of 'material', 'natural' and 'processed'. How would you describe these terms? These teachers used multiple interpretations. For example, there were four main interpretations of what 'material' meant. The most common was to see materials as 'something' to 'make something else'. Is this what you thought? It is, in fact, a fairly limited and exclusive use of the term. A more scientific view would use the term far more inclusively to refer to 'everything you can see, smell, touch or taste' (Schibeci and Hickey 2000a, p. 1164); see above for Driver et al.'s (1994a) description of a 'material'.

Once you have sorted out the difference between objects and materials, then you can ask yourself what are the macroscopic properties – those which are observable – that distinguish the two categories. Materials are distinguished by their intensive properties, which are those that remain the same even when some of the material is removed; examples are hardness, temperature, malleability and states of matter. Objects are defined not only by the material(s) of which they are made, but also by their size and shape (which are examples of extensive properties) (Krnel, Watson and Glazar 1998; Wenham and Ovens 2010). Hence, strength and stiffness are examples of properties of an object but not a material.

Whether you are an early childhood or an upper primary teacher, to have a better understanding of these big ideas can only improve your teaching – this was the view of such teachers in Schibeci and Hickey's (2000a) study. With a clearer understanding of objects, materials and substances, teachers could judge whether the language used in a children's reference source (book, television, website) would assist or confuse future understanding because of the ways in which the terms are used. It would also assist you in the types of conversational remarks or questions you use when interacting with students. As Schibeci and Hickey (2000a, p. 1166) state: 'If teachers do not have scientific models [here, of the nature of objects, materials and substances] to contrast with student models, are they able to foster their students' conceptual change?'

What do you think?

The classification of materials

If students are to learn about materials, then they will need guidance as to how to classify them into categories so that they can appreciate the wide variety of materials that exist and also start to investigate the properties of the different categories. What classifications of materials that make up our world have you encountered? One classification often used by teachers is 'natural' and '(human-)made'. The former (e.g., bone, wool, silk, wood, rocks and minerals) are those produced by natural processes and changes. Of course, sometimes made substances are the same as natural substances (Wenham and Ovens 2010). Every classification scheme has its limitations and this approach often places materials of similar chemical structure into two different groups; for example, leather is not placed with nylon or concrete with granite (Ross 1997). When introducing this or any other classification scheme, teachers need to ask themselves why they are using a particular scheme: reasons may include assisting learners to see that there is often a connection between the type of material and the use(s) to which it is put, or simply to illustrate the wide variety of different materials.

Other classifications of materials do exist. One that emphasises the wide variety of materials (but is not exhaustive) proposes the following categories: air, water, metals, plastics, glass, ceramics, rocks, petroleum and the materials of living things, namely, proteins, carbohydrates, oils/fat/waxes and inorganic substances (Driver et al. 1994a, p. 154). Can you provide some examples? Teachers need to also be conscious of the fact that there are many natural and developed (made) composite materials (e.g., glass-reinforced plastic) that will not readily fit particular categories.

What do you (or your students) know about one or more of these material groupings; for example, metals? Activity 8.3 asks you about your knowledge of plastics by drawing a concept map – see, for example, the concept maps in the online Appendix 5.2, Chapter 6 (see the section 'Students learning about skeletal systems' on p. 233) and the examples in Gallenstein (2005). You could try this approach with any of the material categories from the different classification schemes.

Activity 8.3 A concept map about plastics

Using the terms or phrases 'clothes', 'recycle', 'furniture', 'medicine', 'insulation', 'breakable', 'drinks', 'cars', 'lets heat or electricity through', 'heavy', 'see-through', 'waterproof', 'hard', 'strong', 'container', 'melts', 'saucepan', 'bendy' (Gray et al. 1994), develop a concept map to indicate what you know about plastics.

Add further concept links to include any wider understandings you have about plastics. Check your preconceptions by referring to an appropriate reference – for example, *Children and Plastics* (Horn 1989) or the learning centre at the plastics link in the americanchemistry website (see http://www.americanchemistry.com/s_plastics).

Compare your map with some primary students' maps. What do their maps suggest to you as a teacher about what could be included in a unit of work on plastics?

As teachers, an awareness of other classification schemes for materials may encourage you to view with an open mind the groups that your students propose for their classifications. A novel scheme is Ross' (1997) suggestion that virtually all materials and substances can be placed into five groups: rocklike, lifelike, metal, volatile and saltlike. This scheme could be used from upper primary years onwards. This classification is based on the physical and chemical properties of materials and substances (their elemental make-up and the nature of the forces that are between substances); it can lay a foundation for later studies. Students have to ask themselves four questions:

1. What happens when the material or substance is heated?
2. Does it conduct electricity?
3. Does it dissolve?
4. What happens when you hit it with a hammer?

There are subquestions related to each category. Primary students have little difficulty with the metals, rocklike and lifelike categories; saltlike and volatile are more difficult.

Substances and particles

Your own understanding of the distinction between objects, materials, substances and elements will assist you in helping your students to learn about materials. Activity 8.4 relates to another term used to describe our world, namely 'chemicals', and will encourage you to further clarify your thinking.

Activity 8.4 What is a chemical?

Put a circle around the name of every item in the list below that you think is a chemical or contains chemicals.

detergent	wool	metal	sugar
explosives	your heart	a gum tree	petrol

rubber	prime minister	glass
spider	plastics	sulfuric acid

Discuss your views with a colleague. Compare your responses with those of some secondary students (Dawson 1993) or other teachers (Schibeci and Hickey 2000b). Year 8 results are summarised in Appendix 8.1 (see p. 347).

Activity 8.4 relates to the alternative (non-scientific) conception that chemicals are a special groups of substances. This view probably arises from everyday expressions such as 'the chemicals in our food'. Within the context of thinking about materials, all materials are made of chemicals and chemicals may be equated with the previously mentioned concept of (single, pure) substances.

The concept of a pure substance (or chemical) is a difficult one for young learners and, in many instances, would not be formally introduced at the primary level. This is partly because it relies on ideas that most primary school-age students would find too difficult (e.g., specific melting and boiling points) or too abstract (e.g., that a pure substance is composed of only one kind of particle; that is, a specific element or compound, which they are unable to see) (Johnson 2000, 2002), although this latter view is contested, as mentioned later and in Chapter 9. Over time, an exploration of the properties of substances will lead students to the incredible fact that all materials are made of only 92 stable substances; that is, the periodic table. This is a topic usually left until secondary school, but one primary teacher taught her primary students, at their level, about the uniqueness of the periodic table (see Skamp 1993).

The particulate nature of matter

The question may have arisen in your mind as to how much a primary teacher has to know about the structure of matter. As many primary students use words such as 'atoms' and 'molecules', their teachers need at least a basic appreciation of the substructure of matter. However, the main reason you need to know about the particulate nature of matter is that it underpins most of the ideas primary curricula expect students to learn about materials. Conversations between teachers and learners would be far more meaningful if teachers appreciated this model of matter, as it would influence their scaffolding of student learning. This does not necessarily mean that the particulate nature of matter should be formally introduced to primary students; rather, it emphasises that the particulate nature of matter is the underlying mental model that will inform your interactions with learners on this topic. Depending upon the specific content, context and the age of the students, some teachers may see it as appropriate to help their students think of the world in terms of particles. Suggestions about suitable approaches are mentioned later (see the upcoming section 'Students' conceptions of the structure of matter' on p. 314, and also Chapter 9).

We can conceptualise our material world in several ways. If we think of materials as comprised of substances (chemicals), then these substances can be perceived on three levels: macroscopic, microscopic (nanoscopic) and symbolic (Gabel and Bunce 1994), which can be related as in the three corners of a triangle (Mahaffy 2006):

- The *macroscopic* level is the perceptual or observable level. This is the 'what' of physical and chemical phenomena. Primary students certainly can focus on these features of materials and, where appropriate, a limited range of substances. Such (macroscopic) emphases (for example, the properties of materials and how they may change) form the bases of many of the activities suggested in Chapters 8 and 9.
- The *microscopic* (nanoscopic) level refers to the particulate nature of materials; that is, they are made up of particles that move and have space between them. Some researchers and teachers argue that primary students can probably start to appreciate this model, which explains the behaviour of matter (and hence materials).
- The *symbolic* level describes how substances (or chemicals) can be represented by symbols, and the relationship between chemical symbols and the microscopic and macroscopic world.

Gabel, Samuel and Hunn (1987) report that many students and adults hold non-scientific conceptions about the third level, which certainly is not part of most primary curricula, although primary teachers will find some reference to it in a few topics; for example, carbon dioxide as CO_2 would be mentioned if the greenhouse effect and global warming were being discussed. There are primary level teaching sequences that introduce the representation of materials and substances in a symbolic form (Anonymous 2006; Griffin and Sharp 1998). Alternative conceptions are also common at the particulate level, which is fundamental to an understanding of how substances (chemicals) behave and hence, in turn, how materials behave. To test your understanding of the particulate nature of the material world, try Activity 8.5.

Activity 8.5 Understanding the particulate nature of matter

Gabel, Samuel and Hunn (1987) devised a 14-item nature of matter inventory. One item from the inventory relates to Figure 8.1. Draw your representation of a liquid in the space shown in the figure.

Properties of particles that should be present in representations of matter are listed in Appendix 8.2 (see p. 347). You should analyse your response using them. Not all of the attributes apply to this question, but you should reflect on the list. It represents many features of the particle model of matter.

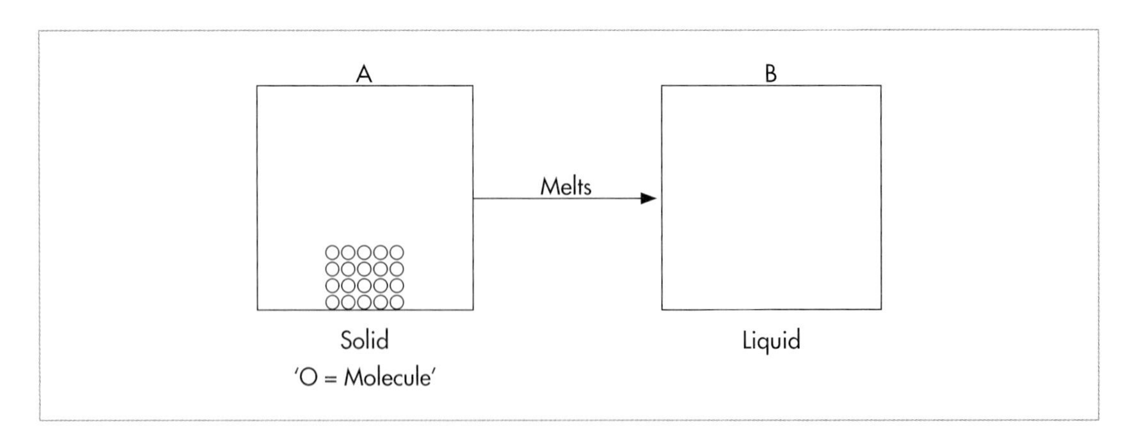

FIGURE 8.1 An item from the nature of matter inventory

Source: Gabel, Samuel and Hunn 1987

If, after completing Activity 8.5, you still feel uncertain about the particulate nature of materials (matter), you can further check your understanding by responding to items asking what happens to particles in matter in a range of contexts (e.g., the size of particles when a solid is pressed): you have to respond 'increase', 'decrease' or 'constant'; see Ozmen and Kenan (2007). You also will find sections in Wenham and Ovens (2010), Stephenson (2000) and del Pozo (2001), and on the website Interactive Library: Explain It with Molecules (see http://www.edinformatics.com/interactive_molecules) useful. More background written for primary teachers about the macroscopic properties of materials is in Farrow (2006), the *Background* booklets in the *Science 5/13* series on metals, wood and plastics (Schools Council 1972, 1973; see the *Children and Plastics* example in this chapter's 'References' section) and the CDs accompanying the *Primary Connections* units listed in Activity 8.1 (see p. 300).

Students' ideas about materials (substances and particles)

Students' conceptions of the nature of materials, their properties, uses and origins, are initially discussed. Attention is then focused on their views of solids, liquids and gases. Finally, what students think about pure substances and the substructure of matter is described. These findings are drawn together by asking if there is a pattern in the development of student thinking about materials. The connection between these studies of students' conceptions about materials and later sections of the chapter describing various teaching approaches is signalled.

The nature of materials, their composition, properties, uses and origins

As described in the introduction to this chapter, early thinkers thought matter was comprised of earth, air, fire and water. What do primary-age students think about the nature of materials? How can teachers access these ideas?

What do students think about materials?

The nature of (objects and) materials

One way to find out your students' ideas on this topic is simply to provide them with a range of objects and ask questions about them; for example, what they are made of, and do they belong to similar categories because of what they are made of. You could also ask students to group the objects and explain their reasons (see Activity 8.2 on p. 301). Many primary students seem to be able to distinguish between an object and the material from which it is made, and usually between the properties of the object compared to the properties of the material from which it is made (Wiser and Smith 2008), although not always (Johnson 2000). Sorsby (1991) tried Activity 8.2 with 10 students (initially, five boys and five girls, aged four to seven) on two different occasions, two years apart, and found that:

- Steel wool caused the greatest identification problems and the most thinking.
- Metal, sand and paper were popular group names, with most students placing sand and brick together, as well as wood and paper because they were made of the same material.
- Plastic (cup and bag) was a grouping used more on the second occasion.
- The apple and potato caused considerable difficulty – the students were unable to give them a group name, resorting to perceptual ideas such as 'apples and potatoes are made of green and yellow or are made of skins ... and hard and soft' (ibid., p. 23), or functional categories, such as what the items are used for (e.g., food).

This last grouping was probably due to the students not being aware of the cellular structure of living things (living things are not the focus of chapters 8 and 9, although all living things are also made of materials; for example, bone and blood, which are themselves composed of substances/chemicals). The above finding about steel wool indicates that upper primary students still may have difficulty appreciating that objects are made of the same material all the way through, in that they may not group powders with chunks of the same material (e.g., iron) (Wiser and Smith 2008).

In a similar but more comprehensive study, Russell, Longden and McGuigan (1991) found that five- to 11-year-olds classified objects based on five criteria:

1. composition (classifying objects according to the material from which they are made)
2. function (what can be done with objects)
3. location (where they are found)
4. perception (perception of an object's characteristics; for example, hard, soft, liquid, heavy, smooth, shiny)
5. made by (e.g., objects are made by humans) (cited in Krnel, Watson and Glazar 1998, p. 262).

There was no obvious age trend, with composition being cited most regularly by five- to seven-year-olds (57 per cent, followed by perception at 43 per cent) and seven- to nine-year-olds (53 per cent, followed by function at 35 per cent), but not the nine- to 11-year-olds (38 per cent), who referred to function more often (53 per cent). Here, composition is concerned with the nature of materials, perception refers to a material's perceptible properties, and function relates more to an object's use.

Teachers need to appreciate that the responses their students give about objects and materials may vary from one context to the next. Reviewing many research studies, Krnel, Watson and Glazar (1998) concluded that the following may influence whether children use object or material criteria to describe objects and materials:

- the age of the students
- the size of the objects
- their familiarity with the objects
- the form of the material (e.g., a powder)
- what the material actually is.

Therefore, the context and manner in which the teacher determines a student's ideas will influence what they uncover (Wiser and Smith 2008); individual conversations will probably reveal much more than writing or drawing elicitation techniques.

Culture can also influence students' conceptions of materials and their use of language. Some Ghanaian children, for example, did not have terms for metal and plastic; rather, they sorted wooden materials into categories such as 'made of tree' and metal materials as 'like a bus'. When asked what objects were made of, they responded as if answering 'Which one is this like?' and 'Where shall we put this one?'. They had a 'real world understanding' of materials in terms of their function: 'What can they be used for?' (Clarke 2006, pp. 15–16).

S&C

Being aware of the influence of culture, context and the mode of elicitation, teachers can listen more attentively to what students are saying and reflect on what responses may mean. A child, for example, may simply state that 'a spoon is a spoon' because they are so familiar with it and do not see it as, say, a metal spoon. More probing may or may not unearth other understandings about the 'nature' of the spoon. Even so, the responses children give to questions in such situations can still be revealing.

The properties of specific groups of materials

As young students are developing a concept of 'material kind', as distinct from kinds of objects, then we can ask what primary students know about the various types of materials and their properties. The Science Processes and Concept Exploration (SPACE) project, although completed in the 1990s, is best known for revealing primary students' ideas about many aspects of the world around them, and its legacy is still influencing pedagogy (Harlen 2007). One of the project's focuses was on students' ideas about materials (Russell, Longden and McGuigan 1991) which are not reported elsewhere. SPACE found that primary students, when applying the above five [object] criteria, used similar categories for classifying materials. This is consistent with primary students having an intuitive theory about materials that is perceptually based: materials are made of 'stuff' if they can be touched or seen. To advance their thinking further, students need to take a 'major epistemological [how we believe we know about something] leap of faith'; that is, to accept 'a mental construct over perceptual evidence' in that, if a material is divided into infinitely small pieces, then it is still there and has, for example, weight (Wiser and Smith 2008, p. 217).

In the SPACE study, 'metals' was a category used more correctly than most others and it could be a useful starting point for teachers when focusing on different types of materials. How could you determine your students' ideas about a particular type of material? Gray et al. (1994) describe the results of using concept maps as an elicitation technique to determine Year 5 students' ideas about plastics (see Activity 8.3 on p. 304). The students were initially assisted by being shown other students' concept maps, and then, using brainstormed words related to 'wood', a concept map was drawn about wood. At this point, the class was provided with the terms and expressions in Activity 8.3. Each term was written on a separate piece of paper. The students then had to (literally) place them on an A3 sheet and link the pieces of paper with lines and words. When satisfied with their effort, the pieces of paper were stuck on the A3 sheet to form their concept map. Analyses of 68 students' maps revealed most students (23) made between five and nine links, several (17) made 15 to 19 links, but eight students had no links to plastic, and many (20) had far fewer links than the teachers had predicted. These students did not link plastics to clothing or recycling and had a limited view of the uses of plastic. Is this what you found when you tried the activity with some students? Now, more than 15 years later, was recycling mentioned?

How are the properties of (objects and) materials determined?

Primary (and older) students sometimes will not be able to identify and/or describe the appropriate properties for comparing the characteristics of particular materials (and objects). In a Year 5 class, after considerable hands-on experiences with various paper types, only one student referred to 'strength' as a property of paper. The teacher argued that this supported the idea that students' language is often 'context specific' (Parker 1995, p. 19). She found that 'strong–weak' is often associated with living things: strong with heavy and weak with light, weak with readily breakable objects while strong was the opposite. When asked what strong and weak might mean when talking about paper, these students referred mainly to observational criteria (thick, hard) and criteria implying a test (hard to rip); sometimes they mentioned functional (wrapping) and compositional criteria. Teachers certainly need to be aware of the meanings their students are ascribing to key words (here, 'strong' and 'weak'), and that students will sometimes not differentiate between meanings of different words such as 'hard' and 'strong'.

There are accepted scientific meanings of material and object properties, such as hardness, plasticity, toughness and strength. Strength, for example, is the force required to break an object (Wenham and Ovens 2010). Teachers would need to decide when it is appropriate to have students investigate particular properties of materials and objects, and how to help them define the characteristic they are comparing. The use of language is clearly a key influencing factor.

Activity 8.6 How do students believe they can determine an object's or material's properties?

Try this task yourself first and then with several students of different ages.

Select two objects from a range provided (for example, a brick, a block of wood) and state how you would decide which was harder and which was stronger. If you are able to get some student responses, determine whether there are any similarities and differences between them and your own ideas. If so, can you categorise the type of response(s) you noted? For example, did you and the students refer only to observation or was experimental testing suggested? Was there evidence of other ways to find answers? Link your responses to the findings of Parker (1995) and Russell, Longden and McGuigan (1991).

As an aside, were you determining the properties of an object or a material? Is it possible to determine the strength of an object and a material? If so, what would you have to do? If not, why not?

For ideas about ways in which students could investigate the properties of materials, see the *Primary Connections* unit 'Material World' (AAS 2008a). It refers to the decomposition, absorbency, tensile strength and thermal insulation properties of materials.

Russell, Longden and McGuigan (1991) found that 38 per cent of primary students (n = 68) would use observation (size, appearance, feel and so on) to determine 'hardness', while 57 per cent thought an empirical test (malleable, breaks when dropped, bends under a load) would be required. Ratios varied across grade levels but both types of responses were always present. As encouraging scientific investigation of materials is a key aim of science teaching, these results suggest students' responses could be used to move in that direction.

The uses and origins of (objects and) materials

How materials are used relates to their properties. In the SPACE project on materials, when students were asked why a particular material was good for a particular use (e.g., wood for furniture), responses were to some extent context-specific (e.g., related to students' familiarity with the material). Functional (e.g., a property such as hardness being related to use) and then manufacturing (e.g., ease of cutting) criteria dominated ahead of aesthetic and economic criteria, which were rarely mentioned.

Activity 8.7 Where do materials come from?

Try the following picture-strip technique (a useful elicitation task) yourself, and then, if possible, with several students as a means of ascertaining your and their preconceptions about the origins of objects and materials.

For a metal spoon and a piece of cotton fabric, draw a series of pictures that would represent the different stages or transformations that had occurred before arriving at the end result of a spoon or a piece of fabric (following the SPACE *Materials* project, pp. 46ff).

Compare and contrast the students' efforts and theirs with yours. If possible, compare with the SPACE *Materials* project picture-strips, or with Qualter, Schilling and McGuigan (1994).

The British students (in the SPACE study) did not know a great deal about the origins of manufactured objects. Only a small number recognised that cotton came from a plant (would students from other countries be more aware?) and only a minority of students seemed to be aware that the origins of metal objects were on – or in – the ground.

Teaching implications

An awareness of the key concepts related to objects and materials and their properties, as well as students' conceptions about them, will guide you in your planning and choice of teaching strategies. Already you will have gleaned various techniques that could be used to elicit students' ideas about materials and you may have thought of some appropriate follow-up teaching strategies. You have probably realised that students' ideas about objects and materials will be enriched by their physical explorations of them, as well as how they interpret the 'linguistic input' by teachers and others (Wiser and Smith 2008, p. 208). In the second main section of this chapter, 'Developing students' ideas about materials' (see p. 319), the strategies associated with activities 1 through 6 in Activity 8.8 (see p. 321) and the follow-up discussion on language use will provide further guidance. Also, the case studies in this chapter's third main section indicate how some teachers have assisted their students to learn about the properties of objects (e.g., their structure) and materials (e.g., their absorbency). When you read these case studies, use the discussion up to this point to guide you in critiquing their conceptual bases and the responses of the students.

Solids, liquids and gases

Solid, liquid or gas refers to the state of a material or substance. A study of solids, liquids and gases is generally a component of most primary science curricula. Here, the macroscopic (surface) properties of solids, liquids and gases will be overviewed (as part of the nature of the materials), while in Chapter 9, changes of state will be considered.

What do students think about solids, liquids and gases?

Even very young students judge solids and liquids to be matter, but many Year 8 students still do not think air is matter. Although such students have been told many times that air is matter, this simply does not make sense to them as they intuitively believe that matter must be perceptible – it must be able to be touched, felt or seen. Consequently, those primary students with an intuitive understanding of matter but who do not 'reject' that gas is matter, will, for example, fragment their understanding by accepting that there are 'unexplained exceptions' or recite that solids, liquids and gases are matter without appreciating what 'matter' is. Interestingly, the concept of weight may or may not help primary students to appreciate whether solids, liquids and gases are matter. This is because many associate weight with size and heaviness and, for example, gases do not align with these perceptual ideas. A few students do, however, start to appreciate that even the smallest quantities of stuff have weight and hence they must be matter (Wiser and Smith 2008, pp. 213, 215).

Solids

Children's ideas about solids, liquids and gases have been more extensively reported than some of the properties of materials described earlier. Many criteria have been used by students to classify materials as solids. Two examples are:

- referring to a property without appreciating the difference between the properties of objects and materials; for example, hard or soft (which is a material or intensive property) and size (which is an object or extensive property) are often used to characterise a solid
- stating what happens when objects or materials are handled; for example, if it can be held it is a solid (Krnel, Watson and Glazar 1998).

Students' understanding of solids also is greatly influenced by everyday usage of the word; namely, if something is solid it will be hard and rigid. Consequently, some children have difficulty classifying soft, malleable and granular materials and rarely refer to them as examples of solids (McGuigan, Qualter and Schilling 1993). Some students may, in fact, see matter as being composed initially of hard and soft materials (and substances), and believe that solids are a subset of hard materials and substances (in a dichotomous categorisation of hollow and solid). Students are here using the word 'solid' as an adjective, not a class of substances (Jones 1984).

Liquids

Liquids, in general, are more readily classified. They are thought to be 'runny', but viscous liquids can cause conceptual difficulty. Students often think about liquids by thinking about water. Water is seen as a prototype upon which knowledge of liquids is often based. In one study, for example, about 90 per cent of children, regardless of age, identified an unknown liquid as water (Krnel, Watson and Glazar 1998). The main characteristic used to distinguish liquids from solids was that they could be poured; water content was also mentioned (Russell, Longden and McGuigan 1991, in Krnel, Watson and Glazar 1998).

Gases

Gases are the least well understood by primary aged students (Wiser and Smith 2008). When asked, students tend not to provide examples of a gas but rather the uses of a gas, or they refer to trigger-word associations (for example, gas flame, the air we breathe, aerosol cans have gas). Gases are sometimes associated with heating or combustion. Research findings vary as to whether students see air as a prototype for gases, as is the case with water and liquids. Some students (up to about age 12) often do not define air as a gas, as gases are regarded as dangerous and combustible (Russell, Longden and McGuigan 1991), while others align gases with air as a prototype (Krnel, Watson and Glazar 1998) and some upper primary students seem familiar with oxygen and carbon dioxide as gases (Liu and Lesniak 2006). Even so, you may find that younger students have difficulty expressing any ideas about the nature of gases.

With specific reference to air, most K-2 students think that an empty container does not contain air (even when it is suggested that the container might not be empty), whereas most seven- to 11-year-olds state that air would be in the container. Properties associated with air include its invisibility, odourlessness, association with movement, temperature, extent and life-sustaining capacity, although these responses definitely are not universal. The last two mentioned could be associated with age, but in many instances, infant students seemed more 'air-aware' (Russell, Longden and McGuigan 1991). In her kindergarten class, Lawther (1994) found that her students made comments supporting the above categories. Some also believed air is found only in a closed container and that it becomes visible when we breathe on a cold day. For other student conceptions about air and suggested activities, see Tytler (2002).

The familiarity of solids, liquids and gases

When given the opportunity to name, draw and label the three states of matter, students tend to be able to give far more examples of solids than liquids and more liquids than gases. The examples students provide often relate to everyday and trigger uses of the words (e.g., washing-up liquid). From another perspective, about half of upper primary Greek students (Years 4 to 6; n = 173) did not recognise the hazardous nature of predominantly solid and liquid household items, detergent being the most stated (10 per cent) – even more adults, in another study, did not recognise household items as hazardous (Malandrakis 2008). Further, it is possible that upper primary students may believe that most solid waste is recycled and food waste is not considered recyclable or useful (Leme, in Cinquetti and de Carvalho 2007). What could be the implications of these studies when your students are learning about the uses of objects and materials?

Scientists' views about solids, liquids and gases

By way of contrast, scientists see solids as having a fixed shape, being able to withstand tension, compression and twist, and unable to flow, whereas liquids do not have a fixed shape, change their shape to fill the bottom of a container, can only withstand compression and can flow. Gases are seen as not having a fixed shape, expanding to fill the entire volume of a container, being easily compressed and able to flow readily. Generally speaking, gases tend to be less dense than solids and liquids (Carlton and Parkinson 1994).

However, as with most branches of science, these descriptions do not always hold and when more complex examples arise in classroom discussion, such as foam, glues and other 'soft matter' (Ogborn 2004), then teachers need to find a balance between meaningful classification and the flexibility that is desirable to empower students to make their own sense of what they are talking about (Goodwin 2002). This is discussed later in this chapter in the section 'Thinking through our reasoning' (see p. 324).

Teaching implications

How do teachers help students learn about the properties of solids, liquids and gases, especially gases and the 'more difficult' examples of these categories, such as foam and glues? You might like to stop and think about what you would do to help your students move forward in their understanding about the three states of matter. Many strategies are suggested in the upcoming section 'Developing students' ideas about materials' (see p. 319).

Substances and particles

The student views presented above all relate to the nature and properties of (objects and) materials from a macroscopic view of matter. Can primary students appreciate the nature of substances and that there is a microscopic/nanoscopic (particulate) interpretation of matter which can explain many of its properties?

Students' conceptions of pure substances

Try the following task yourself and, if you wish, with students in the final year of primary school. Visualise (but if you try it with children, obtain) materials such as air, copper, copper oxide, chocolate, iron ore, iron, water, sugar solution, toffee, oxygen, wood (block), wood (saw dust), rock salt and salt, and ask yourself the question: Are these pure substances? Year 6 primary students ($n = 147$), when asked that question, did not have an understanding of a 'pure substance' in the chemical sense. (You should note that it is more accurate to say 'a pure sample of a substance', as a substance by definition has a fixed identity because of its atomic structure and so it is inappropriate to describe it as 'pure'.) These students tended to think of a pure substance in terms of where it has come from (e.g., Is it 'natural'?) and what has happened to it (e.g., if it had been processed it was not pure). It would seem that thinking about the material world in terms of substances 'is something that children need to learn' (Johnson 2000, p. 735). This has direct implications for activities that are often used at the primary level, such as those relating to mixtures of substances and how teachers introduce chemical change (see Chapter 9).

Students' conceptions of the structure of matter

As primary students do not have an appreciation of the chemical concept of substance, does this indicate they do not have a nanoscopic/microscopic (particulate) view of matter? Most research evidence suggests that primary students think of matter, especially liquids and gases, as being continuous. Solids (e.g., stones) are sometimes thought of as being composed of particles, but not in the atomic or molecular sense – for a review of student ideas, see Hatzinikita, Koulaidis and Hatzinikita (2005), Krnel, Watson and Glazar (1998), and Wiser and Smith (2008). Interviews with seven- to 10-year-olds (like Activity 8.2) suggested three ways they thought about matter. They were classified as:

- 'macrocontinuous children' who thought of matter as continuous
- 'macroparticulate children' who described some materials as comprised of particles, but the particles had 'macro' properties, such as they were the same shape as the actual material
- 'microparticulate children' who gave some responses that suggested aspects of the scientific particulate view of the world.

These children did not use their explanatory framework consistently and often had elements of all three frameworks in their thinking (Nakhleh and Samarapungavan 1999). When students only hold a continuous and static conception of matter, it can be an obstacle to further learning about the nature of matter (Hatzinikita, Koulaidis and Hatzinikita 2005).

As students start to think about particles and matter, they tend to visualise the particles as something *in* matter rather than being the matter itself. This is a 'molecules-in-matter' model and is a very commonly held alternative conception: it is a synthetic model because students are amalgamating their intuitive continuous view of matter with the school-introduced idea of 'particles'. Reasons abound for why this happens – think of textbook diagrams of atoms/molecules shown within, say, a cube, or how language is used, as in 'atoms *in* solids vibrate…' Further, we talk about atoms being microscopic, suggesting they can be seen (and hence why 'nanoscopic' has been added as an alternative term), and even the word 'particles' can lead to images of, say, dust particles. There are numerous other possibly confusing instances. For example, what might be the picture of atoms that could result from teachers asking students to imagine cutting a solid into smaller and smaller pieces? This is a non-scientific 'model' of matter as it does not assist (as a scientific model would) in predicting the behaviour of matter (Wiser and Smith 2008, pp. 219–20).

There are other 'ways of thinking' that students may use which relate to whether matter is conceived to be continuous or particulate or somewhere in between. Some primary students use 'intuitive rules' to explain everyday phenomena. As they get older, they may use the rule 'everything can be divided into two' (an infinite and continuous concept of matter) more than the rule 'everything comes to an end' (a finite concept of matter tending to a particulate view). Awareness that some children may explain their world using intuitive rules can help teachers anticipate children's responses. Other intuitive rules that students may use are 'more of A–more of B', 'same A–same B' and 'less of A–less of B', where A and B represent properties and qualities of objects and materials which are *un*related (Yair and Yair 2004).

Students therefore use various non-scientific ways to try to account for their macroscopic observations of matter (mixing, dissolving and so on). For students to explain the properties of matter using a particulate model requires some appreciation of the nature of 'scientific models'. It is a big thinking shift for primary students to think about the entities within (scientific) models – here, particles (atoms and molecules) – when they (particles) have *different* properties to the realities – here, for example, solids and liquids – that the model is accounting for (Wiser and Smith 2008, pp. 223–4). This has definite implications for teaching – see, for example, the Chapter 8 section 'Narrowing student-generated "scientific" (mental) models' (p. 325) and 'Changes of state: teaching about particles' in Chapter 9 (p. 381).

Finally, when listening to primary students talk about the structure of matter, we need to recognise that they will regularly use scientific labels (e.g., atoms) but be unable to articulate their meanings. Word association tests, used with 11- and 12-year-olds in two countries, in which the particulate model of matter had not been taught, did not find words such as 'solid', 'liquid', 'gas', 'matter', 'element', 'compound', 'chemicals', 'atom' and even 'particle' to be associated with the particulate model of matter. For some words (e.g., 'compound'), students found difficulty in suggesting any associated words. In a country in which the particulate nature of matter was part of the primary curriculum, there was evidence of 'particulate' word associations. (Of interest were some cultural differences: 'atom' in the United Kingdom was associated with 'bomb', but not in other countries) (Maskill, Cachapuz and Kouladis 1997).

Should teachers introduce the particulate nature of matter?

As many books and other resources (and some primary curricula) introduce aspects of the particulate nature of matter (atom, molecule, bond), then what are primary teachers to do? The evidence indicates that primary (and secondary) students find it difficult to accommodate a particulate view of matter. Further, this world view (of matter) develops gradually and over a period of years (Liu and Lesniak 2006; Nakhleh and Samarapungavan 1999; Papageorgiou and Johnson 2005). The emphasis at the primary level, therefore, needs to be on macroscopic examples of the behaviour of materials, but not necessarily to the exclusion of particulate representations (Wiser and Smith 2008). Some young children are capable of fairly complex and scientific reasoning when the learning contexts are supportive (Metz 1995; Tytler and Peterson 2005); it is understandable that students, when scaffolded in appropriate ways, may give indications of thinking in a microparticulate way. Scaffolding strategies are described later.

Teachers clearly need to move carefully in this area. They should do so with a knowledge that most (if not all) of their class has not thought of this way of interpreting their material world. However, as implied, this does not preclude getting students involved in particulate representational and modelling tasks and discussions, especially if they grow out of hands-on observations and investigations described in this and, especially, the next chapter (Skamp 2005, 2008).

R

Approaches, backed by research results, that support the formal introduction of particulate ideas to primary students include:

- a two-level teaching approach – this is when macroscopic and submicroscopic (nanoscopic) levels can concurrently be the focus *but* on a phenomenon-by-phenomenon basis (Lee et al. 1993). This means, for example, that as upper primary students study evaporation from a macroscopic perspective, then discussion may provide an opportunity for particulate ideas to emerge or be introduced, say in role-play, as it applies to this phenomenon in particular
- introducing the idea that particles have an 'ability to hold on to each other' – many 10- to 11-year-olds, using this idea, started to give incomplete particulate explanations for phenomena such as the characteristics of the states of matter, but found it more difficult to apply particulate notions to some phenomena compared to others. These students were

not 'hindered in any way by their exposure' to particulate ideas (Papageorgiou and Johnson 2005, p. 1314)
- using the concept of 'molecule' – this seemed to help seven- to 10-year-olds use particulate ideas, but their use was dependent on the context; for example, they were applied more readily to explain evaporation and rarely for decomposition. This ability to use particulate ideas varied among students. Students, though, still preferred to use explanations of change phenomena (see Chapter 9) based on experiential (that is, their everyday life experiences) rather than particulate thinking (Hellden 2005).

One approach, not accessible to most teachers but interesting because of its results, was created by Novak (2005), who found that, by using specialised audiotutorial strategies with concrete materials and, for example, concept maps, some seven-year-old children understood aspects of the particulate nature of matter; further, the introduction of these ideas assisted later learning in this content area. Studies such as these and others – for example, the impact of students learning about matter, weight, volume and density preceding particulate ideas (Wiser and Smith 2008) – provide evidence that it is probably the pedagogy rather than student characteristics that determine if primary students can benefit from the introduction of particulate ideas.

You may also read some (non-research-based) reports of upper primary students being formally introduced to the structure of molecules; for example, Leisten (2008) and Brown, Rushton and Bencomo (2008). In these instances, concrete molecular models are built; for example, wooden arrangements of methane (CH_4), water (H_2O), ammonia (NH_3) and carbon dioxide (CO_2). This model building is used in conjunction with role-plays and reference to the concept of 'bonding', together with fascinating 'facts' about the molecules they were building and talking about. Positive affective responses resulted. Should you consider using such learning experiences, reflect on how you will explicitly incorporate the nature of mental and physical models (see the Chapter 1 section 'The nature of science: how it works' on p. 5) and listen carefully to students' conversations to determine whether you need to encourage them to move towards mental models that are stepping stones to particulate thinking; that is, away from a 'molecules-in-matter' model.

NOS

In summary, primary teachers must know about the particulate view of matter. If you decide to formally teach about particulate ideas, then you need to be aware of current research findings. Further, you should be conscious that, where success has been reported, then usually very specific teaching approaches were used and the resultant thinking about matter in particulate terms was not consistent across students.

Possible patterns of development in students' ideas about materials

The above summary of students' conceptions suggests that there is not a regular developmental progression of ideas about materials, their nature, properties, uses and origins. This is especially the case for the origins of materials where, in many instances, the only sources of information are secondhand. For uses of materials, students' responses often are context-specific and draw upon experiences they have had with certain products, often in the home or other non-school environment. The nature and properties of materials tend to be described in terms of everyday perceptions of the materials, sometimes using prototypes as a means of generalising. In some instances, there appears to be age-dependent responses, such as a greater familiarity with gases by older students.

Although obvious developmental patterns are difficult to discern, Krnel, Watson and Glazar (1998) have proposed a theory for how students develop a concept of matter. It stresses that young students initially seem to learn about matter by acting on (holding, blowing, pouring) objects and materials. These actions, when applied to particular materials and substances, lead to the development of prototypes (water for liquids, air for gases). When prototypes are combined with each other or with substances with different properties, the concept 'substance' is developed; when actions are combined with prototypes of matter, 'physical changes' (e.g., dissolution) are understood. This suggests the importance of actions (hands-on experiences) in forming concepts, but also indicates a role for teachers in scaffolding understanding by having students discuss their conceptions in relation to any identifiable prototypes (e.g., water) while assisting them to generalise beyond the prototype.

More recent research (Liu and Lesniak 2006, p. 341) argues that there is 'no universal progression' of ideas about matter; rather, student progression is 'multifaceted'. This means that there is a 'unique' progression pattern for each aspect of matter being studied; for example, conceptions of mixing, conceptions of dissolving, conceptions of chemical reaction (also see Chapter 9). To complicate the picture, the conceptual progression related to a phenomenon such as dissolving can vary for different substances. Also, there is much overlap between students of different ages. These findings, according to Liu and Lesniak (2005, 2006), imply that students' conceptual development of matter can be visualised as 'overlapping waves' (ibid., p. 340) in which various aspects of the topic of matter (for example, conservation of matter, physical properties and change, chemical properties and change, structure and composition) are related and overlap. The overlapping suggests that we should not compartmentalise concepts about matter – even so, these studies still recommend that, up to middle primary, students should initially focus on informal learning about matter, with middle primary learning about properties and change involving water and air, and upper primary focusing on the conservation of matter and physical properties and change. Perceiving chemical properties and change also would be appropriate at the upper primary level (Liu and Lesniak 2005, 2006).

How students progress towards a particulate model of matter is problematic. One possible pathway could be from matter as homogeneous and continuous, to particles embedded in continuous matter (maybe like a raisin cake), to particles being the matter but with macroscopic properties, to particles being the matter (the substance) and the way they (that is, the particles) behave collectively describes the properties of matter (Papageorgiou and Johnson 2005). The difficulty with this pathway is that some of its intermediate ideas may be stumbling blocks to developing a particulate view of matter. The evidence suggests that a 'particles-in-matter' model is *not* a useful stepping stone to a particulate understanding, and that students holding this view rarely use it in 'explaining' phenomena. However, students holding models with separate particles, albeit with (non-scientific) macroscopic properties, were able to visualise forces or bonds between the particles. Several researchers encourage teachers to introduce the notion of bonds or forces between particles – Wiser and Smith (2008) detail the evidence for these conclusions. Teachers need to be aware that, when their primary students are talking about the composition of matter, their thinking may contain a mixture of explanatory ideas (about matter) and that the development of these ideas towards a particulate model could be along multiple learning trajectories and may take years.

Developing students' ideas about materials

Appendix 8.5 Websites with ideas for activities related to matter

In this, the second main section of Chapter 8, specific constructivist teaching strategies addressing the alternative conceptions discussed in the first section are the focus. As there are numerous readily available ideas for activities related to this content area (see, for example, the online Appendix 8.5), then what should a constructivist teacher do?

Some guidance has been provided in the previous discussion about primary students' conceptions, although it was noted that there is not an obvious developmental progression of concepts about materials across the primary years. To help students effectively learn about materials, you are encouraged to use ideas and activities you find in resources in particular ways which will be referred to here as 'strategies'. Such strategies can be used at various stages within a lesson and/or across a sequence of lessons, depending upon your purposes. These purposes could be, for example, to elicit, challenge or apply an idea.

Obtaining interest: a first consideration

Engagement with, and interest in, a topic is usually a prerequisite to effective learning (see Chapter 1). Aesthetic experiences (what we find beautiful/ugly and what we find pleasure/displeasure in) are one factor that cannot be ignored when considering how to obtain and maintain interest in ongoing learning in science (Jakobson and Wickman 2008). Students regularly make aesthetic judgements in science lessons. Two students, for example, were using magnifiers to help determine if various materials were solid or liquid: in showing her partner how to look at salt crystals with a magnifier, one student commented, 'It's awesome'. This positive aesthetic experience related to the sorting task as well as to how to use the magnifier; it therefore is not just aesthetic but is coexistent with conceptual learning (ibid., p. 53). How would you introduce and handle aesthetic experiences and responses in order to engage and advance learning? If possible, compare your thoughts with Jakobson and Wickman's (2008, pp. 62–4) advice on how teachers can use aesthetic considerations.

Another key factor determining interest is whether a topic has relevance to an individual (Qualter 1993, p. 315). In a survey of 3400 students, Qualter found that, of 48 statements related to materials, only a couple were among the most popular topics (defined as more than 50 per cent interested), such as 'Why jewellery is made from silver and not aluminium' (55 per cent girls) and 'Why fibreglass and not metal is used for making cars that last' (51 per cent boys). Gender differences were present; for example, 'Why fibreglass and not metal is used for making cars that last' was one of the least popular among girls. To ensure both boys' and girls' engagement in learning about materials and their properties, the choice of topic needs to be considered. Qualter found the least popular topics (disliked by boys and girls) consisted mainly of abstract statements of physical science topics (for example, properties of metals and gases). Those topics most liked by boys and girls contained abstract and application statements, with the latter predominating. Girls oriented more to topics with relevance to social, human or animal needs, while boys chose topics for which personal and social relevance was less apparent. It was concluded that traditional male–female dichotomies should not be applied rigidly and that 'both boys and girls respond to topics which they see as relevant to their interests' (ibid., p. 315).

Qualter's research offers a challenge when teaching about materials. Abstractions (e.g., properties of solids, liquids and gases) fell into the least-liked category, whereas the uses of materials depending upon their properties was more attractive to girls and/or boys, depending upon the perceived relevance of the uses. A lesson sequence focusing on properties of bubbles (Sanderson 2000) and some *Primary Connections* unit sequences (AAS 2005) illustrate this conclusion well (see the case studies later in this chapter). Other possibilities to engage interest are focusing on the environmental aspects of properties of materials and substances; for example, water quality and salinity (Pendlebury 2005; Pressick-Kilborn 2009; Walker, Kremer and Schluter 2007). The importance of context also cannot be underestimated: the supermarket, backyard, household, kitchen and factory are all situations in which materials could be studied. The Primary Upd8 resource (see http://www.primaryupd8.org.uk) provides numerous examples of interesting contexts for learning about properties of materials and their uses, including persuading the local community about alternatives to using plastic for bags and selecting materials to build an on-screen robot.

Many strategies can engage student interest. Concept cartoons showing primary students offering different ideas in familiar surroundings as to whether various objects are solids or what are the properties of particular objects – for example, whether they are insulators or conductors (Keogh and Nalyor 1997) – have encouraged students to offer their own thoughts and have sometimes led to scientific investigations. Theatre performance was an approach used with a group of five to eight-year-olds, centring on a narrative about two young girls in a family situation who were kidnapped by a witch. A series of science investigations incorporated into the narrative were 'played out' on a classroom 'stage'. Examples described related to the properties of bubbles and how to sort objects. These were then followed up in later lessons (Gatt and Gatt 2008). How could you use the performing arts as a means of engaging student interest in materials?

Strategies to facilitate conceptual development

Conceptual development is required to move from an intuitive to a scientific concept of matter. Students will encounter considerable difficulty in crossing and/or linking the macroscopic, microscopic/nanoscopic and symbolic levels. Teachers need to initially be aware of their students' existing ideas. The evidence then suggests that, over time, students need to attend to various aspects of the matter concept, consistent with development resembling overlapping waves. As students are completing activities relating to materials and their properties, teachers need to encourage their reasoning, which can be descriptive and/or explanatory, refer to macro or micro dimensions, and be qualitative or quantitative (Liu and Lesniak 2005, p. 444). With younger children, the emphases will lie towards the former of each of these dichotomies but could overlap with the latter (Wiser and Smith 2008).

Generic strategies related to materials

Various strategies have been advocated to address the issues raised by some of the findings about students' conceptions of materials and their properties (Russell, Longden and McGuigan 1991). These are the focus of Activity 8.8. Twelve activities are suggested that relate to the diversity, properties, uses and origins of materials (1–7) and solids, liquids and gases (8–12). After each activity, a suitable level (lower primary, sometimes referred to as K(R)–2 or 1–3, and upper primary, 3–6 or 4–7) is suggested, but teachers should use their discretion. Each activity illustrates one or more of the following intervention strategies:

1. encouraging students to test their ideas
2. encouraging students to develop more-specific definitions for particular words
3. encouraging students to generalise from one specific context to another through discussion
4. finding ways to make imperceptible changes or features perceptible
5. testing the right idea alongside the students' own ideas
6. using secondary sources
7. discussion with others.

These strategies have been trialled in classrooms and are recommended after teachers have elicited the views of their students about a particular topic.

Activity 8.8 Intervention strategies to change preconceptions about materials

Read through the following descriptions of activities (adapted from Russell, Longden and McGuigan 1991) and decide which one (or more) of the above seven strategies has (have) been used. Justify your response, indicating, where appropriate, why you think the teacher might have used the strategy. Try to suggest a refinement or a variation to the strategy which may further help to achieve its implied purpose.

1. In describing a group of materials using touch in a blindfold game, some of the students seemed to interchange the words 'smooth' and 'soft'. The teacher then asked the students to form sets of smooth and soft objects (lower primary).
2. Children in groups played a guessing game with materials. One student provided descriptive statements about the material (restricting statements to properties, although students sometimes referred to uses). After each descriptive statement, members of the group tried to guess what the material was (lower primary).
3. A teacher provided groups with three materials and asked them to complete a table that required them to list each

material's colour, shape, feel, pattern, smell, weight (heavy or light), size and what happened if it was bent, twisted, pulled, rubbed, scratched and bounced (lower primary).

4 The teacher encouraged her class to raise questions about some supplied materials that they could investigate and they agreed on testing which materials soaked up the most water (the absorbency). They measured the weight of the material before and after. A plan was devised, predictions made and the test carried out with an effort being made to make it fair (for example, the material was immersed for the same time and shaken the same number of times to remove non-absorbed water) (upper primary).

5 Working in groups, students were required to discuss why particular materials were used in certain ways (for example, chalk for writing on the blackboard) (lower or upper primary).

6 Pairs of students compared the uses of paper and plastic straws. They were able to try using them to help them decide which material best suited its use (lower primary).

7 A class was asked to trace a glass window back to its origins. They initially discussed ideas among themselves, then sought information from some firms, watched a video and consulted some references. A series of annotated and sequenced pictures (using students' own words and drawings) was completed by each student to represent their ideas about the formation of the glass window (upper primary).

8 Children were required to classify materials that challenged solid/liquid category boundaries (such as flour) and then studied the materials under a microscope (upper primary).

9 Some students, when asked why a particular substance (for example, Coca-Cola) was a liquid, said it was because you couldn't pour it. They were then provided with some recognised solids, which they crushed and poured, and were asked to reconsider their ideas (lower or upper primary).

10 Children classified liquids according to their own criteria (for example, some drew tables listing liquids under different properties). A few went on to develop overlapping Venn diagrams in which each set defined a property and the liquids were categorised in various sections of the Venn diagram. Both approaches led to discussion among students and with the teacher about the overlap of categories, the clarification of terms and the properties of a liquid (upper primary).

11 A group of students pasted pictures of objects they associated with gases (for example, sprays, gas cookers, oil or gas heaters, fans, a vehicle's exhaust, a football) on a sheet of cardboard. Discussion then occurred as to whether the examples were gases or not (upper primary).

12 A class looked at fizzy drinks, studied the use of an aerosol and blew bubbles in water. They then thought of questions they wanted to ask about gases. Examples included: Can we breathe it? Is it poisonous? Can you see it? Can you see its effects? Is it flammable? (upper primary).

Reflect back on these strategies. Where does their emphasis appear to lie: description or explanation, macro or micro, qualitative or quantitative? Are these emphases appropriate? Why?

These strategies have facilitated conceptual change (Harlen 2007). Investigations (for example, activity 4 above) helped students appreciate the value of empirical tests to distinguish between the properties of different materials, as well as encouraging more thought about the composition of materials. Investigations about properties seemed to be more effective than direct lessons. It was difficult to identify effective strategies to help students make the nature (of a material)–(and its) use connection, but focusing on materials that students regularly use is a starting point. Most students had difficulty tracing objects back to their origins (perhaps because firsthand experiences are often not available) as well as classifying unusual examples of solids such as cotton wool and talc – many examples and probing discussions are still encouraged (e.g., the SPACE *Materials* project).

2-D and 3-D concept maps

Concept maps (see Activity 8.3 on p. 304), which are different to mind maps, promote meaningful learning about matter and change as they require students to specify the links between their ideas, and the quality of these links may be enhanced if the maps are developed in cooperative learning groups as they encourage the exchange of ideas. Concept maps (developed by 'experts') also can be used as scaffolds for learning (Novak 2005). Primary teachers could explore this technique by having students discuss, interpret and suggest changes/additions to teacher-introduced concept maps, which include connections that extend or challenge students' thinking.

With young students, 3-D concept maps using real objects and/or student-created images can replace concept words, and strips of card (with focus words from the students) can be the connections between what is provided and/or images. An example could be using a bottle of water as the central object and providing a range of objects (e.g., bath toys, cup, straw, flotation devices and so on) and asking a focus question such as 'What do you know about water?', and then through discussion create links such as 'wash the dishes' (mop) and 'drink it' (cup). As more ideas evolve, they can be added to the 3-D map which could be left on display. Additional (non-connecting) thoughts could be added in speech bubbles, such as narrative comments like 'helping Dad wash the dishes'. This kinaesthetic and participative approach can effectively relate objects and materials to the students' lives (Howitt 2009).

Language use

Several activities – see Activities 8.3 (p. 304), 8.4 (p. 304) and 8.6 (p. 310) – indicate that language use is integral to facilitating conceptual development and change. Parker (1995; see Activity 8.6) found that talking about and investigating materials were effective strategies in clarifying students' conceptual language – for example, how her students used 'hard' and 'strong' when exploring the properties of paper. The use of surprise as part of the investigation, such as clarifying what the strength of paper means (as most students do not think of paper as having strength), was thought to be conceptually helpful.

Some teachers believe that it is imperative that the correct terminology be introduced. Turner (1997) provides examples related to '[changing] materials'. Researchers urge teachers to use words such as 'objects', 'materials' and 'substances' carefully, although they differ as to whether there should be explicit teaching about such terms (Vogelezang, in Driver et al. 1994b). Papageorgiou and Johnson (2005) clearly drew the distinction between objects and materials before introducing particulate ideas to upper primary students, and this advice was followed in the *Primary Connections* units related to materials (AAS 2008a, 2008b, 2008c). Earlier, it was suggested that the terms 'particles' and 'microscopic' could be misinterpreted as they suggest inappropriate ideas about scale and the actual nature of atoms and molecules that comprise materials (Wiser and Smith 2008). Teachers need to interpret this advice, taking into consideration their students' prior experiences; even so, careful use of conceptual labels, whether introduced formally or otherwise, is strongly encouraged.

Language use is also important in the way teachers interpret students' responses. We may too readily interpret students' responses as 'signs of missing [or incorrect scientific] ideas'. Alternatively we could use students' responses to 'identify what it is they need to learn' (Papageorgiou and Johnson 2005, p. 57). When students say, for example, that they think air (or sometimes oxygen or hydrogen) is what the bubbles are made of when water is bubbling, this could simply mean students 'do not have the idea that water itself can form a body of gas'. Teachers need to be conscious that students' language, apart from indicating their conceptions of how their world works, also may indicate what teachers need to introduce to advance thinking about a phenomenon. It may also be added here that, when students use scientific language, it might not reflect their understanding. From middle primary school onwards, students may commonly use terms such as 'molecule', 'atom' and 'CO_2' but not be clear about their meaning (Liu and Lesniak 2006). An awareness that students' words may not always convey understanding will influence your conversations with them.

Thinking through our reasoning

Having students work in small groups and articulate their reasons is clearly valuable. An example could be classifying materials into particular categories, such as solids, liquids and gases. This strategy can be extended when students disagree about their categories by having them record their thinking. They could, for example, write down their agreements and disagreements and the problems that they have encountered, and then, with teacher encouragement, rethink the task. This encourages metacognitive thinking (thinking about learning how to think) (Levinson 2000). This strategy would also help students to appreciate that definitions in science can have blurred boundaries and hence could be used to explicitly teach about the NOS.

To set up situations that cause disagreement, teachers need a clear task with at least one example that may cause debate (for example, when discussing solids and liquids, steel wool, talc, jelly or even slime or ooze could be used; for a recipe for the last mentioned, see the online Appendix 9.7. Appropriate teacher intervention is recommended to encourage students to think carefully about the consistency of their reasons for classification.

Appendix 9.7
Slime recipe

Thinking through and writing about their ideas (with others) can assist students to clarify their reasoning, provided the teacher has oversight of the concept(s) underpinning the discussion. This was the case in a Year 5 class that had completed some activities related to freezing, melting and condensation and wanted to find out more about solids, liquids and gases. They decided to write a class book about the three states of matter, with each group writing a page. The whole class had to examine the contents of each page and offer comments on the scientific content and how it was written. The teacher, reflecting on the experience, said that it gave her an insight into her students' thinking, and also encouraged students to comment on each other's ideas, requiring them to justify their views (McMahon 1999).

Towards a particulate view of matter

Students do have ideas about what constitutes stuff (matter). There are teaching strategies which may move their thinking towards a particulate model – some of these are outlined below. When teachers make sustained efforts to engage students in constructing new representations to account for the properties of materials, then students appear to make conceptual progress (Wiser and Smith 2008).

Narrowing student-generated 'scientific' (mental) models

R

Contrary to usual expectations, teachers can selectively scaffold primary students' thinking as they generate models to explain the nature of materials in particulate terms. In a lesson sequence (Wiser and Smith 2008), the teacher started with the 'models' that seven- to eight-year-olds had generated after being asked to imagine materials as being made of *discrete* entities (that would be useful in interpreting what they observe). The teacher, through scaffolding, then narrowed the types of models down to ones that embraced certain features consistent with the scientific particulate model, in this instance a 'parts model' but without the 'parts' taking on any name(s). The teacher then had her students manipulate their mental 'models' (of the nature of materials) as a consequence of various physical manipulations of concrete materials – each small group of children had one material: clay, a sponge, water, stones, wood and metal. In other words, the physical manipulation of materials occurred hand-in-hand with the manipulation of mental models. Examples of the physical manipulations included the students' shaping their material into an egg shape and then using a 'breaking method' such as stepping on it. Other breaking methods, such as water and fire, then followed. Ongoing discussions occurred between various small groups and among the class as a whole.

With each physical manipulation, the teacher oriented the students' thinking towards four basic ideas, involving the need to imagine materials as having:

1. discrete parts
2. a very large number of parts
3. bonds between the parts
4. the same number of parts before and after changes.

As students encountered these ideas, the teacher encouraged reflection through imagination (e.g., 'How are the bits joined together?') and asked the students to imagine 'bonds they cannot see'.

In using this approach, even though students may still use non-scientific ideas such as the 'parts' having properties similar to macroscopic entities (e.g., shape), it is argued that it lays down more appropriate stepping-stone ideas that lead towards a scientific particulate understanding. This is because it relates to:

1. a specialised scientific understanding of matter
2. the students' physical manipulations of materials
3. the sociocultural construction of scientific ideas.

Detailed examples of students' diagrams and their comments on their mental model building appear in Acher, Arca and Sanmarti (2007). This approach is a radically different use of classroom practical work in science. It is used not to verify something, but rather the practical work is 'an attempt to use models generated in the classroom and to adapt ways of thinking, ideas and expressions to the formalisation of knowledge itself' (ibid., p. 414).

Activity 8.9 Students generating models

In what ways does the above description relate to 'learning science', 'doing science' and 'learning about science' as discussed in Activity 1.3 (see p. 6)? Try to enumerate how this approach could exemplify various attributes of the NOS.

What would you find challenging about this use of practical activities in your science teaching? How could you prepare for such challenges?

If possible access Acher, Arca and Sanmarti's (2007) report and compare your thoughts with their arguments. Reflect on your role as a teacher when formatively assessing students' development of mental models. How has reading about research in this area influenced your response?

Focusing on more than perceptual responses

To assist children's thinking in moving from a continuous towards a particulate view of matter – possibly through intermediate models which are an amalgam of these two views – some evidence suggests that teachers need to focus not only on students' perceptual responses (e.g., colour, taste, physical state) but also on their particulate-type responses. When there is only a perceptual focus (e.g., 'the aspirin turned from a solid into a liquid' when dissolved in water), students tend to retain a continuous view of matter. However, if teachers focused on students' contributions which suggested changes in

- 'arrangement' – 'the particles of the water and the salt are united and they became one particle'
- 'making' – '[when] the aspirin was broken down ... by the water into particles, it becomes tiny pieces that we cannot see at all'
- 'location' – 'the invisible tiny pieces of the blue stone (copper sulphate) became spread all over the water ... now these pieces circulate in the water',

then students may start to move towards a scientific particulate view (Hatzinikita, Koulaidis and Hatzinikita 2005).

This language focus is similar to the directed scaffolding suggested in Acher, Arca and Sanmarti's (2007) approach to students generating their own models. They emphasise how carefully we have to listen to students' explanations in order to encourage a modification of ideas.

Focusing on volume, mass and density of materials

Interestingly, it has been found that when teachers have upper primary students (i) construct visual models with distinct yet integrated representations of volume, weight and density of materials, and (ii) quantify and measure weight and volume in order to differentiate them from perceptual impressions, while (iii) also engaging them in reasoning about the properties of matter that all materials share, as well as (iv) thought experiments such as continually dividing materials until they are vanishingly small, then more students form abstract ideas that lay the basis for a particulate understanding of matter. These abstract ideas include matter having mass and taking up space, and gases being material (Wiser and Smith 2008). Teachers, therefore, should not feel hesitant in exploring these prerequisite notions with materials as a basis for later particulate thinking.

Teacher-talk

All of the above strategies suggest the importance of teacher-talk in facilitating changes in students' conceptual development. Teacher-talk (not teacher lecturing) has been undervalued. Three types of teacher-talk, which can be used in a variety of contexts, are:

1. tuning-talk – tuning students into a task
2. connecting-talk – helping students make connections; for example, patterns, causes and effects
3. monitoring-talk, which requires students 'to express their understanding so that you can appraise it' (Newton 2002, p. 26; many examples are provided in Part 2).

Your role as a scaffolder of student ideas is partially dependent on your ability to use talk in appropriate ways. You could ask yourself what tuning-talk you would plan to get students interested in plastics, what connecting-talk might be beneficial to help distinguish between objects and materials, and what monitoring-talk you could plan to formatively assess an understanding of solids and liquids.

Activity 8.10 **Intervention and the teacher's role**

Think back over all of the above intervention strategies. What do you think is the appropriate role for a teacher who has a broader knowledge than their students? Is it better to just let discussion flow? Do you act by challenging ideas or simply make suggestions for further discussion? When, at all, do you present the science view? Are there any 'problems' which could arise if the teacher decides to explain science ideas?

If you have already worked through several chapters in this text, try to express your views about what scaffolding student learning means. Provide scaffolding examples related to students' learning about materials (see Harlen 2001b; Newton 2002; also see the online Appendix 1.5).

Appendix 1.5
Scaffolding strategies

Change strategies and materials: a summary

Learners probably have difficulty accommodating some scientific ideas (such as about matter) because they have a high 'learning demand' (Scott, Asoko and Leach 2007). This refers to differences between everyday and scientific ways of knowing about our world. These include, firstly, that scientific 'entities' are rarely used with scientific meanings in everyday language ('atoms' is an example); secondly, scientific explanations are often counterintuitive (atoms have properties which are uniquely their own and do not resemble the macroscopic properties of the materials they are 'explaining'); and thirdly, scientific explanations attempt to generalise across contexts, as is the case with the particulate model of matter. All three features contrast sharply with everyday ways of knowing. Another example would be classifying some materials as solids: this is because the everyday criteria of hard and rigid objects are quite different from the scientific descriptions of solids. Appreciating the concept of learning demand may assist you in empathising with your students and encourage you to be patient with them as you try to facilitate a reduction in the gap between everyday and scientific ways of knowing.

All of the above strategies encourage learners to be active thinkers and would assist in reducing learning demand. They should not be used in isolation but within constructivist lesson sequences (see Chapter 1 and the following case studies). Although learning about some aspects of materials is not straightforward, conceptual development and change is possible with well thought-out approaches.

Constructivist models and case studies

Two case studies are presented. The first, 'Reconstructing ideas about the shape of materials', relates to whether the strength of an object depends on its structure, while the second, 'A comparing investigation', explores the absorbency of fabrics. Each illustrates a schema for sequencing lessons. Following the second case study, some other constructivist planning approaches are mentioned and the importance of sequencing having a conceptual focus is stressed.

The case study 'Reconstructing ideas about the shape of materials' is based on the interlocking steps proposed by the Children's Learning in Science Project (CLISP) (see the Chapter 1 section 'Constructivist teaching models or schemata' on p. 22):

- *eliciting* students' ideas about a concept/idea
- providing experiences that allow students to *explore* the idea/concept and then further experiences that encourage *elaboration and restructuring* of the idea/concept
- opportunities to *apply* the concept/idea are then given
- followed by a *review* of the change in the concept/idea (if any) that is related to the original views held.

The case study 'A comparing investigation' describes how a teacher guided a scientific investigation into the absorbency of fabrics. Scientific investigations (see Chapter 2) are often a phase in constructivist lesson schemata. They can, for example, arise as a consequence of students' questions or could be introduced by teachers as part of, say, the elaboration and restructuring of an idea. This case study could be seen as a sequence within a sequence. It is included because, when studying materials, there are numerous opportunities for students to investigate their various properties.

CASE STUDY

Reconstructing ideas about the shape of materials

Part of learning about materials and their uses could include activities in which students physically change materials in order to use them in various ways; for example, cutting wood for building (ACARA 2011). The following sequence exemplifies how a constructivist conceptual-change teaching model was used to look at the question

'Which shape is strongest?'. The Year 3 students involved in this lesson changed the shape of particular materials to arrive at their solution to this question.

Which shape (of a particular material) is the strongest?

A Year 3 class had been reading *Charlotte's Web*. They were attempting to draw a web for a related activity when the teacher overheard a conversation in which a student commented, 'The hexagon is the strongest so you have to draw it first'. From here, a lesson sequence developed around the above question.

In the first lesson, the teacher asked the class to remember buildings, bridges and other structures in the environment and to try to decide on which shape is used to support them. The class was told that, if they could decide on the supporting shape, then that would probably answer their question of which shape is the strongest. The students were encouraged to discuss their ideas; they did this eagerly and listened to what each other said. They wanted to know the teacher's view, but she resisted the temptation. They then had to write down their thoughts, including their reasons. Several of the students' responses appeared to reflect their school surroundings, especially the buildings. (Original spelling retained in several of the responses.) Those students (eight out of 26) who thought a rectangle was the strongest, wrote reasons such as 'It is the biggest shape', 'It has four corners, four edges, six faces and the middle part is hard' and 'I think the retangol is the strongest because the brick is a retangol and it can hold the beldings'. Those who thought triangles were the strongest (three) included the reason 'It has three sides and it is big' (see Figure 8.2), while several thought it was the semicircle (seven). This last group wrote the comment, 'I believe that the semicircle is the strongest because in the harbour bridge there is a big semicircle and it is a good thing that it has a semicircle because it will hold it in position' (the students live in Sydney). One student thought a sphere was the strongest 'because if you put a sphere on the ground and throw a different shape on it, it will bump off it'. The teacher asked the student to show her exactly what was meant by this statement and a drawing showing objects deflecting off spheres was completed (see Figure 8.3). Five students thought the hexagon was strongest (from the spider study), with one saying, 'It has lots of room. It has six sides', and two nominated a rhombus because 'it looks like a diamond and a diamond is strong and a diamond is a jewel. And that is why I chose rhombus'. The overall class views were represented in a table and graph, which were left permanently displayed in the classroom.

To conclude the lesson, the teacher asked, 'How are we going to find out?' One student suggested the use of Polydron, so a (triangular) pyramid and a cube were constructed immediately. When two of the class attempted to crush them, the cube collapsed first: the class said that pyramids were the strongest. When the teacher brought it to their attention that they hadn't tested the other shapes, this idea was put aside. The class also concluded that Polydron was impractical because of the limitations in the shapes that can be used. At this point, the teacher suggested that the class should be looking at the (2-D) shapes used to construct (3-D) objects instead of the whole shape.

Activity 8.11 Reflections on the introductory lesson

- What do you think the teacher's main purpose(s) may have been in this lesson?
- What techniques were used to achieve the purpose(s)? Are you able to suggest the value(s) of each technique in relation to the overall purpose(s)?
- Are there any comments you would like to make about the nature of the students' ideas? What extra information, if any, do the students' drawings indicate?
- What do you think the students achieved in this lesson?
- What role(s) did the teacher play in this lesson? Why?
- Were there some adjustments you believe the teacher might have to make in the next lesson? What would you do from here?

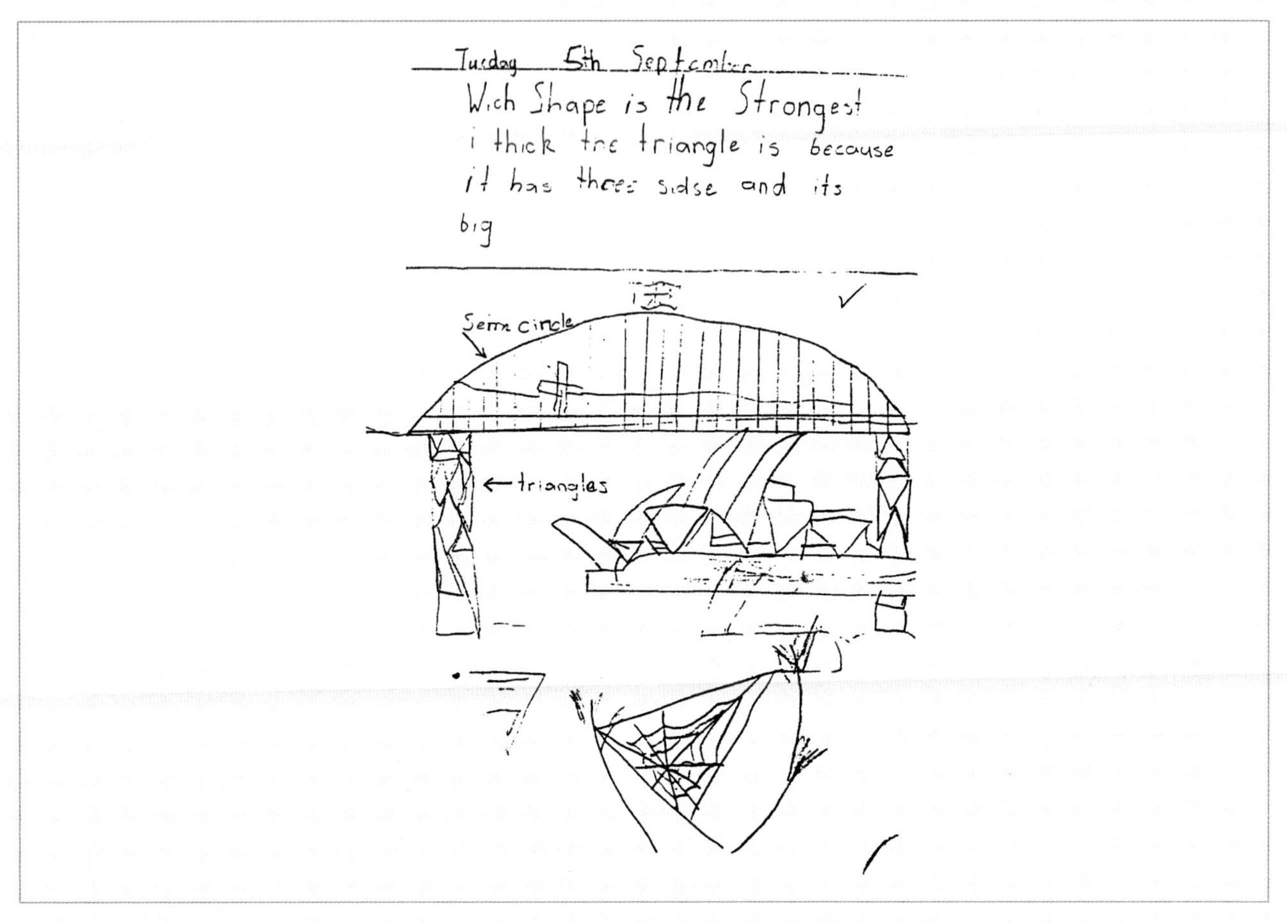

FIGURE 8.2 Children's preconceptions about which shape is the strongest

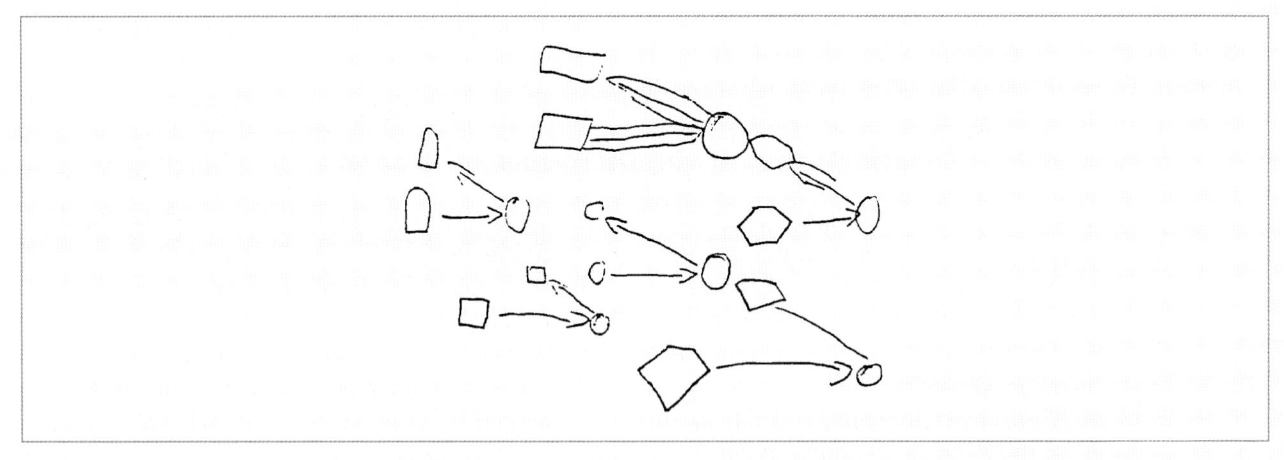

FIGURE 8.3 A child's drawing illustrating why spheres are a strong shape

The second lesson involved the students in testing the strength of various shapes. In the initial class discussion, size was identified as an important factor in determining strength and so size was limited, with each shape being approximately the same in size. In groups of four or five, the students folded a piece of A4 paper in half and then sculpted their shape to be tested. They then stacked dictionaries of equal size and mass on top of the shape until it collapsed. The teacher moved around the room asking questions about how they were setting up their investigations and also challenging their ideas. Using Tunnicliffe's (1988) shape format (see Figure 8.4 for an example), group reports were written and results were reported back to the class. They were recorded for comparison with the earlier views. At this point, only three shapes were thought to be the strongest: cylinder (17), triangle (eight) and rectangle/square/cube (one). The student holding the last-mentioned view wrote her group's report as follows: 'We found out that both shapes we made actually did work. One was a cylinder and a square, but the triangle was the strongest'. The teacher had not expected the introduction of a new shape. Another table was prepared. It listed all the shapes and the number of students who believed the shape was the strongest.

In the third lesson, the teacher encouraged the students to focus on the 2-D shapes left after taking cross-sections of the three remaining 3-D shapes. Asked what shape could be seen after cutting the cube in half, the class agreed it was a square; a similar process was followed for the cylinder and the triangular prism. The students were then given equal lengths of straws and asked to construct the three 2-D shapes. When the task was completed, the students applied pressure on top of the shape and were asked to describe what happened. All of the students found that when pressure was applied to the top of the circle, it was squashed, the square bent sideways and the triangle did not bend. The student who held the view that the square was strongest was asked why the square bent. He said, 'The square needs something to hold it together'. The teacher asked what would happen then and the student replied, 'Then there would be two triangles'. He wrote in conjunction with this lesson: 'The triangle is stronger than the square because the square bends and the triangle doesn't. If you put two triangles together it would make a square and won't bend'. The teacher believed that, at this stage, most of the students had reached the following conclusion, which was expressed in one student's written work, showing diagrams of a square and a triangle with lines showing 'pressure' along the sides: 'I think the triangle is [the] strongest shape because the square bends and the triangle does not'.

Activity 8.12 Reflections on the middle lessons

- What do you believe was the teacher's main purpose(s) in these last two lessons? Think about what she found out in the first lesson. In what way(s) did she respond to the students' ideas? Keep in mind what you believe her overall purpose for the sequence was, and what you know about students' alternative conceptions.
- Compare the teacher's role in these lessons to the first.
- What features of students' ideas have you noticed over the three lessons? How did the teacher consider them?
- Would you have done anything different in the second and third lessons? Why?
- If possible, read about Tunnicliffe's (1988) scheme for assisting students to record their science investigations. What do you believe are the merits of this approach?

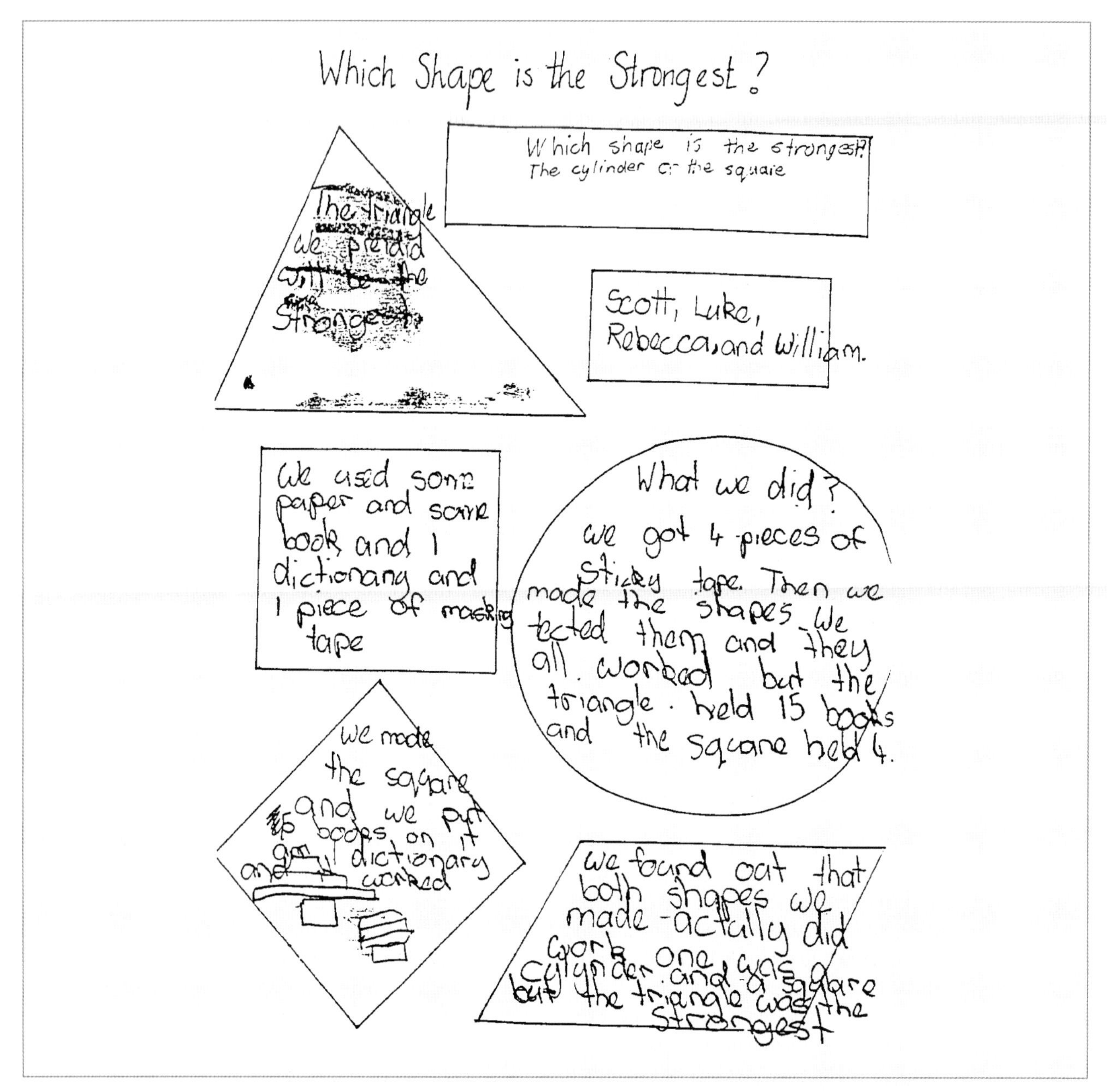

FIGURE 8.4 A student's science report exemplifying Tunnicliffe's (1988) shapes

Although the original intention in the fourth lesson was to allow the students to design and construct a bridge or a tower using their new knowledge, the approaching end of term required the teacher to change plans. Instead, she had the class research in the library to find pictures of structures that clearly showed the strength of the triangle. Several excellent pictures were found of churches, high-rise buildings and other structures that satisfied these requirements. One student said, on viewing the pictures, 'I think the triangle is the strongest because it holds up buildings so they don't collapse'. In response to the question 'Which shape is the strongest?', another wrote, 'I think the triangle is the strongest shape because it can't bend and it's used in most bottoms of buildings and in the

Eiffel Tower and the Harbour Bridge, Opera House, there is heaps of triangles'.

Other responses were 'The boat uses triangles to hold it up sort of like a spider's web is' and 'Look inside the circle is the triangles for support'. The teacher thought the students were using scientific terminology well.

In the final lesson, the students assembled on the floor as an entire group in front of the two tables showing the previously held views. The teacher discussed the charts with the class, had individual students state what they had learnt and asked them to compare what they knew now with what they initially thought; further, they were encouraged to say why they had changed their ideas.

Activity 8.13 **Reflections on the concluding lessons**

- What do you believe were the teacher's main purposes in these last two lessons?
- Contrast the teacher's role in these lessons to the others.
- The fourth lesson could have been expanded into a technological design–make–appraise process (see Chapters 1, 3 and 6). What steps would you have used? Compare them with the steps for this process in Fleer and Jane (2004).

Many other issues concerning this lesson sequence could be productively discussed. What is the implied interpretation of 'strongest'? How did the definition of 'strength' surface? Could it have been clarified further? Were the students really determining the rigidity rather than the strength of the structures? Would noting these differences have made a difference to how the lesson sequence may have progressed (for example, how triangles are used within building structures)?

The question could also be asked, 'What is (are) the conceptual idea(s) underpinning this lesson sequence?' Is the emphasis here more on the uses of objects because of their extrinsic properties – shape and the shape within structures – or the uses of materials because of their intrinsic properties, or some other idea? Also, does this lesson sequence have a scientific and/or a technological focus? This comes back to your understanding of the nature of science and technology, which cannot be expanded on here. Refer back to the nature of science and technology section of Chapter 1; also consult Fleer and Jane (2004) to think more deeply about the nature of technology. This should help you to form an opinion.

Irrespective of these issues, the questions posed in Activities 8.11 to 8.13 relate to a constructivist planning framework the teacher used. You may find it useful.

The constructivist model underpinning this case study

The conceptual focus, from the teacher's perspective, appeared to be that the shape (and rigidity?) of an object affects its strength. What do you think? This teacher based her planning on the CLISP model (see the stages or steps mentioned in the introduction to this case study). Did the purposes you suggested for the various lessons match with the conceptual change stages or steps? Various roles for teachers and students have been proposed during these stages; these are fully outlined in Scott, Dyston and Gater (1987) and, if available, you should compare your responses in the above activities.

Appendix 8.3
A constructivist lesson sequence plan about 'Matter is everything that takes up space' (middle primary)

Activity 8.14 Steps in a constructivist teaching model

What appeals to you about the CLISP approach to planning? Are there any disadvantages?

Choose an idea related to the 'Materials' section of your science syllabus or curriculum framework and develop an outline plan that uses the steps the teacher used in this case study; if possible, try it on a practicum or with your own class and reflect upon the purpose and value of each step. For another case study using this model, see Skamp (1995), which describes Year 1 students restructuring their ideas about evaporation. The online Appendix 8.3 is a student teacher's plan using these steps. The sequence revolved around the concept 'matter takes up space'.

Investigating the properties of materials

When students are studying 'Materials and their uses', opportunities could readily arise for fair-test investigations of the properties of objects and materials, such as their solubility, hardness, conductivity, shininess, strength, stretchiness (tensile strength), malleability, expansion, magnetism, bendability and porosity; for other examples, see ACARA (2011), the *Primary Connections* units in the 'Natural and Processed Materials' strand listed in Activity 8.1 (see p. 300), and the *Science 5/13* (Schools Council 1972, 1973; see the *Children and Plastics* example in this chapter's 'References' section) and the *Teaching Primary Science* (Diamond 1978) series, which include a wealth of suggestions. Although these latter two resources are dated, they are available in many schools. Teachers have reported examples relating to the properties of recycled paper (Nixon 2001); the strength of card (Ruane 1989); the properties of wool (Pattie 1991), fibres (Groves 1986) and fabrics (Griffin and Sharp 1995a); and the best adhesives (Bradley 1994) and sticky tape (Griffin and Sharp 1995b). Such 'tests' can also be completed with different brands of materials or objects. Zeegers (1995) illustrates this with the question 'Is brand X better than brand Z?' using Nike and Reebok shoes to exemplify the fair-testing procedure.

There are also sequencing guidelines that can be used when there is a scientific investigation component within a constructivist lesson sequence. The investigation component may also encompass several lessons. This will especially be the case when primary students are unfamiliar with fair-test investigations and/or the nature of the phenomenon/material(s) being investigated. Before analysing the following case study, you are encouraged to experience the investigation of a material's property (here, the absorbency of fabric) yourself. This will help you to empathise with what you will be asking of your students.

Activity 8.15 Completing your own 'comparing investigation'

Obtain five different types of fabric (e.g., cotton, nylon). Plan, carry out and report on an investigation of the fabrics in order to answer the question 'Which fabrics are best for keeping us dry?'.

On completion, and after discussion with colleagues and/or peers, determine if your approach included general planning (for example, it identified the types of variables in the investigation, such that it was a fair test and considered how a result could be found) as well as specific planning (it decided the number and values of the independent variable to be used and the values of the variables that were kept the same) (see Harlen 2003).

Harlen (1985) calls investigations such as the above 'comparing investigations' and argues that they are suitable for all levels in the primary school (completed with varying degrees of guidance for six-year-olds and up). They all have the same starting point, the question, 'Which X does Y best?'. The following case study illustrates how 30 seven- to eight-year-olds of mixed ability tested the absorbency of a range of fabrics. It is important to note that the teacher did have a general overall plan in mind, which will become evident as you work through the case study, but she did not plan the details of each lesson until the previous one had been completed.

CASE STUDY

A comparing investigation: The absorbency of fabrics

In a brief first lesson, the teacher aimed to set up a problem for investigation. She had planned to spill some water, then seek some suggestions on what to use to mop up the mess. This actually happened earlier than anticipated when water was spilt when a bunch of flowers was being put on the teacher's desk. 'Quick! Get the big red towel!' one student said. The teacher suggested something smaller, and two other students said, 'I've got a tea towel in my bag' and 'Use the rag we keep the [glue] brushes on to dry'. One student got the rag and let it soak up the water. At this stage the teacher invited suggestions for other things that could have been used to mop up the water. Examples were nappies, a mop, a tea towel ('because it dries up dishes'), a kitchen sponge ('because it soaks up into the holes and it's scrunchy'), a towel ('because it dries people'), tissues ('because they soak up things faster') and paper towels ('because they've got little spaces in them to pick up the water'). These were spontaneous reactions, with all the students showing a high level of interest. The teacher now believed she had the basis for an investigation.

In the second lesson, the teacher listed the materials from the first lesson on the board and discussed them with the students, who were still very interested. At this point the teacher produced squares of fabrics: towelling (a washer), flannelette (a blue rag used in the first lesson), polyester/gaberdine, polyester/lawn, organza, taffeta, a synthetic sponge and paper towels. The students were encouraged to feel the materials and discuss them before the teacher told the class they were going to try to find out which fabrics soaked up the water well and which didn't. (The teacher had in mind the establishment of the need for a fair test.) One student said, 'Yes, we'll find out which things that can dry things good and things that can't dry things good'. The students were then asked to choose about five fabrics that would have been good for drying up the water spilt previously. They ruled out the mop as that wouldn't be practical to use on a table, the organza as 'the water would go through the holes and your hands would get wet', and the paper towel 'because it would fall to pieces when it got wet'. The class decided not to include the sponge because they knew from its wide use in their homes that it was a good absorber. Two students still wanted to include a tea towel on the list; they found one made of linen in the lost property box. Interest was aroused when it was explained that there are also cotton tea towels, so a cotton one was included as well. After further discussion and elimination, the following new list was recorded on the chalkboard:

- cotton tea towel
- linen tea towel
- washer (cotton towelling)
- blue rag (flannelette)
- yellow cloth (polyester/lawn)
- brown cloth (polyester/gaberdine).

Each student wrote their predictions for the most absorbent material. The teacher analysed these and found the following preferences:

- towelling (16)
- blue rag (four)
- linen towel (two)
- brown cloth (two).

The least absorbent were thought to be the:

- yellow cloth (13)
- brown cloth (10)
- cotton tea towel (four)
- linen tea towel (one).

Some said the yellow cloth because it was the thinnest. The class decided to plan the investigation in the next lesson.

Activity 8.16 Reflections on introductory lessons

What do you believe were the teacher's purposes in these two lessons? Think about the problem that eventuated and how the teacher kept coming back to it. Can you start to discern a framework for carrying out a scientific investigation which the teacher may have had in mind, and which she is gradually translating into teaching steps?

How did she achieve them? Try to identify the specific steps she took and indicate what you think each of these smaller steps achieved.

After a brief review of what it had done so far, the class formed groups of six and a pile of fabrics was placed in the middle of each group. The following questions were written on the board and the students were asked to focus on them while feeling the fabrics and discussing them:

1. What do we want to find out?
2. How could we find out?
3. What will we need to do it?

The teacher moved from group to group observing and listening, although group work skills still needed refinement (for example, one group spent a lot of time arguing who would be 'boss'). Some of the comments overheard were:

- 'That's good for polishing so it could be good for wiping up water, too [student was rubbing cloth on carpet].'
- 'You're only allowed the same amount of water each time because if you do different lots it will be different each time.'
- 'Yeah, 'cause you have to have the same amount, like 5 mL or 7 mL each time.'
- 'You could get a bucket of some kind.'
- 'You need the same amount each time or it won't do the real thing it should.'
- 'The surface feels kind of funny for drying [feeling the yellow cloth].'
- 'You guys, we're not talking about which is best. We're talking about *how we will do it!*'

Each group chose a scribe to write down ideas. Some of these ideas are shown in Figure 8.5. They were reported back to the class and discussed. Group 1's suggestion was modified to spilling water into a container. A list of equipment needed for the next lesson was made.

In the fourth lesson, a large number of assorted buckets, containers and jugs were made available, as were samples of the six fabrics to be tested. Also, there were textas, paper and rulers. In their previously formed groups, the students decided what equipment they would need and worked happily and busily pouring water into containers and putting cloths in and pulling them out. Their attention was caught by the way the polyester fabrics floated on water and hardly absorbed any. There was much discussion but there seemed to be no real system. The class then reassembled to discuss their findings. At this time a student expanded on a lesson 3 comment about marking the containers to show the water level. He said that a container should be marked at the water level, the cloth then dipped in, then a mark made at the new level and the space between the two marks measured with a ruler. Another student suggested that the cloth be held in the water for 'certain minutes' and another said they would need cloths all the same size to do it properly.

It was consequently decided to try the experiment the next day using two different methods. Firstly, mark the container before and after soaking uniformly sized strips for a set time and compare the results. Secondly, hold uniformly sized fabric strips in a container of water for a set time, then measure the 'dark bit' (wet area) on each cloth. A list of resources was also written.

FIGURE 8.5 Children's initial plans for testing absorbency

Group 1	We are going to spil a bit of water on the desk. Then get the clothe and see which clothe socse up the best.
Group 2	What will we use? (With picture) basin, ice cream bucket, tub How will we find out? spill the water in tub and wipe Spill the cup of water six times in the tub. But first get a ruler to measure the water the same amount each time. Then put a mark there – where you measured water stop where you want it to. Then try out putting the different cloths and metrels in.
Group 3	What will we do? Get a tub and put a little bit of water in it. Then put some of the washer in it the water. Then see if the water will come threo.
Group 4	We think that we shoud make a mark a line on the contana and put some water in it and tip some onto the desk and see which one works the best out of all of them.
Group 5	What will we use? We will put some whater ina cktaner then we wil wip it up witha washa the we will use anater oen.

Source: (Original spelling retained)

Activity 8.17 Cooperative group learning (as in the *Primary Connections* project)

Working in groups is a common strategy in science learning. The *Primary Connections* project (AAS 2005) strongly supports the use of cooperative learning strategies in science classes. Read the introduction to one of the teachers' guides (or another appropriate cooperative group-learning reference) and identify some of the key aspects of cooperative learning that you could try in your classroom.

Locate one of the cooperative learning activities in the teachers' guide for a chosen level and determine how it encourages cooperative learning. If possible, try one of these activities and reflect on its value for enhancing group work. You need to remember that cooperative learning behaviours take a considerable amount of time to develop.

Activity 8.18 Reflecting on the middle lessons

- Distinguish between the teacher's purposes in lessons 3 and 4. What role(s) did the teacher play in each lesson?
- What insights and inferences would you draw from the comments the teacher overheard in lesson 3? How would they help you as a teacher?
- Analyse the group reports in lesson 3. What do they tell you about this class's approach to planning an investigation? Chapter 2 will assist your interpretations.
- The students seemed to make a large mental jump in the discussion at the end of lesson 4. Hypothesise on possible reasons for this. What message(s) do you derive from this as a teacher?

Think about the role of whole-class discussion *after* different groups of students have attempted to solve a problem. Record what sort of issues emerged.

In the fifth lesson, both procedures were tried (partly because the students wanted to try both ways, but also because of a shortage of equipment). Equipment was initially distributed. Two groups attempted the first investigation, which involved filling six clear containers and marking the water level. Each student in the group, one at a time, then lowered his or her fabric strip into a container and held it there while counting to five. When the strip was removed, the new water level was marked. When all members of the group had done this, a ruler was used to measure the difference. A difficulty was that the differences were so slight that they had to be measured in millimetres. The teacher had to direct the measure-and-record step after providing a mini lesson, as the students had no formal experience in measuring small quantities.

The other three groups tried the second procedure. Each group member lowered a different fabric strip into the water in a common container and counted to five while holding it there. They then laid the strip on the desk to measure the length of the wet section, which was easily discernible. Again, however, the teacher had to help because of similar measuring difficulties.

The class then came together to discuss their results, which had been recorded in a two-column table. The results of the first procedure compared favourably, but not so the second, making conclusions difficult to draw. When the class was asked why they thought there were so many different answers, most said it was because they counted to five at different paces and so it was unfair (and some gave examples). It was suggested that the teacher could count, but this was ruled out as she might not always count at the same rate. Eventually, a student suggested a clock.

Discussion of the small variations in the levels was considered but the class was unable to suggest refinements and so the teacher suggested a narrower container. Some students were sceptical about this (maybe they were non-conservers in the Piagetian sense) and so the teacher demonstrated.

The final lesson involved providing the class with the opportunity to apply a fair test using a more structured format. Six students each filled a clear, narrow measuring cylinder to the 200 mL mark. A brief description was given of how liquids can be measured and attention drawn to the markings on the measuring cylinder. Others watched the clock and one student called when to start and finish. Each participant held a fabric strip in the water for 30 seconds, then read out the new water level with teacher assistance. The readings were recorded and two more mathematically able students helped the teacher to calculate and record the differences between the readings (in a table and then enlarged on the board). The results were discussed and the class could clearly see that towelling was the most absorbent (40 mL), while the polyester was the least (2 mL). The fabrics were numbered in order of absorbency and then the class made individual copies of the results sheet and were invited to make comments on the whole investigation (see Figure 8.6).

The teacher commented that some students had done experiments at home. For example, one boy and his mother tested old and new tea towels for absorbency and found that old ones absorbed water better. After the lesson sequence, another student brought a strip of fabric to school and said, 'My mum says this is the most absorbent fabric there is'. The class decided to test it and, with high excitement, observed that for fabric of the same area (as the others), it absorbed 75 mL of water; a separate class report was prepared on this fabric. The class also sent a letter to the boy's mum asking questions about the fabric (it was an artificial chamois). The class went on to investigate manufactured and natural fibres.

FIGURE 8.6 Children's responses at the end of the absorbency lesson sequence

- The excspriment was good. It helptd us to lene abowt absotbening.
- The expeament wasn't fair when we first did it but when we did it agein it we fair and I agre.
- I think it was good fun and a bit unfere becaus some people were counting slower than others. But I think it was still fun.
- The experment was great. Because sience is my greatest intrest and some things were intresting but some of the things were very hard to figer out and Im going to be a sientiste when I grow up.
- The excpriment was great fun I injoed it a lot. Because it helped me to learn wich fabric or twopel was the most absorbent.

Source: (Original spelling retained)

Activity 8.19 Reflections on the concluding lessons

- Suggest the teacher's purposes in these final lessons.
- What role(s) did she play? Could she have made alternative decisions about role(s)?
- What conclusions would you draw from the student comments at the end of the sequence?
- Looking back over the sequence, what learning outcomes do you think the teacher may have had in mind? Did they include NOS outcomes? If so, how; if not what could have been done? If necessary, look at your science syllabus for further guidance. Compare with Appendix 8.4 (see p. 347), which includes the teacher's planned learning outcomes.
- Was there a clear conceptual focus underpinning this sequence? If yes, what was it? If not, what are some actions the teacher could have taken to introduce such a focus? Can you provide some examples?

The case study 'A comparing investigation' uses Harlen and Jelly's (1989) suggestions for sequencing comparing investigations. It clearly indicates that such investigations would rarely occur in a single lesson and that there are steps that teachers could use to guide their planning and implementation. The teacher in this example labelled her steps as:

- setting up the problem
- refocusing the problem
- a workable procedure
- preliminary investigation
- the investigation
- the fair test.

You should be able to relate each of these steps to the above lesson sequence and write a statement that describes the purpose of each step.

Activity 8.19 indicated that, although the emphasis in this sequence was on ways to investigate the properties of fabrics, the teacher still needs to ensure that there is a conceptual focus for the investigation (see Figure 2.1 on p. 65, which emphasises that the concepts involved in an investigation are integral to understanding and using the processes involved). There needs to be 'a conceptual question', 'hypotheses' or 'competing knowledge claims' involved for fair-test investigations to more fully represent what scientific inquiry is about (Tytler and Peterson 2003). Such investigations will then have the potential to advance students' conceptual understanding. In the above comparing sequence, the conceptual focus could have been the weave of the cloth, its thickness or the materials of which it is made and how one or more of these relates to the fabric's absorbency. This principle – that is, ensuring fair tests have a conceptual focus – needs to be applied to any investigation in which you involve students.

Other constructivist models and principles for sequencing lessons

The remaining discussion in this chapter's third section refers to two other examples of how constructivist planning schemata can be applied to teaching and learning about materials. Firstly, the 5E (learning cycle) approach (see Chapters 1 and 3) is illustrated. The second example exemplifies how you can adapt constructivist models for lesson sequencing to suit your own purposes.

The 5E approach

The **e**ngage, **e**xplore, **e**xplain, **e**laborate (apply) and **e**valuate, or 5E, approach is similar to other constructivist planning schemata, except that the 'explain' component is more teacher-directed than in some other approaches (for example, compare with the interactive teaching approach; see Chapters 1 and 4). *Primary Connections* (AAS 2005) is a major curriculum initiative based on the 5Es and is in regular use (Hackling and Prain 2005). It has several units related to materials and objects. Activity 8.20 requires you to analyse the 5E structure of one of these units. If you do not have access to this resource, use another commercial resource that claims it is based on 'the learning cycle' (3E or 5E).

Activity 8.20 *The 5E model in the* Primary Connections *project*

'What's It Made of?' in the *Primary Connections* project (AAS 2008b) has six suggested lessons:

1. go for a school walk, identifying things in the environment and developing a picture map of the school environment
2. help students to identify 'objects' from their school walk and have them discuss what they are made of, how they feel, what other objects are made of the same material and so on; revisit a particular object and complete a worksheet
3. use a 'Feely box' game which requires students to describe various objects, including what they think it is made of (e.g., plastic); have them discuss the uses of particular materials in various objects (e.g., scissors are made of plastic and metal)
4. label various classroom objects with the material(s) they are made from and have the students describe the materials' features as well as seek opinions as to why particular materials were used; have the students make up 'silly stories' that refer to the properties of materials that comprise objects
5. link school/classroom materials to the question 'What happens to different materials when they get wet?' Guide students in carrying out a fair test of different materials; have students make an object for the school environment that is water-resistant
6. ask students to review their learning about objects and materials and how they learned about them.

Read through the actual lesson sequence and discuss with a colleague how the lessons relate to the 5E model by clearly indicating the purpose of each of the five steps and how that purpose is achieved in one or more of the lesson activities. Brief descriptions of the 5Es are in the introductions of all of the *Primary Connections* units (also see Chapter 3).

How has the sequence taken into account research about students' conceptual development of the nature of objects and materials? Being aware of children's thinking about objects and materials (see earlier in this chapter) should influence your reactions to students as you formatively assess them. Can you suggest in what ways?

Indicate how the lesson sequence focuses on generic and science-specific literacy skills; locate examples.

Was learning about the NOS explicit in the sequence? If so, how? If not, how could you include it? Can you relate your thoughts about these NOS considerations to learning from a sociocultural perspective?

Similar questions to those in Activity 8.20 could be asked about the *Primary Connections*' units 'Material World' (AAS 2008a) and 'Package It Better' (AAS 2008c), which are aimed at older students. Lloyd (2007), having taught three *Primary Connections* units, provides insightful comments about her implementation of a draft version of 'Material World' (years 3 to 5) which relates to the properties of materials. Lloyd clearly advocates the value of various attributes of this AAS project, such as the 'word wall' and how the approach assisted students in applying their knowledge of the properties of materials to 'new tasks' examples are provided. This teacher now devises her own 5E sequences.

Teachers who appreciate the reasons for the steps within particular constructivist models for sequencing lessons usually adapt such models (or combinations of steps from them) to create their own constructivist sequences; see Lloyd (2007). Sanderson (2000) also did this when she taught her Year 3 class about the properties of a particular material; namely, detergents. This involved seven lessons and used bubbles as the motivating context. The children were learning how the properties of an object (bubbles and their size) were dependent upon the properties of the materials (here, the concentration and composition of detergent solutions) from which it was made. Sanderson's sequence integrated science and technology tasks: a scientific fair test and technological design activities.

NOS

Initially, students' ideas about bubbles were elicited by asking students to write, draw and label everything they knew about bubbles. Their writing was analysed. Students wrote mainly about the perceptual properties of bubbles (what they could see) and a little about how they move (locational properties). Compositional or functional properties were rare, although one child wrote about the uses of bubbles in fish tanks. Making bubbles with a range of wands was the next activity (an exploration task), which also prompted students to say which was the best bubble wand that they had made. Student questions about bubbles were then collected in the third lesson and some of them were researched in the students' computer time. Several of the students' questions were investigated by testing which concentration of detergent produced the biggest bubbles (a technique for measuring bubble size was shown). This was followed by another investigation, which tested different detergent solutions (with various concentrations of sugar water or glycerine). The penultimate activity had the class making giant bubbles (details from a website) and describing bubbles from inside and out. (A design activity was interspersed with the investigations: students had to design the best bubble wand and justify their choice of materials.) The sequence concluded with a reflective lesson in which the students discussed what they had learnt about bubbles and again were asked to write what they knew. Among other learning outcomes, Sanderson reported that more students referred to the composition of bubbles, an aspect of materials that primary students often overlook. What elements of constructivist models for lesson sequencing are within this unit's description? Can you justify the teacher's activity, strategy and sequence choices from a constructivist perspective? Would you have done anything differently? Why?

Appendix 8.3
A constructivist lesson sequence plan about 'Matter is everything that takes up space' (middle primary)

In the above lesson sequence, Sanderson (2000) focused, in part, on a conceptual understanding of materials; namely, a greater awareness of the material composition of objects (here, bubbles; e.g., composed of detergents and glycerine) and how that affected the size of the objects (bubbles). In the online Appendix 8.3 there is another example in which the teacher focused on the concept 'matter takes up space'. The identification of the conceptual focus of a lesson sequence is critical from a constructivist perspective as we are trying to help students reconstruct their ideas about how their world works. Teachers need to be conscious of the conceptual direction in which they are taking their students, and Lloyd (2007) stressed that this was an outcome of her implementation of several *Primary Connections* units. You are urged to focus on the conceptual outcome(s) students are learning about rather than just what they do. If you do not. your interventions may be unguided and students may not appreciate the aims of their work and, hence, they will not be able to direct their own learning (Harlen 2001a, p. 63). If you have taught some science lessons recently, reflect on them. Can you identify the science conceptual learning that was the focus of the lesson or sequence? If yes, how clear was it in your mind? If not, what could the conceptual learning have been? What differences might it have made to the effectiveness of your lesson(s)?

Science as a human endeavour

The characteristics of science as a human endeavour include primary students being aware of the links between science and culture (e.g., how science ideas have been developed across cultures), the NOS (e.g., the importance of modelling in science) and the influence of science (the effect of science and technology on our lives, as well as how science is used). Each of these can be readily incorporated in lessons focusing on materials.

Science and culture / nature of science: historical and human construction of science ideas

The composition of materials in the world around us has fascinated humankind for several millennia. Students could research how early thinkers explained the nature of matter and this could help students see how humanity (science) constructs mental models to understand their world. Two examples are Empedocles' theory of the four 'elements' of earth, air, fire and water, and Hippocrates' concept of various 'humours' (e.g., melancholic/sad), and how each had various attributes and properties that could 'explain' the observed characteristics of materials in our world. Present-day views from other cultures could be compared, such as the five elements of the Chinese *I Ching* (e.g., metal = Venus) and the Hindus' and Buddhists' 'elements' (wind, fire, earth, water and aether) – in both instances, these relate to beliefs about life. It could be emphasised how it is only in relatively recent times that materials have been envisaged using a 'parts' model, leading to the notion of atoms and molecules (Lievesley 2007). Students could relate their own experiences of mental model building about the composition of matter – see the earlier section 'Narrowing student-generated "scientific" (mental) models' on p. 325 – to these early and current models and reflect on similarities to how the scientific community works.

Related to the composition of matter is the development of the periodic table by Dmitri Mendeleev. You may wish to read how the table's publication was influenced by luck, enthusiasm and social factors, and ask yourself if you could, over time, collect suitable primary level science stories that include these human attributes and elements; see Akeroyd (2007).

The influence of science: socioscientific sustainability issues

SSI

Science needs to be seen within social, cultural, environmental, economic and political contexts. The extraction of materials for human use can exemplify how human stories are interlinked across these contexts. When aluminium is manufactured, it impacts on the environment through various forms of spoliation (e.g., bauxite mining, hydroelectricity) and pollution (waste products including greenhouse gases) but it provides employment and commodities. Students can become involved in 'linking-thinking' tasks (Sterling 2004) by exploring the various ways in which the applications of science affect the lives of people, including children their own age. Although written with secondary students in mind, Levinson (2009) describes how the 'aluminium story' relates to the *catadores de lixo*, the rubbish pickers in Brazilian towns and cities. Gangs of *catadores* would compete for the collection of aluminium drink cans to earn an income from recycling. Through innovative government negotiations with the *catadores*, deals were made that resulted in better health care and education for primary aged children who help in collecting the cans – the story of a 12-year-old is described. Brazil is now the world's second-biggest aluminium recycler. Recycling, of course, has both benefits and drawbacks, including how a reduction in the demand for aluminium in one economy can mean labour losses in another.

Activity 8.21 **Socioscientific issues and linking thinking**

Develop a diagrammatic representation of how the various events in the manufacture of aluminium could be interlinked with social, cultural, environmental, economic and political factors. Suggest how each event could have a human story that impacts on children of primary age. How might you engage primary students in this type of socioscientific issue (SSI)? What might they learn about 'science as a human endeavour'? Compare your thoughts with the *Linking-Thinking* project strategies (http://assets.wwf.org.uk/downloads/linkingthinking.pdf) and the section 'The influence of science, or science in the community: socioscientific issues' in Chapter 1 (see p. 43).

Science in the community: objects, materials and substances

The production and consumption of objects, materials and substances has enormous ramifications in our world. Students can be encouraged to explore how science informs their shared responsibility in the consumption and production of materials and their relationship to maintaining a sustainable environment. This could include actions primary students could take to responsibly dispose of household waste (Malandrakis 2008), reducing the emission of greenhouse gases (Skamp, Boyes and Stanisstreet 2009), finding alternatives to plastic bags (see http://www.primaryupd8.org.uk), exploring recycling (see 'Ideas for science and technology K-6' in the NSW STA newsletter *Science Matters*, 2009, No. 3) and the considered use of water (see http://www.nma.gov.au/exhibitions/water). Initially, awareness could be the main goal, but as students may believe, for example, that they do not have any responsibility for waste control (it is looked after by the 'bin man') or, if they do, then it only involves 'comfort activities', rather than community action, then this is an opportunity for environmental education having a strong overlap with science education in this, and related, areas in the study of 'materials' (Malandrakis 2008). A 'community focus' in such activities/ tasks deserves emphasis as teachers sometimes think mainly of individual, rather than collective, participation by their students when environmental actions are considered (Cinquetti and de Carvalho 2007). An example of how primary students became involved in an action-oriented community project about water quality is outlined by Pressick-Kilborn (2009).

Summary

'Materials' is a topic area about which primary students have many varied but often limited conceptions. These can relate to the properties, uses, origins and ways of testing materials, as well as to whether they are solids, liquids or gases. Ways of helping students develop more scientific ideas about the particulate nature of materials (and, in time, substances) are emerging in the research literature and you are encouraged to include them in your teaching. Numerous constructivist strategies have potential within this content area, including the use of language when teachers are scaffolding student learning about materials. The area also provides excellent opportunities for students to carry out scientific investigations involving fair tests. Obtaining student interest could be a challenge but as the case studies illustrate, this obstacle can be overcome. Science as a human endeavour can be exemplified in many ways in this content strand and can assist in engaging students. The next chapter expands upon the nature of materials by considering how materials (matter) can change.

Concepts and understandings for primary teachers

Below are most of the key conceptual ideas and understandings related to objects, materials and substances with which a primary teacher should be familiar. You should read and interpret this list with an appreciation of its limitations as described in the section 'Concepts and understandings for primary teachers' in Chapter 1 (see p. 47).

Many of the concepts and understandings listed are derived from alternative conception studies and address the (alternative) views that many students and teachers may hold (see Driver et al. 1994a). Emphasis is placed on concepts at the macroscopic level (mostly suitable for primary students) but some key microscopic-level concepts are included. The latter concepts and understandings related to objects and materials require an appreciation of the particulate nature of matter. Where this is the case, an asterisk is used. The particulate explanation related to many of the concepts and understandings is not listed here.

Objects

- All (material) objects are made of stuff (matter) and hence have mass (which students recognise initially as the object's weight).
- Objects are generally made of one or more materials (including liquids; e.g., drops).
- Not all matter need be an object (e.g., sand, air).
- Objects have various properties (these depend upon the material from which they are made, their size and shape). Two object properties are stiffness and strength.

Materials

- Everything tangible (stuff, whether living or non-living) is made of materials (this includes gases).
- All materials ultimately derive from the Earth, including its atmosphere.
- Some materials are fibres (e.g., wool) or can be made into fibres (e.g., glass).
- Materials are made of one or more (pure) substances, although when there are mixtures this may sometimes not be obvious.
- Most materials are mixtures of different substances.
- All materials are comprised of one or more chemicals.

Classification (grouping) of materials

- Materials can be classified in a range of ways for particular purposes. All the classification schemes are human-made and have their strengths and limitations.

Properties and uses of materials

- Each material has its own characteristics; for example, hardness, ability to conduct electricity. This is because of the substances that comprise it; these properties are independent of the material's size and shape.
- Mechanical properties of materials include compressibility, hardness, elasticity, plasticity, brittleness and toughness.
- Some materials are porous in that they contain tiny cavities that can be filled with liquid or gas. A specific property of these materials is absorbency.
- The characteristics of materials are not all obvious or perceptible.
- The uses for each material depend on its properties.
- The properties of a material depend upon the atoms or molecules (or ions) that it comprises and the way these are organised in the material* (if necessary, see elsewhere for the meaning of these terms).
- Materials can be mixed together for particular purposes.
- Materials can often exist in more than one state and can be solids, liquids or gases, depending upon various conditions; for example, temperature, pressure.

Substances

- Substances are unique and pure entities.
- 'Substance' refers to a single pure substance that cannot be broken down into other substances (most students will not be able to appreciate the meaning of 'pure'; for example, having a distinct melting or boiling point).
- Substances may be composed of one or more elements (compounds comprise more than one element held together by chemical bonds). This establishes the substance's unique identity.*

Particles*

- All objects, materials and substances are comprised of particles.
- Atoms and molecules are examples of particles.*
- Particles are very small (of the order of 10^{-9} m).*
- Particles are never still.*
- Particles are different to the materials and substances they comprise. They do not, for example, have the same states, colours or shapes as materials and substances; that is, microscopic properties are different to macroscopic properties.

States of matter: solids, liquids and gases

- Materials can exist in one or more states – solid and/or liquid and/or gaseous forms.

- A material cannot be classified as being in two states simultaneously (e.g., as a solid and a liquid), although its properties may vary under different conditions and so take on the properties of another state.
- When a substance changes from one state to another, it is still the same substance (and its mass does not change, unless something else happens; e.g., an interaction with air occurs).
- Solids may be hard, soft and powdery. They take up a definite amount of space, have their own shape and are not compressible.
- All liquids are runny (to some extent). They take up a definite amount of space and take the shape of the container.
- Most gases are invisible. All spread to fill the available space and are compressible (into smaller spaces). Gases are at least 99 per cent space.
- For a given substance, solids are denser than liquids, which are denser than gases (the exception being water, which, as ice, is less dense).

Water and air

- Water is a liquid substance and has all the usual properties of liquids except that, in solid form, it is less dense than as a liquid. Water has specific freezing and boiling points.
- Water evaporates to form water vapour, which is invisible (although water droplets are visible).
- Air is a material and has all the properties of a material.
- Air is a mixture of several different gases.

Mixtures

- Mixtures are made of a number of substances that have no chemical bonds between them.
- Some mixtures appear homogeneous.

For more detail related to the above concepts and understandings, see, for example, Driver et al. (1994a), Newton (2000) and Wenham and Ovens (2010).

Acknowledgements

The two case studies in this chapter are used with the permission of the two teachers who initially reported on them; namely, Ms Kim Hardy, formerly of Guisne Public School at Macquarie Fields (Which shape is the strongest?), and Ms Patricia O'Hara, of St Augustine's School in Coffs Harbour (The absorbency of fabrics). The lesson sequence plan is used with the approval of a former BEd student at Southern Cross University, Ms Tammy Wilson.

Search me! science education

Search me!

Explore Explore **Search me! science education** for relevant articles on materials. Search me! is an online library of world-class journals, ebooks and newspapers, including *The Australian* and the *New York Times*, and is updated daily. Log in to Search me! through http://login.cengage.com using the access code that comes with this book.

KEYWORDS

Try searching for the following terms:

- particulate nature of matter
- properties of materials
- Solids, liquids and gases

Search tip: **Search me! science education** contains information from both local and international sources. To get the greatest number of search results, try using both Australian and American spellings in your searches: e.g., 'globalisation' and 'globalization'; 'organisation' and 'organization'.

Appendices

In these appendices you will find material that you should refer to when reading Chapter 8. These appendices can be found on the student companion website. Log in through http://login.cengage.com using the access code that comes with this book. Appendices 8.1, 8.2 and 8.4 are included in full below.

Appendix 8.1 Year 8 students' views of what a 'chemical' is

Only detergent, explosives, petrol and sulfuric acid were considered to be chemicals by 100 per cent of the students (n = 25). Plastics and rubber were next (about 85 per cent or more of the sample). Those thought not to be chemicals by more than 50 per cent of the group were a gum tree, sugar, wool, your heart and a spider.

Interestingly, of the 93 students who responded across years 8 to 11, only four thought all the items were chemicals. Even at the primary level, teachers can start to help students appreciate that all matter is made up of chemicals (Skamp 1996).

Appendix 8.2 Attributes of the particulate model of matter

1 *Conservation of particles* – Did you have approximately the same number of particles in your diagrams as were in the original diagram?
2 *Proximity of particles* – Were the particles close to one another for a solid or a liquid? Were they spread out for a gas?
3 *Orderliness of the particle arrangement* – Were solids shown in an orderly fashion and gases shown as disordered?
4 *Location of particles in the container* – Were solids and liquids shown at the bottom of the container? Did liquids go to the sides of the container? Were gases evenly distributed throughout?
5 *Constancy of particle size and shape* – As solids changed to liquids and so on, did the particles remain the same size and shape?
6 *Particle discreteness* – Did you change from a particle model to one that is continuous by adding lines and other fuzzies?
7 *Chemical composition* – Did particle attachments remain the same for physical changes but change for chemical changes?
8 *Arrangement of products* – If a chemical change occurred, were the new products correct? Are particles of the same substance identical?
9 *Bonding* – Are atoms attached to each other in molecules when they should be? (Gabel, Samuel and Hunn 1987)

Appendix 8.3 A constructivist lesson sequence plan about 'Matter is everything that takes up space' (middle primary)

Details of a six-lesson sequence exemplifying the CLISP model are provided in this online appendix.

Appendix 8.4 Teacher's anticipated learning outcomes for the absorbency sequence

- List the ways in which materials could be used for specific purposes.
- Conduct simple tests and describe observations.
- Formulate questions to guide observations and investigations of familiar materials.

Indicators that these outcomes were being reached were that students could:

- compare different materials that can be used for the same purpose (compares materials used to mop up water for their water absorbency)
- plan investigations using resources safely and wisely
- talk about observations and suggest possible interpretations
- identify patterns and groupings in information to draw conclusions
- relate observations and interpretations to other situations
- cooperatively suggest possible improvements to investigations in the light of findings.
- Intended values and attitudes were that students would:
- demonstrate confidence in themselves
- have a positive view of themselves
- persevere with activities to their completion
- work cooperatively in groups
- be curious about the natural and made environments
- gain satisfaction from their efforts to investigate, design, make and use technology.

Appendix 8.5 Websites with ideas for activities related to matter

About 30 annotated websites are listed in this online appendix.

References

Acher, A., Arca, M. & Sanmarti, N. 2007. Modeling as a teaching learning process for understanding materials: A case study in primary education. *Science Education*, 91, pp. 398–418.

Akeroyd, M. 2007. Enthusiasm and luck have roles to play in How science works. *School Science Review*, 89 (326), pp. 93–5.

Anonymous 2006. Communicating science. *Science Matters*, 3, pp. 2–5.

Australian Academy of Science (AAS). 2005. *Primary Connections*. Canberra: AAS.

_____ 2008a. 'Material World': Stage 2 Natural and processed materials. *Primary Connections*. Canberra: AAS.

_____ 2008b. 'What's It Made Of': Early Stage 1 Natural and processed materials. *Primary Connections*. Canberra: AAS.

_____ 2008c. 'Package It Better': Stage 3 Natural and processed materials. *Primary Connections*. Canberra: AAS.

_____ 2009a. 'Change Detectives': Stage 3 Natural and processed materials. *Primary Connections*. Canberra: AAS.

_____ 2009b. 'Spot the Difference': Stage 1 Natural and processed materials. *Primary Connections*. Canberra: AAS.

Australian Curriculum, Assessment and Reporting Authority (ACARA). 2011. K-10 Australian Curriculum: Science. Available at http://www.australiancurriculum.edu.au/Science/Curriculum/F-10 (accessed April 2011)

Bradley, D. 1994. Sticky things. *Investigating*, 10 (3), pp. 15–16.

Brown, T., Rushton, G. & Bencomo, M. 2008. Mighty molecule models. *Science and Children*, 45 (5), pp. 33–7.

Carlton, K. & Parkinson, E. 1994. *Physical Sciences: A Primary Teacher's Guide*. London: Cassell.

Cinquetti, H. & de Carvalho, L. 2007. Teaching and learning about solid wastes: Aspects of content knowledge. *Environmental Education Research*, 13 (5), pp. 565–77.

Clarke, H. 2006. Challenging my assumptions: Young Ghanaian children's 'real world' understanding of materials. *Science Teacher Education*, 46, pp. 14–15.

Dawson, C. 1993. What is a chemical? Some students' views. *Chemeda: Australian Journal of Chemical Education*, 38, pp. 3–8.

del Pozo, R. 2001. Prospective teachers' ideas about the relationships between concepts describing the composition of matter. *International Journal of Science Education*, 23 (4), pp. 353–71.

Diamond, D. 1978. *Introduction and Guide to Teaching Primary Science*. London: MacDonald Educational.

Driver, R., Squires, A., Rushworth, P. & Wood-Robinson, V. 1994a. *Making Sense of Secondary Science: Support Materials for Teachers*. London: Routledge.

_____ 1994b. *Making Sense of Secondary Science: Research into Children's Ideas*. London: Routledge.

Farrow, S. 2006. *The Really Useful Science Book* (3rd edn). New York: Routledge.

Fleer, M. & Jane, B. 2004. *Technology for Children* (2nd edn). Sydney: Prentice Hall.

Gabel, D. & Bunce, D. 1994. Research on problem solving: Chemistry, in D. Gabel (ed.). *Handbook of Research on Science Teaching and Learning*. New York: Macmillan, pp. 301–25.

_____, Samuel, K. V. & Hunn, D. 1987. Understanding the particulate nature of matter. *Journal of Chemical Education*, 64 (8), pp. 695–7.

Gallenstein, N. 2005. Never too young for a concept map. *Science and Children*, 43 (1), pp. 44–7.

Gatt, S. & Gatt, I. 2008. Promoting science through the performing arts. *Primary Science*, 103, pp. 14–17.

Goodwin, A. 2002. Is salt melting when it dissolves in water? *Journal of Chemical Education*, 79 (3), pp. 393–6.

Gray, C., Simpson, L., Sowden, C. & Rodrigues, S. 1994. Children's concept maps about plastics. *Primary Science Review*, 31, pp. 22–3.

Griffin, J. & Sharp, H. 1995a. *Sticky Situations*. Sydney: STA New South Wales and ICI.

_____ 1995b. *Covering Coats*. Sydney: STA New South Wales and ICI.

_____ 1998. Exploring chemical symbols and codes. *Investigating*, 14 (3), pp. 28–30.

Groves, C. 1986. Investigating fibres. *Investigating*, 2 (1), pp. 4–5.

Hackling, M. & Prain, V. 2005. *Primary Connections Stage 2 Trial: Research Report*. Canberra: AAS. Available at http://www.science.org.au/reports/documents/pcreport1.pdf (accessed 10 January 2011).

Harlen, W. 1985. *Teaching and Learning Primary Science*. London: Harper and Row.

_____ 2001a. Research in primary science education. *Journal of Biological Education*, 35 (2), pp. 61–5.

_____ 2001b. Taking children's ideas seriously. *Primary Science Review*, 67, pp. 14–17.

_____ (ed.). 2003. *Primary Science: Taking the Plunge* (2nd edn). Portsmouth, NH: Heinemann.

_____ 2007. The SPACE legacy. *Primary Science Review*, 97, pp. 13–16.

_____ & Jelly, S. 1989. *Developing Science in the Primary Classroom*. Edinburgh: Oliver & Boyd.

Hatzinikita, V., Koulaidis, V. & Hatzinikita, A. 2005. Understanding and explanations concerning changes in matter. *Research in Science Education*, 35, pp. 471–95.

Hellden, G. 2005. Exploring understandings and responses to science: A program of longitudinal studies. *Research in Science Education*, 35 (1), pp. 99–122.

Horn, M. 1989. *Children and Plastics*. London: MacDonald Educational (other Science titles are cited in this text).

Howitt, C. 2009. 3-D mind maps: Placing young children in the centre of their own learning. *Teaching Science*, 55 (2), pp. 42–6.

Jakobson, B. & Wickman, P.-O. 2008. The roles of aesthetic experience in elementary school science. *Research in Science Education*, 38 (1), pp. 45–66.

Johnson, P. 2000. Children's understanding of substances, Part 1: Recognizing chemical change. *International Journal of Science Education*, 22 (7), pp. 719–37.

_____ 2002. Children's understanding of substances, Part 2: Explaining chemical change. *International Journal of Science Education*, 24 (10), pp. 1037–54.

Jones, B. 1984. How solid is a solid? *Research in Science Education*, 14, pp. 104–13.

Keogh, B. & Naylor, N. 1997. *Starting Points for Science*. Sandbach, UK: Milgate House.

Krnel, D., Watson, R. & Glazar, S. 1998. Survey of research related to the development of the concept of 'matter'. *International Journal of Science Education*, 20 (3), pp. 257–89.

Lawther, K. 1994. Children's ideas about air. *Investigating*, 10 (1), pp. 13–14.

Lee, O., Eichinger, D. C., Anderson, C. W., Berkheimer, G. D. & Blakeslee, T. D. 1993. Changing middle school students' conceptions of matter and molecules. *Journal of Research in Science Teaching*, 30 (3), pp. 249–70.

Leisten, J. 2008. What makes children like science? *Primary Science*, 105, pp. 8–11.

Levinson, R. 2000. Thinking through science. *Primary Science Review*, 61, pp. 20–2.

_____ 2009. The manufacture of aluminium and the rubbish pickers of Rio: Building interlocking narratives. *School Science Review*, 90 (333), pp. 119–24.

Lievesley, T. 2007. In our elements. *Primary Science Review*, 96, pp. 5–7.

Liu, X. & Lesniak, K. 2005. Students' progression of understanding the matter concept from elementary to high school. *Science Education*, 89, pp. 422–50.

_____ 2006. Progression in children's understanding of the matter concept from elementary to high school. *Journal of Research in Science Teaching*, 43, pp. 320–47.

Lloyd, M. 2007. Can you teach an old dog new tricks? A teacher's perspective on changing pedagogy using Primary Connections. *Teaching Science*, 53 (3), pp. 27–9.

Mahaffy, P. 2006. Moving chemistry education into 3-D: A tetrahedral metaphor for understanding chemistry. *Journal of Chemistry Education*, 83 (1), pp. 49–55.

Malandrakis, G. 2008. Children's understandings related to hazardous household items and waste. *Environmental Education Research*, 24 (5), pp. 579–601.

Maskill, R., Cachapuz, A. & Kouladis, V. 1997. Young pupils' ideas about the microscopic nature of matter in three different European countries. *International Journal of Science Education*, 19 (6), pp. 631–45.

McGuigan, L., Qualter, A. & Schilling, M. 1993. Children, science and learning. *Investigating*, 9 (4), pp. 23–5.

McMahon, K. 1999. A 'big book' project linking science and literacy. *Primary Science Review*, 59, pp. 7–9.

Metz, K. 1995. Reassessment of developmental constraints on children's science instruction. *Review of Educational Research*, 65 (2), pp. 93–127.

Nakhleh, M. & Samarapungavan, A. 1999. Elementary school children's beliefs about matter. *Journal of Research in Science Teaching*, 36 (7), pp. 777–805.

Newton, D. 2002. *Talking Sense in Science*. London: Routledge and Falmer Press.

Newton, L. 2000. *Meeting the Standards in Primary Science*. London: Routledge and Falmer Press.

Nixon, M. 2001. Exploring the properties of recycled paper. *Investigating*, 17 (2), pp. 21–4.

Novak, J. 2005. Results and implications of a 12-year longitudinal study of science concept learning. *Research in Science Education*, 35 (1), pp. 23–40.

Ogborn, J. 2004. Soft matter: Food for thought. *Physics Education*, 39 (1), pp. 45–51.

Ozmen, H. & Kenan, O. 2007. Determination of Turkish primary students' views about the particulate nature of matter. *Asia-Pacific Forum on Science Teaching and Learning*, 8 (1), pp. 1–15.

Papageorgiou, G. & Johnson, P. 2005. Do particle ideas help or hinder pupils' understanding of phenomena. *International Journal of Science Education*, 27 (11), pp. 1299–317.

Parker, J. 1995. Words on paper. *Primary Science Review*, 36, pp. 18–22.

Pattie, I. 1991. The Golden Fleece – a heritage study. *Investigating*, 7 (4), pp. 2–7.

Pendlebury, J. 2005. Natte Yallock and Tarnagulla primary schools' partnerships in science project: Scientific solutions to a salty problem. *Teaching Science*, 51 (2), pp. 12–13.

Pressick-Kilborn, K. 2009. Steps to fostering a learning community in the primary science classroom. *Teaching Science*, 55 (910), pp. 27–34.

Qualter, A. 1993. I would like to know more about that: A study of the interest shown by girls and boys in scientific topics. *International Journal of Science Education*, 15 (3), pp. 307–17.

_____, Schilling, M. & McGuigan, L. 1994. Exploring children's ideas. *Investigating*, 10 (1), pp. 21–4.

Ross, K. 1997. Many substances but only five structures. *School Science Review*, 78 (284), pp. 79–87.

Ruane, M. 1989. Testing the strength of card. *Primary Science Review*, 11, p. 9.

Russell, T., Longden, K. & McGuigan, L. 1991. *Materials (Primary Science Processes and Concept Exploration [SPACE] project)*. Liverpool: Liverpool University Press.

Sanderson, M. 2000. What's in a bubble? *Investigating*, 16 (3), pp. 21–8.

Schibeci, R. & Hickey, R. 2000a. Is it natural or processed? Elementary school teachers and conceptions about materials. *Journal of Research in Science Teaching*, 37 (10), pp. 1154–70.

_____ 2000b. Primary teachers' conception of chemical. *Investigating*, 16 (2), pp. 33–8.

Schools Council. 1973. *Children and Plastics, Stages 1 and 2 and Background*. London: MacDonald Educational.

Scott, P., Asoko, H. & Leach J. 2007. Student conceptions and conceptual learning in science, in S. Abell and N. Lederman (eds). *The Handbook of Research on Science Education*. Mahwah, NJ: Lawrence Erlbaum Associates, pp. 31–56.

_____, Dyston, T. & Gater, S. 1987. *A Constructivist View of Learning and Teaching in Science*. Leeds: Leeds University, Centre for Studies in Science and Mathematics Education.

Skamp, K. 1993. The periodic table: A place in the primary school. *Chemeda: Australian Journal of Chemical Education*, 37, pp. 7–10.

_____ 1995. Where does the water go? *Investigating*, 11 (3), pp. 10–13.

_____ 1996. Chemistry: Its potential in the primary school. *South Australian Science Teachers' Association Journal*, 96 (1), pp. 15–28.

_____ 2005. Teaching about 'stuff'. *Primary Science Review*, 89, pp. 20–2.

_____ 2008. Atoms and molecules in primary science: What are teachers to do? *The Australian Journal of Education in Chemistry*, 68, pp. 5–9.

_____, Boyes, E. & Stanisstreet, M. 2009. Global warming responses at the primary secondary interface: 1 Students' beliefs and willingness to act. *Australian Journal of Environmental Education*, 25, pp. 15–30.

Sorsby, B. 1991. What are things made of? ... or is the question too hard? *Primary Science Review*, 18, pp. 22–4.

Stephenson, P. (ed.). 2000. *Developing Primary Teachers' Science Knowledge Self Study Materials*. Leicester: SCI Centre.

Sterling, S. 2004. *Linkingthinking*. Dunkeld, Perthshire, Scotland: WWF.

Tunnicliffe, S. D. 1988. 'Writing up' science in the primary years. *Investigating*, 4 (4), pp. 24–5.

Turner, A. 1997. 'We chucked some stuff into a thingy of whatsit and it just went!' *Primary Science Review*, 50, pp. 10–11.

Tytler, R. 2002. Using toys and surprise events to teach about air and flight in the primary school. *Asia–Pacific Forum on Science Learning and Teaching*. Available at http://www.ied.edu.hk/apfslt/v3_issue2 (accessed 17 September 2010).

_____ & Peterson, S. 2003. Tracing young children's scientific reasoning. *Research in Science Education*, 33 (4), pp. 433–66.

_____ 2005. A longitudinal study of children's developing knowledge and reasoning in science. *Research in Science Education*, 35 (1), pp. 63–98.

Walker, M., Kremer, A. & Schluter, K. 2007. The dirty water challenge. *Science and Children*, 44 (9), pp. 26–9.

Wenham, M. & Ovens, P. 2010. *Understanding Primary Science* (3rd edn). London: Sage.

Wiser, M. & Smith, C. 2008. Learning and teaching about matter in grades K-8: When should the atomic-molecular theory be introduced?, in S. Vosniadou (ed.). *International Handbook of Research on Conceptual Change*. New York: Routledge, pp. 205–39.

Yair, Y. & Yair, Y. 2004. 'Everything comes to an end': An intuitive rule in physics and mathematics. *Science Education*, 88, pp. 594–609.

Zeegers, Y. 1995. My 'Nikes' are better than your 'Reeboks'. *Investigating*, 11 (4), pp. 24–6.

9

Materials and change

by Keith Skamp

Changes,
Blackening, destroying,
Smoky, smelly, dirty,
Gobbled up by flames,
Forever.

An eight-year-old's cinquain about the irreversible nature of burning, in Bradley (1996, p. 134)

Introduction

Everyday language introduces young learners to words such as 'metals', 'detergents' and 'acids', the names of common gases such as oxygen and carbon dioxide, and the substances that comprise foods such as iron and carbohydrate. These are all chemicals, another concept word that is also part of everyday language. Primary-school children encounter changes every day in these materials (e.g., food) and substances (e.g., water). They see water and other liquids freeze and boil, solids dissolve in liquids, materials burn, metals tarnish and rust, and foods cook. Recognising the variety of these changes is a stepping stone to seeing the similarities and differences between them and then starting to think and talk about what is happening during such changes. Learning about these changes in the materials and substances in our world and realising that all materials and substances are made of chemicals are important components of the primary school curriculum. This content area is really chemistry for young learners, in which 'chemistry' refers to changes in the material stuff (matter) of this world and, by implication, what matter (stuff) is made of. The study of these changes in matter forms the focus of this chapter. Reference will also be made to what matter is made of, but as this was mentioned in Chapter 8, you will be required to refer back to it.

What role does chemistry content, viewed in this way, have in the primary curriculum? It can:

- provide an opportunity for basic concepts to be developed; for example, melting
- interest and intrigue children – they enjoy learning about chemicals (Skamp 1996a, 1996b)
- help children understand the world around them; this occurs when they investigate and interact with the materials in it
- provide opportunities for inquiry-oriented scientific investigation and for its component skills to develop

- exemplify aspects of the nature of science through children generating and testing representations to explain observations and phenomena
- encourage the development of scientific attitudes
- be taught using equipment that is simple and familiar and so not distract from the phenomenon being studied.

For chemical examples illustrating some of these criteria, see Skamp (1996b).

Although from a scientist's (and perhaps your) perspective, all changes in matter involve some rearrangement of the particles that comprise it, they are commonly divided into either physical or chemical changes. Physical changes in materials may involve changes in form and state; they are considered reversible and do not involve the substance(s)/chemical(s) that comprise the materials changing into a different substance/(s).

Chemical changes, on the other hand, occur when substances in materials react with each other and different substances are formed. Although chemists view these (chemical) reactions as technically reversible, they are usually thought of as irreversible. In the various sections of this chapter, physical changes will always be considered first. This is because primary students have difficulty conceptualising that there is a transformation of substances during a chemical change.

Is there chemistry content in the primary science syllabus/framework you are using? See if you can find references to changes of state (e.g., evaporation, condensation) and reactions and change (e.g., cooking, burning). What you find will be the outcome of Activity 9.1. Examples of chemical content may be found in the content strands of most syllabuses; for an example, see ACARA (2011).

After reviewing your own and students' conceptions about chemical phenomena, this chapter provides numerous suggestions about suitable primary-level chemical learning experiences related to change, as well as advice about teaching the particulate nature of matter. Sequencing lessons using constructivist principles and the encouragement of students to carry out their own chemical investigations is emphasised. Equipment, consumables and safety, and how this content strand can relate to science as a human endeavour, conclude this chapter.

Activity 9.1 Chemistry in your primary science syllabus

Chemical content at the primary level relates to the interaction between, and changes to, materials and substances. This would include the processes of physical and chemical change and factors affecting the rates and character of such changes. Some curriculum guidelines and resources might include a section on the substructure of materials and how this information helps students understand their properties – see, for example, Curriculum Corporation (2006) and the *Primary Connections* unit 'Change Detectives' (AAS 2009b).

Survey your primary science curriculum. What evidence is there that chemical ideas underpin some of the curriculum content? Can you identify key change concepts? Is the substructure of matter (particulate ideas, which may refer to elements, compounds, atoms and molecules) mentioned or implied?

This chapter may broaden your perspective of what primary chemistry involves. Remember to return to this activity at the end of the chapter to determine whether you want to interpret further components of your primary science curriculum as having a chemical basis.

Teachers' ideas about matter and change

The online Appendix 9.1A contains questionnaire items that challenge your beliefs on what happens when matter undergoes change. They refer to boiling, condensation, evaporation, burning, dissolving and melting; see Summers, Palacio and Kruger (1989), and Jarvis, Pell and McKeon (2003).

Appendix 9.1A
Testing teacher beliefs about matter and change

Activity 9.2 What happens when matter undergoes change?

For the first five items in the online Appendix 9.1A, state whether you agree or disagree with the statements associated with the described pictures. If you do not understand the statement, write 'don't understand'; if you do understand it but do not know if it is true or false, write 'not sure'. On completion, discuss your responses with a colleague, and then, if possible, discuss in a group of four. What questions do you need to ask to clarify your understandings?

Appendix 9.1B includes true–false statements about melting, dissolving and evaporation which teachers found to be the most challenging – from 30 questions developed by Jarvis, Pell and McKeon (2003). Repeat the procedure as per Appendix 9.1A to further audit your understanding of physical changes. Provide reasons for your true–false responses.

The online Appendix 9.2 comments on the suitability of particular responses to items in the online Appendix 9.1A. Sufficient information is provided to be definitive about most questions. You may still have to research further to be certain about some of the items: for example, the processes of burning and melting of a candle are not well understood by most people, let alone primary and secondary students (Johnson 2002). This approach is to encourage you to stop and think about each of these statements and hopefully debate them with colleagues. For more details about the first five items, see Kruger and Summers (1989) and Chang (1999); for more information about the true–false items, see Jarvis, Pell and McKeon (2003).

Appendix 9.1B
A further audit of your beliefs about matter and change

Appendix 9.2
Comments on questionnaire items about what happens when matter undergoes change

Your responses to these items will help you recognise your existing ideas about what happens when matter changes. Were your responses perceptual and based only on observable entities? For example, did you confuse steam as being tiny droplets of water with air in between because that is what you can literally see? Did your answers only recall familiar contexts? For example, did you believe that dissolving only refers to solids dissolving in water? Did you merge some scientific ideas? For example, are you clear about the differences between melting and dissolving? What about melting and burning? Did your explanations refer to particulate ideas, which may include reference to atoms and molecules? For example, to explain dissolving, did you refer to both the particles of water and sugar and how they could be arranged? Your understanding related to these concepts should, if necessary, be clarified, and the online Appendix 9.2 will assist your understanding where you are having difficulty. Many of these phenomena – evaporation, condensation, burning – will be the focus of activities in which you will engage children. If you are to assist with children's progressive understanding of change processes, then you do need some appreciation of the major concepts underlying these activities (Jarvis, Pell and McKeon 2003). Are you aware, for

Appendix 9.2
Comments on questionnaire items about what happens when matter undergoes change

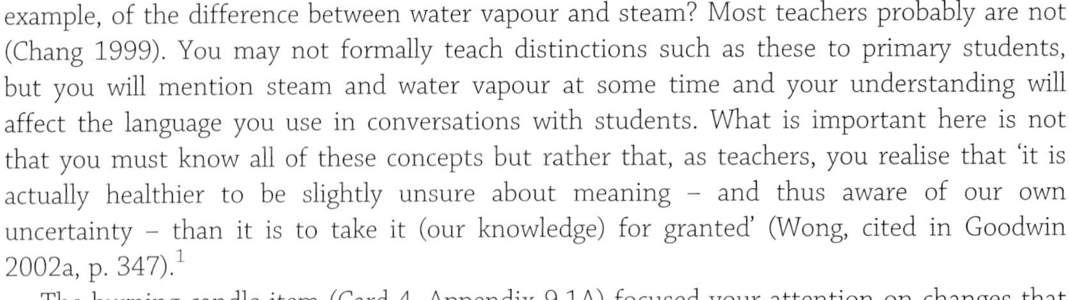

example, of the difference between water vapour and steam? Most teachers probably are not (Chang 1999). You may not formally teach distinctions such as these to primary students, but you will mention steam and water vapour at some time and your understanding will affect the language you use in conversations with students. What is important here is not that you must know all of these concepts but rather that, as teachers, you realise that 'it is actually healthier to be slightly unsure about meaning – and thus aware of our own uncertainty – than it is to take it (our knowledge) for granted' (Wong, cited in Goodwin 2002a, p. 347).[1]

ONLINE

Appendix 9.1A
Testing teacher beliefs about matter and change

The burning candle item (Card 4, Appendix 9.1A) focused your attention on changes that are reversible and those that are not reversible. Melting is a reversible change but burning is not. Burning is an example of a chemical reaction or change. Although this term is possibly problematic (Bucat and Fensham 1995; Palmer and Treagust 1996), the distinction between chemical and physical change is generally considered to be useful in assisting learning about change (Palmer 2003). Fensham (1996) makes reference to many adults misunderstanding what is involved in a chemical reaction. He comments that there is often a lack of recognition of the atomic rearrangements involved in the production of new substances as products from initial ones as reactants (see also Andersson 1986; Bucat and Fensham 1995).

Activity 9.3 The scientist's view of burning (and other chemical changes)

- Discuss with a colleague what you believe to be the currently accepted scientific view of what happens when matter burns. Attempt to write down a set of ideas that encapsulates this view.
- Similarly, share your views about what you believe happens during a chemical change.
- Compare your ideas against the online Appendix 9.3.

ONLINE

Appendix 9.3
Concepts underlying the processes of burning and chemical change

Substances and the particulate nature of matter

The concept of 'substance' is critical to a scientific understanding of physical, and especially chemical, change. Chemical change is defined as a change in the very composition of substances (see the online Appendix 9.3). A substance, from a scientist's perspective, has an identity that is independent of the object or material (of which it is part) and of any particular state; that is, solid, liquid or gas. Substances change into new substances when they chemically react with other substances. The unique identity of a substance is defined by the bonding between its particles and the subsequent structures formed. Each substance, then, by definition, is a unique chemical structure (see the Chapter 8 section 'Substances and particles' on p. 304).

ONLINE

Appendix 9.3
Concepts underlying the processes of burning and chemical change

Does this make sense to you? It would not if you have no appreciation of the substructure of matter; this will, in most instances, depend upon your previous education and reading.

As has been described in Chapter 8 and will be expanded upon in the next section, most primary students do not have an intuitive concept of a chemical substance or the particulate nature of matter (see Activity 8.5 on p. 306). Several researchers (Johnson 1998a, 1998b,

2000; Wiser and Smith 2008) argue that an understanding of how matter is made of particles is a prerequisite for most students to understand the (chemical) concept of 'substance'. Yet chemical change, which is underpinned by the concept of substance and how substances change their unique structures to become new substances, is present in most primary syllabuses. Chemical change is one area where, if your syllabus requires it, it may be advisable to focus on macroscopic ideas rather than the underlying explanations.

There is certainly a case, however, for introducing primary students to physical change and its particulate underpinnings, and this should precede a similar approach with chemical change (Wiser and Smith 2008). The introduction of particulate ideas at the primary level is, for some, still a contentious issue, but evidence is mounting that primary students benefit from being introduced to aspects of the particulate nature of matter in relation to the structure of solids and liquids (see Chapter 8) and what happens during physical changes. This is only the case if non-traditional pedagogy is used, such as students learning through generating and discussing their own representations of matter and what happens when it changes (Tytler, Prain and Peterson 2007; Wiser and Smith 2008). Depending upon your perceptions of your students' understanding of particulate ideas in these two instances, then a decision could be made as to whether to do likewise when macroscopic properties of chemical change are encountered. However, as indicated, be aware that a particulate understanding of chemical change is difficult, even for upper primary students.

Primary teachers, therefore, do need some appreciation of the particulate explanation of chemical (and physical) change. There are many reasons for this, including the following:

- It will assist your confidence when teaching this component of the curriculum, whether it is introduced to students or not.
- You will probably no longer avoid teaching it.
- It will help you select examples of chemical change and ask appropriate (macroscopic) questions of students (for example, about the nature of the new products from a chemical reaction), and, if deemed appropriate, explore students' particulate representations of chemical change.
- There will be some students who will use the language of chemical change and particulate theory, as will the resources they access; for example, the *Primary Connections* unit 'Change Detectives' (AAS 2009b). For teaching to be effective, it will need to be able to respond to these situations.
- It will assist you in making decisions as to whether to formally or informally introduce particulate ideas to your students.
- The research evidence suggests that when the particulate nature of matter has been introduced to young learners using non-traditional approaches, then it appears that it has been useful to later learning (e.g., Novak 2005).
- Your primary syllabus may include particulate interpretations of chemical change and you have no choice.

Depending upon your previous science background, you may require further assistance to understand some of these chemical concepts (for example, substance, chemical reaction, element, compound, bonding). The following references may be useful: Driver et al. (1994b), Farrow (2006), Stephenson (2000) and Wenham and Ovens (2010). Also useful are the lists of concepts and understandings at the ends of chapters 8 and 9, but these are best read after browsing one of the above references.

Students' ideas about matter and change

Many studies have explored students' ideas about matter and change, although relatively few have focused on primary students' ideas about chemical changes. In the summary that follows, physical change will be considered initially. A range of elicitation techniques are implied or illustrated.

Physical change

Melting and freezing

Primary students often describe melting (of a range of solids) using water and ice as a prototype; for example, in many instances, water is said to be the product of melting. It is common for students to say solids 'dissolve' when they are melting. When (secondary and, occasionally, primary) students explain melting by referring to particles, the particles often have macroscopic properties; for example, particles in a solid are represented as cubes (Krnel, Watson and Glazar 1998). In an Australian study (Adams, Doig and Rosier 1991), when Year 5 students were asked to explain how heat melts ice into water (as a consequence of iceblocks being placed into lemonade), only 43 per cent realised that ice and water were the same substance and that heat caused the change. Mechanisms usually were not mentioned; only 5 per cent made any reference to particle ideas.

Activity 9.4 **Students' perceptions of melting and freezing**

Conservation of matter underpins physical and chemical changes. Do upper primary students appreciate this idea? Obtain some children's responses to the following two questions:

- When water freezes, is there a loss in mass, a gain in mass or does the mass stay the same? (You may wish to ask if it is heavier, lighter or the same.)
- When ice melts, is there a loss in mass, a gain in mass or does the mass stay the same? (You may wish to ask if it is heavier, lighter or the same.)

Categorise the answers as loss (lighter), gain (heavier) and same. Interview one student in each category and ask them: 'If you had a cup of water and you put it in the freezer and allowed it to freeze, would it become heavier, lighter or stay the same?' Encourage each child to elaborate on why they gave their answer. They may wish to draw and annotate pictures to show you what they mean.

Analyse your responses in terms of your understanding of what is involved in melting and freezing. Did any students refer to particulate responses? When you have completed this task, compare your responses with those provided in Appendix 9.4 (see p. 394).

What can you suggest students could investigate to advance their thinking about the conservation of mass?

Adapted from Ross and Law 2003, pp. 99–101

Several studies have found that helping primary students appreciate that the mass of materials undergoing physical change does not change, by measuring the weights, can assist them in generating representations of the change in particulate terms (see Wiser and Smith 2008).

Evaporation

Many studies accept that primary (and older) students explain evaporation in a range of ways. For some children water disappears, for others it is 'displaced' (for example, it has gone into something), while some believe it has been 'transformed' (for example, into air) (Krnel, Watson and Glazar 1998). An age-related progression of ideas has been advanced (Bar 1989; Bar and Galili 1994):

- the water disappears (ages five to six)
- the water penetrates solid objects – absorption (ages seven to eight)
- the water 'evaporates' into some container – displacement (ages nine to 10)
- the water evaporates; it is scattered in the sky, clouds, ceiling air – dispersion, transference (ages nine to 11)
- the water goes into air and changes phase (by age 13).

These stages have been related to younger students' emphasis on the descriptive and perceptual aspects of the phenomenon and then (mainly) the different reasoning approaches to the phenomenon that older students use. This sequential progression of ideas, and their implied stability, has been questioned, especially as a means of representing individual students' conceptual growth (Tytler and Prain 2010; Tytler, Prain and Peterson 2007). Students often offer 'more than one interpretation of a phenomenon (such as evaporation), even within the same discussion'. Further, they give different responses, depending upon the context (e.g., a puddle, a water tank, eucalyptus oil) (Tytler 2000, p. 452). In his Australian study, Tytler (2000) found that responses from both Year 1 and Year 6 students ranged across seven identified categories. These were Bar and Galili's (1994), with the addition of other categories such as 'It's just like that' (that is, it always happens, such as puddles dry up) and 'associative thinking' (that is, used as an explanation, such as smell being associated with filling the room). Even so, there were still overall patterns of conceptual development: for example, far more Year 6 students indicated that water went into the (local) air or changed form; Year 1 students used 'Just like that' or associative thinking more (probably because they have fewer life experiences of various change phenomena). Tytler argued that older students appreciated the existence of matter in a more material sense, but emphasised that even Year 6 students used associative ideas. These distinctions may still underestimate the thinking of younger students about evaporation (Varelas, Pappas and Rife 2006). From their longitudinal study of several students' understanding of evaporation from kindergarten to Year 6, Tytler and Prain (2010) found learning pathways to be 'very individual and complex'; they did, though, see a shift from 'simplistic water cycle views to representations of water and alcohol in air', and a slow movement (over years) from a 'global (involving clouds) to a local view of evaporation' (ibid.).

Teachers need to be clear about the difference between evaporation below the boiling point and the phenomenon of boiling, as primary students will encounter both events. It has been claimed that more younger students say water evaporates into the air when it is boiled than when it evaporates (that is, below boiling point). This claim has been questioned as students seem confused about what the bubbles are in boiling water (Do you know? See the online Appendix 9.2). Most primary students (maybe as many as 90 per cent) do not believe that the bubbles in the water are water in the gaseous state, although they know that the water has gone into the air. In fact, many students of all ages think the bubbles that appear during boiling are air, which is often their prototype of a gas. The main problem is that students have real difficulty conceptualising what a gas is, and hence, in the case of evaporation, appreciating that water vapour is a gas (Johnson 1998a; Tytler 2000). These studies and others (Johnson 1998b, 2005) stress that when students use the expression 'into the air' (and other words, such as 'gas', 'oxygen', 'steam' and so on), the word 'air' (and others) may have various meanings for primary students that differ markedly from adults' understandings of the word(s). Johnson (1998a) argued that it is not until most students appreciate the particulate nature of matter that evaporation is really understood as a change of state, although he noted that a small number of Year 6 students did appreciate change of state from a macroscopic perspective and their view was stable across several years.

Appendix 9.2
Comments on questionnaire items about what happens when matter undergoes change

Condensation

Condensation is considered to be more difficult to understand than evaporation as students find it hard to believe that there is always water vapour in the air (Johnson 1998b). Teachers, therefore, may find some students have an earlier understanding of evaporation (Hellden 2005). Responses to condensation tasks (for example, ice in a container and moisture appearing on the outside) suggest that students' answers can be categorised in a similar way to evaporation responses (Adams, Doig and Rosier 1991; Johnson 1998b; Tytler 2000). Examples include 'coldness changes into water' (associative thinking) and 'the water comes through the container' (displacement), but again, context is important – when cold Coca-Cola was used instead of iced water, the latter response was virtually absent (Johnson 1998b). Although access to particle ideas may be the way towards a full understanding of condensation (Johnson 1998b), Tytler (2000) again argues that Year 1 and Year 6 students gave responses across all categories, including water changing from one form to another, although more Year 6 students fell into this category. As with evaporation, individual students often held more than one conception simultaneously, changed explanations depending upon context, and did not necessarily provide more conceptually advanced responses over time.

A key finding in these and other conception studies is that some students use scientific terms but can't demonstrate what they mean by the terms (Johnson 1998b; Liu and Lesniak 2006; Tytler 2000). This is what Presst (cited in Omebe 1987, p. 16) calls ' "verbal wrapping paper" ... which obscures the contents [its meaning] of the parcel [conceptual area]'. Omebe (ibid.) underlines the nature of this surface learning by telling the following story of a class learning arithmetic:

> 'One five is five, two fives are ten, three fives are fifteen, four fives are twenty ...' chanted the class. Walking around the room, the teacher heard one boy repeating, 'Da da dee dah,

Da da dee dah'. 'What are you saying?' the teacher asked. 'Oh, I've learned the tune. It's only the words I don't know yet' was the reply.

Activity 9.5 The water cycle

Many teachers have students (from as young as seven or eight years old) draw the water cycle, often copying it from a book or a chart.

- What do you now think of this practice? Why do you think some teachers persist in using it? Justify your responses.
- Should this practice continue? If not, what experiences, if any, should replace it? Why?

You may find Tytler (2000, pp. 461–2, 464) of some value after thinking through your own response and sharing it with a colleague. He draws the distinction between (young) students explaining evaporation as water going into the air (labelled 'a water cycle' response) compared to water going into the (local) air. He argues that introducing the water cycle to young learners may serve a purpose, but not necessarily one that teachers currently focus on. What do you think this could mean?

The impact of teachers introducing the water cycle in a traditional manner can have long-term effects on students' interpretations of what is happening during the evaporation process; this is illustrated in the same students' responses from kindergarten to Year 6 (Tytler and Prain 2010) – if possible, you should read them. If the traditional way of teaching the water cycle is as a *given* representation, then what could be different ways of engaging students with this concept that illustrates representational learning (see the Chapter 1 section 'Representational and related strategies' on p. 29)? How could you relate your approach to explicitly teaching about the NOS?

How one teacher used literacy connections to encourage Year 2 students' conceptual development about the change processes involved in the water cycle is discussed in Varelas, Pappas and Rife (2006).

Dissolution

Students' explanations of dissolving, such as sugar (the solute) in water (the solvent), to form a solution fall into categories similar to those found with evaporation and condensation. Examples are:

- disappearance of the solute (up to 60 per cent of six-year-olds), which indicates non-conservation of the solute
- displacement, in which students do not indicate that the solute interacts with the solvent
- liquefaction, in which students believe that the solute has become a liquid (students sometimes refer to the sugar 'melting')
- transformation, where students start to talk about the solute breaking up into bits, with some students starting to refer to microscopic explanations.

However, particulate responses, in the molecular sense, are almost non-existent at the primary level (Adams, Doig and Rosier 1991; Driver et al. 1994b; Krnel, Watson and Glazar 1998; Liu and Lesniak 2006) and when they start to occur there are still variations in meaning, such as the solute fills up spaces in the solvent (Selley 1998). Also, conservation of mass during dissolution is late to develop (Driver 1985) – also see Activity 9.4 on p. 356. These findings indicate that children's intuitive reasoning, up to about 10 years old, seems to be dominated by their perceptual experiences. Beyond this age, students tend to argue, as scientists understand, that the sugar is still there (e.g., it is spread out into tiny bits), but particulate (molecular) conceptions of dissolution are slow to appear, even in junior secondary years.

Heating materials

Heating materials can produce physical and chemical changes. There are few reports of students' views about the expansion of solids due to heating. Russell, Longden and McGuigan (1991) found that students of all ages thought a metal rod, when heated, would melt or go soft, with this view being more prevalent in upper primary. More young students thought the metal rod would weigh more, with older students being split between this option and saying there would be no weight change (again, see Activity 9.4 on p. 356). This could be because some students think of heat as a substance (Krnel, Watson and Glazar 1998). When asked to explain why a metal rod expands when heated, teachers gave a wide range of reasons. What would you say? (See Isabelle and de Groot 2008)

As students got older, more of them understood that wool was a combustible material, with about half of all age ranges believing there would be a reduction in the material. With vinegar, there were mixed responses (n = 22, upper primary; n = 23, infants and lower primary): several thought it would become hot but with no other change (infants, 35 per cent; middle primary, 57 per cent; upper primary, 27 per cent), while about one-quarter of upper primary students thought the vinegar would go into the air/evaporate. As students became older, up to about 25 per cent mentioned boiling, bubbling or foaming (Russell, Longden and McGuigan 1991). As with melting and freezing, even upper primary students may still maintain that matter is not conserved across phase changes such as boiling, as the liquids seem to disappear (Wiser and Smith 2008).

Chemical change

An awareness of chemical change probably starts to develop in the upper primary years (Liu and Lesniak 2006). However, most primary students do not intuitively discriminate between physical and chemical change. This has been explained by their non-chemical view of substance and the particulate nature of matter; in fact, students' retention of a continuous and static view of matter (see the Chapter 8 section 'Students' conceptions of the structure of matter' on p. 314) is thought to impede an understanding of chemical change (Hatzinikita, Koulaidas and Hatzinikita 2005). Young students' dependence on the perceptual and directly observable characteristics of change also makes chemical change hard to appreciate. To further complicate matters, many chemical changes involve gases (for example, oxygen in combustion and carbon dioxide as a product of many everyday reactions) and the gaseous state creates many conceptual difficulties for primary learners.

The chemical concept of substance is probably not accessible to most primary students. Even when junior secondary students had completed units of work focusing on materials and substances, but before the concept of chemical change was introduced, there appeared to be no signs of the scientist's view of chemical change in their thinking: '[these] pupils simply did not have the possibility of substances changing into other substances as part of their thinking. They could not entertain the idea of chemical change'. When chemical change was taught, these secondary students still did not recognise the concept. 'Reaction' simply meant a 'change in appearance' and no 12- to 13-year-olds had any idea of the meaning of elements and compounds (Johnson 2000, pp. 730, 732, 733). After several teaching units, most of these secondary students still considered the results of chemical change as a 'mixture of reactants' (Johnson 2002, p. 1050), which is not the chemical view (see the online Appendix 9.3). Johnson concluded that the concept of substance 'needs to be (directly) taught [to secondary students]' (2000, p. 735); he provides evidence that, for example, simply emphasising the macroscopic properties of substances, such as their melting and boiling points, did not necessarily result in the idea of substance being developed (Johnson 2002, p. 1051). It was only when secondary teachers introduced the particulate nature of matter that students started to appreciate chemical change. This study would seem to indicate that most primary students would not be able to intuitively formulate notions of substance. Teachers should note that Rahayu and Tytler (1999) believe that primary students' responses about chemical changes can be interpreted to mean that some of them do start to appreciate the transformation of materials into new materials. However, their research appears to be referring to the transformation of materials and not the transformation of substances.

Appendix 9.3
Concepts underlying the processes of burning and chemical change

What are the implications for primary teachers when they are teaching about chemical change? Their students are probably not going to understand 'substance' as a chemical concept before appreciating the particulate theory of matter. Introducing the particulate nature of matter (see the online Appendix 8.2) at the primary level has been problematic. Most primary students do not naturally see their world in particulate terms. However, there is increasing evidence that some primary students, if taught that their material world is comprised of particles, may start to see their world from a particulate perspective (Acher, Arca and Sanmarti 2007; Hellden 2005; Papageorgiou and Johnson 2005; Tytler, Prain and Peterson 2007; Wiser and Smith 2008; also see the Chapter 8 section 'Students' conceptions of the structure of matter' on p. 314). It must be emphasised, though, that even then, many students usually do not appreciate the chemical view of particles, rather seeing them in other ways; for example, with macroscopic properties (Skamp 1999; see also examples in Chapter 8). However, the conception of matter as 'spaced particles, and only of those particles, and in which they have all the macroscopic properties of a substance, including its state' can be a stepping-stone idea to the scientific particulate model of matter (Wiser and Smith 2008, p. 230). Teachers need to be aware of these findings, and should move carefully in this area. Although macroscopic observations and evidence may not assist in developing the concept of substance, which is fundamental to understanding chemical change, teachers can still lay the foundations for appreciating the everyday occurrence of chemical change (suggestions are provided later; see the section 'Learning about chemical change' on p. 386).

Appendix 8.2
Attributes of the particulate model of matter

Activity 9.6 asks you to check out these conclusions. As a primary teacher who may, at various times, introduce cooking, burning and other chemical reactions into your classroom lessons (e.g., vinegar and baking soda), it is important that you appreciate the lack of accessibility of the concept of chemical change to many primary students. Consequently, you need to ask yourself this follow-up question: If, for example, cooking and burning are in the curriculum and are part of students' everyday life, how, then, do I get them to think about these everyday chemical changes?

Activity 9.6 Determining students' preconceptions of chemical change

Interview two or three primary students about the chemical change between baking soda and vinegar. Initially ask them to predict what would happen, then have them observe the combination of the two substances and ask them to explain what they thought was happening. Use questions to probe their understanding, such as 'Can you tell me what you meant by "X"?' and 'Could you draw what you mean?' (Liu and Lesniak 2006).

Interpret your results in light of the above research about chemical change. What are some implications of the findings for you as a teacher?

Can you suggest some constructivist and representational learning strategies you could use related to chemical change? Compare your thoughts with the teaching suggestions made about chemical change later in this chapter. If you wish to explore children's views about burning (e.g., of a candle, steel wool or paper), see some of the elicitation approaches used in studies such as those of Rahayu and Tytler (1999).

Chemical change, as indicated, is a complex phenomenon; at the particulate level, it not only requires an appreciation that atoms and molecules do not have the same properties as the macroscopic entities they represent, but that atoms are conserved during chemical change but not molecules. Possible ways to introduce these ideas are discussed later in this chapter.

Although there are several studies that have researched secondary students' ideas about specific chemical changes, only a few primary studies have been reported – in Australia, Adams, Doig and Rosier (1991) and Rahayu and Tytler (1999); in Europe, the *Science Processes and Concept Explorations* (SPACE) project, Russell, Longden and McGuigan (1991), Johnson (1998b) and Hellden (2005); and in the United States, Liu and Lesniak (2006). Findings of relevance to primary teachers are summarised next.

Combustion (burning)

Upper primary students (11- to 12-year-olds) may have a 'prototypic view of burning ... based on [their] observations of fibres, matches, splints, etc., burning'. This prototypic view encapsulates ideas such as

> burning involves things going red and a flame appearing
> oxygen (or air) is needed (although its function may not be ... clear)
> things get lighter when they are burnt
> burning drives off the smoke or parts of the material are driven off as smoke
> solid residues or ash are the incombustible bits left behind.
>
> Driver 1985, p. 158

These views are not necessarily scientific and suggest that combustion is a difficult idea for primary students; in fact, Johnson (2002, p. 1053) believes that it 'must be regarded as one of the last things we should expect [primary] students to understand'.

Student explanations of combustion (chemical change) have been classified into five categories by Andersson (1986) and sometimes these have been regarded as hierarchical (Krnel, Watson and Glazar 1998). This scheme is problematic. It may, however, guide teachers in their understanding of, and reactions to, students' responses about the concept of chemical change. Andersson proposed that, initially, students say, 'That's how things happen' (it's just like that), and are unquestioning; then matter is *displaced* and the new substance comes from somewhere else (e.g., smoke from the wood). Thirdly, *modification* may occur when the new substance is the original substance in a new form, and this can be followed by *transmutation*, in which the original substance is considered to be transformed into a completely new substance (not derived from the original reactants, but from some other source; for example, the flame). The scientific view is *chemical interaction*, which is when substances are seen as being composed of atoms of different elements. Chemical interaction involves the dissociation and recombination of atoms from the reactants in the formation of new substances. Later research suggests an additional category may occur after transmutation, namely *transformation of substance*, in which student responses seem to indicate that new substances are formed – '[there is] clear reference [in student responses] to new substances being produced or the reactants used up' (Rahayu and Tytler 1999, p. 201). This, Rahayu and Tytler state, could underpin the scientific idea of a chemical reaction as it indicates that substances have not retained their identity through the change. Further, they found, as did Tytler's (2000) studies of evaporation and condensation, that students at all levels (years 1, 3 and 6) could hold any of the above conceptions of burning except for that of chemical reaction, which was held by less than 1 per cent (n = 362) of Year 6 students. More Year 6 students (approximately 20 per cent) did hold the transformation idea. Again, the context (e.g., burning toast compared to steel wool) seemed a significant factor in the type of explanation given.

Other chemical reactions

There are other types of chemical reactions, although they are not necessarily distinct. Reactions involving decomposition create difficulties for students because often the products are gaseous and cannot be seen (Johnson 1998b). Other factors may also influence their explanations of decomposition. When primary students were introduced to particulate ideas, they did not readily relate them to decomposition (e.g., of leaves), whereas they did so when asked about a burning candle, albeit with incomplete particulate explanations. It appears that when children have rich experiences of phenomena such as leaves on the ground, they will draw on their experiences for explanations rather than formally introduced science ideas (Hellden 2005).

Rust also will be encountered at the primary level and is sometimes considered a different type of chemical reaction (although it is actually slow combustion). Students tend to still see rust as iron, as that is where it came from (Johnson 1998b). They also often believe that rusted nails are lighter – maybe because they look powdery – and, since few secondary students (15-year-olds) appreciate the role of oxygen in rusting (Driver 1985), it would be surprising if any primary students did.

Progression of ideas about physical and chemical change

Students' conceptions of physical and chemical changes (e.g., evaporation, burning) have usually been interpreted in terms of a conceptual progression towards the currently accepted scientific explanation. As described above, the form this progression takes is often assumed to be a hierarchy of increasingly sophisticated conceptions. Being aware of common alternative conceptions is certainly helpful as students' responses can be partially anticipated and suitable activities planned. The difficulty is that there is a wide variability of views about change within the same student population, which suggests that the development of ideas in this area is not 'developmentally inevitable' (Wiser and Smith 2008, p. 216).

This finding and other factors has led to a view of alternative conceptions as 'transcendent' rather than 'real entities'. Consequently, the assumption that each student's conceptual growth follows various 'conceptual trajectories' (for particular concepts) has been questioned (Tytler and Peterson 2005, pp. 68–9; Tytler and Prain 2010). Rather, these trajectories are perhaps better interpreted as generalisations across large numbers of students rather than descriptions for individual conceptual growth.

As teachers focusing on a constructivist view of learning, your interpretation of students' conceptual development may be assisted if you appreciated that their explanations of everyday phenomena are influenced by:

- the context in which the phenomenon is occurring (e.g., evaporation from a puddle)
- the children's views of what they believe are the purposes of particular activities. They may not think that an activity relates to an underlying concept such as condensation and that they should be seeking evidence to check out their thinking about condensation; rather, they may think the activity has other purposes. This has been referred to as an activity's 'epistemological component' (Tytler and Peterson 2005, p. 68; see the Chapter 1 section 'The nature of science: how it works' on p. 5). This refers, in part, to the way children (and teachers) perceive how knowledge may be generated from an activity such as obtaining evidence to support a prediction or supporting an explanation that works across various contexts
- how much particular children perceived activities in terms of their own identity in the classroom (e.g., Are they explorers of ideas?) and in relation to the learning that was happening (e.g., Was the activity seen as predominantly social rather than conceptual?)
- the nature of the questions asked (e.g., by the teacher).

Clearly, there are many factors that may influence students' conceptual growth (see Chapter 1). Students' progress towards a particulate view of matter as a way to interpret physical and chemical changes will, in conjunction with some of the above, be dependent upon:

- their conceptions of the nature of reality – two examples would be that:
  - coldness is seen as a substance or property and is used interchangeably
  - matter, events and properties are not used consistently as entities in explanations.
- the extent of their appreciation of the nature of knowledge – three examples would be how they:
  - perceive what is evidence
  - use the notion of cause and effect
  - appreciate the importance of discriminating between language meanings when describing phenomena (Tytler 2000; Tytler and Peterson 2000; Wiser and Smith 2008).

If conceptual development and change is partially affected by student understanding about the nature of reality and knowledge, then this underscores the importance of students learning about the NOS, and how to think and work scientifically (see Chapters 1 and 2), as these help them to appreciate the role of evidence, the tentativeness of hypotheses, the place of representations and modelling in forming explanations, and so on. Students could, for example, carry out fair tests related to evaporation or cooking – they could vary the amount of ingredients and determine the rate of chemical change. *Primary Connections* units (AAS 2009a, 2009b) provide fair-test exemplars related to chemical and physical change. By completing such investigations, students would be accumulating evidence about physical and chemical changes that may assist their longer-term understanding of these concepts. The factors affecting conceptual growth are so intertwined that, despite the advice provided in this book, it is impossible to provide a route for a way forward – your 'on the job' knowledge of students will be indispensable.

NOS

Several researchers (Johnson 2005; Tytler and Peterson 2005; Tytler and Prain 2010; Varelas, Pappas and Rife 2006; Wiser and Smith 2008) have concluded that there is no set conceptual pathway that each child follows. However, individual children do appear to follow coherent patterns of 'meaning-making' related to the above factors influencing conceptual growth (for example, how they perceive the purpose of activities and their own perceived 'identity'). Longitudinal case studies of the conceptual development of four children related to evaporation (Tytler, Prain and Peterson 2007; Tytler and Prain 2010) indicate how individualistic their learning journeys were. Most expressed their understanding using 'personally meaningful stories' (e.g., evaporation from sand at the beach), associative thinking (e.g., with clouds) and 'expressive' terms (e.g., 'trapped'). Space does not permit further elaboration, but these students' case studies provide a rich insight into the complexity of conceptual development. If you were the teacher of the same students over an extended period of time, you may be able to identify the coherency of some of your students' conceptual responses based on elements such as the child's identity within the class.

Before considering teaching and learning approaches that may help conceptual development in this content area, it needs to be emphasised that students' inability to distinguish between the properties of macroscopic entities and those of the particles (that is, atoms and molecules) that comprise them, will cause real difficulties for most students in advancing their thinking about physical and chemical transformations. This, therefore, is an argument for introducing students to the notion of particles and the 'bonds' between them, as this will help them to start to appreciate that, just because a material displays a certain property (e.g., change of state), there is an alternative explanation for the change apart from the 'particles' themselves changing state. The alternative explanation can refer to the movement of particles and the strength of the 'bonds' between them. Wiser and Smith (2008) argue that these ideas could be applied to thermal expansion, dissolving, melting, freezing and specific examples of mixing, and then extended to evaporation and boiling, although they believe application to chemical change should be left to secondary school.

Elicitation techniques: matter and change

Several elicitation techniques, similar to those referred to in the above and in other chapters, may be used with this topic. Concept cartoons, depicting, for example, students' differing views about how to stop iceblocks from melting or where the water come from on the outside of a cold object, or what happens to sand castles as the tide comes in (melting, dissolving or something else), can be effective (Keogh and Naylor 1997). How a teacher used cartoons such as these to determine Year 3 students' ideas about melting and then, with the aid of a problem-evoking story, helped his students plan and carry out an investigation to further explore what would reduce the rate of melting, has been described (Morris et al. 2007). One different procedure is the logbook, which is especially useful when changes take place over a longer period of time, as is the case with some change phenomena (e.g., rusting). Students are encouraged to draw changes in their logbooks and add explanatory comments. Qualter, Schilling and McGuigan (1994) provide an example related to the evaporation of water from a tank.

A

Developing students' ideas about matter and change

Primary students hold ideas that are vastly different to the accepted scientific view about how matter changes. Research findings are suggesting some general teaching principles. A major focus on physical changes is more appropriate in the early years of primary school, as young students do appear to appreciate that materials can be modified while still retaining their identity; that is, they are the same materials. Chemical change should not necessarily be omitted in the early years, but an emphasis on chemical change would be better left to at least the middle and later primary years, when explicit and carefully considered scaffolding could be used in discussions about the observed changes (Rahayu and Tytler 1999; Wiser and Smith 2008).

How have primary teachers planned lesson sequences about changes in matter that take students' ideas seriously? In the following, teaching suggestions based on constructivist principles are provided. Initially, the focus is on physical change; examples related to chemical change are then described. Physical (reversible) change examples include phase changes (for example, from gas to liquid – condensation; from liquid to gas – evaporation; from liquid to solid – freezing; from solid to liquid – melting; and from solid to gas – sublimation) and dissolution. Chemical change examples refer to cooking and rusting. Some of these examples could lead to dialogue about the particulate nature of matter.

Learning about physical change

M&E

The contexts for students studying physical change can be many. Examples include the backyard, the household, the kitchen, the supermarket and the farm. Suitable activities for use in lesson sequences can revolve around the properties of water and ice, various solutions (e.g., salt, coffee), milk (including making ice-cream), clay, and slime and oozes (Aubussen 1998; Aubussen and Elliott 1997; New Zealand Ministry of Education 1998). These contexts can be planned or incidental; for example, a weather event (Collyer and Ross 2005). Case studies of upper primary investigations related to freezing (Skamp 1991) and sublimation (Skamp 1992) have been reported, as have preschool and middle primary school students' learning about melting (Collyer and Ross 2005; Preston, Dempsey and Benn 2004). All refer to the students' own ideas and investigations of their own questions. Six case studies/descriptions of sequences are included here. The first describes in detail how a lesson sequence commenced with the intention of students investigating the properties of hot and cold water, and how it became a sequence focusing on condensation. The next two are brief accounts of students' development of the concept of melting. Following an outline of further teaching suggestions related to physical change, there is an evaporation lesson sequence which introduces particulate ideas and illustrates students learning through generating and manipulating their own representations (of evaporation). To conclude the section on physical changes, there are lesson sequences based on exploring 'extraordinary substances': the properties and changes associated with 'slime' and dry ice.

CASE STUDY

A condensation sequence

The teacher of a Year 2/3 class with 30 students commenced a lesson sequence with the intention of getting her students to investigate the properties of hot and cold water. She initially showed her class two pictures: one of children playing at a picnic on a sunny day, the other a snow scene with children skating. The students were asked to give their ideas about the differences, especially relating to the water in the pictures. The students seemed clear about the concepts of hot and cold water. Some questions were raised, which the teacher noted on a wall chart. The teacher listened to the students' concepts and asked them questions to help them clarify their ideas. There was not much interaction between the students about their ideas and they did not challenge each other's ideas because they seemed quite clear about the terms.

The second lesson involved exploration of materials and the raising of students' questions about hot and cold water. The students were divided into groups of four and were provided with different containers, hot and cold water, ice, stirrers of different materials, rulers, measuring containers and magnifying glasses. In this lesson, the students handled the materials, observed and then experimented before coming together to report. Each group had been asked to have an observation to relate or a question to ask. Much discussion and organisation occurred within the groups and the teacher encouraged the students to challenge each other's ideas. Some activities observed were

- mixing hot and cold water
- pouring hot water onto ice
- putting ice into hot water
- holding rods of different materials in hot and cold water
- looking at hot and cold water (and many other things) under a magnifying glass
- putting hot water in different containers and feeling the outside
- pouring hot and cold water along the desk.

When the group gathered together, there were dozens of observations – for example, 'When we put the silver spoon in the water it went grey' – but only a few questions to report. To conclude this second lesson, the groups were asked to go to their work areas and try to decide on a group question that they could investigate. However, as they sat down, there was a similar exclamation from all parts of the room. Many students had observed that 'water bubbles' had appeared on their containers. It was immediately discovered that only the cold water containers had water droplets on the outside and that the best examples were on styrofoam cups. The class regrouped. Six of the eight groups wanted to investigate this phenomenon. The other two groups wanted to know, 'Why does the water look grey when we put a silver spoon in it?' and 'Can you boil ice?' With the teacher's help, the students rephrased the most popular question as: 'Where do the drops of water on the cold cup come from?'

During this second lesson. the teacher had circulated among the groups, listening to the discussions and taking quick notes about their ideas. An example noting how students' ideas can be changed through interaction with other students and the teacher is the following:

Neil	The water bubbles aren't really bubbles because they'd have air in them.
Teacher	What do you think is in them?
Neil	Just water.
Teacher	How could we find out?
Neil	I'll show you [finds a pin and pops some of the droplets].

Activity 9.7 What has happened here?

- This teacher was guided by the interactive teaching approach (see Chapter 4) in her planning. What do you think she was trying to achieve in these first two lessons?
- What role(s) did she take on to try to reach these goals?
- Did this teacher experience any difficulties? If so, can you suggest any reasons why this may have been the case?
- What happened in the second lesson? What decision(s) did the teacher make here? Can you suggest what might have been guiding her actions?
- What would you have done?

The other group members were satisfied that the 'bubbles' were indeed full of water.

The students were really interested and showed remarkable perseverance in a long lesson that required a lot of concentration. The teacher was a little disappointed, however, that no-one asked why droplets appeared on the cold containers but not on the hot ones.

In the third lesson, students' ideas were elicited about what they thought was the source of the water droplets on the cold cup. They did this initially in groups, knowing that they had to report to the whole class. The teacher moved around the groups, helping the students clarify their ideas with open-ended questioning and playing a naive role. She was surprised by the assurance with which most students held their beliefs. When the whole class met, the explanations were listed on a wall chart and numbered. After some discussion and merging of ideas, nine acceptable explanations were listed. Several were discarded as being too far-fetched or unrelated to the topic. The nine ideas were:

1. People bumped the desk and spilled the water.
2. (a) The water came through the air holes in the cold cup and mixed with air to form bubbles on the cup.
   (b) The water seeped through the cup.
3. Wind spilled the water.
4. The water is so cold that it freezes the cup and makes water come from inside to outside.
5. People stamped on the floor and bumped the cup and spilled the water.
6. Water could have spilled onto the outside of the cup when we poured it in.
7. When we put the spoon in the cup it could have overflowed.
8. Little drops, like on cobwebs in the cold mornings, could have come onto the cup.
9. It could have been dew on the cup because it was cold. The dew could have come out of the air.

The ninth explanation came from Jayde when she announced, 'I know the answer because my Poppy [father] told me. It comes from tiny bits of water in the air'. (Jayde's father was a secondary science teacher and had told her this. She admitted, though, that she really didn't understand what he meant.) This explanation was amalgamated with one from several other students, that the drops came from 'dew' or 'fog' in the air. Most of the students held their explanations quite strongly. Please read and think about Activity 9.8 before reading further.

Activity 9.8 What would you do now?

- What has the teacher done in the third lesson?
- What, in essence, are the nine ideas the students have proposed?
- How do they relate to what you have read about students' ideas on condensation (and liquids and gases)?
- What would you do now? Why?

The fourth lesson revolved around planning the investigations. The students were asked to form groups supporting one of the nine explanations, which resulted in interest groups of a workable size. The groups were given some time to discuss how they would test their hypotheses; a spokesperson had to explain the plan to the class. A lot of interaction with other students occurred through the questioning and criticism of plans, which resulted in flaws being pointed out and improvements being made. Eventually, the teacher wrote a refined plan to test each proposed explanation. It was numbered to correspond with the explanation and the students who suggested the plan had their names written next to their refined plan.

Activity 9.9 How would you test the students' hypotheses?

For each hypothesis, decide on at least one way you could test the idea. A critical feature of being able to test hypotheses is to identify the variable to be changed so that tests can be planned. For each of the hypotheses the students proposed, can you identify the independent and dependent variables – see Hackling (2005) and Chapter 2 – and hence plan a test of the hypothesis? If you are unable to do this, try using if–then reasoning; namely, if the hypothesis is true, then it could be predicted that ... (and the effect of a variable change is suggested, which could then be tested).
Compare your responses to those the students proposed in the fourth lesson; these are shown in Appendix 9.5 (see p. 395).

NOS

In this fourth lesson, the teacher once again helped students clarify terms and questions as well as acting as a recorder. The students worked very cooperatively to plan their tests and were eager to carry them out immediately. The teacher felt that the tests were, in general, very thoughtful, although no group suggested putting a lid on the cup and observing whether the droplets still appeared (which would have not supported the spilling, overflowing and wind theories).

In the fifth lesson, which was quite short, the students in their groups carried out their tests and observed the outcomes. Prior to testing, the students had a brief whole-class discussion to remind them about fair testing (a concept with which they were now quite familiar). On moving to their groups, there were some accusations of unfair procedures; for example, excessive desk bumping and not observing results carefully. The teacher felt it necessary to call a class meeting to make it very clear that finding that an explanation is not supported is all part of being a scientist.

After the testing, two groups had inconclusive results. For example, the group that used a container that they knew water couldn't come through, still observed droplets and couldn't agree whether this supported their explanation that the water had seeped through the foam of the original cup. The whole class gathered to hear each other's results. It was agreed that six explanations had not been supported, two were inconclusive and one – that the droplets had somehow formed from dew or fog in the air – was most probably right. The students wanted to tick or cross their explanations on the chart during this session.

During this lesson, the teacher was very busy trying to ensure that the students were staying on task, ironing out practical difficulties, providing equipment and wiping up water. There were some practical difficulties in the lesson, but the teacher felt that, for students of this age, the tests were carried out with a fair degree of rigour. (She wondered whether next time she would do the testing one group at a time while the other groups worked at activities not requiring close supervision. This, she felt, would give her more opportunities to interact with the students and to monitor progress.) The teacher also tried to observe and record interactions, which were abundant and interesting. On several occasions, the teacher was able to challenge the students' ideas. For example, a group of students was observing whether water was blown out of a cup with an electric fan:

Dean	It did blow out – there's water on the floor.
Teacher	Did you actually see it blow out?
Dean	No, but it must of [sic].
Teacher	How do you know?
Dean	There's nowhere else it could have come from.
Rachel	Yes, we could have spilled a bit when we poured the water in the cup.
Dean	Oh! Yes!

Activity 9.10 Generating and testing hypotheses

ONLINE

Appendix 4.4 Interactive teaching approach

Appendix 9.6 Dissolution: a pattern-finding investigation

Harlen (1985) has suggested four types of 'investigation' lesson sequences. These can be integrated into constructivist lesson schema such as the interactive teaching approach (see the online Appendix 4.4), depending upon the questions that arise. The questions could lead to a comparing investigation (see the Chapter 8 case study 'A comparing investigation' on p. 335) or a pattern-finding investigation (see the dissolution example in the online Appendix 9.6). In the case study here, the key question led to a testing-hypotheses investigation. (Harlen's fourth investigation type is question raising, exemplified in the dry ice case study found later in this chapter.)

If possible read Harlen (1985, pp. 169–71) and identify the similarities and differences between her 'Generating and testing hypotheses' investigation type and what the teacher did in the third, fourth and fifth lessons of this sequence. How could the teacher have varied her decisions?

This activity stresses that teachers who are aware of different models for sequencing lessons can vary and integrate such approaches to suit their purposes.

The sixth lesson sought the opinion of an 'expert' (Jayde's father). The students knew he was the high-school science master and so he was readily thought of as an expert. He was invited to hear the class's findings. Each group explained their idea and how they had tested it. This turned out to be an unexpected bonus as it further helped in clarifying ideas. To promote discussion, Jayde's father asked the students open-ended questions. The class teacher again played the role of a naive fellow investigator; she observed, listened and asked questions. She felt that this made the students feel secure in not having to know all the answers because even adults have a lot to learn. Jayde's father brought along a kettle and made steam and showed the students how the droplets did come from the water floating in the air. This provoked a new series of questions, including (at last), 'Why do the drops form on cold things and not on hot?'

The final lesson in this sequence had the students critically reflect and report on their findings in written and pictorial form. It was also to assess the extent to which the students' ideas had changed to more-scientifically correct concepts. Working as individuals, the students filled in a proforma relating to their group's testing of their explanation: 'Our aim – to see if …', 'What we did', 'What happened' and 'I think it happened because …' (The teacher felt later that the proforma needed changing, especially in relation to the last unfinished sentence, which didn't apply in all cases.) They were also asked to write their own explanation of where the water droplets on the cold cup came from. Two examples are shown in Figure 9.1.

Using the students' work and formative observation notes written during the sequence, the teacher determined that, at the end of the sequence, 10 students held the same idea with which they had commenced, six changed to another incorrect idea, and 11 changed in the direction of more-scientifically correct ideas. Although the teacher was somewhat disappointed with this outcome, she did comment that many other learning outcomes were achieved; for example, students asked genuine questions and carried out their investigations, and quite a number were critical of their conclusions and considered alternative ideas.

Activity 9.11 Overviewing this interactive teaching approach sequence

Appendix 1.6 Checklists of notes for constructivist teachers

- Try to identify the various steps and the role(s) the teacher played in the above lesson sequence and then compare and contrast them with a description of the steps, as well as the teacher and student roles, in the interactive teaching approach – see, for example, Chapter 4; Fleer, Jane and Hardy (2007, p. 153); and those suggested by Harlen (2009) in the online Appendix 1.6.
- What do you think are the advantages and disadvantages of the sixth lesson?
- During this lesson sequence, what formative assessment procedures did the teacher use? Compare and contrast them with different ways of using observing as an assessment procedure; for example, questioning, dialogue, listening, assessing processes and products (Harlen, Darwin and Murphy 1977).
- Suggest possible interpretations of the results that the teacher found in the final lesson. In particular, refer to the earlier research findings about students' conceptions of condensation. Reflect on how an awareness that each child's conceptual development need not follow a conceptual trajectory but may be influenced by a range of social and other factors, could influence your interpretations.
- Choose a key concept associated with the physical change of materials and develop an outline of a schema for a sequence of lessons based on the interactive teaching approach.

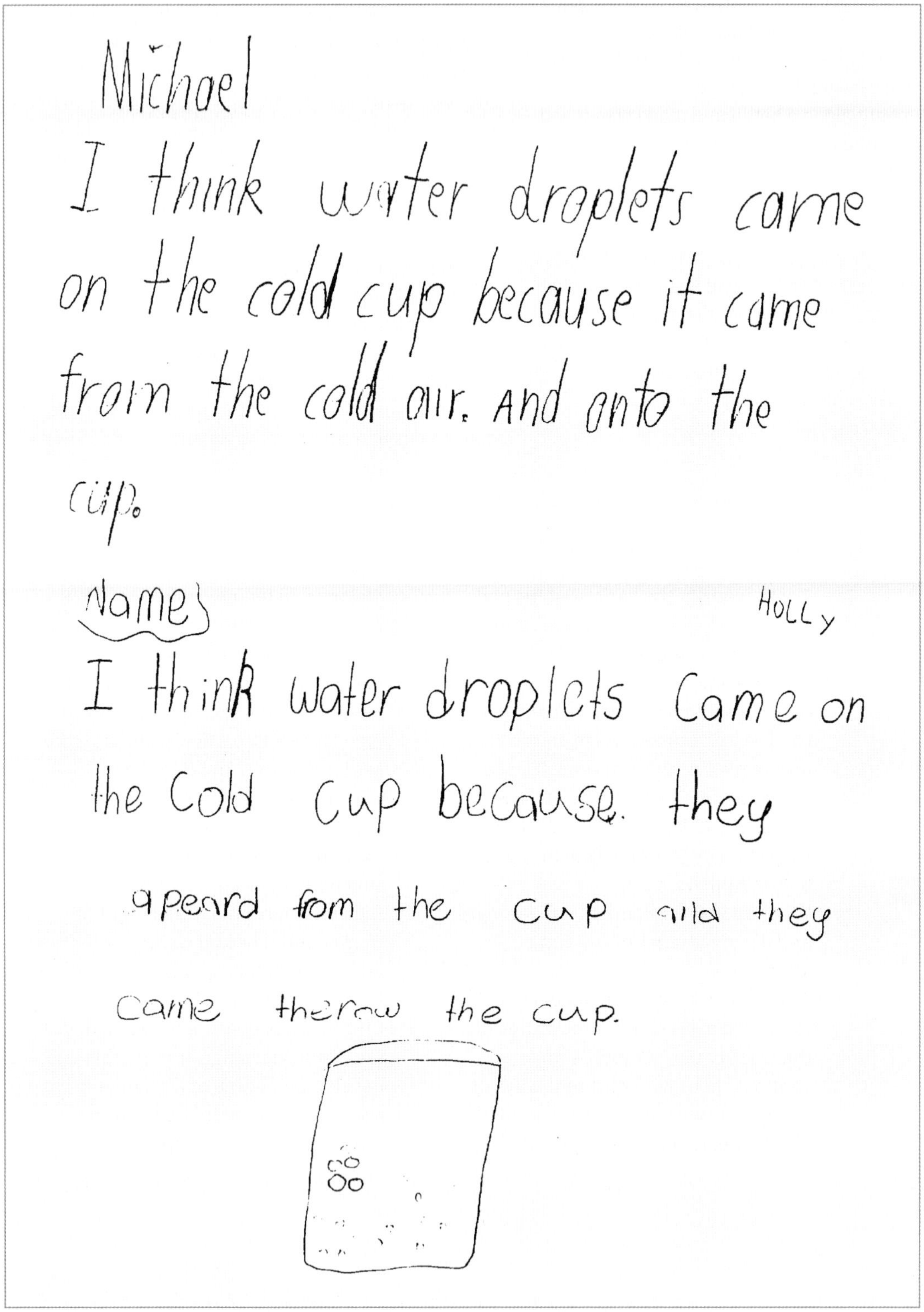

Michael

I think water droplets came on the cold cup because it came from the cold air. And onto the cup.

Name Holly

I think water droplets Came on the Cold cup because. they apeard from the cup and they came therow the cup.

FIGURE 9.1 Two examples of student reports related to the final lesson in the phase change sequence

Changes of state have been common topics in primary science curricula for many years. As the above sequence emphasises, teachers must take into consideration students' ideas. Even then, changes in ideas cannot be assumed because we all firmly hold onto our preconceptions as they have served us well in interpreting our world. This teacher did not consider introducing particulate ideas; it illustrates that conceptual progression can occur without them but if they had been part of these students' thinking, would more students have changed their ideas about condensation? Particulate ideas must be introduced in non-traditional ways to be effective (Tytler, Prain and Peterson 2007; Wiser and Smith 2008) and these will be outlined later.

Another case study of an infants' class learning about evaporation is described in Skamp (1995a). It is an excellent example of how young students strongly believe in their own ideas. One pair of students were convinced that the water on a piece of plastic had gone into the plastic. Even when the teacher encouraged them to cut the plastic to fine threads, they did not readily give up their ideas. This case study revolves around the constructivist sequence proposed by the Children's Learning in Science Project (CLISP; see the Chapter 1 section 'Constructivist teaching models or schemata' on p. 22). It illustrates that there are several approaches a teacher can use which take students' ideas seriously and encourage them to (re)construct ideas.

CASE STUDY

Melting

Melting was the focus for lesson sequences with preschool (Collyer and Ross 2005) and Year 3 students (Preston, Dempsey and Benn 2004). Both were sequences within larger themes: 'Winter and the Arctic' (preschool) and 'People and Substances' (Year 3). Each set the scene and involved the elicitation of students' ideas about melting:

- (for the preschool) There was a discussion of weather in winter, the food we and other animals eat and so on. The role-play area had pretend 'ice' and a life-sized igloo. Using an unexpected event (when it snowed), the children went outside and by playing in the snow familiarised themselves with some of its properties (hot/cold, hard/soft, wet/dry). A mind map was used to record the children's ideas about snow and how it made them feel.
- (for Year 3) The students made observations and gave descriptions of the changing characteristics of chocolate drops, margarine and candle wax as they melted. They were asked to think about how to describe the process of melting.

The preschoolers started to appreciate that snow did not 'go away' but turned into something else, while Year 3 appreciated that heat was required to make things melt and that the materials had not changed, only their physical characteristics.

Teachers of both groups then introduced an intervention to encourage their students to further explore ideas about melting. One was a scientific investigation, the other a design-and-make task which had an investigative component.

- (preschool) Following discussion about using salt on roads, which evolved from looking at books about winter and how people coped, the children investigated whether snow melts faster with salt added. Basic ideas about fair testing were discussed and a very simple activity sheet used, with the headings: 'We will use ...' and 'What I think will happen – I think that ...' The children had to stick pictures of snow, and snow and salt to the sheet to show their planning. It was an example of the ELPS (Experience, Language, Picture,

Symbol) framework for scaffolding children's learning, emphasising language and social interaction (Haylock and Cockburn, cited in Collyer and Ross 2005).

- (Year 3) Following a recap of the previous lesson, the students were asked to think about how they could stop something from melting. They then had to design and make a container to prevent an iceblock from melting. The success of the design was determined by pouring the melted water into containers that could be compared. Elements of fair testing were present (same location, time and so on).

  There was considerable follow-up discussion from both groups.
- (preschool) The children hypothesised, used scientific vocabulary, compared findings and related their investigations to real life, including safety considerations. As only one child had predicted that salt would speed melting, the teacher felt these experiences had been successful – she saw them as stepping stones to children realising that materials do not just 'disappear' and energy is needed for melting.
- (Year 3) The class discussed what features comprised the best designs. To assist discussion and communication of results, digital photographs were taken of the students' 'anti-melt' containers. After transferral to a computer and then an interactive SMART board (SMART Technologies 2003), the teacher and the children matched photos to results. This assisted the teacher to further probe understanding of what had happened.

ICT

For full details of these lesson sequences, see the referenced articles.

Activity 9.12 Identifying constructivist attributes

In the above overviews of two lesson sequences focusing on melting, identify features which you think characterise constructivist learning and constructivist sequencing guidelines.

- How did (or could) each teacher make the lessons relevant to real life? Can you suggest other activities that could have further enhanced real-life connections?
- Discussion and communication helped scaffold and develop these children's ideas. How was this done? What other scaffolds could be used?

If possible, obtain the full descriptions of the above two case studies about melting. Compare your responses with the reflections of the teachers who wrote the articles.

'Engaging science' and engaging students

The above sequences clearly engaged students in their learning. In what ways did they do this? Compare your thinking with the generic ideas about engaging student interest presented in Chapter 1. Sometimes we, as teachers, should consider special ways of motivating student interest, but in conjunction with tested pedagogical principles that will maintain the interest and lead to conceptual progression and an enhancement of inquiry processes. Two examples are provided below that meet these conditions.

M&E

There is evidence for the positive – and sometimes negative – affective (attitudes, feelings or the emotions) value of introducing extraordinary events into the science teaching of primary students – see, for example, Jarvis and Pell (2002) and Shallcross et al. (2006). It is assumed that the unfamiliar, if it produces fascinating effects, will strongly motivate learners to want to find out more. It is also taken for granted that the novelty and fascination of such events will engender a positive attitude towards science. There are probably exceptions to these generalisations, such as the unfamiliar confusing the learner rather than arousing curiosity, but it cannot be denied that young children will not learn if they are bored (Claxton 1989). Affect and aesthetic judgements do influence learning and are linked to conceptual development (Jakobson and Wickman 2008; Osborne, Simon and Collins 2003).

The extraordinary is often remembered as images of real events and phenomena (West and Fensham 1979). Extraordinary and 'engaging science' therefore may have catalyst value in later learning as we apparently store these events in memory and draw upon them in learning; see White's (1988) discussion of 'episodes'. Reflect on your own experiences of chemistry. Do you recall any extraordinary chemical images or episodes? Are the memories positive ones? As a further anecdotal example, one only has to think of the interest shown by primary students – and their teachers and parents – when there are visiting science shows; for example, in Australia, the Science Circus associated with the National Science and Technology Centre in Canberra (see http://www.questacon.edu.au). The chemical demonstrations are always a hit and the materials used are often not difficult to obtain (for example, tomato sauce and cornflour, when looking at the solid/liquid properties of some substances).

Activity 9.13 Extraordinary chemical demonstrations

- Do you believe these teaching approaches have a role? Why?
- In what ways, if any, are they consistent with the approaches to learning science ideas presented in this book?
- If a science show or circus was coming to where you live, how might you integrate some of its program into a constructivist learning sequence?
- Governments spend millions of dollars on science centres. You may have taken your students on a tour of one. Do you think they are effective? If so, in what way(s)?
- Skamp (1995b) may be of interest here: he overviews some research on the impact of interactive science centres.

'Extraordinary' substances: a slime sequence

Appendix 9.7
Slime recipe

Slime and similar materials, those that display properties of solids and liquids, depending upon conditions (e.g., toothpaste), can have a special role in primary science (recipes for PVA slime and others may be found in Aubussen (1998); see also the online Appendix 9.7). Aubussen and Elliott (1997) argue that scientific ways of viewing the world, as often presented in syllabuses and texts, are not always apparent in the everyday world – an example is that there are three distinct states of matter. Materials with plasma-like properties, which are difficult to classify, do not neatly fit these categories. Rather than avoid such (everyday) examples (e.g., snot), they argue that such examples are motivating to learners (also consider using the film *Flubber*) as well as encouraging deeper discussion of what is a solid, a liquid and a gas (see also Levinson 2000). This type of social interaction – emphasising aspects of the social constructivist view of learning – is also important in students' conceptual development.

Based on the above and other considerations, Aubussen and Elliott (1997, p. 151) suggested a sequence for teachers to consider; it combines elements of various constructivist planning models. This sequence exemplifies how you could adapt constructivist teaching schemata. Aubussen and Elliott illustrate the steps in the sequence with slime examples. Their sequence is:

- identify students' experiences with the phenomenon to be studied (e.g., slimy soap)
- have students pose questions about the phenomenon
- encourage students to talk about the phenomenon as they attempt to answer their questions
- facilitate students' exploration and investigation of the phenomenon by observing and manipulating materials; encourage recognition of similarities and differences
- talk with students about relevant scientific views and encourage students to question these views
- support generalisation of ideas where possible (e.g., What about glass that flows?).

In these steps, can you identify aspects of various constructivist schemata for sequencing lessons (e.g., the 5E or interactive approach)?

'Extraordinary' substances: a dry ice sequence

The following lesson sequence revolves around the 'extraordinary' substance dry ice (Skamp 1992). Students find out about the dry ice directly; demonstrations play only a minor role. This provides students with the satisfaction of knowing that they can learn about dry ice themselves.

CASE STUDY

A dry ice sequence

This lesson sequence is based upon Harlen's (1985) question-raising investigation type; an outline is provided below:

Appendix 9.8
Student ideas and questions about the behaviour and properties of dry ice

- The students 'messed around' with the dry ice in a semi-structured way to see what they could find out about it (see the online Appendix 9.8 for some of their observations).
- Comparisons with ice were made (see the responses in the online Appendix 9.8).
- Further directed observations, using a predict-and-test strategy, were made using a range of materials and substances and a very cold mixture (dry ice and acetone).
- The students were then asked what they would like to find out about dry ice using the worksheet in Figure 9.2 (see the online Appendix 9.8 for examples of the questions the students asked).

FIGURE 9.2 Worksheet on dry ice

Finding out more about dry ice

In your groups write down three questions about dry ice that your group would like to find answers to. Before you write down your questions, make sure everyone in the group agrees what the questions should be.

Try to write questions that your group thinks they could find answers to by doing an experiment with simple equipment here in our classroom laboratory.

Question 1

____

____

____

Question 2

____

____

____

Question 3

____

____

____

If you encourage students to raise questions, it is most advisable to address them in some way so that they appreciate that their questions are valued. This can be done by having them interact with materials, demonstrating effects, directly answering factual questions (if you know the answers and believe the students can make sense of them), seeking further clarification of the students' questions – especially the 'how' and 'why' ones – or simply saying 'I don't know' and, if appropriate, guiding the students to reference sources.

Activity 9.14 Reflecting on the dry ice sequence

- Try to identify the steps guiding the dry ice sequence to the stage described above. If possible, compare with Harlen's (1985) question-raising sequence.
  - What strategy has the teacher used in the initial activities? Why do you think this was done?
  - Suggest some reasons why the teacher didn't simply ask for questions from the students to begin with?
  - Can you think of alternative steps or approaches the teacher could have taken? Describe these.
- In the online Appendix 9.8, look at the questions the children raised.
  - What would you do with these questions?
  - Are the questions of different types? In what way(s)?
  - As a teacher, how would you handle 'why' questions?

Appendix 9.8 Student ideas and questions about the behaviour and properties of dry ice

Turning non-investigable questions into investigable or action questions is an important teaching skill. Which of the questions are investigable, in the sense that the students could find answers to them by doing something with materials? How might you help students turn some of the others into investigable questions?

You will find 'Children's how and why questions' in Harlen (1985, pp. 43–4) and 'Handling children's questions' in Harlen and Jelly (1989, pp. 64–5), which provide useful reading about this issue. Other sections in this book also deal with types of questions (for example, Chapters 2, 3, 7 and 10).

With the children's questions about dry ice (in the online Appendix 9.8), the teacher did have a follow-up demonstration/group-experimentation session in which he:

- mixed dry ice with a range of (new) substances
- placed a piece of dry ice in a candle flame
- inserted a lit taper into a beaker with dry ice in it
- placed small pieces of dry ice on large quantities of liquids to see if they bounced – not answering the 'why' question, but turning it into a 'what if' question
- measured the temperature above it ('Why does it steam?')
- put it in an electrical circuit to see if it conducted electricity
- placed it in hot and cold water and compared the effects
- blew through a straw into lime water and compared the effect with placing some dry ice in the lime water.

In activity 9.14 what did you suggest be done with the children's questions? Was it similar to some of the above?

This sequence refers to a range of physical (and some chemical) changes. Identify the physical changes. No reference was made to particulate ideas. If these students had encountered particulate ideas in previous lessons, how could they have been integrated into (or added to) this sequence?

For these students, the dry ice activities were the highlight of a range of science activities undertaken over a year (Skamp 1996a). The 'extraordinary' had a positive affective influence in this instance – see Henderson (1995) for further ideas on 'experiments' that could be used with approaches similar to the above.

What of the long-term effects? This is a difficult question to answer, but from time to time it would seem that it is worth the effort to apply constructivist teaching principles to extraordinary materials and/or events. A point worth bearing in mind is that the extraordinary for students can often be the very ordinary for adults. For example, similar teaching strategies to those used in the dry ice sequence were used by the author with snails as the 'materials'; they received equally enthusiastic responses (see Chapter 6 for responses to other animals). Most very good science teaching can be done with the simplest materials, but the extraordinary, from the student's perspective, has its place.

Changes of state: other teaching considerations

Conceptual development and literacy–science connections

The role of collaborative discussion is now recognised as a key contributor to students' conceptual development of science ideas. Characteristics of this type of dialogue are that children have many opportunities to offer their own ideas, comments and questions, and the teacher and other students can respond to such suggestions in a conversational mode. It requires teachers to become attuned to children's ways of speaking about phenomena. The whole purpose of this dialogue is the 'sense-making' of events and phenomena.

What can encourage such collaborative and conversational dialogue? 'Text' has been argued to be a major catalyst. Text here is interpreted broadly as the 'recounting of events, references to hands-on explorations and connections to prior discourse'. When students make reference to such texts, scientific understanding and use of scientific genres and registers (e.g., third person, general 'you' pronouns, 'if-then' reasoning) can be developed. The students' language may be a hybrid and does not usually use formal scientific registers (Varelas, Pappas and Rife 2006, p. 638).

In one study, Year 2 students engaged in a science literacy unit on states of matter, which emphasised the water cycle. The teacher used in-class and at-home hands-on explorations and related discussion, read-aloud sessions of information books, and other activities such as literature discussion circles with whole-class reports and various writing tasks. All the student discussion that referred to condensation, evaporation and boiling related to some form of text – whether from other media, prior conversations, recall of specific events and so on. Three types of intertextuality were more prevalent in encouraging dialogue about these physical processes. The in-class hands-on experiences (HOE) led to the most science-oriented dialogue, followed by the recounting of generalised events (GE) such as about clouds and fog, and prior classroom talk (PT) about the physical change events. In this research, it was the teacher who initiated dialogue related to HOE and PT, but the students initiated dialogue with GE. Both the teacher and the children were involved in the continuation of dialogue. These conversations often linked back to observations and experiments of the children's, whether through class activities or recall of events, and the teacher would engage her class in 'thinking about "why" – why a particular event happens, what it means, and what would happen in a different case'. She was encouraging a 'theory-data dance' (ibid., p. 656) and hence teaching about the NOS.

NOS

Findings such as these – here, related to students' conceptual development about condensation, evaporation and boiling – and others reported earlier in this chapter (see Tytler and Peterson 2005) are strongly suggesting the importance of collaborative dialogue related to text in helping students think conceptually about everyday phenomena. They also emphasise the scaffolding role of the teacher in encouraging dialogue. These studies are signalling a change of focus in thinking about constructivist learning. Teachers need to be 'centring on what (children) bring rather than what they lack' (Varelas, Pappas and Rife 2006, p. 656) when dialoguing with them.

Further suggestions to advance conceptual progression

Various studies (Driver et al. 1994a; Isabelle and de Groot 2008; Johnson 1998a, 2002; Newton 2002; Ross and Law 2003, Tytler 2000; Tytler and Peterson 2000; Tytler, Peterson and Prain 2006; Tytler, Prain and Peterson 2007; Tytler and Prain 2010; Wiser and Smith 2008) provide specific teaching advice related to physical changes. This advice, for change of state (and other physical change) lesson sequences, varies from general to specific and includes the following:.

- Use a wide variety of contexts related to a phenomenon (e.g., evaporation) and ensure that related phenomena are bracketed together in classroom discourse; for example, 'Why has the puddle disappeared?' and 'Where has the water gone?'. This will help students make links between appropriate phenomena and ideas.
- Include anecdotes about everyday examples of changes of state that provide useful associations to which students can relate, and hence use such experiences to support more plausible explanations. (Can you find any examples in the condensation and other case studies?)
- Refer to examples other than water and ice – students often see these as a prototype and base many of their generalised conceptions upon this prototype. Other examples could include oil, alcohol, wax, chocolate and butter.
- Decide whether the classroom focus will be on a descriptive or explanatory understanding. The former is when children can describe a situation (but not necessarily the underlying reasons for it); the latter is when some reasons for the cause of an event are discussed. An example might be that younger students simply identify the observational differences between chocolate melting and sugar dissolving, rather than identifying the underlying reasons.
- Encourage students to represent their observations and understandings in various forms or modes, and to discuss the links between, and the limitations of, these various representations. These forms include categories such as descriptive (verbal, graphic, tabular), experimental, mathematical, figurative (pictorial, analogous, metaphoric) and kinaesthetic (or embodied). Where appropriate, introduce these forms as teachers but also have students use multimodal representations of their understandings of particular concepts (e.g., dissolving) and investigations relating to these concepts (e.g., the effect of heat on the dissolution rate).
- Appreciate that, on occasion, for conceptually productive student–student and teacher–student dialogue to occur, certain concepts may be required that are not accessible by the student at a particular time. An example would be students needing to understand that, when considering condensation, air is a site for water to exist.
- Consider the use of analogies to assist conceptual understanding. An example could be to build a house with Lego and show how 'solid' it is, and then pull the bricks apart and show the 'pool' of bricks. Help students to draw connections between the analogy and, say, chocolate melting. (The teacher must keep the limitations of analogies in mind; see chapters 5 and 10.)
- Teachers should emphasise that, when materials (and substances) change state, they are still the same material (substance). Avoid language suggesting the solid, liquid and gas forms of the same material (or substance) are three different materials (substances); for example, saying that there are 'solid particles', 'liquid particles' and 'gas particles'. This could be significant because when phase changes look perceptually different, students will often think that there has been a new material formed. It does suggest using phase changes that look similar during transformation (e.g., melting chocolate) before those that do not (e.g., melting wax).

R

- Identify, on the basis of conception studies, appropriate focuses when studying particular phenomena. For example, on what would you focus when introducing the water cycle (see Activity 9.5 on p. 359)?
- If appropriate, use objective rather than perceptual measures, such as weighing before and after phase changes. This can assist in changing inappropriate thinking such as that, because there is more of something (appears to be more of it), it must weigh more.
- Similarities and differences should be a feature of some discussions. Examples of similarities are that gases are matter just like solids and liquids, and that evaporation is occurring all the time, not just when a liquid boils. An example of differences is that melting need involve only one substance and requires heat, whereas dissolving involves two substances and may not need the application of a source of heat.
- Discussion that revolves around water boiling draws student attention to the reduction in the volume of the water, which may assist older students to appreciate that the bubbles are water and that the visible condensed water droplets – steam – do not account for all the water that has boiled away. This may eventually help the small number of those older primary students who start to appreciate the nature of water vapour (that is, water in a gaseous state, which is not the same as steam) to see also that other substances can have a gaseous state.
- As implied in many of these suggestions, use discursive processes (see the Chapter 1 section 'Constructivism and learning through participation in discursive practices' on p. 15) in which there is active classroom conversation where teachers and students respectfully reason and debate ideas and share them aloud with each other.
- Avoid simply 'telling' students an accepted scientific view when it is at odds with their intuitive view – it can confuse more than assist learning. Rather, use some of the suggestions above as other ways forward.

Changes of state: teaching about particles

Whether to introduce students to the particulate nature of matter so that they can start to appreciate a few of its properties is obviously an area in which teachers should move carefully (also see Chapter 8). They should do it realising that most, if not all, of their class has not thought of this way of interpreting the material world. As there is some evidence that speculative discussion can sometimes assist conceptual growth, and as children are natural speculators, then introducing particulate ideas, or extending discussion if students raise such ideas, could be a worthwhile technique to explore (Reynolds and Brosnan 2000).

The formal introduction of particulate ideas, usually but not necessarily at the upper primary level, is being advocated by several researchers. They suggest specific particulate ideas (namely, 'unchangeable particles, existing in a vacuum and held by bonds of varying strength') be taught consistent with the macroscopic experiences that students have encountered and thought about (Wiser and Smith 2008, p. 223). How, then, can teachers effectively do this?

One way forward is to use resources which refer to matter being comprised of particles but that are based on constructivist principles and include elements of a representational learning approach. A series of primary teachers' guides on chemistry topics by Griffin and

Sharp (1995) meets these criteria. It revolves around the learners' questions approach (based on the interactive teaching model). One booklet for infants is called *Strong Colours*. The following exemplifies the approach and indicates how the particle nature of matter is introduced.

The students initially indicate their present ideas about shades of colour by arranging collections (sticks, petals and so on) according to shade and/or telling their teacher about different shades. Extending experiences include activities on light and dark (making lighter and lighter colours; preparing a shade chart), thick and thin (making thinner paints), and strong and weak (cordial strengths where students make a class graph using cup shapes to show which strength they like the best and/or a halve-and-pour activity related to 'Will the colour disappear?'). An extension idea was to make jelly to see if it still sets, depending upon the amount of gelatine used. The authors then encourage students to try to find out answers to their own questions that arise from these activities. To bring the ideas together, a novel simulation game is suggested in which some students play bits of water and others bits of cordial. The weaknesses in the analogy are mentioned, but it is still a useful introduction to particles that is reiterated in later activities in this series. This use of particle role-play has been suggested by others (Skamp 1999; Stringer 2002) as a means of making this idea accessible to students. Clearly, when using such resources/approaches, student learning should be monitored to formatively assess their appreciation of particulate understandings; using knowledge of research findings about their thinking in this area would be an advantage.

R

Generating and testing scientific representations/models

Students can develop their conceptual understanding through the generation and testing of *their own* representations of what is happening in particular phase-change situations. Some writers see this as analogous to scientific (mental) model building and testing. With either interpretation, teachers also can be explicitly teaching about the NOS. Two descriptions of lesson sequences related to evaporation (and in the second, also condensation) illustrate different applications of this approach.

In the first constructivist sequence, the teacher introduced a particle model to Year 5 and students were encouraged to use various representational forms to express their conceptual understanding of some evaporation events (Tytler, Peterson and Prain 2006; Tytler, Prain and Peterson 2007; Tytler and Prain 2010). An overview follows:

- After initially obtaining students' responses to various contexts related to evaporation, condensation and dissolution using elicitation techniques focused on annotated drawings, students completed a worksheet requiring them to draw and annotate, in eight blank boxes, eight stages they had observed in the process of water boiling. A further elicitation discussion followed.
- A physical representation of a particle model – namely, plastic beads in a tray – was then introduced as an idea that may help students think about evaporation. The representation was not presented as *the* way to explain what was happening, but as something students may find useful in understanding what was happening. Discussion ensued, with students being asked to imagine the beads (molecules of water) breaking free of the surface and going into the air. Then a pictorial representation of water boiling was constructed where

water molecules became more spread out as they changed from a liquid into a gas. The discussion was aimed at showing how different representations (pictorial, physical and so on) were all linked and that the representations had limitations.

- Various other evaporation activities were completed over the next two weeks: the level of water in a jug over a week was observed; evaporation of a puddle was noted and students were asked to draw what they thought could be seen through a powerful microscope; and an investigation into the drying of cloth under different conditions was carried out.

Interviews with some students following the sequence showed that particulate ideas as an explanatory model did not expand some students' thinking about the water cycle, while for others it meant new ways of looking at evaporation. About 80 per cent of students used 'molecules' in their explanations of puddles, although their 'molecules' were like 'bubbles' and they 'dissolved' in the air. It is interesting to note that, in a different context, when asked about alcohol evaporating into the air, all students advanced their thinking about the distribution of alcohol molecules in the air. Overall, of nine students who had follow-up interviews, four offered acceptable particulate accounts of alcohol evaporating, while others, for example, hybridised particulate and water-cycle ideas. Also, students displayed various preferences for using visual, verbal and gestural modes and the extent to which they combined them.

R

In the second sequence, also with a Year 5 class, the students constructed, tested, evaluated and revised models of events involving phase changes (Kenyon, Schwarz and Hug 2008). Initially, these students were shown a crudely constructed solar still made from everyday materials. It had collected clear water from a dirty sample and most students were hesitant about where the clear water had come from. After having the students explore some instances of water evaporating (e.g., from a plate) and seeking their responses, the teacher held a whole-class discussion about the purpose of models and the need for them to be tested and shared with others for feedback. The students were then asked to draw diagrams as their first attempt at a model of what was happening – the diagrams had to show what had happened to the water (description) and how it had happened (explanation), and were required to depict a before, during and after sequence. The models were then tested in various ways, such as measuring the weight of the water in an open container over time and using cobalt chloride strips to detect water in the air (e.g., near a humidifier). Students were then asked to evaluate their models using class- and teacher-generated criteria, including the model's consistency with evidence collected from investigations and other scientific ideas, whether important features are shown, and how they are shown (e.g., arrows). Students then revised their diagrammatic representations. At this stage, the students, working in pairs, tested their model against other ideas – they used teacher questions with computer simulations of evaporation and condensation, as well as other information. Further revision of models then ensued before the class compared models in order for it to reach a consensus model, which was then tested against other evaporation phenomena. The process was then repeated for condensation before a return to a discussion of what happened in the solar still. The sequence was spread over several lessons across a number of days.

These two sequences may be compared with each other and with Acher, Arca and Sanmarti's (2007) approach to students generating particulate models, discussed in Chapter 8 (see the section 'Narrowing student-generated "scientific" (mental) models' on p. 325).

Contrast the similarities and differences between the three sequences. In what ways have they engaged students' affective and conceptual interest? What assumptions have they made about students' conceptual development? How has 'representational learning' been integrated into the sequences? Which approach do you prefer? Justify your thinking.

Activity 9.15 Teaching about the substructure of matter

Locate some primary-science teaching resources which introduce the particulate nature of matter; for example, Griffin and Sharp (1995) and the *Primary Connections* unit 'Change Detectives' (AAS 2009b). Identify the ways in which the concept of the particulate nature of matter is introduced and developed: Is it constructivist? Why or why not? Was representational learning used? If so, how; if not, how could it have been? (If you are unable to access these resources, use Jackson (2009), which describes a 5E sequence, with years 3–5 students, using particulate ideas and related to water-changing states.)

If possible, implement a constructivist sequence (perhaps using one of the above resources) that refers to the substructure of matter and attempt to gauge the understanding of your students about the particulate nature of matter. You could, for example, ask individual students to talk to you about the meaning of their written and diagrammatic responses to tasks and/or worksheets (as was the case in the above evaporation sequence).

Did the data you obtained provide evidence that the students appreciated that matter is made up of particles? Remember, you only tried one sequence – such a 'big idea' usually takes considerable time for students to construct: several years – see Wiser and Smith (2008).

A nanoscopic view of particles may still not be appreciated by some primary students, irrespective of the activities used. However, they may form hybrid ideas about particles that could be stepping stones to understanding; see, for example, Johnson (2005). Perhaps, as teachers, we need to more explicitly distinguish students' non-scientific descriptions of particles (such as: they are getting hotter, they are melting) from scientific particulate descriptions (they break free from each other, they vibrate faster). This, for example, could occur in novel ways, such as a creative writing task in which students are imagining they are particles (Ross and Law 2003).

Other advice when teaching about particle ideas

In summary, general advice about introducing particulate ideas with physical changes includes the following:

- Use students' multiple modes of representation of a change in materials as an avenue for uncovering thinking that could lead to clarifying conversations.
- Integrate scaffolded role-play, possibly with appropriate games (e.g., freeze-tag) and songs (e.g., about melting), of particulate distribution and motion, in conjunction with macromolecular observations and investigations; try to ensure the role-plays are consistent with the particulate model of matter.
- If particulate understanding seems to be advancing at some level, then do not necessarily feel that a more complete 'particulate picture' is needed at that time. Remember that some non-scientific views of particles can be stepping stones to later learning.

- When asking students to imagine what could be happening in a physical or chemical change context, interpret their efforts as using the representational resources they have at that time (e.g.,, use of their arms and/or drawings and/or words) as well as suggesting other ways they could represent their ideas.
- Constantly see your role as negotiating meaning with your students through their representations (that is, 'representational negotiation') and use timely scaffolding in this process. This could, for instance, mean asking students to clarify what parts of their representations mean to them, such as squiggly lines or arrows.
- Be explicit about the nature of representation/modelling in science, emphasising that it is one of the ways in which science moves forward in its attempts to explain what is happening in our world. The particulate nature of matter is one of the most powerful representations available to science in that it 'explains' so many phenomena. Students need to grow into its gradual use over many years.
- Introduce the notion of levels of 'ability to hold' between different particles.
- Use, if available, carefully scaffolded multimedia depicting particulate motion. Two key multimedia scaffolds were used with Year 5 students (n = 16) over six lessons. The scaffolds were 'measures' of the differential 'ability to hold' of different substances and their comparative 'energy'. When students manipulated the temperature, then these measures were shown in conjunction with images of substances melting and particle distributions. About 75 per cent of these students progressed towards more-acceptable particulate models of matter. The positive differences compared to a comparison non-multimedia group were minor, but the researcher and teacher-researcher (Papageorgiou, Johnson and Fotiades 2008) still recommended the approach.
- Critique websites depicting particulate properties that are recommended for primary levels (e.g., The Le@rning Federation; see http://www.thelearningfederation.edu.au/for_teachers/sample_curriculum_content/tm_-_science.html) using guidance in this chapter, with the appropriate teacher scaffolding (e.g., teacher questions and/or tasks)

Jackson 2009; Papageorgiou, Johnson and Fotiades 2008; Tytler and his colleagues' various studies

In summary, learning science through representational negotiation is fundamental to conceptual development, as well as assisting in an appreciation of the NOS. Further, it engages students' interest because it is their representations that are the focus of attention and learning. Teachers 'taking on board' this interpretation of learning will be developing a different mindset about concepts and conceptions that could profoundly influence their pedagogy. Concepts become 'ways of thinking that provide the means of explaining a variety of phenomena' and 'conceptions can be understood as a mix of representational accounts of phenomena' (Tytler and Prain 2010). However, be aware that students are not used to thinking about their representations as 'tools for investigating and explaining phenomena' and they will need to see this as 'a way of learning', which studies have shown they can, for it to be effective (Wiser and Smith 2008, pp. 217, 233). Why not compare this view of 'concepts' and 'conceptions' with the view you currently hold and talk it over with a colleague. What could be some of the implications for your teaching?

Learning about chemical change

> The essence of a chemical reaction is not its spectacular display, nor the unexpected outcomes demonstrated in classrooms. Rather, it is the fact that *substances are changed into other substances.*
>
> Bucat and Fensham 1995, p. 60 (emphasis in original); see also the online Appendix 9.3

Appendix 9.3
Concepts underlying the processes of burning and chemical change

Virtually all primary students have difficulty with this idea, which underpins chemical change, because they do not have a chemical concept of substance. Realising this, what types of learning experiences might start to lay the foundation upon which they may later start to construct this proposition? What does your science syllabus or framework include as examples of chemical-change activities and at what level (see Activity 9.1 on p. 352)? One national set of curriculum guidelines (ACARA 2011) suggests that upper primary students could learn about changes to materials that are irreversible and that this can affect their use (examples included burning, rusting, cooking and making cement). Irrespective of your syllabus content, chemical changes will be encountered across the primary years and teachers need to be open to the possible foundational learning they could encourage.

Specific chemical-change activities

Most students enjoy cooking, especially if they get to eat the results, and many teachers include cooking at some time in the experiences of their class (often not for science learning purposes). Teachers could help students to think about cooking as an example of chemical change. Drawing out the differences between raw and cooked food (smell, texture, appearance – inside and out) can be helpful, especially as students may not recognise reactions in everyday situations (Driver et al. 1994a). The use of heat in promoting the formation of new materials can be a focus while being conscious that some students think of heat as a material (substance). Eggs are good examples to use, especially watching an egg being poached, as is making bread and yoghurt (see Sharp and Griffin 1994a, 1994b, 1994c).

An interesting introduction to chemical symbols and codes (for physical and chemical changes) has been described by Griffin and Sharp (1998); teachers trying this suggestion need to appreciate that children often misunderstand what is meant by (shorthand) arrows. Other resources include the *Primary Connections* unit 'Spot the Difference' (AAS 2009a), which introduces changes to foods, including cooking, in the lower primary, while chemical change is more formally encountered in the upper primary unit 'Change Detectives' (AAS 2009b). The following websites provide novel chemical activities: The Le@rning Federation in its 'Science: Natural and Processed Materials' strand (e.g., the kitchen chemistry, mystery substance and chemical science series) (see http://www.thelearningfederation.edu.au/verve/_resources/sci_natl_processed_march10.pdf), Monash University's Primary Science Teaching Resources (e.g., see under chemical activities at http://sciencecentre.monash.edu/ptr/lessons/chem1.html), the CSIRO (http://www.csiro.au/resources/ChemistryActivities.html) and Fizzics Education (http://www.fizzicseducation.com.au/experiments/Kitchen%20Chemistry/chemistry.html).

Activity 9.16 Introducing chemical change (including with particulate ideas)

Read the *Primary Connections* unit 'Change Detectives' (AAS 2009b), especially lessons 5, 6 and 7.

- Identify the ways in which the chemical-change concept was introduced and developed.
- Comment on whether the particulate representation of physical change in lesson 2 (and follow-on lessons) could have been extended to chemical change in the later lessons. Use your knowledge of students' ideas about particles to justify your thinking.
- If you were to extend student thinking about particulate interpretations in the later lessons, what approaches would you take? Why?

Guidelines when teaching about chemical change

Many of the general suggestions for assisting students to develop their ideas about physical change may be applied to learning about chemical change (for example, use of multiple forms of representation of a conceptual idea and collaborative dialogue related to various text types). More specific principles that teachers could use to guide their discussions with students about chemical changes are listed below. They need to be used with discretion because there is no consensus about them; some researchers question whether chemical change should be introduced before particle ideas are accessible to learners (Johnson 2002; Wiser and Smith 2008). The suggestions seen within this context – derived from Driver et al. (1994a), Griffin and Sharp (1998), Johnson (2000), Liu and Lesniak (2005), Papageorgiou and Johnson (2005), Rahayu and Tytler (1999), Ross and Law (2003) and Wiser and Smith (2008) – are as follows:

- Encourage students to think about the products of chemical change as new materials (and substances), rather than as products that are still directly related to the starting materials (for example, try to avoid supporting the idea that rust is still iron) or simply a mixture of the starting materials.
  - This can be assisted by using new names for the products of reactions (e.g., soot, even oxygen and carbon dioxide) as they are words commonly used by many upper primary students. It must be emphasised that there is considerable evidence to show that simply stating that something is the case (for example, 'This is a new material') will not change conceptions. But what is suggested here is not a teacher lecture or recitation, rather an awareness of the role of language in concept development that may have an effect over an extended time.
  - Another idea is to show, if possible, that the products (materials and substances) have different properties to the starting materials, This, of course, is not necessarily straightforward or definitive; for example, in a physical change, liquids and solids of the same material or substance have different observable properties.
  - Where appropriate, indicate that gases have been produced, keeping in mind the difficulty students have in appreciating what a gas is.
  - Sometimes analogies may help, but teachers need to discuss their strengths and limitations (see Chapter 5). For example, to emphasise that new materials are formed,

change a Lego truck into a person's face using the wheels for eyes. The truck is no longer there but the face is, just as when burning occurs, air and wood were there but now there is ash and fumes. For a few students, this may also suggest that matter is conserved even in chemical change.

- Do not overlook focusing student attention on small and slow changes when materials (and substances) interact.
- Help students appreciate that materials (and substances) can be in contact with each other and still not change; this may be contrasted with when change does occur. Be aware that most primary students attribute change to one substance rather than the interactions between substances. Listen for these interpretations and focus on encouraging dialogue related to responses that suggest interactions (also see Chapter 8).
- Remember that noting changes in appearance does not mean that students believe that chemical change has occurred. Students may still think that the same materials (and substances) are present.
- Reiterate that stuff (matter) cannot just be created; it is not that all students will believe you, but it is using language carefully.
- For older students, word equations may highlight new substances being formed.
- Avoid overemphasising that chemical changes are irreversible, while still indicating that new products are formed.
- Do not feel compelled to use everyday examples of chemical change. As students in general find reactions involving chemical change interesting, often because of their magical appearance, contrived instances can sometimes assist understanding more than everyday examples, which can be harder to interpret.
- If particulate ideas are introduced at upper primary level, consider using the idea that particles differ in their 'ability to hold on to each other'.

Equipment, consumables and safety

This issue is clearly the nitty gritty of teaching primary science and obviously relates to activities in all content areas, including those associated with materials and change. Further, as effective primary science teaching requires students to handle materials, then attention must be given to the nature of the equipment and consumables to be used and associated safety concerns. Consequently, some general comments about this topic will be the focus here, although clearly the topic of physical and chemical change will include many activities which require special care as they may involve heating and reactions between substances.

Equipment and consumables

In general, everyday and household materials will suffice for most science activities, but there is a place for gradually introducing specific science equipment at the primary level. Having equipment in primary science allows students to:

- understand that different kinds of equipment have specific uses
- appreciate that classroom equipment often has a real-life equivalent and that skills learnt in equipment use at school can be used elsewhere

- learn that measurement is important in science and equipment can assist in taking measurements (see also the Chapter 2 section 'The role of mathematics in investigations' on p. 73).

A very useful scheme describing student progression in the choice and use of science equipment from preschool to Year 6 has been developed. It provides detailed guidance for teachers across three categories: choosing equipment, using equipment, and safety and care. This developmental listing is highly recommended (Feasey 1998).

In an ideal school situation, science resources need to be wide-ranging, available in suitable quantities, up-to-date, child-friendly, safe, accessible, appropriately stored (and monitored) and well maintained. Feasey (1998) offers practical suggestions on how to make this ideal a reality; she also includes a suggested inventory of equipment and consumables that schools could build up over an extended period.

Safety

Equipment and consumables raise the issue of safety in primary science classes. Most dangers in primary science can be anticipated if teachers:

- stress safety precautions rather than dangers
- check out activities for possible safety issues
- are aware of the background and abilities of the students involved
- realise the impact that excessive movement may cause
- use and teach about effective labelling
- know how to dispose of waste material
- are aware of how to get help in an emergency (Bradley 1996).

Even so, this topic should not be taken lightly. Schools really need a health and safety policy which includes information about 'where we get advice on health and safety; how we deal with risk assessment; how we ensure people are implementing the policy; and how and when we review how successful we are being' (Borrows 2003, p. 18). As risk assessment is the key to safety, it is imperative that teachers consider potential hazards and their associated risks and what control measures need to be in place. Although often a mental process, it always needs to be completed and if necessary noted in program and/or lesson preparation. To encourage you to reflect on safety related to any primary science activities, consider responses to the following questions:

- How much do you know about safety?
- How confident are you when it comes to treating burns and bad cuts, using glue guns and soldering irons, knowing what heat sources to use, using electricity and electrical appliances, safely cutting wood and tin cans, selecting the best containers to store hot water in, and keeping living things in the classroom?
- Are there safety considerations when students are flying hot-air balloons and other such things? Using thin wires? Making things? Testing the strength of shopping bags? Experimenting with mirrors? Seeing how quickly food grows moulds? Measuring temperatures with a thermometer? Cooking? Collecting sun observations? Using toys for force activities? Using household and other chemicals? Investigating plants?

At times, questions about safety do not have clear-cut answers, but you should still think about them – Are safety goggles required at the primary level? When is it safe to mix kitchen chemicals? And so on.

Other safety issues that need to be considered are:

- Where is safety mentioned in your science syllabus? Are there student-learning outcomes that are related to safety? How could these be incorporated into science lessons?
- Is there any legislation about which you or your school should be aware in relation to science activities? (There are requirements about the labelling for chemicals. Other regulations will limit the range of animals that may be kept in classrooms.)
- How can the students in your class be made more safety-conscious? Is your classroom safe? What contributions can students make to classroom safety? Bradley (1996, pp. 136–7) describes how a class of seven-year-olds made their own safety rules, including 'It is not telling tales to let the teacher know that an accident has occurred'.
- What safety issues are involved in outdoor science? Pond activities? Science excursions (Bunyan 1988; Vincent and Wray 1988 – reprinted in ASE 1994)?
- What are the roles of signs and symbols in encouraging safety (ASE 1998)?
- What responsibilities do teachers have for their students' mental safety in primary science? Examples could include sensitivity if investigating human physical characteristics and being aware that handling animal parts may be offensive to students from some cultural backgrounds (Bradley 1996).

Raising these safety issues should encourage you to consider them in relation to your school policy, as well as to develop a resource file related to classroom and outdoor safety. The British Association for Science Education (ASE) has produced excellent resource material for primary teachers and schools (ASE 1990, 1994, 2001). Elliott (1992) also offers sound advice. Other resources, such as Thurmin and Thurmin (1994) and the New South Wales Department of Education and Training (1999), refer to safety issues related to kitchen and other chemicals.

Activity 9.17 Equipment, consumables and safety

Equipment and consumables for teaching primary science can be a problem. As an individual teacher, how will (or do) you handle it? How could (or does) your school organise equipment and consumables? Talk with at least one teacher or colleague from another school to see how he or she and their school handle science equipment and consumables. From your reading – see, for example, Feasey (1999a, 1999b) and Knox (1994) – provide a practical list of hints and suggestions to overcome hurdles you have identified in relation to this area.

Safety is clearly an issue when primary students are using equipment and consumables. What are the basic requirements and considerations you should know about? Suggest some safety procedures that would be standard in your classroom, especially when attempting chemical tasks. How might you involve your students in thinking about safety in science? Is there a place for a primary school safety policy that refers to science? For ideas, see the references within this section, as well as Borrows (1997, 2003) and Mills and Pentland (1997).

Science as a human endeavour

The characteristics of 'science as a human endeavour include primary students being aware of the NOS. In this chapter, the importance of students learning about the role of representations and modelling in formulating explanations about our world (here, especially the particulate nature of matter) has been central. The importance of reasoned discussion and debate within a science classroom learning community has also been stressed. These attributes, when combined, can also help students to appreciate the collaborative NOS. Chemical and physical change also directly relates to the effect of science and technology on our lives through, for example, sustainability issues.

NOS

Science and culture: sustainability issues

Materials in our world can be classified as renewable and non-renewable. The latter are finite resources and their use needs to be monitored for the sake of future generations. The former, such as air and water, are not limited provided we do not pollute them so that they are unusable. When considered from this perspective, 'Materials and change' is a content area that has many environmental/sustainability education links. Does your science syllabus contain references to sustainability? Teachers have a responsibility to assist students to appreciate the environmental consequences of unwise use of resources for our future and, where appropriate, to take action in relation to the conservation of such resources.

The emphasis in this book is primarily on conceptual development. Teachers and students may also have alternative conceptions about environmental concepts associated with materials and change, such as conservation, energy, pollution, recycling, resources, technology and waste – see Mant and Summers (2002) and references therein; see also articles published by the Environmental Education Research Unit, at the University of Liverpool (see http://www.liv.ac.uk/~qe04/eeru). One example would be chemical changes resulting in the production of carbon dioxide and other greenhouse gases which could be related to primary students' beliefs about actions that will affect global warming (Skamp, Boyes and Stanisstreet 2009).

SSI

For teaching resources related to materials and change and the environment, teachers could access articles in journals such as *Green Teacher* (see http://www.encyclopedia.com/Green+Teacher/publications.aspx?pageNumber=1), *Investigating* (now incorporated into *Teaching Science*) and *Primary Science* (formerly *Primary Science Review*), as well as material published by the Gould League (http://www.gould.org.au/index.asp) and the Environment Protection Authority (in Australia, or by analogous State/national and international organisations). Nixon (2001), for example, describes how her primary class learnt about the environmental consequences associated with disposing of the waste materials from making paper. The Primary Upd8 resource (see http://www.primaryupd8.org.uk) includes activities requiring students to investigate the melting of glaciers and rising sea levels.

Clearly, physical and chemical changes can have environmental/sustainability education implications and could provide relevant contexts for helping students consider socioscientific issues (see Chapter 1). Teachers need to be aware of these and take advantage of the potential learning opportunities.

Summary

Chemistry has been a neglected part of some primary science curricula. Children love learning about chemicals, even if in most instances they are household ones. It is unfortunate that more use is not made of such inherent enthusiasm. Although there are conceptual difficulties associated with learning about physical – and especially chemical – changes, teachers can, with careful scaffolding, support stepping-stone understandings of the nature of these processes. This scaffolding should include helping students learn through negotiating their generated multimodal representations of physical- and chemical-change phenomena, and consequently assisting students to learn that representations (modelling) is a way to explore explanations and that this process reflects how science works. Particulate thinking can form an integral part of this learning. This is because the evidence suggests that it is not students' age that hinders their understanding of particulate ideas but the pedagogy that is used. Several case studies in this chapter exemplify the possibilities and it is hoped that you now see the potential of this content area.

Now go back to your responses in Activity 9.1 (see p. 352). What additional components of your primary science curriculum, if any, would you now interpret as having a chemical basis?

Primary teachers can lay the foundation for the development of positive attitudes towards chemistry and growth in the understanding of scientifically acceptable chemical ideas by initially being aware of how primary students think about chemical phenomena. Teachers can then take action to assist the development of physical- and chemical-change ideas in classroom, group and individual discussion, and, where appropriate, challenge or extend them using some of the activities, strategies and procedures suggested in this chapter.

Concepts and understandings for primary teachers

Below are most of the key conceptual ideas and understandings related to materials and change with which a primary teacher should be familiar. You should read and interpret this list with an appreciation of its limitations as described in the section 'Concepts and understandings for primary teachers' in Chapter 1 (see p. 47).

Many of the concepts and understandings listed are derived from alternative conception studies and address the (alternative) views that many students and teachers may hold (see Driver et al. 1994a).

Physical change

Physical change occurs when the appearance of substances and materials change but the actual substances and materials do not. They are usually considered as changes of state, changes of shape and changes caused by mixing. The original substances and materials can be obtained again.

It needs to be noted that it is not always simple to determine if a physical or chemical change has occurred. Distinguishing between difficult cases should not be a focus at the primary level.

Change of state

The nature of solids, liquids and gases was a focus of Chapter 8. Changes of state also involve energy transfer (see Chapter 5) but although mentioned below, to assist teacher understanding, the emphasis here, from a student perspective, is on the nature of the material world. You, as a teacher, do need to appreciate that when energy is transferred (associated with a change) from stuff (matter), it then becomes colder and when energy is transferred (associated with a change) to stuff (matter) it becomes hotter. The matter can be solid, liquid or gas.

The statements below refer to substances. Some materials, which comprise one or more substances (e.g., chocolate and butter), also can change state (that is, melt, boil). It needs to be borne in mind that these changes of the state of materials depend on the substances that comprise the materials. You should also be aware that the most common liquid, water, has properties that are different from all other liquids – it expands when it freezes.

- When energy is transferred to a solid substance it melts to become a liquid (the temperature at which this occurs is its melting point).
- When energy is transferred from a liquid substance it freezes to become a solid (the temperature at which this occurs is its freezing point).
- Each substance has its own temperature for changing from a solid to a liquid (melting) and back again (freezing). This is the same temperature.
- When energy is transferred to a liquid substance it evaporates to become a gas. This is occurring all the time. If enough energy is transferred, the liquid boils and evaporates quickly (the temperature at which this occurs is its boiling point; each substance has a set boiling point at a particular pressure).
- When energy is transferred from a gas substance it condenses to become a liquid.
- Some substances change directly from a solid to a gas and from a gas to a solid. Two examples are iodine and carbon dioxide. This process is called sublimation.

Water cycle

- Water evaporates from leaves and the waters on the earth's surface.
- Water vapour is the presence of water in the air; it is invisible.
- Warm air can hold more water vapour than can cold air.
- When water vapour condenses it becomes water droplets and can be seen as dew and clouds.
- Water droplets collide and join to form bigger drops, which become rain.

Mixtures

- Mixtures involve two or more substances or materials being mixed. The end result is that they often look like one material (e.g., air) but the substances can be separated by various means.
- The substances or materials can be any combination of solid, liquid or gas.

Dissolution

Solutions are when solids, liquids or gases (the solute) dissolve in a liquid (the solvent) to form a homogeneous mixture.

- There is a limit to the amount of substance or material that can dissolve in a given liquid. A solution is saturated if no more substance or material will dissolve in it.
- The temperature of a liquid determines how much solid or gas will dissolve in it. Usually more solid (and less gas) will dissolve in a warmer liquid and vice versa.
- Colloids (e.g., milk) and suspensions have particles of solids still suspended in the solvent.

Chemical change

For an explicit description of chemical change and, in particular, burning, see the online Appendix 9.3.

The particulate nature of matter

The attributes of this model have been detailed in Chapter 8.

ONLINE

Appendix 9.3
Concepts underlying the processes of burning and chemical change

Acknowledgements

Jenny McDonald taught the condensation sequence at Woodburn Central School, New South Wales. The extracts incorporate adapted sections of a subsequent teaching report written by the teacher and are used with her permission.

Search me! science education

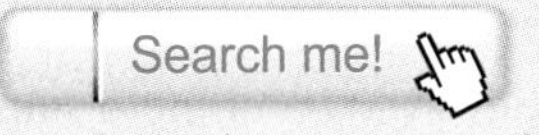

Explore **Search me! science education** for relevant articles on materials and change. Search me! is an online library of world-class journals, ebooks and newspapers, including *The Australian* and the *New York Times*, and is updated daily. Log in to Search me! through http://login.cengage.com using the access code that comes with this book.

KEYWORDS

Try searching for the following terms:

- physical change
- chemical change
- particulate nature of matter

Search tip: **Search me! science education** contains information from both local and international sources. To get the greatest number of search results, try using both Australian and American spellings in your searches: e.g., 'globalisation' and 'globalization'; 'organisation' and 'organization'.

Appendices

In these appendices you will find material that you should refer to when reading Chapter 9. These appendices can be found on the student companion website. Log in through http://login.cengage.com using the access code that comes with this book. Appendices 9.4 and 9.5 are included in full below.

Appendix 9.1A Testing teacher beliefs about matter and change

This online appendix contains five items to challenge your thinking about matter and change.

Appendix 9.1B A further audit of your beliefs about matter and change

Seven statements for you to assess your understanding of dissolving and evaporation are included in this online appendix.

Appendix 9.2 Comments on questionnaire items about what happens when matter undergoes change

Feedback related to the items in Appendix 9.1A.

Appendix 9.3 Concepts underlying the processes of burning and chemical change

This online appendix lists key concepts for teachers about these two ideas.

Appendix 9.4 Year 7 students' responses

Although only a small sample, Ross and Law (2003) reported that all four Year 7 students thought that water would be heavier when it froze and lighter when it melted. Several students in years 9 and 11 also held the same ideas. This may be attributed to students associating 'hard' with 'heavy'. Further interpretations are in Ross and Law (2003).

Appendix 9.5 Children's tests for their hypotheses about why the water was on the outside of the styrofoam cups

1 One person bumps the desk while the other people watch to see if the water spills.
2 a Test a foam cup and a container that we know water can't come through (a tin) to see if water comes through.
  b Measure the height of the water with a ruler before and after the bubbles come.
3 Put the cup near a fan to see if the water spills.
4 Put cold, though not freezing, water in the cup. Do the drops still appear?
5 Ben stamps on the floor near the cup. Does the water still form drops?
6 Pour water into the cup and watch carefully to see if it spills onto the outside of the cup.
7 Put spoons in the cup again and watch carefully to see if the water overflows.
8 Put two cups out and fill with cold water. Put one in a plastic bag. Suck all the air out and tie a knot. After five minutes, look to see if it has drops on it.

Appendix 9.6 Dissolution: a pattern-finding investigation

A lesson sequence exemplifying a pattern-finding investigation related to dissolving time and types of sugar is described in this online appendix.

Appendix 9.7 Slime recipe

Instructions for making slime are provided in this online appendix.

Appendix 9.8 Student ideas and questions about the behaviour and properties of dry ice

Students' descriptions after messing around with dry ice are shown in this online appendix, as well as their comparisons between ice and dry ice. Seventeen of their questions are listed.

References

Acher, A., Arca, M. & Sanmarti, N. 2007. Modeling as a teaching learning process for understanding materials: A case study in primary education. *Science Education*, 91, pp. 398–418.

Adams, R., Doig, B. & Rosier, M. 1991. *Science Learning in Victorian Schools: 1990*. Melbourne: Australian Council for Educational Research (ACER).

Andersson, B. 1986. Pupils' explanations of some aspects of chemical reactions. *Science Education*, 70, pp. 549–63.

Association for Science Education (ASE). 1990, 2001. *Be Safe!* (2nd & 3rd edns). Hatfield, UK: ASE.

_____ 1994. *Safety in Science for Primary Schools*. Hatfield, UK: ASE.

_____ 1998. *Signs and Symbols in Primary Science*. Hatfield, UK: ASE.

Aubussen, P. 1998. Slimes. *Science Education News*, 47 (1), pp. 19–20.

_____ & Elliott, R. 1997. Slime: A solid foundation for liquid thinking. It's a gas. *Science Education News*, 46 (4), pp. 148–52.

Australian Academy of Science (AAS). 2009a. 'Spot the Difference': Stage 1 Natural and processed materials. *Primary Connections*. Canberra: AAS.

_____ 2009b. 'Change Detectives': Stage Natural and processed materials. *Primary Connections*. Canberra: AAS.

Australian Curriculum, Assessment and Reporting Authority (ACARA). 2011. K-10 Australian Curriculum: Science. Available at http://www.australiancurriculum.edu.au/Science/Curriculum/F-10 (accessed April 2011)

Bar, V. 1989. Children's views about the water cycle. *Science Education*, 72, pp. 481–500.

_____ & Galili, I. 1994. Stages of children's views about evaporation. *International Journal of Science Education*, 16 (2), pp. 157–74.

Borrows, P. 1997. Safe as Houses. *Primary Science Review*, 47, pp. 12–13.

_____ 2003. Health and safety. *Primary Science Review*, 79, pp. 18–20.

Bradley, L. 1996. *Children Learning Science*. Oxford: Nash Pollock.

Bucat, B. & Fensham, P. (eds). 1995. *Selected Papers in Chemical Education Research: Implications for the Teaching of Chemistry*. Delhi: IUPAC CTC.

Chang, J. 1999. Teachers college students' conceptions about evaporation, condensation, and boiling. *Science Education*, 83, pp. 511–26.

Claxton, G. 1989. Cognition doesn't matter if you're scared, depressed or bored, in P. Adey (ed.). *Adolescent Development and School Science*. London: Falmer, pp. 155–61.

Collyer, S. & Ross, K. 2005. Snow in the nursery. *Primary Science Review*, 90, pp. 19–21.

Curriculum Corporation. 2006. *National Consistency in Curriculum Outcomes*. Available at http://www.curriculum.edu.au/verve/_resources/SOL_Science_Copyright_update2008.pdf (accessed 19 October 2010).

Driver, R. 1985. Beyond appearances: The conservation of matter under physical and chemical transformations, in R. D. Driver, E. Guesne & A. Tiberghien (eds). *Children's Ideas in Science*. Milton Keynes: Open University Press.

_____, Squires, A., Rushworth, P. & Wood-Robinson, V. 1994a. *Making Sense of Secondary Science: Support Materials for Teachers*. London: Routledge.

_____ 1994b. *Making Sense of Secondary Science: Research into Children's Ideas*. London: Routledge.

Elliott, J. 1992. Safety in primary school science. *Investigating*, 8 (1), pp. 22–3.

Farrow, S. 2006. *The really useful science book* (3rd edn). New York: Routledge.

Feasey, R. 1998. *Primary Science Equipment*. Hatfield, UK: ASE.

_____ 1999a. Getting to grips with resourcing primary science. *Investigating*, 15 (4), pp. 8–14.

_____ 1999b. Choosing, storing and using equipment in the primary science classroom. *Investigating*, 15 (3), pp. 9–13.

Fensham, P. 1996. Public understanding of chemistry – no easy task. *Paper presented at the International Conference of Chemical Education*, Brisbane, July.

Fleer, M., Jane, B. & Hardy, T. 2007. *Science for Children* (3rd edn). Sydney: Prentice Hall.

Goodwin, A. 2002a. Teachers' continuing learning of chemistry: Some implications for science teaching. *Chemistry Education: Research and Practice in Europe*, 3 (3), pp. 345–59.

_____ 2002b. Is salt melting when it dissolves in water? *Journal of Chemical Education*, 79 (3), pp. 393–6.

Griffin, H. & Sharp, H. 1995. *Our Colourful World*. Sydney: STANSW Publications.

_____ 1998. Exploring chemical symbols and codes. *Investigating*, 14 (3), pp. 28–30.

Hackling, M. 2005. *Working Scientifically*. Perth: Education Department of Western Australia.

Harlen, W. 1985. *Teaching and Learning Primary Science*. London: Harper & Row.

_____. 2009. Teaching and learning science for a better future. *School Science Review*, 90 (333), pp. 33–42.

_____, Darwin, A. & Murphy, M. 1977. *Match and Mismatch: Raising Questions*. Edinburgh: Oliver & Boyd.

_____ & Jelly, S. 1989. *Developing Science in the Primary Classroom*. Edinburgh: Oliver & Boyd.

Hatzinikita, V., Koulaidis, V. & Hatzinikita, A. 2005. Understanding and explanations concerning changes in matter. *Research in Science Education*, 35, pp. 471–95.

Hellden, G. 2005. Exploring understandings and responses to science: A program of longitudinal studies. *Research in Science Education*, 35 (1), pp. 99–122.

Henderson, D. 1995. Slime time. *Investigating*, 11 (4), pp. 4–5.

Isabelle, A. & de Groot, C. 2008. Alternate conceptions of preservice elementary teachers: The Itakura method. *Journal of Science Teacher Education*, 19, pp. 417–35.

Jackson, J. 2009. H_2O and you. *Science Activities*, 46 (1), pp. 3–6.

Jakobson, B. & Wickman, P.-O. 2008. The roles of aesthetic experience in elementary school science. *Research in Science Education*, 38 (1), pp. 45–66.

Jarvis, T. & Pell, A. 2002. Effect of the Challenger experience on elementary children's attitudes to science. *Journal of Research in Science Teaching*, 39 (10), pp. 979–1000.

_____, Pell, A. & McKeon, F. 2003. Changes in primary teachers' science knowledge and understanding during a two year in-service program. *Research in Science and Technological Education*, 21 (1), pp. 17–42.

Johnson, P. 1998a. Children's understanding of changes of state involving the gas state, Part 1: Boiling water and the particle theory. *International Journal of Science Education*, 20 (5), pp. 567–83.

_____ 1998b. Children's understanding of changes of state involving the gas state, Part 2: Evaporation and condensation below the boiling point. *International Journal of Science Education*, 20 (6), pp. 695–709.

_____ 2000. Children's understanding of substances, Part 1: Recognizing chemical change. *International Journal of Science Education*, 22 (7), pp. 719–37.

_____ 2002. Children's understanding of substances, Part 2: Explaining chemical change. *International Journal of Science Education*, 24 (10), pp. 1037–54.

_____ 2005. The development of children's concept of substance: A longitudinal study of interaction between curriculum and learning. *Research in Science Education*, 35 (1), pp. 41–62.

Kenyon, L., Schwarz, C. & Hug, B. 2008. The benefits of scientific modelling. *Science and Children*, 46 (2), pp. 40–4.

Keogh, B. & Naylor, N. 1997. *Starting points for science*. Sandbach, UK: Milgate House.

Knox, M. 1994. Let's organise the science store. *Investigating*, 10 (3), pp. 26–9.

Krnel, D., Watson, R. & Glazar, S. 1998. Survey of research related to the development of the concept of 'matter'. *International Journal of Science Education*, 20 (3), pp. 257–89.

Kruger, C. & Summers, M. 1989. An investigation of some primary teachers' understanding of changes in materials. *School Science Review*, 71 (255), pp. 17–27.

Levinson, R. 2000. Thinking through science. *Primary Science Review*, 61, pp. 20–2.

Liu, X. & Lesniak, K. 2005. Students' progression of understanding the matter concept from elementary to high school. *Science Education*, 89, pp. 422–50.

_____ 2006. Progression in children's understanding of the matter concept from elementary to high school. *Journal of Research in Science Teaching*, 43, pp. 320–47.

Mant, J. & Summers, M. 2002. Teaching sustainable development – Why? What? How? *Primary Science Review*, 75, pp. 16–19.

Mills, J. & Pentland, D. 1997. Practising safe science in the physical world. *Investigating*, 13 (3), pp. 12–13.

Moorehouse, C. 1987. Planning science lessons with process skills in mind. *Primary Science Review*, 3, pp. 7–9.

Morris, M., Merrit, M., Fairclough, S., Birrell, N. & Howitt, C. 2007. Trialling concept cartoons in early childhood teaching and learning of science. *Teaching Science*, 53 (2), pp. 42–5.

New South Wales Department of Education and Training. 1999. *Chemical Safety in Schools*. Sydney: New South Wales Department of Education and Training (DET).

New Zealand Ministry of Education. 1998. *Making Better Sense of the Material World*. Wellington: New Zealand Ministry of Education.

Newton, D. 2002. *Talking Sense in Science*. London: Routledge and Falmer Press.

Nixon, M. 2001. Exploring the properties of recycled paper. *Investigating*, 17 (2), pp. 21–4.

Novak, J. 2005. Results and implications of a 12-year longitudinal study of science concept learning. *Research in Science Education*, 35 (1), pp. 23–40.

Omebe, E. 1987. Evaporation: Drying up or just soaking in. *Primary Science Review*, 3, pp. 13–16.

Osborne, J., Simon, S. & Collins, S. 2003. Attitudes towards science: A review of the literature and its implications. *International Journal of Science Education*, 25 (9), pp. 1049–79.

Palmer, W. 2003. A study of teaching and learning about the paradoxical concept of physical and chemical change. PhD thesis. Perth: Curtin University of Technology.

_____ & Treagust, D. 1996. Physical and chemical change in textbooks. *Research in Science Education*, 26 (1), pp. 129–40.

Papageorgiou, G. & Johnson, P. 2005. Do particle ideas help or hinder pupils' understanding of phenomena. *International Journal of Science Education*, 27 (11), pp. 1299–317.

_____, Johnson, P. & Fotiades, F. 2008. Explaining melting and evaporation below boiling point: Can software help with particle ideas? *Research in Science and Technological Education*, 26 (2), pp. 165–83.

Preston, C., Dempsey, A. & Benn, S. 2004. Team teaching in primary science: How good could that be? *Teaching Science*, 50 (2), pp. 9–13.

Qualter, A., Schilling, M. & McGuigan, L. 1994. Exploring children's ideas. *Investigating*, 10 (1), pp. 21–4.

Rahayu, S. & Tytler, R. 1999. Progression in primary school children's conceptions of burning: Towards an understanding of the concept of substance. *Research in Science Education*, 29 (3), pp. 295–312.

Reynolds, Y. & Brosnan, T. 2000. Understanding physical and chemical change: The role of speculation. *School Science Review*, 81 (296), pp. 61–6.

Ross, K. & Law, E. 2003. Children's ideas about melting and freezing. *School Science Review*, 85 (311), pp. 99–102.

Russell, T., Longden, K. & McGuigan, L. 1991. *Materials (Primary Science Processes and Concept Exploration [SPACE] project)*. Liverpool: Liverpool University Press.

Selley, N. 1998. Alternative models of dissolution. *School Science Review*, 80 (290), pp. 79–84.

Shallcross, D., Harrison, T., Wallington, S. & Nicholson, H. 2006. Reaching out to primary schools: The Bristol ChemLabs experience. *Primary Science Review*, 94, pp. 22–9.

Sharp, H. & Griffin, J. 1994a. *Food and Cooking*. Sydney: STANSW Publications.

_____ 1994b. *Rumble, Gurgle and Fizz*. Sydney: STANSW Publications.

_____ 1994c. *Hot Bread*. Sydney: STANSW Publications.

Skamp, K. 1991. Investigating 'cold things'. *Chemeda: Australian Journal of Chemical Education*, 32, pp. 37–9.

_____ 1992. The role of the 'extraordinary': Investigating 'very cold things'. *Chemeda: Australian Journal of Chemical Education*, 34, pp. 37–9.

_____ 1995a. Where does the water go? *Investigating*, 11 (3), pp. 10–13.

_____ 1995b. Chemistry and the Ontario Science Centre. *Chemeda: Australian Journal of Chemical Education*, 42, pp. 19–23.

_____ 1996a. Chemistry: Its potential in the primary school. *South Australian Science Teachers' Association Journal*, 96 (1), pp. 15–28.

_____ 1996b. Elementary school chemistry: Has its potential been realised? *School Science and Mathematics*, 96 (5), pp. 247–54.

_____ 1999. Are atoms and molecules too difficult for primary school children? *School Science Review*, 81 (295), pp. 87–96.

_____ Boyes, E., & Stanisstreet, M. 2009. Global warming responses at the primary secondary interface: 1 Students' beliefs and willingness to act. *Australian Journal of Environmental Education*, 25, pp. 15–30.

SMART Technologies 2003. Available at http://www.smarttech.com (accessed 15 September 2010).

Stephenson, P. (ed.). 2000. *Developing Primary Teachers' Science Knowledge: Self Study Materials*. Leicester: SCI Centre.

Stringer, J. 2002. All change. *Child Education*, February, pp. 42–4.

Summers, M., Palacio, D. & Kruger, C. 1989. *Primary School Teachers and Science Project*. Oxford: Oxford University Department of Educational Studies.

Thurmin, R. & Thurmin, R. 1994. Testing kitchen chemicals. *Investigating*, 10 (4), pp. 11–13.

Tytler, R. 2000. A comparison of Year 1 and Year 6 students' conceptions of evaporation and condensation: Dimensions of conceptual progression. *International Journal of Science Education*, 22 (5), pp. 447–67.

_____ & Peterson, S. 2000. Deconstructing learning in science: Young children's responses to a classroom sequence on evaporation. *Research in Science Education*, 30 (4), pp. 339–55.

_____ & Peterson, S. 2005. A longitudinal study of children's developing knowledge and reasoning in science. *Research in Science Education*, 35 (1), pp. 63–98.

_____, Peterson, R. & Prain, V. 2006. Picturing evaporation: Learning science literacy through a particle representation. *Teaching Science*, 52 (1), pp. 12–17.

_____ & Prain, V. 2010. A framework for re-thinking learning in science from recent cognitive science perspectives. *International Journal of Science Education*, 32 (15), pp. 2055–78.

_____, Prain, V. & Peterson, S. 2007. Representational issues in students' learning about evaporation. *Research in Science Education*, 37 (3), pp. 313–31.

Varelas, M., Pappas, C. & Rife, A. 2006. Exploring the role of intertextuality in concept construction: Urban second graders make sense of evaporation, boiling and condensation. *Journal of Research in Science Teaching*, 43 (7), pp. 637–66.

Wenham, M. & Ovens, P. 2010. *Understanding Primary Science* (3rd edn). London: Sage.

West, L. & Fensham, P. 1979. What is learning in chemistry?, in Chemical Education: A view across the Secondary–Tertiary Interface. *Proceedings of the Royal Australian Chemical Institute Chemical Education Division Conference, Gippsland Institute of Advanced Education*, pp. 162–9.

White, R. 1988. *Learning Science*. Oxford: Blackwell.

Wiser, M. & Smith, C. 2008. Learning and teaching about matter in grades K-8: When should the atomic-molecular theory be introduced?, in S. Vosniadou (ed.). *International Handbook of Research on Conceptual Change*. New York: Routledge, pp. 205–39.

Endnote

1 In two articles, Goodwin (2002a, 2002b) poses questions such as 'Is salt melting when it dissolves in water?' He argues that we can sometimes draw too strong a delineation between polarised classificatory concepts such as solids and liquids and chemical and physical change. Of course, with primary classes it is important to be sufficiently certain about various categories and processes. Even so, if the circumstances arise where legitimate questions are asked about such categories and processes, it can be 'empowering' for students to discuss such matters (2002a, p. 395).

10

Our place in space

by Keith Skamp

Space isn't remote at all. It's only an hour's drive away if your car could go straight upwards

Sir Fred Hoyle, *London Observer*, 1979

Introduction

This chapter focuses on students finding out about the universe in which they live. Most primary science syllabuses include content related to the Earth and space. The International Year of Astronomy (IYA) (2009) highlighted the vast array of activities that could engage students at all levels. An IYA web search would readily uncover many articles and sites which teachers could access as resource material – see, for example, *Primary Science* (vol. 108, 2009), D. Smith (2009) and Australian Science Teachers' Association (ASTA; 2004, 2009). In this chapter, attention is directed towards the nature and movements of the sun, Earth, moon, other planets and the stars. This is because these celestial bodies are directly observable. Space travel and its spin-offs are briefly mentioned.

Learning about our place in space has the potential to change students' views about the way they perceive themselves and humanity in general. On the day after humankind set foot on the moon, a newspaper editorial said:

> only the most fearful mind would insist that (humanity) should rest content in ignorance of the nature of the universe; that a better understanding of the cosmos in which (we) live would not deepen (our) respect for (our) world or the miracle of (our) own existence (*Sydney Morning Herald*, 22 July 1969).

Further, with the realisation that there is a high probability of other life in the universe (see, for example, SETI at http://www.seti.org), students may see that the uniqueness of humanity is problematic – they certainly ask questions about life elsewhere in the universe (Strange and Fullam 2009). Space exploration also has utilitarian value. Students could research some of these spin-offs; for example, new ways of studying and protecting the environment, disaster control, weather forecasting and understanding human spatial behaviour (Reynolds 2009; A. Smith 2009). There is a danger, though, that we have become 'used to what is completely extraordinary'; for example, the 'technical difficulty of capturing images' such as the beautiful pictures of Saturn's rings transmitted from the Cassini spacecraft, which was launched in 2004 (*New York Times* editorial, 26 April 2009). The awesome nature of space and space travel must always be obvious when you teach this content area.

Primary students love to learn about space – its vastness, the sense of the unknown, our fragile place within it. It is a topic liked by boys and girls (Sharp et al. 1999) and a common primary school theme. The challenge for teachers is usually not one of motivation as the area is intrinsically fascinating; rather, it is one of avoiding making it a dry, sterile topic through an overuse of library and Internet research activities and the rewriting of specific information from these sources. Of course, secondhand sources have a role, provided students' minds are engaged and they are required to think about the information they are reading or accessing. However, space is sometimes thought of as an area of science virtually devoid of hands-on investigating potential. This need not be the case.

Activity 10.1 Teaching about the Earth in space: your existing ideas

Before reading any further, think about how you would teach early childhood and primary-age students about the sun, Earth, moon, other planets and stars. Develop a mind map that encapsulates your current thoughts. In the centre of your mind map write 'Teaching about space to five- to 13-year-olds'. Then write your ideas around this circled caption. Connect thoughts where you think it is necessary. Ensure you focus on teaching ideas, not content. Share and discuss your mind map with another teacher, making additions if necessary. Retain it for the last activity in this chapter.

Now that you have thought about how you could teach primary students about our place in space, you might wonder how this chapter will encourage you to think further about your teaching approaches. It will use the following structure. Firstly, you will be asked to reflect upon your existing ideas/conceptions about this topic. Secondly, ways to elicit primary students' ideas about celestial phenomena will be overviewed and a summary of the ideas they hold will be provided; the question as to whether there is a progression in conceptual development across the primary years will be addressed. Thirdly, the main focus of the chapter – ways to teach about celestial phenomena from a constructivist perspective are introduced. After initially presenting research evidence that constructivist sequences related to astronomy are effective, the following teaching principles are identified and illustrated:

- Experiential concepts (what you can actually experience) are critical; explanatory concepts should not replace them.
- Scientific investigations (as in Chapter 2) and firsthand observations should be integral to learning about space.
- Teacher- and student-generated representations can be significant determinants in students' reasoning and learning about astronomy and the NOS.

These principles and the suggested activities can be applied across the primary years. Some activities are more suitable for older learners and these are discussed separately. As models and secondhand experiences are often used in this content area, there are sections that specifically consider these approaches and suggest effective ways of using them. These strategies need to be combined into lesson sequences; hence, suggestions are made about how to plan such sequences and examples are provided. The chapter concludes by commenting on how this content area can address 'science as a human endeavour'.

Teachers' ideas about astronomy

Before teaching any topic, it is advisable to reflect on your own background. What preconceptions about celestial bodies and events do you bring to the classroom? Many children, as well as adults, have shaky understandings about day and night, seasons, the 'changing' shape of the moon, the stars and the planets. The following activity will challenge your ideas about some of these areas. It takes on learning value when your mind interacts with the statements or, better still, also with the ideas of others.

Activity 10.2 Our existing ideas about space

Read the following statements and decide if you agree (A) or disagree (D) with each of them. It is important you provide reasons for your responses. You may, of course, be uncertain (U). If possible, do the task with a colleague or friend and with some everyday resources at hand (e.g., a torch, different-sized balls). Compare your thoughts and talk through your opinions together.

- Summer and winter are caused by the Earth swinging backwards and forwards 23.5 degrees from the vertical.
- The sun is our nearest star.
- The planets that we can see are the ones that give out the most light.
- The planets are closer to the Earth than are the stars.
- The phases of the moon are caused by the shadows the Earth casts on it.
- We can see the moon because of the light reflected off it from the sun.
- Day and night are caused by the sun moving to the other side of the Earth.
- The moon takes about a week to go around the Earth.
- Although the moon spins on its axis, we only see one side of it.
- People can take large steps on the moon because there is no gravity.
- It is safe to look at the sun with sunglasses.
- The Earth goes around the sun in a circle.
- Our solar system is in the centre of the Milky Way galaxy.
- There is probably life on other stars.
- The sun is made up of gases.
- The universe is contracting.
- Red giants, white dwarfs and black holes are stages in the life cycle of stars.
- The morning and evening star is a planet.
- Stars have points.
- Stars appear to move around a particular point.
- Scientists are agreed on how the universe began.
- Planets in the solar system are made of rocks and dirt.
- Comets are balls of rocks and fire.
- The tides are caused by the gravitational pull of the sun on the Earth.

If you would like to determine your understanding on related content, try the multiple-choice items in Trumper (2001a, 2001b, 2003) or the astronomy items on the Science Belief survey (Stein, Larrabee and Barman 2008; and see http://www.citejournal.org/vol6/iss3/science/article1.cfm).

You were probably uncertain about and/or disagreed with your colleague on some of the items. Why do you think you were asked to talk through the task with a colleague or friend? By doing so, your ideas had to interact with those of another person, which probably forced you to be active in your thinking. As it is believed that learning is an active mental and usually social process, you opened up the opportunity for more learning to occur.

The purpose of Activity 10.2 was to highlight that we all hold alternative (non-scientific) conceptions about a range of everyday phenomena. Twenty per cent of Americans, for example, think that the sun orbits the Earth, and 17 per cent believe the Earth revolves around the sun once a day, while 45 per cent think the sun is not a star (Reneke 2005). Teachers are no exception (Mulholland and Ginns 2008; Stein, Larrabee and Barman 2008); they need to be aware of the fact that they may also hold alternative conceptions.

You can be confident the young learners in your class will hold many non-scientific views. Before you assist students to learn about some of these phenomena, ensure you have reflected on your own understandings. Effective ways of doing this would be learning with colleagues, using models, taking and interpreting observations of celestial bodies (e.g., moon and stars), devising multiple and multimodal representations, discussing conceptual explanations of phenomena, and so forth. However, if these opportunities do not arise, you should take heart from research which suggests that primary teachers who are aware of how students learn constructively were 'triggered' through classroom student interactions to improve their understanding of astronomy as they taught this topic (Ackerson 2005). If you are a teacher who appreciates the characteristics of learners' ideas about how their world works, and that learners need to engage in constructivist and representational learning strategies such as testing their ideas and models and discussing them with peers and teachers (see Chapter 1) in order to modify them, then you, like the teachers in the research study, will be triggered to read more to improve your understandings about the topic as it progresses.

Simply reading about the content area you are to teach will not necessarily ensure your understanding (Ackerson 2005), but it may be the only option you have. You are strongly encouraged to access chapters related to this content area which have been written especially for primary teachers. Examples include Farrow (2006), Wenham and Ovens (2010) and the CDs accompanying *Primary Connections* units (e.g., AAS 2006); also, *Primary Science* (*PS*) has articles called 'Wobbly Corner' which help with explanations included in primary syllabus outcomes but which are not straightforward, such as phases of the moon (*PS*, vol. 108, 2009). At the end of this chapter is a list of key concepts and understandings with which you should familiarise yourself. You must be aware that these statements are not to be considered as propositions that can simply be 'taught' (told) to students. Rather, these often abstract statements need to be linked to real and/or simulated situations, such as various observations or the results of investigations. Even then, the resultant conceptual growth may only be a 'stepping stone' to these statements (Harlen 2009). If you would like to read further about responses to Activity 10.2, see the online Appendix 10.1 (Skamp 1994).

Appendix 10.1 Some alternative conceptions held by some preservice primary teachers about astronomical phenomena

Students' ideas about celestial phenomena

Primary students hold many ideas about the nature of celestial bodies and events that are not in accord with currently accepted scientific opinion. Apart from motivating student interest in a topic, being aware of, and usually eliciting, your students' ideas about these phenomena is a necessary first step in teaching a sequence of lessons. It is the 'starting point in providing support for [the] progression of student ideas' (Millar and Murdoch 2002, p. 29).

Eliciting students' ideas about space

Teachers need to be aware of their students' ideas if they wish to assist them to learn. Without such knowledge, it will be difficult to challenge or progress their thinking or link appropriate activities and/or other ideas (for example, from the teacher or a secondhand source such as a website, book or chart) to where their students are at. Further, we know students' ideas – and our own – can be resistant to change.

To make it more difficult for teachers, students can say one thing and mean another; for example, Vosniadou, Skopeliti and Ikospentaki (2005) found that many Year 1 and Year 3 students who held non-spherical views about the shape of the Earth agreed that the Earth was actually spherical when provided with a globe. However, several of these same students gave responses to other questions related to the shape of the Earth which showed that they really held a non-spherical view despite their earlier 'correct' responses. We need to be aware that the presence of various cultural artefacts or tools (e.g., globes, physical models, drawings) and even the language the teacher uses may influence the responses that students provide; that is, their responses are often context-dependent. 'Context', here, clearly is interpreted in several ways. This can make it problematic as to how to interpret students' responses in the presence of an artefact and teachers must take this into consideration (Brewer 2008; Ehrlen 2008).

Finding out what our students *think* is therefore not necessarily straightforward. However, there are sufficient studies of students' ideas to indicate, despite some ambiguity in interpretation, the most common ways primary students think about celestial bodies and phenomena. Obviously the more we know about our students' ideas, the more we can use appropriate techniques and strategies to try to help them move their ideas towards those which are more scientifically acceptable.

A

How can we elicit students' ideas about astronomy topics? There are numerous procedures teachers can try (Naylor, Keogh and Goldsworthy 2004; White and Gunstone 1992). Different approaches will probably suit some students more than others (Millar and Murdoch 2002), so teachers need to consider the motivational value of the technique and its suitability for particular children. Students' responses to open-ended rather than closed-ended questions may provide more accurate representations of their thinking (Brewer 2008). The following is an open task. A Year 3 teacher (McGrogan 1992) gave her class the words 'sun', 'moon', 'stars', 'space' and 'Earth' and asked them to write down everything they knew about these words – 'emphasising that spelling was unimportant on this occasion and that it was their ideas [she] was interested in' (ibid., p. 28).

Activity 10.3 **Using students' writing about space to elicit ideas**

Try the above concept-writing exercise with a few middle and upper primary students. Compare it with your own attempt at the task and the responses McGrogan reported (see below). Identify those conceptions the students held that you believe needed further discussion, challenging or expanding. For each alternative or limited conception, suggest a possible activity or strategy you could try. Why did you choose the activity or strategy? Compare your suggestions with later sections in this chapter.

Drawings, concept maps, true–false cards and concept cartoons, all with associated dialogue, are elicitation techniques you could try. You may wish to obtain some student responses using one or more of these approaches as well.

Responses reported by McGrogan included: 'The Sun shines during the day and the Moon at night', 'The Sun goes around the Earth' and 'I estimate it would take two days to get to the Moon'. She had the students share their ideas and collected the scripts. They were invaluable in guiding the way the teacher interacted with the students and selected activities over the next six weeks. McGrogan then returned the sheets; the students' reactions were varied. These students then had to complete the same exercise on the back of the original. The changes were obvious (to the teacher and the students) and examples are given in the article. It is worth noting that this teacher said it gave her 'more satisfaction than any other exercise in her 18 years teaching' (ibid.). Other teachers have reported similar experiences. Kang and Howren (2004, p. 32) thought that having their Year 2 students compare past and current ideas was 'the most critical part' of the culminating activities of a unit related to the movement of objects in the sky. The importance of using the students' ideas in helping them learn cannot be overstated.

Many teachers and researchers have used drawings to elicit students' ideas. The derived information may be limited (Brewer 2008) unless accompanied by annotations and discussion (Hubber and Haslam 2009). There are many examples you can access:

- Baxter (1989, 1991), Osborne (1994) and Sharp (2000) determined infants' to 16-year-olds' ideas about time and distance, planets, Earth and gravity, day and night, phases of the moon and the seasons.
- Marsh, Willimont and Boulter (1999) and Sharp (2000) provide examples related to the solar system.
- Kang and Howren (2004) accessed Year 2 students' conceptions of the movement of objects in space.

Other elicitation techniques that have been used with astronomy concepts at the primary level are concept cartoons (Keogh and Naylor 2000a, 2000b), selected multiple-choice items requiring reasons for the selection of the response (for example, use items in Diakidoy and Kendeou (2001) provided reasons were sought for the responses selected), concept maps (Sharp et al. 1999) and true–false picture cards (Kibble 2002a). The last mentioned is a set of cards with a sketch describing an aspect of space with an accompanying statement. For example, six cards for day and night are:

1. At night the sun moves behind a hill.
2. Night happens because a cloud covers the sky.
3. At night the moon covers the sun.
4. The sun goes around the Earth once each day.
5. The Earth goes around the sun once each day.
6. The Earth spins around once each day.

Other examples are provided for gravity, seasons and phases of the moon. These cards can be used in various ways (e.g., a card game) but it is the discussion that the cards generate that is most significant. Conversational and argumentative (the latter distinguishing, for example, 'claim' and 'evidence' statements) discussion between peers and with teachers (with such aids) can assist children to alter their ideas. This can be contrasted with simply challenging student ideas (that is, cognitive conflict) through demonstrations, simulations, model building and fair-test investigations. The distinction is important as sometimes the impression can be given that cognitive conflict is the only way to help students modify existing ideas (revisit 'Modifying learners' ideas' in Chapter 1 and Appendix 1.7).

This emphasis on talk is consistent with social constructivist premises, which emphasise the importance of discussion as a means of rethinking existing conceptions. Of course, hands-on and observational activities are still significant, but sometimes are not essential to assist students to modify or change their ideas when (discussion) strategies such as these are used. You are encouraged to try some of these elicitation techniques; however, use them in such a way that you encourage discussion among a group of children about their thinking on a celestial concept and note how the peer group discussion evolves and what interventions, if any, you believe you could make as a constructivist teacher.

Activity 10.4 Sources of students' alternative and limited conceptions

Alternative and limited conceptions derive, in part, from the many experiences we encounter. We might either accept others' non-scientific conceptions or explain the experiences with our own intuitive (rather than accepted scientific) understandings. Read the nursery rhyme 'Twinkle twinkle little star' and identify possible statements that could be sources of alternative conceptions.

Devise a series of questions related to phrases or statements in the nursery rhyme to pose to children with different levels of understanding. Your purpose is to create peer-group discussion about their conceptual understanding of stars and other celestial phenomena.

Can you find any other children's literature that could be used in similar ways?

It is worthwhile reflecting on the impact of literature, newspapers and other media on children's conceptual development. When asked to draw pictures of the moon, students and adults often draw waning rather than waxing moons. What cultural influences may be responsible for this observation? Analyses of children's literature depicting the moon and moon phases reveal that inappropriate images are often used and could be a source of alternative conceptions. How might you use such cultural factors to encourage conceptual development? How might you use children's fiction with inaccurate images and text to advance conceptual development? Compare your thoughts with some interesting ways forward suggested by recent research – see Appendix 10.2 (p. 442), Trundle and Sackes (2008), Trundle and Troland (2005) and Trundle, Troland and Pritchard (2008).

Primary students' ideas about astronomy concepts

Numerous research studies, across many different countries, have documented primary students' ideas about space. Brewer (2008) analysed many studies which described students' views about the shape of the Earth, the day/night cycle, the seasons and the phases of the moon. Students (and adults) develop different models and theories about these celestial phenomena and the frequency with which these models are held varies across age ranges. The age at which each model was most commonly held is shown below. In reading these summaries, be aware that some primary students, across a wide age range, may hold any one of the described models, even if it is most commonly reported with adults.

- *The Earth's shape* – The common models fell into two groups: those with gravity acting 'down' and those with gravity acting towards the centre of the Earth. The 'gravity down' models were mainly a flat Earth (highest frequency was with five- to six-year-olds; 23 per cent), a disc (six-year-olds; 25 per cent), a hollow Earth in space (11-year-olds; 30 per cent) and a sphere in space (11-year-olds, 37 per cent). The 'gravity to the centre' models were either a flattened sphere in space (six-year-olds; 9 per cent) or a sphere in space (11-year-olds; 60 per cent). Interestingly, many students hold a 'dual earth' model (seven- to eight-year-olds; 53 per cent); that is, they hold concurrent conceptions of the Earth's shape – for example, as a spherical object in space and the flat place where we live.
- *The day/night cycle* – The common models were sun behind clouds (six- to seven-year-olds; 10 per cent), sun (way) out in space (eight-year-olds; 5 per cent), sun (goes) down to other side (eight-year-olds; 11 per cent), sun revolves around the Earth (eight-year-olds; 16 per cent); sun behind moon (adults; 4 per cent), Earth rotates while sun and moon are fixed (adults; 16 per cent), and Earth rotates, sun is fixed and the moon revolves around the Earth (adults; 68 per cent).
- *Seasons* – The common models were sun out more in summer (eight-year-olds; 39 per cent), sun turns hotter in summer (seven-year-olds; 18 per cent), snow makes winter colder (seven-year-olds; 16 per cent), the Earth spins (adults; 11 per cent), the Earth is in an elliptical orbit around the sun (adults; 22 per cent), tilt plus distance (or no mechanism) (adults; 11 per cent) and full tilt (fixed axis and spread of heat) (adults; 6 per cent).
- *Phases of moon* – The common models were clouds block the moon (six-year-olds; 14 per cent), night blocks moon (six-year-olds; 4 per cent), the Earth spins (10-year-olds; 7 per cent), something blocks moon (adults; 6 per cent), eclipse (adults; 50 per cent) and scientific model (adults; 12 per cent).

Other studies have reported students' ideas about different celestial bodies and phenomena. An extensive but dated Australian study (Adams, Doig and Rosier 1991) reported Year 5 students' (n = 583) views about the above phenomena, as well as the following: the nature of the sun (most common response, 55 per cent, was that the sun is hot and uninhabitable, but did not indicate how its structure differed from the Earth); why the moon shines (43 per cent thought light comes from the sun but more than half of these did not appreciate it was reflected light); the moon's atmospheric conditions (58 per cent thought there was no appreciable atmosphere or oxygen); and gravity (responses indicated only 1 per cent understood the relationship between mass and gravity). Many later studies, from a range of countries, generally support such findings, namely that many primary students hold a wide range of alternative conceptions about many topics referred to in primary science syllabi – for example, Baxter (1989, 1991), Osborne (1994), Reneke (2005), Sharp et al. (1999), Trumper (2003) and Webb and Morrison (2000). Students' conceptions about gravity, atmosphere and the planets (Palmer 2001; Webb and Morrison 2000) and the movement, formation and life of stars (Dove 2002; Finegold and Pundak 1990) have also been published.

Not all studies, though, are consistent. Osborne (1991, 1994), for example, suggested that in some middle to upper primary classes, up to 50 per cent of students may hold a scientific conception of gravity and the Earth, but maybe as few as 10 per cent of upper primary students might be able to correctly draw the (apparent) path of the sun across the sky. Interestingly, Osborne found that up to 67 per cent of upper primary students could order cards (depicting locations, including astronomical bodies) in the correct order of distance from their school, but only 10 per cent were able to suggest what the absolute distances might be.

For this topic, students from different societies may hold views influenced by their cultures and religions. This is not always the case, with students from different nations and across indigenous cultures often holding similar conceptions. However, some studies report models and beliefs apparently unique to a society and culture (Brewer 2008). Some primary students in India constructed spherical Earths supported by water consistent with local cosmological beliefs, while some Samoan children depicted the earth as a horizontal ring, possibly related to the 'circular organization of space in their everyday environment' (Vosniadou et al. 2005, p. 325). Mohapatra (1991) reported that about 15 per cent of the secondary students in his Indian study, which investigated the causes and associated effects of solar eclipses, believed that the basic reason for solar eclipses was the devouring of the sun by Rahu (in Hindu myth, the demon who causes eclipses); significant numbers also held particular societal views about practices such as food consumption during eclipses (for example, 72 per cent of 13-year-olds agreed that food should not be consumed during an eclipse). In a study involving Australian Aboriginal students, most described models of day and night similar to those held by Euro-American students but with more supernatural and animistic responses (Fleer 1997). With many primary teachers having students from a range of cultures in their classes, an awareness that such conceptions could exist will heighten a teacher's sensitivity to students' views.

S&C

In summary, there is only a relatively small set of models students use to explain celestial phenomena. Variations across cultures are minor. Some important considerations need to be kept in mind when reflecting on these summaries of students' ideas. Firstly, primary students may not apply a consistent interpretation of a phenomenon from one context to the next. In one study it was found that only 27 per cent of 10- to 11-year-old students (n = 26) used their model of how gravity worked in a consistent manner (Webb and Morrison 2000; see also Palmer 2001). Secondly, and related to the first caveat, students will sometimes hold fragments of scientifically correct conceptions and alternative conceptions simultaneously. An example related to gravity would be that some Year 6 students hold one or more of the following three correct conceptions – gravity acts downwards on falling and stationary objects, as well as objects moving vertically upwards – while also holding one or more inappropriate conceptions (for example, gravity does not act on falling objects) (Palmer 2001). What teaching implications would you draw from such results? Space does not allow further expansion but you are encouraged to reflect on them because, for example, there may be links between students' concurrent alternative and scientifically correct conceptions that teachers could use to facilitate students' reasoning about phenomena (Palmer 2001, p. 703; see also the online Appendix 10.3).

Appendix 10.3
Ways to handle correct conceptual fragments

Possible patterns of development in students' celestial ideas

Is there a pattern to students' conceptual development about astronomy concepts? Several studies suggest developmental progressions but there is not definitive agreement. Baxter (1989) proposed three categories:

1. Students' early notions are based on *observable features*.
2. Intermediate notions involve the *motions of astronomical bodies* but not the accepted scientific ones.
3. Usually at a later time, these are sometimes changed to the accepted views.

Baxter (1989, p. 508; 1995) postulated that, depending upon the way we think about the sun, the Earth – see Brewer's (2008) analysis – moon and stars, various phases in the development of ideas can be identified:

> The first phase is characterised by a static view of the Earth. This is frequently drawn saucer-shaped, North being 'up' and South being 'down'. Any changes in celestial bodies are caused by familiar and near objects like hills and clouds.
>
> The second phase is characterised by a round Earth but the naive idea of 'up' and 'down' still persists. The Earth is commonly thought of as central and static. Celestial bodies can move to cause observed phenomena but their movement is represented as 'up', 'down', 'right' or 'left'.
>
> In the third phase the same notions about the Earth and gravity still persist. However, celestial bodies are now seen to move in orbits, though this motion is Earth-centred.
>
> The fourth phase is that of the present heliocentric view and its associated gravitational ideas.

Related to Baxter's phases, some researchers have labelled these 'initial models' (derived from, and consistent with, observations; for example, day is replaced by night), 'synthetic models' (in which scientific and everyday information is integrated; for example, the Earth goes around the sun, the sun moves in space and so on) and 'scientific models'. Children up to about Year 2 usually have initial models that sometimes are replaced by synthetic and scientific models during later primary years. Brewer (2008) provides several examples of this progression (see the online Appendix 10.4). A challenge for teachers is implicit in this progression, because it appears that, as students encounter, at school and elsewhere, 'scientific' models for day and night and so on, many tend to form models that mix initial and scientific ideas – namely, synthetic rather than scientific models. If you are aware that your students may be developing synthetic models, then this may influence your interactions with them.

Appendix 10.4
Examples of hypothesised conceptual progressions

Others writers (e.g., Finegold and Pundak 1990; Wyn Davies 2002) refer to the following four related frameworks, which may assist in understanding how students' ideas are developing:

1. *prescientific* – a view of the universe with humanity at its centre and the universe as a flat, infinite space
2. *geocentric (Earth-centred)* – the universe is seen as having changes in the vicinity of the Earth, which is surrounded by a finite sky with heavenly bodies such as the sun, moon, planets and the stars
3. *heliocentric (sun-centred)* – the sun is seen as surrounded by the Earth, moon, planets and stars, and these are thought to be in the centre of the universe, which is fixed and unchanging except in our solar system
4. *sidereal (star-centred)* – our solar system is only one of many in the universe, which is constantly changing with various suns, moons, earths, planets, red giants, white dwarfs, galaxies, black holes, neutron stars and other bodies.

Interspersed with these frameworks can be precausal (offering reasons but not explaining how a phenomenon occurs) and intuitive (for example, based simply on what children see or hear) thinking.

These theories of conceptual progression may assist in a general way; you should not, however, interpret them as indicating that individual students' conceptual progression is linear across time. Individual conceptual development related to a particular concept may be haphazard and very much context-dependent (see Chapter 1). One teacher mapped how the explanations of the phases of the moon of individual Year 6 students in her class changed as a consequence of a unit of work on space. Of 32 children, none had a scientifically correct idea before the unit, while 10 did afterwards; of the others, some did not change their ideas (e.g., phases are due to clouds), while others changed from one alternative conception to another and still others started to develop partially correct conceptions (Sharp et al. 1999).

Activity 10.5 The development of astronomy ideas

- Using an elicitation technique, select a concept associated with the Earth, sun, moon and/or stars and obtain the responses of 10 students across a range of primary grades. Analyse their responses in terms of one or more of the suggested frameworks for conceptualising the development of ideas about astronomy. Share your responses with others.
- In what ways, if any, were your responses consistent with some of the research findings? What do you think are some of the implications of your findings for your teaching of this topic at different year levels?
- If the opportunity arises to teach a sequence of lessons related to an astronomy concept, then elicit the existing conceptions of your class in relation to the concept. At the end of the sequence, repeat the elicitation task and map the changes in your individual students (for example, follow Sharp et al. 1999). Reflect on the results. Draw some inferences about the nature of (constructivist) learning.
- Obtain Finegold and Pundak's (1990) multiple-choice questionnaire and attempt it. Then discuss your responses with others. What framework do you predominantly use? How could knowledge of these frameworks assist you as a primary teacher?

These studies provide background for teachers about to involve their students in learning about one or more astronomy concepts. They touch upon most of the space content and learning outcomes which would be included in all primary science syllabuses. Further, these studies also suggest how teachers might commence a learning sequence as well as obtain some indication of conceptual growth and change in their students.

Children's questions about space

One approach to helping students change their alternative conceptions is to have them explore, investigate and/or research their own questions (see, for example, Chapter 4). How do you get students to ask questions that interest them about the Earth, sun, moon, planets and stars? There are many techniques (see, for example, Chapter 2) although, on this topic, you may find that the questions arise spontaneously from the content and activities you introduce to your class. When keeping records of moon positions, Year 4 students, for example, asked questions about the movements of the moon, such as why it always appeared to move in a particular direction (Trundle, Atwood and Christopher 2007a). A lower primary teacher encouraged parents/carers to help their children keep a structured observational moon journal in which they had to keep moon observations noting date, day of week and time of day. Three statements on each page evoked many questions: 'What I saw' (students would draw a response), 'What I think' and 'What I wonder'. The approach could be used with any celestial body(ies) (Roberts 1999). What strategies or techniques do you know that would encourage students to ask questions about the world around them (e.g., question boxes)?

M&E

Developing students' ideas about astronomical objects and events

Teachers aware of hypothesised developmental phases and frameworks, and research findings about students' ideas, will better appreciate the views their students may hold about celestial phenomena. Many primary students will use intuitive and precausal thinking and hold initial and synthetic, rather than scientific, models to explain celestial observations and phenomena. This background will assist teachers in their interaction with students as they provide opportunities for them to move their conceptual thinking and possibly frameworks towards a sun-centred and then star-centred universe, and their reasoning patterns from the intuitive to the causal.

How learning celestial ideas occurs

How then can we assist students to move forward in their conceptual development and formation? The question does not have a straightforward answer. Several views have been proposed to explain how students learn astronomy concepts (Ehrlen 2009). Each is reflected in the pedagogical approaches described later in this chapter. Students' conceptual growth in this area might be explained by:

(a) their movement from an (Piagetian) egocentric perspective – this would be shown by their ability to imagine how their surroundings look from space (see Activity 10.12 on p. 424, which concerns the use of different perspectives to help understand moon phase observations)

(b) their differentiation of different conceptual frameworks (e.g., initial, synthetic, scientific) for explaining phenomena – this would be evident if students could distinguish between a commonsense framework based on everyday experience (e.g., the Earth is flat) and a scientific framework (e.g., planetary Earth in space)

(c) the influence of the discourse that occurs around an idea and any associated cultural tools (e.g., globes, maps, computer simulations) that are used (in teaching)

(d) the situation in which the learning occurs being significant to the student – this is because students will make judgements as to what type of explanation is 'activated' (in their thinking) depending upon the context.

Each of these theoretical perspectives on learning emphasises context, whether it be the:

(a) physical context
(b) conceptual context
(c) sociocultural context
(d) or simply different contexts per se.

Each of these theoretical perspectives on learning emphasises context, whether it be the physical context as in (a), the conceptual context (b), the socio-cultural context (c), or simply different contexts per se (d). Interestingly, when six- to eight-year-olds were interviewed about a satellite photo of the Earth, it appeared they gave responses that reflected (a) to (c) depending upon the situation (in the interview); that is, (d) (Ehrlen 2009). Hence, these views on learning need not be mutually exclusive – as Harlen (2009) explicates (see 'Constructivist teaching approaches' in Chapter 1), they can complement each other. Teachers taking this complementary position will probably make wiser judgements about how to understand and facilitate students' conceptual learning; for example, they may be less likely to misinterpret a student's comments as an alternative conception when it is actually the student making 'an alternative judgment of what explanation is appropriate in the situation' (Ehrlen 2009, p. 290). Further, such teachers will probably make better teaching decisions about what to do next.

Constructivist approaches in astronomy: How effective are they?

Upper primary students taught over an extended period in a traditional, teacher-centred and textbook manner about astronomical concepts (e.g., equator, axis, orbit) and explanations (e.g., day/night, seasons) could recall textbook definitions and explanations two months after instruction but had reverted to everyday (non-scientific) ideas four years later (Kikas 1998). In a similar but shorter sequence, students could not even recall the introduced ideas two months later, while in another study comparing traditional approaches with a constructivist sequence (using collaborative problem-solving activities and whole-class discussion), the latter students showed considerable conceptual growth relative to the class using the textbook-style instruction (Diakidoy and Kendeou 2001). Similar findings were reported from a study comparing traditional (n = 650; years 4 to 6) and constructivist approaches (n = 750; years 3 to 6) to learning about shadows, the sun's apparent path over time, day and night, and the seasons (Ward, Sadler and Shapiro 2007/2008), as well as a review of several research studies investigating how effectively students learned about explanations for phases of the moon (Kavanagh, Agan and Sneider 2005). The contrasting teaching methods were suggested to explain these results; for example, in the traditional approaches, teachers did not link learning to the students' everyday knowledge, and information was not discussed but memorised. These studies indicate that approaches based on constructivist principles are more effective than direct instruction. They also support activity-based methods (e.g., taking celestial observations and using models, both advocated in constructivist sequences described later) as being superior to teacher telling and using textbook 2-D images.

As teachers, we need to acknowledge these conclusions. For students to modify their existing ideas about celestial concepts and explanations, they must be given the opportunity to question their beliefs and to interact with different ideas. Depending upon the learning context – see Ehrlen's (2009) summary – students may see the 'new' ideas as being more fruitful, plausible and explanatory than their existing ideas (revisit 'Does constructivist teaching work?' in Chapter 1).

To facilitate conceptual growth relating to celestial concepts and explanations, numerous specific teaching suggestions have been proposed. This is the main emphasis of the remainder of this chapter. Many opportunities to 'think and work scientifically' (see Chapter 2), including building competencies in discursive practices of science such as interpreting, generating and manipulating representations of celestial phenomena, and using science-specific literacies, are included in the following activities and strategies.

Learning experiences about astronomy

This major section initially suggests ways to engage students in wanting to learn about our place in space. It then emphasises the difference between experiential and explanatory concepts and the consequent pedagogical implications. This consideration leads into the nature of scientific investigations about space suitable for students at different levels in the primary school. The use of models and representational learning in assisting conceptual growth in this area are then explored together with a reminder about the importance of dialogical conversation. Finally, ways to effectively employ secondhand experiences consistent with constructivist principles are overviewed.

Engaging students' interest in astronomy

If students are engaged with the content of a lesson or a theme, then they will generally focus their attention on learning. Orienting students to a topic and engaging their interest is the first step in sequencing lessons for constructivist learning, but retaining interest is an ongoing consideration. Generic suggestions are given in Chapter 1 and you should revisit them; for example, listening to students and taking seriously their ideas about how to describe or explain celestial bodies and phenomena. Some specific approaches that teachers have used to engage their students' interest in space are:

- linking with other curriculum areas; examples include:
  - *visual arts* – using colour and texture to express feelings about what it might be like to make a journey through space (Strange and Fullam 2009); preparing a mural of the

solar system, including a current space probe, while also discussing the accuracy of the mural during preparation and on completion (Paige and Whitney 2008); drawing pictures to illustrate science facts and have students sharing the end products and talking about them (Bilderbeck 2008)
  - *numeracy* – exploring the concept of shape while keeping a record of moon observations
  - *literacy* – having students use their persuasive writing skills to 'market a planet', asking themselves what would prospective buyers or explorers need to know about the planet (Comstock 2008), or writing their own media story such as about the discovery of a new planet.
- relating space events to everyday life; for instance, through news items such as on whether Pluto is a planet or not (Jarman and McClune 2007)
- organising an astronomy 'conference' where the presentations are given by primary students; for ideas, see Comstock (2008) and Walker (2009)

M&E

- incorporating field trips (e.g., to a planetarium or a major telescope), ensuring they are well integrated with in-school tasks
- linking with the community in person or electronically, both locally (e.g., through an amateur astronomer or club) and/or further afield, such as with schools in another hemisphere (e.g., to explore seasonal differences) or even by talking to astronauts (see Figure 10.1 on p. 420) (Thomas, Christie and Leverment 2009)
- encouraging kinaesthetic engagement with celestial concepts (Slater, Morrow and Slater, in Plummer 2009)
- using children's fiction and non-fiction literature while being cautious about its scientific accuracy, and possibly linking the content and images with hands-on investigations and computer simulations (Trundle and Sackes 2008)
- organising a 'Space Week' at your school where each grade level could focus on one astronomy concept; this can create talk and sharing across grades, as well as provide the opportunity for the display of completed investigations, all of which can encourage enthusiasm across a whole school (Comstock 2008)
- focusing on, and posing questions about, the validity of space-related films (Hugman and Placing 2007).

Experiential and explanatory concepts and activities

There are prerequisites for the development of explanatory models of the solar system (e.g., about the cause of day and night). Younger students need to appreciate daily and seasonal changes and have an elementary understanding of time. Many students under the age of seven may not know a day consists of 24 hours, a week of seven days, the number of days in a month and so on. To establish an appreciation of these times and sequences, construction of time lines (the events of a day, week, year, my lifetime, the history of a phenomenon or object such as communication and even the age of the universe) can assist (Osborne 1991).

Other ways to lay the foundations for later understanding of celestial explanatory models is to revolve learning experiences around the key principle of students making observations outside during the day and in the evening. Observational astronomy is essential as many students lack familiarity with the movements of celestial bodies. Plummer (2009), for example, reported that many Year 1 students do not appreciate the similar directions in the rising and setting of the sun and moon and their apparent paths across the skies, and more than half of Year 6 students believed stars never move or only do so at the end of the night. Changes in the position, path, shape and appearance of the sun and moon initially, then, with older students, the stars (and maybe, on suitable occasions, the planets), can be noted. The use of written records for observations (logbooks, hour-by-hour or day-by-day charts and seasonal wall charts) are a real aid in helping students see the changes that occur over short and longer time intervals. When appropriate, simple measurements can be introduced.

These suggestions support the view that young learners (and maybe older learners on topics with which they are unfamiliar) should be exposed to sufficient 'experiential concepts' before, or in conjunction with, explanatory concepts and/or models (Lucas and Cohen 1999). Experiential concepts are those that learners can directly experience; for example:

- Shadows change in length and shape during the day and this would seem to be related to the apparent movement of the sun.
- The sun appears to rise in the east and set in the west and appears to move in a regular pattern across the sky from day to day.

Can you list some experiential concepts that students could determine from observations of the moon and/or stars? Explanatory concepts (for example, the constancy of the tilt of the Earth on its axis as it revolves around the sun can explain the occurrence of the seasons), in general, should not precede experiential concepts: in this instance, for example, noting the duration of daylight in summer and winter and the pattern by which the duration of daylight varies.

Being too rigid in applying these guidelines is not advisable. Observing, describing and finding patterns in moon shapes, for example, is often thought to be more developmentally appropriate for young learners, than explaining such apparent changes (check to see if this is what your syllabus suggests). However, when taught using a scaffolded inquiry-oriented focus that included predicting moon shapes and appearance patterns, taking and analysing actual observations over many weeks, weekly data sharing and discussion sessions, most Year 4 (including special needs) students displayed a competent understanding of the causes of moon phases. For these students, a desire to 'explain' was a natural concomitant of investigating moon shapes and patterns (Trundle, Atwood and Christopher 2007a).

Activity 10.6 Experiential shadow stick activities

Set up a shadow stick at home with a large sheet of white cardboard. What questions could you ask yourself or students prior to taking observations? Keep a shadow record over the day, accurately noting length and times. What questions could you ask about the record during, and at the end of, the day? For further ideas about shadow sticks and related activities, see Aldridge (2002), George (1992a), Kibble (2002b) and the 'Science as a human endeavour' section later in this chapter (p. 435).

Various questions could be asked about shadow stick activities. Can you devise some that clearly relate to experiential concepts? What about some that require explanatory concepts or models? Some example questions are:

- Is the shadow getting longer or shorter?
- What does this tell us about the apparent motion of the sun?
- If we were on Mars, would we have shadows like this and would we be able to see the same things happening? (adapted from George 1992a)

Would these questions require experiential or explanatory understandings?

As a teacher, what other short- and long-term observations or investigations with shadow sticks can you think of? What preconceptions might such activities challenge? For example:

- How might you find the 'north' direction?
- What is the direction of the midday shadow?
- When is the shadow the shortest?

Also, can you suggest similar types of activities in relation to the movement and shape of the moon? On what types of concepts and/or models would they be focusing?

Other examples of possible outdoor activities are listed next. They place initial emphasis on experiential concepts; some would be suitable for younger primary learners. It is most important that, integrated with the use of these activities, teachers provide 'sustained emphasis on *discussion and reflection* [as this will] encourage the growth of coherent understanding rather than the assimilation of individual fragments of knowledge' (Osborne et al. 1994, p. 113, emphasis added; Barnett and Morran 2002).

- The sun's apparent movement across the sky can be observed and recorded in 2-D representations at different times of the year. Observations at the equinoxes and in between provide contrasting findings. Upper primary students could be challenged by introducing 3-D representations for their experiential observations; computer simulations of the sun's path at their location on Earth could assist this task (see http://engnet.anu.edu.au/DEpeople/Andres.Cuevas/Sun/SunPath/SunPath.html) (Rowlands 2005). It is also relatively simple to take photographs of the sun across the sky during a six-month period using a pinhole camera (Quinnell 2009; see http://www.pinholephotography.org and http://www. solargraphy.com).
- Shadow stick activities also can assist students to appreciate the sun's apparent movement across the sky. They can be used in the same location but in different months and seasons to appreciate the sun's altitude changes. A novel idea, if it can be arranged, is to exchange shadow stick observations with students from a different hemisphere (perhaps by email).
- Moonrise can be recorded on particular evenings (Kepler 1990) and, if obtainable, be compared, by older students, to Earthrise (access website images of the Earth rising as seen from the moon or space).
- A record of the moon's (apparent) shape over the period of a month can be noted and entered onto a monthly chart (perhaps a large class wall chart), which can highlight the sequence of changes that occur (George 1992b). With younger students, a cheese and biscuits activity involving covering the moon with appropriate amounts of cheese spread to indicate its (apparent) different shapes and in what order they occur can reinforce observations (Kepler 1990).
- A more sophisticated but still reasonably straightforward activity is to record the location of the moon relative to a fixed landscape location over a period of a month and, where possible, indicate its angular separation in the sky from the sun. Hand measurements can

be used as the measure of separation between the sun and the moon; for further ideas, see the Elementary Science Study (1968) volume *Where Was the Moon?*

This list suggests that there are numerous experiential activities that can be undertaken by observation of the skies. As Schaaf (1990, p. 24) says, 'All we need to discover things new and beautiful ... is to look up, be alert and be patient'. He then describes observational activities related to moon maps, features of each lunar phase, shades of bright and dark on the moon, the size of the moon, the thinnest moons in twilight and day, earthshine on the moon, differences between planets and stars (twinkling), colour variation and twinkling of stars (seeing), visibility of the Milky Way and others. Many can be done in one or two nights; others take longer. Teachers can use such resource activities to help students answer their own questions about astronomical objects and events, as well as to challenge students' existing ideas.

Osborne (1991, p. 9) offers the following advice when involving students in astronomical observations of the type just described: 'Observations must be directed and the salient features discussed and explored before the activity if pupils are to see what we would like them to discern'. This directed scaffolding (see the online Appendix 1.5) could follow after students have reported on their unaided and exploratory observations. An example would be having students identify the shapes of major constellations (say, with a ready-made constellation viewer – see the upcoming section 'Mainly upper primary experiences' on p. 418) and teachers then providing instructions for parents or carers as to where to direct their children's attention in the evening sky.

Appendix 1.5
Scaffolding strategies

Activity 10.7 Selecting and assessing learning outcomes for lower primary space activities

If primary students up to Year 2 were engaged in some of the above activities, or variations of them, what could be some of the important conceptual learning outcomes? Can you discern whether they are experiential or explanatory concepts and understandings? Compare your responses with the appropriate learning outcomes in your primary science syllabus or framework. (Many syllabuses and related documents – e.g., Curriculum Corporation 2006 – do not explicitly state concepts and understandings as at the end of this chapter. When this is the case, you need to make professional judgements as to what are appropriate conceptual expectations from the available learning outcomes, statements of learning and so on that are in education system documents.)

Suggest some ways in which you could formatively assess such outcomes. Decide on a specific concept that you think might be appropriate for lower primary and plan a specific formative assessment process/task (see types of formative assessment in Chapter 1). If you obtained responses from a student using your assessment procedure, how might you judge the quality of the conceptual learning? Could you develop an assessment rubric for your specific learning outcome? (The learning outcome, for example, might require students to draw and talk about their understanding of the apparent movement of the sun across the sky.) You could use categories such as 'confused', 'incomplete/ inaccurate', 'partial understanding' and 'complete understanding', or devise your own categories. Young and Guy (2008) report how they used a rubric with the moon phase task associated with Activity 10.12 (see p. 424).

Scientific investigations about space

For lower and upper primary levels

Many activities about space can become scientific investigations if investigable questions are posed by the teacher or the students. An investigable question is one which requires students to take observations and/or use hands-on materials in order to find an answer. Investigating in this sense does not mean looking up a secondhand source, such as the Internet, a book or a CD-ROM (these are better termed 'research' activities; see Chapter 2). Investigable questions are, therefore, to be distinguished from questions that do *not* lead to students doing (that is, either handling materials or taking field observations). Investigable questions are sometimes referred to as 'productive questions' or 'action' questions (Harlen 2001). These may be better descriptors, as 'investigating' sometimes has been given a specific meaning, namely fair testing, and this may not be applicable in some scientific investigations of celestial bodies (see the Chapter 2 section 'Fair-test investigations' on p. 69). Many teachers are familiar with the value of productive or action questions in other science topic areas, but may feel that space does not lend itself to such questions, which can lead to direct observations or hands-on investigations.

Activity 10.8 Devising productive, action or investigable questions about space

Using the above description of productive, action or investigable questions, devise suitable questions related to the sun, moon, Earth and stars that you believe students could find answers to by using field observations and/or manipulating physical models. State why your questions are productive, action or investigable in terms of the way they are worded. (The questions must require students to handle materials or observe real events and phenomena. For example, consider how a student might respond to a 'what if' question compared to a 'why' question.)

Action questions tend to suggest that certain science processes (e.g., observing or predicting) will be used to answer them. This is usually determined by the way they are phrased and their content. Into what categories do your productive, action or investigable questions fall? Are they testing or perhaps design-and-make questions? Why? (A 'testing' question, for example, usually would require a fair test to be carried out to find a response.) Or are they, for example, observing, predicting or hypothesising questions that might *not* result in an investigation or a testing activity? Why might this be the case? (Hint: see the Chapter 2 section 'Activities to develop thinking and working scientifically' on p. 67.) If this is the case, they could, though, with teacher input, still lead to students planning and carrying out their own scientific tests. What could a teacher do to encourage this to happen when these more limited types of questions were the initial focus?

Questions about space that could lead to observations and/or investigations

The following productive questions about the sun, moon, planets and stars clearly suggest the firsthand observations and the type of hands-on investigations (mostly with models) that would be required. Where appropriate, there is considerable value in having students make, and act out, predictions based on their prior knowledge first (Plummer 2009).

The movement of the sun and shadow sticks

- What (apparent) path does the sun take across the sky?
- Is there a noon shadow? Does it change from day to day?
- Are there evening shadows?
- Predict where the shadow will be in X minutes. Were you right?
- When are shadows the shortest? The longest?
- How can shadow records be used as clocks? (How could you design and build a sundial? MacLennan (1992, pp. 156–8) and Kibble (2002b) provide some suggestions.)
- How can you find the direction from a shadow record?
- How do shadow records change over the year? What happens on special days (shortest and longest days, equinoxes)?

The shape and position of the moon

- How can a map be drawn of the moon? What would you need to draw it? How does it change over time?
- How does the moon's shape change over a month? Two months?
- What path does the moon take across the sky?
- How does the moon change its position in relation to the sun? How can you find out?

The stars and planets

- Do the stars move? Are the stars in the same place from night to night? Week to week? Month to month?
- Do all stars look the same? If not, how do they differ?
- Which patterns can you see in the stars? How could you find out more about these patterns?
- Can direction be found using the stars?
- Is it possible to see planets with the naked eye? In what ways can they be observed to be different from stars? How do they move?

There are numerous ways such questions could arise in a classroom. One Year 1/2 teacher used a story with a concept cartoon that revolved around the question, 'Will grandma stay in the shade all afternoon under the tree or will the children need to wake her up and move her at some point?' After initial discussion around the cartoon, the students devised a hands-on way to investigate the problem. What might they have done? What are some questions you could use to help the students reflect on what they found out? What types of explanations might enter the class discussions? How could you address these? Compare your thinking with the case study reported in Morris et al. (2007), which also provides the sources of concept cartoons and associated stories.

Clearly, depending upon previous experiences the students may have had, teachers could, in some instances, encourage their students to plan scientific investigations in order to find answers to these questions, as just described. For some questions, the importance of variables could be stressed. Students could plan their investigation (experiment) so that only one variable is changed at a time: a *fair test* – see the Chapter 2 section 'Fair-test investigations' (p. 69) or Hackling (2005) for information on how to assist students to plan fair tests. This can also happen when simulations or models are used to explore students' ideas about reasons for events such as day and night.

An example of such an investigation would be simulating day and night using a light source, such as a strong torch for the sun and, say, a rough-surfaced white soccer ball for the Earth, and changing particular variables associated with the Earth to determine possible effects. Examples could include its movement (stationary and circular), direction of movement (up and down, circular with spin and no spin), the speed of the spin and other variables that may be elicited from the students' questions or previous ideas. For a 'seasons' investigation, variables could be tilt and no tilt, distance between the sun and the Earth and so on. For eclipse investigations, similar variables could be tried.

Encouraging discussion around models by changing particular variables that evolve from the students' own ideas can only assist students to clarify their thinking about how they explain events such as the seasons, day and night, moon phases and other astronomical phenomena. For ideas on how to sequence lessons based on this approach, refer to the 'Question, Identify, Gather, Interpret, Model, Predict, Demonstrate' (QIGIMPD) steps (see the upcoming section 'Planning teaching sequences for conceptual change or development concerning celestial phenomena' on p. 432).

Activity 10.9 Observing or investigating the sky as a teacher

It has been strongly suggested that teachers get their students involved in actual observations and investigations of the (day and) night sky. One way to appreciate the impact this can have on your students and to see it from your students' perspectives is to undertake a real investigation or set of observations yourself (as described above; that is, not just research a topic) in order to answer some question(s) to which you do not know the answer(s). An example follows.

Take observations of the moon and/or a planet, say, Venus, for one month. Make an entry in a diary for each observation. What patterns can you find in your observations? Try to make sense of them. As a challenge, try to determine if there is any pattern to:

- the moon (and/or planet) and sun movements
- the moon and/or planet movements
- the moon (and/or planet) and star movements (see also Activity 10.10 on p. 419).

What did you learn about observational investigations of the sky from completing this task? How will it influence the type of tasks (e.g., problems to solve) that you might give to lower and upper primary students? How could such 'investigations' fit into a constructivist teaching sequence (compare your responses with descriptions later in this chapter)?

When preservice teachers engaged in extended observations and analysis of moon observations, together with scaffolded psychomotor modelling of moon–Earth–sun relationships, there was a marked improvement in their appreciation of these relationships to explain a range of phenomena. Various characteristics of constructivist learning were present. Compare the features of your teaching sequence in this activity with the characteristics of the sequence in Trundle, Atwood and Christopher (2007a, 2007b), which are based on Vosniadou (2003) (see the online Appendix 10.7).

Appendix 10.7 Some key characteristics of teaching for effective conceptual development

Mainly upper primary experiences

For older students, some observations could focus on the stars and planets. The use of a star wheel or map can initially assist teachers to locate key stars, constellations and planets (for availability, see Activity 10.10; see also http://www.ozskywatch.com or http://www.heavens-above.com). For teachers not familiar with identifying constellations, it is important to realise the large expanses of sky that most constellations cover and to have a mental picture of the images that ancient astronomers and astrologers saw in the heavens – see, for example, Floyd (2002b), Kerrod (2002) and Ridpath (2006). In Australian schools, teachers could focus on the Southern Cross and the two pointers and the Saucepan (part of Orion). Scorpio, once recognised, is usually not forgotten. Apart from the satisfaction of being able to recognise some key stars and constellations, it gives students some sense of finding their way around the heavens. It also can lead to a greater awareness of the movement and nature of the stars. Students can, for example, observe how stars in a particular constellation appear to move over an hour and whether all stars in a constellation look the same. The view held by some students that the sun is not a star can also be challenged.

Activity 10.10 Identifying planets, stars and constellations

You really cannot teach confidently about this part of the curriculum unless you can readily identify some key celestial bodies. Obtain a copy of a star chart or map suitable for your area and over the period of a week attempt to locate the constellations mentioned above (if in the Southern Hemisphere). Note that at certain times of the year you may not be able to observe all the constellations.

Are there any planets visible at the time of observation? You will find this information in some newspapers, science education journals such as *Teaching Science* (e.g., Forma 2008), annual Sky Guide publications (e.g., Lomb 2009) and on various websites (e.g., http://www.sydneyobservatory.com.au/observations-blog/?cat=10). Ensure that you can identify Venus (the morning and evening 'star'). Other possibilities can be found on websites (e.g., see http://outreach.atnf.csiro.au/education/teachers/viewing, which outlines naked-eye, binocular and small-telescope observations).

By completing this activity, a whole range of other teaching possibilities will occur to you because you have had a go.

Apart from using a strong torch on a clear night away from city lights, there are several ways to assist students to identify constellations. A constellation viewer, made using cans, heavy-duty paper, a pencil, rubber bands, scissors or similar materials (ASTA 2009; Revell 1994), can help students appreciate the constellation patterns. For other Australian constellation activities, see Floyd (2002b).

Planets, apart from Venus, usually are not as readily identified. However, as the solar system is such a popular primary project topic, every effort should be made with upper primary students to assist them to see some planets with the naked eye or a pair of binoculars (and with the help of an interested parent or carer). Special events with the planets can be a means of gaining interest. The Sydney Observatory website (see Activity 10.10) indicates when these occur and how to observe them (also see the CSU Remote Telescope website at http://black-hole.mit.csu.edu.au/telescope). For most of the year at least, one planet is usually visible in the early evening sky (Floyd 2002a). Such observations could be supported by aids such as a student-made 'flip-book' which can show how a planet moves against a background of stars (Floyd 2003).

Observations with the naked eye raise the question of what is the role of optical telescopes in primary science. Although there may be difficulty for younger learners in appreciating what they are seeing, many upper primary students would obtain a real thrill from seeing Saturn with its rings or the colours of Jupiter; the smallness of the image with amateur telescopes may limit the satisfaction for some students. When teachers plan astronomy nights with a local amateur astronomer, they need to ascertain exactly what can be seen and weigh up the learning benefits of the experience. In some instances, the benefits are obvious. Primary students at one school used a telescope to sight an orbiting space station; they spoke with NASA astronaut Jeff Williams on the space station (see Figure 10.1). Always be on the lookout for what may be possible. At one (secondary) school in England, students met astronauts, astronomers and space shuttle pilots (Jones 2009). What might be possible in your country?

Some primary schools can now access real-time astronomy using high-powered telescopes via the Internet. Student responses to seeing the images and receiving them by email have been very positive (Anonymous 2004); for example, see Floyd (2003).

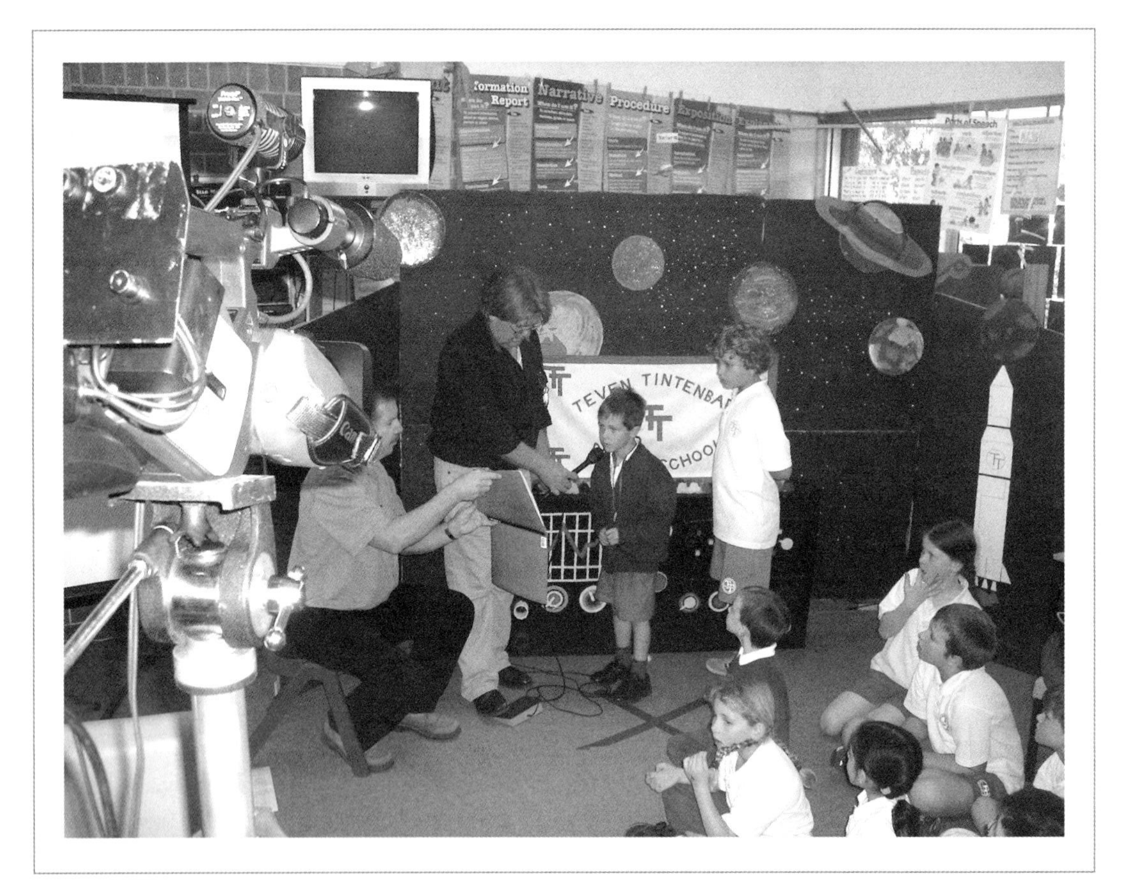

FIGURE 10.1 Students talking with NASA astronaut Jeff Williams in 2006

Source: *Northern Star*, 10 August 2006

Teachers in England have reported how primary students accessed the Bradford Robotic Telescope (see http://www.telescope.org) and requested images of the moon, planets, stars and galaxies; interactive whiteboards were also used with these images to further engage the students. Baruch, Machell and Norris (2005) describe how one Year 6 class used such images to find sites of the Apollo moon landings and study craters and other moon features.

Planetariums can speed up the movement of celestial bodies and assist students in visualising celestial phenomena. If learning at a planetarium links verbal descriptions with non-verbal representations (visual, auditory, haptic and kinaesthetic), then it may be more effective. When year 1 and 2 students' body motions were linked with visual images and verbal descriptions to connect to target concepts, such as the sun's apparent path across the sky, then the students showed significant improvements in their understanding of the movement of the sun (and moon and stars). This was, in general, independent of grade and gender. An excursion to a planetarium is recommended for upper primary-age students, although, as described, it can be of value to younger students (Foo 2009; Plummer 2009). It is imperative that you discuss with planetarium staff the celestial events to be covered, as sometimes these marvellous teaching aids can be used to introduce concepts far beyond what we know most students appreciate about space; such a visit then loses the interest of the learner. Ideally, the visit should be integrated with in-school experiences before and after, and, if possible, be an interactive session (Plummer 2009). Portable planetariums, which are inflatable and accommodate small numbers of students, may be available for teacher use in your State, Territory or country; for example, see Ould (2009). Another alternative is to access the computer program Stellarium (see the online Appendix 10.5; ASTA 2009). It shows the sky as seen from a particular location, date and time, and can be used as if you were at a planetarium. Hughes (2008) provides a range of suggestions on how to use the program.

Appendix 10.5 Software, websites and webcams

Activity 10.11 Selecting and assessing learning outcomes for middle and upper primary space activities

If middle and upper primary students were engaged in some of the above activities, what could be some of the important conceptual learning outcomes? Compare your responses with those in your own primary science curriculum. How could you formatively assess them? See Activity 10.7 (p. 415) for guidance and other ideas.

Explanatory concepts, representations and conceptual formation

The importance of experiential concepts and associated hands-on and observational learning has been stressed. It has also been noted that it is natural for students, even very young learners, to offer explanations for what they are observing. Their 'explanations' often go beyond describing experiential concepts as they try to explain why things are happening; that is, explanatory concepts. In this section, how teachers can use physical models to assist students to move towards accepted scientific explanatory modes related to astronomy concepts will initially be explored. Physical models are a way of 'representing' a concept or understanding. There is now an emerging view that engaging students in generating their own 'representations' of phenomena is, in part, how they develop their conceptual understanding (revisit 'Representational and related strategies' in Chapter 1). This wider view of learning, through representation, is then developed.

Accepting explanatory models related to the solar system

Many primary students have considerable difficulty accepting certain astronomical models, such as those that explain night and day, the seasons, and the fact that we see only one side of the moon. However, at upper primary levels, up to half of the students might internalise some of these explanatory models (Osborne 1994; Webb and Morrison 2000), and with appropriate scaffolding even more, and younger, students may understand the causes of some celestial phenomena; for example, moon phases (Trundle, Atwood and Christopher 2007a) and the movement of the sun, moon and stars (Plummer 2009). Strategies that could help are those that require students to see events from another perspective, and drawing objects from a range of viewpoints, which can be exemplified by having students play the roles of celestial bodies but also observing the role-plays from the 'outside'. A number of astronomy role-playing activities – for example, various planets revolving around the sun, zodiac star patterns through a year, eclipses, and moon rotation and phases – are described in van den Berg (2000).

Physical models can assist understanding

Understanding celestial phenomena can, at times, be made more difficult as 2-D diagrams are often inadequate and detract from learning. Find a typical textbook moon-phase

FIGURE 10.2 Students displaying physical models and launching a model of a 'rocket' (from a unit of work on space during which they spoke to a NASA astronaut – see Figure 10.1)

Source: *Northern Star*, 10 August 2006

diagram, for example, and ask yourself if it distinguishes moon phases from solar and lunar eclipses – clear 2-D moon-phase diagrammatical representations are difficult to locate; two examples are ASE (1990) and Stringer (1999). Consequently, using 3-D models to assist understanding in this area is often recommended; for example, see 'eclipses' in Allen (1999).

You need to appreciate that using such models can clarify or confuse your students' learning. In one study, using a globe with six- to eight-year-olds did not assist all of them to articulate a scientific view of the Earth as a planetary sphere; unintended consequences can result, such as students developing dual models of the shape of the Earth – see Brewer's (2008) summary in the earlier section 'Primary students' ideas about astronomy concepts' on p. 405). There are several reasons why using models may not result in the desired understanding. It could be because the 'visual representation does not always give students access to that which is being represented' (in the example here, a globe representing the Earth in space). It may also be that students are not aware of, or do not differentiate between, the different 'speech genres' (e.g., scientific or everyday) that may be associated with the use of a physical model. In this example, with the globe, a student may respond with 'geographical (everyday) speech' related to the nations or topographical features depicted on the globe, but the teacher and other students may use speech (and a scientific genre) that is referring to the 'model' (here, a globe) and the 'actuality' (here, planet Earth) that it represents. Teachers need to be conscious of the 'visual language' students are using in association with an artefact employed to help them understand a scientific idea (Ehrlen 2008, pp. 237–9).

Another consideration, especially with primary students, relates to how students interpret the relationship between physical scientific models and reality. There appear to be three different levels of understanding the relationship:

- level 1 – where models are seen as either toys or copies of reality; possibly up to 60 per cent of Year 7 students see models in one of these two ways
- level 2 – where constructed models are seen as having a particular purpose in mind
- level 3 – the scientist's perspective, where the model is a tool for developing and testing ideas (Ehrlen 2008, p. 224, summarising Grosslight, Unger and Jay; Ehrlen also points out that most primary students appreciate 'picture-world' relationships in visual representations but not the intentions of the producer of a picture until they are about 11–14 years).
- So what advice is there for teachers using physical models?
- Firstly, ensure your students have had observational or hands-on and/or everyday experiences of the phenomenon (for example, they have kept a record of apparent moon shapes over time) and do not substitute the (secondhand) model for the (firsthand) experience.
- Secondly, do not uncritically accept a particular physical model for use with your students.
- Thirdly, be aware of possible ways in which your students may focus on aspects of the model that do not relate to the understanding(s) (sometimes called 'target ideas') you are trying to help them learn (Frazier 2003). The above example related to the globe and the position and shape of the Earth emphasises the value of this advice.

Teachers can help their students analyse physical models by using three questions: 'What makes the model an effective model, that is, how does it resemble that which it represents? What makes the model an insufficient model, that is, how does it differ from that which it represents? How could we improve the model?' (Eichinger 2005).

With more and more teachers (and students) having access to computer simulations, it needs to be recognised that everything portrayed on a computer is a model. Hence, hands-on experiences should preferably precede computer models and students should be encouraged to analyse computer models using the above three questions.

The following activity relates to a well-known use of a physical model for explaining moon phases and encourages you to think about the conceptual learning implications of the model. As an extension activity, you could ask yourself what is the assumed perspective for viewing the moon phases system in Activity 10.12. What would the moon phase system look like from the moon, the sun, another planet or out in space? Responding to such questions will check your understanding.

Activity 10.12 Physical moon phase (and other) models

Firstly, keep a moon-observation record for an extended period (see Activity 10.9 on p. 418). Then, assume that you are going to use, in an appropriate constructivist lesson sequence, a model to illustrate moon phases – for example, an overhead projector or strong light source (sun), a person (Earth) holding a stick with a white polystyrene ball (moon) on it and a piece of plasticine on the ball. Instructions may be found in many references (e.g., ASTA 2009). Identify some of the strengths (how might it help clarify alternative conceptions) and potential difficulties (how might it add to, or strengthen, alternative conceptions) of using such a model to help students understand moon phases.

More generally, can you now itemise some obvious teaching implications of using models to assist student understanding? DeBoo and Asoko (2000) provide a list of benefits and potential difficulties of using models and analogies. As an additional task, try to apply some of this advice to the static and linear scale (distance and size) model that is often used to depict the solar system (see also the upcoming section 'Secondhand experiences: a popular example (planets)' on p. 427).

How does the implied and other pedagogical advice from this activity relate to students learning about the NOS?

Activity 10.12 is an 'inside the system' model (illustrating moon phases). Try this extension activity: have a colleague take your place and view the moon from outside the system. What do you notice? This is an example of using different perspectives to help students learn. Young and Guy (2008) provide further details and introduce the novel teaching idea of a 'self-shadow' (a shadow on an object blocking the light and on the opposite side of the light source) compared to the commonly experienced 'cast shadow'. In what ways could you use these two types of shadows when having students explore moon-phase models?

Teacher- and student-generated multimodal representations

There is increasing evidence that student learning is enhanced if new ideas are encoded in multiple ways (for background to this, see Chapter 1). This can be done if teachers use and encourage students to use different modalities (verbal and non-verbal) to represent phenomena. Non-verbal modalities include movement (kinaesthetic), touch (haptic), visual and auditory. As many celestial concepts depend on understanding phenomena in three dimensions, this approach is considered to be especially helpful (Hubber and Haslam 2009).

Several principles can guide the use of multiple representations using different modes (Hubber and Haslam 2009). These are that teachers:

- identify the main ideas and concepts underpinning a topic (e.g., cause of day and night) and the most common alternative conceptions primary students may hold. This will assist in guiding students' refinement of their representations.
- ensure that the representation(s) are explicitly critiqued by the students, as all representations are limited – for suggested questions above, that is, Eichinger (2005).
- insist on students manipulating their models (whether mental, 2-D or 3-D) by, for example, varying the variables within the model and reasoning out the effects. An example of how this could be guided is the QIGIMPD approach (see the upcoming section 'Planning teaching sequences for conceptual change or development concerning celestial phenomena' on p. 432).

R

Activity 10.13 Using multiple representations to develop celestial concepts

For a middle to upper primary class, select an astronomy concept (e.g., the cause of eclipses) and devise how multiple and multimodal representations (see the Chapter 1 section 'Representational and related strategies' on p. 29) could be used by you as the teacher, as well as a few representational tasks you could ask the class to complete. At what points in a lesson sequence could they be used? Justify your selections. Ensure you refer to some personal and sociocultural constructivist characteristics.

If feasible, try out some of your ideas with a class or group of students. Have a colleague keep a record of some of the students' comments (written and verbal) and actions (including gestures) as they engage in representational learning. Reflect on your colleague's and your own observations in light of any research and theoretical readings you have encountered about this approach to learning.

As with Activity 10.12, how does the implied and other pedagogical advice from this activity relate to students learning about the NOS?

There are many responses to this activity. Did you, for example, use representations in any of the following ways?

- *To elucidate the nature of science* – assisting upper primary students towards an appreciation that models (whether 2-D or 3-D) can only partially represent a concept/understanding and that this is how scientists view models. You could ask your students in what ways does a globe represent and not represent the Earth. Models could also help students derive evidence about scientific decisions. How, for example, might a model be interrogated to have a class discussion as to whether Pluto should be called a planet or not? This would be using models as in science (as per the level 3 understanding of the relationship between physical scientific models and reality; see the earlier section 'Physical models can assist understanding' on p. 422).
- *To challenge students' thinking and to facilitate reasoning* – This could be done, for example, by having students kinaesthetically act out

NOS

their existing ideas about a phenomenon, such as the length and position of shadows over a day. Observations could then be collected and their thinking re-enacted. Open ended dialogue would be integral to both instances.

- *To represent and then re-represent an understanding* – Were different modes used to represent an idea? Was, for example, a diagrammatic representation used to explain the seasons, re-represented in 3-D?
- *As a means of formative and summative assessment* – If you tried activities 10.7 (p. 415) or 10.11 (p. 421) with some students, what were some instances of formative assessment to assist learning (see Chapter 1)?

You should compare your decisions and insights from Activity 10.13 with features that underpinned a 'representational astronomy' sequence taught to Year 8, but which could be readily applied to primary levels (Hubber and Haslam 2009; see the Chapter 1 section 'Representational and related strategies' on p. 29).

A

Representing and re-representing ideas in different modes is a fundamental way of learning science that embraces many constructivist characteristics. It is important to appreciate that representations are 'tools for thinking [communicating and reasoning about concepts] rather than summarised pieces of knowledge [ends in themselves] to be learned' (Hubber and Haslam 2009, p. 20). This is probably different to the way many teachers have thought about representations (e.g., textbook diagrams, 3-D models) and is certainly different to how primary students envisage models used in learning (see the discussion of levels 1 to 3 in the earlier section 'Physical models can assist understanding' on p. 422).

Discussion and dialogue

Encouraging peer and teacher–student conversations is now believed to be integral to progressing conceptual understanding and has been emphasised several times in this book. 'Dialogic' conversations, where the classroom talk is more 'searching, reciprocal and cumulative' and is focused on the use of evidence, is a more effective way of assisting conceptual learning than 'question–answer–tell routines' (Alexander 2004; Harlen 2009). To have children talk about their astronomy ideas usually will also assist engagement. Sort cards and concept cartoons (Keogh and Naylor 2006) can be very helpful. An example of the latter could be a cartoon depicting several students looking at the moon at night and each offering a different reason for the moon's apparent change in shape.

It is critical that students of all ages be able to offer their ideas about celestial events without ridicule or censure. Often, further discussion, done sensitively, or related activities can help students to see that there could be other explanations, as when a young learner shared his idea that a 'moon monster' ate a little of the moon every night (Foo 2009). Integral to this style of dialogue is teachers having the mindset that learning occurs 'from the inside out'; that is, from where the students' thinking is, rather than from the 'outside in', or imposed by the teacher. Students need to be encouraged to be open about their ideas with their peers and adults/teachers, and to consciously think and reflect upon them, often through participation in practical or other activities (Collins and Simpson 2007).

Firsthand and secondhand experiences in astronomy

Three teaching principles have been emphasised about the types of strategies to use when teaching about space:

1 the importance of taking firsthand observations (and planning and conducting hands-on or field-based investigations, even if it be via models and simulations)
2 where possible, using the students' own ideas as the stimulus for the investigations and other means of learning
3 incorporating multiple and multimodal representational learning into the selected range of tasks.

Also, students need to do some of the science learning outside, which, for many students, is motivation in itself; it emphasises that science does not just occur indoors. Another principle then follows. Secondhand experiences support, but should not replace, these firsthand observations or hands-on (albeit at times simulated) activities, except when phenomena are inaccessible in other ways. This fourth principle is important, especially when the topic is space. This is because secondhand experiences are often used at the expense of the first two principles.

Secondhand experiences: a popular example (planets)

Secondhand data do fulfil a significant role when students learn about celestial objects not readily accessible by unaided observation or at any particular time (e.g., some planets, meteorites and comets) or simply are not observable (e.g., black holes). If teaching a lesson sequence about, say, the planets – a very popular primary school topic – then engaging interest and eliciting students' ideas is again an important initial step. Also, where possible, it is helpful to determine the questions about the planets to which students may wish to find answers (as in the interactive teaching approach discussed in Chapter 4). Although their questions may be 'researchable' rather than investigable (see Chapter 2), they are still important because such questions provide an indication of what the students already know and where they wish to head next (Tasker 1992). From this point, an adapted version of an interactive sequence could be used: after questions have been categorised, students (older learners in particular) could be asked how they could find answers to their questions. Access to reference material can help students acquire interesting facts and understandings about the planets that are related to their questions. It may not be hands-on, but it is still related to their ideas and there are many teaching activities associated with such secondhand experiences, including, for example, writing postcards from the planets (Aldridge 2002). Secondhand resources about the planets are extensive and include 'big books', space resource centres in large cities, CD-ROMs and the Internet (e.g., see the Nine Planets website can be found at http://nineplanets.org).

Activity 10.14 **Secondhand space experiences for primary students**

Locate the names of some secondhand resources that primary teachers could use to assist learning about the Earth and space. Some suggestions are provided to help you get started. Prepare a sheet for other teachers that lists the resource and an annotation about its content and usefulness. Ensure that you indicate some examples of specific conceptual idea(s) to which it could relate and suggest how it could be used constructively in a lesson sequence. Can you find some examples of:

- posters and charts – see websites associated with observatories, museums and science teacher associations; for example, ASTA (see http://www.asta.edu.au)
- software and Internet sites, including webcams – see, for example, ASTA (2004, 2009), Hollow (2006), McCall, Pickford and Hart (2002), and http://www.webcamworld.com; see also the online Appendix 10.5
- songs (e.g., Wilson and Wisbey 2000)
- CD-ROMs (e.g., associated with the CSU remote telescope; see http://black-hole.mit.csu.edu.au/telescope)
- IMAX films (for example, *Hubble 3D*; *Magnificent Desolation: Walking on the Moon 3-D* and *Space Station 3-D*)
- DVDs (documentaries and movies; for 'authenticity' see *Apollo 13* – Hugman and Placing (2007) list where to find excerpts related to various celestial concepts – some could be adapted for primary levels)
- fiction and non-fiction books
- magazines (e.g., see the *Double Helix* science club and *Helix* and *Scientriffic* magazines; visit http://www.csiro.au/products/DoubleHelixClub.html) and comics
- games, including computer games
- models
- museums and similar institutions (e.g., an observatory, such as the radio telescope at Parkes, New South Wales)
- satellite images
- any other type of secondhand experience you can think of.

Appendix 10.5
Software, websites and webcams

Secondhand experiences can be used in multiple ways, such as eliciting or challenging non-scientific views, having dialogic discussions about representations, and seeking students' questions (e.g., Wilson 2002a). With student questions, for example, rather than directly answering them, teachers could turn them into investigable/productive questions and guide students in seeking answers from hands-on tasks and other sources (Kallery 2000).

On the topic of planets, student questions may result in the construction of scale models of the planets and the distances between them; these are worthwhile activities if students appreciate scale (see Wilson 2002b; see also Appendix 10.6 on p. 443) and teachers assist them in appreciating the limitations of representations (see the earlier discussion about physical models). The construction of the planets can become a class task, with different groups preparing scale models of a particular planet or planetoid (for scale measurements, see http://www.exploratorium.edu/ronh/solar_system). Distances between planets can be emphasised using the schoolgrounds, which will underline the enormity of the distances involved. Include the sun (in two dimensions, as it will be very large) to stress the contrast in size between it and the planets. Such a model, together with role-play, may help to clarify many alternative conceptions students have about the solar system (Marsh, Willimont and Boulter 1999). While outside, teachers could ask 'What do you want to learn' or 'What new wonderings do you want to investigate' about the planets. Your task will be to turn questions such as 'How do the planets move around the sun?' or 'Where are the moons around various planets' into investigations that can lead to actions (here, the role-play of planetary and moon motions). You might also categorise some questions as explanatory questions about which older students can make predictions and then research them, such as 'Why do planets have moons?' – see Schuster (2008) for more examples. In relation to solar system models, it is also possible to have students construct scale models of the planets in their actual location at any particular time; see Floyd (2002c).

Reflective use of secondhand experiences

The above discussion suggests that, when appropriate, students can learn science from non-practical experiences (books, DVDs, posters, computer simulations and so on) and that these can be catalysts for firsthand activities. Lund (1999) describes how science understandings developed during timetabled literacy hours. Just as reflection on learning is critical to the development of ideas related to firsthand experiences (Barnett and Morran 2002), learning experiences using secondhand sources require reflective, not receptive, interaction with the resource. Osborne (1991) refers to the literature on directed activities related to text (DART) that highlights techniques such as discussion of the text and articulating one's own understanding of it. Specific tasks can include sentence reordering, text completion, table construction, diagram labelling, creating a (space) dictionary, faxing or emailing a message to a younger student or friend, even creating a book (Bentler 1988; Lund 1999). Blending science and literacy and assisting students to use a range of generic and science-specific literacies in responding to problems and tasks about space can assist their conceptual development (see 'Literacy-science connections' in Chapter 1). These techniques have been emphasised here to stress that, for many students, simple note-taking and project work probably do not cause sufficient reflectivity for students to modify their ideas.

Information and communication technology

There is an increasing literature on how primary teachers have used ICT to help students to learn space concepts. Some suggestions have been integrated into earlier sections of this chapter; for example, students accessing real-time images from telescopes. Brief details of further examples are provided below; you are encouraged to access the suggested references. You need to apply WISE and other principles (see the Chapter 1 section 'Information and communication technology (ICT)' on p. 39) to determine if these are, in the contexts in which you intend to employ them, effective uses of ICT.

Spreadsheets

Students used an Excel spreadsheet to calculate their weight on different planets. It stimulated one upper primary student to query why his weight on Earth was similar to that on the larger planet Uranus (Ball 1999).

Multimedia software

Such software can be 'content-specific' (e.g., a teacher PowerPoint program) or 'content-free' e.g., students preparing their own multimedia PowerPoint slide show on what they have learnt). In Australia, the Le@rning Federation has developed learning objects which teachers can access. These are 'chunks of digital material – for example, graphics, text, audio, animation, interactive tools – specifically designed to engage and motivate student learning' (see http://www.thelearningfederation.edu.au/default.asp). As Davison et al. (2004, p. 9) comment, 'they support rather than replace hands-on activities and authentic inquiry'. There

ICT

is an 'Earth and Beyond' catalogue of objects. O'Conner (2003) has argued that content-free software is more empowering for students. What do you think?

Digital photographs/images

One Year 2 teacher used real-time images of the Earth taken from space (see the NASA solar system simulator at http://space.jpl.nasa.gov) every two hours to prepare two PowerPoint presentations: one was simply showing the sequences of images, while the other repeated the sequence but with arrows showing where the students lived. With strategic questioning, the teacher created a discrepant event for her students, as they did not believe that the Earth moved (Kang and Howren 2004). A Year 5 teacher reported that the use of Earth and moon images from space, when displayed on an interactive whiteboard, made them 'much more real' and evoked 'awe' and 'wonder'; the whiteboard was also used by the students to search for information related to space tasks (Ferris 2005).

More generally, using photographs of your class learning science appears to act as a later stimulus in encouraging whole-class discussion of what was happening and what students were thinking at the time (Lias and Thomas 2003). This could be applied to almost any science context.

Electronic communication

Primary school students have spoken directly to astronauts (e.g., through the NASA International Space Station link-up program; also see Figure 10.1 on p. 420). When organised, it can inspire many hours of model building, investigation, research and discussion. In one school, students devised questions to ask the astronaut, such as 'Which way does a compass point in space?', 'Can you grow plants in space?' and 'What inspired you to become an astronaut?' (Terracini 2006).

Communicating by email with scientists widens young students' views of science and scientists, not to mention their English skills (reading, writing, listening and talking) (Hafford and Meadows 1999). This use of ICT provides additional opportunities for talking with others, including peers; it extends classroom discourse. It is another way of helping 'hands-on science' to become 'minds-on' science.

Slow-motion animation (slowmation)

Slowmation is a narrated animation that can be created by students (and teachers) and played in slow motion at two frames per second to explain a science concept. It integrates aspects of clay animation, object animation and digital storytelling, and encourages students to design a multimodal representation of their learning (see http://www.slowmation.com.au). Many celestial examples are included on this website, such as day and night, and lunar phases. There is also a suggested four-phase teaching approach to using slowmation and guidelines for teachers using the software.

Sophisticated cognitive tools

Some software allows interactive student thinking and testing of ideas. The software Knowledge Integration Environment (KIE) was used by Year 5 students to determine which crop (regular earth plants or Astroplants, a fast-food plant) is best suited for accompanying a NASA scientist on a space flight mission. This project integrated hands-on tasks with

Web-based ones and included student–scientist interactions, making predictions, and planning and testing various scenarios by manipulating variables. There was strong evidence of conceptual growth (Williams and Linn 2002). Hopefully, in time, such software will be more readily available.

Websites

Teachers are strongly encouraged to use reviewed websites to make effective use of their own and their students' time. Reviews may be found in most professional journals for teachers of primary and middle school science (K-8); for example, *Primary Science*, *Science and Children* and *Teaching Science* – in connection with the last mentioned, see ASTA (2004, 2009) for extensive lists of websites on astronomy topics for students. As 2009 was the International Year of Astronomy, many articles in these journals that year were astronomy-based and referred to numerous websites. Other annotated lists of relevant astronomy websites for primary students may be found through organisations such as observatories (e.g., see http://www.sydneyobservatory.com). Although dated, Slater et al. (2001) describe NASA's educational Internet resources, as well as websites that are most useful for teachers of astronomy topics; for example, astronomy activities classified into various levels (see http://btc.montana.edu/ceres – most of the sites are still current). The online Appendix 10.5 lists further websites. For effective student use of websites, teacher scaffolding is strongly recommended (see the online Appendix 1.5).

Appendix 10.5
Software, websites and webcams

Appendix 1.5
Scaffolding strategies

Activity 10.15 **ICT and space topics**

Identify three uses of ICT, with specific examples, that you think will assist primary students to constructively learn about space concepts. Ensure the examples relate to your science syllabus. Identify the possible advantages and disadvantages of the uses of ICT you have listed – for example, will they assist in understanding the NOS, further develop skills in thinking and working scientifically, and help representational learning?

Find a teacher or colleague who has incorporated ICT into their teaching to assist student learning about space. Compare their views about the use of ICT to some of the critiquing principles you read about in Chapter 1 (for example, WISE). If possible, attempt to teach a lesson in a 'space' sequence that involves the use of ICT. Reflect on and then compare your experiences with the teacher's views and the literature. As a consequence of one or more of these tasks, determine your current position on the role of ICT in teaching about celestial concepts.

The complexity of conceptual development: a gentle reminder

Various principles and strategies have been suggested to assist conceptual growth related to astronomy concepts. It is appropriate to reiterate a point from Chapter 1 here – namely, that even when these approaches are used, conceptual growth for individual students will still be influenced by many contextual factors. For example, you may be tempted to use a cognitive conflict situation with your students (e.g., a 3-D model that demonstrates the seasons and that is in conflict with alternative conceptions held by your class). While such approaches have had some success, you should recall that students may not necessarily change their ideas. Barnett and Morran (2002) reported that students changed their ideas as a consequence of time being provided for evaluation and reflection of their evolving ideas on concepts (here, for example, seasons). Sometimes there may simply need to be more teacher–student and student–student (conversational/dialogic) talk about classroom tasks (e.g., representations) and what they mean (Newton 2002). This implies that group and class discussion of ideas must be valued. Strategies such as learning journals or thinking books also have a role in this evolving view of conceptual change. Teachers still need to be aware of the existing (alternative) ideas of their students to effectively engage in these conversations (also see the online Appendix 10.3). To paraphrase Osborne (1991, p. 10), 'good teaching practice [will devise first and secondhand ways for students to encounter] all the evidence to the contrary and [do this] several times [and in a variety of modes using a range of literacies] ... in a form that engages children [in thinking and reasoning]'. Such advice, integrated with the above comments about the importance of discussion, reflection and specific scaffolding strategies (see the online Appendix 1.5), indicates a way forward for teachers. These comments all reinforce the following advice about student learning (for teachers who are worried that students did not 'get' the 'right answer'):

Appendix 10.3
Ways to handle correct conceptual fragments

Appendix 1.5
Scaffolding strategies

> The idea that children might go away believing the 'wrong answer' seems very prevalent among teachers, and causes much anxiety. I think that the anxiety is misplaced. Accepting a child's ideas as positive contributions, and building on them, is more likely to lead the child to question her thinking, than is giving her an answer which she cannot accept intellectually, but must be right because the teacher says so.
>
> Gibson 1992, p. 7

Planning teaching sequences for conceptual change or development concerning celestial phenomena

The constructivist strategies and techniques described in this chapter may be used or adapted by teachers to help students modify, where appropriate, their existing conceptions towards those that are more scientifically acceptable. As Gibson's (1992) quotation above emphasises, the intention is to help students further develop their own ideas rather than simply the teacher (or a resource) telling (transmitting) the idea to the learner. The latter is often ineffective. The former process, you now realise, can be complex. It will include students engaging in a social learning context, mentally interacting with the ideas they and others already have as well as with those they encounter via teacher-planned experiences, generating and manipulating their own and others' representations of the focus concept(s), and being metacognitive about their conceptual learning. This section describes various lesson sequences that some teachers and researchers have suggested meet these expectations.

Before commencing a sequence of lessons, you need to be clear about the conceptual understanding(s) around which your sequence of activities is planned; for example, an explanation of the phases of the moon. This would also entail consideration of the prerequisite concepts that may be required to study the focus idea, such as the reflection of light if students were to study why we see celestial objects. This decision will guide your selection of activities or tasks and your reading about children's preconceptions related to the celestial concept(s) on which you are focusing. Having made these decisions, you can then apply some sequencing principles.

In general, you must first engage your students' interest and this may happen concurrent with or prior to eliciting some of your students' ideas about the focus celestial concept(s) or event(s). You should limit the conceptual focus of the sequence; for example, to be only on the phases of the moon or the description and interpretation of shadow records. Consider getting your students to write down and, if appropriate, draw and annotate their ideas about the event. This will enable the students to come back to these views at a later time. It would be useful to have groups of students discuss the views that other individual children hold. The advantage of using small-group work is that every child gets an opportunity to share ideas with at least two other students. Of the views that evolve, one or more could become the focus of activities or tasks that follow; for example, students might take structured observations to test their predictions or, if that has been done, they could generate and test their mental models (representations) with simple materials such as torches and polystyrene balls or an interactive computer simulation. Dialogue must continue through these and the next phases. Application of what has been learned reinforces understanding. Finally, reviewing what has been learned (by revisiting starting ideas) and how it was learned helps students to be metacognitive in their learning.

A specific adaptation of these steps can be used if the teacher's intention is to have students test their mental models. The QIGIMPD (investigating with models) approach requires students to have had some learning experiences related to the celestial phenomenon they are trying to explain before using the approach. MacIntyre, Stableford and Choudry (2002) provide an example related to seasons:

R

- *Question (Q)* – What is the cause of the seasons on planet Earth?
- *Identify (I)* alternative notions and ideas; for example, the sun is further away from the Earth in the winter.
- *Gather (G)* and *Interpret (I)* evidence or data (sometimes from recollections of experiences; for example, the midday sun is low in the sky during winter and high in the sky in summer).
- *Model (M)* the notions or ideas (explanations) and look to see if they are consistent with the range of suggested evidence (add other bits to student models, if need be, in order to observe the evidence). This modelling step is achieved with a matrix that has explanations or alternative ideas along one side and patterns, evidence or data along the other. If the explanation fits the evidence, a tick is placed in the box; otherwise, a cross is used. The explanation that fitted most or all of the evidence is accepted as the most plausible, or more scientific, explanation. Teachers who have used this approach have found that

considerable teacher input is required to assist the students to test all their explanations. As a consequence of completing such a task, students are usually able to:

- *Predict (P)* events using their accepted model.
- Learning is reinforced by having students *Demonstrate (D)* their understanding using a 3-D model with an audience.

This QIGIMPD sequence places the students' own astronomy ideas at the centre of the activities – they are testing their ideas. The model-and-predict steps have students involved in representational learning. For another example, related to day and night, see MacIntyre, Stableford and Choudry (2002).

Another constructivist sequence involves using the 5E steps suggested by the *Primary Connections* project (AAS 2006; see http://www.science.org.au/primaryconnections; also see Chapter 1). Redman (2001) describes numerous activities appropriate to each of the 5E steps for a sequence about the moon. An example related to each of the steps follows:

- **E**ngage (write about what would happen if the moon did not exist)
- **E**xplore (record the time and shape of the moon over a few weeks)
- **E**xplain (using models, develop a way to explain moon phases)
- **E**laborate (research and/or investigate the uses and types of satellites)
- **E**valuate (write an updated version of what would happen if the moon did not exist).

R

NOS

A

Activity 10.16 The *Primary Connections* 'Spinning in Space' unit

Obtain a copy of this major Australian curriculum project unit, which focuses on the nature and movements of the sun, Earth and moon, shadows, and the cause of day and night. If it's unavailable, locate a published unit in your country that relates to similar content and is based on the 5E sequencing principles, or the three-stage learning cycle (see Chapter 1). For the unit, answer or perform the following:

- In what ways does this unit engage and retain student interest?
- Develop an overlapping Venn diagram in which you label two circles 'personal constructivism' and 'sociocultural constructivism'. Identify the characteristics of these views of learning in the unit and include them in appropriate sections of the diagram. Justify your choices with a colleague.
- How does the unit exemplify learning through multiple modalities and various forms of representation? Provide specific examples.
- What attributes of the NOS can you locate in the unit? Are they made explicit to the student?
  If so how; if not, how could this happen?
- What formative assessment strategies are used to assist in conceptual growth?

Appendix 4.4 Interactive teaching approach

An alternative but related set of steps is the interactive teaching approach (see Chapter 4, especially the steps in the online Appendix 4.4). The following sequence uses this approach and was suggested for a space topic involving a camp or a sleepover (Curnow and Paige 2006):

- *Teacher preparation* – Prior to commencing the sequence, check for visibility and location of the moon, planets and constellations on the designated night; prepare helpers

(e.g., parents) by organising a community viewing night (prior to the camp) with a local amateur astronomer or astronomical society in a larger city.
- *Exploration* – Elicit students' prior conceptions (examples of concept maps, role-plays and prediction tasks are described). Involve students in exploratory experiences: examples suggested are observations of the night sky with possible teacher focus questions; connecting students with how Indigenous Australians use the night sky; prior to the camp, have students act out role-plays or simulations, such as designing their own constellations.
- *Children's questions* – A visit to a planetarium could be one means of evoking questions from students about stars, planets and other celestial bodies (examples of student questions are listed).
- *Specific investigations* – These are in response to possible questions raised by students. Three possibilities are described: determining the shape and the movement of constellations; using a telescope to determine whether both planets and stars twinkle; and investigating how the size and shape of moon craters vary by using binoculars (or a telescope) and moon maps.
- *Reflection* – Students revisit prior conceptions, (e.g., mind maps) and alter and/or add more detail.

In summary, you need to decide what your purposes are for a lesson sequence and develop a sequence using models like the above as a guide. The key issue is that you appreciate why you are using the steps in the first place. If you understand that, then you can create your own justifiable sequence.

Science as a human endeavour

The characteristics of 'science as a human endeavour' include primary students being aware of the links between science and culture (e.g., the historical nature of science ideas; contributions of Indigenous knowledge), the NOS (the human construction of ideas), the contributions of scientists (science-related careers and a nation's contribution to science), and the influence of science (the effect of science and technology on our lives, as well as how science is used). Each of these can be readily incorporated into lessons focusing on space. This section provides some examples.

Science and culture/nature of science: historical and human construction of science ideas

Astronomy has fascinated humanity since ancient times. With their intrinsic interest in this topic, students could be encouraged to research how early thinkers explained celestial events. The Babylonians kept accurate records of the movement of the sun, moon and planets as far back as 400 BCE. Mental models used to explain observations, a hallmark of science, date back to this time. How the Babylonians and Greeks used models to explain the moon's phases assuming the earth was flat or spherical, but at the centre of the universe, exemplifies how our observations and beliefs influence our mental models. For upper primary students, access to some of these records (Kavanagh, Agan and Sneider

2005; Redman 2001; Solomon 1983) may help them to appreciate that science knowledge can be tentative – relevant attributes of the NOS, such as humanity's desire to explain observations using models, need to be explicitly drawn out through various tasks and open-ended discussion (see 'The nature of science: how it works' in Chapter 1).

Activity 10.17 We (humanity) construct scientific concepts: the case of Pluto

Ask some upper primary students if they know why Pluto was in the news a couple of years ago. Depending upon their response, ask them why they think there was a problem with Pluto as a planet and what their opinion is of the decision. Can they remember where they found out about Pluto? When 9–11-year-olds were asked this question three to four months after the news story, about 50 per cent were able to provide relevant answers (Jarman and McClune 2007).

Primary teachers need to make the NOS explicit. How might you use the case of Pluto to illustrate the social nature of science? What other attributes of science could be made explicit using this case study? What about the role of observation and the changing nature of evidence, and their relationship to constructing ideas? What tasks/activities could upper primary students complete to facilitate their appreciation of some of these nature-of-science characteristics (other than you simply telling the students)?

For some further ideas about relating the case of Pluto to 'science as a human endeavour', see Schibeci (2007), Wilkinson (2009a, 2009b) and the Primary Upd8 resource (see http://www.primaryupd8.org.uk/activity.php?actid=37). What other modern space news stories could you use in similar ways?

NOS

The influence of science: the effect of science and technology on our lives

Investigating the nature and movements of celestial bodies has fascinated and influenced humanity since ancient times and continues to do so today. When teachers engage learners in shadow stick activities, as described earlier (see Activity 10.6 on p. 413), this can be an opportunity for students to research the history of sundials from Babylonian times to the present. If you accessed Merson (1989, pp. 38–9), O'Toole (1999, p. 21) and also the Wikipedia entry on sundials (http://en.wikipedia.org/wiki/History_of_sundials), you could use them to encourage students to speculate on their uses over the ages, such as for predicting the change of seasons, the development of more accurate calendars and their relationship to people's beliefs (see ASTA 2009; Kibble 2002b, 2008).

ICT

Today, space exploration in particular opens up possibilities for technological activities related to rockets, vehicles and equipment for lunar and space travel (ASTA 2004; Fowler 1994; Fuller 1992). Forbes (2006) involved her Year 4 students in a simulation of the NASA Mars Exploration Rover missions featuring the spacecraft *Spirit* and *Opportunity* (see http://marsrovers.nasa.gov/overview) – the students worked on a 'Mars Mission' webquest devised by the teacher. She showed her students how to use Lego Mindstorms' iconic ROBOLAB software programming system (see http://en.wikipedia.org/wiki/Lego_Mindstorms). These students quickly became adept problem-solvers and programmed their wheeled robots to initially navigate a coloured floor tile and then a simulated Martian terrain littered with rocks and 'craters'. The unit of work involved the planning and building of a colony on Mars (see Figure 10.3). This is an example of technology as a problem-solving process of designing, making and appraising (Fleer and Jane 2004). Older students could be asked how being engaged in space exploration might be affecting our lives – materially, emotionally, intellectually and spiritually. (The NASA website, http://www.thespaceplace.com/nasa/spinoffs.html, could be used as a resource.)

Science and culture: cosmology and Indigenous knowledge

When students are learning about the sun, Earth, moon and stars, teachers can integrate Indigenous interpretations in a holistic and/or thematic, rather than a tokenistic, way (Michie 2002). Such interpretations can be used to help students appreciate how Indigenous peoples have a close relationship with the Earth; Indigenous stories also can be used as a motivation to learn about the Earth in space (Caduto and Bruchac 1989). One story tells of how the moon had been stolen from the Indigenous Kalispel people of Idaho, who needed it to return to provide them with light. The Coyote took this role for a period of time because he said he could do a better job than the Yellow Fox, and because he would be able to look down and see all that was happening on Earth. From the moon the Coyote spied on the Kalispel and constantly interfered in their lives, which partly explains the legendary snoopiness attributed to the coyote. This story illustrates the connections of these Indigenous people to the animal and celestial worlds. Telling this story to young students would help them to appreciate the different ways the Kalispel perceived their world and could then provide motivation for exploring various features of the moon, such as how we see it, its surface, phases, eclipses and its temperature. On a related theme, Kibble (2008) describes how, through story and diorama building, students could explore the relationship between standing stones and stone circles and the apparent movement of the sun across the sky.

Australian Aboriginal dreaming stories could be used in a similar manner. For example, *There's an Emu in the Sky* (Malcolm 1995; also see http://www.questacon.edu.au/html/aboriginal_astronomy.html) includes a unit about space that refers to some Aboriginal interpretations of astronomical phenomena, while Floyd (2002b) offers teaching suggestions related to Indigenous views about constellations. Michie provides specific websites and possible activities at http://members.ozemail.com.au/~mmichie/astronomy.htm and a CSIRO website shows Aboriginal rock carvings that reflect various astronomical features (see http://www.atnf.csiro.au/research/AboriginalAstronomy). For comprehensive coverage of the topic, see the website for the Australian Aboriginal Astronomy project (http://www.warawara.mq.edu.au/aboriginal_astronomy). Also available are Indigenous star charts and a tape, *Aboriginal Sky Figures*, produced by the Australian Broadcasting Corporation (c. 1998). Such interpretations can be shared with students in other countries, as occurred about constellations between schools in South Africa and England (Thomas, Christie and Leverment 2009).

By using Indigenous perspectives in this way, teachers help students to explore how other cultures can contribute to our understanding of our physical and biological world. Such an approach also indicates how a study of the Earth in space can have spiritual significance in

S&C

FIGURE 10.3 Year 4 students programming and testing their model Mars Exploration Rover

the sense that the Aboriginal or First Nation peoples were connected to the physical world (universe) around them. It was not a case of the Earth belonging to humanity but humanity belonging to the Earth (from Chief Seattle's speech, in Caduto and Bruchac 1989, p. 4).

Contributions of scientists: astronomy- and space-related careers

Upper primary students usually know whether they like studying science or not, and by this stage may even have made some career choices (Osborne 2008; Ramsay, Logan and Skamp 2005). Space fascinates boys and girls (Webb and Morrison 2000), but fewer women tend to work in this field, although this is changing with women now comprising about 40 per cent of research degree enrolments in astronomy at some universities. Introducing primary students, especially girls, to the range of occupations associated with space exploration may be beneficial. Female role models could assist: Roberta Bondar (author in 1994 of *Touching the Earth*) was a Canadian astronaut; Annie Cannon, one of America's best-known astronomers, catalogued and classified over 275 000 stars (Wydoga 1993); Ruby Payne-Scott was the first female radio astronomer in Australia, if not the world, and worked throughout the 1940s and 1950s; and Penny Sackett helped discover a cool planet outside our solar system (ASTA 2009). Male role models include Andy Thomas, an Australian astronaut (ASTA 2004). An aware teacher could take advantage of girls' and boys' interest in space by referring to the nature of the work involved and the stories of some of the people, including their childhood ambitions and extracurricular activities.

M&E

You may also be able to locate information about your country's contribution to the development of astronomy. Australia, for example, is a world leader, partly because of its unique position in the Southern Hemisphere and the clear skies that prevail in several locations. A useful resource is the bicentennial (1988) issue of the *Australian Science Teachers Journal*, 34 (2), which contains information about 200 years of Australian astronomy; also see ASTA (2004, 2009).

Summary

Overall, this chapter stresses that, when students learn about space, a topic which they love, it need not be research work in the library, but rather can revolve around the students' existing ideas, real observations, investigable questions and representations of their thinking. Ideas for activities on this topic are readily available, but this chapter emphasises that it is the way in which the teacher uses such activities that is important – where possible, activities need to be related to the students' ideas and questions, but in all instances they must be of relevance to their interests and lives. Examples have been given to illustrate the types of ideas that students will probably hold and how to use the questions they might ask; typical investigable/productive questions that teachers might use have also been discussed. Issues such as ways to use mental and physical models and their relationship to representational learning have been considered, as has the role of secondhand experiences and the place of ICT. Several examples of how this topic exemplifies science as a human endeavour were provided.

Activity 10.18 Reflections on teaching about the Earth in space

There are numerous activities available on websites and in other resources that teachers could use to teach this topic. Use the content in this chapter to devise a list of 'constructivist-oriented' criteria that could be used to evaluate the potential learning effectiveness of activities found in these resources.

Select a key celestial concept and locate a website that describes some learning activities (not necessarily ICT-related) associated with the concept. Critique the activities using your criteria indicating their strengths and weaknesses and, if necessary, suggest ways that they could be improved from a constructivist perspective.

Finally, look back at your initial mind map in Activity 10.1 (see p. 399). Reflect on the same question and, in another colour, add further connections. Try to identify the three main preconceptions about teaching this topic that you have either replaced, had supported or expanded. Share your thoughts with a colleague.

Concepts and understandings for primary teachers

The major topics primary syllabuses would refer to in relation to the content of this chapter are:

- the Earth in space (day and night and the changing year, especially the seasons)
- the Earth, moon and sun (including phases of the moon and eclipses)
- the solar system
- stars and the universe.

Below are most of the key conceptual ideas and understandings related to these topics with which a primary teacher should be familiar. Read and interpret this list with an appreciation of its limitations as described in the section 'Concepts and understandings for primary teachers' in Chapter 1 (see p. 47).

In broad terms, the basic conceptual ideas and understandings towards which we should be encouraging primary students to move are as follows:

- The sun, moon and stars move relative to the Earth in regular, repeated patterns.
- Changes in the apparent position of the sun in the sky are connected with night and day, and seasonal changes in the weather (Harlen 1985).

More specifically, these broad ideas would encompass the following, adapted in particular from Driver et al. (1994).

The Earth

- The Earth is spherical.
- 'Down' refers to the centre of the Earth (in relation to gravity).

Day and night

- Light comes from the sun.
- Day and night are caused by the Earth turning on its axis. (It should be noted that 'day' can refer to a 24-hour time period or the period of daylight; the reference being used should be made explicit to students.)
- At any one time, half of the Earth's sphere is in sunlight (day) and half is in darkness (night).

The changing year

- The Earth revolves around the sun every year.
- The Earth's axis is tilted 23.5 degrees from the perpendicular to the plane of the orbit of the Earth around the sun. The Earth's tilt is always in the same direction.
- As the Earth revolves around the sun, its orientation in relation to the sun changes because of its tilt.

The seasons are caused by the changing angle of the sun's rays on the Earth's surface at different times during the year (due to the Earth revolving around the sun). The seasons are not caused by the Earth being farther away (from the sun) in winter than in summer, or by the fact that the Earth's orbit is not exactly circular. The radiation from the sun in winter has to, firstly, pass through more of the Earth's atmosphere; secondly, warm more of the Earth's surface; and thirdly, has a shorter time in which to warm the Earth's surface. For all three reasons, it is not as warm in winter. (This explanation is not always easy for students to understand because some of it involves inverse proportion: for example, more surface area with the same heat source means less heat overall.)

The Earth, moon and sun

- The Earth, moon and sun are part of the solar system, with the sun at the centre.
- The Earth orbits the sun once every year.
- The moon orbits the Earth in one lunar month (about 28 days). The moon is the Earth's only natural satellite.
- The moon turns on its axis at a rate that means we always see the same face.
- The moon orbits the Earth at an angle to the plane in which the Earth and sun are located.

The phases of the moon and eclipses

- The moon is visible because it reflects light from the sun.
- The sun always illuminates half of the moon's sphere.
- The moon appears to change shape (its phases) each month because we see different amounts of the illuminated surface of the moon at different times each month due to the relationship between the positions of the Earth, sun and moon at any particular time.
- The phases of the moon occur in a regular pattern.
- Eclipses occur in two ways: when the Earth lies between the sun and the moon and casts a shadow – full or partial – over the moon (that is, a full or partial lunar eclipse), or when the moon lies between the Earth and the sun and casts a shadow – full or partial – over part of the Earth (that is, a full or partial solar eclipse). These occur irregularly.

The solar system and stars

- Stars emit light. The sun is a star. The sun emits light.
- The sun is the centre of the solar system and is the only body in the solar system that emits light.
- The sun is the solar system's main source of energy.
- The planets orbit the sun. Some planets, other than Earth, have their own moons (natural satellites).
- The planets are great distances from the Earth, but relatively much closer than the stars, apart from the sun.

The universe

- The solar system is only a small component of one particular galaxy, the Milky Way, which is made up of millions of stars.
- Even the nearest stars (apart from the sun) are gigantic distances away compared to the planets.
- The universe (which is everything that exists) is comprised of countless galaxies. Our galaxy, the Milky Way, is not the centre of the universe.

Teachers should appreciate that this content area also includes understandings about gravity. Further, there are difficulties associated with learning about the Earth in space because of the sizes and distances involved and the three-dimensional nature of many of the understandings. Appreciating the meaning of scale and the limitations of models and simulations to scaffold student development of ideas related to size, distance and movement of the Earth, moon, sun and stars needs to be remembered.

For a related list of concepts, readers could view the astronomy concepts listed for years K–4, 5–8 and 9–12 in the American National Research Council Science Education Standards (Slater et al. 2001; see also http://www.nap.edu).

Concepts and understandings can sometimes be found in reliable curriculum resources; for example, the *Primary Connections* unit 'Spinning in Space' (AAS 2006).

Acknowledgements

Anne Forbes, a lecturer at Australian Catholic University, Sydney, taught students from Epping North Public School, NSW the lessons related to the Mars Rover Mission and provided the images. The images of the students talking to the astronaut and their associated astronomy studies are from Teven-Tintenbar Public School, NSW.

Search me! science education

Search me!

Explore **Search me! science education** for relevant articles on our place in space. Search me! is an online library of world-class journals, ebooks and newspapers, including *The Australian* and the *New York Times*, and is updated daily. Log in to Search me! through http://login.cengage.com using the access code that comes with this book.

KEYWORDS

Try searching for the following terms:

- astronomy and children
- physical models and astronomy
- misconceptions and astronomy

Search tip: **Search me! science education** contains information from both local and international sources. To get the greatest number of search results, try using both Australian and American spellings in your searches: e.g., 'globalisation' and 'globalization'; 'organisation' and 'organization'.

Appendices

In these appendices you will find material related to our place in space that you should refer to when reading Chapter 10. These appendices can be found on the student companion website. Log in through http://login.cengage.com using the access code that comes with this book. Appendices 10.2 and 10.6 are included in full below.

Appendix 10.1 Some alternative conceptions held by some preservice primary teachers about astronomical phenomena

Alternative conceptions related to phases of the moon, the shape of stars, the nature of comets, the gravity on the moon, the visibility of the morning and evening star, the tides and the composition of planets are listed, together with the accepted scientific view.

Appendix 10.2 Use of misrepresentations in books to encourage conceptual development

When children's books have misrepresentations of celestial events, consider:

- asking probing questions that raise awareness of the issue
- completing an inquiry by taking observations and modelling and comparing with the book
- using a scientifically accurate non-fiction book
- pairing together the use of fiction and non-fiction books (Trundle and Sackes 2008; Trundle, Troland and Prichard 2008).

Appendix 10.3 Ways to handle correct conceptual fragments

Two approaches are briefly outlined.

Appendix 10.4 Examples of hypothesised conceptual progressions

Examples of progressions are given for the shape of the Earth, explanations for day and night, and the seasons.

Appendix 10.5 Software, websites and webcams

A selection of software, websites and webcams are listed. Some need additional computer programs.

Appendix 10.6 A scale task

To appreciate the distance to the stars, mathematically able upper primary students could use a scale calculation. The nearest star to Earth, other than the sun, is Alpha Centauri, which is 8.4 light years away. Using a scale of 1 metre to represent the distance from the Earth to the sun (149.6 million kilometres), and assuming that light travels at 300 000 kilometres/second, students could determine how many kilometres it is to Alpha Centauri (Answer: 531.2 kilometres).

Appendix 10.7 Some key characteristics of teaching for effective conceptual development

Critical features of constructivist astronomy sequences for students to advance their conceptual understanding – as advocated by, for example, Vosniadou (2003) – are described.

References

Ackerson, V. 2005. How do elementary teachers compensate for incomplete science content knowledge? *Research in Science Education*, 35, pp. 245–68.

Adams, R., Doig, B. & Rosier, M. 1991. *Science Learning in Victorian Schools: 1990*. Melbourne: Australian Council for Educational Research (ACER).

Aldridge, C. 2002. Reach for the stars! *Primary Science Review*, 72, pp. 20–2.

Alexander, R. 2004. *Towards dialogic teaching: Rethinking classroom talk*. York: Dialogos.

Allen, P. 1999. Make your own eclipse demonstrator. *Primary Science Review*, 58, pp. 8–9.

Anonymous 2004. Bringing the cosmos into the classroom. *Science Education News*, 53 (4), pp. 179–81.

Association for Science Education (ASE). 1990. *Earth and Space*. Hatfield, UK: ASE.

Australian Academy of Science (AAS). 2006. 'Spinning in Space'. *Primary Connections*. Canberra: AAS.

Australian Science Teachers Association (ASTA). 2004. *Out of this World: Investigating Space*. Canberra: ASTA.

_____ 2009. *Astronomy: Science without Limits*. Canberra: ASTA.

Ball, S. 1999. How big's Uranus? *Primary Science Review*, 56, pp. 25–6.

Barnett, M. & Morran, J. 2002. Addressing children's alternative frameworks of the moon's phases and eclipses. *International Journal of Science Education*, 24 (8), pp. 859–79.

Baruch, J., Machell, J. & Norris, K. 2005. Watching the moon from Tenerife. *Primary Science Review*, 88, pp. 7–9.

Baxter, J. 1989. Children's understanding of familiar astronomical events. *International Journal of Science Education*, 11, pp. 502–13.

_____ 1991. A constructivist approach to astronomy in the national curriculum. *Physics Education*, 26, pp. 38–45.

_____ 1995. Children's understanding of astronomy and Earth sciences, in S. Glynn & R. Duit (eds). *Learning Science in the Schools*. Mahwah, NJ: Erlbaum.

Bentler, S. A. 1988. Using writing to learn about astronomy. *Reading Teacher*, 41 (4), pp. 412–17.

Bilderbeck, N. 2008. Astronauts or astrnust? Learning about space by illustrating it. *Primary Science*, 103, pp. 14–17.

Bondar, R. 1994. *Touching the Earth*. Canada: Key Porter.

Brewer, W. 2008. Naïve theories of observational astronomy: Review, analysis and theoretical implications, in S. Vosniadou (ed.). *International Handbook of Research on Conceptual Change*. London: Routledge, pp. 155–204.

Brown, D. 1992. Using examples and analogies to remediate misconceptions in physics: Factors influencing conceptual change. *Journal of Research in Science Teaching*, 29 (1), pp. 17–34.

Caduto, M. J. & Bruchac, J. 1989. *Keepers of the Earth*. Saskatoon, SK, Canada: Fifth House.

Collins, R. & Simpson, F. 2007. Does the moon spin? *Primary Science Review*, 97, pp. 9–12.

Comstock, D. 2008. A week for space: Set aside time to explore our place in the universe. *Science and Children*, 46 (1), pp. 46–9.

Curnow, P. & Paige, K. 2006. The cosmic story board: Practical ways of using the night sky. *Teaching Science*, 52 (2), pp. 36–9.

Curriculum Corporation. 2006. *Statements of Learning for Science*. Available at http://www.curriculum.edu.au/verve/_resources/StmntLearning_Science_2008.pdf (accessed 19 October 2010).

Davison, J., Kenny, J., Johnson, J. & Fielding, J. 2004. Use learning objects to bring an exciting new dimension to your classroom. *Teaching Science*, 50 (1), pp. 6–9.

DeBoo, M. & Asoko, H. 2000. Using models, analogies and illustrations to help children think about science ideas. *Primary Science Review*, 65, pp. 25–7.

Diakidoy, I. & Kendeou, P. 2001. Facilitating conceptual change in astronomy: A comparison of the effectiveness of two instructional methods. *Learning and Instruction*, 11, pp. 1–20.

Dove, J. 2002. Does the man in the moon ever sleep? An analysis of student answers about simple astronomical events: A case study. *International Journal of Science Education*, 24 (8), pp. 823–34.

Driver, R., Squires, A., Rushworth, P. & Wood-Robinson, V. 1994. *Making Sense of Secondary Science: Support Materials for Teachers*. London: Routledge.

Editorial. 1969. The wonder beyond Apollo. *Sydney Morning Herald*, 22 July.

Ehrlen, K. 2008. Children's understanding of globes as a model of the Earth: A problem of contextualising. *International Journal of Science Education*, 30 (2), pp. 223–40.

_____ 2009. Understanding of the Earth in the presence of a satellite photo. *European Journal of Psychology of Education*, XXIV (3), pp. 281–92.

Eichinger, J. 2005. Using models effectively. *Science and Children*, 42 (7), pp. 43–5.

Elementary Science Study (series). 1968. *Where Was the Moon?* St Louis: McGraw Hill.

Farrow, S. 2006. *The Really Useful Science Book*. (3rd edn). London: Falmer Press.

Ferris, E. 2005. Barely able to see a star. *Primary Science Review*, 88, pp. 10–12.

Finegold, M. & Pundak, D. 1990. Students' conceptual frameworks in astronomy. *Australian Science Teachers' Journal*, 36 (2), pp. 76–83.

Fleer, M. 1997. A cross-cultural study of rural Australian Aboriginal children's understandings of night and day. *Research in Science Education*, 27, pp. 101–16.

_____ & Jane, B. 2004. *Technology for Children* (2nd edn). Sydney: Prentice Hall.

Floyd, P. 2002a. Beyond Earth: Astronomy. *Investigating*, 18 (1), pp. 32–4.

_____ 2002b. Beyond Earth: Patterns in the sky. *Investigating*, 18 (2), pp. 37–9.

_____ 2002c. Earth and space science events and teaching opportunities in 2002. *Science Education News*, 51 (1), pp. 24–30.

_____ 2003. *Science Education News*, 52 (1), pp. 39–42.

Foo, J. 2009. The wonders of the night sky come alive at Glasgow Science Centre planetarium. *Primary Science*, 108, pp. 31–4.

Forbes, A. 2006. Personal communication. Email, 22 August.

Forma, R. 2008. The sky. *Teaching Science*, 54 (2), pp. 54–5.

Fowler, B. 1994. More 'space' in the classroom. *Science and Children*, 32 (1), pp. 40–1, 55.

Frazier, R. 2003. Models. *Science & Children*, 40 (4), pp. 29–33.

Fuller, J. 1992. Investigating some problems of space exploration through technology. *Investigating: Australian Primary Science Journal*, 8 (2), pp. 6–7.

George, M. 1992a. Our star 1: Watching shadows. *Investigating: Australian Primary Science Journal*, 8 (2), pp. 2–3.

_____ 1992b. Keeping track of the moon's phases. *Investigating: Australian Primary Science Journal*, 8 (2), pp. 24–5.

Gibson, F. 1992. Learning about sky and space: A story and some lessons we might draw. *Primary Science Review*, 25, pp. 6–7.

Grayson, D. 1994. Concept substitution: An instructional strategy for promoting conceptual change. *Research in Science Education*, 24, pp. 102–11.

Hackling, M. 2005. *Working Scientifically*. Perth: Education Department of Western Australia.

Hafford, A. & Meadows, J. 1999. Electronic mail provides positive role models. *Primary Science Review*, 57, pp. 20–2.

Harlen, W. 1985. *Teaching and Learning Primary Science*. London: Harper & Row.

_____ (ed.). 2001. *Primary Science: Taking the Plunge* (2nd edn). Portsmouth, NH: Heinemann.

_____ 2009. The Presidential Address 2009: Teaching and learning science for a better future. *School Science Review*, 90 (333), pp. 33–42.

Helix Double Helix, The, magazine of the Double Helix Science Club, Canberra.

Hollow, R. 2006. Back to basics astronomy: Pedagogical modelling for day and night. Paper presented at the Science Teachers' Association of New South Wales Annual Conference, December.

Hubber, P. & Haslam, F. 2009. The role of representation in teaching and learning astronomy. Paper presented at the Australasian Science Education Research conference, Geelong, July.

Hughes, S. 2008. Stellarium: A valuable resource for teaching astronomy in the classroom and beyond. *Science Education News*, 57 (2), pp. 83–4.

Hugman, A. & Placing, K. 2007. Movies in the classroom. *Science Education News*, 56 (2), pp. 86–7.

Jarman, R. & McClune, B. 2007. Do children really take note of science in the news? *Primary Science Review*, 103, pp. 10–14.

Jones, D. 2009. Carleton Community High School walking with astronauts. *School Science Review*, 90 (333), pp. 67–72.

Kallery, M. 2000. Making the most of questions and ideas in the early years. *Primary Science Review*, 61, pp. 18–19.

Kang, N. & Howren, C. 2004. Teaching for conceptual understanding. *Science and Children*, 42 (1), pp. 28–32.

Kavanagh, C., Agan, L. & Sneider, C. 2005. Learning about phases of the moon and eclipses: A guide for teachers and curriculum developers, *Astronomy Education Review*, 4 (1), pp. 19–52.

Keogh, B. & Naylor, S. 2000a. *Concept Cartoons in Science Education*. Cheshire: Millgate House Publishers (see also Investigating, 17 (1), pp. 47–8).

_____ 2000b. Teaching and learning in science using concept cartoons. *Investigating*, 16 (3), pp. 10–14.

_____ 2006. Access and engagement for all, in V. Wood-Robinson (ed.). *ASE guide to secondary science education*. Hatfield, UK: ASE, pp. 133–9.

Kepler, L. 1990. Date with science. *Science & Children*, 28 (1), pp. 69–70.

Kerrod, R. 2002. *The Book of Constellations* Sydney: Allen & Unwin.

Kibble, B. 2002a. Misconceptions about space? It's on the cards. *Primary Science Review*, 72, pp. 5–8.

_____ 2002b. Simple sundials. *Primary Science Review*, 72, pp. 23–5.

_____ 2008. Shadows and stone circles: A storyline. *Primary Science*, 103, pp. 33–6.

Kikas, E. 1998. The impact of teaching on students' definitions and explanations of astronomical phenomena. *Learning and Instruction*, 8 (5), pp. 439–54.

Lias, S. & Thomas, C. 2003. Using digital photographs to improve learning in science. *Primary Science Review*, 76, pp. 17–19.

Lomb, N. 2009. *2010 Australian Sky Guide*. Haymarket, NSW: Powerhouse Publications.

Lucas, K. & Cohen, M. 1999. The changing seasons: Teaching for understanding. *Australian Science Teachers' Journal*, 45 (4), pp. 9–17.

Lund, A. 1999. Blast off to space. *Primary Science Review*, 72, pp. 4–6.

MacIntyre, B., Stableford, J. & Choudry, H. 2002. Teaching for conceptual understanding in astronomy: Using an investigating with models approach. *Investigating*, 18 (1), pp. 6–10.

MacLennan, G. 1992. *Generate, Create, Investigate*. Brisbane: Jacaranda.

Malcolm, C. (ed.). 1995. *There's an Emu in the Sky*. Melbourne: Curriculum Corporation.

Marsh, G., Willimont, G. & Boulter, C. 1999. Modelling the solar system. *Primary Science Review*, 59, pp. 24–6.

McCall, D., Pickford, M. & Hart, C. 2002. Internet resources for primary science, *Investigating*, 18 (1), pp. 35–8.

McGrogan, S. 1992. Assessing children's learning about Earth in space. *Primary Science Review*, 23, pp. 28–9.

Merson, J. 1989. *Roads to Xanadu: East and west in the making of the modern world*. Sydney: Child & Associates/ABC.
Michie, M. 2002. Why Indigenous science should be included in the science curriculum. *Australian Science Teachers' Journal*, 48 (2), pp. 36–40.
Millar, L. & Murdoch, J. 2002. A penny for your thoughts. *Primary Science Review*, 72, pp. 26–9.
Mohapatra, J. K. 1991. The interaction of cultural rituals and the concepts of science in student learning: A case study on solar eclipse. *International Journal of Science Education*, 13 (4), pp. 431–8.
Morris, M., Fairclough, S., Birrell, N. & Howitt, C. 2007. Trialling concept cartoons in early childhood teaching and learning of science. *Teaching Science*, 53 (2), pp. 42–5.
Mulholland, J. & Ginns, I. 2008. College MOON Project Australia: Preservice teachers learning about the moon's phases. *Research in Science Education*, 38 (3), pp. 385–99.
Naylor, S., Keogh, B. & Goldsworthy, A. 2004. *Active Assessment: Thinking, Learning and Assessment in Science*. Cheshire: Millgate House Publishers.
Newton, D. 2002. *Talking Sense in Science*. London: Routledge and Falmer Press.
O'Conner, L. 2003. ICT and primary science: Learning 'with' or learning 'from'. *Primary Science Review*, 76, pp. 14–16.
Osborne, J. 1991. Approaches to the teaching of AT16 – the Earth in space: Issues, problems and resources. *School Science Review*, 72 (260), pp. 7–15.
_____ 1994. Coming to terms with the unnatural: Children's understanding of astronomy. *Primary Science Review*, 31, pp. 19–21.
_____ 2008. Engaging young people with science: Does science education need a new vision? *School Science Review*, 89 (328), pp. 67–76.
_____, Black, P. J., Wadsworth, P. & Meadows, J. 1994. *Science Process and Concept Exploration Research Report: The Earth in Space*. Liverpool: Liverpool University Press.
O'Toole, J. M. 1999. *History and Nature of Science: A Broad Introduction*. Sydney: Five Senses Education.
Ould, S. 2009. Kids love space. *Primary Science*, 108, pp. 8–11.
Paige, K. & Whitney, J. 2008. Vanishing boundaries: Making the connections between art and science. *Teaching Science*, 54 (1), pp. 42–6.
Palmer, D. 2001. Students' alternative conceptions and scientifically acceptable conceptions about gravity. *International Journal of Science Education*, 23 (7), pp. 691–706.
Plummer, J. 2009. Early elementary students' development of astronomy concepts in the planetarium. *Journal of Research in Science Teaching*, 46 (2), pp. 192–209.
Quinnell, J. 2009. Making a solar graph. *Primary Science*, 108, pp. 12–14.
Ramsay, K., Logan, M. & Skamp, K. 2005. Primary students perceptions of science and scientists and upper secondary science enrolments. *Teaching Science*, 51 (4), pp. 21–6.
Redman, C. 2001. Moon rise, moon set. *Investigating*, 17 (1), pp. 22–7.
Reneke, D. 2005. What star is that? *Science Education News*, 54 (2), p. 61.
Revell, M. 1994. Simply space. *Investigating*, 10 (4), pp. 7–10.
Reynolds, R. 2009. Weather satellites and their use in weather analysis and forecasting: Some examples. *School Science Review*, 90 (333), pp. 83–90.
Ridpath, I. 2006. *Astronomy*. New York: DK Publishing.
Roberts, D. 1999. The sky's the limit. *Science and Children*, 37 (1), pp. 33–7.
Rowlands, M. 2005. Developing understanding of the sun's movement across the sky. *School Science Review*, 86 (316), pp. 69–78.
Schaaf, F. 1990. Seeing the sky: 100 projects, activities and explorations in astronomy. *Science Activities*, 27 (1), pp. 24–36.
Schibeci, R. 2007. Pluto is a planet: True or false? *Teaching Science*, 53 (1), pp. 44–5.
Schuster, D. 2008. Take a planet walk: Upper elementary students are challenged to create and explore a more accurate model of the solar system. *Science and Children*, 46 (1), pp. 42–5.
Sharp, J. 2000. The Earth and beyond, in P. Stephenson (ed.). *Developing Student Teachers' Subject Knowledge and Understanding*. Leicester: SCI Centre.
_____, Bowker, R., Mooney, C. M., Jeans, R. & Grace, M. 1999. Teaching and learning astronomy in primary schools. *School Science Review*, 80 (292), pp. 75–86.
Skamp, K. 1994. Determining misconceptions about astronomy. *Australian Science Teachers' Journal*, 40 (3), pp. 63–7.
Slater, T., Beaudrie, B., Cadtiz, D., Governor, D., Roettger, E., Stevenson, S. & Tuthill, G. 2001. A systematic approach to improving K-12 astronomy education using NASA's internet resources. *Journal of Computers in Mathematics and Science Teaching*, 20 (2), pp. 163–78.
Smith, A. 2009. Beyond SatNav: Using the global positioning system as a tool for the behavioural sciences. *School Science Review*, 90 (333), pp. 99–104.
Smith, D. 2009. Celebrate the International Year of Astronomy. *The Science Teacher*, 76 (1), pp. 60, 62–4, 66.
Solomon, J. 1983. *Space, Cosmology and Fiction*. Hatfield, UK: ASE.
Stein, M., Larrabee, T. & Barman, C. 2008. A study of common beliefs and misconceptions in physical science. *Journal of Elementary Science Education*, 20 (2), pp. 1–11.
Strange, D. & Fullam, C. 2009. Through the eyes of a great astronomer. *Primary Science*, 108, pp. 5–7.
Stringer, J. 1999. Once in a lifetime: The total solar eclipse. *Primary Science Review*, 58, pp. 4–7.
Tasker, R. 1992. Effective teaching: What can a constructivist view offer? *Australian Science Teachers' Journal*, 38 (1), pp. 25–34.
Terracini, A. 2006. Beam us up, Jeffrey! *Northern Star*, 17 August.
Thomas, R., Christie, S. & Leverment, S. 2009. From the opposite sides of the Earth. *Primary Science*, 108, pp. 27–30.
Trumper, R. 2001a. A cross-age study of junior high school students' conceptions of basic astronomy concepts. *International Journal of Science Education*, 23 (11), pp. 1111–23.
_____ 2001b. Assessing students' basic astronomy conceptions from junior high school through university. *Australian Science Teachers' Journal*, 47 (1), pp. 21–9.
_____ 2003. The need for change in elementary school teacher training: A cross-college age study of future teachers' conceptions of basic astronomy concepts. *Teaching and Teacher Education*, 19, pp. 309–23.
Trundle, K., Atwood, R. & Christopher, J. 2007a. Fourth grade elementary students' conceptions of standards-based lunar concepts. *International Journal of Science Education*, 32 (5), pp. 595–615.
_____, Atwood, R. & Christopher, J. 2007b. A longitudinal study of conceptual change: Preservice elementary teachers' conceptions of moon phases. *Journal of Research in Science Teaching*, 44 (2), pp. 303–26.
_____ & Sackes, M. 2008. Sky observation by the book: Lessons for teaching young children astronomy concepts with picture books. *Science and Children*, 46 (1), pp. 36–9.
_____ & Troland, T. 2005. The moon in children's literature. *Science and Children*, 43 (2), pp. 40–3.
_____, Troland, T. & Pritchard, T. 2008. Representations of the moon in children's literature: An analysis of written and visual text. *Journal of Elementary Science Education*, 20 (1), pp. 17–28.
van den Berg, E. 2000. Role-playing in astronomy. *School Science Review*, 81 (296), pp. 125–9.
Vosniadou, S. 2003. Exploring the relationships between conceptual change and intentional learning, in G. Sinatra and P. Pintrich (eds). *Intentional Conceptual Change*. Mahwah, NJ: Lawrence Erlbaum Associates, pp. 377–406.
_____, Skopeliti, I. & Ikospentaki, K. 2005. Reconsidering the role of artifacts in reasoning: Children's understanding of the globe as a model of the Earth. *Learning and Instruction*, 15, pp. 333–51.

Walker, R. 2009. 'Out of this world': A science conference for primary schools. *Primary Science*, 108, pp. 15–17.

Ward, R., Sadler, P. & Shapiro, I. 2007/2008. Learning physical science through astronomy activities: A comparison between constructivist and traditional approaches in grades 3–6. *Astronomy Education Review*, 6 (3), pp. 1–19.

Webb, L. & Morrison, I. 2000. The consistency of primary children's conceptions about the Earth and its gravity. *School Science Review*, 81 (296), pp. 99–104.

Wenham, M. & Ovens, P. 2010. *Understanding Primary Science* (3rd edn). London: Sage.

White, R. & Gunstone, R. 1992. *Probing Understanding*. London: Falmer Press.

Wilkinson, J. 2009a. The new solar system. *Teaching Science*, 55 (4), pp. 32–5.

_____ 2009b. *Probing the Solar System*. Canberra: CSIRO.

Williams, M. & Linn, M. 2002. WISE Inquiry in fifth grade biology. *Research in Science Education*, 32 (4), pp. 415–36.

Wilson, H. 2002a. Bright ideas time. *Primary Science Review*, 72, pp. 13–15.

_____ 2002b. A crowded view of space. *Investigating*, 18 (2), p.62.

_____ & Wisbey, L. 2000. Spotlight on space. *Primary Science Review*, 61, pp. 23–4.

Wydoga, L. J. 1993. Women in science. *Science Teacher*, 60 (8), pp. 24–7.

Wyn Davies, R. 2002. There's a lot of learning about the Earth in space. *Primary Science Review*, 72, pp. 9–12.

Young, T. & Guy, M. 2008. The moon's phases and the self shadow: A new perspective on the teaching of the phases of the moon to upper-elementary students. *Science and Children*, 46 (1), pp. 30–5.

11

Our planet Earth

by Christine Preston, Beverley Jane and Graham Crawford

Introduction

Nothing makes you more aware of the Earth we live on than noticing what's under your feet: 'Ouch! The sand is hot! I need to find some shade to cool my feet, I forgot my thongs, and I can't believe how hot the sand can get in summer!' You may remember running across gravel in bare feet as a child, without a care in the world. Noticing what is under their feet, what it looks like, smells like and even tastes like, often comprises a child's first geological learning experiences. It is where students' thinking starts; young children view the Earth on a much smaller scale than adults. This is where we need to start to develop knowledge about what our Earth is made of, leading into the dynamic processes that shape and change the Earth.

Planet Earth formed about 4.6 billion years ago; this gargantuan time span, known as 'Geological Time', is hard to imagine. For much of this time (80 per cent) there was no life on Earth. Amazingly, scientists have discovered that Australia has very ancient rocks formed during this time. To geologists these are not dusty old rocks that hurt if you stub your toe on them; they are a window through which we can view Earth as it was near the very beginning. Did you know that Australia has some of the oldest crystals on Earth? With such rich history right on our doorstep, how can we fail to excite and enthuse our students to want to learn about our planet Earth? This chapter is about what comprises the Earth (not including the atmosphere, which is dealt with in Chapter 12), from the small stuff (dust and dirt) to the processes that shaped and change our Earth. The chapter is divided into four sections, each framed around *an initial* major question.

Firstly, what are teachers', students' and scientists' ideas about soils? Knowledge of students' ideas about soil, sand and dust from teaching experiences and research are outlined and compared with those of teachers and scientists. Constructivist learning experiences are described and suggestions for encouraging students' questions are included to help guide you in teaching about soils.

Secondly, what are teachers', students' and scientists' ideas about rocks and minerals? Like the first section, this one contrasts adults' and students' ideas, this time about rocks and minerals. We also describe teaching approaches to aid students in developing the skills of 'observing and thinking geologically' and enhancing their conceptual understandings about rocks and minerals.

Thirdly, how can you develop students' ideas about geological processes? The large- and small-scale processes that have shaped and changed the Earth over time are included in this section. Learning about rock formation and changes includes the processes of volcanism, weathering and erosion, and metamorphosis. Related topics – earthquakes, dinosaurs and other fossils – are also discussed to help guide you in treating these common interest areas scientifically.

Finally, why is 'science as a human endeavour' significant in learning about our planet Earth? Science and culture and the nature of science are focused on through the historical and human construction of science ideas. Contributions of scientists in geology-related careers, the influence of science and the effect of science and technology on our lives are addressed using mining as an example. Socioscientific issues are raised through consideration of different cultural constructions of the Earth and its origins, including Indigenous views of the land.

Throughout the chapter, you will reflect on what you know about the topics and also the views of scientists and students. The teaching approaches suggested illustrate constructivist principles and focus on the methods and concepts that Earth scientists use when they are investigating their environment.

What are teachers', scientists' and students' ideas about soils?

The term 'soils' is used here collectively (as it is by scientists) to refer to all components and types of soil. Consistent with research that strongly suggests that learning starts from the ideas, skills and concepts of the learner, your prior knowledge is again our starting point.

Your initial ideas about soils as teachers

All of us are familiar with soils, perhaps because our childhood experiences involved eating it, digging in it, making mud pies out of it or gardening in it. But what do you really know about soil? Find out by doing Activity 11.1.

Activity 11.1 Record what you already know about soils

- Think for a few minutes and then jot down whatever comes to mind when you think of soil.
- Now write your answers to the questions:
  - What is soil made of?
  - Where does soil come from?
  - How does soil get there?
  - Why is soil important?
  - Who or what needs soil?

Primary teachers' willingness to teach about science topics is influenced by their own background knowledge. Few students have the opportunity to study geology in their senior secondary years and it is often lacking from the primary science repertoire (except for students' favourite topics like dinosaurs and volcanoes). This doesn't have to be the case. There are many opportunities for teaching about Earth's materials and the geological processes that form them. The next activity helps you to see what concepts primary students could learn about soils (and rocks).

Activity 11.2 What and when should we teach students about soil and rocks?

Take a look at your science syllabus or standards/outcomes statement and do the following:

1. List any explicit references to the inclusion of soil, rocks, minerals, geological processes and so on.
2. Now read the syllabus or standards/outcomes statement again and list other opportunities for teaching about these topics in connection with other science concepts.

The second approach may make for more meaningful learning, allowing you to make links between concepts and view the Earth as an interconnected system; for example, this approach has been applied below to the 'Science understandings' strand of the Australian national curriculum (ACARA 2011) at the kindergarten level.

SCIENCE UNDERSTANDING CONTENT DESCRIPTIONS	TEACHING OPPORTUNITIES
Living things – features and basic needs of humans and other familiar living things.	Basic needs of plants (soil), animals that live in soil (worms), non-living things (rocks, soil, water).
Everyday materials – sorting objects based on features (size, shape, texture and colour).	Sorting rocks (rough vs smooth, dark vs light, big vs small, heavy vs light).
Movement – ways in which objects of different shapes and sizes move.	Investigating and comparing tracks left in the sand or soil by different things that move.

This example illustrates how the topics included in this chapter can be integrated with other science concepts to make learning about them more meaningful. Learning about soils in isolation will not help students make

Expanding your ideas about soils as teachers

Soil is full of living things. In Activity 11.3 you can look at soil under a microscope and see that it is moving with different kinds of algae, fungi, nematodes, mites and springtails. Microscopic animals fertilise the soil by mixing water and air into the soil as they feed on other organisms, and when they die they enrich it with their dead bodies: 'Geologists have drilled down 4 kilometres below the surface of the Earth and even found micro organisms in solid rock' (Suzuki and Vanderlinden 1999, p. 50).

Activity 11.3 Sensory experience: getting your hands dirty

Half fill a large jar or a small sealable plastic bag with soil from your garden or a schoolground.

1. Work on several layers of newspaper.
2. Never pour soil or mud into the sink. Rather, collect these residues in a bucket for disposal outside.
3. Dirty hands and sharing information are encouraged.
4. Use your senses of sight, feel and smell to find out about your soil. Describe it.
5. Look at the soil under a magnifying glass and a stereo microscope. Use labelled sketches and written descriptions to record the various bits the soil is made up of according to the shapes, sizes (in millimetres), colours and lustres and how they stick together. You can use the table below to name your bits if you have access to a mini grid slide and a monocular microscope.

SIZE	NAME
More than 2 mm	Gravel or stone
0.2–2.0 mm	Coarse sand
0.02–0.2 mm	Fine sand
0.002–0.02 mm	Silt
Less than 0.002 mm	Clay

6. Compare your soil with that of the members of another group by noting similarities and differences in your descriptions.
7. Record the properties of soil that you all found most useful when making these comparisons.

Activity 11.4 Reflecting on your experiences doing Activity 11.3

- What were the advantages and disadvantages of working with soil? What were the advantages and disadvantages of working in groups for this activity?
- Indicate a year level and then write down the sorts of things you would do in order to overcome the disadvantages when you are helping students to investigate soils.
- Through handling and examining soils you know more about soil than you did before. Is this new knowledge or have you simply remembered more? Go back to the ideas about soil that you recorded in Activity 11.3 and add what you now know.

This form of recorded self-assessment is also a good method of assessing what students have learnt from an activity. It also helps reinforce the important principle that, in addition to having a responsibility for their own learning, students also have a big part to play in assessing that learning (Baird and Northfield 1992; Harlen 2009).

Activity 11.5 Manipulating mud – scientifically

A

Place a dessert-spoon of soil in the palm of your hand. Add water drop by drop until the soil is thoroughly wet but there is no excess water on your hand. Record what happens when you try to:

- use your finger to press the soil out flat on your palm
- roll the soil into a long, thin worm
- mould the soil into a tiny bowl
- make the soil into a cube.

Scientists' ideas about soil

Soil scientists operating in field locations describe soils as sands, loams or clays, according to whether they fall apart (sand), hold together in blocks but do not roll into a thin worm (loam), or roll out well to a thin worm (clay) when moistened and then manipulated, as you have just done. This property of a soil is known as its 'texture'. If some gravel, sand or silt-sized grains occur in a soil that rolls out like clay, it is called 'gravelly', 'sandy' or 'silty' clay. Table 11.1 shows a common range of textures, with descriptions of how these soils feel when moistened and worked between your fingers.

TABLE 11.1 Field classification of soil texture

SAND	WILL NOT HOLD TOGETHER	FEELS VERY GRITTY
Loamy sand	Just holds together	Feels gritty
Clayey sand	Holds, but will not roll out	Feels gritty
Loam or silty loam	Holds, but breaks up when pressed	Feels gritty and silky
Sandy loam	Just holds together	Feels gritty
Sandy clay	Rolls out into thick worms	Feels gritty
Silty clay	Rolls out but cracks when bent	Feels silky
Clay	Rolls out into long, thin worms	Feels slippery

Activity 11.6 Uses of soil

How do the properties of soils influence the many and varied uses of them?

Add three more drops of water to the moist soil you were manipulating and attempt to roll, mould and cube your soil. Record the implications of what you have discovered for people who make:

- pottery
- bricks
- deep trenches

As for other topic areas, we need to find out what our students already know before they engage in planned learning experiences. Before reading any further, stop and think for a few minutes about what ideas you may expect students in your class (or students of a particular age group) to have about soils.

What do students know about soils?

Students in middle primary have fairly simple ideas about what soil is made of and what we can do with it. For example, Year 4 students' answers to questions about soil included the role of worms, awareness of compost, soil types and the needs of plants.

Question – What is soil?
'Worms business.'
'It's dirt and germs mixed together.'
'It has got fruit peels and compost in it, my Granny does this.'
'There are different types, dry sandy soil, proper soil, compost soil.'
'Worms get into soil and eat the fruit scraps and dead leaves.'

What does soil do?
'It helps the plants grow.'
'Well it's in gardens and it helps seeds grow.'

Students in upper primary are aware that soil is not homogeneous. They know it is made up of organic material (both animal and plant) with inorganic components (grains of sand/rock), that different soils vary in colour and muddiness, and that the wetness or dryness of a particular soil changes over time (Russell, Longden and McGuigan 1993). Many students understand that soils can be eroded by water, that plants grow in soil, and that there is water in ('wet') soil. However, few believe that there is water in 'dry' soil or list air as a component of soil (Brass and Duke 1994).

It is not surprising that most students see soils as having existed 'forever' or occurring as the result of outside agents (brought in by gardeners, rain or erosion), because the rates at which soils develop from parent rock are so imperceptibly slow. The process where rain soaks through the topsoil, slowly breaks down the mineral particles and leaches the more soluble products into the subsoil, takes thousands of years to produce any significant soil profile (vertical variation in soils). So, although many students may be aware of the layers of soil they have seen in road cuttings or beside steps on a path to the beach, they have little idea about the changes that produced this profile. In addition, the countless millions of bacteria necessary to break down organic litter into the humus that makes a soil fertile are invisible.

For students to include the above abstract ideas within their concept of 'soil', they must have concrete learning experiences compatible with what they already know. Try giving students in small groups samples of soil, as well as straws, gauze fabric, water, hammers, boards and magnifying glasses. Allow the students to conduct their own investigations to immediately answer questions they may have about the different soils. At the end of the investigation, the teacher can ask each child to record (in a drawing, labelled diagram or with words) what they found about the soil samples.

Year 5 students' responses at the end of a science unit where they investigated a microhabitat within their schoolgrounds revealed the learning of specific ideas about soils:

> **Question – What do you know about soil that you didn't know before?**
> 'Soil has a lot of stuff in it.'
> 'The plant matter always floats on the top.'
> 'Sand is part of soil.'
> 'Our dirt was very dry, that's why we have a red layer.'
> 'I never knew that soil had so many layers.'

The above responses relate to the experiment 'Sorting Soil with Water', during which the students placed some soil and water in a plastic bottle, shook it and waited for it to settle. In time, the soil sorted according to particle sizes and the students used rulers to measure the depth of the different layers. They did drawings of the soil and water in the bottles, showing the layers with measurements and written descriptions of where they found their soil (see Figure 11.1 for one child's drawing). For example:

> *Description 1* It was damp when we found it. We saw grass around it and it had a little bit of sand in it. It was moist and thick.

> *Description 2* There was [sic] lots of rocks, weeds, tanbark, leaves, grass and sand. We also found lots of clay and moss.

Encouraging students' questions about soils

From the above activities, you now know that all soils contain water, air and organic material. You also found that all soils (sands, silts and clays) contain mainly rounded fragments of different sizes. The explorations and reflections you have undertaken have equipped you to pose many questions about soils and to devise investigations that will help you test your ideas. When students have completed similar explorations and reflections, they, too, will come up with questions to pursue and will have ideas about how to carry out their investigations. Students' questions are essential for engagement as it taps into their personal interests. Encouraging students to design their own practical investigations to answer their questions or test out their ideas helps them develop their scientific thinking skills. The case study below illustrates how a question from a student can lead to purposeful investigation.

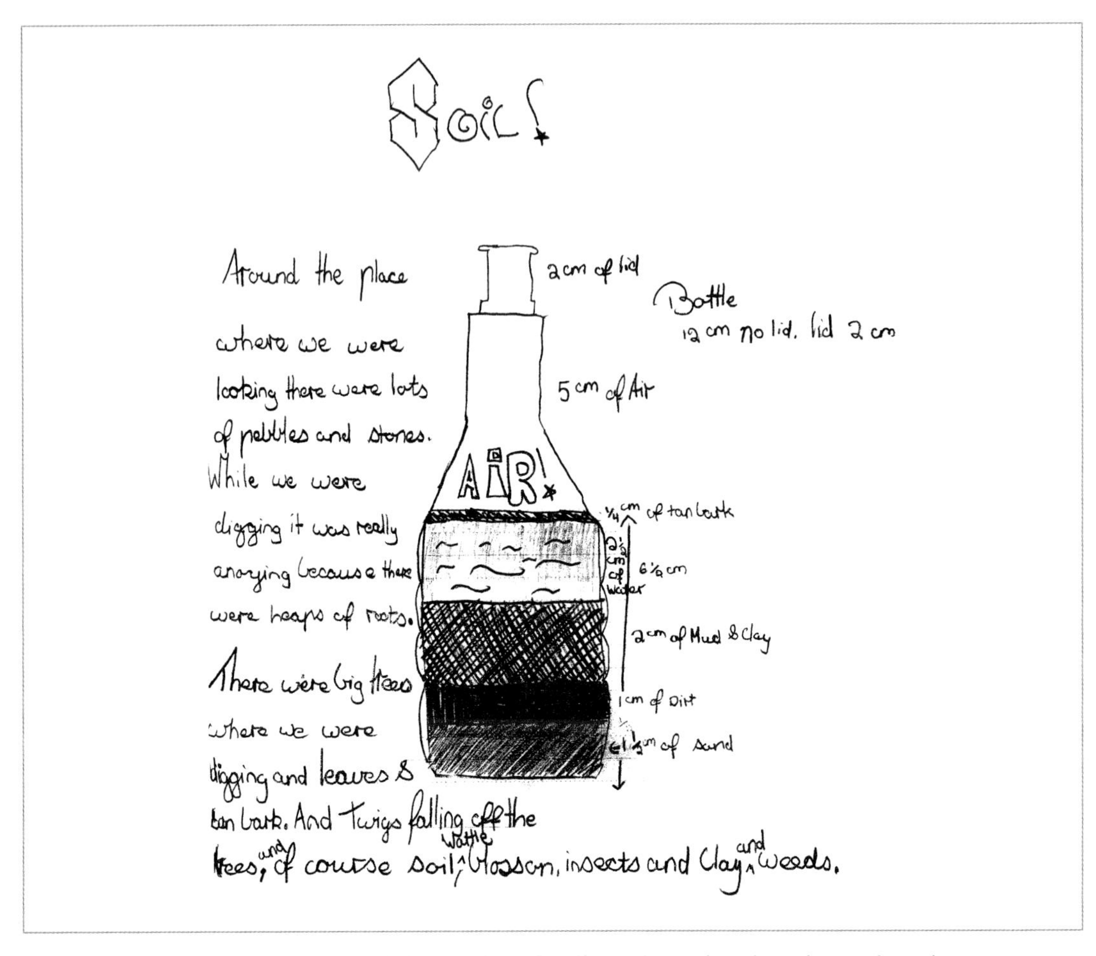

FIGURE 11.1 Work sample of child's results of soil sorting, showing the various layers

CASE STUDY

Students investigating soils

A teacher brought soil from different locations into her Year 5 classroom. She encouraged the students to carry out the activity of manipulating mud scientifically and to classify the several types of soil. After completing the activity, one child asked if she could investigate the question, 'Why does rain soak into sandy soil, but it makes puddles on top of clay?' Her teacher accepted the question, praised her for making careful observations and asked for ideas about what the answer might be. When the girl replied that sandy soil might 'let water leak through it faster', the teacher wrote on the board the question 'Does water leak through a sandy soil more quickly than it does through a clayey soil?' and asked if there was anyone else who also wanted to know the answer to this question. Two students did, so they joined the original questioner and discussed how they might answer the question.

NOS

The students quickly established that they knew where there were patches of sandy soil and clayey soil in the schoolyard. They decided to 'pour water on them and watch'. By raising questions such as 'How much water do you think you should use?', 'Should it always be the same amount?' and

'How will you stop it running off over the surface?' (Pattie 1993), the teacher helped them clarify their ideas and also introduced them to the importance of fair testing.

After some trial and error, the students decided to push cans (opened at both ends) the same distance into the different soils, and then pour the same amount of water into each can, noting which lot of water disappeared first. Subsequent modifications, such as using a stopwatch to measure soak-in times at several locations, produced some interesting data that they entered in table format on the classroom computer. They also located each test site on a schoolyard grid that the class had developed for the computer during an earlier mathematics investigation. They used colours to shade around sites with similar soak-in times and in this way produced a very effective soak-in time map.

The table and map were displayed, along with the apparatus used and an outline of the students' method, as they presented their findings to the class. They also displayed a quite detailed account titled 'Our thinking' (for the teacher). Their teacher often asked them to record their thinking as they went along because she found this a very useful technique for assessing their learning. At any point in the investigation, the teacher had access to a record of invaluable insights into the outcomes of the learning experiences the students had all enjoyed and, by asking a few follow-up questions of individuals, quickly compiled a comprehensive record of who had learnt what and at what level.

The question that this particular group investigated was fairly typical of the sort of questions that other groups had raised. The teacher cheerfully pointed out that, when she was preparing for the topic, she had 'not even thought of most of the questions the students came up with'. Some of the other groups' questions are listed in Activity 11.7. The questions show that the students had used their library time well and/or knew someone who was a keen gardener.

Eliciting students' questions about soils

Some students may be hesitant to ask questions. By listening sensitively to their discussions, asking probing (usually open) questions, observing them during the exploratory activities, and reading their reports (especially their tentative generalisations and explanations), you can facilitate further learning experiences relevant to their needs.

Activity 11.7 Reflecting on students' questions about soils

For each question below, reflect on how you would help the child answer it.

Record whether each question would best be addressed by:

- direct personal investigation
- further exploration to clarify and refine the question
- predict–observe–explain activities (White and Gunstone 1992) devised by the teacher
- referring to information from books, DVDs, the Internet and so on
- inviting experts (e.g., parents) to school to share their expertise.

All of the above are excellent learning and teaching strategies, so it is a matter of selecting which one you think is best, rather than using the 'right one'.

The students' questions are:

- Is there water in dry soil or air in moist soil?
- What is 'humus' and where does it come from?
- What makes some soils acid? Will plants grow in acid soil?
- How can farmers stop soil erosion?
- How can we find out the size of the grains in our soil?
- What happens to soil that is washed away by erosion?
- What do earthworms do to soil?
- How much water can different soils hold and what is in the spaces before the water runs into them?

Students' learning experiences about soils

Research shows that, when learners have opportunities to handle Earth's materials as soon as the topic is begun, they offer a tapestry of ideas richer than their prior knowledge and show far more interest in the properties and origins of rocks, minerals and soils than is the case when initial experiences are with notes, tables, charts and/or films (Paige and Chartres 2002; Russell et al. 1993).

Soil activities in field locations

Teachers and researchers alike agree that observing and investigating soils where they occur, rather than simply taking samples back to the classroom, is a powerful way of extending students' learning about soil. Apart from helping students link their knowledge of soils to what they know about their local environment, fieldwork also enables them to see relationships between different soils.

When students look for similarities and differences in soils found in different locations, and examine the 'soil profile' (see Figure 11.2) observable in road cuttings, quarries, coastal cliffs and river banks, they construct a more extensive and coherent concept of soil. Research into how soil profiles develop also helps students to get a better idea of the slowness of the process of soil development, especially when compared to the rapid rate at which the topsoil can be eroded (see Figure 11.3). When students see that the roots of the vast majority of plants penetrate only the thin layer of topsoil, they begin to appreciate that when it is

FIGURE 11.2 Soil profile showing the various layers (horizons)

removed by erosion in Australia, the land will have limited plant growth for the hundreds of years necessary to build that topsoil again.

Activity 11.8 Identifying the layers (horizons) in a soil profile

Label the photograph above (Figure 11.2) to identify the following layers:

- leaf litter
- humus-rich topsoil (loam)
- leached layer (clayey sand)
- clay with minerals from leached layer
- broken rock fragments.

Upper primary school students may work out that the downward percolating rainwater, which breaks down the minerals in the topsoil (aided by organic acids from decaying plant and animal debris), also leaches nutrients out of the topsoil, so that soils will eventually become fragile and less fertile. This has occurred with many of the soils covering the old, stable and largely arid landmass of Australia. It was not a serious problem before European settlement, because the plants and animals had evolved with the soils. During the many thousands of years they have lived in Australia, Indigenous people have respected and adapted to the land. All of this changed with the clearing of natural vegetation and the advent of European farming methods and crops. The grazing of hard-hoofed animals on land that had evolved with a soft-footed fauna destroyed the soil structure and stripped away much of the remaining vegetation cover. The soil nutrients were quickly exhausted, and the wind and water removed a great deal of the topsoil. Fortunately, Australian soil scientists have learnt, and continue to learn, how to restore the health of our soils. Indigenous people's voices are being heard, listened to, and their special understanding of the land recognised. Scientists can learn more about preserving the land by talking with them. The idea of a fragile 5- to 10-centimetre-thick layer of topsoil providing most of the food consumed by the world's human population is a motivating starting point for studies about ecological and social and sustainability issues, such as water use and land degradation.

Soil activities in the classroom

S&C

The role of earthworms and other soil dwellers in mixing the layers of topsoil and providing a steady supply of digested organic material to the soil is one which can readily be investigated in the classroom by setting up a wormery (refer to the Chapter 6 section 'Concepts and understandings for primary teachers' on p. 242). The constant removal of the organic material placed on the soil surface, and the slow colour change in the topsoil, are observable indicators of how important earthworms are to healthy soils. The big book *Earthworms* (Pigdon and Woolley 1989) is a good resource to have in the classroom for the students to refer to during this activity.

An example of a sequence of lessons focusing on the topic of soil was planned and implemented by two preservice primary teachers with a group of students. The sequence began by finding out the students' ideas about soil, and then moved to making soil in which they germinated bean seeds. Firstly, the students observed some soil and then discussed what

FIGURE 11.3 These photographs show the 'danger' sign has collapsed after strong waves pounded the cliff – evidence that the coastal cliffs at Mornington are constantly being eroded

they thought soil was, what makes soil and how soil helps plants to grow. Secondly, the students made their own soil by mixing ingredients in a big mixing bowl. Thirdly, they planted some bean seeds in their soil to test it. They were excited to see if the beans would grow in their soil. Once they observed their bean plant, the students said that they had learnt:

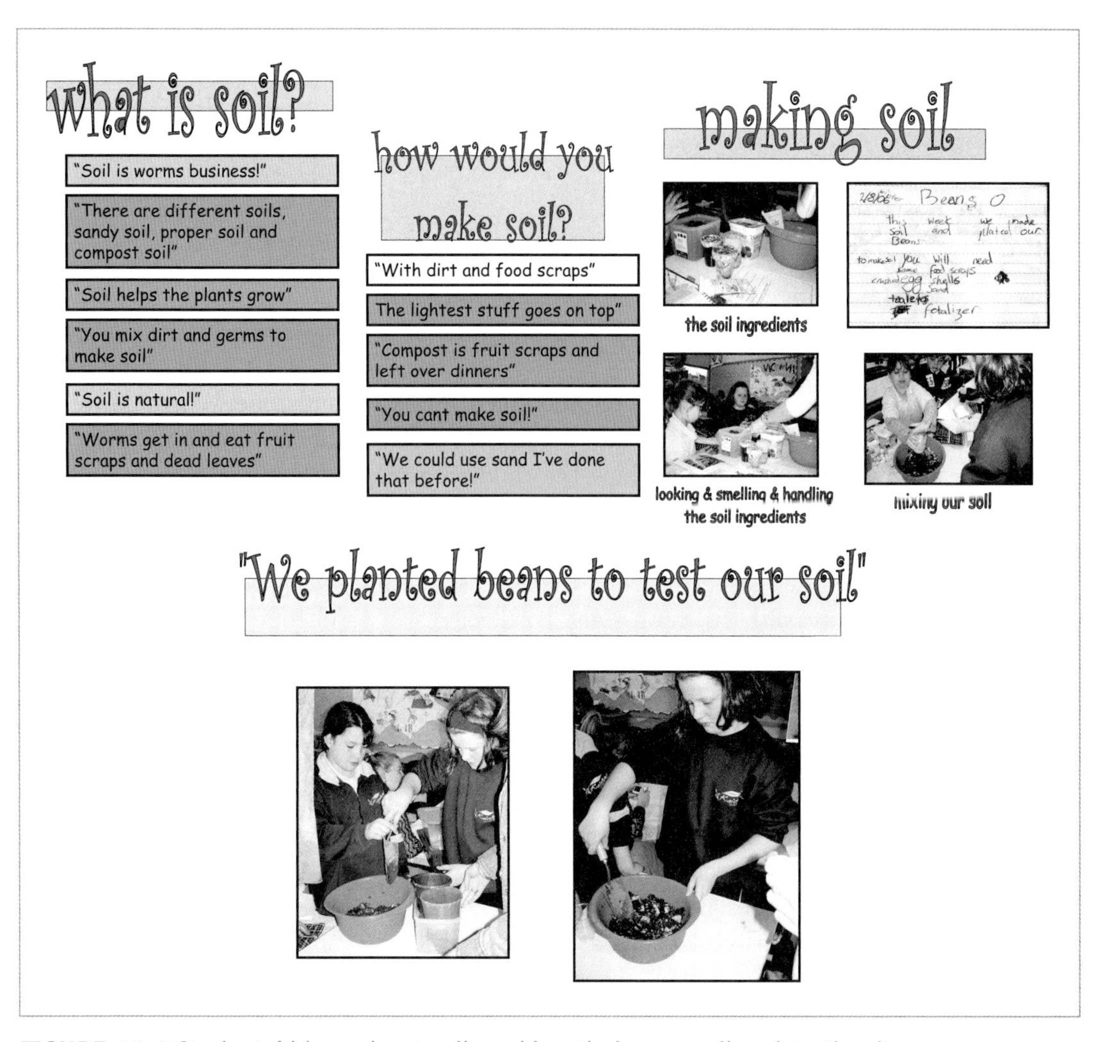

FIGURE 11.4 Students' ideas about soil, making their own soil and testing it

- 'that plants need sunshine and warmth to grow'
- 'that soil isn't dirt. It's got heaps of food in it for plants'
- 'our soil was successful because our bean plant is growing and I know how to make soil now'.

A similar activity, growing radishes in different types of soil, was designed to teach young students about the role of soil and organic matter. Students probably know soil is made up of sand, silt and clay (inorganic components from weathering and erosion) but few know about organic matter (organic material from the decomposition of plants and animals) (Piotrowski et al. 2007). Children were asked thought-provoking questions: 'What is soil made of?'; 'Where does it come from?'; 'Why do plants need soil to grow?'; 'What's the difference between a good soil for plants and a bad soil?' They then predicted and tested which soil mixture would grow the biggest radish. Some useful websites cited include: the US Natural Resources Conservation Service web page 'Soil Education' (http://soils.usda.gov/education), Soil Science Education (http://soil.gsfc.nasa.gov) and the National Science Teachers Association's soil ecology lessons (http://www.nsta.org/elementaryschool/connections/200710SoilEcologyLessons.pdf). An interesting Australian case of soil scientists providing evidence in forensic investigations is included towards the end of this chapter (see the section 'Contributions of scientists: geology-related careers' on p. 489).

Exploring what students know about sand

What is the difference between sand and soil? Do you know? Christine finds out what her kindergarten students know about sand by getting them to make some in the classroom. Here is a segment from the lesson plan:

1. **Introduction** – Show children a pile of sand and a chunk of sandstone. Where do we find lots of sand? How is this sand made? Demonstrate how rubbing sandstone with your finger makes bits of sand fall off. Show the children a piece of basalt. Do you think this rock will make sand? Demonstrate rubbing basalt, showing that bits do not fall off. Why not? Ask children, 'Do you think we could make some sand from rocks?'
2. **Lesson activity** – The teacher explains the process, noting safety. Students wear safety goggles and work in pairs, crushing some small pieces of sandstone into smaller pieces. The pieces are placed into a mortar and pestle (you may have one at home in the kitchen) and ground until they become sand. Children complete sentences in their science record book: Sand is made from *rocks*. Rocks wear away to make sand or *dirt*. Some rocks are *soft*, some are *hard*. Students also draw a picture of the sand they made.
3. **Conclusion** – What have we made? Is this how sand is made naturally? What things might break up sandstone around here? Look at the colour of your sand. Is everyone's sand the same colour? Why not? Explain this is why sand and soil at different locations look different: some beaches are white, others are yellow and some can even be black.
   - *Note for teachers* – Soft rocks such as partly weathered sandstone are required (easily sourced in the Sydney region, where sandstone is one of the major rock groups forming the Sydney basin). With some pre-preparation (gently tapping small pieces with a hammer while wearing safety glasses), children will easily be able to grind them down (see Figure 11.5).

Answers to kindergarten students' questions before and after investigating rocks and sand gives some insights into young children's thinking about geology concepts. Responses to children's questions are shown in Table 11.2.

Allowing students to play with sand (wet or dry) can enable them to learn through sensory activities and trial and error. They can use sand to build structures (like they may do at the beach) and explore the properties of sand and how they change when the sand is wet. Ashbrook (2010b) describes some simple activities for engaging younger students in hands-on tasks with sand and water: 'What about dust then?'; 'What do you think dust is and where does it come from?' Students may think sand, silt, clay and dust have no relationship to soil or rocks. Many students think that the dust that comes from rubbing two stones together are not rocks because they are too small (Eberle and Keeley 2008).

FIGURE 11.5 Kindergarten children making sand

TABLE 11.2 Children's ideas about sand

BEFORE THE UNIT – WHAT IS SAND?	AT THE END – WHAT DID YOU LEARN ABOUT SAND?
'Sand is little tiny pebbles, little tiny broken-up rocks.'	'When you look through a big magnifier, it looks really different. You can see small shells better.'
'Sand comes from beaches.'	'Sand from different beaches looks different.'
'Sand is from a volcano that exploded a long, long time ago – like fifty hundred years ago. When it got old from the volcano, it got cold and shrunk and then ... sand.'	'It's made from little pieces of shells. Some looked like crystals, and they were different colours.'
'Sand is something that is at the sea, and it came from the water. People get it from the beach and bring it to other places.'	'The water rubbed the sand smooth.'

Source: Ogu and Reynard Schmidt, 2009

Another interesting question is that of what happens to soil. Does it always stay the same or does it change? Is it easily transported? Where does it go if the wind blows loose soil away? Questions such as these are great discussion starters and can be used to alter the direction or focus during a geology-based topic.

Can soil become a rock?

When you pose this question to most primary school students, you are likely to be reminded that rocks are very hard and sharp, whereas soil is soft and yielding. However, if you had felt clay that had been baked hard in the heat of a scorching outback sun, as some students have done, then you might, after due consideration, concede that it could be possible, if the sun made the soil very, very hot, that the soil could be hard (Symington, Biddulph and Osborne 1987).

As a follow-up to the sand-making activity described earlier, students can use the sand (or dirt) they made to make a rock. Students simply mix sand with a small amount of glue (or preferably clay), stirring it with a paddle-pop stick in a plastic takeaway container. When the mixture is ready, take a handful and roll it into a ball or squash it flat (depending on what type of rock they want to make). Place the rocks on newspaper or a plastic sheet to dry. Discuss whether this is how rocks are made in real life. What 'glues' the grains together in nature? Where and how does this happen?

Figure 11.6 shows a sketch drawn by a Year 2 student before beginning the topic 'What is under our feet?' The class was asked to sketch what they might encounter if they dug down as far as possible beneath a grassed area in the schoolground. The somewhat haphazard distribution of the various materials was typical. However, some students showed two or three horizontal layers in the materials, but did not show the Earth as being spherical or even curved.

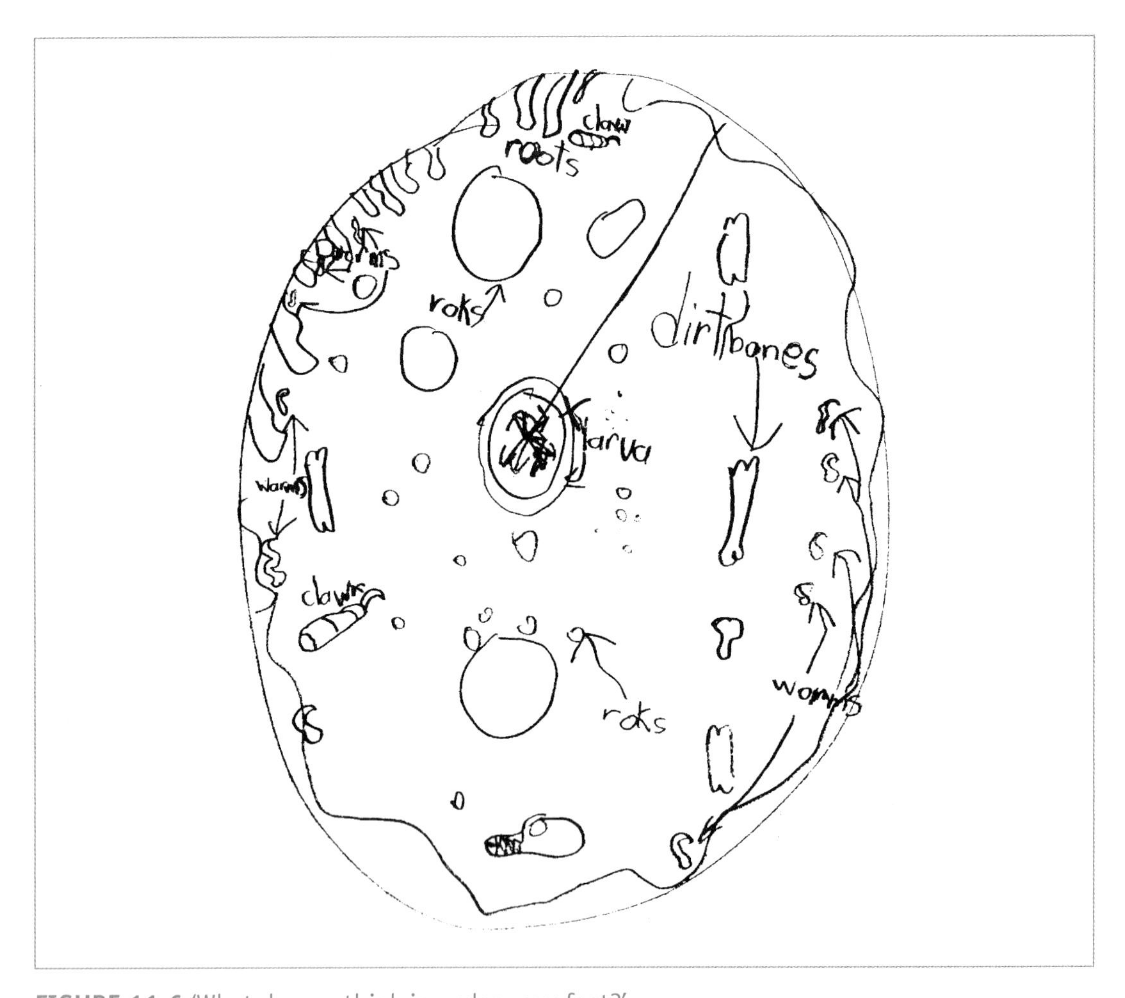

FIGURE 11.6 'What do you think is under your feet?'

The experience of most primary school students with regard to Earth's materials leads them to believe that the Earth's crust consists mainly of soil, with odd bits of rock embedded in it and occasional very big bits of rock sticking up to form mountains (Happs 1985; Russell, Longden and McGuigan 1993). If that is where students are starting, there seems little point in us, as their teachers, deciding to start somewhere else. We must provide our students with learning activities that will broaden and enhance their experience of soils and rocks while at the same time challenging their current explanations in a supportive manner. Having students explore a wide range of rocks in the classroom and where they occur naturally seems a promising option (Symington 1986). Raising questions about how mountains form and awareness of the geological processes that shape and change Earth's surface seems a logical teaching sequence and has guided the organisation of this chapter.

What are teachers', students' and scientists' ideas about rocks and minerals?

In this second main section of Chapter 11, the focus moves onto rocks and minerals and, as in the previous section on soils, after considering teachers', students' and scientists' ideas about rocks and minerals, various constructivist teaching and learning suggestions are explored.

Your ideas about rocks

What we think rocks are, and where we think they have come from, will largely determine what each of us learns about them. Therefore, we need to reflect on our ideas about rocks and their formation. One strategy is to discuss our ideas with several colleagues while handling examples of materials we each think are rocks. This helps us clarify our thinking about rocks and also reveals the range of ideas held by other people.

Activity 11.9 **Exploring rocks**

Preparation for this exploratory activity requires you and one or two colleagues acting as co-explorers, to accumulate 20 or so rocks. Please collect them in environmentally respectful ways.

The easiest way to get the rocks you need is to ask each co-explorer to collect five to 10 rock samples. Other useful sources of rocks are stonemasons, quarries and suppliers of building stone (dimension stone) and stone tiles (quarry tiles). Visits to such places can yield not only rock samples, but also information about which properties make particular rocks useful for specific purposes.

When you and your colleagues have acquired the necessary numbers of rocks, each person should decide whether each sample is really a rock and explain why they think that. As you do so, record the criteria you used to make your decisions. Did you use the same set of criteria each time? Do you think there are degrees of 'rockness' or is it a simple yes/no decision? It often helps to make a pile of the samples that everyone is sure is a rock and list the qualities that make their qualification so obvious.

What you have done so far is to use properties that these samples have in common in order to include them in – or exclude them from – the class of objects we call rocks. We can now use other properties of these rocks to sort them into groups and subgroups, always keeping in mind that the working set of properties of 'rockness' must be constantly subjected to review in the light of further examples.

Activity 11.10 Differences between rocks

From among the 20 or so rocks with which you have been working, choose eight that look least like each other. Divide them into two groups of four according to whether they have some particular property. Record the property you used. Examine the way in which others have sorted their rocks into two groups. Work out the property they used. Check with them to see if you agree. Record these distinguishing properties.

Having explored these properties, record what you think are some useful properties to use when sorting rocks into groups. Look again at your group of rocks and divide each group of four into subgroups of two. Record the property you used in each case. Further divide each subgroup so that you now have every rock in a class of its own. Record the properties you used to do this. Arrange your rock samples into a key like the one in Figure 11.7, recording the distinguishing properties you used in each set of brackets.

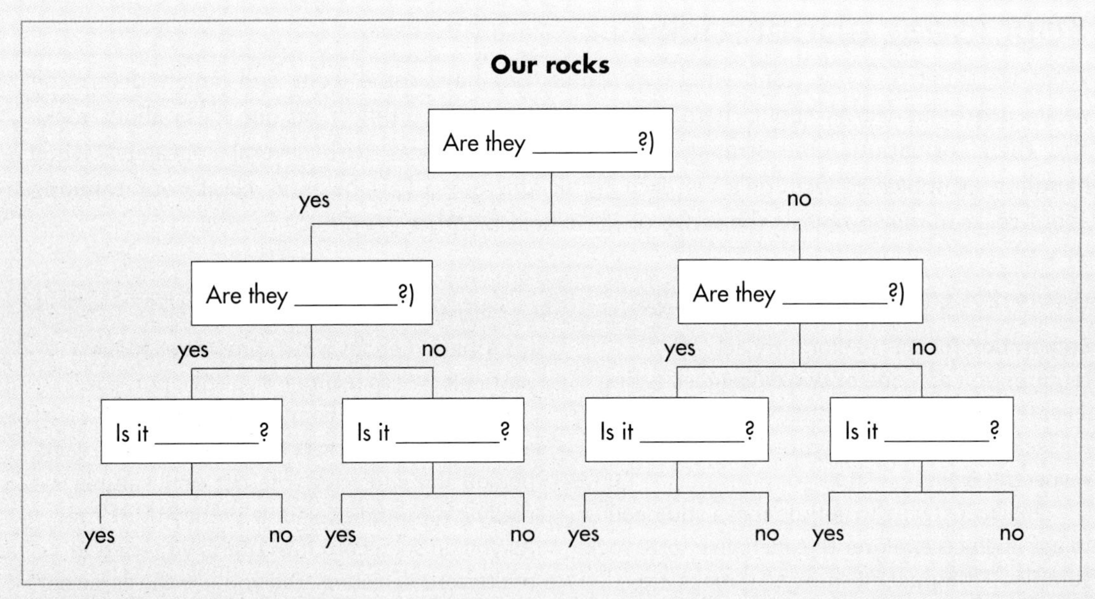

FIGURE 11.7 Dichotomous key

By using the properties of these rocks to produce eight different classes, you have classified your rocks. In the process you have constructed a dichotomous key, which is exactly the sort of process geologists often use to construct their rock (and mineral) identification systems. Scientists use dichotomous keys in many branches of the natural sciences in order to help them to construct a better understanding of the materials they are working with. Keep the rocks you have accumulated and add some others that you think are different from the ones you already have, as you shall be using them in a later part of this chapter.

As you have seen from your previous examination of some rocks, they are usually made up of bits stuck together. These bits are either crystals or grains. Crystals have flat, shiny faces, straight edges and sharp corners, while grains have had their corners, edges and faces worn away and so appear dull and rounded. The crystals and grains that make up most rocks may be seen by using a magnifying glass or with your naked eye.

Can minerals exist on their own; that is, can you have a whole specimen that is crystalline and composed of only one mineral? Of course you can, and these are not only often very pretty but quite alluring to girls. This represents a good way of getting girls into geology – work with the pretty stuff to gain their interest (if it is not already there). Figure 11.8 shows two examples of minerals and a Year 6 girl showing Christine part of her extensive crystal collection. If you have a high school or university nearby and you can borrow a mineralogy microscope, this is a wonderful way to open students' eyes about what is inside rocks. The Australian Museum website (see http://australianmuseum.net.au/Minerals) provides resources for teachers about minerals and includes a photo gallery of mineral specimens.

Note that some shops and stall holders at outdoor markets sell crystals with claims of medicinal qualities and fortune bringing. Examining some of these claims may be a useful way of reinforcing the evidence-based NOS. See the CSIRO website for a crystal-growing activity (see http://www.csiro.au/resources/crystal-grower-activity.html).

Activity 11.11 Examining the bits

- With a magnifying glass to help you see the bits, divide your rocks into three groups – those made of crystals (crystalline), those made of grains (granular) and those in which the particles are too fine to be seen even with a magnifier.
- Run the tips of your fingers across the rocks made of very fine particles. Place those that feel smooth and have broken along fairly flat surfaces into the granular group and those that feel rough and have broken along fairly uneven surfaces into the crystalline group. (Visually impaired students can often carry out this test with an acuity that amazes their peers.)
- Divide the crystalline rocks into those in which the crystals show some alignment in a particular direction and those in which the crystals show random orientation. The feel test is helpful in making this distinction.
- Draw up a table with three columns headed 'Crystals (not aligned)', 'Crystals (aligned)' and 'Grains' and list your rocks here. You may need a fourth column for, as we have seen, every system of classification of natural objects has its limits. Primary school students usually find 'Don't fit' an appropriate heading for this column.

FIGURE 11.8 Minerals samples, quartz and calcite, and a Year 6 girl showing Christine some of her crystal collection

Teachers' and other adults' ideas about rocks and minerals

The research base for adults' (and children's) understanding of geology concepts and processes is not as extensive as in other areas of science. While an understanding of rocks and how they form is fundamental to explaining Earth's history and processes, research into adult geology students and preservice teachers reveals they have an incomplete understanding, or alternative conceptions, about Earth science concepts. Adults' ideas about what rocks are, how they form/change and natural disasters generally differ from those of geologists (Bulunuz and Jarrett 2010; Dal 2009; Kortz and Murray 2009; Kusnick 2002).

An investigation into preservice primary teachers' conceptions of rock formation revealed three misconceptions about sedimentary rocks:

- Rocks are formed by being broken down into smaller pieces or by sediments sticking together at the bottom of rivers.
- Rounded pebbles or rocks found near rivers must be sedimentary rocks.
- Sedimentary rocks are formed through natural disasters such as earthquakes or volcanic eruptions.

These preservice teachers did not consider geologic processes as responsible for forming different categories (igneous, sedimentary and metamorphic) of rocks (Kusnick 2002).

Similarly, in a French study of preservice teachers' conceptions about rock formation (n = 24), very few teachers gave a scientifically acceptable explanation of the rock-forming process (9 per cent), while many equated the term 'rock' with 'stone' (59 per cent) or 'related rock' to a category (e.g., sedimentary) (32 per cent). Other misconceptions about rock formation held by these teachers included the idea that lithification (sedimentary rock formation) involves pebbles growing or different minerals melting together, and that rocks are formed in situ, not involving transportation processes. Only a third of these preservice teachers referred to the changing effects of atmospheric conditions (36 per cent), linked processes to natural disasters (32 per cent), or included human intervention through transportation of particles (32 per cent) (Dal 2009).

Research with adult geology students showed they view rocks as static, unchanging objects rather than being part of a process (rock cycle) where they form and change over time. They also confused igneous, sedimentary and metamorphic rock types and failed to discriminate the differences between how they form (Kortz 2009, cited in Kortz and Murray 2009).

The prevalence of alternative conceptions adults hold about science concepts, including 'rocks', has led to deeper research examining the underlying factors involved in forming these ideas and their influence on the development of further understanding. It has been argued that students (of all ages) retain deeply held beliefs called 'conceptual prisms' that interfere with their understanding of how rocks form (Kusnick 2002). The prisms include:

- the use of common language compared to scientific language in describing what is a rock
- an understanding of scales of space and time
- the notion that landscapes are forever, hence Earth is stable and unchanging
- the prevalence of the idea that humans play a major (rather than minor) role in rock formation (ibid., p. 301).

Building on this work (and that of other researchers), Kortz and Murray (2009) sought to determine the underlying conceptual causes of students' alternative conceptions about rocks and their formation that act as barriers to their understanding. While the purpose of this chapter does not allow a full discussion of this research, their findings will be used in the next section as recommendations for teaching children about geology concepts. The conceptual barriers identified in this study and one example of a related alternative conception are presented in Table 11.3 to reinforce that, as a result of these conceptual barriers, students cannot form scientifically correct mental models of rock formation (Kortz and Murray 2009, p. 312).

Such insights into adults' thinking may help us to develop more-effective methods for teaching concepts such as rocks and their formation at the primary level, which may reduce or prevent these conceptual barriers from forming or persisting into later learning.

TABLE 11.3 Critical barriers and examples of resulting alternative conceptions

CRITICAL BARRIER	EXAMPLE ALTERNATIVE CONCEPTION
Deep time	A 'long time' is at most a thousand years
Changing Earth	Features on Earth do not appear or disappear
Large spatial scale	Layers in rocks are the same as layers in the Earth
Bedrock	A rock forms as a hand sample
Materials	Magma turns into a black rock and black rocks were magma (that is, black = igneous)
Atomic scale	Minerals form separately, then come together to form rocks
Pressure	Pressure to form rocks is caused by things like heat, water, faults and air

Activity 11.12 Classifying the rocks in each column

- Arrange the rocks in each column of your table (from Activity 11.11; see p. 465) into three (vertical) groups of coarse, medium and fine.
- Display your rock table, attaching labels to show how you have grouped the rocks. Compare your display with that of other groups and discuss any differences.

We have used the properties of crystals or grains and size of particles because they are usually obvious features of the surface of rocks and they are geologically significant features used by geologists. However, there is a further reason for using them in that these properties also give insights into the origin of rocks. Russell, Longden and McGuigan (1993) found that students do not regard the origin of rocks as a relevant issue and so very rarely suggest that this is a way of classifying them. Nor is it being suggested here that you should do so when working with your students. However, it should be noted that geologists use the origin of rocks as the fundamental criterion in classifying them. Two reasons have been proposed for teaching children about igneous, sedimentary and metamorphic rocks: it provides the required background for looking at rocks and minerals through a geologist's lens (Fries-Gaither 2008) and links the observable properties of rocks and minerals to their formation (Ford 2005). Further discussion of children's ideas about rock formation is in the next section.

Scientist's ideas about rocks and minerals

Alternative conceptions about geology concepts frequently stem from language usage. For example, a 'rock' in everyday language refers to a single, particular specimen ('That's a rock'); however, a geologist would use 'rock types' (e.g., 'metamorphic rock'), reserving the term 'clast' for a single specimen (Fries-Gaither 2008). Geologists define rocks as the solid aggregations of minerals that make up the Earth's crust.

Properties geologists use to identify rocks with include colour, the minerals present and patterns (e.g., layers and arrangements of crystals) (Ford 2005). When geologists view rocks, they pay attention to characteristic properties that help them to identify the specimens and use them to gain a picture of the past environmental conditions under which they were formed. Geologists do not list physical properties of rocks or minerals in isolation; they draw from background knowledge of rock types, their formation and mineral characteristics, knowing which properties of a specimen to attend to and which to ignore (Ford 2005). Geologists describe as 'coarse' rocks in which the crystals or grains are over 2 millimetres across. Those finer than this but that can still be seen with the naked eye (0.2 to 2 millimetres) are described as 'medium', and those in which the crystals or grains can be seen only with the aid of a magnifying glass or microscope are said to be 'fine'.

Geologists use the term 'mineral' to describe rocks or parts of rocks that are inorganic in origin, crystalline and made of one chemically distinct substance. Quartz, for example, is a mineral made up of silicon dioxide, while pyrite ('fool's gold') is iron sulphide and gypsum is calcium sulphate. There are over 2000 different minerals making up the Earth's crust, but only about 200 of them are well known, either because they have important roles in forming common rocks or they are economically important. Geologists use the unique properties of minerals to identify them, including hardness (harder minerals scratch softer ones), cleavage (breakage along the surface), fracture (the way a mineral breaks), streak (the colour left by scraping a sample on a white tile), lustre (how the surface of the mineral reflects light), colour (most minerals show characteristic colours, especially when viewed in thin section) and specific gravity (a standard measure of density). Not all of these properties are always obvious in minerals present as components of rocks, so 'the chemical compositions of minerals are the basis for the main classification of the mineral kingdom' (Press and Siever 1982). As primary students have difficulty in appreciating the uniqueness of chemical substances (see chapters 8 and 9), then we will refer to physical rather than chemical properties for activities for students in the next section.

Students' ideas about rocks and minerals

Students have some definite ideas about what rocks are, what they look like and where they are found. Even though some of these ideas are rather fuzzy, students cling to their views with great tenacity. Several studies have found students' ideas about the nature and origin of rocks, and the rates at which geological processes occur, were very different from those of geologists (Dove 1998; Eberle and Keeley 2008; Fries-Gaither 2008; Happs 1982, 1985; Russell, Longden and McGuigan 1993). Most students agree that rocks are solid, but other conceptions of a rock depend on generic properties: colour (grey), shape (jagged), texture (rough), shininess (dull) and size (big). A reliance on general characteristics overlooks geologically important properties, while the use of everyday language (bumpy, rough, dirty, heavy, crumbly) shows primary students lack knowledge of more correct scientific terms (Ford 2005; Happs 1982). Young students may describe 'big chunks' as rocks but 'little bits' as stones, even if they are made of the same stuff. Children commonly think that if something is a 'pebble', 'stone' or 'crystal', then it is not a rock. This should be kept in mind when planning learning experiences with Earth's materials.

A review of research in Europe and the United States into children's understanding of: Earth's materials, structures, process and geological time confirmed that 'children develop their own, mostly non-scientific, understanding of Earth science concepts before instruction. Children describe and interpret them in everyday terms that are familiar to them' (Dal 2007, p. 252). These conceptions are often different from those of Earth scientists.

An investigation into Year 3 students' descriptions of rocks and minerals identified common descriptors for rocks and minerals, as shown in Table 11.4.

TABLE 11.4 Children's descriptors for rocks and minerals

MINERALS	ROCKS
Feel/texture (rough, sharp, smooth)	Feel/texture (bumpy, rough, sharp)
Colour (white, clear, pink, green)	Colour (white, grey, tan)
Lustre (shiny, sparkly)	Shape (shaped like [various])
Shape (cube, rectangle)	

Source: Ford 2005, p. 284

Children's use of non-scientific criteria to classify rocks (colour, shape and size) shows they lack an understanding of geologically important properties that are more helpful in identification. This means that children are learning to observe general properties; 'they are not observing geologically, nor learning how observation and rock and mineral properties frame one another in the geosciences' (Ford 2005, p. 291). Children as young as those in kindergarten have some definite ideas about rocks before learning about them in formal settings. For example, before a unit investigating rocks and sand, children's answers to the question 'What is a rock?' included: 'A rock is a little pebble', 'Rocks are mostly hard things, and God made them and dropped them', 'I think a rock is a part of the ground, but it broke ...' and 'Yes, rocks could have been little pieces of dirt from the ground that got really old – that got really smooshed together and then got hard'. Listening to what children say, not just the words they use, can give you as the teacher useful starting points for teaching. For example, what does 'smooshed together' mean, and how does this student (and her classmates) think this could happen?

Sources of alternative conceptions about minerals and crystals also arise from misuse of these terms in everyday language. When children are asked to observe and describe a set of rocks and minerals, the term 'mineral' is almost never used, and the word 'crystal' is applied to anything that sparkles or shines (within a rock or a whole specimen) (Fries-Gaither 2008; Happs 1982).

The following account reminds us that our understanding and use of terminology is not always viewed in the same way by our students. Some years ago, a Year 6 class, their teacher and a researcher developed and evaluated a unit of work about rocks and minerals. The students and the teacher had been thoroughly involved in the program and results from the various instruments used to assess the students' learning in the key concept areas were most impressive. During the final workshop, one of the most enthusiastic students in the group approached the teacher and researcher to ask whether the topic of Earth's materials would be continued next week. When the teacher said they were starting a new topic, the child seemed disappointed. The researcher, sensing that new data were about to unfold, asked, 'What else would you like to do?' 'Well', came the reply, 'when we started learning about rocks and minerals, you asked us to bring in any we had at home and I didn't have any. But I've got lots of stones I've collected, and now that we've learnt so much about rocks and minerals I thought we might be going to learn about stones, too'. As a teacher, how would you respond in this situation?

In contrast to the specific, but inconsistent, usage of properties to determine whether something is a rock, research has shown that primary students have only vague ideas about where rocks came from ('underground') and how old they were ('old') (Russell, Longden and McGuigan 1993). Many children think rocks were just there and always had been, presumably since the Earth was formed, and do not relate the properties (that they so assiduously describe) to the reasons for these properties; namely, the conditions under which they were formed (Ford 2005). Ford argues the need to connect properties of minerals and rocks with the processes that form them (which involves deeper understanding) in order to identify and not merely describe them. She also notes that the processes for identifying rocks are not the same as those for minerals. The same criteria are not used (which few learners realise).

Finding out students' ideas about rocks and minerals

Although research can provide some very helpful general ideas about what students are likely to think, it is still important to collect specific information about the prior knowledge of each particular group of students with whom we work. You will recall that when we began thinking about rocks, you spent quite some time handling materials while discussing whether or not you thought they were rocks. This learning experience had a clear purpose, but it also allowed free exploration of both the various rock samples and the ideas that they helped to generate. In this way, you shared your prior knowledge (even ideas you were not sure about) and you came to an understanding about what you and your colleagues saw as the meaning of the term 'rock'. This greatly facilitated further learning.

Some methods you could use to elicit students' prior knowledge are as follows:

- Visit a rocky beach, a rocky riverbed, an old (but safe) quarry, stone buildings or a cemetery and have students make field notes about their observations.
- Read or make up a story about a group of (possibly Stone Age) people who were confronted by a problem, such as where all the different rocks they encountered or used in their daily lives had come from. The class discusses and suggests solutions to this problem.
- Show DVD/video clips of rocky mountains, coastal cliffs or other rocky landforms to initiate discussion to find out what students know and want to know about these landforms and rocks. Display these ideas as classroom posters.
- Ask students to make a labelled sketch showing what they think they would encounter if they went outside and dug downwards as far as they could go. Older students can apply a scale indicating the relative depth of the various features.
- Ask students to make a sketch of a rock or rocky place they know, and then label it, indicating what they know about these rocks (Hayes, Symington and Martin 1994).
- Ask students to bring in for display the rocks they have collected or bought in a tourist shop and tell the class what they know about these specimens.

- Ask thought-provoking questions – What are rocks? What do you know about rocks? Where do they come from? Are all rocks the same? Do rocks change? What are they made of? What do you notice about these two minerals? What do you think will happen if we try to scratch one with the other?
- Display a few ordinary and not so ordinary rocks for the students to examine. Suggest that, in pairs, students write letters to each other about what they have observed about these rocks and any others they have seen recently.
- Help students prepare concept maps to show what they think are the most important things they know about rocks.
- Display and discuss the sentences that students write when they associate the following groups of words:
  - rocks, mountains, valleys
  - beaches, waves, rocks
  - soil, rocks, change
  - size, rocks, sharp
  - jagged, hills, boulders.
- Provide students with some interesting rocks to handle – for example, a large piece of pumice that students will assume is heavy and become interested in when they discover it is not. How can it be so big and not be heavy? (See Figure 11.9.)

Formative assessment probes were used by a Year 5 teacher to elicit the conceptual rules students used to form initial ideas about rocks (as well as matter). The use of probes can

FIGURE 11.9 Kindergarten children holding light and heavy rocks of similar size

promote deeper conceptual understanding by helping to solidify students' ideas (Eberle and Keeley 2008). In the study, the teacher used a specially designed probe to determine whether students thought that a rock must have certain properties (size, shape and texture) to be called a rock. The teacher used students' responses as a stimulus for small-group discussion to further refine their ideas. Students were then engaged in a demonstration in which a small rock was presented and all agreed it was a rock. After breaking the rock into smaller and smaller pieces in front of the students, the students had to reconsider their ideas. This study showed that students 'lacked an understanding that the formation and composition of a rock is what makes it a rock, rather than the properties of size or texture of a rock' (ibid., p. 52). Later, the teacher used another probe about matter and related this back to learning about rocks, which provided an opportunity for students to connect learning about the two areas, leading to a deeper understanding rather than isolated pieces of knowledge (see the Chapter 8 section 'The particulate nature of matter' on p. 305, where a similar approach is described). You are encouraged to access this paper as it provides a range of teaching strategies for addressing these issues.

Developing primary students' ideas about rocks and minerals

A

When developing primary students' ideas about rocks and minerals, we need to value and build on students' prior knowledge, follow their interests, and guide them to learn how to observe geologically. So, where might we start? Zeegers (2002) suggests that we start small, explaining: 'One student's interest in rocks collected during the school holidays inspired a unit on the nature and uses of rocks' (ibid., p. 15). In pairs, students shared their ideas about rocks, stones, pebbles and minerals, and then chose a secret rock that they later had to identify. This activity proved to be a challenging one that generated many questions for investigation.

Try showing younger students your pet rock; explain that you know this rock so well that it could be placed among other rocks and you would be able to pick it. Ask a student to test this out. Discuss why the students thought you could find your rock. Give each student a rock and a magnifying glass. Explain that they need to look at it very carefully so that they will be able to find it when we put all the rocks back together again. Students draw their rock and write words that they can use to describe it. Students put all their rocks into a box. Shake the box to mix up the rocks. One child at a time comes out to try to find their rock. Discuss what features made it hard or easy to find their rock. Discuss the different features of rocks. Ask students to tell you what rocks are made of. What are the bits like? Give students a selection of rocks and ask them to find one that is made of crystals and one that is made of grains. Ask students which of these rocks is the hardest? Why do they think this? Where do they think these rocks came from? To conclude, show students a selection of different materials and ask them to vote on whether they think each is a rock and why they think so. Conclude with a definition of a rock (from the students' viewpoint). This kind of activity ensures you are starting from the ideas your students have.

A similar activity to the one above was used with upper primary students but focused on making observations and guiding students in writing an essay based on observations of their rock. This study incorporated assessment (guided by a rubric with a grading scale) which was used for mid-unit evaluation (Lark, Kramp and Nurnberger-Haag 2008). Another way to introduce a rock-sorting activity is to use *Dave's Down-to-Earth Rock Shop* (Murphy 2000) because it provides an excellent review of the visible characteristics and formation for sorting rocks, as well as the hardness test.

Anecdotal and research evidence indicates that students often lose interest in materials making up the Earth as they progress through the primary years unless they spend time investigating these materials and are also encouraged to continue asking the sorts of questions that interest them (Happs 1985). To prepare for practical activities, you will need to borrow, purchase or in other ways accumulate sets of rocks and a select number of local field sites where rocks may be observed. Once more, students are an invaluable source of such information and materials; other teachers, parents and guardians, nearby secondary schools and your State, Territory or country's science teachers associations are also very helpful sources, as are the stonemasons, building stone suppliers and quarry operators mentioned in Activity 11.9 (see p. 463).

Learning experiences where students observe rocks and minerals require teacher guidance so the students learn to distinguish between geologically relevant and irrelevant properties. Ford (2005) recommends helping students form explanations about their observations rather than write simple descriptions. This enables students to learn to classify rocks and minerals in the same way as geologists. Have students try to link observable properties to how the samples formed and write their own explanations, justifying these with evidence from their rocks. This emphasis moves students more towards geologically authentic inquiry (Ford 2005). The challenge (albeit difficult at the primary level) is to get students to relate their observations to the reasons for their existence (geological formation processes) and try to understand that the Earth as they know it today is different to when the rocks they are observing were formed.

The overall message here is that teaching geological concepts to primary students is not as straightforward as *letting them loose* among a rock collection. It requires the careful, deliberate choice of learning experiences with clear aims, underpinned by a desire to involve children in observing and thinking geologically. We will look more at developing students' ideas about rocks and minerals in the context of the processes that form and change them in the next section.

How can you develop students' ideas about geological processes?

It may seem strange to separate the concepts of soils, rocks and minerals from the processes that form them; this is just how children, as opposed to geologists, tend to think about them. We are not suggesting they should be taught in this way, but this presentation is simply due to the structural organisation of the chapter. In fact, this artificial separation may serve as a reminder that our role as teachers is to make explicit the links between concepts and processes, helping children make connections in order to develop deeper understandings. Within this third section of Chapter 11, the concepts of rock types, fossils, natural disasters and Earth structure are integrated. This approach has been taken for two reasons. These concepts and processes are integrally linked and it has been argued that 'children's ideas about concepts such as "crystal", "volcano", "rock", and "Earth", reveal broader patterns of understanding that provide a more informative guide for teachers to address than the consideration of such ideas in isolation' (Dal 2007).

The rock cycle

The idea of rocks at the Earth's surface being slowly uplifted while at the same time being worn down by agents of weathering and erosion, with the products being transported, deposited, buried and hardened into rock again, is simply not linked to primary students' experience. Similarly, the idea that some of these sediments may be pushed so far down into the Earth's crust that they are melted by the intense heat they encounter, and reappear later to make volcanoes or to become the molten core of new mountain ranges, seems further removed from reality than stories about dragons and fantasy lands. We should not be too surprised at this, for until about 200 years ago no-one seems to have considered it worth writing about except in the most speculative terms (see the upcoming section 'Science as a human endeavour' on p. 488).

The research described above confirms the experience of many teachers that, for primary school students, the fundamental geological concept of the 'rock cycle' has little or no meaning in terms of their daily observation of their environment. Further rock and mineral activities that focus on descriptions that only ignore important properties linked to their formation, retard the development of deep understanding about geological processes. A study of Year 6 students after they had completed a geology unit showed they knew about the major rock types and could explain the formation of igneous and sedimentary rocks. Despite this, they could not relate rocks present in their local environments to their formation (Ford 2005).

For many students, the idea that rocks even have an origin is confusing, for it implies that rocks were once something else and for most students that just does not make sense. It is important that you try to appreciate the enormity of geological time if you are to establish for your students a framework of learning compatible with the overall geological knowledge structure you want them to build.

Learning about rock types and how they form

The words 'often', 'most', 'probably' and 'usually' appear frequently in discussions about the classification of rocks, soils and minerals. This is because we are dealing with naturally occurring materials. When putting them in groups, we do so in order to make better sense of the properties and structures we observe that they have. Therefore, all such classifications are human constructs that have been modified and added to as scientists try to make them more encompassing. There are no right classification systems, only those that are considered, at present, to be most useful. The same process of classification is being used when a child says 'These rocks are pretty and those are ugly' and when a geologist says 'These rocks are igneous and those are sedimentary'. The difference lies in the sophistication of the concepts used and the degree to which the statements help us make sense of the world of rocks. The articles by

Fuller (2002) and Pyle (2003) in *Investigating* provide information for teachers about rocks and how they can be classified.

At the primary level, you are faced with finding ways of enabling students to make sense of their geological environment given their developmentally constrained world views. Ford (2005) suggests 'giving children access to more authentic geological concepts', which, with a little imagination, using supplies from the kitchen or art cupboard, you could develop some intriguing practical activities around rock formation. Think about melting and hardening (linked to physical and chemical changes) and the formation of metamorphic rocks. What could you use to simulate the action of a stream rolling rock fragments around, making them smooth? Consider what sorts of substances children may be able to squish together (like rocks and minerals under the ground). These kinds of activities may help students to form deeper geological concepts. The challenge is to hold student interest as you develop a deep understanding of the underlying geological processes.

The following section aims to point out observable features of the three main rock types that teachers can use to link to how they were formed and incorporate topics of interest such as volcanoes, earthquakes and fossils (including dinosaurs) that may help to maintain students' engagement.

Igneous rocks and volcanoes

When we look at the crystals that make up an igneous rock (see Figure 11.10), we notice that there are groups of crystals that look similar to each other, but are different from the other crystals in the rock. The similar crystals are each made of the same mineral. For example, in granite, the lighter-coloured minerals are usually feldspar, the dark crystals that can be removed in flakes are mica, and the glassy-looking crystals that do not break along flat surfaces are quartz. The rocks you placed in the 'Crystals (not aligned)' column in Activity 11.11 (see p. 465) are almost certainly of igneous origin; that is, they were formed by the cooling of liquid

FIGURE 11.10A Granite (coarse crystals)

FIGURE 11.10B Basalt (fine crystals)

magma at various depths in the Earth's crust. Each crystal is made of a particular mineral. In general, the coarser the crystals, the deeper in the crust the rock formed (see Figure 11.10). Figure 11.10A shows the coarse crystals in granite and Figure 11.10B shows the fine crystals in basalt. These are both igneous rocks formed from molten magma – Why the difference? Are there any other features that may suggest how these rocks were formed?

Research involving Year 8 students showed they held alternative conceptions about molten lava, rock formation and crystals, saying, for example, 'These heavy rocks are full of holes; there are bits of black and yellow stones' and 'While forming (the basalt), it picked up small stones on the way, which then integrated themselves into the rock' (Dal 2007). These alternative conceptions suggest that students think of minerals not as integral rock components but inclusions (like chocolate chips in cookies). They also show a lack of understanding about just how hot the lava is and that it is actually molten rock. The observation that basalt is 'full of holes' provides an opportunity for open-ended questions such as: What might have caused these holes? What might explain why granite doesn't have similar holes? The holes or 'vesicles' in basalt are due to gas bubbles in lava, which gives a key to its origin.

Two activities that can be done in a primary classroom and give students concrete examples to relate to, are:

- making crystals (from Alum, obtainable from a pharmacy), allowing half the crystals to take more time to cool (at room temperature) and making the rest cool quickly (placing them on ice) to see if this makes a difference to the size of the crystals
- making honeycomb, allowing some to set before the mixture boils and some after (take care with boiling liquids).

FIGURE 11.11 Fuji san, a cone-shaped active volcano

Ensure you discuss the limitations of these 'models' and how they are similar yet different to the natural processes (see Chapter 1).

When students think about a volcano, they probably visualise an active volcano, such as Krakatoa or Mount Etna – one that is erupting and throwing out hot lava that flows down to the sea, covering everything in its path. In a class discussion, students might ask, 'Why do people choose to live close to a volcano?' Such a question may have been prompted by an awareness of movies that take viewers through a terrifying experience of an active volcano erupting and the town's inhabitants' desperate plight in trying to leave the scene. Mount Etna, situated 30 kilometres from Catania, Sicily's second-largest city, last erupted in July 2001 (Pfeiffer 2003). In Japan, many people live near Fuji san (Mt Fuji), which is one of the most impressive cone-shaped active volcanoes in the world (see Figure 11.11). But what does all this have to do with how rocks form?

Many hands-on activities involving volcanoes include making something that 'spurts' stuff out the top of a model volcano to simulate an eruption. While this is exciting and interesting to students, it does little to engage them in thinking about the under-surface processes that cause volcanoes to erupt. A more meaningful model is shown in Figure 11.12, where students in Chris' Year 5 classes used red or orange balloons to simulate molten magma and plasticine to represent the Earth's surface. One student gradually blew up the balloon, with others observing changes in the Earth's surface (mountain formation) and eventually eruption (when the magma forces cracks or holes in the surface). While not as dramatic, this activity leads students to think more deeply about the processes involved in volcano formation and eruption rather than simply focusing on above-surface 'fireworks'.

Volcanoes give us clues about the past. The volcano discovery trail that runs from Victoria to South Australia takes students through a surprising journey of the landscape of Australia's

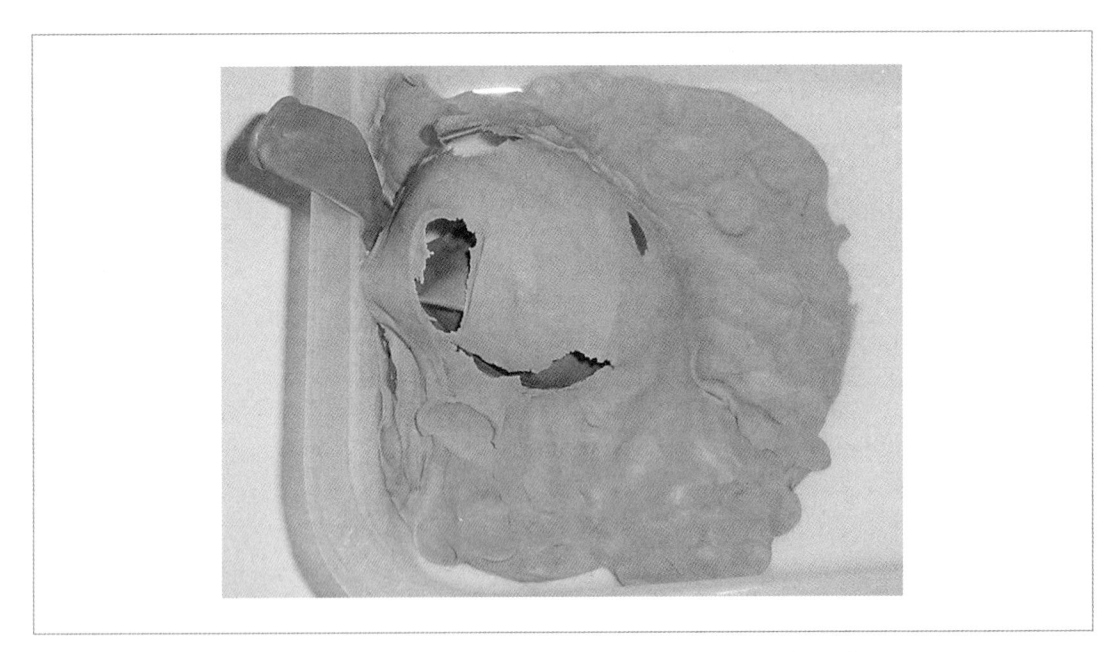

FIGURE 11.12 Model volcano eruption focusing on the cause not the effect

youngest volcanoes, all of which are dormant. The discovery trail experience encourages students to learn to recognise volcanic features and understand the natural events that create volcanoes.

Though students' interest in volcanoes is more to do with the theatrical displays of eruptions, geologists' interests lie in the formation of new rock and physical changes to the landscape. Often, we only hear about the disaster an active volcano can cause, but volcanoes also have a crucial role in maintaining the Earth's environmental balance. On cooling, the gases and lava flung from the volcanoes form new islands and continental crust. Formations such as lava caves provide sanctuary for wildlife. Volcanoes also provide materials for use in building and construction. In the past, basalt (bluestone) was extensively used for buildings, paving and roads, and red scoria for driveways. Students could examine scoria, which forms when escaping gases form bubbles in the lava as it sets. As part of a field trip, students could also inspect the sites of past volcanic action to identify the effect of erosion of volcanic products. Internal volcanic structures that have been exposed by erosion can be seen at the Warrumbungle Mountains in New South Wales, the Glass House Mountains in Queensland, and at Shiprock in New Mexico (Mason 2002).

Sedimentary rocks, fossils and dinosaurs

Meanwhile, 'weathering' of surface rocks occurs due to heat, water, ice and wind. These forces break down the rocks into small pieces called sediments that may be changed into sedimentary rock or, over a longer time, form the mineral basis for soils. Sedimentary rocks hold interest for children due to their inclusion of fossils, the most exciting, of course, being dinosaur bones. Some geologists, called palaeontologists, find sedimentary rocks of interest for the same reasons, but others are more interested in the processes of weathering (breaking down of rocks into bits called sediments), erosion (transportation of sediments to other places) and deposition (laying down of sediments), leading to new rock formation.

In Activity 11.12 (see p. 468), the rocks you placed in the 'Grains' column are of sedimentary origin; that is, they were formed when particles derived from pre-existing rocks were stuck together by chemical cement and/or pressure from overlying sediments, and hardened into new rocks like sandstone and limestone (see figures 11.13 and 11.14). Sedimentary rocks are usually composed of layers that were deposited by wind or water. Limestone is a sedimentary rock that forms from shell remains and has a high calcium content. All rocks have originated in one of these two ways: the cooling of molten magma or the laying down of sediments over time.

As you have no doubt discovered, we cannot always be sure exactly what shape the particles making up a rock are, nor how they are arranged, even when we examine them under a magnifying glass – hence the 'Don't fit' column. A common example of such a rock type is quartzite, which, although it is sedimentary in origin, has often been changed so much by heat and pressure that it is difficult to see the sand-sized grains of which it is composed or the layers in which these grains were originally deposited. We can resolve many such difficulties by cutting a thin wafer of the rock and examining this under a powerful microscope. This technique makes it much easier to establish the rock's origin and its history, as well as identifying the minerals of which it is made.

Observable features of sedimentary rocks that can be linked to their origin include their layers (see Figure 11.14) or the inclusion of a conglomeration of various-sized bits and pieces.

FIGURE 11.13 Sedimentary rock

FIGURE 11.14 Sedimentary rocks showing layers (Mornington, Victoria)

An example of an open-ended question you might ask students is: 'Why do you think this rock (sandstone) has layers and this rock (conglomerate) doesn't? To help answer this question, your students could sprinkle different colours of sand into a jar of water, allowing each colour to settle before adding the next. Ask questions like: How is this like the way sandstone forms? Where might the different-coloured sand (sediments) have come from?

Linking sedimentary rocks to fossils and, indeed, news events like new fossil dinosaur finds is a good way of linking students' learning to their interests, but we must be sure of our learning aims. Is the topic about dinosaurs or sedimentary rocks?

Learning about dinosaurs

As most students seem to be naturally interested in dinosaurs, it is a popular context for science in primary schools. This is especially so for infants classes where students often come with amazing prior knowledge about dinosaurs, including accurate identification and correct pronunciation of their names. While this topic is not hands-on in the same way it can be for living things, learning about dinosaurs 'develops critical thinking and introduces animal diversity and the relations between body form and function' (Ashbrook 2010a). Ashbrook reports that students love to make dinosaur models that are bigger than themselves, and the paper includes some useful references and Internet resources. Wilkinson (2000), who works with the Queensland Museum, keeps track of fossils. She is aware that many students ask the question 'Can I find a dinosaur?', so she takes students on fossil-hunting trips. During these trips, students learn the importance of carefully recording the details related to the fossils, which are numbered and later added to the collections. Use of a field notebook can help stimulate students' curiosity, and provide a place for recording careful observations using scientific drawings and descriptive writing (Morris 2010).

Most primary-school students would not have the opportunity to participate firsthand in a dinosaur fossil hunt, but they can search the Internet to access *The Fossil Files* (see http://www.newscientist.com), which outlines The Paleobiology Database project that provides details about approximately 30 000 fossil collections. The project organisers aim to provide online information concerning every fossil that has been found on Earth. It is intended that this database will record the history of biodiversity, provide insight into the role environmental changes have played in shaping life on Earth so far, and predict their possible effect in the future (Sohn 2003).

M&E

Another secondary resource for science learning about dinosaurs is museums; many have dinosaur skeletons as exhibits. The Australian Museum website features a section on dinosaurs with lots of ideas for teachers, including photos from the dinosaur exhibit that you can enlarge and view on an interactive whiteboard, and a photo gallery of Australian dinosaurs and fossils (see http://australianmuseum.net.au/Dinosaurs-and-their-relatives). If you lack resources for teaching about dinosaurs, you can borrow *Dinosaurs in a Box* (special kits designed for schools and available from the Australian Museum). There are also interactive CD-ROM packages that deal with aspects of the Earth, including *Learn about Dinosaurs* (Wings for Learning) and *Dinosaur Database* (South Australia Department of Education and Students' Services).

ICT

Many students will have seen films that involve dinosaur reconstructions, such as *Jurassic Park* (1993), *Walking with Dinosaurs* (2000), *The Ballad of Big Al* (2000) and *Dinotopia* (2002). We can begin with the students' experiences of viewing these films to elicit questions about dinosaurs. Viewing small sections of such movies may be useful as motivators to engage students, but viewing the entire movie may degrade to nothing more than 'edutainment'.

A source of students' prior learning about dinosaurs is often the reading of books, of which there are a plethora about dinosaurs. But how good are these books? An analysis of books used in North American school classrooms on four topics, including dinosaurs, revealed they were inconsistent in matching curriculum goals and elements of scientific literacy (Schroeder et al. 2009). Hence, although many books on dinosaurs may be readily available, teachers need to carefully select those that provide accurate representations of science and the NOS.

NOS

Watson, Gregson and Webb (2003) question whether dinosaurs should be taught 'as part of science'. After analysing the data in their study, they concluded that the understandings and processes associated with the topic of dinosaurs might not sit well with science education intentions. Analysis revealed that the students in the study had poor understandings of the size of dinosaurs and their behaviour. Furthermore, the students tended to relate these understandings to their own experiences. With a partner, take some time to reflect on and discuss the science learning and teaching implications of the findings of this study, which are presented below.

The study showed that:

- students were interested in learning about dinosaurs
- the main scientific ideas students recalled were:
  - the size of dinosaurs
  - the time frame in which they lived
  - how dinosaurs reproduced
  - what dinosaurs ate.
- students do not have accurate knowledge or understandings about the scientific information they recalled
- most students did not identify additional information they wanted to know about dinosaurs
- the main source of students' information about dinosaurs was television or the movies, with school and teachers the least cited source of information.

Watson, Gregson and Webb 2003, p. 8

In contrast to the above findings, a study by Murray and Valentine-Anand (2008) describes an eight-month 'dinosaur immersion' for preschool children. The children were read books, examined artefacts, drew pictures and watched videos that helped them 'develop science-process skills, such as observation, measurement, and communication, and taught their teachers how to help young students learn and to find the answers to their own questions and love science' (ibid., p. 36). Learning involved students drawing and explaining their theories about how dinosaurs became extinct, during which three main ideas emerged:

1 'A rock fell down from the sky and crushed the dinosaurs.'
2 'The T-rex ate all the dinosaurs then there was nothing left to eat so he died.'
3 'A volcano erupted and poured on the dinosaurs and they died.'

The most popular theory was the falling-rock theory. The teachers explained that no-one really knows for sure how dinosaurs became extinct and that research is going to discover this (Murray and Valentine-Anand 2008), providing an opening for emphasising attributes of the NOS (see the Chapter 1 section 'The nature of science: how it works' on p. 5).

Apart from dinosaurs, fossils in general are frequently a source of interest for primary-age students. Fossils appeal to the young and the old, possibly due to the excitement of discovering something that has been hidden in the rocks; it's a bit like finding buried treasure! When you question kindergarten children about rocks, they may mention fossils without prompting; for example: 'If you see a rock that looks like it has a living creature in it, it is a fossil' (Ogu and Reynard Schmidt 2009, p. 13). The aspects of fossils that students find difficult to comprehend is that plants and animals that lived in the past differ to those alive today, and that they lived in quite different conditions on the Earth's surface. Again, you need to consider carefully what geological concepts and processes you want students to learn about as they interact with fossils. The Australian Museum also has a section on fossils on its website, with lots of great photo galleries (see http://australianmuseum.net.au/Fossils), and there is also a 'Museum in a Box' for fossils (see http://australianmuseum.net.au/Museum-in-a-Box). The easy-to-read book illustrated with stunning photographs of fossil plants, by Mary White (1998), would make an excellent reference book when teaching about fossils.

The following case study shows how students' interests were used as a starting point for a unit of study.

CASE STUDY

Students learning about longer-term changes in Antarctica, dinosaurs and fossils

When Mr M taught a 10-week science unit on the topic 'Change' to a Year 2 class, he knew that these students also were fascinated by eggs. He introduced the unit by exploring the Emperor penguin egg and life cycle before moving on to examining longer-term changes in Antarctica. Working in groups, the students participated in a predict–observe–explain (POE) activity by estimating and measuring relevant temperatures and then making deductions. The students accessed the Internet for up-to-the-minute information and observations of Antarctica, viewed episode 4 ('Polar Dinosaurs'; 5 minutes) of the movie *Walking with Dinosaurs* (1999), interacted with computer programs, examined maps of ancient Gondwanan scenes (White 1990) and made judgements about what changes have occurred and how these have affected Antarctica over the longer term. Observations of the students involved in these activities revealed that they were highly motivated throughout the sequence of lessons. When students were asked about their favourite activity for this topic, most indicated that they enjoyed the DVD *Walking with Dinosaurs* the best. This is not surprising given that watching videos is a common early-childhood experience and many students are visual learners. Mr M's teaching approach is consistent with a constructivist view of learning because when planning the activities, he took into account the students' prior interests. He moved them forward by encouraging participation in activities (such as exploring in the sandpit and viewing video) that he knew they were familiar with and enjoyed.

In the later lessons, the students examined fossils and talked about how these formed. The students carefully observed the preserved remains of crinoids and corals and compared these with remains of today's creatures (sea urchin tests, corals, shells). Mr M fostered the links between learning domains by setting a technology task that required the students to design and make their own fossil-digging tools, which they later used to search for plaster fossils in the school sandpit (see figures 11.15A and 11.15B). When digging up the fossils, one girl said, 'When we get all the fossils we'll be able to make something by putting everything together'. This comment reveals her thinking – that the fossils she found were parts of a more-complex whole. She naturally went the next step and combined her fossil with some of the fossils her peers had found. She was excited after they had rearranged the parts: 'We've put it together. It's a plant!'

In the next lesson, Andrew expected the students to mark where they found their fossils on a simple map of the sandpit. As an extension for those students who were capable, more-accurate mapping techniques were encouraged, such as depth of find and measurements from side borders. In following lessons, students attempted to identify individual fossils from their activity and also deduce the environment from which the fossils originally came.

The remaining lessons in the sequence were designed to build on the students' ideas about ancient Antarctica and fossils, and provided opportunities for the class to investigate dinosaurs. Students shared what they already knew about dinosaurs and were provided with posters, pictures and information on what dinosaurs actually were, what separated them from other animals, and when they lived. From there, the students were given the option as to which direction they took next. Andrew thought that some students might like to examine the rise and change of different plants over the period of dinosaur extinction to the present (65 million years), the development of birds from reptiles, the speciation of marsupials, or some of the more-recent geomorphologic changes in the nearby coastal areas. Students could refer to *The Age of the Dinosaurs* (Catchpole 2003) and *Dinosaurs of Australia and New Zealand and Other Animals of the Mesozoic Era* (Long 1998).

Reflect on the case study by identifying what constructivist learning and teaching characteristics are present or could be inserted into the sequence.

FIGURE 11.15A Students fossil digging in the school sandpit

FIGURE 11.15B A tool the children made to uncover fossils

Metamorphic rocks

So what of metamorphic rocks? What does the word itself mean and where did it arise from? Once formed, igneous and sedimentary rocks that are subsequently buried deep in the Earth's crust, and so subjected to heat and pressure, will change in form and so are called 'metamorphic' ('changed form') rocks; for example, sandstone changes to quartzite and limestone forms marble. Interestingly, research into secondary students' understanding of science concepts revealed common confusion between the metamorphic processes of rock formation and metamorphosis in biology (King 2008). If you use the word 'metamorphosis' when teaching primary students about animal life cycles, this is something to keep in mind. For adults, these two processes are totally different, but for children the words are so similar a connection is logical.

What observable features give clues as to the origin of metamorphic rocks? Metamorphic rocks usually show crystals (formed by heat) that are lined up (due to directed pressure) and so they are the ones in the 'Crystals (aligned)' column. Metamorphic rocks are crystalline and often appear as wavy stripes and bands (see Figure 11.16). The properties of

FIGURE 11.16 Metamorphic rock

metamorphic rocks vary widely; for example, splits along flat, hard surfaces (e.g., slate); shows a lustrous, hard, attractive surface (e.g., granite); can be carved or shaped easily (e.g., marble, alabaster). The chemical composition of limestone, dolomite, rock salt and coal is what makes them useful. As you seek to extend your knowledge about rocks, try to keep in mind specific types of rocks you have seen used in various settings and how these rocks have been used. This helps you to set your learning in a familiar context – an excellent way to enhance learning.

From Chris' teaching experience, hands-on investigations using familiar materials helped to develop students' understanding of the processes involved in rocks being altered by the two main metamorphic agents of heat and pressure. A few examples are outlined below:

- heating soft lollies (e.g., red and green frogs) in the sun and laying them on top of each other before squishing them together. Cutting through the model rocks shows how they have been changed, revealing wobbly rather than straight layers like in sedimentary rocks
- subjecting one end of a layer cake to pressure and discussing the changes that occurred. This shows that parts of rock beds can be metamorphosed while others are left unchanged
- melting large, layered marshmallows in a fire and observing the margin between the heated end and the other end. Discuss what happened to the heated end. What type of rock is it now and what might cause this to happen? An issue here is that children have no trouble believing that lollies can melt but will still find it hard to believe that rocks (let alone huge areas of rock) can melt.

Discuss the limitations of these 'modelling' activities with students so they understand these are representations of the real process and as such are not exactly what happens (see Chapter 1).

Note that any activities involving food must be strictly controlled to avoid problems with dietary requirements and food allergies.

Dynamic changes in planet Earth

Some barriers to students' understanding about rock formation stem from the view that the formation of rocks is not a 'normal event' of the changing Earth, hence they think an unusual process such as a volcano or an earthquake is needed for this to happen (Kortz and Murray 2009). Natural disasters are events that impact on humans and their built environment. Some earthquakes can cause enormous damage, even loss of life, for example the 2011 Christchurch, New Zealand earthquake, whereas others are barely felt. As the Earth's continents are always moving, the edges of these plates grind together and earthquakes can occur. Students (and adults) who view Earth as static and unchanging can have difficulty in understanding that earthquakes are caused by moving continental plates. In order to facilitate an understanding that rocks form as a result of changes to the Earth, there needs to be a focus on what rocks can tell us about Earth in the past (Kortz and Murray 2009).

Again, primary children at first may be interested in the violent, catastrophic nature of earthquakes, but from a geological perspective they are a natural part of living on the planet. Research into adults' and students' ideas about earthquakes shows some common trends:

- Earthquakes are always bad, cause destruction, injury and confusion, and why they happen is unknown (Year 3 students; Hubenthal, Braile and Taber 2008). Earthquakes are a bad event, mostly in relation to humans (a majority of K-2, Year 6 and Year 8 students (37 out of 40) (Lacĭn Sĭmsek 2007). In the latter study, no scientific explanation for the cause of earthquakes was provided and many students (regardless of age) thought it was an 'act of God'.
- The movement of tectonic plates is associated with earthquakes (Hubenthal, Braile and Taber 2008). Teachers' lack of knowledge of Earth's sections and composition contributes to a lack of understanding of the causes of earthquakes (Hubenthal, Braile and Taber 2008; King 2000).

The above findings suggest that, when teaching about earthquakes, the two focuses should be their cause and addressing the *incorrect* idea that they are always devastating. The major alternative conception to be addressed is that it is not simply plate movement (the Earth's tectonic plates are moving all the time) but plate movement in different directions that causes earthquakes. When this happens, there is shearing, shuddering and tearing, and the massive forces that are created when the plates eventually slip send shock waves through the planet. The *Primary Connections* initiative includes a curriculum resource unit called 'Earthquake Explorers' that incorporates factual and narrative texts (AAS 2009).

A positive 'spin' on earthquakes is to focus on the science and technology of building design for human protection in earthquake-prone areas. Participation in a technological activity, such as making and testing buildings designed to withstand earthquakes, could assist students in making meaningful links between science, mathematics, engineering and technology. Two projects involving Year 5 students helped them learn about how building codes protect people who live in earthquake-prone areas (Gilstrap, Sheldon and Schimmoeller 2010) and about structures that isolate the building from the shaking ground (Maltese 2009); these projects also enabled them to work like scientists and learn in the context of real-world situations.

You could extend students' ideas about earthquakes by looking at linked occurrences, such as a tsunami. As a teacher you can provide activities to demonstrate how a tsunami works and develop understandings by addressing questions such as:

- What is a tsunami?
- How do you tell a tsunami is going to happen?
- Where is a tsunami likely to happen?
- Can you stop a tsunami?
- Do you have to be directly in line with a tsunami to be in danger?
- How does a shoreline affect a tsunami?
- Would it be safe to be in a bay when a tsunami strikes?
- What is the seiche effect?

Geological processes and geological time

Why were people so slow to begin to investigate and understand the everyday workings of the Earth's crust? Why is it that students who are fascinated by rocks, stones and crystals seem to have made so little progress in their thinking about where these things came from, how they got there and how long it all took to happen? It seems reasonable to assume that we do not seek such explanations because they only make sense if viewed in the context of events that have occurred over a timescale so vast that it is calibrated in millions of years. Who among us has any real understanding of the claim that the Earth is 4.6 billion years old? We can reflect with curiosity about the oldest rocks making up the crust of continents being over 4000 million years old and the oldest rocks making up the crust under the oceans being less than 300 million years old because we can compare 4000 and 300, but when we try to conceptualise what a million years is, we have great difficulty.

SSI

Watson, Gregson and Webb (2003) found that eight- to 12-year-old students (n = 30 across three schools) had poor understandings of geological time frames. In fact, their interviews revealed that the students' statements demonstrated a limited view of any large time frames. This is not surprising given that research with adult geology students has shown they think a long time is 'about a thousand years' (Kortz and Murray 2009). Yet some kindergarten students will tell you, 'Rocks are very old. Some rocks are millions of years old' (Ogu and Reynard Schmidt 2009). Young children may know that rocks can be millions of years old but if you ask them how old their teacher is, they will probably say about 100, hence they lack understanding of time scales. At the primary level, this is a fact we just have to keep in mind.

Integrating learning about rocks and minerals across the curriculum

Integration and learning – for example, blending literacy and science – was discussed in Chapter 1 (see the section 'Science and literacy' on p. 38). Here, examples are outlined of how learning about rocks and minerals can engage and motivate students through bringing science together with art and literacy skills:

- Reading-to-learn in science enabled students to learn to read information texts critically and develop an understanding of science concepts. An active reading process, where students turn headings into questions and annotate the text as they find the answers, was used to develop Year 6 students' understanding of how mechanical weathering changes rocks. Such learning experiences can aid the development of higher-order thinking skills and provide opportunities for assessing students' comprehension of the science concepts involved (Wardrip and Tobey 2009).

- A novel way of integrating art and literature with science involved analysing, improving and creating cartoons to communicate science content. Exciting texts on minerals, crystals, volcanoes, earthquakes, glaciers, caves, dinosaurs and other fossils relating to students' interests were used to gain information that was then used to make original cartoons communicating science explanations. This approach 'was successful in motivating underachieving students to read science books and practice communicating the information by drawing humorous cartoons' (Sallis, Rule and Jennings 2009, p. 27).
- Kindergarten children were supported in inquiry-based learning investigations of rocks and sand which integrated learning across the curriculum. Children drew and painted while comparing and contrasting rocks and sand; they also designed natural compositions with small rocks and stones; used descriptive language, created and re-created landscapes, wrote stories, built rock museums, sang, danced and designed. This range of strategies catered for different learning styles (Ogu and Reynard Schmidt 2009).
- Food items were used to make 'model rocks' and help special-needs students to develop meaningful understandings of the words 'metamorphic', 'sedimentary' and 'igneous'. The use of constructivist principles challenged these students but also provided them with concrete simulations of the different rock-formation processes which supported their learning (Prestwich, Sumrall and Chessin 2010).

Another example on integrating literacy skills involved students making field observations, creating a journal entry and drawing field sketches, and then using these to construct poems. The authors stated that 'field note poetry forges interdisciplinary connections by combining knowledge of poetry structures and science content' that enables students to 'qualitatively express developing knowledge' that 'facilitates learning and engage students' (Jackson, Dickinson and Horton 2010, p. 31).

Science as a human endeavour

This final section focuses on why this topic is significant in learning about planet Earth. As you will see, there are many opportunities related to 'Our Planet Earth' to develop students' understanding of science as a human endeavour, such as the historical and human construction of science ideas and the contributions of scientists to society. Some examples of these are developed below.

Science and culture: historical and human construction of science ideas

For as long as humans have been on the planet, they have wondered about natural events and posed ideas to explain them. Considering students' intrinsic interest in earthquakes and volcanoes, they could be encouraged to research how early thinkers explained natural disasters. This could help students to see how humanity (science) constructs mental models to understand its world and to trace how such models have changed over time and hence help them to appreciate some attributes of the NOS.

Up until the 18th century, philosophers sought natural explanations for individual phenomena, such as volcanoes, fossils and meteorites. In the past, change was linked to catastrophic events like eruptions, earthquakes and storms because people believed they generated sufficient power to make these huge changes possible. It was not until 1785, when James Hutton read his dissertation 'concerning the system of the Earth, its duration and stability' (Hutton 1788) to the Royal Society of Edinburgh, that the idea that the sediments composing sedimentary rocks may have been derived from weathering and erosion of the rocks they overlie was taken seriously. Fifty years later, Charles Lyell expanded the idea of rock recycling into the science of geology.

NOS

An interesting example of how the personalities and egos of scientists can sometimes get in the way of scientific discoveries is how two wealthy rival scientists became obsessed with being the first to discover new dinosaur species. The tale of these two men and their antics demonstrates how conflicts and errors in science are overcome and corrected by the scientific community (Clary, Wandersee and Carpinelli 2008).

Contributions of scientists: geology-related careers

Children's learning in primary school can have a long-lasting effect and contribute to later career choices, though geology is still an area where women are underrepresented. Contributions of science to the community may stimulate girls' interests in geoscience-related careers, as they tend to develop a social conscience early on in life.

The following are two examples of ways that soil scientists have contributed to society:

- How much water is needed to produce food? Scientists working for the Australian Commonwealth Scientific and Industrial Research Organisation (CSIRO) used precision weighing systems to reveal that 1 kilogram of oven-dried wheat grain requires 715 to 750 litres of water for production, compared with 1550 litres for paddy rice and 1650 to 2200 litres for soya beans. This information may encourage students to ask questions about issues relating to soil and water use.
- CSIRO's Dr Rob Fitzpatrick and his soil scientists teamed up with police forces, governments and other groups to help fight crime and solve environmental disasters. This crime-solving success led to the establishment of The Centre for Australian Forensic Soil Science (CAFSS). CAFSS is the first worldwide network of forensic and soil scientists. It has many partners and can be found at CSIRO Land and Water in Adelaide, South Australia. So far, CAFSS has helped out on more than 50 criminal and environmental forensic investigations. So, criminals beware! These scientists will get the dirt on you! For more information, visit the following link at the CSIRO website: http://www.csiro.au/science/The-Dirt-On-Dirt.html.

The influence of science: the effect of science and technology on our lives – mining

Mineral resources and products derived from them are vital to the economy of many countries, including Australia. Whole communities are built around the mining industry, with towns throughout history having been formed and deserted in relation to mining activities. The underground rescue of the trapped miners Brant Webb and Todd Russell after the Beaconsfield mine disaster in Tasmania, Australia in May 2006 raised local awareness of the dangers of working in mines and the effect such tragedies have on the local community. A recent issue of *Primary Science* (no. 112) featured articles on industrial alchemy, which the editors suggest may be a way of linking the sciences in students' learning, as modern industries work across science disciplines. Articles in this issue discuss lead mining (Peacock 2010), tin mining and smelting (Gardner 2010), today's alchemists (Serret 2010), iron manufacture (Weatherly 2010), students' perceptions of industry (Waller 2010) and civil engineering in primary schools (Brown and Strong 2010). This issue of the journal could make a useful resource for comparing and contrasting industry in your local area.

Many people do not realise the extent to which scientific research and development processes underpin all mining activity, producing the scientific and technological knowledge necessary to support and advance this huge industry. As our knowledge of our impact on the environment continues to increase along with new scientific discoveries, leading to a reduction in our reliance on fossil fuels, mining priorities are changing. Primary-age children in the future will be involved in making decisions about mining, the environment and alternative energy sources that are currently controversial.

Whether mining occurs close to or away from major cities, it inevitably disturbs delicately balanced ecosystems, including arid and fragile environments. Students can research legislation (including environmental management plans and rehabilitation requirements) that has been put in place to reduce environmental impact.

As the world's supply of fossil fuels is limited and is being used up at a fast rate by developed countries, the search for alternative materials and ways to harness energy is happening in earnest. Advances in scientific research, including solar-cell technology, biofuels, wind turbines, wave generators and nuclear power, may provide alternatives and lead to changes in current mining practices.

The controversial nature of uranium mining and export due to the possibility of other countries using it to make nuclear weapons' (Lang 2004, p. 10), and debate over whether it is a viable solution, continue. The recipient of the 2006 Australian Peace Prize, Helen Caldicott (2006), argues convincingly that nuclear power is not the solution to global warming or declining fossil fuels. In contrast, environmental scientist Tim Flannery (2006) makes strong claims that nuclear power could offer Australians a cleaner conscience, as currently they are the worst greenhouse gas polluters in the world. This is an excellent topic for engaging students in socioscientific issues in the classroom. An example of how such an issue from the Earth sciences was implemented at the primary level is described by Dolan, Nichols and Zeidler (2009). It related to mechanical erosion and its impact on beach erosion in Florida. Students participated in a hands-on activity to help them understand the concept of mechanical erosion before debating the moral question of whether the country should 'continue to purchase beach sand to fix the beaches or use crushed glass as a new alternative?' Students' arguments included the effect on nesting sea turtles and human safety, raising questions such as 'What happens if I bury someone in the sand? and 'What happens if it gets in my eyes?' (ibid., pp. 20–1).

Socioscientific issues: cultural constructions of Earth and Indigenous views of mining

The majority of understandings related to a changing Earth put forward in this chapter are from a perspective that has sometimes been labelled 'Western Modern Science' (WMS; see the Chapter 1 section 'Science and culture: Indigenous and cultural knowledge' on p. 42). As teachers, we should also be aware that there are science topics in which an Indigenous perspective may be different to WMS. Indigenous people often hold views of the origins of the Earth that are different to the accepted scientific view. For example, in the Sioux culture it is believed that, in the beginning, everything was in the mind of Wakan-Tanka. All things were to exist only as spirits that moved about in space seeking somewhere to manifest themselves. Finally, they came to the Earth after travelling to the sun, where it was too hot for creation to begin. At that time, the Earth was covered with water and there was no dry land or life. Then a great burning rock rose up, making the dry land appear and forming clouds from steam so life could begin. The rock is called Tunka-shila, 'Grandfather Rock', which is the oldest rock. Therefore, rocks must be respected. This story shows awareness of basic geological concepts: 'As in this first story, science tells us that over four billion years ago, the hot, molten rock of the Earth's crust gave off water vapour that formed clouds and rain' (Caduto and Bruchac 1989, p. 57).

We must recognise that, for Indigenous people, rocks have a special meaning. Rocks are considered to be guides and hold stories about the many people who have rubbed the rock. Indigenous elder Frans (Sinatra and Murphy 1999, pp. 19, 23) talks about the land and rocks: 'See the face in that cloud, the face in that tree, the rock you can see him talking? Nobody owns the land ... The black sands along the trail are rich in minerals: Our function is to look after it. We're only passing through'.

Australian Indigenous tribal art tells stories about the land, creation and supernatural transformations (Isaacs 1984). By looking at rock art from various parts of Australia, we can glimpse how the Indigenous communities interacted with the land. People from different areas of Australia created different styles of rock engraving. The rock paintings, particularly in caves in New South Wales and Western Australia, resemble creation ancestors. Many paintings are also powerful communicators of Australian Aboriginal history, including the massacres that occurred in the past. Arnhem Land's rock paintings provide historical evidence of changing weapons and animal species at different times. Australian literature, such as *The Timeless Land* (Dark 1941), describes the conflict between the Aborigines and white settlers in the earliest days of European settlement in Australia. Dark's historical novel was the first attempt to recall events of the failed conquest of the land from the perspective of Indigenous people.

When teaching Earth science, it is important that, as teachers, we have an appreciation of the way Indigenous people connect with the land. The significance of the land for Indigenous people is that the land is the giver of all life, shelter, food and protection, and is the dwelling place of the spirits. The spirits are associated with totem animals that must be looked after, as they are regarded as teachers. This means that although animals are a source of food, there are certain sacred sites that belong to a particular animal of the Dreamtime. The Pannemukeer people of Tasmania value their tribal totem, the possum, for its diversity, agility and versatility. When the people experience hard times, they sing to the possum for help. The possum enters the Dreaming because it is connected to the stars. Experiences of possums getting into the roofs of homes and causing what non-Indigenous urban students might recall as havoc would be a contrast to the view of possums held by some Indigenous students.

Because of their special relationship with the land, Indigenous peoples have identified specific issues of concern that relate to the process of mining. Using library and archival records and, if possible, by gaining information directly from community elders, try to establish an understanding of the reasons for their concern. Do some yarning with several Indigenous people and listen to their stories. In so doing, honour the cultural protocols for that particular community.

In Australian Indigenous communities, the elders own the knowledge, which is not readily available to non-Indigenous people. Some elders do speak out, such as Eileen Wani Wingfield, a Woman Senior Elder from Coober Pedy, South Australia who opposes any form of mining, but particularly uranium mining and nuclear activities in Australia. In resisting change due to mining, she describes the detrimental effects on water supply in the following way:

> They let off two nuclear bombs – Emu Junction in the fifties and Maralinga in the sixties. Now they want to bring the nuclear dump here. It's too close to the underground water – that's what we don't want: That next generation are going to have a hard time with the land if the government don't pull up. What I'm saying is, Mother Nature put life here to hunt, gather, be a part of. This mining lot came in and ruin everything. We want nothing to do with this way. Mother Nature is Mother Nature. If the whitefella stands still, not blowing things up, not mining the Earth, the land will stay as it is. When it's opened up, people will suffer – like they are. Whitefellas never stop and think, they're really spoilt.

Wingfield, in McConchie 2003, p. 21

Science and culture interact to influence personal and community choices. The above is an example of the consideration of Aboriginal cultural perspectives on decisions about resource use and sustainable management of the environment (ACARA 2011).

Summary

Our planet Earth as we know it today has been shaped and changed by continuous geological processes. Soil profiles, the study of rocks and minerals, and the analysis of fossil-bearing rock strata can be used by Earth scientists to interpret the Earth's history. The use of scientific evidence to piece together puzzles explaining the past and predicting future changes to our planet represents a stimulating part of the primary curriculum that should not be overlooked.

In this chapter, geology concepts and processes have been used as the basis for discussing constructivist learning experiences through which your students can explore their physical surroundings and develop a scientific understanding of the dynamic processes at work on our planet Earth. Throughout the chapter, ways to elicit prior knowledge and develop primary school students' understanding of ideas and topics, including soils, rocks and minerals, rock formation and change, fossils, volcanoes and earthquakes, were discussed – to review your understanding of Earth science terms and concepts, refer to Appendix 11.2 (see p. 495). Key findings from research into primary students' understanding of the above topics were also briefly included, where available. Pedagogic strategies, especially hands-on investigations, careful observations and questioning that encourage students to think geologically were introduced – for a review of the key constructivist learning and teaching strategies discussed, refer to Appendix 11.1 (see p. 495). Examples of how this topic exemplifies science as a human endeavour were presented at the end of this chapter.

Earth scientists are intrigued by the secrets held in rock strata and fossils, marvel at the tremendous forces unleashed by volcanoes and in earthquakes, and are dazzled by the intricate beauty of crystal structures. With the right stimulation, your students can also become captivated with learning about our planet Earth.

Concepts and understandings for primary teachers

Below are most of the key conceptual ideas and understandings related to 'the changing Earth', with which a primary teacher should be familiar. You should read and interpret this list with an appreciation of its limitations, as described in the section 'Concepts and understandings for primary teachers' in Chapter 1 (see p. 47).

The major topics that primary curricula would refer to in relation to this chapter are:

- soils – their nature and origins, and the identification of soil types
- how rocks form and their classification
- volcanoes, tsunamis and earthquakes (how they are measured)
- change as a result of human impact (e.g., salinity problems).
- Concepts and understandings related

to these topics include:

- The Earth is covered with rocks, soil, ice and water.
- Rocks are made up of one or more minerals that can form crystals.
- Natural rocks are made in different ways.
- Some natural rocks form quickly while others take a long time to solidify.
- Igneous rocks arise from volcanic action. Because granite cooled below the Earth's surface, it has large crystals compared to basalt, which cooled above the Earth's surface. Pumice is cooled lava froth.
- Rocks can vary in shape, colour, texture, mineral content and markings.
- Some rocks are harder than others; varying hardness determines their use.
- Rocks slowly change by wearing away (weathering).
- Rocks can be eroded and form soil that can be sand, clay or mud.
- The eroded material can be deposited in layers to form sedimentary rocks, such as sandstone, limestone and mudstone.
- Rocks that we find at a particular place may have been made elsewhere.
- When sedimentary rocks are subjected to heat and pressure, they form metamorphic rocks, such as marble, quartzite and slate.

- Some rocks contain prints or filled parts of old plants and animals (fossils).
- The mining industry extracts minerals from rocks and changes the environment.
- Crumbled rocks form part of soil.
- Soil scientists describe soils as sand, clay or loam according to whether they fall apart (sand), hold together in blocks but do not roll into a thin worm (loam), or roll out well to a thin worm (clay) when moistened in your hand.
- Soil can be made up of crumbled rock, air, water and remains of dead plants and animals.
- Earthquakes are ground-shaking events, the result of strain built up in rocks due to movements in the Earth. When the rocks break, energy is suddenly released, causing vibrations or tremors called 'earthquakes'.
- Most earthquakes occur below the boundaries of well-defined zones or plates.
- Seismologists measure earthquake magnitude using instruments called seismographs.
- Volcanoes are gaps in the Earth's crust through which hot rocks and lava spill out onto the Earth's surface, and which shoot gases and dust into the upper layers of the atmosphere. Volcanoes can be active or inactive.
- A tsunami is a destructive wall of water driven by an undersea tremor. It is a giant wave caused by an undersea earthquake.
- The impact of human activity on the Earth is producing environmental problems, such as salinity, which is the result of an elevated watertable.
- Plants that have the ability to absorb water are called 'halophytes'.

Acknowledgement

We wish to acknowledge that the case study about dinosaurs and fossils was provided by, and used with the permission of, Andrew Mains.

Search me! science education

Search me!

Explore **Search me! science education** for relevant articles on our planet Earth. Search me! is an online library of world-class journals, ebooks and newspapers, including *The Australian* and the *New York Times*, and is updated daily. Log in to Search me! through http://login.cengage.com using the access code that comes with this book.

KEYWORDS

Try searching for the following terms:

- Rocks and minerals
- Children and scientists

Search tip: **Search me! science education** contains information from both local and international sources. To get the greatest number of search results, try using both Australian and American spellings in your searches, e.g. 'globalisation' and 'globalization'; 'organisation' and 'organization'.

Appendices

In these appendices you will find material related to our planet Earth that you should refer to when reading Chapter 11. These appendices are provided in full below and can also be found on the student companion website. Log in through http://login.cengage.com using the access code that comes with this book.

Appendix 11.1 Teaching strategies in primary Earth science

ONLINE

The key constructivist learning and teaching strategies employed in this chapter have included:

- valuing the prior knowledge of learners
- exploring aspects of the topic using concrete materials
- having learners share their observations, questions and tentative explanations
- generating questions for investigation
- devising fair tests to investigate these questions
- sharing the ideas and explanations that arose from the investigations
- introducing ideas selected from previous scientific work
- assessing the ideas produced and the learning that you believe occurred.

Record three or four specific strategies used in this chapter that you believe are important for you when you teach Earth science to primary school students. Give reasons for your choices.

Appendix 11.2 Your understanding of Earth science

Construct a concept map using as many as you can of the key Earth science terms used in this chapter. Twenty are listed below as suggestions, but add any others you wish to use. Pay particular attention to developing propositions that make cross-links between the longitudinal (vertical) propositions.

crystals	volcanoes	rocks	minerals	grains
earthworms	crust	water	soils	(soil) texture
igneous	change	(soil) profile	sedimentary	time
earthquakes	air	magma	leaching	plants

When you have completed your concept map, find a colleague with whom you can compare and discuss the concepts you each hold about 'Our Planet Earth'.

References

Ashbrook, P. 2010a. If you were a dinosaur. *Science and Children*, 47 (5), pp. 19–20.

_____ 2010b. Building with sand, *Science and Children*, 47 (7) pp. 17–18

Australian Academy of Science (AAS). 2009. 'Earthquake Explorers'. *Primary Connections*. Canberra: AAS.

Australian Curriculum, Assessment and Reporting Authority (ACARA). 2011. K-10 Australian Curriculum: Science. Available at http://www.australiancurriculum.edu.au/Science/Curriculum/F-10 (accessed April 2011)

Baird, J. & Northfield, J. 1992. *Learning from the PEEL Experience*. Melbourne: Monash University Press.

Brass, K. & Duke, M. 1994. Primary science in an integrated curriculum, in P. Fensham, R. Gunstone and R. White (eds). *The Content of Science*. London: Falmer Press.

Brown, M. & Strong, A. 2010. Civil engineering in primary schools. *Primary Science*, 112, pp. 25–8.

Bulunuz, N. & Jarrett, O. 2010. The effects of hands-on learning stations on building American elementary teachers' understanding about Earth and space science concepts. *Eurasia Journal of Mathematics, Science & Technology Education*, 6 (2), pp. 85–99.

Caduto, M. & Bruchac, J. 1989. *Keepers of the Earth: Native Stories and Environmental Activities for Students*. Saskatoon, SK, Canada: Fifth House.

Caldicott, H. 2006. *Nuclear Power Is Not the Answer to Global Warming or Anything Else*. Carlton: Melbourne University Press.

Catchpole, H. 2003. The age of the dinosaurs. *The Helix*, 88, pp. 10–15.

Clary, R., Wandersee, J. & Carpinelli, A. 2008. The great dinosaur feud: Science against all odds. *Science Scope*, 32 (2), pp 32–40.

Dal, B. 2007. How do we help students build beliefs that allow them to avoid critical learning barriers and develop a deep understanding of geology? *Eurasia Journal of Mathematics, Science & Technology Education*, 3 (4), pp. 251–69.

_____ 2009. An Investigation into the understanding of Earth sciences among students' teachers. *Educational Science Theory and Practice*, 9 (2), pp. 597–606.

Dark, E. 1941. *The Timeless Land*. London: Collins.

Dolan, J., Nichols, B. & Zeidler, D. 2009. Using socioscientific issues in primary classrooms. *Journal of Elementary Science Education*, 21 (3), pp. 1–12.

Dove, J. 1998. Students' alternative conceptions in Earth science: Review of research and implications for teaching and learning. *Research Papers in Education*, 13, pp. 183–201.

Eberle, F. & Keeley, P. 2008. Formative assessment probes. *Science and Children*, 45 (5), pp. 50–4.

Flannery, T. 2006. No nukes. *Good Weekend*, 5 August, pp. 22–6.

Ford, D. 2005. The challenges of observing geologically: Third grader's descriptions of rock and mineral properties. *Science Education*, 89, pp. 276–95.

Fries-Gaither, J. 2008. Common misconceptions about rocks and minerals. *Beyond Penguins and Polar Bears*, (6), pp. 14–19. Available at http://beyondpenguins.nsdl.org/issue/column.php?date=September2008&departmentid=professional&columnid=professional!misconceptions (accessed 8 June 2010).

Fuller, M. 2002. Do you know? What is the difference between a rock and a mineral? *Investigating*, 18 (3), p. 31.

Gardner, R. 2010. Cornish tin mining and smelting. *Primary Science*, 112, pp. 10–13.

Gilstrap, T., Sheldon, P. & Schimmoeller, P. 2010. Shake it up. *Science and Children*, 47 (6), pp. 32–5.

Happs, J. 1982. Some aspects of student understanding of rocks and minerals. University of Waikato: Science Education Research Unit Working Paper.

_____ 1985. Regression in learning outcomes: Some examples from earth sciences. *European Journal of Science Education*, 7, pp. 431–3.

Harlen, W. 2009. Teaching and learning science for a better future. *School Science Review*, 90 (333), pp. 33–42.

Hayes, D., Symington, D. & Martin, M. 1994. Drawing during science activity in the primary school. *International Journal of Science Education*, 16 (3), pp. 265–77.

Hubenthal, M., Braile, L. & Taber, J. 2008. Redefining earthquakes. *The Science Teacher*, 75 (1), pp. 32–6.

Hutton, J. 1788. *Theory of the Earth*. Transcripts of the Royal Society, Edinburgh, 1 (2), pp. 209–304.

Isaacs, J. 1984. *Australia's Living Heritage: Arts of the Dreaming*. Sydney: Lansdowne Press.

Jackson, J., Dickinson, G. & Horton, D. 2010. Rocks and Rhymes. *The Science Teacher*, 77 (1), pp. 27–31.

King, C. 2000. The Earth's mantle is solid: Teachers' misconceptions about the Earth and plate tectonics. *School Science Review*, 82, pp. 57–64.

_____ 2008. Geoscience education: An overview. *Studies in Science Education*, 44 (2), pp. 187–222.

Kortz, K. & Murray, D. 2009. Barriers to college students learning how rocks form. *Journal of Geoscience Education*, 57 (4), pp. 300–15.

Kusnick, J. 2002. Growing pebbles and conceptual prisms: Understanding the source of student misconceptions about rock formation. *Journal of Geoscience Education*, 50 (1), pp. 31–9.

Lacĭn Sĭmsek, C. 2007. Children's ideas about earthquakes. *Journal of Environmental & Science Education*, 2 (1), pp. 14–19.

Lang, R. 2004. *Resources: Australian Focus on Issues*. Alexandria: Watts Publishing Australia & New Zealand.

Lark, A., Kramp, R. & Nurnberger-Haag, J. 2008. My Pet Rock. *Science and Children*, 45 (5), pp. 24–7.

Long, J. A. 1998. *Dinosaurs of Australia and New Zealand and Other Animals of the Mesozoic Era*. Sydney: University of New South Wales Press.

Maltese, A. 2009. Shake, rattle and hopefully not fall. *Science and Children*, 46 (8), pp. 40–3.

Mason, D. 2002. Volcanoes: Molten rock at the Earth's surface. *Investigating*, 18 (2), pp. 32–5.

McConchie, P. 2003. *Elders*. Cambridge: Cambridge University Press.

Morris, K. 2010. Thinking like a geologist. *Primary Science*, 114, pp. 30–2.

Murphy, S. 2000. *Dave's Down-to-Earth Rock Shop*. New York: HarperCollins.

Murray, M. & Valentine-Anand, L. 2008. Dinosaur extinction, early childhood style. *Science and Children*, 46 (4) pp. 36–9.

Ogu, U. & Reynard Schmidt, S. 2009. Investigating rocks and sand: Addressing multiple learning styles through an inquiry-based approach. *Young Children*, 64 (2), pp. 12–18.

Paige, K. & Chartres, M. 2002. Using sensory trails to develop environmental awareness. *Investigating*, 18 (2), pp. 26–9.

Pattie, I. 1993. Soil. *Investigating*, 9 (2), pp. 9–12.

Peacock, A. 2010. Pipes, petrol, paint and pewter: The rise and fall of lead. *Primary Science*, 112, pp. 5–8.

Pfeiffer, T. 2003. Mount Etna's ferocious future. *Scientific American*, 288 (4), pp. 40–7.

Pigdon, K. & Woolley, M. 1989. *Earthworms*. Toronto: Modern Curriculum Press.

Piotrowski, J., Mildenstein, T., Dungan, K. & Brewer, C. 2007. The radish party. *Science and Children*, 45 (2), pp. 41–5.

Press, F. & Siever, R. 1982. *Earth* (3rd edn). San Francisco: W. H. Freeman and Company.

Prestwich, D., Sumrall, J. & Chessin, D. 2010. Science rocks! *Science and Children*, 47 (7), pp. 86–8.

Pyle, C. 2003. A first look at rocks. *Investigating*, 19 (1), pp. 30–1.

Russell, T., Longden, K. & McGuigan, L. 1993. *Rocks, Soil and Weather* (Primary SPACE project). Liverpool: Liverpool University Press.

Sallis, D., Rule. A. & Jennings, E. 2009. Cartooning your way to student motivation. *Science Scope*, summer, pp. 22–7.

Schroeder, M., McKeough, A., Graham, S., Stock, H. & Bisanz, G. 2009. The contribution of trade books to early science literacy: In and out of school. *Research in Science Education*, 39, pp. 231–50.

Serret, N. 2010. The alchemist of today. *Primary Science*, 112, pp. 15–16.

Sinatra, J. & Murphy, P. 1999. *Listen to the People, Listen to the Land*. Melbourne: Melbourne University Press.

Sohn, E. 2003. The fossil files. *New Scientist*, 179 (2409), pp. 32–5.

Suzuki, D. & Vanderlinden, K. 1999. *You are the Earth*. St Leonard's, NSW: Allen & Unwin.

Symington, D. 1986. Upper primary students' knowledge of rocks. *Investigating*, 2 (1), pp. 9–11.

_____, Biddulph, F. & Osborne, R. 1987. Research into students' prior knowledge: Some implications for teaching science? *Investigating*, 3 (3), pp. 14–15.

Waller, N. 2010. It's not all flat caps and chimneys, you know! *Primary Science*, 112, pp. 21–3.

Wardrip, P. & Tobey, J. 2009. How does mechanical weathering change rocks? *Science Scope*, 32 (5), pp. 25–9.

Watson, K., Gregson, R. & Webb, C. 2003. A dig at dinosaurs: Real science education or not? Paper presented at the 34th Annual Conference of the Australasian Science Education Research Association, Melbourne, July.

Weatherly, M. 2010. Coalbrook: The birthplace of industry. *Primary Science*, 112, pp. 17–19.

White, M. E. 1990. *The Flowering of Gondwana*. Princeton, NJ: Princeton University Press.

_____ 1998. *The Greening of Gondwana* (3rd edn). East Roseville, NSW: Kangaroo Press.

White, R. & Gunstone, R. 1992. *Probing Understanding*. London: Falmer Press.

Wilkinson, J. 2000. Dinosaur hunting in Queensland. What many students would like to do ... find a dinosaur: What is it really like? *Investigating*, 3 (16), pp. 15–17.

Zeegers, Y. 2002. Is a rock the same thing as a stone? *Investigating*, 18 (2), pp. 14–16.

12

Weather and our environment

by Christine Preston and Graham Crawford

Introduction

Weather happens all around us every day (whether we like it or not); in fact, it is most notable when we don't like it! At other times the different elements of the weather are simply taken for granted, and why not? After all, we can't change it – or can we? Unfortunately, we can and we are, often without realising it. Our weather is so explicably linked to our environment that things humans do that change the environment (in the long term) can also change the weather. Environmental change is now a significant socioscientific issue and primary teachers need to be able to engage students in learning about its causes and consequences.

We all know that air and water, driven by the vast energy output of the sun, produce the rain, winds, clouds and temperatures, which we sum up as 'the weather'. We also know how much our way of life is influenced by these phenomena. In fact, weather so dominates our thinking about the Earth's atmosphere that we can sometimes forget that it also contains the air we breathe and provides the water we drink. The atmosphere also provides the blanket that prevents the Earth's surface being unbearably hot or cold and protects us from ultraviolet and other high-energy rays that would rapidly destroy most of Earth's life forms if they struck the surface. In view of all that the atmosphere does for us, it really is surprising that we don't treat it with greater respect. Learning to recognise our environment for what it is – a life-support system for our planet – requires the development of deep understanding of scientific concepts.

This chapter is divided into the following four sections, each framed around a main question. Firstly, what are teachers' and students' ideas about weather? Students' ideas about evaporation and condensation, cloud formation and rain, weather and climate, the ozone layer, global warming and the greenhouse effect are discussed. Students' questions about weather are also raised to help determine *what* to teach about weather.

Secondly, how can you develop students' ideas about weather using a constructivist and representational approach? Various techniques and strategies, consistent with these perspectives, provide examples of *how* to teach students about weather. Ways to build on students' existing knowledge, challenge their ideas, link science to everyday situations and make use of resources to aid the explanation of weather-related science concepts are presented.

Thirdly, where should teaching and learning about weather be situated? Teaching outdoors, integrating learning across subjects and linking knowledge and understanding about weather to related topics of environmental change are discussed as ways to engage students and help them see the relevance of science to their everyday lives. This section emphasises *where*, physically, structurally and conceptually, students' learning about weather and our environment should take place.

Fourthly, *how* can weather as a topic exemplify science as a human endeavour? The nature of science and the contributions of scientists are illustrated by profiling the work of a prominent Australian scientist in measuring past, and predicting the future impacts of human-driven climate change. Finally, Australian Aboriginal perspectives on weather provide an example of the link between science and culture.

What are teachers' and students' ideas about weather?

Before reading any further, think about what you know about the weather. The following activity will help you to reflect on your own ideas about weather and how confident you would be explaining these concepts to students. Developing a personal knowledge hierarchy is helpful as it can be used to guide the prioritisation of your own learning.

Activity 12.1 Developing a personal knowledge hierarchy about the weather

Earlier chapters in this book have introduced you to many different ways of eliciting prior knowledge. Now is the time to put that knowledge into practice by reflecting on what you know about weather. Firstly, make a list of all of the things you know about the weather. Next, complete the following table by rating your confidence (based on your own knowledge and understandings) in being able to explain each idea to students. Those items gaining the highest score become the highest priorities for you to learn more about prior to teaching weather topics. Use a highlighter to identify the items representing the ideas you need to learn most about (you will use these in the next activity).

	YOUR CONFIDENCE IN EXPLAINING THIS IDEA		
IDEAS YOU MAY HAVE TO TEACH ABOUT WEATHER	VERY CONFIDENT	NOT VERY CONFIDENT	NOT AT ALL CONFIDENT
	1	2	3
What the atmosphere is and how it helps us to survive			
Formation of water vapour in the air			
Air pressure and its effects on weather			
What we can use to tell us about the weather			

IDEAS YOU MAY HAVE TO TEACH ABOUT WEATHER	YOUR CONFIDENCE IN EXPLAINING THIS IDEA		
	VERY CONFIDENT	NOT VERY CONFIDENT	NOT AT ALL CONFIDENT
	1	2	3
Clouds and rain formation			
Forms of precipitation other than rain			
How wind forms and changes			
The difference between weather and climate			
What causes the seasons on the Earth			
How global warming impacts on the Earth			
The cause of the hole in the ozone layer			
The difference between weathering and erosion			
What the greenhouse effect is and what it does			
Components of the weather and how we measure them			
How weather affects people and other living things			

Try this activity now and then later when you have addressed your learning priorities about weather. Comparing your 'before' and 'after' views is a very effective way for you to assess what you have learnt. It will work with your students, too.

You probably found that you knew quite a bit about the weather; for example, that it changes (how rapidly will depend on where you live and the seasons); that temperature, wind strength, wind direction, sunshine, clouds and rain are all components of weather; and that maps can be drawn to show patterns that enable the weather to be predicted or forecast. Other terms, such as 'relative humidity', 'dew point', 'high', 'low', 'cold front', even 'barometric pressure' and 'isobar', may have scored a mention, and, of course, anyone who has spent time in primary classrooms knows that the students are always restless when a strong wind is blowing. Perhaps you also recorded the effect that particular kinds of weather have on your activities or on the activities of others in your community. When you thought about your ability to explain some of the phenomena associated with weather to students, you may have found your knowledge to be incomplete.

For all our awareness about how the weather affects our daily lives and its impact on industries such as fishing, agriculture, livestock, fruit and vegetables, viticulture and tourism, most of us really don't know much about what causes the changes in the weather we experience. In this, our views are not unlike those of most primary school students. The sheer complexity of weather tends to make us focus on what we experience rather than why it happens. This means that, if we are to move from an awareness of weather to some understanding of it, we need to simplify what we are trying to make sense of.

The traditional scientific approach to such a task is to deconstruct this complex phenomenon of weather to its major components, such as temperature, wind speed, wind direction, rain and clouds, which is probably what you began with in Activity 12.1. We can then examine our understanding of what each component is and what causes it to change. This is also a useful strategy to use when working with primary school students. Note that 'deconstructing' means breaking weather down into component concepts or chunks to make understanding each idea more straightforward for students. The accuracy of scientific ideas must not be lost in this process of deconstruction. While we may focus on teaching one aspect of weather at a time, it is important to also convey how the components can affect each other and contribute to changes and patterns in the weather. This will assist students to put this learning in the context of the bigger picture of weather and climate in the world around them. Such interlinking of ideas is not intuitive to many students; the concepts can be forgotten without a framework to attach the concepts to.

Activity 12.2 Primary science curriculum and learning about weather

In the previous activity, you made a list of what you know about weather and your confidence in explaining weather-related ideas to students. Now you will use your primary and junior science syllabuses to find out what weather concepts you are expected to teach and what students will learn more about in secondary school. Take a copy of your current syllabus (practising teachers) or the one you are most likely to use when you begin teaching (preservice teachers). Find the section that includes weather and the environment.

- Make a list of topic areas, phrases and words that indicate teaching content; for example, seasons, daily environmental changes, energy.
- List all of the major concepts and generalisations that will need to be included in a teaching program addressing the syllabus outcomes; for example, hot and cold, temperature, cloud formation, cloud types, wind direction, solar energy and heat energy.
- Using the scoring criteria from the previous activity, rate each of the items in your list according to your confidence (based on your level of knowledge and understanding) in teaching about these ideas.
- Reflect on your ratings and make a list starting with all of the ideas that scored a 3 (in this and the previous activity). Repeat for the ideas that scored a 2. You now have a personal learning hierarchy for weather that can guide your preparation for teaching regarding background knowledge. Use this hierarchy to set yourself some goals and time lines for improving your knowledge and confidence to teach these ideas to students.

It is all very well to be aware of what you will need to teach students about weather, but what if you are not confident about your understanding of the concepts involved? To successfully engage with students in meaning-making through constructivist practices, such as engaging in teacher–student talk in the context of collaborative inquiry (Hodson and

Hodson 1998), you will need to have a sound understanding of the concepts you intend to teach. Your studies at university may not have included formal instruction in Earth science; therefore, you will need to access information and explore your own understanding of science concepts prior to teaching about them. Without such preparatory learning, you may inadvertently add to students' alternative conceptions (Sewell 2002).

It is not unusual for adults to have alternative conceptions about key weather processes, such as the water cycle. A recent study of Turkish university science students, using student-generated drawings and interviews, revealed that one-fourth of students held alternative conceptions. These included the following notions: the process of evaporation of water from the Earth is only determined by the sun; the amount of water in the biosphere is gradually declining due to the melting of glaciers and global warming; underground water cannot be drunk unless it's first purified, since it is polluted; and living things cannot exploit water in the seas and oceans since it is salty (adapted from Candek 2009, p. 870).

Many of you will find that you need to increase your content knowledge of weather prior to teaching about it. Smith (1997) showed that primary teachers are able to increase their content knowledge through focused reading. So where do you go to find information to read up on? Many Internet sites now provide background information linked to sample classroom activities. An excellent resource available for teachers is the site provided by the Australian Bureau of Meteorology (see http://www.bom.gov.au/lam). This website has links to available publications and a glossary providing definitions of common weather terms, as well as experiments and activities designed for students and teachers and which include interactive animated models, lesson plans and student worksheets. The *Primary Connections* unit 'Weather in My World' also includes a DVD with background information for teachers on weather topics (AAS 2005). Locally, search out resources developed by regional education centres with materials linked to specific syllabuses. The Riverina Environmental Education Centre (NSW), for example, developed e-learning resources on energy and climate change (Boylan 2008); they are available to teachers and students at the REEC website (http://www.reec.nsw.edu.au).

Secondary science textbooks and specialist science teachers in high schools are also a good source of help. Teaming up with a colleague from the local high school to engage in active discourse – asking questions, describing your synthesis of new ideas with your current knowledge, and talking about what students will learn in secondary school – would be an excellent way of assessing your understanding of key concepts and may help you avoid the trap of believing the content of primary science activity books, which, in the experience of these authors, is often oversimplified, misleading and sometimes grossly incorrect. A cautionary note is that textbooks may not include up-to-date information informed by recent research on climate change issues. As scientific knowledge has a temporary status and should not be accepted as unquestionable truth (see the online Appendix 1.1), then this means that some ideas change as we become better informed by research, so look out for updates in the media, professional journals such as *Teaching Science*, and key science publications like *Nature* (also see http://www.nature.com). The CSIRO website has a section with current information titled 'Understanding Climate Change' (see http://www.csiro.au/science/Changing-Climate.html).

Appendix 1.1
What do you think science is?

Eliciting students' ideas about weather

Research into constructivist approaches to teaching must begin with identifying students' ideas and views (Harlen 2009; Hodson and Hodson 1998; Tytler 2002), including alternative conceptions (Sewell 2002). Prior to, or in conjunction with, teaching a unit of work, a range of strategies can be used to elicit students' existing conceptions. These can be used as a basis for planning lessons to enhance students' conceptual development. Four general strategies for finding out students' existing scientific ideas include:

1. questioning (teacher or student prompted)
2. visual elicitation (poster, spider diagram, flow chart)
3. concept cartoons (mentioned in previous chapters)
4. play; for example, where the teacher observes what students already know what to do with equipment (Knight 2009).

Specific elicitation strategies for finding out students' preconceptions about weather could include the following:

- Questioning – Ask students (individually or as a class) to respond verbally to questions: Where do puddles go when it stops raining? Why is it only sometimes windy?
- Using puppets – Introduce a puppet that is confused about the greenhouse effect and engage the students in sharing their ideas to help the puppet; for example, see the PUPPETS project (Belohlawek, Keogh and Naylor 2010).
- Brainstorming – Ask the class to tell you what they already know about weather. Using ICT resources, such as SMART board technology (SMART Technologies 2003), the teacher and/or student can write on the interactive whiteboard's large touch-sensitive surface with electronic pens. The writing is converted into digital/electronic data and can be saved as a notebook computer file. The results of a brainstorm can be recorded in this way prior to commencing a unit of work. This information can then be recalled for review and discussion at the end of the unit. It could also be uploaded to the school website for student reference at home and to share the data with teacher colleagues.
- Drawing – Have the students draw visual representations of their understanding of a particular concept or process. Draw what is inside clouds. Use a drawing to show how clouds change when it is going to rain.
- Testing – Use a diagnostic testing method for formative assessment. This can take the form of a before-and-after written test.
- Creating a glossary – Provide students with a list of weather terms and ask them to write definitions for those with which they are already familiar, to create a glossary. Students can add to the definitions as they develop an understanding of the listed terms.
- Interviewing – Conduct interviews with a small group of students to find out their ideas (Abdullah and Scaife 1997). Have a list of questions; for example: What do you know about cyclones? What makes the wind so strong? Why don't we get cyclones in Sydney?
- Sharing stories – Invite students to share stories about their experiences with particular weather conditions: Has anyone been to the snow? Can you tell me what it was like?
- Letter writing – Have students write a letter to a pen friend in a different country describing what the weather is like (including daily changes) where they live.

- Record keeping – Organise your students to keep a written and/or pictorial record of the weather for a week or so, and then discuss these records with the class.
- Explaining weather changes – Ask your students to make a series of annotated drawings about what they think happens when the weather changes – e.g., from sunny to stormy, or when the wind direction changes from west to south-east – and then use these picture strips (Qualter, Schilling and McGuigan 1994) for group discussion.
- Reading and problem solving – Read the storybook *Ducks in the Flow, Where Did They Go?* This book models and promotes student inquiry and is supported by hands-on activities that help students learn about ocean currents. The story and activities are available free on the Internet (see http://www.windows.ucar.edu/ocean_education.html) (Coffman et al. 2009).

Activity 12.3 Eliciting students' ideas about the weather

- Select a grade level that is relevant to you – for example, Year 1, if you are going to, or are already, teaching this age group.
- Plan a lesson segment that will enable you to collect students' ideas about a specific weather concept; for example, cloud formation.
- If possible, implement this lesson segment with a class and evaluate the success of the strategy in determining students' ideas. Alternatively, share this plan with a colleague or fellow student and ask them to evaluate its usefulness as a strategy for elicitation of students' ideas about the selected concept.
- Compare your findings with relevant literature that has reported students' ideas (see below).

Students' ideas about weather

Students are conscious of how the weather affects their lives. Even very young children are keen to dress their dolls and action figures in appropriate clothes for the day or season. Figure 12.1 shows simple dolls made from pipe cleaners – the children selected from a variety of fabric scraps to decide the most appropriate clothes for the seasons. They build up their own ideas about weather from personal experiences, explanations offered by adults and other students, information presented in television programs, and from children's books. These sources often provide inadequate scientific information that is not linked with students' prior knowledge.

Children will record assiduously the day's weather on a chart, doing so by drawing or sticking on symbols well before they can write the words. These children tend to think that the weather, although changeable from day to day, is constant within the day. This false conception is reinforced by having a once-a-day weather observation and recording. The group who record their observations at 9 a.m. are often very puzzled when they find their observations do not match with those of another group who take theirs at 1.30 p.m. They usually doubt that this time difference is relevant and seem to believe that entire days are rainy, sunny, windy or cloudy. If groups within a class make their observations at different times on the same day, it will challenge this view.

All this indicates that, although younger students are often acute weather observers, they are more likely to perceive (and record) their observations as a series of isolated and

FIGURE 12.1 Examples of 'seasons dolls' dressed by kindergarten children for summer and winter

unrelated events rather than as snapshots from which to construct a continuing story. The learning experiences we provide for them, therefore, must seek to challenge this episodic view of the weather and encourage a picture based on continuity and pattern. A good example of this type of activity can be found in 'Weather in My World' (resource sheet 3), an Early Stage 1 unit in *Primary Connections*, an Australian national resource that promotes scientific literacy (AAS 2005; also see http://www.science.org.au/primaryconnections).

As with their ideas about many other phenomena, young children are not much interested in abstract and general perceptions of weather but perceive it in the more personal terms of the impact it is having on them – now. In similar fashion, their explanations of weather phenomena are usually self-centred, perceptual, associative and unconnected to other events happening around them, so it is not unusual to find that they associate rain with God crying or someone (television weather presenters and astronauts are popular choices) turning on a tap in the clouds or the clouds starting to sweat (Bar 1989; Tytler and Prain 2010).

Many researchers have investigated students' understanding of the weather and its causes and effects. Students' ideas have been reported about evaporation and condensation (for a review, see the Chapter 9 section 'Students' ideas about matter and change' on p. 356), cloud formation and rain (Dove 1998; Henriques 2002; Russell, Longden and McGuigan 1993; Saçkes and Flevares, 2010), the ozone layer (Fisher 1998; Henriques 2002; Potts, Stanisstreet and Boyes 1996), the greenhouse effect (Fisher 1998; Henriques 2002), the weather (Dove 1998), weather and climate (Spiropoulou, Kostopoulos and Jacovides 1999), and the water cycle and its relationship to water in environmental systems (Covitt, Gunckel and Anderson 2009). The research has focused more on students' incomplete knowledge or misunderstandings about weather. An analysis of this research has led to a growing body of knowledge about students' alternative conceptions relating to weather, which has been summarised by Henriques (2002). To provide an overview of the ideas young children hold about weather, a summary and related discussion of some of the research findings is presented next. For additional details of students' alternative conceptions about weather, try to obtain a copy of the Henriques (2002) paper.

Evaporation and condensation

Research into students' thinking about the concepts of evaporation and condensation, including their role in the water cycle, has already been discussed in Chapter 9. This section looks at these concepts only as they relate specifically to weather and our environment. Sparked by recent emphases on rich learning tasks aimed at promoting deep understanding and sophisticated thinking by students, Covitt, Gunckel and Anderson (2009) assessed upper elementary and high school students' understanding of water in environmental systems. In contrast to previous studies on students' ideas about the water cycle, which focused on the identification of alternative conceptions about phase change (again, see Chapter 9), their study was based on students' reasoning about water in environmental systems.

Most primary students showed water underground in human-made containers, with only a few showing groundwater in spaces and cracks between rocks and soil. They knew water came to houses in pipes but did not make connections between natural and human-engineered systems. Whether landfill (garbage dumps) could pollute well water was mostly explained by above-ground processes: 'It can. Some of it might fall in a well' and 'Yes, a landfill can pollute [sic] the water by getting trash into the water' (Covitt, Gunckel and Anderson 2009, pp. 46–7). These results show that primary students think about pollution as macroscopic trash and so do not consider microscopic/particulate processes such as solutions (see Chapters 8 and 9); neither do they consider what happens below the ground. Probing of where water comes from before it gets to your home and where it goes afterwards revealed knowledge of house water coming from natural sources but rarely (20 per cent) was there a reference to water treatment before supply, and less (12 per cent) after the water left the house (ibid., p. 48). Overall, the researchers concluded that students' understandings about water were incomplete and unconnected, lacking deep understanding. Students were unable to trace water and other materials through human and natural systems. This, they argue, is necessary for people to understand and act 'responsibly on environmental issues related to maintaining and protecting water quality for all life systems on Earth' (ibid., p. 49).

Cloud formation and rain

While primary-age students relate clouds to rain, they often have difficulty explaining what clouds are made from, how they form and their relationship to rain. Students' ideas about cloud composition include the notions that clouds are made of:

- snow, cotton or something soft (Saçkes and Flevares 2010)
- stones, earth, smoke or steam; cold, heat, fog, snow or night; sponges that hold water (Dove 1998; Henriques 2002)
- dust particles, water vapour rather than water droplets, and ice crystals (Henriques 2002).

Students' explanations of how clouds form include:

- clouds are made or moved by God
- empty clouds are refilled by liquid water from the sea

- clouds are formed by vapours from kettles or the sun boiling the sea
- clouds come from somewhere above the sky (Dove 1998; Henriques 2002).

A majority of students across the entire five to 11 years age range saw clouds as having a key role in determining the weather (Russell, Longden and McGuigan 1993). The common argument ran along the lines that, in the absence of clouds, the weather was sunny, but if clouds appeared it became cloudy. Further, if the clouds became heavy, rain would fall. Occasionally, even children as young as five or six years would include in their drawings indications of water being 'sucked' (straws were often included) or 'blown' (by the wind) up from the sea into clouds, but the clouds were already there to act as temporary containers for the water. When the water load became 'too heavy', the clouds were seen to 'leak' or to develop 'holes' or 'burst', which in turn determined how heavy the rainfall would be.

Rain formation is also frequently attributed to the actions of God, angels or people in the sky. Students' explanations about the mechanisms of rainfall include thinking that rain is made by God, angels crying, men in clouds throwing buckets of water or using sprinklers, smoke that melts, heat that makes clouds melt or sweat, or clouds splitting or colliding (Dove 1998; Henriques 2002). Four- to six-year-olds think that rain: comes from between clouds, is water that comes from clouds, is produced when clouds collide, is made by trees shaking their roots, comes from the sky, or that God creates rain (Saçkes and Flevares 2010). They also hold different views about what happens to water after it rains, some reckoning that water goes underground after falling or collecting into puddles; that, when rainwater goes underground, it ceases to exist; that it drains into holes or cracks in the ground; or that it's dried up by God after it rains (ibid.).

Notable in the above conceptions of rain by young children are the variety of ideas that you, as a teacher, may have to deal with during classroom dialogue. The important thing to remember is that, for the children, these ideas are neither right nor wrong but represent their 'true' (current) understandings built from observations and experience.

We must remember that students do have prior knowledge of the weather and it may be quite sophisticated. Figure 12.2 shows a picture strip produced prior to studying the topic by a Year 3 student to illustrate how he thought clouds form and how they produce rain. The idea that water in some way rose from the sea to provide the clouds with the water that would later become rain became increasingly prevalent as the age of the students increased. At the same time, there was a steady increase in the number of drawings showing that this water formed the clouds rather than simply filling them. Very few students showed any understanding that the cooler air temperature found where the clouds are forming had a role in this process. We should not be surprised at these findings, however, as previous research with older students (Osborne and Cosgrove 1983) and primary school teachers (Kruger and Summers 1989) indicated a similar lack of understanding regarding links between lower temperatures and the condensation of water vapour. This may, in turn, indicate an alternative conception that atmospheric temperatures increase rather than decrease with elevation, despite the very familiar photographic evidence of snow-capped mountains, even near the equator.

It seems that students' ideas about cloud formation are based on their understanding of the water cycle (Henriques 2002) or (for younger children) on what adults have led them to

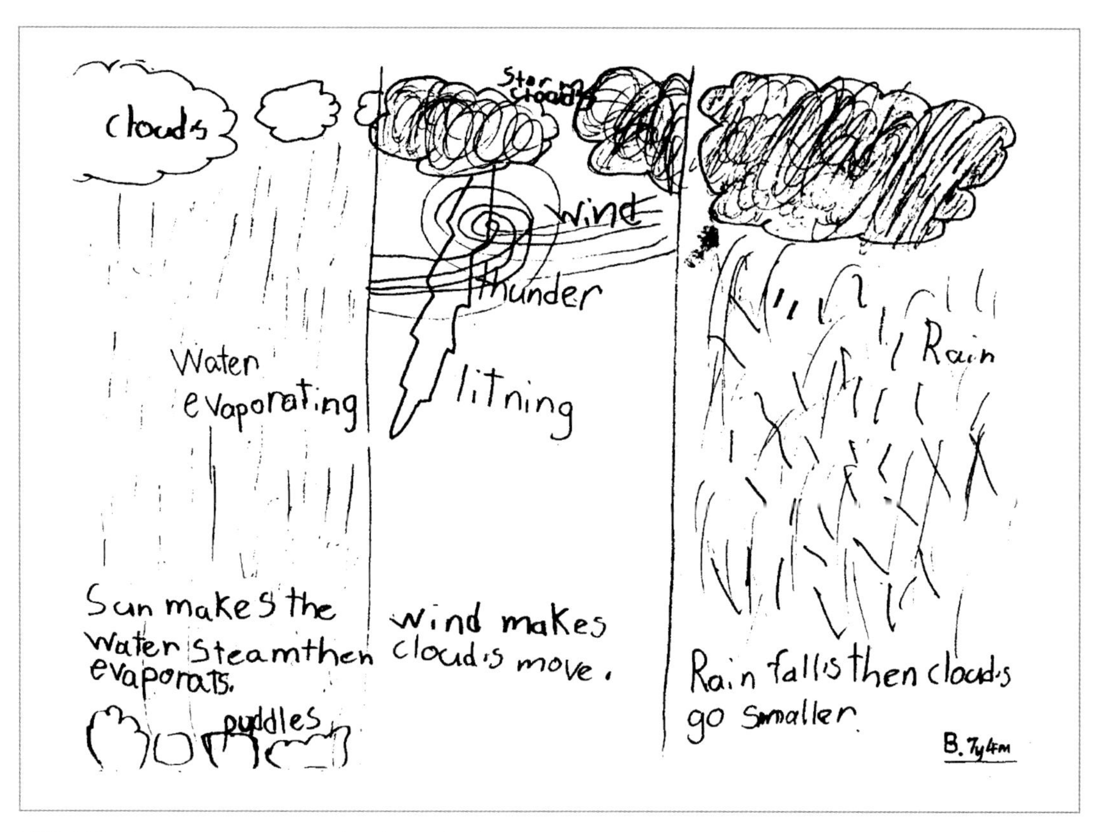

FIGURE 12.2 Evaporation, clouds and rain

believe in the absence of personal observations. Even students who had previously studied the water cycle rarely linked evaporation with phenomena such as clouds, rain and fog (Russell, Longden and McGuigan 1993). The scientists' grand explanatory overview presented as the water cycle does not appear to impinge greatly on the explanations students give for the changes they observe in the weather (Symington and Symington 1983). These findings complement the studies and further discussion set out in Chapter 9.

In an attempt to explain that clouds are formed when water vapour changes into water droplets that are visible, and happen under certain atmospheric conditions, Christine involves children in her classes in making a cloud in a bottle. Figure 12.3 shows visible fog (actually a small cloud) that can be made to appear and disappear in a bottle due to pressure changes when the balloon sealing the top of the bottle is pushed and pulled up and down. This experiment takes a little time to perfect and takes keen eyesight and concentration to see the cloud appear, but once seen it is very convincing. The idea is to demonstrate that the process can be controlled, showing that clouds don't magically appear. If there is no water vapour or no small particles (such as smoke) for the water to condense around, or the conditions are not right, clouds will not form. This fact sometimes explains the lack of rain, and why, at times, cloud seeding will not work due to insufficient water vapour to form rain-producing clouds. Figure 12.3 also shows children making their own clouds. Active discussion about the

FIGURE 12.3 Teacher demonstrating how to make a cloud form in a bottle and children making clouds themselves

experiment is needed to help children understand how the model is similar to, or different from, reality.

Weather and climate

A survey of Greek primary school and middle school students sought to determine students' conceptions about weather and climate (Spiropoulou, Kostopoulos and Jacovides 1999). The authors' study supported earlier research findings that students confused the meanings of the terms 'weather' and 'climate', and that most students associated climate with yearly weather data. Students' knowledge about weather and climate was mostly non-scientific, concentrated on elements that can be perceived using the senses, and did not include abstract ideas such as air pressure and humidity. Students' personal experiences of the weather seem to dominate over formal learning in students' thinking (see Tytler and Prain 2010).

The ozone layer, global warming and the greenhouse effect

Environmental change involving climate change has become such a significant societal issue that primary-age students can't help but become interested in the subject. Previous research revealed that upper primary students (ages 12 to 13) have an awareness, but an incomplete understanding, of the scientific concepts related to ozone depletion, the greenhouse effect and global warming – see Fisher (1998), Henriques (2002) and Potts, Stanisstreet and Boyes (1996). A study of preservice science teachers' understandings of ozone-layer depletion revealed they had insufficient knowledge as well as alternative conceptions about this environmental issue. Paralleling the above results for students, these teachers also confused ozone layer depletion with global warming (Bahar, Bağ and Bozhurt 2008). Clearly, if we are going to teach about these issues there needs to be greater emphasis on deep understanding. This means that primary teachers need to develop a more sophisticated understanding themselves to be able to design thought-provoking activities to support students' thinking about the role of the ozone layer in sustaining life on Earth.

Boylan's (2008) survey to establish what primary students understand about energy sources and climate change issues concluded that 'many students were not clear about how the key environmental concepts of climate change, greenhouse emissions and global warming are different from each other' (ibid., p. 12). This study highlights the importance of eliciting prior knowledge and developing responsive teaching programs capable of facilitating deep understanding of the concepts. Investigations into Year 7 students' conceptions of global warming and climate change showed a lack of deep understanding, especially the connection between the greenhouse effect and global warming (Shepardson et al. 2009). These students underestimated the impacts of global warming and climate change. They thought temperature, precipitation and only local weather events or conditions would be affected. They rarely considered the increased frequency, variability and severity of weather events, and did not consider widespread changes (ibid., p. 562). While these are complex scientific concepts to teach primary age students, a deeper understanding of the greenhouse effect is crucial to developing an understanding of the impacts of global warming and its explicit connections with weather and climate change – clearer, stepping-stone ideas can form a foundation at this stage.

SSI

Students' questions about weather

The previous section showed that researchers have investigated students' understanding of the weather and its causes and effects. Much of this research has focused on students' explanations of particular weather components rather than examining which components of weather the students thought were important. In addition to finding out students' ideas about the topics you intend to teach, it is also beneficial to determine their questions and interests. Starting from what students want to know can be a powerful way to engage their interest. Some methods for determining students' questions and interests include:

- group consensus – allowing students time to discuss with others in small groups what they want to know about the weather, decide on some common interests, record these on butcher's paper and then report back to the whole class
- an interest brainstorm – for example, introduce the topic 'people and weather', ensuring that students know what the title means, then have them tell you the things they are interested in learning about the weather
- a secret poll – students individually write down questions they have about the weather, and place these in a box for the teacher to read later on and select questions for class activities
- inviting an expert – for example, weather reporter to the school. Students discuss and prepare in advance questions that they would like to ask about the weather; alternatively, visit a site where there is a resident expert, such as a scientific officer, ranger or horticulturalist, whom the students can interview.

To provide a specific example of the last suggestion, a visit to a local university's climate lab spiced up a Canadian Year 4 unit on weather. The outing enabled students to connect to scientists, showing them they were real people who collected and analysed weather data from a weather station similar to the one that happened to be installed on the roof at their school. The teacher used the visit as an opportunity to find out what questions the students wanted to know about weather. A list of questions was compiled for students to ask the scientists, which included the following:

- How do the northern lights start up?
- What makes thunder and lightning?
- Do tornadoes spin left or right or both?
- How are water spouts formed?
- How strong can a tornado be?
- Can a hurricane be over category 5?
- How many times does lightning strike in one hour?
- Why don't tornadoes come here?
- Why are there no hurricanes here?
- Why don't we have lots of snow here? (Weaver and Mueller 2009)

Asking students what questions they have about weather can be a useful way for teachers to find out aspects they don't understand or interest them about the weather. Types of questions students have asked include the following:

- What are clouds made of and is this the same for all clouds?
- If wind is just air moving, what makes it move?
- How do they know what the weather will be?
- How do clouds change shape?
- What happens to puddles when the sun shines?
- How does pollution change the weather?
- I think clouds are made of water because it rains, but I don't know why the water drops making white clouds don't just fall down too.
- How do we get rainbows?

Helping students develop better questions

Although students have questions they want to ask about the weather, these questions are not always asked in a form that will help the students make better sense of their world. Part of the teacher's role, therefore, is to help the students clarify, modify and extend their questions. Often, this can be done by suggesting that the students examine the assumptions underlying the question. For example, an appropriate response to a question such as 'Why does it always rain at lunchtime and not when we are inside?' can be along the lines of 'I haven't noticed that. How do you think we could find out if that does always happen?'

Some questions, however, are stated so vaguely that they require considerable discussion with the student before you can identify what question is really being asked. 'Why does it rain?' and 'Why do we have weather?' are examples of such questions and, like so many 'why' questions, in their current form they are unanswerable through classroom activities. The question 'Why does it rain?' can be addressed by asking students if they think we could make rain in the classroom. This can be answered by a hands-on activity with concrete materials – a kettle, a sheet of cold glass – as used to demonstrate the water cycle. Used in this context, the demonstration perhaps has greater relevance because its purpose is to answer a question of interest posed by a student rather than to demonstrate a process related to weather phenomena that occurs naturally in their environment.

Developing students' ideas about weather using a constructivist approach

When students' alternative conceptions are known, teaching and learning activities can be planned that not only challenge naive ideas, but also present new information that extends thinking and/or provides cognitive conflict. The ineffectiveness of traditional teaching in changing Greek students' preconceptions about weather and climate (Spiropoulou, Kostopoulos and Jacovides 1999) showed that the direct teaching of concepts does not guarantee students' learning and understanding. Stepans and Kuehn (1985, p. 47) point out that 'merely teaching a concept is no guarantee that students learn it' – an assertion that is still applicable today. The second aspect of 'challenging students cognitively' is required to bring about a change in students' ideas (although challenge alone may not do that, as discussed in Chapter 1). Failing to take into account students' ideas when planning instruction means that alternative conceptions are unlikely to be changed or, worse, others could be developed because of the learning experiences.

How can you use what you have learnt from the research literature to help students develop more scientific understandings about weather? Having first identified the ideas, views and questions your students hold about weather, you then use this knowledge about your students' thinking to plan and implement lessons that will further develop their understanding.

While varied teaching approaches are advocated by researchers and teachers, depending on their specific constructivist emphasis, they have some common elements. You will need to develop learning experiences that:

- take students' existing knowledge into account
- provide learning experiences to extend and/or challenge students' ideas
- offer relevant alternative ideas
- encourage students to reflect on their understandings
- link scientific explanations to everyday situations.

Before you start planning, you need to ensure that you won't add to alternative conceptions by increasing your own personal understanding of the concepts. You also need to be aware of what alternative conceptions are commonly held by students. As you have seen, this book provides a good starting point for both of these.

Much of the research into students' understanding of weather summarises findings about students' alternative conceptions and states that this has implications for teaching. What is lacking are specific examples of how teachers can use this knowledge for changing students' ideas about weather. The following section provides a brief discussion of several of the general constructivist teaching elements listed above and illustrates these with examples for developing and changing students' ideas about weather. As you read through these examples, try to identify which of the teaching elements are emphasised in the teaching scenarios.

Students' existing knowledge

Researchers state that knowledge of students' existing ideas, including alternative conceptions, should be used in the planning of, and as a starting point for, your lessons (Sewell 2002; Tytler 2002). Knowing that some young children think that the clouds are moved by humans pushing them, and that some children in her kindergarten class also thought this, the teacher in the following conversation directly challenged a child and asked for supporting evidence.

Teacher	We have just been outside and we saw clouds moving across the sky. Can anyone tell me how clouds move?
Child	Clouds can't move by themselves, so there is a little man in the sky that pushes them around.
Teacher	Have you ever seen a person pushing a cloud?
Child	No.
Teacher	Why not? If you can see the clouds, why can't you see the person who is pushing them?
Child	Because they are invisible.
Teacher	How do you know that they are there?
Child	I know because daddy said so.
Teacher	OK. Well, how does the person get up there?
Child	(Shrugs; looks puzzled and frustrated).
Teacher	Do you think there may be a different explanation for how clouds move?
Child	Maybe.
Teacher	Let's do some activities to see if we can find out and then we will continue our discussion.
Child	OK.

Asking a child to provide evidence for their ideas is a way of testing their level of understanding. It is also a way of reinforcing thinking scientifically, where evidence is sought to support a hypothesis or theory. A child who cannot explain how their idea works or provide evidence that it is true is put in a situation where they may need to reflect on their understanding.

Activity 12.4 Using an awareness of students' alternative conceptions in teaching

1 Read the above author's account of using commonly held alternative conceptions to aid teaching and learning.
2 List how she has applied knowledge of research findings and of her students' alternative conceptions.
3 Evaluate this technique by listing any possible advantages or disadvantages.
4 Suggest other strategies that you could apply in this situation.

For further ideas that may assist your thinking about these tasks, see Chapter 1.

Challenging students' ideas

In order for children to change their views, they need to have cause to question their existing ideas. When children are presented with information that is in conflict with their own ideas, they will either reconstruct their thinking or modify or reject the new views. Many young children have the idea that clouds are made of cotton wool or smoke, probably because that is what they look like to them. The suggestion that clouds are made from cotton or other substances might also result from teacher or adult descriptions of clouds or from art projects where cotton wool has been used to represent clouds (Dove 1998). We should remember that our teaching experiences could lead students to learn other than what we are seeking to teach them.

Teachers may assume that young children have already made connections between clouds and rain. However, research shows that preschool and kindergarten children 'believe that rain is water, but they do not relate clouds with rain and many believe rain simply comes from the sky' (Saçkes and Flevares 2010, p. 9). These authors recommend teaching students about the content of clouds and emphasising the relationship between the change in the colour of the clouds and rainfall. Children may not have had any experiences to suggest that clouds could be made of water droplets so small that they can be suspended in the sky. As children cannot observe water vapour evaporating and then condensing in the sky to form clouds, they think rain just falls magically from the sky; they may not even relate rainfall to the presence of clouds.

Here is a teaching situation in which the teacher attempted to challenge these ideas and lead the children towards a preliminary understanding of scientists' views:

Teacher	What are clouds made of?
Child 1	Cotton wool.
Teacher	No, they are not made of cotton wool. How could it stay up there? What are clouds made of?
Child 2	Rain.
Teacher	Well, you are sort of correct. We call it rain when it falls from the clouds. What is rain?
Child 3	Bits of water.
Teacher	Yes. So what must be in clouds if water falls from them when it rains?
Child 2	Water?
Teacher	Yes, that is correct. Do all clouds rain?
Child 4	No. Grey clouds rain, white clouds don't rain.
Teacher	OK. Do all clouds have water in them?
Child 4	No. Grey clouds have water in them, white clouds don't.
Teacher	Well, actually, all clouds have water in them but not all of them rain. Now we are going to do an activity to see what happens in clouds to make it rain.

Stepans and Kuehn (1985, p. 47) state that the sharing of views by students in a classroom setting 'exposes them to many different views at the same time as it helps the teacher become aware of the way [these] children perceive the world'.

The 'activity' mentioned by the teacher above is described in the next section.

Presenting alternative ideas

New ideas need to be presented in a way that is relevant to the student or they will simply reject it. Successful methods to assist students in changing alternative conceptions include the use of visual activities, collaborative conversation, small-group discussion with other students and the creation of concept maps (Sewell 2002). The scenario below describes a practical activity followed by a collaborative discussion with a group of three kindergarten children.

Clouds and rain: using a visual model

Children are shown how to work in pairs to hold a plastic plate vertically over a sink and spray the plate with water. First, a small amount of water is sprayed onto the plate, forming small water droplets that stay stuck to the plate. Then more water is sprayed onto the plate until the droplets run off the plate in the manner of rain. The teacher discusses the children's observations and relates them to the making of a visual model to show what clouds are made of and what makes it rain:

Teacher What happened to the small water drops on the plate?
Child 1 They stayed on the plate.
Teacher Yes, they did. What happened when you sprayed more water onto the plate?
Child 2 The drops all joined up and ran down the plate.
Teacher Excellent observation. Why did the big drops fall down?
Child 3 They were too heavy to stay there.
Teacher Yes, very good. Now, did you realise that this is what happens in clouds?
Child 1 Clouds are made of water?
Teacher Yes.
Child 2 But how does it stay up in the air without falling down?
Teacher We just saw that in our activity. Clouds are made of tiny little water drops (like on the plate in the beginning) that can stay up in the air.
Child 1 Oh! And when they get bigger and heavier the drops fall and this is rain.
Teacher Yes, just like what happened when we sprayed lots of water onto the plate. If we wanted to show this as a display model, what would we show inside the cloud?
Child 3 Little water droplets.
Teacher Yes, we could have some clouds that are just little water drops and don't make rain. Then we could have some other clouds that will make rain. What will they have in them?
Child 1 Bigger water drops.
Teacher Yes, that is correct, and we could put that cloud over our model rain drops that we already have in our display.

The resulting model is shown in Figure 12.4. In this photograph, the smaller clouds can be seen to contain small, light-grey water droplets, while the larger clouds contain larger, darker water drops. Raindrops can be seen falling from the large clouds, but not from the smaller clouds.

FIGURE 12.4 Rain and clouds visual model made by kindergarten children

Activity 12.5 Using a model for young children's ideas about clouds and rain

- Consider the following views of scientists:
  - Visible clouds are tiny water droplets suspended in the air.
  - When water droplets in clouds join together and become too big to stay suspended, they fall as rain.
- Read the author's account above of using models to aid teaching and learning.
- List how she has applied knowledge of research to children's alternative conceptions.
- Evaluate this technique by listing any possible advantages or disadvantages.
- Suggest other strategies that you could apply in this situation.

The practical activity above provided the children with concrete evidence about the behaviour of small and large water droplets. As children experienced this activity firsthand, the evidence was plausible. While children cannot see what clouds are made of, they have experienced water droplets falling from them, so it makes sense that clouds are composed of rain. It also makes sense, then, that if clouds contain water that is not falling, the droplets must be small and for some reason are able to stay up there and not fall as rain. The collaborative discussion enabled the teacher and the children to co-construct a new level of understanding about cloud composition. Like all models, this one has its limitations and where appropriate these need to be discussed with students (see chapters 1 and 10). What is not explained is that, although the cloud droplets are tiny before they join to form larger rain drops, the reason they remain suspended in the air is not due to their light weight (liquid water is heavier than air despite the droplet size). It is due to the heat energy that is released when the water changes from gas to liquid, warming the air between the water droplets and consequently changing the air pressure that keeps them up there (see http://amasci.com/amateur/clouds.html). This explanation is best left until secondary school, when students learn about energy in relation to changes of state.

Discussions linked to the earlier activity on water evaporation helped children understand how the water got into the air to form the clouds. Active discourse with the children included asking for evidence to support their casual explanations (Saçkes and Flevares 2010). For example, when children explained that water from puddles goes up into the air, the teacher asked, 'How could that be? Do we see water drops whizzing past us like raindrops in reverse?' A resounding 'No!' from the children then resulted in the search for a plausible answer: 'Could it be that the water is invisible?' Teacher drawings like the one shown in Figure 12.5 facilitated collaborative discussions with the children. Note that the small circles are water drops, not particles (explanations involving particles should be left until later primary classes – see chapters 8 and 9, as well as Tytler, Peterson and Prain (2006) and Tytler, Prain and Peterson (2007), who explored the use of a particle representation in developing students' understanding of evaporation. The problem of dealing with abstract concepts such as evaporation can be aided by a multimodal teaching approach using a range of resources,

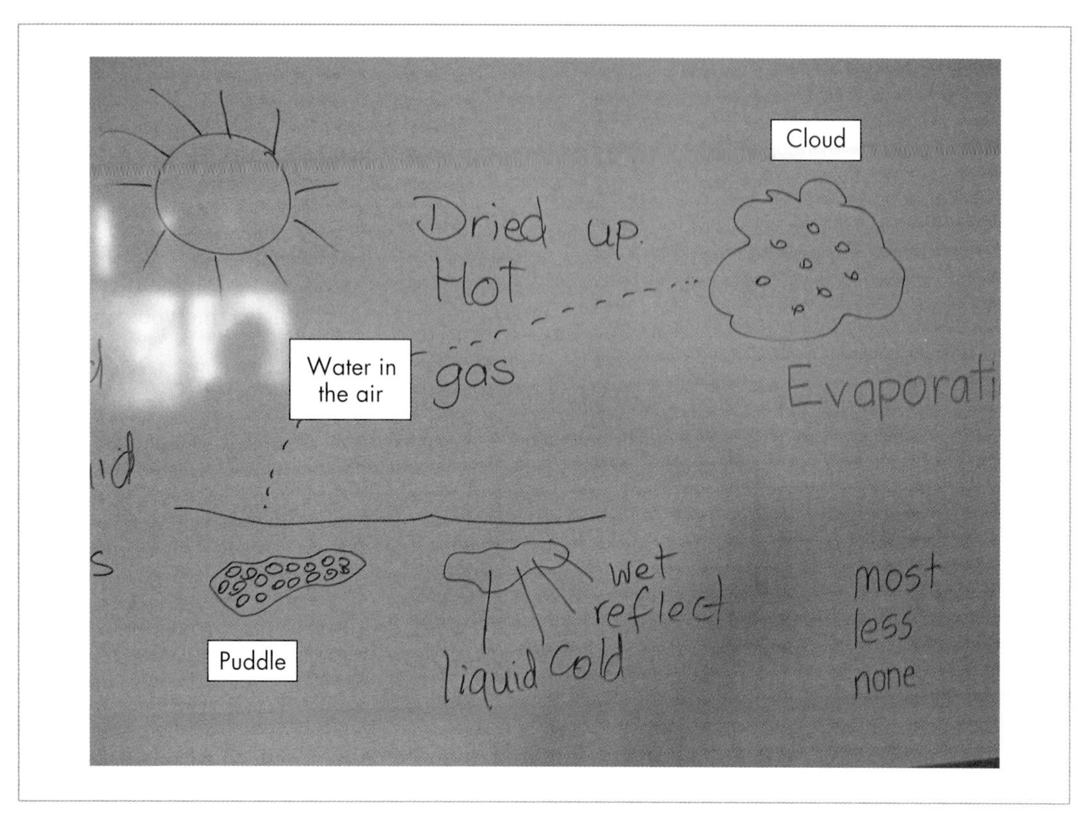

FIGURE 12.5 Teacher's white board drawings

Appendix 12.3
Once I was a water drop – narrative

including children's books such as *Where Do Puddles Go?* (Tunkin 2004) and narratives (see the online Appendix 12.3) in conjunction with children's drawing, writing, hands-on activities, active discussion and verbal explanations.

Children accepted and held onto the view that all clouds are made of water droplets throughout the nine-week topic, and could still explain this when questioned after a three-week vacation and a few weeks of learning about a new topic. Figure 12.6 shows two labelled diagrams drawn by kindergarten children showing invisible water vapour going up from a puddle to form a cloud, along with a darker cloud that is raining.

The long-term retention of these children's understandings about weather was reassessed the following year (Year 1) when they studied weather and its effects on people. A child's work sample shown in Figure 12.7 demonstrates conceptual understanding. Annotated drawings are an effective method of assessing students' conceptual understanding.

This account supports the need for a spiralling curriculum in which students' ideas about concepts are checked and challenged through regular and recurrent teaching and learning experiences. Isolated lessons are less effective in changing students' preconceived ideas than is a planned sequence of topics or units designed to build on previous learning experiences with the aim of continual conceptual development.

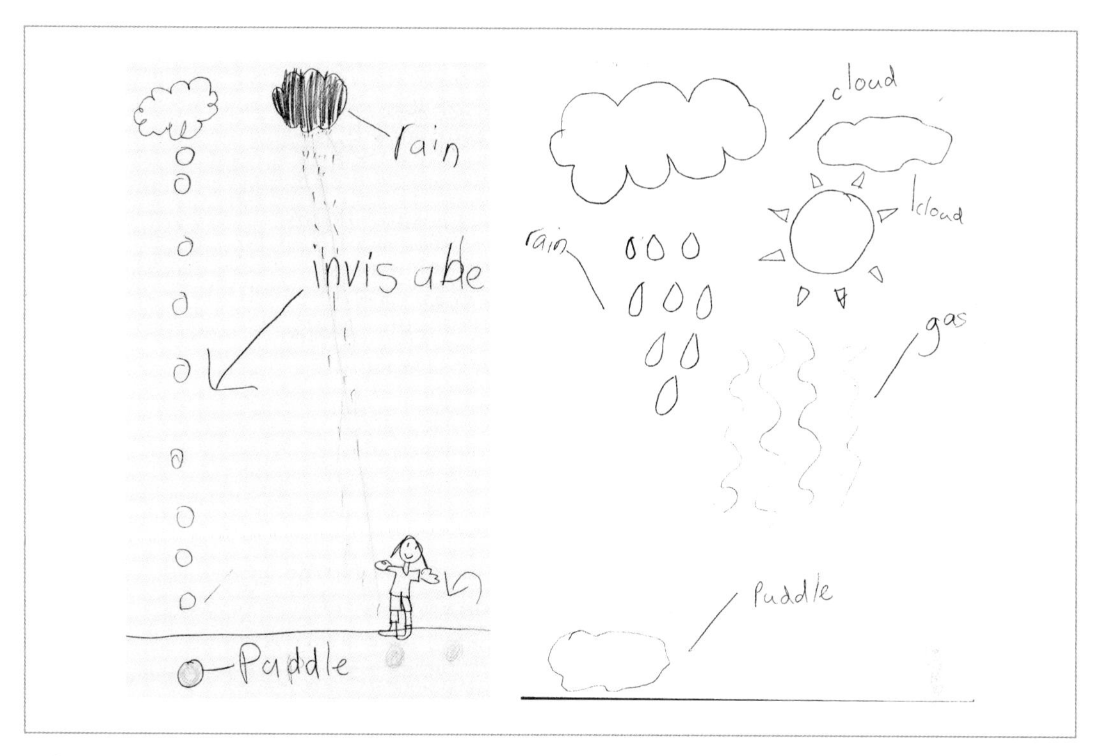

FIGURE 12.6 Kindergarten children's labelled drawings explaining cloud formation and rain

Evaporation, clouds and rain: acting it out

Teachers have devised many effective and enjoyable learning experiences that give concrete support to students' tentative ideas about evaporation and condensation. A group of students miming the behaviour of water particles during evaporation as the faster-moving particles escape from the liquid to form water vapour is a memorable way of clarifying and reinforcing their tentative ideas.

The students may also be asked to show in mime how water vapour particles, moving more slowly at the lower temperatures found up in the sky, will join together when they collide (gently) and condense to form visible water droplets (or clouds). Similarly, they can demonstrate how small water droplets in white clouds can, on further cooling, join together on collision and form bigger drops (making the clouds darker) and, finally, form raindrops.

We must realise that students' alternative conceptions can arise as a result of poor teaching practices, including the incorrect use of analogy and confusion over closely related concepts (Dove 1998). A possible source of the alternative conception that evaporation is just like boiling could be due to water vapour demonstrations being done at school using a kettle (Henriques 2002). Students see what we call steam rising from the kettle and, when asked what steam is, often reply that it is water vapour (which is actually invisible). What they are seeing just above the opening of the kettle is water that has condensed, which is, in fact, a

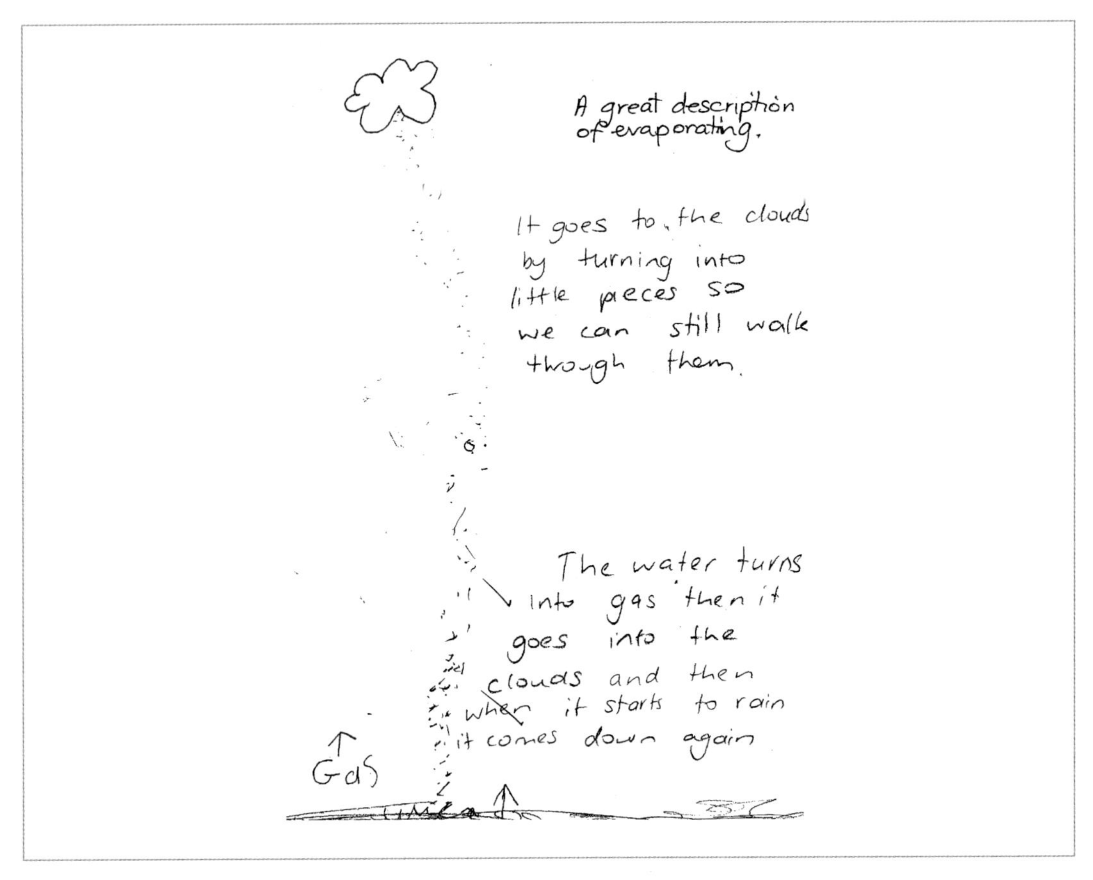

FIGURE 12.7 Child's annotated drawing explaining evaporation

tiny cloud. The use of the everyday term 'steam' to name the visible condensed water not only confuses children but also misses a more accurate teaching opportunity. If using this demonstration, children should be encouraged to look for any signs of water vapour (seen as an invisible space) between the opening of the kettle and the cloud. As teachers, we need to be aware of the limitations of the models or demonstrations we use and make these explicit to the children.

Students' alternative conceptions can result from misrepresentation of science concepts in the community. Take Figure 12.8, for example. This photograph, taken by Christine, clearly shows a highly visible white substance escaping from the chimney of a coal-generated power plant. From your earlier readings, you will know that this white stuff is certainly NOT water vapour (remembering that water vapour is invisible). An attempt by the company running the plant to allay people's concerns by stating that it is water and not smoke going from the chimney into the atmosphere has resulted in an obvious misrepresentation of science. Amusingly, the graffiti on the left-hand side of the sign reads, 'Climate change is happening' (also see the Chapter 9 section 'Evaporation' on p. 357).

FIGURE 12.8 Incorrectly labelled cloud at the top of a chimney

Linking science to everyday situations

When students can make use of their learning in science to explain experiences and observations in the world around them, they begin to see the value of science. If science can help explain how things work or why they happen, it is useful knowledge to have. The challenge for primary school teachers is to develop stimulating activities that help students relate scientific activities to everyday life. Put another way, 'When students are being taught about weather and climate they need to have the opportunities to use stimulating scientific methods and reasoning which bring science and everyday life together' (Spiropoulou, Kostopoulos and Jacovides 1999, p. 58). One way of achieving this is to set their knowledge in a familiar context. This is outlined below.

Setting students' ideas in a familiar context

Richer ideas and questions will be produced if you provide an opportunity for the students to explore not only the materials they will use, but also to set their knowledge in a context with which they are familiar. This familiar context helps them to draw together all they know about the topic. A practical way of setting the context involves studying weather maps. Weather maps are one of the most familiar ways in which science represents ideas; namely, diagrams. They appear on nightly television weather reports and in daily newspapers.

While most people can probably recognise a weather map, many do not know how to read them. Listening to the weather reporter referring to a weather map to explain the day's weather and the next day's forecast does not necessarily lead to an understanding of how to read these specialised diagrams. Everyone could benefit from learning how to read weather maps. Some members of the community, such as pilots, farmers and fishermen, rely on this skill for their work. Others, including sailors, skiers and outdoor adventurers, rely on reading weather maps for their personal safety. Being able to read a weather map may simply help you decide on the best day to plan a picnic, a barbecue or an outdoor excursion.

Researchers who have investigated the interpretation of scientific diagrams have concluded that students need to be taught the skills involved, including the recognition of conventional symbols (Lowe 1986, 1993). Recent research focusing on the literacies of science points to the need for teachers to assist students in learning how to use representations such as diagrams and illustrations, to help them understand science concepts (Tytler, Peterson and Prain 2006; Tytler and Prain 2010). A good starting point is to consider common weather map symbols – what each represents and what they can tell us about the weather. You can do this by collecting some weather maps from your local newspaper or printing some from the Bureau of Meteorology website (see http://www.bom.gov.au/info/weathmap/showwhat.htm). Begin by asking students to extract the different symbols from the map and use the key to name them. Use the information in Table 12.1 to explain the meanings of the symbols used in weather maps and how analysis of the changes in symbols between subsequent weather maps can be used to make weather predictions. A basic explanation of weather map symbols is provided. Students will learn more about weather maps in middle school when they study the concept of air pressure. At this point, how highs and lows form and their effect on other aspects of the weather will be more understandable.

R

TABLE 12.1 Common weather map symbols and what they mean

SYMBOL	MEANING	WHAT IT TELLS US ABOUT THE WEATHER
Isobars	Air pressure	Isobars are lines joining areas of equal air pressure and can be used to determine wind strength. Isobars close together indicate strong winds; the further apart the isobars are the weaker the wind will be.
Pressure zones	High and low	Concentric isobars show zones of high or low pressure. Warm air rises, while cold air sinks. Warm rising air forms low pressure zones, while cold sinking air forms high pressure zones. Highs and lows can be used to predict the weather. Lows usually bring clouds and rain, while highs usually indicate fine weather.
Front	Cold and warm	These form when large air masses of different temperature containing different amounts of water vapour hit each other. The boundary between warm and cold air is called a front. Cold fronts usually mean gusty winds followed by clouds and heavy but brief rain, whereas warm fronts usually bring lighter rains that last longer.
Wind	Wind direction and speed	Special shaped arrows show the direction of the wind and are read from the point they originate to the end of the arrow. The more barbs there are on the end of the arrow the stronger is the wind.
Rainfall	Rain in the last 24 hours	Rain is shown by shaded areas with darkest shading indicating the highest rainfall.

Source: Bureau of Meteorology (http://www.bom.gov.au/info/weathmap/weathmap.htm)

Following direct teaching about weather map symbols, students could be involved in a range of activities to further their understanding of how to use weather maps. Examples of strategies that teachers use are listed below.

- Students observe, measure and record at least two components of the weather as they contribute to a class set of observations and records carried out three or four times a day for a school week (Biddulph and Biddulph 1993). The results could be used by students to draw their own weather maps. Involving students in the construction of their own diagrams is an effective way for students to become familiar with the conventions used and improve their comprehension of like diagrams.
- Upper primary students can combine this activity with making a display sheet of three or four successive weather maps and weather forecasts clipped from the previous day's newspaper, such as those in Figure 12.9. (Please note: this requires careful instruction. Parents do not take kindly to discovering a see-through section when they first open their morning paper; the explanation 'I need it for school' does not win gold stars for the teacher either.) Each student can record their personal description of the weather for that day, and then the next morning place the previous day's forecast next to this description. Group discussions of the differences between the two as perceived by different students are usually animated and productive.
- Record the evening television weather report on whichever channel provides the most comprehensive weather data relating to that day and for the forecast period. Play the recording the next day, pausing frequently to allow the students to discuss in groups various segments of the presentation, in particular:
  - the report of that day's weather, which is based on records of local measurements of maximum and minimum temperatures, rainfall, cloud cover or sunshine, wind strength and relative humidity
  - the forecast of weather, which is based on measurements taken throughout and beyond the country, of air pressure (notably highs and lows), wind speed and direction,

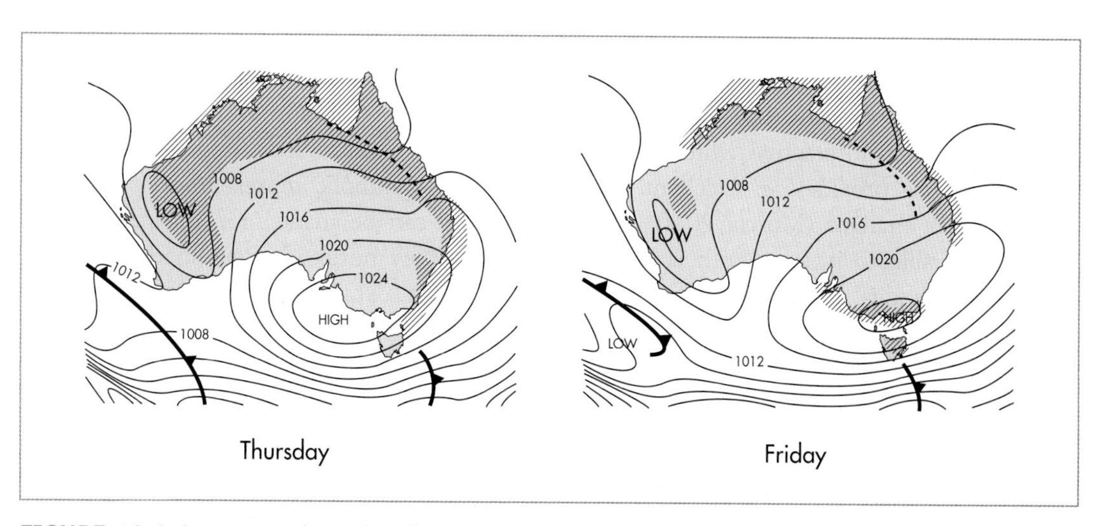

FIGURE 12.9 A prognostic and a diagnostic chart

cloud type and direction of movement, and upper atmosphere conditions derived from balloon probes and satellite imagery.

Russell and his colleagues (Russell, Longden and McGuigan 1993) found that many children thought that the forecast was a description of the weather in the places 'we get our weather from' rather than a prediction based on measurements of the various weather parameters. As this has an element of truth to it, children who hold this view can be invited to test its accuracy over a period of two weeks or so.

A number of recent studies recommend teaching scientific phenomena such as the water cycle in the 'context of their effects on daily lives of humans' (Candek 2009, p. 87) in order to help 'students develop a richer understanding of water systems' (Covitt, Gunckel and Anderson 2009, p. 32). Not only does such an approach lead to increased scientific literacy, it is vital in developing 'big picture' views on environmental systems and the interconnectedness of science with our everyday lives. These researchers argued that 'a fundamental understanding of water in environmental systems is essential in helping citizens reason effectively about how human actions impact natural and environmental systems' (ibid., p. 50).

Explaining the weather

The purpose of this next section is to discuss certain aspects of the weather to help you develop ways of explaining them to children. As you work through this section, stop and reflect on ideas that challenge your previous understanding.

Explaining changes in temperature

We are all aware that the temperature of our environment changes, but what causes this? (You may wish to record your current knowledge about this before you read further.)

Immediate changes

Before you can start to explain changes in the weather, you need to be sure that students are aware of such changes. Just because it happens all around them, don't take for granted that your students have a) really taken notice or b) tried to use their science knowledge to understand observations that we (as adults) may think are important. A recent article considering what interactive whiteboard technology offers multimodal representations in primary science provides an excellent example of using symbolic representations to show daily weather changes (Murcia 2010). Such an activity can be made more powerful when children are able to interact with it during the day as they notice changes in the weather, reinforcing the idea that weather does not necessarily remain stable even for a day's weather. These provide really good examples of effective teaching, making the most of modern technology.

R

Involving students in the regular collection of weather data also presents opportunities for teaching about the importance of organising data in ways that we can make sense of the information. One way of doing this is to turn the problem over to the students by setting a scenario. For example, tell the students that a weather reporter from the local TV station has heard they have been collecting weather information and has asked if they could provide a report on last week's weather. Point out that the data they have collected is all over the place: How can it be organised so that this person can understand their data? This presents a science problem for them to solve together (Coskie and Davis 2009). This kind of activity presents a realistic situation that stimulates students and involves them in problem solving.

Short-term changes may be produced by clouds moving between the sun and us. They absorb and reflect some of the sun's light and heat, which was falling on the part of the Earth where we are located. While such changes can temporarily reduce our enjoyment of being at the beach or on a picnic, they cause little immediate change in the air temperature. Daily temperatures are, therefore, recorded in the shade in order to discount these momentary and frequent fluctuations. If the increase in cloud cover remains for some time, however, particularly if it is accompanied by a colder air mass (cold front) moving in, the temperature of the surrounding air (ambient temperature) will also be reduced.

Changes through the day and the year

In addition to these sporadic, short-term changes, there are more gradual daily and seasonal temperature changes. Despite their different timescales, both types of change are caused by much the same mechanism; namely, changes in the intensity (or concentration) of the sun's heat falling on a particular location.

In the middle of the day, the sun is more nearly overhead than at any other time, which means that the heat (or infra-red) rays from the sun are falling on a smaller area of the Earth's surface. In the early morning and late afternoon, the sun's rays approach at an angle to the Earth's surface and so are spread over a greater area.

In similar fashion, because of the tilt of the axis on which the Earth rotates, the midday sun is more directly overhead in the Southern Hemisphere during the months of December, January and February than it is during June, July and August, when the Earth is on the other side of its annual revolution around the sun. Hence, locations in the Southern Hemisphere receive a higher concentration of the sun's energy during these summer months and our average temperatures are therefore higher.

Figure 12.10 may make sense to you, but many primary students do not have a clear understanding of 'area'. To them, the idea that a given 'beam' of the sun's rays will produce a greater intensity of heat (and therefore rise in temperature) where it is distributed over a smaller area will not make much sense.

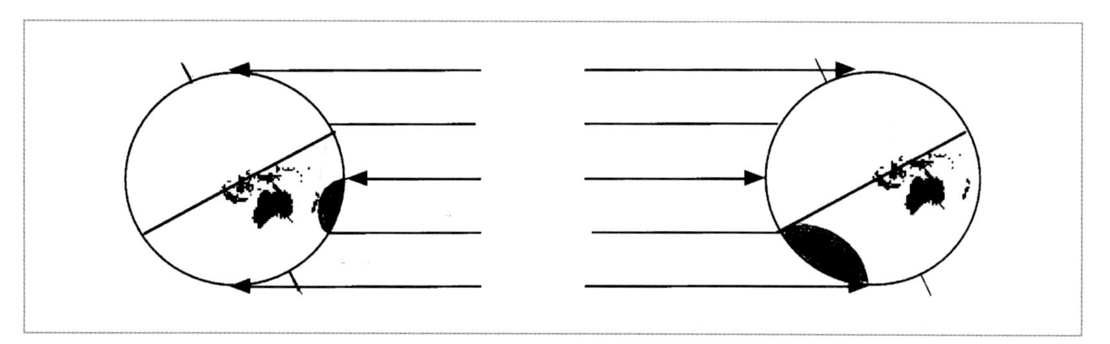

FIGURE 12.10 Diagrams showing the sun's rays falling on Earth

Activity 12.6 Explaining by illustration and analogy

To illustrate this relationship in a qualitative sense, shine a torch beam on a flat surface (or, even better, the surface of a large sphere) from a distance of about 50 centimetres in a darkened room. You will observe that, when the torch beam is shining directly downwards on the surface, the circle of light produced is smaller than when the beam is directed at an angle to the surface.

To advance from this illustration to the idea that there is a greater intensity of heat when the sun is directly overhead, think in terms of the analogy of equal quantities of jam being spread over each of the two light patches. You will probably sense intuitively that the light (or heat or jam) is spread more thinly over the bigger area.

Other causes of temperature variations

Heat from the sun produces minimal increase in the temperature of the Earth's (gaseous) atmosphere as it passes through it, but warms the surfaces of the (solid) continents and (liquid) oceans when it strikes them. Because the atmosphere is heated almost entirely by its contact with the continents and oceans, atmospheric temperature decreases with elevation. Thus, high mountains are snow-capped and water vapour condenses out of rising moist air to form the multitudinous tiny water droplets we call clouds.

Significant long-term temperature reductions are believed to occur when huge amounts of fine dust from repeated volcanic eruptions or the impact of a massive meteorite is suspended in the atmosphere. Variations in the percentage of carbon dioxide and other greenhouse gases in the atmosphere are considered to be the cause of the most significant long-term variations in world temperature (and climate) (IPCC 2007). Rapid expansion in the numbers of photosynthesising green plants is thought to have been the major cause of the Earth's ice ages, as the effectiveness of the Earth's greenhouse blanket was reduced by the removal of carbon dioxide. Conversely, when plant numbers dwindled, carbon dioxide levels rose, as did the Earth's temperature. This is why environmentalists are so concerned about the rising level of greenhouse gases in our atmosphere over the past 50 years.

The significance of global warming due to the enhanced greenhouse effect as an environmental issue means that students are likely to encounter these topics via the media and from the conversations of adults (Mintz 2006; Skamp, Boyes and Stanisstreet 2009). As primary-age students will already have embedded notions about this topic, they might ask questions in class or raise it as a topic of interest. While the sophistication of these phenomena and the need for conceptual understanding of energy, heat and change has led to their inclusion in secondary science syllabuses, instances may arise in the primary classroom in which a teacher is asked to answer questions or facilitate a discussion about the greenhouse effect, global warming or the hole in the ozone layer.

Activity 12.7 Thinking about global warming and ozone depletion

This activity will help you consider and reflect on your personal knowledge of global warming and ozone depletion. It will be most effective if done with a colleague.

1. Write your conceptions of:
   - global warming and its causes
   - ozone layer depletion and its causes.
2. Comment on:
   - the impact of global warming and ozone layer depletion on ecological systems
   - In what way, if any, are global warming and ozone depletion different?
3. Compare and discuss your answers to these questions with a colleague.
4. Assess your level of understanding of these ideas by referring to a summary of common alternative conceptions and scientists' ideas (Henriques 2002).

While students can be the source of their own alternative conceptions, teachers can also contribute to their development. Research into preservice teachers' understanding of environmental issues reveals 'insufficient knowledge as well as misconceptions regarding ozone layer depletion' (Bahar, Bağ and Bozhurt 2008). The oversimplification of concepts, the imprecise and inconsistent use of language by teachers, the incorrect use of analogy and confusion over closely related concepts can result in the formation of alternative conceptions by students (Dove 1998; Lee et al. 2007). Referring to global warming as the greenhouse effect can be misleading because the increase in temperature in a greenhouse is partly due to reduced movement of the air (Dove 1998). Winds are still active near the surface of the Earth, despite the increase of greenhouse gases in the atmosphere. When using analogies or models to help explain ideas, their limitations also need to be noted and discussed: 'If the points at which the analogy breaks down are not made clear to the students, the analogy can introduce more alternative conceptions, instead of clarifying concepts as intended' (Venville and Donovan 2006, p. 18). Research by Lee et al. (2007) noted that students were often confused by the analogy of the greenhouse effect and an actual greenhouse. Their study showed advancements (post-instruction) in Year 5 students' scientific understanding about the greenhouse effect, its causes and consequences, but students remained confused about the difference between the enhanced greenhouse effect and ozone layer depletion (ibid., p. 124). To assist your understanding of the differences between the 'greenhouse effect' and the 'hole in the ozone layer', scientists' views are summarised below.

Greenhouse effect

When heat energy from the sun hits the Earth's surface, much of the energy rebounds into space. Some of the radiating heat bounces off the ozone layer and stays close to Earth, making the planet warmer than it would be if it had no atmosphere. This natural process makes Earth warm enough for life as we know it to exist. Increased atmospheric gases – carbon dioxide, methane, chlorofluorocarbons (CFCs) – prevent more heat than normal from rebounding into space, which results in an increase of surface temperatures. It is feared that significantly increased temperatures on Earth could lead to rising sea levels as ice at the poles melts. Dramatic changes in weather and climate conditions, such as violent thunderstorms, droughts, cyclones, heatwaves and so on, may also occur (Henriques 2002), and there are also predictions that some low-lying islands will disappear (Peacock 2007).

Ozone depletion

While ozone depletion is technically not an example of temperature change, it is included here as many students associate greenhouse warming with ozone-layer depletion and think that an effect of ozone depletion is increased temperatures (Boylan 2008; Fisher 1998). A thick layer of the Earth's atmosphere is made of a special gas called 'ozone'. This ozone layer acts as a protective blanket above the Earth, blocking potentially harmful ultraviolet (UV) radiation from the sun from entering the Earth's atmosphere. Specific chemicals, such as CFCs, react with ozone and destroy it. This has led to the thinning of areas of the ozone layer, especially above the poles. These thinner areas, popularly titled 'the hole in the ozone layer', let greater amounts of UV radiation through to the Earth's surface. Increased UV radiation levels can damage living things due to the effects of radiation on biological tissues.

Activity 12.8 **Do you drive a greenhouse?**

This activity is best done in collaboration with two or three car-driving colleagues.

- Record the air temperature outside and inside your car when you first leave it in a car park, preferably on a sunny day. Take these temperatures again at about noon. After noting the noon temperatures, leave a window down for about 10 minutes and see what difference this makes. Repeat these procedures later in the afternoon. Compare your results with those of your colleagues.
- Devise some variations in the procedure; for example, the direction the windscreen faces, placing a reflector shield (or rug) across the windscreen (inside/outside), or covering the roof with a white sheet.
- Record your findings in a table.
- Add an explanation to show how the windows of your car are acting like the Earth's atmosphere with regard to the transmission of the sun's heat.

Reflect on your understandings of the greenhouse effect.

Discuss any limitations of this activity as an analogy for demonstrating the greenhouse effect to upper primary students.

Note that the two ideas outlined above are, in the main, different phenomena – the hole in the ozone layer is not responsible for the greenhouse effect, nor does ozone-layer depletion increase greenhouse warming. As Mintz (2006, p. 2) states:

> Although there are points of interrelation between the two phenomena, they are in essence conceptually separate processes. Teachers who, in response to students' interest, suggest that they are one and the same thing risk instilling misconceptions that may persist into secondary schooling and even further into adult life.

The fact that each process resides somewhere in the Earth's atmosphere and cannot be directly observed by children and adults, and that CFCs can be a contributor in both cases, is a likely source of confusion. The teaching of these ideas to primary-age students requires strategies that can make these phenomena more concrete and clearly distinguish between them. The Australian Greenhouse Office has produced some resource booklets that contain practical examples of how daily human activities contribute to global warming, and include suggested ways we can reduce this; for example, *Global Warming Cool It!* is available from http://www.portstephens.nsw.gov.au/files/217567/File/Globwarm_Cool_It.pdf.

FIGURE 12.11 Child experiencing wind from a fan

Given the complexities of the concepts involved in the above environmental issues, it is not the primary teacher's aim to develop a complete understanding of how these phenomena happen. Your role should be geared more towards raising students' awareness of the consequences of these changes for weather and our environment. Developing an interest in and concern for the long-term impacts of these issues may encourage children to learn more about them in future education.

SSI

Explaining the wind

Sensing the wind

Students are not quite sure about wind. It is a bit like a riddle: 'You can feel it but you can't touch it; you can see what it does but you can't see it; it carries smells but you can't smell it; and it makes sounds but it is not a sound. What is it?'

One way to introduce your students to the idea of wind as moving air is to have them individually stand in front of a fan (see Figure 12.11) and ask them to explain what they felt. Other ways to make air move can also be investigated, such as flicking the pages of a book in front of a student's face. The idea that heat can make air move can be introduced by children hanging streamers (at a safe height) above a heater and observing them move by themselves as the air below them is heated and rises. Acting out hot air rising and cold air sinking, creating air currents (which we feel as wind), can help students to understand how winds are formed.

Activity 12.9 **Is wind really only air that moves?**

Record in brief outline three or four other activities that you would use to help students (of whichever age group you choose) to understand that wind is moving air. Bubbles, balloons, bags, blowing paper and bending trees may be useful pieces of equipment to begin with. For each activity, include the symbols of an ear, eye, nose, hand (skin) and/or tongue to indicate the sense(s) that you are using to collect the information. Exchange your ideas with a colleague. Is your colleague convinced? If so, team up to try out your most convincing activities with a child.

TABLE 12.2 Modified Beaufort scale of wind force (speed)

BEAUFORT NUMBER	DESIGNATION	DESCRIPTION	WIND SPEED (KM/H)
0	Calm	Smoke rises vertically	Less than 1
1	Light air	Smoke drifts but no leaf movement	1–5
2	Slight breeze	Wind felt on face; leaves rustle	6–11
3	Gentle breeze	Leaves and twigs moved constantly	12–19
4	Moderate breeze	Loose paper and small branches moved	20–28
5	Fresh wind	Small trees sway	29–38
6	Strong wind	Large branches move	39–49

A

Students also have difficulty in reasoning about wind direction. An effective way to teach this concept is through an investigation designed to explain how windvanes work. For an engaging hands-on activity, see Koballa's (2008) description of assessing Year 4 students' understandings about the operation of windvanes. The article follows the 5E model and includes an assessment rubric for teachers to use.

Wind strength and wind speed

Primary school students are often confused when adults, including television weather presenters, use the terms 'wind strength' and 'wind speed' interchangeably. This conceptual confusion will usually be overcome when the students use the Beaufort scale of wind strength (see Table 12.2). This scale is based on our observation of the effects of winds of different strengths, but has also been correlated with wind speeds. The Beaufort scale ranges from 0 (calm) to 12 (hurricane), but it is the 0–6 range that is of most use to us. It is particularly valuable for use with primary school students because the wind strength scale is descriptive (and so they can readily identify with it); it is also directly linked with a quantitative measure of the more abstract but synonymous wind speed.

Activity 12.10 **Measuring wind speed**

Do you think you understand all this? Of course you do – but just to make sure, perhaps you should try it out for yourself. Thinking about what you are doing is always a great way to learn.

- Make a simple anemometer (a device used to measure wind speed) based on one of the designs featured in a primary science activity book.
- Calibrate it using the modified Beaufort scale provided in Table 12.2.
- Use it to compare the wind speed at ground level and on top of the tallest building you have access to (or a hill if you have one handy).
- Note the maximum and minimum wind speeds recorded over a 10-minute period.
- Record your results along with what you learnt about wind speed/strength, any difficulties you encountered and any ideas you developed about what specific learning outcomes you would be looking for when your students carried out these activities.

Explaining changes in atmospheric water content

Water in the air: evaporation, condensation and clouds

Read the following conversation between a teacher and a student:

Teacher	What do you think has happened to the water?
Student	It evaporated.
Teacher	What do you mean by 'evaporated'?
Student	It disappeared.
Teacher	OK, but where has it gone?
Student	It hasn't gone anywhere, it just disappeared.

Conversations such as this between teachers and their primary school students once stopped after the second line. The student had said the magic word. By continuing the dialogue, we are now aware that the student's understanding of these magic words very rarely matches the teacher's. In fact, they are more likely to obstruct rather than assist our students in coming to an understanding of evaporation of water as the process by which water particles at the surface of liquid water change into water vapour (gas) and move into the spaces between the gas particles in the air above the water's surface.

A case study of conceptual change regarding evaporation is described in Skamp (1995); further discussion of evaporation and condensation may be found in Chapter 9 (see the section 'Students' ideas about matter and change' on p. 356). In order to refresh your memory about water in the air, you may find it useful to try the following activity, which attempts to address the problem of making perceptible the imperceptible changes involved in evaporation.

Activity 12.11 Now you see it, now you don't

For this activity you will need three to five noses, a dark background, an aerosol can of low-allergy fly spray or room freshener or, even better, an old-fashioned perfume spray, and a pump-action water atomiser of the type used by hairdressers.

Stand with the dark background to your right, the observers attached to the noses to your left at 2-metre intervals and all facing your left shoulder. Point the aerosol can directly forward and ask the observers to record as much as they can of what they see in front of the nozzle when you press the button, then to indicate when they smell the insect spray, room freshener or perfume. Discuss their observations.

Repeat the procedure using the water atomiser, but this time ask the observers to focus entirely on what they see and where they think the water has gone.

Join with your observers to make statements on what this tells you about:

- water droplets
- water vapour
- air
- how smells travel
- evaporation
- the mixing of gases
- whether clouds are made of water droplets or water vapour
- how clouds move
- anything else that you think is worth recording.

CASE STUDY

A 5E constructivist lesson sequence

Applying learning about working scientifically (see Chapter 2) will involve using investigation work to facilitate conceptual learning. Although investigating is usually listed as a separate outcome in syllabus documents, integrating investigations into the conceptual outcomes of the curriculum enables the concepts to provide the learning context (see Figure 12.1 on p. 504). A useful model for structuring investigations and sequencing lessons that is grounded in constructivist perspectives is the 5E instructional model (AAS 2005; Hackling 1998). The application of this model to a weather topic is outlined below.

Investigating how we stay dry in the rain

Phase 1: Engage

This lesson sets the context, raises questions and elicits students' existing beliefs. Teacher comes to class dressed in a raincoat and holding an umbrella; water is dripping off her. Teacher props the umbrella on the desk and takes off the raincoat, explaining that she had to go through the sprinkler on her way to class, or complaining about the rain outside. Children are asked to explain why the teacher chose to wear these clothes. What properties of the materials stopped her from getting wet, and why? The teacher allows children to feel the umbrella on the top and bottom surface and describe what they observe. The teacher asks if they can explain their observations. Children record their observations and explanations on a worksheet and draw an umbrella and a raincoat. Teacher asks children to share their ideas with the class as the teacher makes a summary of their ideas about the materials and how they work onto a sheet of cardboard. Students' ideas are compared and discussed.

Phase 2: Explore

This lesson involves investigation work in which students gain firsthand (and, where possible, concrete) experience of the phenomena of interest. In small groups, children investigate a range of materials to test their suitability of use for making waterproof garments. Teacher asks children to suggest how they can test the materials to decide which type of material is the best for stopping water from getting through it. Teacher and children discuss how they could investigate this, making sure the test is fair and can lead to a plausible conclusion. Once decided, children conduct the test (for example, see Figure 12.12) while the teacher monitors and listens to the students talk about what happens and why. Children record and share their findings (see the Chapter 8 case study 'A comparing investigation' on p. 335).

Phase 3: Explain

This lesson involves drawing students' beliefs from the 'Engage' lesson, concepts introduced by the teacher, or from text reading. These are used to construct explanations for the 'Explore' phase. Children discuss the findings of

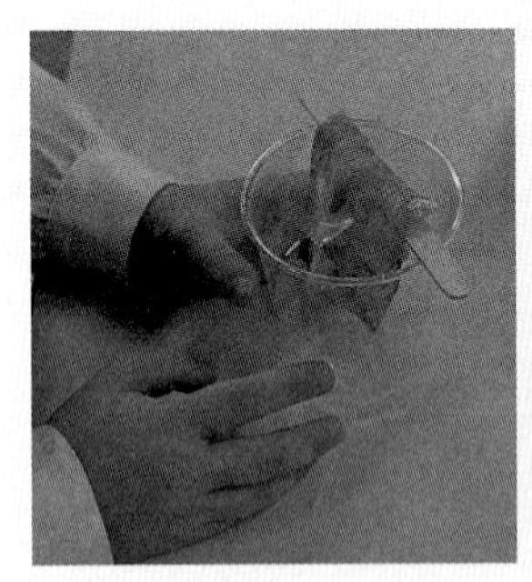
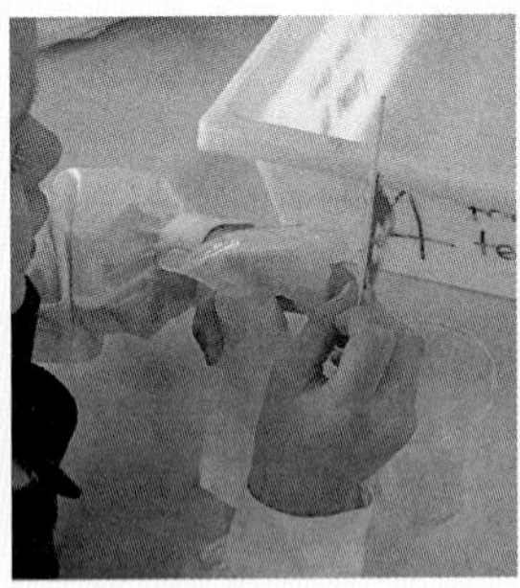
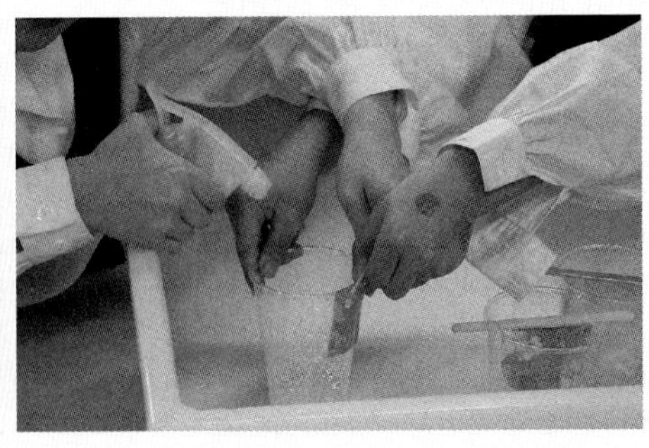

FIGURE 12.12 Method children decided on to test how waterproof fabrics were

their investigation and compare their ideas. The teacher introduces terms and definitions of relevant concepts, such as 'porous', 'absorbent' and 'water repellent'. Children jointly construct explanations for why the raincoat and the umbrella kept the teacher dry, drawing on observations and data from their investigation.

Phase 4: Elaborate

This lesson involves students applying conceptions developed in the 'Explain' lesson to new contexts, thus extending and integrating their learning. The teacher brings in various items of clothing that are labelled 'showerproof' and 'waterproof'. Children are asked to observe the materials and explain reasons for the differences in the labels. Children use their knowledge to compare the degree to which the items of clothing are waterproof. The teacher also shows students something that is not waterproof and asks them to suggest how they could stop water going through it. Discuss various coatings or treatments that can be applied to a range of materials to make them waterproof.

Phase 5: Evaluate

This lesson provides an opportunity for students and the teacher to assess developed conceptions and compare them to their beliefs at the 'Engage' phase. Children are given a second copy of the worksheet completed in lesson one and asked to redo the exercise in the light of their learning from previous lessons. The teacher asks children to share their ideas and again makes a summary on cardboard. The old and the new summary of ideas are shown simultaneously and children are asked to compare them and to identify where their ideas have changed and what led them to these changes.

Hackling's (1998) booklet (also see http://www.eddept.wa.edu.au) is a good resource for information on types and examples of investigations, scaffolding tools to guide and support students' (and teachers') investigations, cooperative learning strategies for effective investigation work, and assessing students' investigative skills (also see Chapter 2). A further source of lesson ideas constructed around the 5E instructional model is the *Primary Connections* unit 'Weather in My World' (AAS 2005).

Where should teaching and learning about weather be situated?

Weather is a theme which provides teachers with the opportunity to use the outdoor environment and integrate across subject areas, as well as explore related topics such as environmental change. Each of these areas can be ways of engaging student interest, which is usually a prerequisite to conceptual learning (see the Chapter 1 section '"Hot" conceptual considerations: student engagement' on p. 25). These three areas are explored in this section.

Teaching outdoors

Teaching outdoors can help students contextualise learning about weather and the environment. Even brief outings can help engage students and enhance their learning; for example:

- taking young children outside to see how wet the ground is in different parts of the playground after rain
- making daily visits to a weather station to collect data
- visiting the Bureau of Meteorology to find out about weather forecasting
- going on a trip to the ski fields to experience snow conditions firsthand.

Outdoor activities can be used to introduce a unit of work – for example, a visit to the weather station on Lord Howe Island by students from the local primary school could lead to learning about how weather data are collected and used to aid fishermen, tourist operators and visiting sailing craft in planning their daily activities. Placed in the middle of a unit of work, an outing could form an integral part of the topic whereby students develop their knowledge and practise their skills before applying them on location. Students might visit a local dairy farm to measure the wind speed in a variety of locations to determine the best place to install wind turbines to generate electricity. Alternatively, an outing can be placed at the end of a unit of work to consolidate students' learning and as an opportunity for assessment. Students could go to a location such as the Forest of Tranquillity in Gosford, New South Wales, where they are able to measure and compare the climatic conditions in the open parkland to those in the rainforest and communicate their findings through a scientific report that can be assessed by the teacher.

Other considerations for teaching about weather outdoors

Students usually find outdoor-learning activities fun; they learn by being there and doing things that they cannot experience from using books, doing experiments at school, watching videos or using CD-ROMs or computers. As most students enjoy being outside and exploring the natural environment, learning activities that occur outdoors are motivating (Melancon 2000). This natural motivation can be channelled into meaningful learning (Purdie 2000). The inquisitive nature of young children is fostered when they are given the opportunity to explore their environment (Conezio and French 2002).

M&E

While indoor teaching during science lessons is frequently aimed at the development and practice of skills, there are certain skills that require environments or facilities located only outside the classroom. To make the measurement of wind speed meaningful, for example, it needs to be done outside, where the wind is a component of the natural environment.

Outdoor teaching is not only about learning content and skills; it is also about attitude development. Outdoor teaching experiences, such as field trips, can be very exciting, stimulating and memorable learning occasions that enhance learning outcomes. While students may forget the details of things learnt on field trips, they may long remember the experience itself (Dale 1995). Outdoor activities are often used to promote positive attitudes towards the environment (Ramsey-Gassert 1997; Skamp 2002). However, students do not adopt pro-environmental behaviours simply by being involved in outdoor activities. The act of taking children outdoors for learning does not guarantee success. The key is to ensure that such experiences are planned with outcomes in mind. The achievement of environmental education learning outcomes requires careful planning and explicit teaching of action-oriented processes.

Technology can be used to enhance outdoor teaching through capturing images, sounds, data and actions that can be used for post-visit experiences. For example, students can:

- record sounds to play back at school to see if they can remember the activities associated with those sounds
- take photos with a digital camera to use in the construction of an account of the day when they are back at school
- act like a film crew, recording the outing for part of a documentary to promote their school to the community
- use data loggers to measure various aspects of the physical environment.

ICT

The above suggestion of using digital cameras on field trips was implemented on an excursion to the zoo with Year 2 students. Students created an open-ended question to investigate and then used digital cameras in a focused way to capture and record directed observations before creating a digital story to share their findings with the teacher (for assessment) and their family. Questions students investigated included: 'I wonder if all baby animals look like their mother?', 'I wonder what a zoo keeper does?' and 'How do new animals get into the zoo?' (Davidson 2009). Upon returning to the school, the students used the computer lab to assemble the photos they took into a book and adding captions using iPhoto (software for Macintosh computers), providing opportunities for integrating both ICT and literacy into the culmination of the science unit on animals. Students could take photos on a weather-related excursion such as a visit to a local weather station or a brief trip outside the classroom to capture cloud types.

If you have not taken students outside for science learning activities, you may find the tips from Christine's firsthand experiences of working in an environmental education centre and as a classroom teacher of young children a useful guide (see the online Appendix 12.4).

Appendix 12.4 Tips for teaching outdoors

Activity 12.12 Outdoor teaching and weather

Outdoor activities, including field trips, can be memorable experiences for children but hard work for the organising teacher(s). What are your thoughts about teaching outdoors?

Reflect on your own schooling. Did you ever go on excursions or venture out of the classroom for lessons?

- Make a list of the times you can remember having outdoor lessons.
- Were they a positive experience for you?
- What do you remember about them?
- Compare your memories with at least two colleagues to see if they had similar or different experiences.

As a teacher, how will (or do) you incorporate outdoor teaching in your teaching program?

Identify the advantages and disadvantages of taking children for weather-related outdoor experiences (whether it be the schoolgrounds, a weather station or some other location).

Integrating weather across the curriculum

Integration and learning – for example, blending literacy and science – was discussed in Chapter 1 (see the section 'Integration and learning' on p. 37). Here, examples are outlined of how learning about weather can have enhanced relevance through bringing science together with generic and science-specific literacy skills and also technology (interpreted in several ways).

Weather and literacy skills

Figure 12.13 provides an example of readily available stimulus material that can be used to develop writing, talking and listening, drawing and analytical skills.

Activity 12.13 Using a weather extract from a local newspaper

- List the ways you could use Figure 12.13 as a constructivist science teaching resource while also noting where writing, talking and listening, drawing and analytical skills are used.
- Include reference to your Science and English syllabus and list the learning outcome(s) students could achieve using this resource as a stimulus.

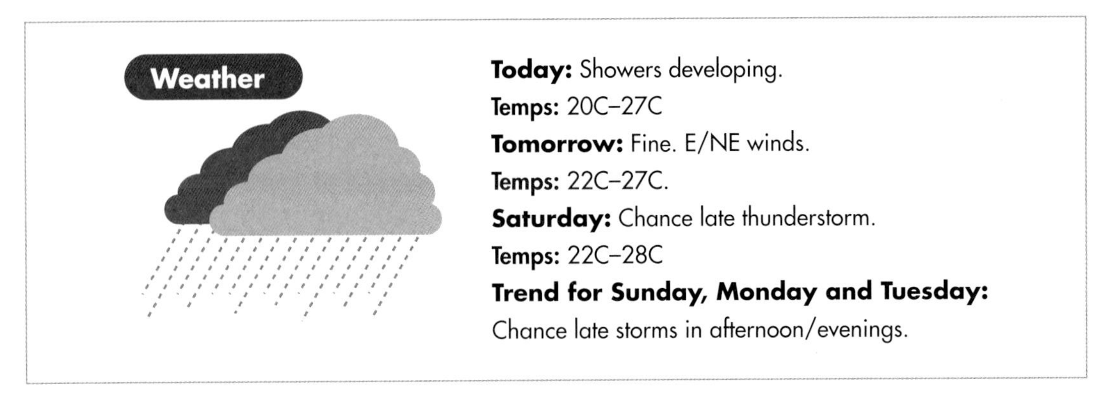

Weather

Today: Showers developing.
Temps: 20C–27C
Tomorrow: Fine. E/NE winds.
Temps: 22C–27C.
Saturday: Chance late thunderstorm.
Temps: 22C–28C
Trend for Sunday, Monday and Tuesday:
Chance late storms in afternoon/evenings.

FIGURE 12.13 Newspaper clipping of a daily weather forecast

Appendix 12.2 Examples of cross-curriculum activities using an extract from a local newspaper

Appendix 12.3 Once I was a water drop – narrative

A variety of cross-curriculum activities can be developed around this one stimulus (see the online Appendix 12.2 for examples). After the completion of some of these activities, students could be asked to list what else they want to know or learn about weather. Text types such as narratives, procedures and explanations can be used to enhance learning in science. To aid kindergarten children's understanding of the complexities involved with water evaporating from the ground and travelling invisibly into the air before forming clouds and falling as rain, Christine wrote a story called 'Once I Was a Water Drop' (see the online Appendix 12.3) to help children link the concepts of evaporation to cloud formation and rain. Other examples of integrating related to a weather theme include the following:

- Singing songs such as *Incy Wincy Spider* that many children would remember from preschool can be a fun way of activating students' prior knowledge and relating learning about rain to everyday experiences.
- Involving older students in the debate about alternative energy sources and their long- and short-term effects on the environment can help to highlight the implications of science on society.
- Students can use their creative abilities in poster art of digital imagery using software applications such as Kidpix or Adobe Photoshop to create an advertisement urging people to conserve and recycle water during drought periods.
- Constructing outdoor shelters with different roof materials and investigating their suitability in different weather conditions could help develop experimental design skills and data collecting, recording and reporting.
- Exploring the symmetry, symmetrical design and hexagons in snowflakes can help develop mathematics concepts (Ellison 2009).
- Ocean science literacy can be combined with artistic representation of microscopic organisms to enhance art skills (Haley and Dyhrman 2009).

Books and stories can be used to enhance students' understanding of science concepts. Research for the BASICS (Books and Stories in Children's Science) project reveals that stories can support children's learning by providing 'a context for the science concepts, a shared experience for exploring current ideas and understanding, lesson structure, real life problems, real characters who the children can connect with and strive to help and opportunities for developing overall and scientific language' (McCullagh, Walsh and Greenwood 2010). Christine uses a big book, *Who Cares about the Weather?* (Berger 1992), to introduce the topic of weather to her kindergarten class.

Weather and technology

Technology as a subject embraces many aspects, two of which are:

1. the use of technological equipment
2. the design, make and appraise (DMA) process (see chapters 1, 3 and 6).

As we have seen already, weather can greatly influence the things we do and so we are well aware of its major components – wind, rain, temperature and sunshine or cloud cover. We can, and often do, observe these by direct sensory experience, but to measure them we need the help of technology. This should not be seen as a problem in primary classrooms, but as a wonderful opportunity to integrate two areas of the curriculum.

Integrating science and technology can also be an effective way to stimulate student engagement. An article by Claymier (2009, p. 36) describes a project where Year 6 students 'work for a fictitious power company as they explore electricity and build working wind turbines'. This project provided a real-world connection where students answered questions such as 'What percentage of the total electricity produced in the United States today comes from wind turbines?' and 'How do wind farms affect wildlife in the area?' (ibid., p. 38). To meet the challenge, students had to design and test a wind turbine that would produce an electric current and it involved them learning about magnetic fields, electric motors and energy conversions. Learning in the context of a real-life situation engages students by providing them with the 'need to know' as they work towards building a working model that functions due to scientific principles.

M&E

Students can also be involved in designing, making and appraising their own equipment to collect data for investigations. Christine had her Year 1 children make simple rain gauges to enable them to see how they worked. After making the rain gauges, the children went outside and tested them under a variety of artificial rain conditions (e.g., the teacher using a hose). The children observed that less water was collected in their rain gauges when it was 'sprinkling rain' compared to when it was 'pouring rain'. Some of the children wanted the rain (hose) to be placed directly over their rain gauge rather than the gauge collecting the drops that fell where it was positioned. This enabled the teacher to restate the purpose of the rain gauge to sample the amount of water falling on average across the surface of the ground rather than just in one spot. Some children made incidental observations about the effects of sprinkling rain as wetting the grass and soaking into the soil, compared with pouring rain when localised flooding was seen, as the children had to step back to avoid getting their shoes wet by water forming small streams as it ran to lower ground. The children were delighted that they were able to take their rain gauges home to show their families. Parents and carers came back with stories about the gauges being erected at home along with rain record charts being stuck to the refrigerator and of children explaining how they worked to anyone prepared to listen.

Integrating engineering is also possible in weather topics. For example, Year 3 students in an inclusive classroom learnt about the core processes of engineering to design windmill blades. Students were guided by the Engineering is Elementary (EiE) steps – Ask, Imagine, Plan, Create and Improve – in a unit titled 'Catching the Wind: Designing Windmills'. In this unit, students learned about position, force, motion and energy using 'inquiry based science instruction to help students understand these basic physics concepts prior to engagement in the engineering design process' (Lottero-Perdue, Lovelidge and Bowling 2010).

NOS

Interdisciplinary tasks such as learning about extreme weather conditions in order to design and construct buildings that students test to see if they can withstand simulated hurricane-scale wind and rain can provide rich learning experiences for students. Sterling (2010, p. 51) reported that students were involved in 'doing science like a scientist' and 'started to understand the complexity and interdisciplinary nature of science'.

Weather and environmental change

Weather topics for lower primary students focus on what weather is and how it affects humans. Topics for upper primary are more globally oriented and involve the effects of weather on the natural and built environment.

To assist students to relate slow changes in our environment to the causal agents of weathering, such as air, water, and heat and light from the sun, learning experiences could include:

- direct observations of natural and built structures in our physical environment
- hands-on investigations to make the imperceptible perceptible
- examining environmental change using secondary sources.

Changes in our physical environment

We see every day the long-term results of slow changes in our physical environment – peeling paint, rusting iron, crumbling stone buildings and decaying timber – caused by the weather. Similarly, features of the natural environment, such as sandstone cliffs, deep gorges, wide rivers, limestone caves and soil, are the results of slow, long-term effects of the weather. How can you help your students understand the role of the weather in such changes? One method could involve a trip outside the classroom in which the students go on a hunt for any examples they can find of long-term changes. The students could use written descriptions, labelled sketches or take photographs to record their ideas about:

- what it is like now
- what they think it was like originally
- what agents may have caused it to change
- how the changes may have happened
- how long they think it has been there for
- what it might look and feel like 100 years from now.

With available technology, students could create a slide show to demonstrate the changes involved.

Making the imperceptible perceptible

Many of the environmental changes caused by interactions between air, water and the Earth's crust are imperceptibly slow. How can you make the imperceptible perceptible for your students? Rocks are so hard that it is very difficult to envisage their slow breakdown by agents of weathering and erosion. One way to overcome this is to have students make their own rocks (for example, with plaster of Paris), which they could subject to experimentally controlled, artificial weathering processes, then observe and discuss their effects. Follow-up activities could involve students theorising about:

- how blocks of rock are changed to rounded boulders by rivers and the action of waves
- why soils and sand in different locations are different in colour
- what shape they would expect grains of beach sand to be and where they think the grains come from
- how they think rounded Australian landforms, such as the Bungle Bungles, Uluru, the Snowy Mountains or Mt Coot-tha, were formed (see the Chapter 11 section 'Learning about rock types and how they form' on p. 475).

Another way to help students observe the progress of slow changes is to use technology such as a digital camera to record these changes over time. Taking a photograph of the same deciduous tree as it changes with the seasons can assist students in relating the changes to the time of the year and to see the cyclic pattern of change. The same technique could be used to study erosion in areas of high wear in the playground; for example, a grassy hill above playground equipment. Take a photo at the beginning of the year (when grassed areas are mostly covered) and then take snapshots every week over a one-month period (or longer if needed). Studying the photographs may help students to identify the causes of the observed changes in the environment and lead to a realisation of their personal impact on the environment.

ICT

Examining environmental change using secondary sources

Some of the environmental changes caused by interactions between air, water and the Earth's crust are extraordinarily complex. For this reason, firsthand investigation of these interactions often involves computer modelling, radiometric analysis, satellite imaging and other high-technology techniques not available in primary schools. This is where secondary sources of information are valuable, particularly with the development of multimedia CD-ROMs, television science programs and the Internet. The humble research project has been given a new and exciting life in primary schools as students draw on these highly accessible sources of information to build new concepts and refine their skills of drawing relevant information from these new sources as well as the more traditional print media. Primary students' interest in such tasks is enhanced by the social, human dimension, which frequently arises in projects looking at environmental change (see the online Appendix 12.1 for suggested topics for primary school projects).

Appendix 12.1 Suggested topics for primary school projects

Science as a human endeavour

The characteristics of science as a human endeavour include making primary students aware of the NOS (scientists work by asking questions, and solving problems; science helps us to understand our world and can be used to make predictions), the contributions of scientists (e.g., Australian scientists are currently making a significant contribution to scientific understanding in various fields) and the links between science and culture (e.g., contributions of Indigenous knowledge). Each of these can be readily incorporated into lessons focusing on weather. This section provides some examples.

The nature of science: asking questions, solving problems and making predictions

In the introduction of this chapter, the following statement was made: our weather is so explicably linked to our environment that things humans do that change the environment (in the long term) can also change the weather. This link is so strong that changes to the weather or climate (climate change) can also change the environment. This revelation may lead you (and your students) to ask questions like: How do we know that? What will happen in the future? Does humanity-caused environmental change have serious consequences? Is there anything we can do about it? These are the sorts of questions that scientific research can help us to answer by providing evidence. Equipped with evidence from scientific research communities, country and world organisations can make informed decisions leading to sustainable living practices.

A synopsis of current research being conducted by an Australian professor and his research team provides an example of 'science as a human endeavour' in action related to the effects of climate change on the Great Barrier Reef.

Christine has been lucky enough to be part of a scientific research project that is asking questions about the impacts of increasing greenhouse gases on our oceans. These questions are:

1. What does increased carbon dioxide levels in the atmosphere do to sea water?
2. Do changes in sea water affect the growth of corals?
3. Are changes in coral growth likely to endanger major ecosystems like the Great Barrier Reef?

These questions stem from the fact that changes in one part of the Earth's environment, the atmosphere, also affect other parts, including the oceans. As explained previously, the atmosphere and oceans are inextricably linked and together they play a major role in controlling the Earth's climate. Do you and your students (present or future) know how the enhanced greenhouse effect impacts on the ocean and the things that live there? Professor Malcolm McCulloch (from the University of Western Australia) and his team of scientists are researching how increasing carbon dioxide in our oceans is leading to the process known as 'ocean acidification', and they are also looking at its potential impacts on coral reefs, including the Great Barrier Reef.

Carbon dioxide in the atmosphere can mix with and dissolve in sea water. It is in the surface waters where carbon dioxide (and other gases) is readily absorbed. The problem with carbon dioxide is that when it dissolves it makes an acid (called carbonic acid). So, as the amount of carbon dioxide in the atmosphere increases, so does the amount absorbed and dissolved in sea water. The process results in an overall increase in acidity, lowering the overall pH of sea water. This becomes a problem when too much carbon dioxide is added to the atmosphere and the oceans, which has been happening rapidly since industrialisation. In fact, over 40 per cent of human-made carbon dioxide has been taken up by the oceans, which means that if the oceans didn't absorb carbon dioxide, then the impact of global warming and climate change would be far greater. Too much carbon dioxide mixing with the water will increase the acidity of sea water which lowers the pH to levels that will likely threaten many marine organisms. What you probably know about marine organisms (including corals) is that many have shells or exoskeletons made of calcium carbonate. When corals grow, they absorb carbonates (and other elements) in the surrounding sea water to build their skeletons, a process called calcification. The pH level of sea water affects the 'carbonate saturation state' that controls this process. If the pH becomes too low, coral won't be able to make the skeleton that is vital to its existence.

Malcolm's research involves taking long core samples from large corals that are very old (often over 200 years). Figure 12.4 shows how big these corals are and the underwater drilling process. As these corals have been living for hundreds of years and started growing before greenhouse gas emission started to increase dramatically from human activity, they provide an excellent record of changes in coral growth and sea water composition, important information that is not available from historical recordings. Changes in the environment, such as temperature, pH and degradation of land from farming, can all be recorded by changes in sea water, which are reflected in coral skeletons. The concentrations of some of the elements and the ratios of their isotopes can fluctuate with changes in the environment. From the techniques that the scientists use, they can accurately determine the ocean temperatures and ocean pH as the coral skeleton grew. Figure 12.15 shows Malcolm and Christine examining freshly drilled coral cores.

It is speculated that rate of coral growth as well as density (calcification) is declining due to increasing atmospheric carbon dioxide levels that are reducing the pH of sea water and carbonate ion concentration in sea water, which inhibits the ability of corals to calcify. What this means is that if carbon dioxide levels in the atmosphere continue to increase, there may

FIGURE 12.14 Dr Stuart Fallon and Christine drilling a large coral

FIGURE 12.15 Malcolm and Christine examine freshly drilled coral cores

come a time when corals can no longer grow. This is what makes the work of scientists like Malcolm significant, because scientific evidence obtained by his research can be used to predict the future of major reef communities such as the Great Barrier Reef, including their possible demise. This kind of scientific research helps quantify the real impacts that humans are having on the environment, from local and regional issues to global effects such as climate change. These studies can hopefully help us better manage our environment and live more sustainably, in harmony with the rest of the planet.

Contributions of scientists: an Australian Earth science example

Malcolm McCulloch (see Figure 12.15) is a Professor in the School of Earth & Environment at the University of Western Australia, and is a WA Premier's Fellow. He has received a number of prestigious awards, including fellowships of the Australian Academy of Science (2004), the Geological Society of Australia (2007), the Geochemical Society (2008) and the American Geophysical Union (2002). In 2009, he was awarded the Jaeger Medal by the Australian Academy of Sciences for 'Career Research Excellence in the Earth Sciences' and his contributions to the study of solid earth and environmental research. Malcolm has also recently been elected Fellow of the prestigious Royal Society (London; see http://www.news.uwa.edu.au/201005252519/international/uwa-scientist-and-premiers-fellow-elected-royal-society). His work on the effects of human-driven climate change on corals, outlined above, is just one of the important contributions Malcolm has made to our understanding of climate and the environment. Raising students' awareness of the contributions being made by scientists in their own State, Territory or country helps them realise that these are real people doing real jobs of importance to society, just next door to them. This approach may prevent students putting scientists on a pedestal and help them realise science is a career they could aspire to, with many branches and fields of interest to follow.

Science and culture: Australian Aboriginal perspectives

A focus on Indigenous Australian perspectives about weather can assist students to appreciate that life practices of different cultures can be governed by their respect for, and being at-one with, their surrounding environment. Clarke (2009) describes how Aboriginal Dreaming gives order to their world and acknowledges a common origin for people and the environment, and he provides considerable insight into Indigenous perspectives on weather, based on historical records and ethnographic fieldwork. In brief, to encourage you to read further, Clarke refers to:

- Europeans viewing specific weather components as having independent causes, compared with Aboriginal people who believe that natural phenomena (rain, fire and wind) are integrally linked
- Aboriginal people's knowledge of weather as mythical and expressed in song
- ancestors and spirits being viewed as weather makers, but people also playing a role by offering totems and performing rituals, and using weather-making 'tools' such as crushed snails, 'rain-sticks' and 'rain stones'
- 'weather-making' places where rituals were performed.

Indigenous people's seasonal calendars

Another example of the difference in ways of knowing between cultures relates to the seasons. In Australia, Indigenous people have lived in harmony with the land for more than 40 000 years. In the past, their survival depended on understanding the sequence of significant natural events, especially the life cycles of useful plants and animals (Reid 1995). Their calendars were directly related to local sequences of natural events. While some of these tribal calendars are still in use in northern Australia, those in the south have not persisted.

The four seasons currently in use in mainstream Australia no longer seem to fit the climatic conditions the people are experiencing. This situation is not surprising, given that it was the British who introduced the seasons of equal length to the new colony. The continued adherence to a calendar designed for the European climate is evidence of how much out of touch we are with the land (Jane 2001). In recent times, there have been moves to design more appropriate and useful calendars for the various regions. Suggestions include six seasons for the south and five for the north.

Leaders of the Indigenous Weather Knowledge project are listening to and valuing Indigenous Australians' voices. For the first time, Indigenous people who own the knowledge will control what information is included. Bodkin, a traditional D'harawal Aboriginal descendant who has inherited years of weather wisdom, said:

> In Aboriginal society one of the laws is you never tell another person's knowledge. The knowledge runs through women. The women have most things to do with plants so the seasons are most important to them anyway, in terms of food and medicines.

The project team aims to record Indigenous Australians' knowledge of weather patterns and long-term environmental changes. Indigenous knowledge is site-specific. There are clear signs that the seasons are changing, possibly due to global warming. In the future, it is highly likely that Australia will have new seasons that are determined with the help of Indigenous people sharing their knowledge.

As teachers of science, we can question the appropriateness of the European seasons for Australia by organising a classroom debate on the topic, 'It is better to refer to the seasons in the way they do in northern Australia: "the wet", "the dry" and "the build–up" '. To prepare arguments for and against, see 'The lost seasons' at http://www.abc.net.au/science/features/indigenous/default.htm to find out the latest developments and the views concerning the seasons that the elders are willing to share. Students can follow the seasonal calendar of the Gundjeihmi-speaking people of the Murrumburr clan in the story book *Walking with the Seasons in Kakadu* by Lucas and Searle (2003).

Summary

Environmental change encompasses a substantial part of the primary curriculum, involving not only science but also linking scientific approaches with those used in other curriculum areas; notably, society and environment, and technology and mathematics.

In this chapter, the weather has been used as the main vehicle for discussing the development of constructivist learning experiences through which your students can explore, research and investigate changes in the local and global environment. The term 'weather' has been interpreted broadly and links have been drawn between weather and the associated ideas of climate, weathering, landscape and seasons in order to address a wide range of environmental changes. How plants, animals (including humans) and non-living materials at or near the Earth's surface respond to these changes has also been addressed.

Pedagogic strategies, especially learning outdoors, that encourage primary students to challenge their existing concepts and develop new concepts while thinking and working scientifically have been introduced. Key findings from research into primary students' concepts about environmental change have also been briefly included. Several examples of how this topic exemplifies science as a human endeavour were provided.

It's our planet. We can live within its natural capabilities, if we can learn what they are and respect them – OR, we can live it to death! What choices will our children make?

Concepts and understandings for primary teachers

The major topics that primary syllabuses would refer to in relation to the content of this chapter are:

- weather and climate, especially the seasons
- components of the weather (wind, rain, snow, clouds and temperature)
- the effects of weather on people, other living things and the environment (weathering and erosion)
- global weather effects (greenhouse effect, global warming, the hole in the ozone layer).

Below are most of the key conceptual ideas and understandings related to weather with which a primary teacher should be familiar. You should read and interpret this list with an appreciation of its limitations, as described in the section 'Concepts and understandings for primary teachers' in Chapter 1 (see p. 47).

Many of the concepts and understandings are derived from alternative conception studies and are adapted, in particular, from Henriques (2002) and Driver et al. (1994).

Air

- Air is a material and exists; it is all around us.
- Air is a mixture of gases, including oxygen, nitrogen, water vapour, carbon dioxide and others.

- Water vapour is water in the state of a gas.
- Air, like other types of matter, has mass and takes up space.

Air pressure

- Air exerts a pressure that acts in all directions, not just 'down'.
- Pressure is when force is exerted on a particular area; the smaller the area on which the force is exerted, the greater the pressure.
- Low air pressure results when there are fewer air particles in a particular space.
- High air pressure results when there are more air particles in a particular space.

Atmosphere

- The atmosphere is a layer of air that surrounds the Earth.
- The air gets thinner as the distance from the Earth increases; so does the air pressure and temperature.
- The atmosphere makes the Earth warmer than it would be without an atmosphere.
- There is a layer of ozone in the upper atmosphere; it is called the ozone layer and it blocks out damaging ultraviolet radiation. This ozone is good.
- The hole in the ozone layer is an area of the upper atmosphere that has lower-than-expected levels of ozone.
- Ozone near the Earth's surface is a major part of smog. This ozone is bad.

The water cycle and water in the air

- Water can evaporate from plants, animals, puddles, soil and other ground surfaces, and from oceans, lakes, rivers and streams.
- The water cycle comprises liquid water being evaporated (from any of the above sources) and water vapour condensing in the clouds, then falling to Earth as rain.
- Water vapour in the air is invisible.
- Condensation forms when water vapour (gas) in the air changes into liquid due to cooling.
- Humidity is a measure of the amount of water vapour in the air.
- Humid air is less dense than dry air, moist air contains more water particles, and water has a lower molecular weight than dry air.

Clouds

- Clouds form under certain conditions, such as when:
  - more water vapour evaporates from the Earth into the atmosphere than condenses on the Earth
  - there are dust, smoke or other particles suspended in the air
  - water vapour condenses onto particles in the air.
- Clouds are made of tiny water droplets or ice crystals.
- Clouds float in the air and are moved by the wind.
- There are different types of clouds and not all clouds produce rain.

Rain

- Rain is liquid water that falls from clouds.
- Rain occurs when the water drops in a cloud get too large to stay in the sky and so fall due to gravity.
- Tiny raindrops are spherical.
- The shape of raindrops falling through the air is affected by surface tension and pressure.

Frost, snow, sleet, hail and dew

- Frost forms when water vapour in the air changes directly from gas to solid (no liquid stage) on a very cold surface; for example, grass.
- Snow is formed when ice crystals from clouds fall to the ground without melting.
- Sleet occurs when ice crystals melt as soon as they hit the ground.
- Hail is solid ice chunks that fall from clouds.
- Dew is formed when water vapour in the air condenses into water drops on cold surfaces.
- Mist consists of tiny water droplets suspended low in the air and is actually a low cloud.

Wind

- Wind is moving air.
- Winds on the Earth's surface flow from areas of high to low pressure.
- Winds at high altitude are affected by air currents.

- Winds form due to some areas of the Earth's surface being heated more than others.
- Hot air masses rise while cold air masses fall, creating the movement of air (wind).

Weather

- Weather is the state of the atmosphere at any given moment.
- The main weather elements are temperature, air pressure, humidity and wind.
- Weather determines the climate of a region.
- The temperature of the day is affected by the time of year, location on Earth, altitude, winds, cloud cover and precipitation.

Climate

- Climate is a combination of the long-term atmospheric conditions characteristic of a particular place.
- Climate is determined by observing the patterns in weather elements over a period of 30 or more years.

Seasons

- The climate in different parts of the Earth changes with the seasons.
- Seasonal climate change is due to the tilt of the Earth on its axis as it orbits the sun.

Global warming and the greenhouse effect

- Global warming is the phenomenon that makes the surface of the Earth and the lower atmosphere hotter.
- The greenhouse effect is the phenomenon that makes the Earth warmer than it would be without an atmosphere.
- The greenhouse effect occurs because the atmosphere radiates some of the heat energy reflected off the Earth's surface back to Earth, thereby making the Earth's surface and the lower atmosphere warmer.
- A greenhouse gets warmer because it traps the heat energy inside and it all stays there due to a lack of convection.
- The natural phenomenon of the greenhouse effect is not a bad thing because life as we know it could not exist without the atmosphere keeping the Earth much warmer than it would be without one.
- The enhanced greenhouse effect due to increased gases in the atmosphere results in more heat being radiated back to Earth than normal.

Weather and its effects on the Earth's surface

- Weathering is when parts of the Earth's surface (rocks) are broken down by the weather (wind, rain, temperature changes) or by chemical interactions (for example, effect of slightly acidic rain water or ground water on particular types of rocks, such as limestone).
- Weathering is usually a very slow process.
- Weathering is a surface process; it only occurs when the surface is exposed to the elements of the weather.
- Rocks can be worn away by water and wind-borne rock fragments.
- Water can enter rocks through cracks and pores.
- Rocks can be cracked by trapped water expanding when it freezes and by plant roots as they grow and push through the soil or between rocks.

Erosion

- Erosion is the movement of weathered fragments away from the parent rock.
- Weathered rock moves due to gravity or moving air, water and ice.

Acknowledgement

Photos of children and some work samples provided courtesy of Abbotsleigh Junior School, Sydney.

Search me! science education

Search me! Explore **Search me! science education** for relevant articles on weather and our environment. Search me! is an online library of world-class journals, ebooks and newspapers, including *The Australian* and the *New York Times*, and is updated daily. Log in to Search me! through http://login.cengage.com using the access code that comes with this book.

KEYWORDS

Try searching for the following terms:

- Children and weather
- Children and rain

Search tip: **Search me! science education** contains information from both local and international sources. To get the greatest number of search results, try using both Australian and American spellings in your searches: e.g., 'globalisation' and 'globalization'; 'organisation' and 'organization'.

Appendices

In these appendices you will find material related to weather and the environment that you should refer to when reading Chapter 12. These appendices can be found on the student companion website. Log in through http://login.cengage.com using the access code that comes with this book.

Appendix 12.1 Suggested topics for primary school projects

Brief outlines of some suggested topics for primary school projects relating to the weather are provided.

Appendix 12.2 Examples of cross-curriculum activities using an extract from a local newspaper

Some examples of cross-curriculum activities designed around using an extract from a local newspaper are listed.

Appendix 12.3 Once I was a water drop – narrative

This appendix comprises a narrative titled 'Once I was a water drop', which makes use of this writing genre to aid young children in understanding the abstract concepts of evaporation and precipitation in the context of a common classroom experiment.

Appendix 12.4 Tips for teaching outdoors

Some tips are given for teaching outdoors, including preparation before leaving the classroom, while teaching outdoors and follow-up back in the classroom after the outing are provided.

References

Abdullah, A. & Scaife, J. 1997. Using interviews to assess children's understanding of science concepts. *School Science Review*, 78 (285), pp. 79–84.

Australian Academy of Science (AAS). 2005. 'Weather in My World'. *Primary Connections*. Canberra: AAS.

Bahar, M., Bağ, H. & Bozhurt, O. 2008. Pre-service science teachers' understandings of an environmental issue: Ozone layer depletion. *Ekoloji*, 18 (69), pp. 51–8.

Bar, V. 1989. Children's views about the water cycle. *Science Education*, 73 (4), pp. 481–500.

Belohlawek, J., Keogh, B.& Naylor, S. 2010. The PUPPETS project hits WA. *Teaching Science*, 56 (1), pp. 36–8.

Berger, M. 1992. *Who Cares about the Weather?* New York: Newbridge Educational Publishing.

Biddulph, F. & Biddulph, J. 1993. *Clouds, Rain and Fog*. Hamilton, VA: Applecross.

Boylan, C. 2008. Exploring elementary students' understanding of energy and climate change. *International Electronic Journal of Elementary Education*, 1 (1), pp. 1–15.

Candek, O. 2009. Science students' misconceptions of the water cycle according to their drawings. *Journal of Applied Sciences*, 9 (5), pp. 865–73.

Clarke, P. 2009. Australian Aboriginal ethnometeorology and seasonal calendars. *History and Anthropology*, 20 (2), pp. 79–106.

Claymier, R. 2009. Breezy power: From wind to energy. *Science and Children*, 46 (9), pp. 36–40.

Coffman, M., Eidietis, L., Gardiner, L., Hatheway, B., Henderson, S. & Rutherford, S. 2009. Need an upgrade from 'Once upon a time ...'? Try this storybook. *Teaching Science*, 55 (4), pp. 45–9.

Conezio, K. & French, L. 2002. Science in the preschool classroom: Capitalizing on children's fascination with the everyday world to foster language and literacy development. *Young Children*, 57 (5), pp. 12–18.

Coskie, T. & Davis, K. 2009. Organizing weather data. *Science and Children*, 46 (5), pp. 52–4.

Covitt, B., Gunckel, K. & Anderson, C. 2009. Students' developing understanding of water in environmental systems. *The Journal of Environmental Education*, 49 (3), pp. 37–51.

Dale, S. 1995. Science and children field trips. *Investigating: Australian Primary Science Journal*, 11 (3), pp. 16–19.

Davidson, S. 2009. A picture is worth a thousand words. *Science and Children*, 46 (5), pp. 36–9.

Dove, J. 1998. Alternative conceptions about the weather. *School Science Review*, 79 (289), pp. 65–9.

Driver, R., Squires, A., Rushworth, P. & Wood-Robinson, V. 1994. *Making Sense of Secondary Science: Support Materials for Teachers*. London: Routledge.

Ellison, T. 2009. Snowflake symmetry. *Science and Children*, 47 (4), pp. 58–60.

Fisher, B. W. 1998. There's a hole in my greenhouse effect. *School Science Review*, 79 (288), pp. 79–99.

Hackling, M. 1998. *Working Scientifically: Implementing and Assessing Open Investigation Work in Science*. Perth: Education Department of Western Australia.

Haley, S. & Dyhrman, S. 2009. The Artistic Oceanographer program. *Science and Children*, 46 (8), pp. 31–5.

Harlen, W. 2009. Teaching and learning science for a better future. *School Science Review*, 90 (333), pp. 33–42.

Henriques, L. 2002. Children's ideas about weather. *School Science and Mathematics*, 102 (5), pp. 202–15.

Hodson, D. & Hodson, J. 1998. From constructivism to social constructivism: A Vygotskian perspective on teaching and learning science. *School Science Review*, 79 (289), pp. 33–41.

Intergovernmental Panel on Climate Change (IPCC). 2007. Fourth Assessment Report: Climate Change. Geneva: IPCC. Available at http://www.ipcc.ch/publications_and_data/ar4/syr/en/spms5.html (accessed 25 May 2010).

Jane, B. 2001. Accessing spirituality and sustainability: Letting go of a Western science worldview, in S. Gunn and A. Begg (eds). *Mind, Body & Society: Emerging Understandings of Knowing and Learning*. Melbourne: University of Melbourne, Department of Mathematics and Statistics, pp. 21–8.

Knight, R. 2009. Starting with what they know. *Primary Science*, 109, pp. 22–4.

Koballa, T. 2008. The point of it all: Exploring variations in wind vane design. *Science and Children*, 45 (6), pp. 32–5.

Kruger, C. & Summers, M. 1989. The 'materials' interviews: Detailed findings. *Primary School Teachers and Science Project*, Working Paper no. 6. Oxford: PSTS Project.

Lee, O., Lester, B., Ma, L., Lambert, J. & Jean-Baptiste, M. 2007. Conception of the greenhouse effect and global warming from diverse languages and cultures. *Journal of Geoscience Education*, 55 (2), pp. 117–25.

Lottero-Perdue, P., Lovelidge, S. & Bowling, E. 2010. Engineering for all. *Science and Children*, 47 (7), pp. 24–7.

Lowe, R. 1986. The scientific diagram: Is it worth a thousand words? *Australian Science Teacher's Journal*, 32 (3), pp. 7–13.

_____ 1993. Scientific diagrams: How well can students read them?, in B. Fraser (ed.). *Research Implications for Science and Mathematics Teachers*, 1, pp. 14–19.

Lucas, D. & Searle, K. 2003. *Walking with the Seasons in Kakadu*. Crows Nest, NSW: Allen & Unwin.

McCullagh, J., Walsh, G. & Greenwood, J. 2010. Books and stories in children's science. *Primary Science*, 111, pp. 21–4.

Melancon, M. 2000. Nature transects: Examining outdoor habitats to assess students' science-process skills. *Science and Children*, 38 (3), pp. 34–7.

Mintz, J. 2006. 'Global warming, what's that?' Student teacher understanding of the science related to sustainability issues. *Science Teacher Education*, 46, pp. 2–3.

Murcia, K. 2010. Multi-modal representations in primary science: What's offered by interactive whiteboard technology. *Teaching Science*, 56 (1), pp. 21–7.

Osborne, R. & Cosgrove, M. 1983. Children's conceptions of the change of state of water. *Journal of Research in Science Teaching*, 20 (9), pp. 825–38.

Peacock, A. 2007. Rising sea levels: Truth or scare? *Primary science review*, 98. pp. 4–6.

Potts, A., Stanisstreet, M. & Boyes, E. 1996. Children's ideas about the ozone layer and opportunities for physics teaching. *School Science Review*, 78 (223), pp. 57–62.

Purdie, C. 2000. Going out for science and other Key Learning Areas too: Lend us a bus, put on some walking shoes, have a group of enthusiastic students and we'll learn lots. *Investigating: Australian Primary and Junior Science Journal*, 16 (4), pp. 11–13.

Qualter, A., Schilling, M. & McGuigan, L. 1994. Exploring children's ideas. *Investigating: Australian Primary and Junior Science Journal*, 10 (1), pp. 21–4.

Ramsey-Gassert, L. 1997. Learning science beyond the classroom. *The Elementary School Journal*, 97 (4), pp. 433–51.

Reid, A. 1995. *Banksias and Bilbies: Seasons of Australia*. Melbourne: Gould League.

Russell, T., Longden, K. & McGuigan, L. 1993. *Rocks, Soil and Weather (Primary SPACE project)*. Liverpool: Liverpool University Press.

Saçkes, M. & Flevares, K. 2010. Four-to-six-year-old children's conceptions of the mechanisms of rainfall. *Early Childhood Research Quarterly*, Vol. 25, issue 4, pp. 536–46.

Sewell, A. 2002. Constructivism and student alternative conceptions. *Australian Science Teachers' Journal*, 48 (4), pp. 24–8.

Shepardson, D., Niyogi, D., Choi, S. & Charusombat, U. 2009. Seventh grade students' conceptions of global warming and climate change. *Environmental Education Research*, 15 (5), pp. 549–70.

Skamp, K. 1995. Where does the water go? *Investigating: Australian Primary and Junior Science Journal*, 11 (3), pp. 10–13.

_____ 2002. Learnscapes, science and technology teachers and the curriculum. *Australian Science Teachers' Journal*, 48 (1), pp. 8–14.

_____, Boyes, E. & Stanisstreet, M. 2009. Global warming responses at the primary secondary interface: 1. Students' beliefs and willingness to act. *Australian Journal of Environmental Education*, 25, pp. 1–17.

SMART Technologies. 2003. Available at http://www.smarttech.com (accessed 15 September 2010).

Smith, R. G. 1997. 'Before teaching this I'd do a lot of reading': Preparing primary student teachers to teach science. *Research in Science Education*, 27 (1), pp. 141–54

Spiropoulou, D., Kostopoulos, D. & Jacovides, C. P. 1999. Greek children's alternative conceptions on weather and climate. *School Science Review*, 81 (294), pp. 55–9.

Stepans, J. & Kuehn, C. 1985. Children's conceptions of weather. *Science and Children*, 23 (1), pp. 44–7.

Sterling, D. 2010. Hurricane proof this! *Science and Children*, 47 (7), pp. 48–51.

Symington, D. & Symington, S. 1983. The not so simple cycle. *Australian Science Teachers' Journal*, 29 (1), pp. 34–8.

Tunkin, D. 2004. *Where Do Puddles Go?* Melbourne: Reed International Books Australia, Rigby Literacy–Science Titles.

Tytler, R. 2002. Teaching for understanding in science: Constructivist/conceptual change teaching approaches. *Australian Science Teachers' Journal*, 48 (4), pp. 30–5.

_____, Peterson, S. & Prain, V. 2006. Picturing evaporation: Learning science literacy through a particle representation. *Teaching Science*, 52 (1), pp. 12–17.

_____ & Prain, V. 2010. A framework for re-thinking learning in science from recent cognitive science perspectives. *International Journal of Science Education*, 32 (15), pp. 2055–78.

_____, Prain, V. & Peterson, S. 2007. Representational issues in students' learning about evaporation. *Research in Science Education*, 37 (3), pp. 313–31.

Venville, G. & Donovan, J. 2006. Analogies for life: A subjective view of analogies and metaphors used to teach about genes and DNA. *Teaching Science*, 52 (1), pp. 18–22.

Weaver, A. & Mueller, A. 2009. Partners in learning. *Science and Children*, 46 (8), pp. 36–9.

Index

T